MW00674880

de Gruyter Expositions in Mathematics 15

de Gruyter Expositions in Mathematics

1 The Analytical and Topological Theory of Semigroups, *K. H. Hofmann, J. D. Lawson, J. S. Pym (Eds.)*

2 Combinatorial Homotopy and 4-Dimensional Complexes, *H. J. Baues*

3 The Stefan Problem, *A. M. Meirmanov*

4 Finite Soluble Groups, *K. Doerk, T. O. Hawkes*

5 The Riemann Zeta-Function, *A. A. Karatsuba, S. M. Voronin*

6 Contact Geometry and Linear Differential Equations, *V. R. Nazaikinskii, V. E. Shatalov, B. Yu. Sternin*

7 Infinite Dimensional Lie Superalgebras, *Yu. A. Bahturin, A. A. Mikhalev, V. M. Petrogradsky, M. V. Zaicev*

8 Nilpotent Groups and their Automorphisms, *E. I. Khukhro*

9 Invariant Distances and Metrics in Complex Analysis, *M. Jarnicki, P. Pflug*

10 The Link Invariants of the Chern-Simons Field Theory, *E. Guadagnini*

11 Global Affine Differential Geometry of Hypersurfaces, *A.-M. Li, U. Simon, G. Zhao*

12 Moduli Spaces of Abelian Surfaces: Compactification, Degenerations, and Theta Functions, *K. Hulek, C. Kahn, S. H. Weintraub*

13 Elliptic Problems in Domains with Piecewise Smooth Boundaries, *S. A. Nazarov, B. A. Plamenevsky*

14 Subgroup Lattices of Groups, *R. Schmidt*

Orthogonal Decompositions and Integral Lattices

by

Alexei I. Kostrikin
Pham Huu Tiep

Walter de Gruyter · Berlin · New York 1994

Authors

Pham Huu Tiep
Hanoi Institute of Mathematics
P.O. Box 631
10000 Hanoi, Vietnam

Present address:
Institute for Experimental Mathematics
University of Essen
Ellernstraße 29
D-45326 Essen, Germany

Alexei I. Kostrikin
Department of Mathematics
MEHMAT
Moscow State University
119899 Moscow GSP-1, Russia

1991 Mathematics Subject Classification: 11H31, 17Bxx, 20Bxx, 20Cxx, 20Dxx, 94Bxx

Keywords: Orthogonal decompositions, Euclidean lattices, finite groups, Lie algebras

∞ Printed on acid-free paper which falls within the guidelines of the ANSI to ensure permanence and durability.

Library of Congress Cataloging-in-Publication Data

Kostrikin, A. I. (Alekseĭ Ivanovich)
Orthogonal decompositions and integral lattices / by
A. I. Kostrikin, Pham Huu Tiep.
p. cm. – (De Gruyter expositions in mathematics ; 15)
Includes bibliographical references and index.
ISBN 3-11-013783-6
1. Lie algebras. 2. Orthogonal decomposition. 3. Lattice theory.
I. Pham Huu Tiep, 1963– . II. Title. III. Series.
QA252.3.K67 1994
512′.55–dc20 94-16850
CIP

Die Deutsche Bibliothek – Cataloging-in-Publication Data

Kostrikin, Aleksej I.:
Orthogonal decompositions and integral lattices / by
Aleksei I. Kostrikin, Pham Huu Tiep. – Berlin ; New York :
de Gruyter, 1994
(De Gruyter expositions in mathematics ; 15)
ISBN 3-11-013783-6
NE: Tiep, Pham Huu:; GT

Printed in Germany.
Typeset with LATEX: D. L. Lewis, Berlin. Printing: Gerike GmbH, Berlin.
Binding: Lüderitz & Bauer GmbH, Berlin. Cover design: Thomas Bonnie, Hamburg.

Preface

The present book is the result of investigations carried out by algebraists at Moscow University over the last fifteen years. It is written for mathematicians interested in Lie algebras and groups, finite groups, Euclidean integral lattices, combinatorics and finite geometries. The authors have used material available to all, and have attempted to widen as far as possible the range of familiar ideas, thus making the object of Euclidean lattices in complex simple Lie algebras even more attractive. It is worth mentioning that orthogonal decompositions of Lie algebras have not been investigated before now, for purely accidental reasons. However, automorphism groups of the integral lattices associated with them could not be investigated properly until finite group theory had reached an appropriate stage of development.

No special theoretical preparation is required for reading and understanding the first two chapters of the book, though the material of these chapters enables the reader to form rather a clear notion of the subject-matter. The subsequent chapters are intended for the reader who is familiar with the basics of the theories of Lie algebras, Lie groups and finite groups, and is more-or-less acquainted with integral Euclidean lattices. As a rule, undergraduates receive such information from special courses delivered at the Faculty of Mathematics and Mechanics of Moscow University. In any case, it is essential to have in mind a small collection of classic books on the above-mentioned themes: [SSL], [SAG], [Gor 1], [Ser 1] and [CoS 7].

Our teaching experience shows that the material of Part I and some chapters from Part II can be used as a basis for special courses on Lie algebras and finite groups. The material on integral lattices enrich the lecture course to a considerable extent. The interest of the audience and the readers grows, due to the great number of concrete unsolved problems on orthogonal decompositions and lattice geometry. In connection with integral lattice theory, we mention here a comprehensive book [CoS 7] and an interesting survey [Ple 3], where one can find much information contiguous with our book.

It is a pleasure to acknowledge the contributions of the many people from whose insights, assistance and encouragement we have profited greatly. First of all we wish to express our thanks to Igor Kostrikin and Victor Ufnarovskii, who were among the first to investigate orthogonal decompositions, and whose enthusiasm has promoted the popularisation of this new research area. Their impetus was kept up by the concerted efforts of K. S. Abdukhalikov, A. I. Bondal, V. P. Burichenko and D. N. Ivanov, to whom the authors are sincerely grateful. We are particularly indebted to D. N. Ivanov and K. S. Abdukhalikov: Chapter 7 is based on the results of D. N. Ivanov's C. Sc. Thesis, and the first five paragraphs of Chapter 10 are taken from K. S. Abdukhalikov's C. Sc. Thesis. Some brilliant ideas came from A. I. Bondal and V. P. Burichenko. We would like to thank A. V. Alekseevskii, A. V. Borovik, S. V. Shpektorov, K. Tchakerian, A. D. Tchanyshev and B. B. Venkov, who have made contributions to progress in this area of mathematics. A significant part of the book has drawn upon the Doctor of Sciences Thesis of the second author. We are indebted to our colleagues W. Hesselink, P. E. Smith and J. G. Thompson for a number of valuable ideas mentioned in the book.

Our sincere thanks go to Walter de Gruyter & Co, and especially to Prof. Otto H. Kegel, for the opportunity of publishing our book. We would also like to express our gratitude to Prof. James Wiegold for his efforts in improving the English. We are grateful to Professor W. M. Kantor for many valuable comments.

The authors wish to state that the writing of the book and its publication were greatly promoted by the creative atmosphere in the Faculty of Mathematics and Mechanics of Moscow University.

The present work is partially supported by the Russian Federation Science Committee's Foundation Grant # 2.11.1.2 and the Russian Foundation of Fundamental Investigations Grant # 93–011–1543. The final preparation of this book was completed when the second author stayed in Germany as an Alexander von Humboldt Fellow. He wishes to express his sincere gratitude to the Alexander von Humboldt Foundation and to Prof. Dr. G. O. Michler for their generous hospitality and support.

A. I. Kostrikin
Pham Huu Tiep

Table of Contents

Introduction

The theory of finite-dimensional simple Lie algebras over the field $\mathbb{C}$ of complex numbers, which is used intensively in many different branches of mathematics, has been developed in such a way that the absence of ready answers to concrete questions that arise is sometimes perceived as an irritating obstacle. Meanwhile, the cases of this kind bear witness to the fundamental nature of the subject, but not to the gaps in its elucidation. As is shown by experience, finite-dimensional simple Lie algebras over $\mathbb{C}$, just as with classical linear groups, are an inexhaustible source of new problems, new approaches to old questions and interesting interpretations of known facts.

It is well known that the dimension $\dim_{\mathbb{C}} \mathcal{L}$ of any complex simple Lie algebra $\mathcal{L}$ is divisible by its rank $\text{rank}\mathcal{L} = \dim_{\mathbb{C}} \mathcal{H}$ (here $\mathcal{H}$ is a Cartan subalgebra). Considered from the viewpoint of root systems, this fact seems at first sight to be a casual one. An internal explanation of this property is contained in the following simple statement.

As vector space, every complex simple Lie algebra $\mathcal{L}$ can be decomposed into a direct sum of Cartan subalgebras $\mathcal{H}_i$:

$$\mathcal{L} = \mathcal{H}_0 \oplus \mathcal{H}_1 \oplus \ldots \oplus \mathcal{H}_h \tag{0.1}$$

(here, h is the Coxeter number).

It is also well known that the Killing form

$$K(x, y) = \text{Tr}(\text{ad}x.\text{ad}y)$$

is non-degenerate on $\mathcal{L}$. The same holds for its restriction $K|_{\mathcal{H}}$ to any Cartan subalgebra $\mathcal{H}$. Let us try to combine these two facts by making all components $\mathcal{H}_i$ of decomposition (0.1) pairwise orthogonal with respect to the form K:

$$K(\mathcal{H}_i, \mathcal{H}_j) = 0 \text{ for } i \neq j. \tag{0.2}$$

We shall call decompositions of this kind *orthogonal decompositions* (abbreviated as OD). We demonstrate the simplest example: $\mathcal{L} = sl_2(\mathbb{C})$ is the Lie algebra of type A_1 with the decomposition

$$\mathcal{L} = \left\langle \begin{pmatrix} 1 & 0 \\ 0 & -1 \end{pmatrix} \right\rangle_{\mathbb{C}} \oplus \left\langle \begin{pmatrix} 0 & 1 \\ -1 & 0 \end{pmatrix} \right\rangle_{\mathbb{C}} \oplus \left\langle \begin{pmatrix} 0 & 1 \\ 1 & 0 \end{pmatrix} \right\rangle_{\mathbb{C}}.$$

The decomposition thus constructed possesses the remarkable property of multiplicativity:

$$[\mathcal{H}_i, \mathcal{H}_j] \subseteq \mathcal{H}_k \tag{0.3}$$

for all i, j and some $k = k(i, j)$. A decomposition of the form (0.1), with properties (0.2) and (0.3), will be called a *multiplicative orthogonal decomposition* (abbreviated as MOD).

A more interesting example of an MOD for the Lie algebra of type E_8 was discovered by J. G. Thompson while constructing the sporadic simple group $F_3 = Th$. In this construction, Thompson started from the two following facts. Firstly, F_3 has exactly one irreducible ($\mathbb{Q}$-valued) character χ of degree 248. Secondly, F_3 has a 2-local maximal subgroup $D = \mathbb{Z}_2^5 \circ SL_5(2)$ which is the unique non-split extension of $\mathbb{Z}_2^5$ by $SL_5(2)$ (the Dempwolff group), and which can be embedded in the complex Lie group $\mathcal{G} = E_8(\mathbb{C})$ (as an irreducible subgroup in the adjoint representation of $\mathcal{G}$). Thompson showed [Tho 2] that D fixes some MOD of the Lie algebra $\mathcal{L}$ of type E_8. Moreover, as it is shown by Thompson and P. E. Smith [Tho 3], [Smi 1, 2], D preserves some $\mathbb{Z}$-module $\Lambda \subset \mathcal{L}$, which, being equipped with the Killing form K is an even unimodular Euclidean integral lattice with automorphism group $\mathrm{Aut}(\Lambda) = \mathrm{Aut}_{\mathbb{Z}}(\Lambda) = \mathbb{Z}_2 \times F_3$. Recall that a lattice Λ equipped with a form B is said to be *even* if $B(x, x) \in 2\mathbb{Z}$ for all $x \in \Lambda$, and *unimodular* if $\det B = \pm 1$. Thompson's nontrivial example and a series of other examples, which we shall not dwell on here, lead us to the following questions.

Question 1. *Does every complex simple Lie algebra $\mathcal{L}$ possess an OD? If so, how many are there up to* $\mathrm{Aut}(\mathcal{L})$*-conjugacy?*

Suppose now that $\mathcal{L}$ admits some orthogonal decomposition $\mathcal{D}$. Suppose also that the complex Lie group $\mathrm{Aut}(\mathcal{L})$ contains a finite subgroup G fixing $\mathcal{D}$ (that is, permuting the components $\mathcal{H}_i$) and acting irreducibly on $\mathcal{L}$. Consider the set $\mathcal{M}_G$ consisting of all G-invariant $\mathbb{Z}$-modules $\Lambda \subset \mathcal{L}$ such that:

(i) Λ is a full module, that is, $\Lambda \otimes_{\mathbb{Z}} \mathbb{C} = \mathcal{L}$;

(ii) the restriction $K|_\Lambda$ is an integral positive definite form on Λ.

Question 2. *What can one say about the set $\mathcal{M}_G$, and what properties do Euclidean lattices in $\mathcal{M}_G$ possess? In particular, what are their automorphism groups?*

In posing this question, the hope is expressed of obtaining a "Lie" realization of some finite simple groups, and also of finding interesting new lattices. As will be seen, these hopes have proved to be well-founded, especially in the part related to constructing new Euclidean lattices, even unimodular ones. The realizations à la Thompson for a few of the sporadic finite simple groups are also obtained.

The systematic investigation of the above-mentioned questions, undertaken by the authors and their colleagues in Moscow University, has shown that the restriction of Question 1 to the class of MODs is too limited. Meanwhile, the constructions of ODs have turned out to be much more attractive, both in themselves and from the viewpoint of lattice theory. Quite quickly, the first part of Question 1 was crystallized in the form of the following conjecture, which is still open.

Conjecture 0.1. *A simple Lie algebra $\mathcal{L}$ over $\mathbb{C}$ possesses an OD if and only if $\mathcal{L}$ is not of the following types: A_n, where $n+1$ is not a prime power; and C_n for $n \neq 2^m$.*

The existence of ODs for Lie algebras of all types, except those mentioned above, was established quite some time ago (in about 1983). The exceptional nature of types A_n and C_n will be seen from different viewpoints in many of the chapters in this monograph. From the practical point of view, only the decompositions $\mathcal{D}$ whose automorphism groups

$$\mathrm{Aut}(\mathcal{D}) = \{\phi \in \mathrm{Aut}(\mathcal{L}) \mid \forall i, \exists j, \phi(\mathcal{H}_i) = \mathcal{H}_j\}$$

act irreducibly on $\mathcal{L}$ are of interest. Such decompositions will be called *irreducible orthogonal decompositions* (abbreviated as IODs). The solution of Question 1 is a more-or-less complete success only in the class of IODs. The following conjecture seems very plausible at the present time.

Conjecture 0.2. *A complex simple Lie algebra $\mathcal{L}$ admits an IOD if and only if $\mathcal{L}$ is of one of the following types: A_n, where $n = p^m - 1$ and B_n, where $n = [(p^m-1)/2]$, for some prime-power p^m; C_2; D_n, $n = 6, 14$ or $n = 2^m$; G_2, F_4, E_6 and E_8.*

Relaxing the irreducibility condition to the assumption that the permutation action of $\mathrm{Aut}(\mathcal{D})$ on components $\mathcal{H}_i$ of decomposition $\mathcal{D}$ is transitive we arrive at the notion of a *transitive orthogonal decomposition* (abbreviated as TOD). The following conjecture is fundamental for the class of TODs:

Conjecture 0.3. *A complex simple Lie algebra $\mathcal{L}$ possesses a TOD if and only if $\mathcal{L}$ is of one of the following types: A_n, where $n+1$ is primary; B_n, C_{2^m}, D_n, G_2, F_4, E_6, E_8.*

A comparison of Conjectures 0.2 and 0.3 shows that the class of TODs occupies a place intermediate between the ODs and IODs; moreover, the class of TODs is essentially wider than the class of IODs. Note also that the Lie algebra of type E_7 has an intransitive OD that agrees with the Cartan decomposition of $\mathcal{L}$ with respect to a standard Cartan subalgebra. Such *root orthogonal decompositions* (RODs) exist precisely for the types A_1, B_{2m-1}, D_{2m}, G_2, E_7, E_8. The information on the different classes of ODs mentioned up to this point is presented in the following scheme, where arrows show the corresponding embeddings:

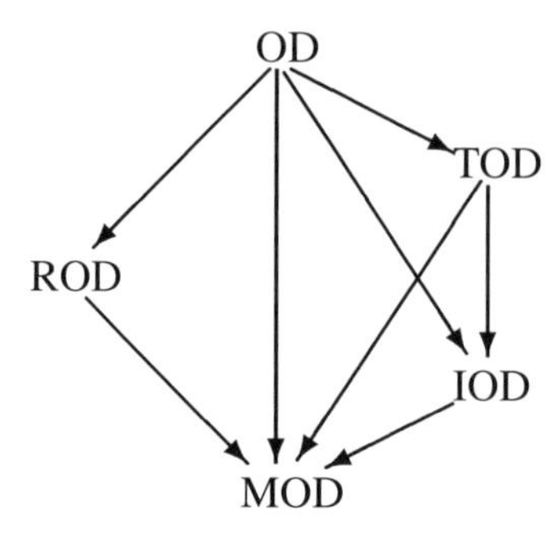

From the very beginning, an extraordinary analogy between the list of Lie algebras possessing TODs and the list of complex Lie groups having *Jordan subgroups* in the sense of A. V. Alekseevskii (for the precise definition, see Chapter 3) has been observed. The normalizers of Jordan subgroups usually represent maximal finite subgroups in Lie groups (see [Ale 2], [Bor 2, 4]), and local maximal subgroups in finite groups of Lie type (see [Asch 1], [Bor 2], [CLSS]). We shall see that the analogy mentioned above is not accidental. Namely, the presence of a Jordan subgroup in the Lie group $\mathcal{G} = \mathrm{Inn}(\mathcal{L})$ implies the existence of a TOD in a complex Lie algebra $\mathcal{L}$. It appears that a proof of the converse implication will be achieved only with great difficulty. As we shall show, even the corresponding result for a narrower class of IODs is far from being a triviality.

The investigation of ODs in the general case touches upon various questions of combinatorics and geometry. It is not excluded that a proof of our basic Conjecture 0.1 might be obtained using the ideas of such mathematical disciplines as algebraic geometry and finite projective planes. Up to now, we are only able to say the following. If Lie algebras of types A_n, $n \neq p^m - 1$; C_n, $n \neq 2^m$ possess some (hypothetical) orthogonal decomposition $\mathcal{D}$, then the group $\mathrm{Aut}(\mathcal{D})$ is highly likely to be trivial (compare this with the situation in the case of a projective plane of order 10 before undertaking use of a Cray!).

Although the problem of describing ODs for complex simple Lie algebras is fascinating in itself, our original aim is to use an existing orthogonal decomposition $\mathcal{D}$ to construct reasonably interesting integral Euclidean lattices in Lie algebras. To ensure that the task can be carried out in practice, the most natural approach is to assume finiteness of the $\mathbb{Z}$-similarity classes of lattices that are invariant under the group $G = \mathrm{Aut}(\mathcal{D})$. By Ivanov's theorem, the latter condition is equivalent to the requirement that $\mathcal{D}$ be an IOD. Following Thompson, we consider the slightly wider class of G'-invariant lattices, where G' is any subgroup of G acting irreducibly on $\mathcal{L}$. The task of producing a complete classification of all these lattices, although it is algorithmically solvable, is in fact quite cumbersome even in the case of Lie algebras of small rank. A complete description of invariant lattices is obtained for the Lie algebras of types G_2, F_4 and A_{p-1}, p a prime. The description for the Lie algebra of type A_{p^m-1} is near completion.

The task of describing automorphism groups for all invariant lattices (without factual description of such lattices, but sometimes using the classification of finite simple groups) is in a much more advanced stage of development. As an example, consider the Lie algebra $\mathcal{L}$ of type G_2 with non-split group $G = \mathrm{Aut}(\mathcal{D}) = \mathbb{Z}_2^3 \circ SL_3(2)$. It is clear that the description of the corresponding G-invariant lattices in $\mathcal{L}$ is equivalent to the classification of the $\mathbb{Z}$-representations of G of degree 14. It can be checked immediately that, up to $\mathbb{Z}$-similarity and duality, there exist just six G-invariant lattices Λ_k, $0 \le k \le 5$. Here Λ_0 is the root lattice of type $7G_2$, and $\Lambda_0 \supset \Lambda_k \supset 2\Lambda_0$ for $k < 5$. Furthermore, Λ_5 lies in Λ_0 but does not contain $2\Lambda_0$. Direct calculations show that $\mathrm{Aut}(\Lambda_k)$ is an imprimitive linear group provided that $k < 5$: $\mathrm{Aut}(\Lambda_k) \subseteq D_{12} \wr \mathbb{S}_7$, whereas $\mathrm{Aut}(\Lambda_5)$ acts primitively on $\mathcal{L}$ and, moreover, $\mathrm{Aut}(\Lambda_5) = \mathbb{Z}_2 \times G_2(3)$. The appearance of the Chevalley group $G_2(3)$ is not unexpected, since $G \hookrightarrow G_2(3)$ and $G_2(3)$ has an integer-valued irreducible complex character of degree 14.

This example leads us to the validity of the following general alternative, which we shall call $(\mathcal{A})$. Let Λ be an invariant lattice and set $\mathbb{G} = \mathrm{Aut}(\Lambda)$. Then one of the following statements holds:

(i) $\mathbb{G}$ acts imprimitively on $\mathcal{L}$, and then $\mathbb{G}$ has a well-described composition structure;

(ii) $\mathbb{G}$ acts primitively on $\mathcal{L}$ and $\mathbb{G}/Z(\mathbb{G})$ is an almost simple group, that is $L \lhd \mathbb{G}/Z(\mathbb{G}) \subseteq \mathrm{Aut}(L)$ for some non-abelian simple group L.

Further investigations show that alternative $(\mathcal{A})$ does in fact hold for all invariant lattices associated with ODs. Almost simple automorphism groups of invariant lattices are listed in the table below.

$\mathcal{L}$	$\dim \Lambda$	Λ	$\mathbb{G} = \mathrm{Aut}(\Lambda)$
G_2	14		$\mathbb{Z}_2 \times G_2(3)$
F_4	52		$\mathbb{Z}_2 \times L_4(3) \lhd \mathbb{G} \subseteq \mathbb{Z}_2.L_4(3).\mathbb{Z}_2$
E_6	78		$\mathbb{Z}_2 \times \Omega_7(3) \lhd \mathbb{G} \subseteq \mathbb{Z}_2.\mathrm{Aut}(\Omega_7(3))$
E_6	78		$\mathbb{Z}_2 \times Fi_{22} \lhd \mathbb{G} \subseteq \mathbb{Z}_2.\mathrm{Aut}(Fi_{22})$
E_8	248	even unimodular	$\mathbb{Z}_2 \times F_3$
E_8	248		$\mathbb{Z}_2 \circ L_4(5) \lhd \mathbb{G} \subseteq \mathbb{Z}_2.\mathrm{Aut}(L_4(5))$
A_2	8	even unimodular	$\mathbb{Z}_2 \circ O_8^+(2)$
A_4	24	even unimodular	$\mathbb{Z}_2 \circ Co_1$
B_3	21		$\mathbb{Z}_2 \times Sp_6(2)$
D_4	28	odd unimodular	$\mathbb{Z}_2 \times \Omega_8^+(2) \lhd \mathbb{G} \subseteq \mathbb{Z}_2.\mathrm{Aut}(\Omega_8^+(2))$

From the viewpoint of integral lattice theory, invariant lattices which are even unimodular are of special interest. The point is that in spite of the abundance of such lattices (for example, in dimension 32 there are at least 8.10^7 even unimodular Euclidean lattices), most of them have trivial automorphism groups (see [Ban]).

Moreover, very few explicit constructions of these lattices are known. Lattices in small dimension are studied in detail in the beautiful book mentioned above [CoS 7]. In the general case, apart from root lattices, there are only a few infinite series of even unimodular Euclidean lattices known at present. Namely, 1) the Barnes-Wall series [BaW] in dimension 2^{2d+1}, $d > 1$; 2) lattices of dimension $p+1$ ($p \equiv -1 (\mathrm{mod}\ 8)$ a prime) and $2(p-1)$($p \equiv 1(\mathrm{mod} 4)$ a prime) with isometries of order p (see, for instance, [Que 1,2]); 3) the Gow-Gross series [Gow 2], [Gro] in dimension $p^{2m}-1$, p a prime; 4) the Gross series [Gro] in dimension $2p^2(p^2-1)$, p a prime; and 5) the Gow series [Gow 3] of dimension 2^n, where $n = m^2 - 1$ or $n = 2m^2 + 2m - 1$, m a natural number. Quite recently, a new series of even unimodular lattices in dimension $2(p^n - 1)$, where $p \equiv 1(\mathrm{mod}\, 4)$ is a prime, is discovered in [Tiep 20].

Orthogonal decompositions of Lie algebras of type A_{p^m-1} lead us to a new series of even unimodular Euclidean lattices in the same dimension $p^{2m}-1$ as the Gow-Gross lattices, but with poorer automorphism groups. Unlike the series 3), 4) and 5), which were discovered implicitly, in our case one can say quite a lot about the lattices: automorphism groups, minimal vectors and so on.

Now we draw attention to the content of the book which consists of two parts. Part I comprises Chapters 1 - 7 and is devoted to orthogonal decompositions of complex simple Lie algebras as well as of associative algebras. Every chapter is accompanied by a very detailed commentary containing references to original works, sometimes unpublished. Contiguous questions are also discussed, parallel constructions described and unresolved problems stated. We hope that these commentaries will be useful for everyone who would like to make a contribution to this research field.

In Chapter 1 the only known construction of ODs of special form (the so-called *J-decompositions*) for Lie algebras of type A_{p^m-1} is described. This construction is based heavily on symplectic spreads, which appear widely in finite geometry. The classification theorem for irreducible J-decompositions is proved. The exceptional J-decomposition for $p^m = 27$ presented here corresponds to the Hering exceptional translation plane of order 27 [Her 1]. Note that, provided that $p = 2$ and $m > 2$, the automorphism group of the decomposition realizes the unique non-split extension of $\mathbb{Z}_2^{2m}$ by $SL_2(2^m)$. It is shown that, for $p^m \leq 5$, the J-decompositions constructed give the unique OD of the Lie algebra of type A_{p^m-1}; however, for $m \geq 2$, there is no uniqueness even in the class of J-decompositions. The question on the uniqueness of orthogonal pairs of Cartan subalgebras (or orthogonal pairs of maximal abelian $*$-subalgebras in von Neumann algebras) is discussed here; note that it is of interest in its own right. We also suggest a very curious, though not too elaborate, approach to Conjecture 0.1 using representations of Hecke algebras.

Chapter 2 is devoted to orthogonal decompositions of the remaining types of classical Lie algebras. Among them, type C_n is nearest to type A_n, in the sense that all known ODs are J-decompositions. On the other hand, types B_n and D_n

distinguished themselves by having a variety of different ODs. The construction (of the so-called *E-decompositions*) that is most natural for them is related to the skew-symmetric realizations of these algebras; it leads at once to the known combinatorial question of 1-factorizations of complete graphs, or, in other words, of 1-parallelisms. (Besides, we prefer to talk about admissible partitions). All irreducible E-decompositions of Lie algebra of type D_n are classified in this chapter.

Chapter 3 begins with a common construction of TODs using Jordan subgroups. This construction is especially useful for studying ODs of exceptional Lie algebras. Furthermore, we describe the construction of ROD, which gives in particular the only OD that we know for the Lie algebra of type E_7. The uniqueness theorem for MODs is proved in this chapter.

Beginning in Chapter 4, we set out to prove Conjecture 0.2. The first result we cover is the finiteness theorem motivating interest in the class of IODs. After this we give the general outline of the classification of IODs for Lie algebras with special Coxeter number. This class includes all exceptional Lie algebras, and also the algebras of types A_{p-2}, $C_{(p-1)/2}$, C_p, D_p (p a prime) and D_4. In Chapters 5 and 6, the same scheme is applied to Lie algebras of types A_n and B_n, n an arbitrary natural number. Note that we use here some consequences of the classification of finite simple groups: the validity of Schreier's conjecture and the description of doubly transitive permutation groups. Applying this to type A_n we obtain the expected final result: the number of conjugacy classes of IODs is

$$N = \begin{cases} 0 & \text{if} \quad n+1 \text{ is not a prime-power,} \\ 1 & \text{if} \quad n+1 \text{ is a prime-power and } n \neq 26, \\ 2 & \text{if} \quad n = 26. \end{cases}$$

As for types C_n and D_n, the full picture for IODs remains rather vague.

Chapter 7 is written following the works of Ivanov. Its position in the book is a rather isolated one, although in the idealogical plan it springs immediately from the OD problem. The question is that of the so-called *orthogonal decompositions of semisimple associative algebras*. Roughly speaking, a system $\mathcal{D} = \{\mathcal{B}_i\}$ of semisimple subalgebras, each containing the unit $1_{\mathcal{A}}$, of a semisimple matrix algebra $\mathcal{A}$ forms an OD if $\mathcal{A}$ is the linear span of subalgebras $\mathcal{B}_i$ in $\mathcal{D}$, and any two subalgebras $\mathcal{B}_i$, $\mathcal{B}_j \in \mathcal{D}$ are orthogonal with respect to the trace $\mathrm{Tr} = \mathrm{Tr}_{\mathcal{A}}$ of the regular representation in the sense that $\mathrm{Tr}xy = 0$ for all $x \in \mathcal{B}_i$, $y \in \mathcal{B}_j$ with $\mathrm{Tr}x = \mathrm{Tr}y = 0$. It is clear that $\mathcal{B}_i \cap \mathcal{B}_j = < 1_{\mathcal{A}} >_{\mathbb{C}}$. Furthermore, if $\mathcal{D} : \mathcal{L} = \bigoplus_{i=1}^{n+1} \mathcal{H}_i$ is an OD of the Lie algebra of $n \times n$ matrices with zero trace, then the system $\{\mathcal{B}_i\}$ of associative commutative subalgebras given by $\mathcal{B}_i = < 1_{\mathcal{A}} >_{\mathbb{C}} \oplus \mathcal{H}_i$, forms an OD of the associative algebra $\mathcal{A} = M_n(\mathbb{C})$. It turns out that in addition to ODs of Lie algebras, this new notion containes familiar topics such as finite affine planes, Hadamard matrices and finite split groups. In spite of the fact that none of these topics receives full elucidation, the fresh point of view produces new, interesting and challenging questions.

Part II of the book is devoted to integral Euclidean lattices and their automorphism groups.

The model example of an invariant lattice in the Lie algebra of type G_2 mentioned above is studied in detail in Chapter 8. The following chapter deals with the much more interesting case of the Lie algebras of type A_{p-1}, p an odd prime. It is shown that there exist exactly $\frac{(p-1)(2p+1)}{2}$ $\mathbb{Z}$-similarity classes of invariant lattices, and of these $(p-1)([p/4]+\frac{1}{2})$ are even unimodular lattices of dimension p^2-1. In particular, if $p=3$ we obtain the root lattice of type E_8, while for $p=5$ we obtain the remarkable Leech lattice. Most of the unimodular invariant lattices in the Lie algebras of type A_{p-1} are new and have no roots. Note that if $p \to \infty$, then the minimal length of the vectors in some of these lattices grows like $\sqrt{p}$. It should be observed that there exists an interesting connection between projections of invariant lattices into one component $\mathcal{H}_i$ of the OD and the Mordell-Weil lattices that were constructed recently by N. Elkies (see Chapter 14 on this topic).

The results of Chapter 9 can be generalized in part to the case of Lie algebras of type A_{p^m-1}. In the first part of Chapter 10, we follow Abdukhalikov and indicate the algorithm classifying invariant lattices in these Lie algebras, and give a construction of even unimodular lattices without roots of dimension $p^{2m}-1$. In some sense, these lattices remind ones of the Barnes-Wall lattices: their automorphism groups are $\mathbb{Z}_2 \times \mathbb{Z}_p^{2m}.Sp_{2m}(p)$. Furthermore, we prove a general reduction theorem which facilitates a description of the automorphism group of any invariant lattice in the given Lie algebra.

In Chapter 11 we consider invariant lattices associated with MOD of Lie algebras of types B_{2^m-1} and D_{2^m}. Here the partition of the set of all such lattices into *root-* and *nonroot-type* introduced by Burichenko has turned out to be very effective. Namely, a lattice Λ with $\mathbb{G} = \mathrm{Aut}(\Lambda)$ has root-type if Λ contains a $\mathbb{G}$-invariant sublattice Λ' which is a root lattice in the classical sense. Otherwise, Λ is of nonroot-type. We are convinced that this approach will be useful in describing invariant lattices associated with general ODs for Lie algebras of types B_n and D_n. In this chapter we also consider the relation between $\mathbb{Z}$-forms and invariant lattices in Lie algebras.

Chapters 12 and 13 are devoted to invariant lattices in Lie algebras of the exceptional types F_4, E_6 and E_8. The questions treated in these chapters have many overlaps with the investigations of R. L. Griess (Moufang loops), Thompson and Smith (constructing the sporadic simple group F_3), B. N. Cooperstein, A. M. Cohen, M. W. Liebeck and others (the exceptional embeddings $L_4(3) \hookrightarrow F_4(2)$, $\Omega_7(3) \hookrightarrow Fi_{22} \hookrightarrow {}^2E_6(2)$ etc.). Here, the decisive ingredients are the above-mentioned alternative $(\mathcal{A})$, and a technique from the representation theory of finite groups.

In the final Chapter 14, we survey recently-discovered constructions of integral lattices. There we have in mind the results of Gow and Gross (lattices associated with minimal representations of the finite groups $Sp_{2n}(p)$ and $2\mathbb{A}_n$), Elkies and Gross (the Mordell-Weil lattices), Burichenko (the Steinberg module) and the second author (a description of the automorphism groups of the lattices arising here).

We do not claim that our bibliography is exhaustive, especially the part that concerns lattices. This is because literature on this theme is covered in detail in [CoS 7]. The same is true of combinatorics and finite geometries.

Part I

Orthogonal decompositions of complex simple Lie algebras

In spite of its simplicity, the notion of orthogonal decompositions of complex simple Lie algebras turns out to be a very fruitful one. On the one hand, when fully realized, it leads to interesting combinatorial problems. On the other hand, quite substantial connections between ODs and other areas of mathematics, including integral lattices and finite simple groups, are uncovered almost immediately.

It is interesting to note that the removal of the orthogonality property makes our decomposition problem trivial, as is shown by the following statement.

As vector space, every complex simple Lie algebra $\mathcal{L}$ can be decomposed as a direct sum of Cartan subalgebras.

In fact, it is sufficient to check that, for every subspace $V \subset \mathcal{L}$ of codimension $\geq n = \operatorname{rank}\mathcal{L}$, there exists a Cartan subalgebra $\mathcal{H}$ that intersects V trivially. To this end, we consider the projective space $\mathbb{P}\mathcal{L}$ of dimension $\dim \mathcal{L} - 1$, its subspace $\mathbb{P}V$ and the algebraic subvariety $\mathcal{M} \subset \mathbb{P}V \times \mathbb{P}\mathcal{L}$ given by

$$\mathcal{M} = \{(\bar{v}, \bar{x}) \mid [v, x] = 0; v \in V, x \in \mathcal{L}\}$$

(here $[v, x]$ is the Lie product in $\mathcal{L}$). It is well known that the elements of $\mathcal{L}$ that are regular in the sense of Lie algebra theory, form a set that is open in the Zariski topology; its centralizer has dimension n ($n-1$ under projectivization). So a standard argument (see, for example, [Shaf, Chapter 1, §6.2, proof of Theorem 7]) on the fibres of a regular transformation – in the case under discussion, the projection of $\mathcal{M}$ into $\mathbb{P}V$ – shows that

$$\dim \mathcal{M} = \dim \mathbb{P}V + n - 1 < \dim \mathbb{P}\mathcal{L}.$$

Hence, there exists a regular element $x \in \mathcal{L}$ such that $\bar{x}$ is not contained in the image of $\mathcal{M}$ under the projection into $\mathbb{P}\mathcal{L}$. The required Cartan subalgebra is $C_{\mathcal{L}}(x)$.

A visual variant of this argument in the case of Lie algebras of type A_n is contained in §1 of [KKU 2]. Note also that the assertion proved above gives an internal explanation of the well known fact that the dimension of a simple Lie algebra is divisible by its rank.

Chapter 1

Type A_n

In [PaZ], which is devoted to generalizations of the Pauli matrices and graduations of the Lie algebra of type A_n, it is stated with reference to [KKU 2] that the existence problem for ODs for these Lie algebras is solved there. In fact this is far from the truth. In this chapter, we explain some of currently known constructions of ODs, the beginning of which was simply stated in the paper [KKU 2] mentioned above. The problem formulated there remains open, as before:

Problem. *Prove that the Lie algebra of type A_n admits an OD if and only if $n = p^m - 1$ for some prime p and for some natural number m.*

1.1 Standard construction of ODs for Lie algebras of type A_{p^m-1}

We start with the simplest case $m = 1$. In the Lie algebra $\mathcal{L}$ of type A_{p-1} (p a prime), we introduce the following matrices:

$$D = \mathrm{diag}(1, \varepsilon, \varepsilon^2, \ldots, \varepsilon^{p-1}),\ P = \begin{pmatrix} 0 & 0 & \ldots & 0 & 1 \\ 1 & 0 & \ldots & 0 & 0 \\ 0 & 1 & \ldots & 0 & 0 \\ & & \ldots & & \\ 0 & 0 & \ldots & 1 & 0 \end{pmatrix},$$

where ε is a primitive p-th root of unity. It can be checked immediately that $P^b D^a = \varepsilon^{-ab} D^a P^b$, and so

$$[J_{(a,b)}, J_{(c,d)}] = \varepsilon^{-ad}(\varepsilon^{ad-bc} - 1) J_{(a+c,b+d)}, \tag{1.1}$$

where $J_{(a,b)} = D^a P^b$ and matrix commutation is standard: $[A, B] = AB - BA$. Here and below, it is convenient to interpret the indices a, b as elements of the prime field $\mathbb{F}_p = \mathbb{Z}/p\mathbb{Z}$. From the commutation rule (1.1) and the obvious trace formula

$$\mathrm{Tr} J_{(a,b)} = 0, \ \forall (a,b) \neq (0,0),$$

it follows that the subspaces

$$\mathcal{H}_k = \mathcal{H}_{(1,k)} = \langle J_{(a,ka)} \mid 1 \leq a \leq p-1 \rangle_{\mathbb{C}}, \ 0 \leq k \leq p-1,$$

$$\mathcal{H}_\infty = \mathcal{H}_{(0,1)} = \left\langle P, P^2, \dots, P^{p-1} \right\rangle_{\mathbb{C}},$$

are commutative subalgebras of $\mathcal{L}$ and are pairwise orthogonal with respect to the Killing form $K(A,B) = \mathrm{Tr}(\mathrm{ad}A.\mathrm{ad}B) = 2p.\mathrm{Tr}AB$. On the other hand, all the elements $J_{(a,b)}$ are ad-semisimple in $\mathcal{L}$; therefore, these subspaces are in fact Cartan subalgebras. This assertion also follows from the non-degeneracy of the restrictions $K|_{\mathcal{H}_k}$ and the following general statement:

Lemma 1.1.1. *Let $\mathcal{L}$ be a complex semisimple Lie algebra and $\mathcal{H}$ a subalgebra. Then the following conditions are equivalent:*

(i) *$\mathcal{H}$ is a Cartan subalgebra;*

(ii) *$\mathcal{H}$ is a soluble subalgebra such that the restriction $K|_{\mathcal{H}}$ of the Killing form to $\mathcal{H}$ is non-degenerate, and the dimension of $\mathcal{H}$ is exactly* $\mathrm{rank}\mathcal{L}$.

Proof. The implication (i) $\Rightarrow$ (ii) is obvious. Now assume that $\mathcal{H}$ is a subalgebra satisfying condition (ii). Then in fact $\mathcal{H}$ is abelian. For, according to Lie's Theorem, $\mathcal{H}$ acts triangularly on $\mathcal{L}$:

$$\mathrm{ad}x = \begin{pmatrix} * & & & & \\ & * & & & \\ & & * & & \\ & & & * & \\ 0 & & \dots & & * \end{pmatrix}, \ x \in \mathcal{L}.$$

In particular,

$$\mathrm{ad}[x,y] = \begin{pmatrix} 0 & & & & \\ & 0 & & & \\ & & 0 & & \\ & & & 0 & \\ 0 & & \dots & & 0 \end{pmatrix}$$

for all $x, y \in \mathcal{H}$. But in that case, for any $z \in \mathcal{H}$ the operator $\mathrm{ad}z.\mathrm{ad}[x,y]$ also has zero diagonal, whence

$$K(z,[x,y]) = \mathrm{Tr}(\mathrm{ad}z.\mathrm{ad}[x,y]) = 0.$$

From the non-degeneracy of $K|_{\mathcal{H}}$ it follows that $[x,y] = 0$. Thus $\mathcal{H}$ is abelian.

Recall (see [Bour 2]) that every element $x \in \mathcal{L}$ can be written as the sum $x = x_s + x_n$ of its ad-nilpotent part x_n and its ad-semisimple part x_s; moreover, there exists a polynomial $f(t) \in \mathbb{C}[t]$ such that $\mathrm{ad}x_s = f(\mathrm{ad}x)$. We show next that $K(x, y) = K(x_s, y_s)$ for all $x, y \in \mathcal{H}$. Indeed, by the commutativity of $\mathcal{H}$ the operators $\mathrm{ad}x$ and $\mathrm{ad}y$ commute with each other. In particular,

$$[\mathrm{ad}x_s, \mathrm{ad}y_n] = [\mathrm{ad}y_s, \mathrm{ad}x_n] = [\mathrm{ad}x_n, \mathrm{ad}y_n] = 0$$

on the one hand. On the other hand, by their definition, the operators $\mathrm{ad}x_n$ and $\mathrm{ad}y_n$ are nilpotent. Hence, the operator $\mathrm{ad}x_s.\mathrm{ad}y_n$ is also nilpotent, so $K(x_s, y_n) = \mathrm{Tr}(\mathrm{ad}x_s.\mathrm{ad}y_n) = 0$. Similarly, $K(y_s, x_n) = K(x_n, y_n) = 0$. Therefore, we have

$$K(x, y) = K(x_s, y_s) + K(x_s, y_n) + K(y_s, x_n) + K(x_n, y_n) = K(x_s, y_s),$$

as stated.

Now choose some basis $\{x^1, \ldots, x^m\}$ of the complex space $\mathcal{H}$, where $m = \dim \mathcal{H} = \mathrm{rank}\mathcal{L}$. From the non-degeneracy of the form $K|_{\mathcal{H}}$ it follows that the matrix $(K(x^i, x^j))_{1 \le i,j \le m}$ is not singular. But, as we have shown above, in that case the matrix $(K(x_s^i, x_s^j))$, where x_s^i is the ad-semisimple part of x^i, is also non-singular. In particular, the m elements $x_s^1, \ldots, x_s^m$ are linearly independent. Furthermore, by the commutativity of $\mathcal{H}$ we have also that

$$0 = [\mathrm{ad}x_s^i, \mathrm{ad}x_s^j] = \mathrm{ad}[x_s^i, x_s^j]$$

for all i, j. The adjoint representation of the semisimple algebra $\mathcal{L}$ on itself is faithful, so that $[x_s^i, x_s^j] = 0$ for all i, j. Thus, the complex space $\mathcal{H}_s = < x_s^1, \ldots, x_s^m >_{\mathbb{C}}$ is an abelian subalgebra of dimension m consisting of ad-semisimple elements. It is well known (see [Bour 2]) that in that case $\mathcal{H}_s$ is a Cartan subalgebra and $C_{\mathcal{L}}(\mathcal{H}_s) = \mathcal{H}_s$. But $\mathcal{H}$ obviously centralizes $\mathcal{H}_s$ and $\dim \mathcal{H} = \dim \mathcal{H}_s$, so that in fact $\mathcal{H} = \mathcal{H}_s$. □

So we have arrived at an orthogonal decomposition:

$$\mathcal{L} = \mathcal{H}_\infty \oplus \mathcal{H}_0 \oplus \ldots \oplus \mathcal{H}_{p-1}.$$

All these arguments are based on the primality of the integer p. Replacing p by a composite number n, and regarding $\varepsilon^n = 1$ we do not even obtain a decomposition of $\mathcal{L}$ into a direct sum of Cartan subalgebras, because generally speaking the subspaces defined above have non-trivial intersections.

Suppose now that m is an arbitrary natural number, p a prime, and set $q = p^m$. Consider the plane $W = \mathbb{F}_q \oplus \mathbb{F}_q$ as a $2m$-dimensional space over $\mathbb{F}_p$ equipped with the following symplectic form:

$$< u \mid u' >= \mathrm{Tr}_{\mathbb{F}_q/\mathbb{F}_p}(u \circ u'),$$

where $u \circ u' = \alpha\beta' - \alpha'\beta$ for $u = (\alpha, \beta)$, $u' = (\alpha', \beta')$; $\alpha, \beta, \alpha', \beta' \in \mathbb{F}_q$. For this form on W there exists a symplectic basis $\{e_1, \dots, e_m, f_1, \dots, f_m\}$ such that

$$< u \mid u' >= \sum_{i=1}^{m} (a_i b_i' - a_i' b_i)$$

for $u = \sum_{i=1}^{m} (a_i e_i + b_i f_i)$, $u' = \sum_{i=1}^{m} (a_i' e_i + b_i' f_i)$. In view of Witt's Theorem, one can suppose that the first component $\mathbb{F}_q$ of W is generated (over $\mathbb{F}_p$) by $e_1, \dots, e_m$, and the second component by $f_1, \dots, f_m$. In this basis set

$$J_u = J_{(a_1,b_1)} \otimes J_{(a_2,b_2)} \otimes \dots \otimes J_{(a_m,b_m)}, \tag{1.2}$$

identifying a vector u with the collection of coordinates:

$$u = (a_1, \dots, a_m, b_1, \dots, b_m).$$

In our terminology the set $\{J_u \mid 0 \neq u \in W\}$ forms a *J-basis* of the Lie algebra $\mathcal{L}$ of type A_{q-1} as a $\mathbb{C}$-space. The operations on matrices J_u are performed by exactly the same rules as in the case of the matrices $J_{(a,b)}$ (see (1.1)):

$$J_u^k = \varepsilon^{\frac{k(k-1)}{2} B(u,u)} J_{ku}, \; J_u J_{u'} = \varepsilon^{B(u,u')} J_{u+u'} \tag{1.3}$$

$$[J_u, J_{u'}] = \varepsilon^{B(u',u)} \{\varepsilon^{<u|u'>} - 1\} J_{u+u'},$$

where

$$B(u, u') = -\sum_{i=1}^{m} a_i' b_i. \tag{1.4}$$

By (1.3), the matrices of the form $\varepsilon^s J_u$ form an extraspecial group $\hat{H}$ of order p^{2m+1} under the usual product. In particular,

$$J_u J_v J_u^{-1} = \varepsilon^{<u|v>} J_v. \tag{1.5}$$

This remark will be required later.

Now it is easy to see that the subspaces

$$\mathcal{H}_\alpha = \left\langle J_{(\lambda,\lambda\alpha)} \mid \lambda \in \mathbb{F}_q^* \right\rangle_{\mathbb{C}}, \alpha \in \mathbb{F}_q^*,$$

$$\mathcal{H}_\infty = \left\langle J_{(0,\lambda)} \mid \lambda \in \mathbb{F}_q^* \right\rangle_{\mathbb{C}}$$

of $\mathcal{L}$ are Cartan subalgebras and the decomposition

$$\mathcal{L} = \mathcal{H}_\infty \oplus \left(\oplus_{\alpha \in \mathbb{F}_q} \mathcal{H}_\alpha\right) \tag{1.6}$$

is an orthogonal decomposition for $\mathcal{L}$. To this end, it is sufficient to remark that

i) $< (\lambda, \lambda\alpha) \mid (\lambda', \lambda'\alpha) >= 0$,

ii) the trace $\mathrm{Tr} J_{(\alpha,\beta)} J_{(\gamma,\delta)}$ is not zero if and only if $(\gamma, \delta) = (-\alpha, -\beta)$ (see (1.3)), and then apply Lemma 1.1.1. Henceforth, we shall call a decomposition like (1.6) a *standard OD* of the Lie algebra $\mathcal{L}$ of type A_{q-1}.

1.2 Symplectic spreads and *J*-decompositions

The standard construction described in §1.1 suggests us that it is appropriate to consider all the ODs of special form in the Lie algebra $\mathcal{L}$ of type A_{q-1}.

Definition 1.2.1. An orthogonal decomposition

$$\mathcal{L} = \sum_{i=1}^{q+1} \mathcal{H}_i, \tag{1.7}$$

where the Cartan subalgebras $\mathcal{H}_i$ admit bases consisting of elements J_u of the form (1.2):

$$\mathcal{H}_i = \langle J_u \mid u \in U_i \rangle_{\mathbb{C}} \,,$$

is said to be a *J-decomposition*.

Definition 1.2.2. A collection

$$\pi = \{W_i \mid 1 \le i \le p^m + 1\} \tag{1.8}$$

consisting of subspaces W_i of dimension m of the space $W = \mathbb{F}_p^{2m}$ and satisfying the condition $W = \bigcup_{i=1}^{p^m+1} W_i$ is called a *spread.* If in addition W is endowed with a non-degenerate symplectic form, under which each W_i is totally isotropic, then π is called a *symplectic spread.*

We would like to show that there exists a bijective correspondence between J-decompositions and symplectic spreads. Namely, the J-decomposition (1.7) corresponds to the spread (1.8), where

$$W_i = U_i \cup \{0\}. \tag{1.9}$$

Obviously $W = \cup_i W_i$. Furthermore, from (1.3) and the commutativity of the $\mathcal{H}_i$, it follows that $< u \mid u' >= 0$ for all $u, u' \in W_i$; that is, the subspace $< W_i >_{\mathbb{F}_p}$ is totally isotropic on the one hand. On the other hand, the dimension of any totally isotropic subspace in W does not exceed m, and $|W_i| = p^m$. Hence W_i coincides with $< W_i >$, and so it is a maximal totally isotropic subspace of W.

Conversely, if the subspaces W_i from (1.9) form a symplectic spread, then it is obvious that (1.7) is a J-decomposition (again use the formulas (1.3) and Lemma 1.1.1).

Definition 1.2.3. By the *kernel* $K(\pi)$ of a spread π of the form (1.8) we mean the set

$$K(\pi) = \{\varphi \in \mathrm{End}(W) \mid \forall i, \varphi(W_i) \subseteq W_i\}.$$

Lemma 1.2.4. *The kernel $K = K(\pi)$ is a field of order p^l, where $1 \le l < 2m$ and l divides $2m$. Furthermore, every automorphism ψ of the spread π is a K-semilinear map; that is,*

$$\exists \sigma \in Gal(K/\mathbb{F}_p),\ \forall f \in K,\ \forall u \in W,\ \psi(f(u)) = f^{\sigma}\psi(u).$$

The space W is a $\frac{2m}{l}$-dimensional space over K and π is a spread of the K-space W.

Proof. First we show that K is a division ring. To this end, it is enough to show that $U = \mathrm{Ker} f = 0$ for every $f \in K\backslash\{0\}$. Suppose that $U \neq 0$ for some $f \in K\backslash\{0\}$. Fix a vector $u \in U\backslash\{0\}$, and consider an arbitrary element W_i such that $0 \neq f(W_i)$. Let $v \in W_i$ be such that $f(v) \neq 0$ and let $u + v \in W_j$. Then $W_i \ni f(v) = f(u+v) \in f(W_j) = W_j$; that is, $0 \neq f(v) \in W_i \cap W_j$, $i = j$, $u = (u+v) - v \in W_i$. Consequently, $f(W_i)$ can be non-zero for only one index $i = i_0$. Moreover, we have also shown that $u \in W_{i_0}$ for every $u \in U\backslash\{0\}$. In other words, $U \subseteq W_{i_0}$, $U = \mathrm{Ker}(f|_{W_{i_0}})$. So,

$$\begin{aligned}\dim W &= \dim \mathrm{Ker} f + \dim \mathrm{Im} f = \dim U + \dim f(W) = \\ &= \dim \mathrm{Ker}(f|_{W_{i_0}}) + \dim f(W_{i_0}) = \dim W_{i_0},\end{aligned}$$

a contradiction.

Hence, K is a finite division ring. By the Molin-Wedderburn Theorem, K is a field of order p^l for some l, $1 \le l \le 2m$. Now note that $K^* = K\backslash\{0\}$ is a normal subgroup of $G = \{\psi \in GL(W) \mid \psi(\pi) = \pi\}$. Therefore, for $\psi \in G$, the map $\sigma : f \in K \mapsto \psi f \psi^{-1}$ (restricted to K^*) is an automorphism of the group K^* and an automorphism of the field K, that is, $\sigma \in Gal(K/\mathbb{F}_p)$. Clearly, for $f \in K$, $u \in W$ we have $\psi(f(u)) = \psi f \psi^{-1}\psi(u) = f^{\sigma}\psi(u)$, that is, the map ψ is K-semilinear.

Furthermore, we turn W into a vector space over K by setting $f.u = f(u)$ for $f \in K$, $u \in W$. Then

$$2m = \dim_{\mathbb{F}_p} W = \dim_{\mathbb{F}_p} K.\dim_K W = l.\dim_K W.$$

In particular, l divides $2m$. Since multiplication by $f \in K$ fixes every element W_i of the spread π, it follows that W_i is a K-subspace of W and that π is a spread of the K-space W. In particular, $2m/l = \dim_K W > \dim_K W_1 \ge 1$, $l < 2m$. □

Definition 1.2.5. A spread π is said to be *desarguesian* if $|K(\pi)| = p^m$.

Clearly, the spreads corresponding to the decompositions described in §1.1 are desarguesian. The converse assertion is also true. To formulate it, we set

$$\begin{gathered} Sp^{\pm}(W) = \{\varphi \in GL(W) \mid \\ \exists \varepsilon = \pm 1, \forall u, v \in W, < \varphi(u) \mid \varphi(v) >= \varepsilon < u \mid v >\}, \end{gathered} \tag{1.10}$$

$$\mathrm{Aut}(\pi) = \{\varphi \in Sp^{\pm}(W) \mid \forall i, \exists j, \varphi(W_i) = W_j\}.$$

Lemma 1.2.6. *Any two desarguesian symplectic spreads are $Sp(W)$-conjugate. If π is a desarguesian symplectic spread, then*

$$\mathrm{Aut}(\pi) = (SL_2(p^m).\mathbb{Z}_m).\mathbb{Z}_\upsilon,$$

where $\upsilon = \frac{2}{\gcd(2,p)}$.

Proof. Consider two arbitrary desarguesian symplectic spreads $\pi = \{W_i\}$ and $\pi' = \{W_i'\}$.

1) Since $K = K(\pi)$ has dimension m over $\mathbb{F}_p$, it follows from Lemma 1.2.4 that π is a spread of the 2-dimensional K-space $W = K \oplus K$, that is, π has the form

$$\{W_\infty, W_\lambda \mid \lambda \in K\},$$

where $W_\infty = \{(0, u) \mid u \in K\}$, $W_\lambda = \{(u, \lambda u) \mid u \in K\}$. The same holds for π' with $K' = K(\pi')$: $\pi' = \{W_\infty', W_\lambda' \mid \lambda \in K'\}$. By the Noether-Skolem Theorem, a field isomorphism $K \simeq K'$ can be extended to an automorphism of the central simple algebra $\mathrm{End}_{\mathbb{F}_p}(W)$, that is, to conjugation by some element $f \in GL(W)$. So, up to $GL(W)$-conjugacy, we can suppose that $K = K'$. Fix two vectors $u \in W_0 \backslash \{0\}$, $v \in W_\infty \backslash \{0\}$. Without loss of generality, one can assume that $u \in W_0'$, $v \in W_\infty'$, $u + \lambda v \in W_\lambda'$, $\lambda \in K$. Then $W_\infty' = \{\alpha' v \mid \alpha' \in K'\} = \{\alpha v \mid \alpha \in K\} = W_\infty$. Similarly, $W_\lambda' = W_\lambda$ for every $\lambda \in K$. Hence, there exists an operator $A \in GL(W)$ such that $\pi' = A(\pi)$.

2) Suppose that $\pi = \{W_\infty, W_\lambda \mid \lambda \in K\}$ is a symplectic spread (over $\mathbb{F}_p$) of the space $W = K \oplus K = \mathbb{F}_p^{2m}$ with respect to non-degenerate symplectic forms $< * \mid * >_1$ and $< * \mid * >_2$ on W. We shall show that we can find an element $\vartheta \in K^*$ such that $< u \mid v >_2 = < \vartheta u \mid v >_1$ for all $u, v \in W$. Without loss of generality, one can suppose that

$$< u \mid v >_1 = \mathrm{Tr}_{K/\mathbb{F}_p}(u \circ v),$$

where $u \circ v = \alpha\delta - \beta\gamma$ for $u = (\alpha, \beta)$, $v = (\gamma, \delta)$, $\alpha, \beta, \gamma, \delta \in K$. Provided that $\lambda \in K$, the vectors u and λu lie in one and the same subspace W_μ, so that $<u \mid \lambda u>_2 = 0$. Similarly, $<v \mid \lambda v>_2 = 0$, $<u + v \mid \lambda(u + v)>_2 = 0$.

Consequently,

$$0 = < u + v \mid \lambda u + \lambda v >_2 - < u \mid \lambda u >_2 - < v \mid \lambda v >_2 =$$

$$= < u \mid \lambda v >_2 - < \lambda u \mid v >_2;$$

in other words, $< u \mid \lambda v >_2 = < \lambda u \mid v >_2$.

For the fixed element $w = (0, 1) \in W_\infty$, consider the $\mathbb{F}_p$-linear functional $f : \alpha \in K \mapsto <(\alpha, 0)|w>_2$. Obviously, one can find $\vartheta \in K$ such that $f(\alpha) = \mathrm{Tr}_{K/\mathbb{F}_p}(\vartheta\alpha)$. So, for $u = (\alpha, \beta)$, $v = (\gamma, \delta)$ we have

$$< u \mid v >_2 = < (\alpha, 0) + (0, \beta) \mid (\gamma, 0) + (0, \delta) >_2 =$$

$$= < (\alpha, 0) \mid (0, \delta) >_2 - < (\gamma, 0) \mid (0, \beta) >_2 =$$

$$= < (\alpha, 0) \mid \delta w >_2 - < (\gamma, 0) \mid \beta w >_2 =$$

$$= < \delta(\alpha, 0) \mid w >_2 - < \beta(\gamma, 0) \mid w >_2 =$$

$$= f(\alpha\delta - \beta\gamma) = \mathrm{Tr}_{K/\mathbb{F}_p}(\vartheta(\alpha\delta - \beta\gamma)) = \mathrm{Tr}_{K/\mathbb{F}_p}(\vartheta u \circ v) = < \vartheta u \mid v >_1,$$

as stated.

3) We return to symplectic spreads π, $\pi' = A(\pi)$ of the space W with the symplectic form $< * \mid * >$, and introduce a new symplectic form $< u \mid v >' = < Au \mid Av >$. Then π is a symplectic spread with respect to $< * \mid * >$ and $< * \mid * >'$. Hence, in accordance with 2), there exists $\vartheta \in K^*$ such that $< Au \mid Av > = < \vartheta u \mid v >$. In 2) we have also shown that $< \vartheta u \mid v > = < u \mid \vartheta v >$. So it is easy to see that the operator B defined on $W = K \oplus K$ by the rule $B((\alpha, \beta)) = (\vartheta\alpha, \beta)$ preserves π, and also that $< Bu \mid Bv > = < \vartheta u \mid v >$. In particular, $< Au \mid Av > = < Bu \mid Bv >$, that is, $AB^{-1} \in Sp(W)$. Moreover, $AB^{-1}(\pi) = A(\pi) = \pi'$. In other words, π and π' are $Sp(W)$-conjugate.

4) Finally, we compute the group $\mathrm{Aut}(\pi)$ for a desarguesian symplectic spread π. According to the result proved above, it is enough to find this group for the spread $\pi = \{W_\infty, W_\lambda \mid \lambda \in K\}$ of the space $W = K \oplus K$ with the form $< u \mid v > = \mathrm{Tr}_{K/\mathbb{F}_p}(u \circ v)$. By Lemma 1.2.5, any $\psi \in \mathrm{Aut}(\pi)$ is contained in the group $\Gamma L_2(q) = GL_2(q).Gal(\mathbb{F}_q/\mathbb{F}_p)$ of all K-semilinear transformations of the space $W = K \oplus K$, $K = \mathbb{F}_q$. Conversely, $\Gamma L_2(q)$ fixes π. So,

$$\mathrm{Aut}(\pi) = \Gamma L_2(q) \cap Sp^{\pm}(W) = (SL_2(q).\mathbb{Z}_m).\mathbb{Z}_\upsilon. \qquad \square$$

We are interested firstly in symplectic spreads which correspond to groups $\mathrm{Aut}(\pi)$ that are transitive on $W \setminus \{0\}$. Obviously, any desarguesian spread satisfies this condition. A classification of such spreads is given in the following statement.

Theorem 1.2.7. *Up to $Sp^{\pm}(W)$-conjugacy, there exists precisely one symplectic spread with group* $\mathrm{Aut}(\pi)$ *transitive on $W \setminus \{0\}$. The case $p^m = 27$ is exceptional, when there exists one further spread, with group* $\mathrm{Aut}(\pi) = SL_2(13)$.

Proof. i) Set $A = \mathrm{Aut}(\pi)$. Let $\bar{W}$ be the translation group of the space W. Considering the natural permutation representation of the group $\bar{A} = \bar{W}.A$ on W, we see that $\bar{A}$ acts doubly transitively on W. Such groups $\bar{A}$ – the so-called *affine 2-transitive permutation groups* – are classified by Hering [Her 2] and Liebeck [Lie 2]. In this connection, A belongs to one of the following classes.

(a) Infinite classes.
(a1) $A \subseteq \Gamma L_1(r) = GL_1(r).Gal(\mathbb{F}_r/\mathbb{F}_p)$, $r = p^{2m}$;
(a2) $A \rhd SL_a(r)$, $r^a = p^{2m}$, $a > 1$;
(a3) $A \rhd Sp_{2a}(r)$, $r^a = p^m$, $a > 1$;
(a4) $A \rhd G_2(r)'$, $p^m = 2^m = r^3$.

(b) Extraspecial classes. The group A normalizes an extraspecial subgroup R, where either $p^m = 5, 7, 11, 23$ and $R = 2_-^{1+2} = Q_8$, or $p^m = 9$ and $R = 2_-^{1+4} = D_8 * Q_8$.

(c) Exceptional classes.
(c1) $p^m = 9, 11, 19, 29, 59$, $A \rhd SL_2(5)$;
(c2) $p^m = 4$, $A = \mathbb{A}_6$ or $A = \mathbb{A}_7$;
(c3) $p^m = 27$, $A = SL_2(13)$.

Since the symplectic spread is unique when $m = 1$, we shall assume henceforth that $m > 1$. Set $l = \dim_{\mathbb{F}_p} K(\pi)$. Our argument will be based on the transitivity of A on the elements W_i of the spread π.

ii) Here we show that π is desarguesian in case (a1). Set $S = GL_1(r)$, $B = Gal(\mathbb{F}_r/\mathbb{F}_p) \simeq \mathbb{Z}_{2m}$ and $H = A \cap S$. Note that $p^{2m} - 1$ divides $|A|$, so if $|H| = (p^{2m} - 1)/t$, we have that t divides $\gcd(2m, p^{2m} - 1)$. Since A acts transitively on the elements W_i of π, the normal subgroup H acts $\frac{1}{2}$-transitively on π, that is, with orbits of the same length, say d, $d \mid (p^m + 1)$. But H is cyclic, so that the subgroup H_1 of index d in H is unique. Hence, H_1 fixes every element $W_i \in \pi$. Thus $H_1 \subseteq K(\pi)^*$, which implies the divisibility of $p^l - 1$ by $(p^{2m} - 1)/td$. Recall that $d \mid (p^m + 1)$, $t \mid \gcd(2m, p^{2m} - 1)$, $1 \le l < 2m$, $l \mid 2m$. It is now easy to show that we always have $l = m$, with the single exception $p^m = 9$, $t = 4$, $d = 10$, $l = 1$. But in this exceptional case, A cannot act transitively on $W \backslash \{0\}$. So $l = m$ and π is desarguesian.

iii) Assume now that one of cases (a1), (a2) or (a3) holds. Then $A \rhd S$, where $S = SL_a(r)$, $Sp_{2a}(r)$ or $G_2(r)'$. Again, S acts $\frac{1}{2}$-transitively on π with orbits of length d, $d \mid (p^m + 1)$. Note that $d > 1$, because the equality $d = 1$ means that $S \subseteq K(\pi)^*$ and S is soluble. Thus, S has a subgroup $S_1 = St_S(W_1)$ of index d, $d > 1$, $d \mid (p^m + 1)$. Since d and p are coprime, by the Borel-Tits Theorem [BoT], [Sei 1], S_1 is contained in some maximal parabolic subgroup P of S. If $S = G_2(r)'$ we have

$$(S : P) = \frac{r^6 - 1}{r - 1} = \frac{p^{2m} - 1}{p^{m/3} - 1} > p^m + 1.$$

If $S = Sp_{2a}(r)$, $a > 1$ then

$$(S : P) = \prod_{k=1}^{j} \frac{r^{2(a-j)+2k} - 1}{r^k - 1}$$

for some j, $1 \leq j \leq a$. Hence $(S : P) \geq p^{3m/2} > p^m + 1$. Finally, if $S = SL_a(r)$, $a > 1$ then

$$(S : P) \geq \frac{r^a - 1}{r - 1} = \frac{p^{2m} - 1}{p^{m/a} - 1}.$$

But $p^m + 1 \geq d = (S : S_1) \geq (S : P)$, so we obtain the unique possibility $S = SL_2(q)$, $q = p^m$, $d = q + 1$, $S_1 = P$. In this case, we can identify the space W with the natural S-module $\mathbb{F}_q \oplus \mathbb{F}_q$. Because W_1 is invariant under

$$S_1 = P = \left\{ \begin{pmatrix} a & b \\ 0 & a^{-1} \end{pmatrix} \mid a, b \in \mathbb{F}_q, a \neq 0 \right\},$$

we must have $W_1 = \{(u, 0) \mid u \in \mathbb{F}_q\}$. Making S act on W_1, one can see that the spread π has the form $\{W^\infty, W^\lambda \mid \lambda \in \mathbb{F}_q\}$, where $W^\lambda = \{(u, \lambda u) \mid u \in \mathbb{F}_q\}$, $W^\infty = \{(0, u) \mid u \in \mathbb{F}_q\}$. But in that case, the kernel $K(\pi)$ contains the multiplications by scalars from $\mathbb{F}_q$, that is, $|K(\pi)| \geq p^m$; hence π is desarguesian.

iv) Direct calculations show that when $p^m = 9$, every symplectic spread is desarguesian. In particular, cases (b) and (c1) can be eliminated. Furthermore, for $p^m = 4$, the groups $\mathbb{A}_6$ and $\mathbb{A}_7$ have no subgroups of index 5, so case (c2) is also impossible.

It remains to consider the case (c3): $p^m = 27$, $A = SL_2(13)$. We mention some properties of A:

α) A acts transitively on $W \backslash \{0\}$, $W = \mathbb{F}_3^6$;

β) A is a maximal subgroup of $Sp_6(3)$. Moreover, $Sp_6(3)$ has two conjugacy classes of maximal subgroups of type $SL_2(13)$, which are permuted by the outer automorphism $X \mapsto BXB^{-1}$, where $B = \begin{pmatrix} E_3 & 0 \\ 0 & -E_3 \end{pmatrix} \in Sp_6^{\pm}(3)$;

γ) every subgroup of index 28 in $SL_2(13)$ is conjugate to the subgroup

$$Q = \left\{ f_{a,b} = \begin{pmatrix} a & b \\ 0 & a^{-1} \end{pmatrix} \mid b \in \mathbb{F}_{13}, a \in \mathbb{F}_{13}^{*2} \right\}.$$

Consider the element $\varphi = f_{1,1}$ of order 13 in Q. As $13^2 \nmid |Sp_6(3)|$, we can suppose that φ lies in the maximal parabolic subgroup

$$\left\{ \begin{pmatrix} X & Y \\ 0 & {}^tX^{-1} \end{pmatrix} \mid X \in GL_3(3), Y \in M_3(\mathbb{F}_3), X.{}^tY = Y.{}^tX \right\}$$

(matrices are written in a symplectic basis $\{e_1, e_2, e_3, f_1, f_2, f_3\}$). Setting $\mathbb{F}_{27}^* =< \vartheta >$, one can assume that $\varphi|_U$ is multiplication by ϑ^2, where $U = <e_1, e_2, e_3>_{\mathbb{F}_3}$. Then the operator φ has eigenvalues ϑ^2, ϑ^6, ϑ^{18} on U; while on $\bar{U} =< f_1, f_2, f_3 >_{\mathbb{F}_3}$, φ has eigenvalues ϑ^{-2}, ϑ^{-6}, ϑ^{-18}. Note that $Q =< \varphi, \psi >$, where $\psi = f_{4,0}$. Since $\psi\varphi\psi^{-1} = \varphi^3$, we have

$$\psi : \mathrm{Ker}(\varphi - \vartheta^2.1_W) \to \mathrm{Ker}(\varphi - \vartheta^{18}.1_W) \to$$

$$\to \mathrm{Ker}(\varphi - \vartheta^6.1_W) \to \mathrm{Ker}(\varphi - \vartheta^2.1_W),$$

that is, $\psi(U) = U$. Thus $Q(U) = U$.

In fact, we have shown that if U' is a Q-invariant maximal totally isotropic subspace of W, then U' is U or $\bar{U}$. Now, making A act on U we obtain the collection $U = W_1, W_2, \dots, W_t$ of t (different) maximal totally isotropic subspaces, where $t \le 28 = (A : Q)$. Because $\cup_{i=1}^t W_i \setminus \{0\}$ is an A-orbit of $W\setminus\{0\}$, and A is transitive on $W\setminus\{0\}$, we must have that $\cup_{i=1}^t W_i = W$. But in that case, $26.28 = |W\setminus\{0\}| \le 26t$, so $t = 28$. Thus $\pi = \{W_1, \dots, W_{28}\}$ is an A-invariant symplectic spread. In addition, the element $f_{2,0}$ of A interchanges U and $\bar{U}$. It is possible to show that $\mathrm{Aut}(\pi) = A$. Since desarguesian spreads have automorphism group $(SL_2(27).\mathbb{Z}_3).\mathbb{Z}_2$ when $p^m = 27$, and $L_2(13)$ cannot be embedded in $L_2(27)$, it follows that π is not desarguesian. We have also established the $Sp_6^{\pm}(3)$-conjugacy of all $SL_2(13)$-invariant spreads. □

Definition 1.2.8. A symplectic spread $\pi = \{W_i\}$ of a space W is said to be *irreducible* if $\mathrm{Aut}(\pi)$ acts transitively on $W\setminus\{0\}$; and *transitive* if $\mathrm{Aut}(\pi)$ is transitive on the set $\{W_i\}$.

Theorem 1.2.7 gives an example of an irreducible non-desarguesian spread. An example of a transitive spread which is not irreducible is contained in the following statement.

Theorem 1.2.9. *Suppose that $r = 2^{2n+1} > 2$. There exists a transitive non-desarguesian symplectic spread π of the space $W = \mathbb{F}_2^{4(2n+1)}$ with automorphism group $\mathrm{Aut}(\pi) = \mathrm{Aut}(Sz(r)) = Sz(r).\mathbb{Z}_{2n+1}$, where $Sz(r) = {}^2B_2(r)$ is the Suzuki twisted simple group.*

Proof. The embedding of $S = Sz(r)$ in $Sp_4(r)$ may be described as follows (see [Lun]). Identify W with a 4-dimensional space over $\mathbb{F}_r$ with basis $\{e_1, e_2, e_3, e_4\}$, and define its symplectic form via the matrix $\phi = \begin{pmatrix} 0 & 0 & 0 & 1 \\ 0 & 0 & 1 & 0 \\ 0 & 1 & 0 & 0 \\ 1 & 0 & 0 & 0 \end{pmatrix}$. Then $S = < B, \phi >$, where B is generated by the matrices

$$\begin{pmatrix} 1 & 0 & 0 & 0 \\ a & 1 & 0 & 0 \\ a^{1+\theta} + b & a^\theta & 1 & 0 \\ a^{2+\theta} + ab + b^\theta & b & a & 1 \end{pmatrix}, \begin{pmatrix} c^{1+\theta^{-1}} & 0 & 0 & 0 \\ 0 & c^{\theta^{-1}} & 0 & 0 \\ 0 & 0 & c^{-\theta^{-1}} & 0 \\ 0 & 0 & 0 & c^{-1-\theta^{-1}} \end{pmatrix},$$

where a, b and c run over $\mathbb{F}_r$, and θ is the automorphism of the field $\mathbb{F}_r$ given by $\theta(x) = x^{2^{n+1}}$. The subgroup B has index $r^2 + 1$ in S, and the representation of S on S/B is doubly transitive, so that $S = B \cup B\phi B$.

It is clear that B fixes the maximal totally isotropic subspace $W_2 = < e_3, e_4 >$. We show next that $\pi = \{s(W_2) \mid s \in S\}$ is a symplectic spread. Since B is maximal in S and $W_1 = \phi(W_2) =< e_1, e_2 >\neq W_2$, it follows that $B = St_S(W_2)$ and $|\pi| = r^2 + 1$. Furthermore, for $s \in S \backslash B$ we can write $s = b_1 \phi b_2$ for some $b_1, b_2 \in B$, and obtain that

$$W_2 \cap s(W_2) = W_2 \cap b_1 \phi b_2(W_2) = W_2 \cap b_1(W_1) = 0.$$

Hence,

$$|\bigcup_{U \in \pi} U| = 1 + (r^2 + 1)(|U| - 1) = |W|,\ W = \bigcup_{U \in \pi} U,$$

as stated. The outer automorphism ψ of S is induced by the Frobenius automorphism $x \mapsto x^2$ of $\mathbb{F}_r$, and it fixes π. Thus $\mathrm{Aut}(\pi) \supseteq \mathrm{Aut}(S)$. In fact, one can show that $\mathrm{Aut}(\pi) = \mathrm{Aut}(S)$.

Obviously, π is a transitive spread. Suppose that π is irreducible. Then by Theorem 1.2.7, π is desarguesian and $\mathrm{Aut}(\pi) = SL_2(r^2).\mathbb{Z}_{2(2n+1)}$ on the one hand. On the other hand, the simple group $S = Sz(r)$ cannot be embedded in $SL_2(r^2)$. Hence π is not irreducible. □

From what follows it will be clear that we have in fact listed all transitive symplectic spreads with insoluble automorphism groups.

Definition 1.2.10. Let $\mathcal{V} = (P(\mathcal{V}), L(\mathcal{V}), \in)$ be an incidence structure, where $P(\mathcal{V})$ is a finite set of points, $L(\mathcal{V})$ a finite set of lines and $\in$ is the incidence relation "a point lies on a line". Every pair (a, l) with $a \in P(\mathcal{V})$, $l \in L(\mathcal{V})$, $a \in l$, is called a *flag*. Assume that $\mathcal{V}$ satisfies the following axioms:

(i) every line contains at least two points;
(ii) every point belongs to at least two lines;
(iii) every pair of different points is contained in exactly one line.

Then $\mathcal{V}$ is called a *finite linear space*. If in addition $\mathrm{Aut}(\mathcal{V})$ acts transitively on the set of flags, then $\mathcal{V}$ is said to be a *flag-transitive space*.

The following lemma is obvious.

Lemma 1.2.11. *Suppose that $\pi = \{W_i\}_{i=1}^n$ is a nontrivial (that is, $n > 1$) spread of a space W. Let $\mathcal{V}$ be an incidence structure, where points are vectors of W and lines are affine subvarieties $W_i + u$, $u \in W$. Then $\mathcal{V}$ is a finite linear space. Moreover, transitivity of the spread π implies flag-transitivity of the space $\mathcal{V}$.* □

Proposition 1.2.12. *Suppose that $W = \mathbb{F}_p^{2m}$ is a symplectic space, $\pi = \{W_i\}_{i=1}^n$ a symplectic spread, and G_0 a subgroup of* $\mathrm{Aut}(\pi)$ *with transitive action on $\{W_i\}_{i=1}^n$. Then one of the following statements holds.*

1) *G_0 is soluble and* $G_0 \subseteq \Gamma L_1(p^{2m})$*;*
2) *π is desarguesian;*
3) $p^m = 27$, $G_0 = SL_2(13)$*;*
4) $p^m = r^2$, $r = 2^{2l+1}$ *($l \geq 1$),* $Sz(r) \lhd G_0 \subseteq \mathrm{Aut}(Sz(r))$.

Moreover, in cases 2) – 4)*, the corresponding spread exists and is unique up to conjugacy.*

Proof. Let $\mathcal{V}$ be a finite linear space corresponding to π (see Lemma 1.2.11). Then $G = \bar{W}.G_0$ is a flag-transitive automorphism group of the space $\mathcal{V}$, where $\bar{W}$ denotes the translation group of W. Furthermore, $\mathcal{V}$ has $N = p^{2m}$ points and each line of $\mathcal{V}$ contains $k = p^m$ points. According to a fundamental result [BDDKLS], in this case one of the conditions 1) – 4) is satisfied. Note that here we have excluded a translation plane of order 9, which is constructed via a near-field (see [Lun] or [Hall 1]). This is because, as mentioned in the proof of Theorem 1.2.7, every symplectic spread of $\mathbb{F}_3^4$ is desarguesian. The uniqueness of π in cases 2) – 4) is well-known (see [Lun]). □

There are many known examples of transitive symplectic spreads with soluble automorphism groups [Kan 4, 5]. These examples suggest that such spreads may be too numerous to be classified. Transitive spreads π such that $\dim_{K(\pi)} W = 4$ have been studied in [BaE 1, 2].

1.3 Automorphism groups of J-decompositions

In the preceding paragraph, with each J-decomposition $\mathcal{D}$ of the Lie algebra $\mathcal{L}$ of type A_{p^m-1} we associated a certain symplectic spread $\pi = \pi_{\mathcal{D}}$ of the symplectic space $W = \mathbb{F}_p^{2m}$. Our aim is to prove that the group $\mathrm{Aut}(\mathcal{D})$ can be expressed in terms of $\hat{H}$ (see §1.1) and $\mathrm{Aut}(\pi)$. Recall that $\mathrm{Aut}(\mathcal{L}) = \mathrm{Inn}(\mathcal{L}). < T >$, where $\mathrm{Inn}(\mathcal{L}) \simeq PSL_{p^m}(\mathbb{C})$ is the group of all inner automorphisms of $\mathcal{L}$, and T is the outer automorphim $X \mapsto -{}^tX$. (In the case of the Lie algebra of type A_1, the automorphism T is inner). Set $Z = Z(SL_{p^m}(\mathbb{C})) \simeq \mathbb{Z}_{p^m}$, $H = \hat{H}Z/Z \simeq \hat{H}/Z(\hat{H}) = \mathbb{Z}_p^{2m}$,

$$\mathcal{A}(J) = \{\varphi \in \mathrm{Aut}(\mathcal{L}) \mid \forall u \in W, \exists v \in W, \varphi(< J_u >_{\mathbb{C}}) = < J_v >_{\mathbb{C}}\},$$

$$\mathcal{G}(J) = \mathcal{A}(J) \cap \mathrm{Inn}(\mathcal{L}).$$

Lemma 1.3.1. *The following equalities hold:*
(i) $\mathcal{G}(J) = H.Sp_{2m}(p)$*;*
(ii) $\mathcal{A}(J) = \mathcal{G}(J). < T >$.

Proof. The embedding of H in $PSL_{p^m}(\mathbb{C})$ via matrices J_u, described in §1.1, occurs in the works of many authors (see, for example, [Sup 1]). In [Sup 1] it is shown that

$$\mathcal{G}(J) = N_{PSL_{p^m}(\mathbb{C})}(H) = H.Sp_{2m}(p).$$

The extension of H by $Sp_{2m}(p)$ is split when $p > 2$. More precisely, H consists of the maps

$$\hat{f} : J_u \mapsto \varepsilon^{f(u)} J_u, \tag{1.11}$$

where $f \in W^* = \mathrm{Hom}_{\mathbb{F}_p}(W, \mathbb{F}_p)$, and $Sp_{2m}(p)$ is identified with the set of transformations

$$\hat{\varphi} : J_u \mapsto \varepsilon^{(B(\varphi(u),\varphi(u))-B(u,u))/2} J_{\varphi(u)}, \tag{1.12}$$

where $\varphi \in Sp(W)$.

If $p = 2$, then the extension $H.Sp_{2m}(p)$ is non-split in general. This happens if $m > 2$ (see below). Furthermore, equality (ii) follows from the fact that the action of T takes the form

$$T(J_u) = -\varepsilon^{B(u,u)} J_{\bar{u}},$$

in the basis $\{J_u\}$, where

$$\bar{u} = (a_1, \ldots, a_m, -b_1, \ldots, -b_m)$$

if $u = (a_1, \ldots, a_m, b_1, \ldots, b_m)$. □

Clearly, every automorphism $\varphi \in \mathcal{A}(J)$ maps J-decompositions into J-decompositions. It is interesting to note that the converse assertion is also valid.

Lemma 1.3.2. *Every automorphism (inner automorphism, respectively) of the Lie algebra $\mathcal{L}$ that maps a J-decomposition to a J-decomposition, is contained in $\mathcal{A}(J)$ (in $\mathcal{G}(J)$, respectively).*

Proof. Suppose that $\varphi \in \mathrm{Aut}(\mathcal{L})$ maps some J-decomposition $\mathcal{D}^1 : \mathcal{L} = \oplus_{i=1}^n \mathcal{H}_i^1$ into a J-decomposition $\mathcal{D}^2 : \mathcal{L} = \oplus_{i=1}^n \mathcal{H}_i^2$.

1) Firstly, suppose that $\varphi \in \mathrm{Inn}(\mathcal{L})$. Using the transitivity of $Sp(W)$ on the set of maximal totally isotropic subspaces of W, we can assume that $\mathcal{H}_1^1$ is the standard Cartan subalgebra

$$\mathcal{H}_0 = \{\mathrm{diag}(\lambda_1, \ldots, \lambda_{p^m}) \mid \lambda_i \in \mathbb{C}, \sum_i \lambda_i = 0\}.$$

Suppose that $\varphi(\mathcal{H}_0) = \mathcal{H}_i^2$ for some i. As above, we can find $f \in \mathcal{G}(J)$ such that $f(\mathcal{H}_i^2) = \mathcal{H}_0$. Here, f maps $\mathcal{D}^2$ into a new J-decomposition $\mathcal{D}^3 : \mathcal{L} = \oplus_{i=1}^n \mathcal{H}_i^3$. Consider an arbitrary element $J_u \notin \mathcal{H}_0$. Then

$$f\varphi(J_u) = \sum_{J_v \in \mathcal{H}_j^3} \lambda_v J_v$$

for some $\mathcal{H}_j^3 \neq \mathcal{H}_0$. Our aim is to show that exactly one coefficient λ_v differs from zero. Since $f\varphi(\mathcal{H}_0) = \mathcal{H}_0$, we have $f\varphi(X) = FXF^{-1}$ for every $X \in \mathcal{L}$, where F is a certain monomial matrix. In particular, $f\varphi(J_u) = FJ_uF^{-1}$ is also a monomial matrix not lying in $\mathcal{H}_0$. On the other hand, all matrices $J_v \in \mathcal{H}_j^3$ are monomial, and for different v the nonzero coefficients of these matrices are disposed in different places. So only one coefficient λ_v can be nonzero.

Now for $J_u \in \mathcal{H}_0$, choose two matrices $J_{v_1}, J_{v_2} \notin \mathcal{H}_0$ such that $J_u = \lambda J_{v_1}.J_{v_2}$, $\lambda \in \mathbb{C}$. Then the matrix $f\varphi(J_u) = \lambda f\varphi(J_{v_1}).f\varphi(J_{v_2})$ is again a multiple of some matrix J_w. This means that $f\varphi \in \mathcal{G}(J)$, whereas by choice f is contained in $\mathcal{G}(J)$, so that $\varphi \in \mathcal{G}(J)$.

2) Suppose now that φ is not inner. Then $T\varphi \in \mathrm{Inn}(\mathcal{L})$, and $T\varphi$ maps $\mathcal{D}^1$ into the J-decomposition $T(\mathcal{D}^2)$. So, by 1), we have $T\varphi \in \mathcal{G}(J)$ and $\varphi \in \mathcal{A}(J)$. □

For what follows set

$$H^+ = \begin{cases} H. < T > & \text{if} \quad p = 2, \\ H & \text{if} \quad p > 2. \end{cases}$$

Proposition 1.3.3. *For the J-decomposition $\mathcal{D}$ corresponding to a spread π, the following formula holds:*

$$\mathrm{Aut}(\mathcal{D}) = H^+.\mathrm{Aut}(\pi).$$

Proof. By Lemma 1.3.2, the group $G = \mathrm{Aut}(\mathcal{D})$ is contained in $\mathcal{A}(J) = H^+.Sp^{\pm}(W)$. Furthermore, the subgroup H^+ consists precisely of those elements of G that fix every line $< J_u >_{\mathbb{C}}$, $u \in W$. Finally, the factor-group G/H^+ is obviously just $\mathrm{Aut}(\pi)$. □

The following assertions are immediate consequences of Proposition 1.3.3 and Lemma 1.3.2.

Corollary 1.3.4. *The standard orthogonal decomposition $\mathcal{D}$ of the Lie algebra $\mathcal{L}$ of type A_{p^m-1} constructed in §1.1 has automorphism group*

$$\mathrm{Aut}(\mathcal{D}) = \mathbb{Z}_p^{2m}.(SL_2(p^m).\mathbb{Z}_m).\mathbb{Z}_2.$$

The last factor $\mathbb{Z}_2$ is absent in the case $p^m = 2$. □

Corollary 1.3.5. *A J-decomposition $\mathcal{D}$ of the Lie algebra $\mathcal{L}$ of type A_{p^m-1} is an irreducible (transitive, respectively) OD if and only if the corresponding symplectic spread $\pi = \pi_{\mathcal{D}}$ of the space $W = \mathbb{F}_p^{2m}$ is irreducible (transitive, respectively). A classification of irreducible (transitive, respectively) J-decompositions of $\mathcal{L}$ up* *to* Aut($\mathcal{L}$)*-conjugacy is equivalent to a classification of irreducible (transitive, respectively) symplectic spreads of W up to* $Sp^{\pm}(W)$*-conjugacy.* □

Combining Corollary 1.3.5 with Theorem 1.2.7 we obtain:

Corollary 1.3.6. *Every irreducible J-decomposition of the Lie algebra $\mathcal{L}$ of type A_{p^m-1} is standard. The unique exception is the case $p^m = 27$, where there exists (up to* Aut($\mathcal{L}$)*-conjugacy) just one more irreducible (non-standard) J-decomposition $\mathcal{D}$ with automorphism group*

$$\mathrm{Aut}(\mathcal{D}) = \mathbb{Z}_3^6.SL_2(13). \qquad \square$$

Note (see Chapter 4) that Corollary 1.3.6 remains valid if we exchange the words "J-decomposition" in the statement by "orthogonal decomposition" !

From Corollary 1.3.4, it follows that, for a standard decomposition $\mathcal{D}$ of the Lie algebra $\mathcal{L}$ of type A_{p^m-1}, we have $\mathrm{Inn}(\mathcal{D}) = \mathbb{Z}_p^{2m}.\Sigma L_2(p^m)$, where $\mathrm{Inn}(\mathcal{D}) = \mathrm{Aut}(\mathcal{D}) \cap \mathrm{Inn}(\mathcal{L})$, and

$$\Sigma L_2(p^m) = SL_2(p^m).Gal(\mathbb{F}_{p^m}/\mathbb{F}_p)$$

consists of all special $\mathbb{F}_{p^m}$-semilinear transformations of the space $W = (\mathbb{F}_{p^m})^2$.

Theorem 1.3.7. *The extension* $\mathrm{Inn}(\mathcal{D}) = \mathbb{Z}_p^{2m}.\Sigma L_2(p^m)$ *is non-split if and only if* $p = 2$ *and* $m > 2$.

Proof. 1) The fact that $\mathrm{Inn}(\mathcal{D})$ is split over $\mathbb{Z}_p^{2m}$ for $p > 2$ follows from the explicit formulas (1.11) and (1.12). Henceforth, we shall suppose that $p = 2$ and $q = 2^m$. Then all J_u's are of order 4. But according to Lemma 1.3.2 every $\Psi \in \mathrm{Inn}(\mathcal{D})$ permutes the lines $< J_u >_{\mathbb{C}}$. Therefore $\mathrm{Inn}(\mathcal{D})$ consists of the elements

$$\Psi_{f,\varphi} : J_u \mapsto i^{f(u)} J_{\varphi(u)},$$

where $i = \sqrt{-1}$, f is a functional on W with values in $\mathbb{Z}_4 = \mathbb{Z}/4\mathbb{Z}$, and the map φ acts by the rule

$$\varphi(\alpha + \beta\vartheta) = a\alpha^{2^n} + b\beta^{2^n} + (c\alpha^{2^n} + d\beta^{2^n})\vartheta \tag{1.13}$$

with $0 \leq n \leq m-1$, $a, b, c, d \in \mathbb{F}_q$, $ad - bc = 1$, $\alpha, \beta \in \mathbb{F}_q$. Here (see also §2.1), we identify the space W with $\mathbb{F}_{q^2} = \mathbb{F}_q(\vartheta)$. Furthermore, for $u = \alpha + \beta\vartheta$ and $v = \gamma + \delta\vartheta$ we have

$$< u \mid v >= \mathrm{Tr}_{\mathbb{F}_q/\mathbb{F}_2}(\alpha\delta - \beta\gamma),\ B(u, v) = \mathrm{Tr}_{\mathbb{F}_q/\mathbb{F}_2}(\beta\gamma).$$

We agree upon the following convention: the complex number i^a with $a \in \mathbb{F}_2$ is equal to i if $a = 1$ and 1 if $a = 0$. Observe that

$$i^a i^b = i^{a+b}(-1)^{ab};\ a, b \in \mathbb{F}_2. \tag{1.14}$$

2) Consider the new basis $\{J'_u\}$ of the $\mathbb{C}$-space $\mathcal{L}$ obtained by setting $J'_u = i^{B(u,u)}J_u$. Then every automorphism $\Phi \in \mathrm{Inn}(\mathcal{D})$ has the form

$$\Phi = \Phi_{h,\varphi} : J'_u \mapsto (-1)^{h(u)} J'_{\varphi(u)},$$

where φ and h satisfy (1.13) and the following condition:

$$h(u+v) = h(u) + h(v) + C(u, v) + C(\varphi(u), \varphi(v)) \tag{1.15}$$

with $C(u, v) = B(u, u).B(v, v) + B(v, u) + <u|v> (B(u, u) + B(v, v))$. In fact, since Φ is an automorphism of the associative algebra $M_q(\mathbb{C})$, the identity $(J'_u)^2 = J_0$ obtained by using (1.14) implies that Φ can be written in the form $\Phi = \Phi_{h,\varphi}$: $J'_u \mapsto (-1)^{h(u)} J'_{\varphi(u)}$ for some h and φ. We clarify when the map $\Phi = \Phi_{h,\varphi}$ is an inner automorphism of $\mathcal{L}$. By (1.14), we have

$$J'_u.J'_v = i^{B(u,u)}J_u.i^{B(v,v)}J_v = (-1)^{B(u,u).B(v,v)+B(u,v)}.i^{B(u,u)+B(v,v)}J_{u+v} =$$

$$= (-1)^{B(u,u).B(v,v)+B(u,v)}.i^{B(u,u)+B(v,v)}(-1)^{B(u+v,u+v)}i^{B(u+v,u+v)}.J'_{u+v} =$$

$$= (-1)^{C(u,v)} i^{<u|v>} J'_{u+v}.$$

Similarly,

$$\Phi(J'_u).\Phi(J'_v) = (-1)^{h(u)+h(v)+C(\varphi(u),\varphi(v))} i^{<\varphi(u)|\varphi(v)>} J'_{\varphi(u)+\varphi(v)}.$$

The identity $\Phi(J'_u.J'_v) = \Phi(J'_u).\Phi(J'_v)$ obviously implies (1.15).

3) Fix the following notation for elements of $SL_2(q)$:

$$\varphi_t = \begin{pmatrix} t & 0 \\ 0 & t^{-1} \end{pmatrix}, \ \psi_b = \begin{pmatrix} 1 & 0 \\ b & 1 \end{pmatrix}, \ \omega = \begin{pmatrix} 0 & 1 \\ 1 & 0 \end{pmatrix}.$$

Furthermore, let $\rho : \alpha + \beta\vartheta \mapsto \alpha^2 + \beta^2\vartheta$ be the map defined by the Frobenius automorphism of $\mathbb{F}_q$. It is not difficult to see that the following equalities hold:

$$B(\varphi_t(u), \varphi_t(v)) = B(\rho(u), \rho(v)) = B(u, v), \ B(\omega(u), \omega(v)) = B(v, u). \tag{1.16}$$

4) Assume now that $m > 2$ and $\mathrm{Inn}(\mathcal{D})$ is split over $\mathbb{Z}_2^{2m}$: there exists a map $\varphi \mapsto \hat{\varphi}$ that realizes an embedding $SL_2(q) \hookrightarrow \mathrm{Inn}(\mathcal{D})$. We can write

$$\hat{\varphi}(J'_u) = (-1)^{h_\varphi(u)} J'_{\varphi(u)}$$

for $\varphi \in SL_2(q)$ and $h_\varphi : W \to \mathbb{F}_2$. Then the identity $\hat{\psi}(\hat{\varphi}(J'_u)) = \widehat{\psi\varphi}(J'_u)$ implies that

$$h_{\psi\varphi}(u) = h_\varphi(u) + h_\psi(\varphi(u)). \tag{1.17}$$

We can suppose that $h_\varphi(u) = 0$ for any $\varphi \in D$, where $D = \{\varphi_t \mid t \in \mathbb{F}_q^*\} \simeq \mathbb{Z}_{q-1}$ is the subgroup consisting of the diagonal elements of $SL_2(q)$. In fact, from

(1.15) and (1.16), it follows immediately that $h_\varphi \in W^* = \mathrm{Hom}_{\mathbb{F}_2}(W, \mathbb{F}_2)$ for every $\varphi \in D$. Furthermore, if the collection $\{h_\varphi \mid \varphi \in SL_2(q)\}$ realizes an embedding $SL_2(q) \hookrightarrow \mathrm{Inn}(\mathcal{D})$, and $l \in W^*$, then the collection $\{h'_\varphi = h_\varphi + l + \varphi \circ l\}$, where $(\varphi \circ l)(u) = l(\varphi(u))$, also embeds $SL_2(q)$ in $\mathrm{Inn}(\mathcal{D})$, because h'_φ satisfies (1.15) and (1.17). Setting $l = \sum_{\mu \in D} h_\mu$, for $\varphi \in D$ we have

$$h'_\varphi(u) = h_\varphi(u) + \sum_{\mu \in D} h_\mu(u) + \sum_{\mu \in D} h_\mu(\varphi(u)) =$$

$$= h_\varphi(u) + \sum_{\mu \in D} h_\mu(u) + \sum_{\mu \in D} (h_{\mu\varphi}(u) + h_\varphi(u)) = (|D| + 1)h_\varphi(u) = q.h_\varphi(u) = 0,$$

as required.

Furthermore, from (1.15) it follows that, for $b \in \mathbb{F}_q$, we have

$$\hat{\psi}_b(J'_{\beta\vartheta}) = (-1)^{g_b(\beta)} J'_{\beta\vartheta}$$

for some functional $g_b \in \mathrm{Hom}_{\mathbb{F}_2}(\mathbb{F}_q, \mathbb{F}_2)$. In fact, $g_b = 0$. To see this, note that, for $t \in \mathbb{F}_q^*$ we have

$$\hat{\psi}_{t^2}(J'_{\beta\vartheta}) = \hat{\varphi}_t^{-1} \hat{\psi}_1 \hat{\varphi}_t(J'_{\beta\vartheta}) = (-1)^{g_1(t^{-1}\beta)} J'_{\beta\vartheta},$$

so that $g_{t^2}(\beta) = g_1(t^{-1}\beta)$. In particular, $g_{s^2+t^2}(\beta) = g_1((s+t)^{-1}\beta)$ for all $s, t \in \mathbb{F}_q^*$, $s \neq t$. On the other hand, from (1.17) it follows that

$$g_{s^2+t^2}(\beta) = g_{s^2}(\beta) + g_{t^2}(\beta) = g_1(s^{-1}\beta) + g_1(t^{-1}\beta) = g_1((s^{-1} + t^{-1})\beta).$$

Hence, $((s+t)^{-1} + s^{-1} + t^{-1})\beta \in \mathrm{Ker} g_1$ for all $\beta \in \mathbb{F}_q$ and $s, t \in \mathbb{F}_q^*$, $s \neq t$. Since $m > 2$, there exist $s, t \in \mathbb{F}_q^*$ such that $s \neq t$ and $(s+t)^{-1} + s^{-1} + t^{-1} \neq 0$. But in this case $\mathrm{Ker} g_1 = \mathbb{F}_q$, $g_1 = 0$, and $g_b = 0$ for all b, as claimed.

Recall that we have identified $W = \mathbb{F}_q \oplus \mathbb{F}_q$ with $\mathbb{F}_{q^2} = \mathbb{F}_q(\vartheta)$. For the vector $v = 1 \in W$ choose $\psi = \psi_\alpha$ such that $\mathrm{Tr}_{\mathbb{F}_q/\mathbb{F}_2}\alpha = 1$. Suppose that $\hat{\psi}(J'_1) = \lambda J'_{1+\alpha\vartheta}$, $\lambda \in \mathbb{C}^*$. Then

$$J'_1 = \hat{\psi}^2(J'_1) = \lambda \hat{\psi}(J'_{1+\alpha\vartheta}) = \lambda \hat{\psi}(iJ'_1.J'_{\alpha\vartheta}) = i\lambda \hat{\psi}(J'_1).\hat{\psi}(J'_{\alpha\vartheta}) =$$

$$= i\lambda^2 J'_{1+\alpha\vartheta}.J'_{\alpha\vartheta} = -\lambda^2 J'_1.$$

Hence $\lambda^2 = -1$, on the one hand. On the other hand,

$$J_0 = \hat{\psi}(J_0) = \hat{\psi}(J'_1.J'_1) = \hat{\psi}(J'_1).\hat{\psi}(J'_1) = \lambda^2 (J'_{1+\alpha\vartheta})^2 = \lambda^2 J_0,$$

which implies that $\lambda^2 = 1$. This contradiction shows that the group $\mathrm{Inn}(\mathcal{D})$ is non-split over $\mathbb{Z}_2^{2m}$ if $m > 2$.

5) Finally, we define a section $\mathcal{S} : \Sigma L_2(q) \to \mathrm{Inn}(\mathcal{D})$, which is in fact an embedding $\Sigma L_2(q) \hookrightarrow \mathrm{Inn}(\mathcal{D})$ if $m \leq 2$. By the Bruhat decomposition, it is

enough to define the values φ^- of $\mathcal{S}$ on elements $\varphi \in \{\varphi_t, \psi_b, \omega, \rho\}$. To this end, we set

$$\varphi_t^-(J'_u) = J'_{\varphi_t(u)}, \; \omega^-(J'_u) = (-1)^{B(u,u)} J'_{\omega(u)}, \; \rho^-(J'_u) = J_{\rho(u)},$$

$$\psi_b^-(J'_u) = (-1)^{Q(\alpha\sqrt{b}) + \mathrm{Tr}\alpha\beta.\mathrm{Tr}\alpha\sqrt{b} + \mathrm{Tr}(b\beta+\alpha\sqrt{b})} J'_{\psi_b(u)},$$

where $u = \alpha + \beta\vartheta$ and the quadratic form $Q : \mathbb{F}_q \to \mathbb{F}_2$ satisfies the condition $Q(\alpha+\beta) = Q(\alpha) + Q(\beta) + \mathrm{Tr}\alpha\beta + \mathrm{Tr}\alpha.\mathrm{Tr}\beta$. (Here, the suffix $\mathbb{F}_q/\mathbb{F}_2$ is omitted from the notation of the trace $\mathrm{Tr}_{\mathbb{F}_q/\mathbb{F}_2}$). Relations (1.15) and (1.16) show that the maps φ_t^-, ω^-, ρ^- and ψ_b^- so constructed really are contained in $\mathrm{Inn}(\mathcal{D})$. □

Corollary 1.3.8. *The group $\mathcal{G}(J) = \mathbb{Z}_p^{2m}.Sp_{2m}(p)$ is non-split over $\mathbb{Z}_p^{2m}$ if $p = 2$ and $m > 2$.* □

To conclude this section, we mention the following fact. Let $W = \mathbb{F}_q^n$ be the natural $\mathbb{F}_q$-module for $SL_n(q)$. Then the second cohomology group $H^2(SL_n(q), W)$ is trivial except in the cases $(n, q) = (2, 2^m)$, $m > 2$; $(3, 2)$; $(5, 2)$; $(3, 3^m)$, $m > 1$; $(3, 5)$ and $(4, 2)$, when it is $\mathbb{F}_q$ (see [Sah], [Avr]). In the first three exceptional cases a (unique) non-split extension of W by $SL_n(q)$ is realized as the automorphism group of some OD of the Lie algebra of type A_{2^m-1} (as we just convinced ourselves), G_2 and E_8 (see Chapter 2), respectively.

1.4 The uniqueness problem for ODs of Lie algebras of type A_n, $n \leq 4$

Examples of ODs for Lie algebras of type A_n, $n \leq 4$, have been constructed in §1.1. In fact the following result holds.

Theorem 1.4.1. *A Lie algebra $\mathcal{L}$ of any of the types A_1, A_2, A_3 and A_4 possesses an orthogonal decomposition, that is unique (up to conjugacy).*

Sketch of the proof. In principle, the problem of classifying ODs can be solved using the following scheme. As all Cartan subalgebras are conjugate, and the orthogonality property is preserved by conjugation, one can assume that one component of the OD is the subalgebra $\mathcal{H}_0$ consisting of the diagonal matrices. Thus, the problem is reduced to extending $\mathcal{H}_0$ to an orthogonal decomposition and showing that any two such decompositions are conjugate under the stabilizer of $\mathcal{H}_0$ in $\mathrm{Aut}(\mathcal{L})$. To this end, we first investigate the set $C = \{\mathcal{H}\}$ of all Cartan subalgebras orthogonal to $\mathcal{H}_0$. A parametrization of $\mathcal{C}$ which is minimal in some sense is an essential part of this proof. Secondly, a decisive step consists of analysing the relations between the parameters that follow from the pairwise orthogonality condition for certain members of $\mathcal{C}$.

We make some general remarks concerning a general $\mathcal{H} \in \mathcal{C}$.

1) Every diagonal matrix is orthogonal to $\mathcal{H}$.

2) Every matrix $A \in \mathcal{H}$ has zero principal diagonal.

3) For any matrix $A \in \mathcal{H}$ and any polynomial $f(t)$, the matrix $f(A)$ has equal coefficients on the principal diagonal.

4) In $\mathcal{H}$ there exists an invertible matrix X such that $X^{n+1} = E_{n+1}$ (the unit matrix of degree $n+1$) and $\mathcal{H} = < X, X^2, \dots, X^n >_{\mathbb{C}}$.

5) A nonzero matrix from $\mathcal{H}$ cannot have a zero row nor a zero column.

6) The subalgebra $\mathcal{H}$ admits a basis $\{A^{(k)}, 2 \le k \le n+1\}$ satisfying the following conditions:

(i) the first row of $A^{(k)}$ contains exactly one nonzero element; it is equal to 1 and stands in the k-th position;

(ii) the k-th column of $A^{(k)}$ contains exactly one nonzero element; by (i), it is equal to 1 and stands in the first position;

(iii) the j-th row of $A^{(k)}$ coincides with the k-th row of $A^{(j)}$.

Only assertion 6) requires proof. From 5) it follows that there exists a basis $\{A^{(k)}, 2 \le k \le n+1\}$ with property (i). To prove (iii), we use the commutativity of $\mathcal{H}$: by (i), the j-th row of $A^{(k)}$ is equal to the first row of the product $A^{(j)}A^{(k)}$, whereas $A^{(j)}A^{(k)} = A^{(k)}A^{(j)}$, so this first row coincides with the k-th row of $A^{(j)}$. To prove (ii), it remains to note that for $j \ge 2$ the j-th element of the k-th column of $A^{(k)}$ is the k-th element of the j-th row of $A^{(k)}$, and by (iii) the latter is equal to the k-th element of the k-th row of $A^{(j)}$, and therefore it is zero by 2).

Applying these elementary remarks to the case $n \le 3$, we obtain the following parametrization.

Lemma 1.4.2. *Every subalgebra $\mathcal{H} \in \mathcal{C}$ has a basis of the form indicated below.*

(i) *If $n = 1$, then*

$$\mathcal{H} = \left\langle A^{(2)} = \begin{pmatrix} 0 & 1 \\ a & 0 \end{pmatrix} \right\rangle_{\mathbb{C}}$$

for some $a \neq 0$.

(ii) *If $n = 2$, then*

$$\mathcal{H} = \left\langle A^{(2)} = \begin{pmatrix} 0 & 1 & 0 \\ 0 & 0 & a \\ ab & 0 & 0 \end{pmatrix}, A^{(3)} = \begin{pmatrix} 0 & 0 & 1 \\ ab & 0 & 0 \\ 0 & b & 0 \end{pmatrix} \right\rangle_{\mathbb{C}}$$

for some a, b, $ab \neq 0$.

(iii) *If* $n = 3$, *then* $\mathcal{H} =< A^{(2)}, A^{(3)}, A^{(4)} >_{\mathbb{C}}$, *where*

$$A^{(2)} = \begin{pmatrix} 0 & 1 & 0 & 0 \\ ab & 0 & c_3 & bc_2/d \\ bc_2 & 0 & 0 & a \\ c_3d & 0 & b & 0 \end{pmatrix}, \quad A^{(3)} = \begin{pmatrix} 0 & 0 & 1 & 0 \\ bc_2 & 0 & 0 & a \\ ad & c_1 & 0 & c_2 \\ bc_1 & d & 0 & 0 \end{pmatrix},$$

$$A^{(4)} = \begin{pmatrix} 0 & 0 & 0 & 1 \\ c_3d & 0 & b & 0 \\ bc_1 & d & 0 & 0 \\ bd & bc_1/a & c_3d/a & 0 \end{pmatrix};$$

here $abd \neq 0$ *and at least two of* c_1, c_2, c_3 *are equal to* 0.

Because of the cumbersome nature of the calculations, we omit the proof of Lemma 1.4.2 (and also of some following lemmas), referring the reader for details to [KKU 4]. □

If $n = 4$, the following statements hold for $\mathcal{H} \in \mathcal{C}$.

Lemma 1.4.3. *Suppose that* $\mathcal{H}$ *contains at least one monomial matrix. Then* $\mathcal{H}$ *admits a basis* $\{A^{(i)}, 2 \leq i \leq 5\}$, *where* $A = A^{(2)}$ *is monomial,* $A^5 = \lambda E_5$ *for some* $\lambda \neq 0$, *and all the matrices* $A^{(i)}$ *are proportional to powers of* A. □

Lemma 1.4.4. *Assume that* $\mathcal{H} = \mathcal{H}_2$ *does not contain monomial matrices. Then, up to simultaneous conjugation by a diagonal matrix and multiplication by a constant, its basis matrices* $A^{(i)}$ *have the following form:*

$$A^{(2)} = \begin{pmatrix} 0 & 1 & 0 & 0 & 0 \\ 2 & 0 & 1 & 1 & 1 \\ 1 & 0 & 0 & 1 & 1 \\ 1 & 0 & 1 & 0 & 1 \\ 1 & 0 & 1 & 1 & 0 \end{pmatrix}, \quad A^{(3)} = \begin{pmatrix} 0 & 0 & 1 & 0 & 0 \\ 1 & 0 & 0 & 1 & 1 \\ 2\lambda^2 & \lambda^3 & 0 & \lambda^2 & \lambda^2 \\ \lambda & \lambda^2 & 0 & 0 & \lambda \\ \lambda & \lambda^2 & 0 & \lambda & 0 \end{pmatrix},$$

$$A^{(4)} = \begin{pmatrix} 0 & 0 & 0 & 1 & 0 \\ 1 & 0 & 1 & 0 & 1 \\ \lambda & \lambda^2 & 0 & 0 & \lambda \\ 2\lambda^2 & \lambda^3 & \lambda^2 & 0 & \lambda^2 \\ \lambda & \lambda^2 & \lambda & 0 & 0 \end{pmatrix}, \quad A^{(5)} = \begin{pmatrix} 0 & 0 & 0 & 0 & 1 \\ 1 & 0 & 1 & 1 & 0 \\ \lambda & \lambda^2 & 0 & \lambda & 0 \\ \lambda & \lambda^2 & \lambda & 0 & 0 \\ 2\lambda^2 & \lambda^3 & \lambda^2 & \lambda^2 & 0 \end{pmatrix},$$

where $\lambda^2 + \lambda - 1 = 0$. *Every Cartan subalgebra* $\mathcal{H}' \in \mathcal{C}$ *admitting no basis consisting of monomial matrices (briefly: non-monomial Cartan subalgebra) may be obtained from* $\mathcal{H}_2$ *via conjugation by a diagonal matrix.* □

Having parametrized basis matrices for $\mathcal{H} \in \mathcal{C}$, we can begin the proof of Theorem 1.4.1. For example, suppose that $n = 2$ and let $\mathcal{L} = \oplus_{i=0}^{3}\mathcal{H}_i$ be an orthogonal decomposition of the Lie algebra $\mathcal{L}$ of type A_2. According to Lemma 1.4.2, one can suppose that

$$\mathcal{H}_1 = \left\langle \begin{pmatrix} 0 & 1 & 0 \\ 0 & 0 & 1 \\ 1 & 0 & 0 \end{pmatrix}, \begin{pmatrix} 0 & 0 & 1 \\ 1 & 0 & 0 \\ 0 & 1 & 0 \end{pmatrix} \right\rangle_{\mathbb{C}},$$

$$\mathcal{H}_2 = \left\langle \begin{pmatrix} 0 & 1 & 0 \\ 0 & 0 & a \\ ab & 0 & 0 \end{pmatrix}, \begin{pmatrix} 0 & 0 & 1 \\ ab & 0 & 0 \\ 0 & b & 0 \end{pmatrix} \right\rangle_{\mathbb{C}},$$

$$\mathcal{H}_3 = \left\langle \begin{pmatrix} 0 & 1 & 0 \\ 0 & 0 & c \\ cd & 0 & 0 \end{pmatrix}, \begin{pmatrix} 0 & 0 & 1 \\ cd & 0 & 0 \\ 0 & d & 0 \end{pmatrix} \right\rangle_{\mathbb{C}}.$$

Because of the orthogonality condition $K(\mathcal{H}_1, \mathcal{H}_2) = 0$, we have $a = b$ and $1 + a + a^2 = 0$. If ε is a primitive cube root of 1, either $a = b = \varepsilon$ or $a = b = \varepsilon^2$. Similarly, $c = d \in \{\varepsilon, \varepsilon^2\}$. But $K(\mathcal{H}_2, \mathcal{H}_3) = 0$, so we can suppose that $a = \varepsilon$, $c = \varepsilon^2$. Thus the decomposition is uniquely defined.

The same arguments are applicable to the cases $n = 1$ and $n = 3$. Now suppose that $n = 4$. Firstly, by using Lemma 1.4.4, one shows (and the main difficulty of this case lies here!) that every component of the decomposition $\mathcal{L} = \oplus_{i=0}^{5}\mathcal{H}_i$ is a monomial subalgebra. The monomial matrix $A^{(i)}$ (see remark 6)) is the product of some diagonal matrix and the permutation-matrix $\pi(A^{(i)})$. If $\mathcal{H}$ is a monomial Cartan subalgebra orthogonal to $\mathcal{H}_0$, then it is easy to check that $\pi(\mathcal{H}) = \{E_{n+1}, \pi(A^{(i)}) \mid 2 \leq i \leq n+1\}$ is an abelian subgroup of order $n+1$ of the symmetric group $\mathbb{S}_{n+1}$. In our case $n = 4$, the group $\pi(\mathcal{H})$ is the cyclic group $\mathbb{Z}_5$. For arbitrary n, the following statement holds:

Lemma 1.4.5. *Suppose that $\mathcal{H}$ is a monomial Cartan subalgebra of the Lie algebra of type A_n that is orthogonal to the standard Cartan subalgebra $\mathcal{H}_0$. Then $\pi(\mathcal{H})$ is a transitive regular abelian permutation group of degree $n+1$.*

Proof. According to remark 3), for $\pi \in \pi(\mathcal{H})$ every permutation π^i is trivial or else has no fixed point. Thus, π is a product of disjoint cycles of the same length $d = d(\pi)$. In particular, if π fixes some point, then $d = 1$ and π is trivial. But $|\pi(\mathcal{H})| = n+1$, so $\pi(\mathcal{H})$ is a regular transitive permutation group (of degree $n+1$). □

Lemma 1.4.6. *Suppose that $\mathcal{H}$ and $\mathcal{H}'$ are monomial Cartan subalgebras of the Lie algebra of type A_n that are orthogonal to the standard Cartan subalgebra $\mathcal{H}_0$. If the set*

$$\pi(\mathcal{H}).\pi(\mathcal{H}') = \{\sigma.\sigma' \mid \sigma \in \pi(\mathcal{H}), \sigma' \in \pi(\mathcal{H}')\} \subset \mathbb{S}_{n+1}$$

contains some permutation fixing exactly one symbol, then the subalgebras $\mathcal{H}$ *and* $\mathcal{H}'$ *are not orthogonal.*

Proof. As we mentioned in the proof of Lemma 1.4.5, neither $\pi(\mathcal{H})$ nor $\pi(\mathcal{H}')$ contains permutations with exactly one fixed point. Therefore, if $\sigma.\sigma'$ fixes precisely one symbol, where $\sigma \in \pi(\mathcal{H})$ and $\sigma' \in \pi(\mathcal{H}')$, then $\sigma \neq 1$ and $\sigma' \neq 1$. So $\sigma = \pi(A)$ and $\sigma' = \pi(A')$ for some $A \in \mathcal{H}$, $A' \in \mathcal{H}'$. But in that case the principal diagonal of AA' has exactly one nonzero coefficient and $K(A, A') \neq 0$. □

Corollary 1.4.7. *If* $\mathcal{H}$ *and* $\mathcal{H}'$ *are two monomial Cartan subalgebras of the Lie algebra of type* A_4 *that are orthogonal to* $\mathcal{H}_0$ *and to each other, then* $\pi(\mathcal{H}) = \pi(\mathcal{H}')$.

Proof. By Lemma 1.4.5, one can suppose that $\pi(\mathcal{H}) =< (1, 2, 3, 4, 5) >$ and $\pi(\mathcal{H}') =< (1, 2, x, y, z) >$, where $\{x, y, z\} = \{3, 4, 5\}$. The required statement follows from the following equalities:

$$(1, 2, 4, 3, 5)(1, 2, 3, 4, 5) = (1, 4)(2, 5)(3);$$

$$(1, 2, 3, 5, 4)(1, 2, 3, 4, 5) = (1, 3)(2, 5)(4);$$

$$(1, 2, 5, 4, 3)(1, 2, 3, 4, 5)^4 = (1, 4)(3, 5)(2);$$

$$(1, 2, 4, 5, 3)(1, 2, 3, 4, 5)^3 = (1, 5)(2, 3)(4);$$

$$(1, 2, 5, 3, 4)(1, 2, 3, 4, 5)^3 = (2, 3)(4, 5)(1).$$ □

As we have just shown, all subalgebras $\mathcal{H}_i$ with $i \neq 0$ are monomial and have the same group $\pi(\mathcal{H}_i)$. Permuting rows and columns if necessary we can suppose that $\pi(\mathcal{H}_i) =< (1, 2, 3, 4, 5) >$, that is, $\mathcal{H}_i =< A, A^2, A^3, A^4 >_{\mathbb{C}}$ for some $A = \mathrm{diag}(1, a, b, c, d).P$, where P is the permutation matrix $(1, 2, 3, 4, 5)$. For brevity, we denote such a Cartan subalgebra by $\mathcal{H}(a, b, c, d)$. Direct calculations show that the subalgebras $\mathcal{H}(a, b, c, d)$ and $\mathcal{H}(a', b', c', d')$ are orthogonal if and only if one of the following conditions holds:

(i) $a'/a = d'/d = 1$, $b'/b = c/c' = \alpha$;

(ii) $a'/a = \varepsilon^k$, $b'/b = \varepsilon^{2k}$, $c'/c = \varepsilon^{3k}$, $d'/d = \varepsilon^{4k}$;

(iii) $a'/a = \alpha.d'/d$,

where $\alpha + \alpha^{-1} + 3 = 0$, $1 \leq k \leq 4$ and ε is a primitive 5-th root of 1.

After establishing this fact, it is not difficult to show that (up to conjugacy) the Lie algebra of type A_4 has a unique orthogonal decomposition

$$\mathcal{L} = \oplus_{k=0}^{4} \mathcal{H}(\varepsilon^k, \varepsilon^{2k}, \varepsilon^{3k}, \varepsilon^{4k}) \oplus \mathcal{H}_0.$$

This completes the proof of Theorem 1.4.1. □

After proving Theorem 1.4.1, it is natural to treat the case of the Lie algebra $\mathcal{L}$ of type A_5. Unfortunately, all attempts to prove or to disprove Conjecture 0.1 for $\mathcal{L}$ undertaken so far have met with no success. In view of Theorem 1.4.1, an essential part of the argument confirming the validity of this conjecture for type A_5 is contained in the following statement.

Proposition 1.4.8. *The Lie algebra $\mathcal{L}$ of type A_5 has no monomial OD.*

Sketch of the proof. It would be boring to go into details, which often require the direct checking of a great number of special cases. Instead, we highlight the most important ideas in the proof, which are to a considerable extent typical for Lie algebras $\mathcal{L}$ of type A_{n-1} for arbitrary n.

1. Let $\mathcal{L} = \oplus_{i=0}^{n}\mathcal{H}_i$ is a monomial OD, that is, $\mathcal{H}_0$ consists of diagonal matrices and $\mathcal{H}_i$ possesses a monomial basis $\{M_{i,1}, \dots, M_{i,n-1}\}$, $1 \le i \le n$. According to Lemma 1.4.5, when $n = 6$ the group $\pi(\mathcal{H}_i)$ is a cyclic regular transitive permutation group of degree 6. Using Lemma 1.4.6 and checking special cases directly, one can show that $K(\mathcal{H}_i, \mathcal{H}_j) \neq 0$ if the groups $\pi(\mathcal{H}_i)$ and $\pi(\mathcal{H}_j)$ are different.

2. If all groups $\pi(\mathcal{H}_i)$ are the same and are cyclic, then the orthogonality condition $K(\mathcal{H}_i, \mathcal{H}_j) = 0$, $i \neq j$, leads to the following system of equations:

$$\sum_{s=0}^{n-1} x_s x_{1+s} \cdots x_{k+s} = 0,\ 0 \le k \le n-2 \qquad (*)$$

(indices i of variables x_i are taken modulo n: $i = 0, 1, \dots, n-1$).

Without loss of generality, one can suppose that

$$M_{1,1} = \mathrm{diag}(a_0, a_1, \dots, a_{n-1})P,\ M_{2,1} = \mathrm{diag}(b_0, b_1, \dots, b_{n-1})P,$$

$$M_{3,1} = \mathrm{diag}(c_0, c_1, \dots, c_{n-1})P;$$

where P is the permutation-matrix $(0, 1, 2, \dots, n-1)$ and $a_i b_i c_i \neq 0$. The conditions

$$K(\mathcal{H}_1, \mathcal{H}_2) = K(\mathcal{H}_2, \mathcal{H}_3) = K(\mathcal{H}_3, \mathcal{H}_1) = 0$$

lead to three systems of type $(*)$ with variables $x_i = a_i/b_i$, $y_i = b_i/c_i$, $z_i = c_i/a_i$, $0 \le i \le n-1$. In particular, $x_i y_i z_i = 1$. Hence, to establish the existence of three pairwise orthogonal subalgebras $\mathcal{H}_i$, $1 \le i \le 3$, it is necessary to find three solutions $(x_0, \dots, x_{n-1})$, $(x'_0, \dots, x'_{n-1})$ and $(x''_0, \dots, x''_{n-1})$ of the system $(*)$ such that $x_i x'_i x''_i = 1$, $0 \le i \le n-1$.

3. If $n = 6$, it turns out that $x_4 = \alpha x_1$, $x_5 = \beta x_2$ and $x_6 = \gamma x_3$, where (α, β, γ) is one of the following 19 triples that are obtained by permuting the components

in triples:

$$(-1,-1,-1),\ (1,-2+\sqrt{3},-2+\sqrt{3}),\ (1,-2-\sqrt{3},-2-\sqrt{3}),$$

$$(-1,-1-\sqrt{3}-\sqrt{3+\sqrt{12}},-1-\sqrt{3}+\sqrt{3+\sqrt{12}}),$$

$$(-1,-1+\sqrt{3}-i\sqrt{3+\sqrt{12}},-1+\sqrt{3}+i\sqrt{3+\sqrt{12}}).$$

Similarly, $x'_4 = \alpha' x'_1$, $x'_5 = \beta' x'_2$ and $x'_6 = \gamma' x'_3$; $x''_4 = \alpha'' x''_1$, $x''_5 = \beta'' x''_2$ and $x''_6 = \gamma'' x''_3$; where $(\alpha', \beta', \gamma')$ and $(\alpha'', \beta'', \gamma'')$ belong to the same set of triples. But it follows from the condition $x_i x'_i x''_i = 1$ that $\alpha\alpha'\alpha'' = 1$, $\beta\beta'\beta'' = 1$ and $\gamma\gamma'\gamma'' = 1$. However, no three triples from the 19 triples described above satisfy this condition. Hence we cannot even find three pairwise orthogonal Cartan subalgebras $\mathcal{H}_1$, $\mathcal{H}_2$, $\mathcal{H}_3$ (each of which is, by definition, orthogonal to the standard $\mathcal{H}_0$). □

1.5 The uniqueness problem for orthogonal pairs of subalgebras

Up to now, we have dealt only with sets consisting of $n+1$ Cartan subalgebras of the Lie algebra $\mathcal{L}$ of type A_{n-1} that are pairwise orthogonal. Meanwhile, orthogonal pairs of Cartan subalgebras are also worthy of attention in connection with, firstly, the uniqueness problem for OD and, secondly, the requirements of algebraic topology and geometry.

The notion of a *commutative square* of von Neumann algebras introduced by Popa plays an essential role in recent works [GHJ], [Pop 1, 2, 3, 4], [Sun]. We recall the definition in a special case. For given $n \geq 2$, consider the von Neumann algebra $\mathcal{N} = M_n(\mathbb{C})$ of complex $n \times n$ matrices with normalized trace $\tau(X) = \frac{\mathrm{Tr}X}{n}$.

Definition 1.5.1. Let $\mathcal{A}$ and $\mathcal{B}$ be $*$-subalgebras of $\mathcal{N}$ (that is, subalgebras of the associative algebra $\mathcal{N}$ that contain the unit $1_{\mathcal{N}}$ of $\mathcal{N}$ and are closed under the involutory antiisomorphism $* : X \mapsto^t \bar{X}$). We say that the diagram

$$\begin{array}{ccc} \mathcal{B} & \subset & \mathcal{N} \\ \cup & & \cup \\ \mathcal{A}\cap\mathcal{B} & \subset & \mathcal{A} \end{array}$$

is a *commutative square* if $\tau(XY) = 0$ for all $X \in \mathcal{A}$, $Y \in \mathcal{B}$ such that $\tau(XZ) = \tau(YZ) = 0$ whenever $Z \in \mathcal{A}\cap\mathcal{B}$. In particular, if $\tau(XY) = \tau(X)\tau(Y)$ for all $X \in \mathcal{A}$, $Y \in \mathcal{B}$, then $\mathcal{A}\cap\mathcal{B} = <1_{\mathcal{N}}>_{\mathbb{C}}$ and the subalgebras $\mathcal{A}$, $\mathcal{B}$ form a commutative square which is an *orthogonal pair* in the sense of [Pop 1]. Correspondingly, orthogonal pairs $(\mathcal{A}, \mathcal{B})$ and $(\mathcal{A}', \mathcal{B}')$ are said to be *conjugate*, if there exists an automorphism of the von Neumann algebra $\mathcal{N}$ mapping $\mathcal{A}$ into $\mathcal{A}'$ and $\mathcal{B}$ into $\mathcal{B}'$.

In the first place we are interested in orthogonal pairs $(\mathcal{A}, \mathcal{B})$, where $\mathcal{A}$, $\mathcal{B}$ are maximal abelian $*$-subalgebras of $\mathcal{N}$. It is easy to give an example of such a pair for any n. Let ε be a primitive nth root of 1, set $D = \mathrm{diag}(1, \varepsilon, \varepsilon^2, \dots, \varepsilon^{n-1})$, and let P be the permutation matrix $(1, 2, \dots, n)$. Then the pair $(\mathcal{A}_0, \mathcal{A}_1)$, where $\mathcal{A}_0 = < E_n, D, \dots, D^{n-1} >_{\mathbb{C}}$, $\mathcal{A}_1 = < E_n, P, \dots, P^{n-1} >_{\mathbb{C}}$ and $E_n = 1_{\mathcal{N}}$ is the unit matrix of degree n, is orthogonal. The pair $(\mathcal{A}_0, \mathcal{A}_1)$ will be called *standard.*

In [Pop 1] the following conjecture is formulated:

Conjecture 1.5.2. *Any orthogonal pair of maximal abelian $*$-subalgebras of $\mathcal{N}$ is conjugate to the standard pair.*

We shall consider Conjecture 1.5.2 together with the following:

Conjecture 1.5.3. *Any orthogonal pair of Cartan subalgebras of the Lie algebra $\mathcal{L}$ of type A_{n-1} is conjugate to the standard pair $(\mathcal{H}_0, \mathcal{H}_1)$.*

Here $\mathcal{H}_i = \mathcal{A}_i \cap \mathcal{L}$, $i = 0, 1$ and $\mathcal{L}$ is realized as the Lie algebra of all matrices in $\mathcal{N}$ with zero trace. The interconnection between Conjectures 1.5.2 and 1.5.3 becomes clear after the following result is established.

Lemma 1.5.4. 1) *A $*$-subalgebra $\mathcal{A}$ of $\mathcal{N}$ is maximal abelian if and only if $\mathcal{A} = X\mathcal{A}_0X^{-1}$ for some unitary matrix X.*

2) *Let $(\mathcal{A}, \mathcal{B})$ be an orthogonal pair of maximal abelian $*$-subalgebras of $\mathcal{N}$. Then the pair $(\mathcal{A} \cap \mathcal{L}, \mathcal{B} \cap \mathcal{L})$ is an orthogonal pair of Cartan subalgebras of $\mathcal{L}$.*

Proof. 1) Suppose that $\mathcal{A}$ is an abelian $*$-subalgebra of $\mathcal{N}$, and take $A \in \mathcal{A}$. Then the linear operator A is normal in the sense that $AA^* = A^*A$, since $A^* \in \mathcal{A}$ and $\mathcal{A}$ is abelian. Thus $\mathcal{A}$ consists of normal operators commuting with each other. Hence, $\mathcal{A}$ can be reduced to diagonal form, that is, $\mathcal{A} \subseteq X\mathcal{A}_0X^{-1}$ for some unitary matrix X. In particular, $\mathcal{A}$ is maximal if and only if $\mathcal{A} = X\mathcal{A}_0X^{-1}$.

2) From 1) it follows that $\mathcal{A} \cap \mathcal{L} = X\mathcal{A}_0X^{-1} \cap \mathcal{L} = X(\mathcal{A}_0 \cap \mathcal{L})X^{-1} = X\mathcal{H}_0X^{-1}$. Thus $\mathcal{A} \cap \mathcal{L}$ (and $\mathcal{B} \cap \mathcal{L}$) are Cartan subalgebras of $\mathcal{L}$. Furthermore, for $A \in \mathcal{A} \cap \mathcal{L}$, $B \in \mathcal{B} \cap \mathcal{L}$ we have $K(A, B) = 2n^2\tau(AB) = 2n^2\tau(A)\tau(B) = 0$. □

Having investigated the Lie algebras of type A_n, $n \leq 4$ we see that Conjectures 1.5.2 and 1.5.3 cannot both hold for large n. But following Popa, who first raised these questions, we have formulated the conjectures in the positive form.

Using the transitive action of $\mathrm{Aut}(\mathcal{N})$ ($\mathrm{Aut}(\mathcal{L})$, respectively) on the set of maximal abelian $*$-subalgebras (the Cartan subalgebras, respectively), we may suppose that the given orthogonal pair contains the standard maximal abelian $*$-subalgebra $\mathcal{A}_0$ (the standard Cartan subalgebra $\mathcal{H}_0$, respectively). The simplest tool for studying orthogonal pairs with fixed first component $\mathcal{A}_0$ ($\mathcal{H}_0$, respectively) are the so-called *GH-matrices*, that is, matrices X such that $X^{-1} = \hat{X}$, where $X = (x_{ij})$, $\hat{X} = (\hat{x}_{ij})$, $\hat{x}_{ij} = 1/(n.x_{ji})$. The following assertion holds:

Lemma 1.5.5. 1) *A pair* $(\mathcal{H}_0, \mathcal{H} = X\mathcal{H}_0X^{-1})$ *is orthogonal if and only if* X *is a* GH*-matrix.*

2) *A pair* $(\mathcal{A}_0, \mathcal{A} = X\mathcal{A}_0X^{-1})$*, where* X *is a unitary matrix, is orthogonal if and only if* X *is a* GH*-matrix.*

Proof. 1) The orthogonality of the subalgebras $\mathcal{H}_0$ and $\mathcal{H}$ is equivalent to the requirement that the principal diagonal of every matrix $B = X(E_{jj} - E_{kk})X^{-1}$, where $1 \le j, k \le n$ and

$$E_{jj} = \mathrm{diag}(0, \ldots, 0, \underbrace{1}_{j}, 0, \ldots, 0)$$

is zero; that is, $x_{ij}.\hat{x}_{ji} - x_{ik}.\hat{x}_{ki} = 0$ for all i, j, k. Here, we set $X^{-1} = (\hat{x}_{ij})$. Summation over k gives us $nx_{ij}.\hat{x}_{ji} = \sum_{k=1}^{n} x_{ik}.\hat{x}_{ki} = 1$, that is $\hat{x}_{ji} = 1/(n.x_{ij})$.

2) It is sufficient to note that the orthogonality of the pair $(\mathcal{A}_0, \mathcal{A})$ is equivalent to that of the pair $(\mathcal{H}_0, \mathcal{H} = \mathcal{A} \cap \mathcal{L})$. □

Clearly, a unitary matrix X is a GH-matrix if and only if $|x_{ij}| = 1/\sqrt{n}$. Such matrices will be called *UGH-matrices*. The standard pairs $(\mathcal{A}_0, \mathcal{A}_1)$ and $(\mathcal{H}_0, \mathcal{H}_1)$ correspond to the so-called *standard UGH-matrix*

$$X_n = \frac{\varepsilon^{(i-1)(j-1)}}{\sqrt{n}},$$

where $\varepsilon = \exp(2\pi i/n)$.

Roughly speaking, the task of classifying orthogonal pairs up to conjugacy in $\mathcal{L}$ (in $\mathcal{N}$, respectively) is equivalent to that of enumerating double cosets $PXP = \{aXb \mid a, b \in P\}$ of GH-matrices (UGH-matrices, respectively), where $P = St_{GL_n(\mathbb{C})}(\mathcal{H}_0)$ is the group of monomial matrices ($P = St_{U_n(\mathbb{C})}(\mathcal{A}_0)$ is the group of monomial unitary matrices, respectively). In fact, the pairs $(\mathcal{A}_0, (aXb)\mathcal{A}_0(aXb)^{-1})$ and $(\mathcal{A}_0, X\mathcal{A}_0X^{-1})$ are conjugate:

$$(\mathcal{A}_0, (aXb)\mathcal{A}_0(aXb)^{-1}) = (\mathcal{A}_0, aXb\mathcal{A}_0b^{-1}X^{-1}a^{-1}) =$$

$$= (a\mathcal{A}_0a^{-1}, aX\mathcal{A}_0X^{-1}a^{-1}) = a(\mathcal{A}_0, X\mathcal{A}_0X^{-1})a^{-1}.$$

The first (and also the last, as will be seen!) argument confirming the validity of Conjectures 1.5.2 and 1.5.3 is contained in the following statement:

Lemma 1.5.6. *Conjectures* 1.5.2 *and* 1.5.3 *hold if* $n \le 3$.

Proof. It is sufficient to show that any orthogonal pair $(\mathcal{A}_0, \mathcal{A} = X\mathcal{A}_0X^{-1})$ ($(\mathcal{H}_0, \mathcal{H} = X\mathcal{H}_0X^{-1})$, respectively) is conjugate to the standard pair $(\mathcal{A}_0, \mathcal{A}_1)$ (to $(\mathcal{H}_0, \mathcal{H}_1)$, respectively). Multiplying the matrix X on the left and on the right by suitable matrices, one can suppose that

$$X.\sqrt{n} = \begin{pmatrix} 1 & 1 & \dots & 1 \\ 1 & & & \\ \vdots & & * & \\ 1 & & & \end{pmatrix}.$$

When $n = 1$, the statement is trivial.

When $n = 2$, suppose that $X.\sqrt{2} = \begin{pmatrix} 1 & 1 \\ 1 & a \end{pmatrix}$. Since $\hat{X}.\sqrt{2} = \begin{pmatrix} 1 & 1 \\ 1 & 1/a \end{pmatrix}$ and $X^{-1} = \hat{X}$ it follows that $a = -1$. Thus, X is just the standard UGH-matrix X_2.

When $n = 3$, suppose that $X.\sqrt{3} = \begin{pmatrix} 1 & 1 & 1 \\ 1 & x & y \\ 1 & z & t \end{pmatrix}$. Since $X^{-1} = \hat{X}$, we have $(x, y, z, t) = (\varepsilon, \varepsilon^2, \varepsilon^2, \varepsilon)$ or $(x, y, z, t) = (\varepsilon^2, \varepsilon, \varepsilon, \varepsilon^2)$, where $\varepsilon = \exp(2\pi i/3)$. Conjugating X by the matrix $\begin{pmatrix} 1 & 1 & 1 \\ 1 & 0 & 1 \\ 1 & 1 & 0 \end{pmatrix}$ (if necessary), we obtain

$$X = \frac{1}{\sqrt{3}} \begin{pmatrix} 1 & 1 & 1 \\ 1 & \varepsilon & \varepsilon^2 \\ 1 & \varepsilon^2 & \varepsilon \end{pmatrix} = X_3.$$

In any case, there exists only one double coset PXP of GH-matrices. □

The following lemma shows that Conjecture 1.5.3, generally speaking, is stronger than Conjecture 1.5.2.

Lemma 1.5.7. *For $n = 5$, Conjecture* 1.5.2 *holds but Conjecture* 1.5.3 *does not.*

Proof. 1) Consider first Conjecture 1.5.3. Clearly, the standard pair $(\mathcal{H}_0, \mathcal{H}_1)$ consists of monomial Cartan subalgebras. Meanwhile, Lemma 1.4.4 gives us an example of an orthogonal pair $(\mathcal{H}_0, \mathcal{H}_2)$, where $\mathcal{H}_2$ is not monomial. Thus the pairs $(\mathcal{H}_0, \mathcal{H}_1)$ and $(\mathcal{H}_0, \mathcal{H}_2)$ are not conjugate, because the stabilizer $P = St_{GL_n(\mathbb{C})}(\mathcal{H}_0)$ consists of monomial matrices and the outer automorphism $T : X \mapsto -{}^tX$ fixes the pair $(\mathcal{H}_0, \mathcal{H}_1)$. Hence Conjecture 1.5.3 is false.

2) Now consider an orthogonal pair $(\mathcal{A}_0, \mathcal{A})$ of maximal abelian $*$-subalgebras of $\mathcal{N}$. Set $\mathcal{H} = \mathcal{A} \cap \mathcal{L}$.

Suppose that $\mathcal{H}$ is not monomial. By Lemma 1.4.4, there exists a diagonal matrix $X = \mathrm{diag}(1, a, b, c, d)$ such that $\mathcal{H} = X\mathcal{H}_2X^{-1}$, where $\mathcal{H}_2$ is as described in Lemma 1.4.4. So, $\mathcal{A}$ contains the matrices

$$Y = XA^{(2)}X^{-1} = \begin{pmatrix} 0 & a^{-1} & 0 & 0 & 0 \\ 2a & 0 & a/b & a/c & a/d \\ b & 0 & 0 & b/c & b/d \\ c & 0 & c/b & 0 & c/d \\ d & 0 & d/b & d/c & 0 \end{pmatrix},$$

$$Y^* = \begin{pmatrix} 0 & 2\bar{a} & \bar{b} & \bar{c} & \bar{d} \\ 1/\bar{a} & 0 & 0 & 0 & 0 \\ 0 & \bar{a}/\bar{b} & 0 & \bar{c}/\bar{b} & \bar{d}/\bar{b} \\ 0 & \bar{a}/\bar{c} & \bar{b}/\bar{c} & 0 & \bar{d}/\bar{c} \\ 0 & \bar{a}/\bar{d} & \bar{b}/\bar{d} & \bar{c}/\bar{d} & 0 \end{pmatrix}.$$

The matrix YY^* has first row $(1/a\bar{a}, 0, 0, 0, 0)$, and Y^*Y has first row

$$(4a\bar{a} + b\bar{b} + c\bar{c} + d\bar{d}, 0, \frac{2a\bar{a} + c\bar{c} + d\bar{d}}{b}, \frac{2a\bar{a} + b\bar{b} + d\bar{d}}{c}, \frac{2a\bar{a} + b\bar{b} + c\bar{c}}{d}).$$

Since $abcd \neq 0$, we have $YY^* \neq Y^*Y$ and $\mathcal{A}$ is not abelian, a contradiction.

Hence $\mathcal{H}$ is monomial. By Lemma 1.4.3, one can suppose that $\mathcal{H} =< A, A^2, A^3, A^4 >_{\mathbb{C}}$, $\mathcal{A} =< E_5, A, A^2, A^3, A^4 >_{\mathbb{C}}$, where $A = \begin{pmatrix} 0 & & & & e \\ a & 0 & & & \\ & b & 0 & & \\ & & c & 0 & \\ & & & d & 0 \end{pmatrix}$, $abcde \neq 0$. Rescaling A, we may assume that $abcde = 1$. Furthermore, $\mathcal{A}$ contains the matrix

$$AA^* = \mathrm{diag}(e\bar{e}, a\bar{a}, b\bar{b}, c\bar{c}, d\bar{d}).$$

So $a\bar{a} = b\bar{b} = c\bar{c} = d\bar{d} = e\bar{e} = 1$. In particular, the matrix $X = \mathrm{diag}(1, a, ab, abc, abcd)$ is unitary. But $X\mathcal{A}_0X^{-1} = \mathcal{A}_0$, $XAX^{-1} = P$, so the pairs $(\mathcal{A}_0, \mathcal{A})$ and $(\mathcal{A}_0, \mathcal{A}_1)$ are conjugate. □

In [HaJ] the following invariant is introduced for an orthogonal pair $(\mathcal{A}, \mathcal{B})$ of maximal abelian $*$-subalgebras of $\mathcal{N}$:

$$N^*(\mathcal{A}, \mathcal{B}) = \{X \in \mathcal{B} \mid XX^* = E_n, X\mathcal{A}X^* = \mathcal{A}\}.$$

By analogy, we introduce the following invariant for an orthogonal pair $(\mathcal{H}, \mathcal{K})$ of Cartan subalgebras of $\mathcal{L}$:

$$N(\mathcal{H}, \mathcal{K}) = \{X \in \hat{\mathcal{K}} \mid \det X \neq 0, X\mathcal{H}X^{-1} = \mathcal{H}\},$$

where $\hat{\mathcal{K}} =< \mathcal{K}, E_n >_{\mathbb{C}}$. In fact, it is more convenient to work with the following "truncated" invariants:

$$I^*(\mathcal{A}, \mathcal{B}) = N^*(\mathcal{A}, \mathcal{B})/Z(U_n(\mathbb{C})),\ I(\mathcal{H}, \mathcal{K}) = N(\mathcal{H}, \mathcal{K})/Z(GL_n(\mathbb{C})).$$

Clearly, all the invariants introduced are groups (under the usual product). Note that every two conjugate (ordered) orthogonal pairs have isomorphic invariants. For pairs that are conjugate under $\mathrm{Aut}(\mathcal{N})$ or $\mathrm{Inn}(\mathcal{L})$, this remark is obvious. Assume now that ordered orthogonal pairs $(\mathcal{H}, \mathcal{K})$ and $(\mathcal{H}', \mathcal{K}')$ are conjugate under the automorphism $T : X \mapsto -{}^tX$. Then for $X \in N(\mathcal{H}, \mathcal{K})$, $A \in \mathcal{H}$, we have $T(X) \in \widehat{\mathcal{K}'}$, $\det T(X) = (-1)^n \det X \neq 0$,

$$T(X)T(A)T(X)^{-1} = -{}^tX\,{}^tA\,{}^tX^{-1} = -{}^t(X^{-1}AX) = T(X^{-1}AX) \in$$

$$\in T(X^{-1}\mathcal{H}X) = T(X^{-1}(X\mathcal{H}X^{-1})X) = T(\mathcal{H}) = \mathcal{H}'.$$

Thus $T(X) \in N(\mathcal{H}', \mathcal{K}')$. An isomorphism between $N(\mathcal{H}, \mathcal{K})$ and $N(\mathcal{H}', \mathcal{K}')$ is given by the map $X \mapsto {}^tX^{-1} = -T(X)^{-1}$.

Concerning the standard pairs $(\mathcal{A}_0, \mathcal{A}_1)$ and $(\mathcal{H}_0, \mathcal{H}_1)$, the following statements hold.

Lemma 1.5.8. 1) *The ordered pairs $(\mathcal{A}_0, \mathcal{A}_1)$ and $(\mathcal{A}_1, \mathcal{A}_0)$ ($(\mathcal{H}_0, \mathcal{H}_1)$ and $(\mathcal{H}_1, \mathcal{H}_0)$, respectively) are conjugate.*

2) *The standard pair $(\mathcal{A}_0, \mathcal{A}_1)$ has invariants $N^* = S^1.\mathbb{Z}_n$ and $I^* = \mathbb{Z}_n$. The standard pair $(\mathcal{H}_0, \mathcal{H}_1)$ has invariants $N = \mathbb{C}^*.\mathbb{Z}_n$ and $I = \mathbb{Z}_n$.*

Here $S^1 = \{z \in \mathbb{C} \mid z\bar{z} = 1\}$.

Proof. 1) Suppose that the matrices D and P are written in an (orthonormal) basis $\{e_0, \ldots, e_{n-1}\}$. Define a new (orthonormal) basis

$$\{f_j = \frac{1}{\sqrt{n}} \sum_{i=0}^{n-1} \varepsilon^{ij} e_i \mid 0 \leq j \leq n-1\}.$$

Then we have

$$D(f_j) = \frac{1}{\sqrt{n}} \sum_{i=0}^{n-1} \varepsilon^i \varepsilon^{ij} e_i = f_{j+1},\ P^{-1}(f_j) = \frac{1}{\sqrt{n}} \sum_{i=0}^{n-1} \varepsilon^{ij} e_{i-1} = \varepsilon^j f_j.$$

Thus, the pair (P^{-1}, D) of operators in the basis $\{f_j\}$ plays the same role as the pair (D, P) in the basis $\{e_j\}$. Hence, the ordered pairs $(\mathcal{A}_0, \mathcal{A}_1)$ and $(\mathcal{A}_1, \mathcal{A}_0)$ are conjugate. The same holds for the pairs $(\mathcal{H}_0, \mathcal{H}_1)$ and $(\mathcal{H}_1, \mathcal{H}_0)$. In view of this statement, we can disregard the order of components in standard pairs in what follows.

2) Note that $N^*(\mathcal{A}_0, \mathcal{A}_1) = N(\mathcal{H}_0, \mathcal{H}_1) \cap U_n(\mathbb{C})$. Now suppose that $X \in N(\mathcal{H}_0, \mathcal{H}_1)$, and fix n different complex numbers $\lambda_1, \dots, \lambda_n$ such that $\sum_i \lambda_i = 0$. Then the operator XAX^{-1}, like $A = \operatorname{diag}(\lambda_1, \dots, \lambda_n) \in \mathcal{H}_0$, has spectrum $\{\lambda_1, \dots, \lambda_n\}$, and reduces to diagonal form in the same basis $\{e_1, \dots, e_n\}$. Because of the fact that the λ_i are pairwise different, X must permute the lines $< e_j >_{\mathbb{C}}$, that is, X is monomial. But $X \in \hat{\mathcal{H}}_1 = \mathcal{A}_1 =< E_n, P, \dots, P^{n-1} >$, so that the monomiality of X means that $X = xP_j$ for some $x \in \mathbb{C}^*$, $0 \le j \le n-1$. Conversely, xP_j is contained in $N(\mathcal{H}_0, \mathcal{H}_1)$. Hence, $N(\mathcal{H}_0, \mathcal{H}_1) = \mathbb{C}^*.\mathbb{Z}_n$, $N^*(\mathcal{A}_0, \mathcal{A}_1) = S^1.\mathbb{Z}_n$. □

In the general case, let $(\mathcal{H}_0, \mathcal{H} = A\mathcal{H}_0A^{-1})$ be an orthogonal pair and suppose that $X \in N(\mathcal{H}_0, \mathcal{H})$. Recall that we have $A^{-1} = \hat{A} = (\hat{a}_{ij})$. Since $X \in \mathcal{H} = A\mathcal{H}_0\hat{A}$, we can write X in the form $X = A(\sum_{i=1}^n g_i E_{ii})\hat{A} = \sum_i g_i A E_{ii} \hat{A}$ for some $g_i \in \mathbb{C}^*$. Note that the matrices E_{ii}, $1 \le i \le n$ (in other words, minimal projectors of $\mathcal{A}_0$) exhaust all elements of $\hat{\mathcal{H}}_0$ with spectrum $\{1, 0, \dots, 0\}$. So acting by conjugation on $\hat{\mathcal{H}}_0$, X permutes these matrices: $XE_{kk}X^{-1} = E_{\pi(k),\pi(k)}$ for some permutation $\pi \in \mathbb{S}_n$. Hence,

$$\sum_j g_j E_{jj} \hat{A} E_{kk} A = \hat{A} X E_{kk} A = \hat{A} E_{\pi(k),\pi(k)} X A = \sum_j \hat{A} E_{\pi(k),\pi(k)} A g_j E_{jj}.$$

In particular, the (r, s)-coefficients of the first and last matrices coincide: $g_r \hat{a}(r, k) a(k, s) = \hat{a}(r, \pi(k)) a(\pi(k), s) g_s$, where a_{ij} and $\hat{a}_{ij}$ have been written in the form $a(i, j)$ and $\hat{a}(i, j)$. From this and the equality $\hat{a}(i, j) = 1/(na(j, i))$, the first basic equation follows:

$$\frac{g_r}{g_s} = \frac{a(k, r).a(\pi(k), s)}{a(k, s).a(\pi(k), r)}, \tag{1.18}$$

(for all k, r, s). Furthermore, note that $X^l \in N(\mathcal{H}_0, \mathcal{H})$ for all l and

$$X^l = A(\sum_j g_j E_{jj})^l \hat{A} = A \sum_j (g_{jj})^l E_{jj} \hat{A}, \; X^l E_{kk} X^{-l} = E_{\pi^l(k),\pi^l(k)}.$$

Applying (1.18) to the element X^l, we obtain the second basic equation:

$$\left(\frac{g_r}{g_s}\right)^l = \frac{a(k, r).a(\pi^l(k), s)}{a(k, s).a(\pi^l(k), r)}, \tag{1.19}$$

(for all k, l, r, s).

Equations (1.18) and (1.19), which were first established in [HaJ], allow us to define the invariants N, N^*, I and I^*. But first we make the following simple remark:

Lemma 1.5.9. *Suppose that $A \in M_m(\mathbb{C})$ and $B \in M_n(\mathbb{C})$ are GH-matrices (UGH-matrices, respectively). Then $A \otimes B$ is a GH-matrix (a UGH-matrix, respectively) of degree mn.*

Proof. Consider the (r, s)-coefficient C_{rs} of the matrix $C = A \otimes B$, where $r = (i-1)n + k$, $s = (j-1)n + l$, $1 \le i, j \le m$, $1 \le k, l \le n$. Clearly, it is $a_{ij}b_{kl}$. Hence,

$$(\hat{A} \otimes \hat{B})_{rs} = \hat{a}_{ij}.\hat{b}_{kl} = (mna_{ji}b_{lk})^{-1} = (mnC_{sr})^{-1} = \hat{C}_{rs},$$

that is, $\hat{A} \otimes \hat{B} = \hat{C}$. Thus, provided that A and B are GH-matrices, we have

$$C\hat{C} = (A \otimes B)(\hat{A} \otimes \hat{B}) = A\hat{A} \otimes B\hat{B} = E_m \otimes E_n = E_{mn},$$

and C is a GH-matrix. If in addition both A and B are UGH-matrices, then $|C_{rs}| = |a_{ij}b_{kl}| = 1/\sqrt{mn}$. □

Using this tensor construction, we obtain the following statement:

Lemma 1.5.10. *Suppose that n is divisible by the square of some prime number p. Then both Conjectures 1.5.2 and 1.5.3 are false.*

Proof. Assume as above that X_m is a standard UGH-matrix of degree m. Set $A = X_p \otimes X_p \otimes X_{n/p^2}$, $\mathcal{H} = A\mathcal{H}_0A^{-1}$, $\mathcal{A} = A\mathcal{A}_0A^{-1}$. By its definition, A is a UGH-matrix; moreover, $(\sqrt{n}a_{ij})^{n/p} = 1$ for all i, j. Setting $l = n/p$ in (1.19) and preserving the notation $a(i, j) = a_{i,j}$, we obtain:

$$\frac{a(\pi^l(k), s)}{a(k, s)} = \frac{a(\pi^l(k), r)}{a(k, r)}$$

for all k, r, s. Now assume that $t = \pi^{n/p}(k) \neq k$. Then $a(t, r)/a(k, r) = a(t, 1)/a(k, 1)$ for every r, that is, $a(t, r) = \alpha.a(k, r)$ for some $\alpha \in \mathbb{C}^*$. Recall that A is a UGH-matrix, so that we have:

$$0 = \sum_r a(t, r)\overline{a(k, r)} = \alpha \sum_r a(k, r)\overline{a(k, r)} = \alpha,$$

a contradiction. So $\pi^{n/p} = 1$.

The correspondence $X \mapsto \sigma$, where $XE_{kk}X^{-1} = E_{\sigma(k),\sigma(k)}$, is in fact a homomorphism $N(\mathcal{H}_0, \mathcal{H}) \to \mathbb{S}_n$ ($N^*(\mathcal{A}_0, \mathcal{A}) \to \mathbb{S}_n$, respectively) with kernel $Z(GL_n(\mathbb{C}))$ ($Z(U_n(\mathbb{C}))$, respectively). Consequently, the identity $\pi^{n/p} = 1$ proved above means that the exponents of $I(\mathcal{H}_0, \mathcal{H})$ and $I^*(\mathcal{A}_0, \mathcal{A})$ divide n/p. Meanwhile, the exponents of $I(\mathcal{H}_0, \mathcal{H}_1)$ and $I^*(\mathcal{A}_0, \mathcal{A}_1)$ are equal to n. Hence, the pairs $(\mathcal{H}_0, \mathcal{H}_1)$ and $(\mathcal{H}_0, \mathcal{H})$ ($(\mathcal{A}_0, \mathcal{A}_1)$ and $(\mathcal{A}_0, \mathcal{A})$, respectively) are not conjugate. □

In [HaJ], strongly regular graphs are used to disprove Conjecture 1.5.2 in the case $n \equiv 1 \pmod 4$, $n > 5$ and n a prime.

Definition 1.5.11. A graph with n vertices having no loops or multiple edges and satisfying the following conditions:

(i) exactly k edges issue from every vertex;

(ii) any two different adjacent vertices have exactly λ adjacent vertices in common;

(iii) any two non-adjacent vertices have exactly μ adjacent vertices in common; is called a *strongly regular graph* (s.r.g.) with parameters (n, k, λ, μ).

Example 1.5.12. Let p be a prime, $p \equiv 1 \pmod 4$. Then the *Paley graph* (with elements of the field $\mathbb{F}_p$ as vertices, where $a, b \in \mathbb{F}_p$ are adjacent if and only if $a - b$ is a nonzero quadratic residue) is an s.r.g. with parameters $(p, (p-1)/2, (p-5)/4, (p-1)/4)$.

Example 1.5.13. The *Petersen graph*

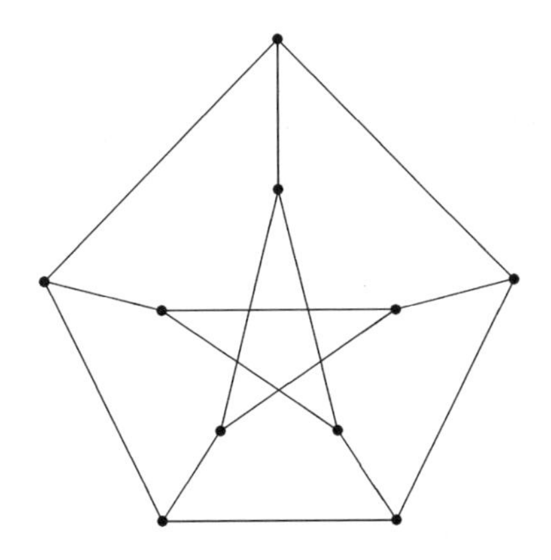

is an s.r.g. with parameters $(10, 3, 0, 1)$.

If A is the incidence matrix of an s.r.g. with parameters (n, k, λ, μ), then by [BaI],

$$A^2 = kE_n + \lambda A + \mu(J - E_n - A),\ AJ = JA = kJ,$$

where J is the matrix all of whose coefficients are equal to 1. Suppose now that $\mu = \lambda + 1$ and set $R = E_n + \beta A + \beta^{-1}(J - E_n - A)$ for some β with $|\beta| = 1$. Then for $b = \beta + \beta^{-1}$ we have

$$RR^* = E_n(2 - b - (b^2 - 4)(k - \mu)) + J((b^2 - 4)(k - \mu) + b - 2 + n).$$

If we choose β so that

$$(b^2 - 4)(k - \mu) + b - 2 + n = 0, \tag{1.20}$$

then $RR^* = nE_n$, $R_{ij} = 1$ if $i = j$ and $R_{ij} \in \{\beta, \beta^{-1}\}$ if $i \neq j$, that is, $\frac{1}{\sqrt{n}}R$ is a UGH-matrix.

Set $b = -\frac{1}{2}$ for the Petersen graph and $b = \frac{2}{1+\sqrt{p}}$ for the Paley graph. Then equation (1.20) is satisfied; moreover, in both cases there exists β such that $|\beta| = 1$, $\beta + \beta^{-1} = b$. Furthermore, if $n \neq 5$ then β is not a root of unity. Indeed, suppose that $\beta^N = 1$ for some N. Then β and b are algebraic integers. In particular, $b \neq -\frac{1}{2}$. If $b = \frac{2}{1+\sqrt{p}}$, the element b of the field $\mathbb{Q}(\sqrt{p})$ has norm $\frac{2}{1+\sqrt{p}} \cdot \frac{2}{-1+\sqrt{p}} = \frac{4}{p-1}$, so that b can be an algebraic integer only in the case $n = p = 5$ (if $p = 5$, then in fact $\beta = \exp(2\pi i/5)$). We have just proved the following lemma:

Lemma 1.5.14. *Suppose that $n = 10$ or n is a prime with $n \equiv 1 \pmod 4$ and $n > 5$. Then there exists a symmetric UGH-matrix $B = (b_{ij})$ of degree n with the following properties:*

$$\sqrt{n} b_{ij} = \begin{cases} 1 & \text{if } i = j, \\ \beta \text{ or } \beta^{-1} & \text{if } i \neq j, \end{cases}$$

where $|\beta| = 1$ but β is not a root of unity. □

We describe one more construction of UGH-matrices. Suppose that p is a prime, $p \equiv 3 \pmod 4$, and η is a quadratic character of the field $\mathbb{F}_p = \{0, 1, \ldots, p-1\}$:

$$\eta(x) = \begin{cases} 0 & \text{if } x = 0, \\ 1 & \text{if } x \text{ is a nonzero quadratic residue,} \\ -1 & \text{otherwise.} \end{cases}$$

Consider the matrix $A = (\eta(i-j))_{1 \leq i,j \leq p}$. Then A has the following properties:
(i) $A = \bar{A} = -{}^tA = -A^*$ (since $\eta(-1) = -1$);
(ii) $AJ = JA = 0$;
(iii) $AA^* = pE_p - J$ (because, the (i,j)-coefficient of AA^* is

$$\sum_k \eta(i-k)\eta(j-k) = \sum_k \eta((i-k)(j-k)) = \sum_{x \in \mathbb{F}_p} \eta((x-i)(x-j)).$$

The latter sum is $p\delta_{ij} - 1$, by [LiN]).

Now choose $\alpha, \beta, \gamma \in \mathbb{C}$ so that the matrix $B = \alpha A + \beta E_p + \gamma J$ is a UGH-matrix. For definiteness we add the condition $\alpha + \gamma = \beta + \gamma = \frac{1}{\sqrt{p}}$. Then $\alpha = \beta = \frac{\sqrt{p}-i}{p+1}$, $\gamma = \frac{i\sqrt{p}+1}{(p+1)\sqrt{p}}$; furthermore $B = (b_{ij})$, where

$$b_{ii} = \beta + \gamma = \frac{1}{\sqrt{p}},$$

$$pb_{ij}b_{ji} = p(\alpha\eta(i-j) + \gamma)(\alpha\eta(j-i) + \gamma) = p(\gamma + \alpha)(\gamma - \alpha) = \sqrt{p}(\gamma - \alpha) = \delta$$

if $i \neq j$ and $\delta = (1 - p + 2i\sqrt{p})/(p+1)$. Suppose that δ is a root of unity. Then $-(\delta + \bar{\delta}) = 2(p-1)/(p+1)$ is an algebraic integer, so that $p = 3$. Conversely, if $p = 3$ then $\delta = \exp(2\pi i/3)$. Thus we have proved:

Lemma 1.5.15. *Suppose that p is a prime, $p \equiv 3 \pmod 4$, $p > 3$. Then there exists a UGH-matrix $B = (b_{ij})$ of degree p with the following properties: $\sqrt{p}b_{ii} = 1$, $pb_{ij}b_{ji} = \delta$ for $i \neq j$, where $|\delta| = 1$ but δ is not a root of unity.* □

We are now able to prove the main result of this section.

Theorem 1.5.16. 1) *Conjectures* 1.5.2 *and* 1.5.3 *hold for $n \leq 3$.*

2) *Conjecture 1.5.2 holds for $n = 5$ and, perhaps, for $n = 15$; but is false for all other $n \geq 4$.*

3) *Conjecture* 1.5.3 *is false for all $n \geq 4$.*

Proof. 1) Assertion 1) was proved above, as well as part of assertions 2) and 3) (see Lemmas 1.5.6 and 1.5.7). Now assume that $n \geq 4$. If n is divisible by the square of some prime p, then both Conjectures 1.5.2 and 1.5.3 are false, by Lemma 1.5.10. Suppose that $n = p_1 p_2 \ldots p_t$, where $p_1 < p_2 < \ldots < p_t$ are primes in ascending order, $n > 6$ and $n \neq 15$. Then n is divisible by an integer q, where $q = 10$ or q is prime and $q \geq 7$: $n = mq$. According to Lemmas 1.5.14 and 1.5.15, there exists a UGH-matrix $B = \{b_{ij}\}$ of degree q with the following properties:

(i) $\sqrt{q}b_{ii} = 1$;

(ii) if $i \neq j$, the number $qb_{ij}b_{ji}$ has modulus 1 but is not a root of unity.

Set $A = B \otimes X_m$, $\mathcal{H} = A\mathcal{H}_0 A^{-1}$, $\mathcal{A} = A\mathcal{A}_0 A^{-1}$, where X_m is a standard UGH-matrix of degree m. By Lemma 1.5.9, A is a UGH-matrix. Recall that every $X \in N(\mathcal{H}_0, \mathcal{H})$ satisfies equations (1.18) and (1.19). We split the set $\Omega = \{1, 2, \ldots, n\}$ up into p subsets $\Omega_i = \{(i-1)m + 1, \ldots, im\}$, $1 \leq i \leq p$.

Firstly, assume that the permutation $\pi \in \mathbb{S}_n$ corresponding to some element $x \in N(\mathcal{H}_0, \mathcal{H})$ does not fix some component, Ω_i say, of the partition $\Omega = \Omega_1 \cup \ldots \cup \Omega_p$, so that $\pi(\Omega_i) \cap \Omega_j \neq \emptyset$ for some i, j, $1 \leq i \neq j \leq p$. Thus, one can find $k = (i-1)m + r$, $1 \leq r \leq m$ such that $\pi(k) = (j-1)m + s$, $1 \leq s \leq m$. By (1.18), we have

$$\frac{g_{\pi(k)}}{g_k} = \frac{a(k, \pi(k)).a(\pi(k), k)}{a(k, k).a(\pi(k), \pi(k))}.$$

Furthermore,

$$a(k, k) = b_{ii}x_{rr} = \frac{x_{rr}}{\sqrt{p}}, \; a(\pi(k), \pi(k)) = b_{jj}x_{ss} = \frac{x_{ss}}{\sqrt{p}},$$

$$a(k, \pi(k)).a(\pi(k), k) = b_{ij}b_{ji}x_{rs}x_{sr},$$

so that

$$\frac{g_{\pi(k)}}{g_k} = \frac{(pb_{ij}b_{ji}).(x_{rs}x_{sr})}{x_{rr}x_{ss}}.$$

Here the (r, s)-coefficient x_{rs} of the standard UGH-matrix X_m is an m-th root of unity, and $pb_{ij}b_{ji}$ is not a root of unity. Hence, $g_{\pi(k)}/g_k$ is not a root of unity.

Meanwhile, if in (1.19) we take the order of the permutation π in $\mathbb{S}_n$ in place of l, we obtain that $(g_{\pi(k)}/g_k)^l = 1$, a contradiction.

Hence, for any $X \in N(\mathcal{H}_0, \mathcal{H})$, the permutation π fixes every component Ω_i of the partition $\Omega = \Omega_1 \cup \ldots \cup \Omega_p$. In other words, if $k = (i-1)m + u$, $1 \le i \le p$, $1 \le u \le m$, then $\pi(k) = (i-1)m+v$, $1 \le v \le m$. Furthermore, set $r = (j-1)m+z$, $s = (l-1)m+t$ for some j, l, z, t with $1 \le j, l \le p$, $1 \le z, t \le m$. Then, by (1.18) we have

$$\frac{g_r}{g_s} = \frac{b_{ij}x_{uz}.b_{il}x_{vt}}{b_{il}x_{ut}.b_{ij}x_{vz}} = \frac{x_{uz}x_{vt}}{x_{ut}x_{vz}}.$$

In particular, $(g_r/g_s)^m = 1$ for all r, s. Setting $l = m$ in (1.19), we obtain that

$$\frac{a(\pi^m(k), s)}{a(k, s)} = \frac{a(\pi^m(k), r)}{a(k, r)}$$

for all k, r, s. Using the arguments in the proof of Lemma 1.5.10, we come to the conclusion that the exponents of $I(\mathcal{H}_0, \mathcal{H})$ and $I^*(\mathcal{A}_0, \mathcal{A})$ divide $m = n/p$. Hence, the pairs $(\mathcal{H}_0, \mathcal{H})$ and $(\mathcal{H}_0, \mathcal{H}_1)$ ($(\mathcal{A}_0, \mathcal{A})$ and $(\mathcal{A}_0, \mathcal{A}_1)$, respectively) are not conjugate.

2) Here we assume that $n = 6$. In [Ufn 2] the following UGH-matrix A of degree 6 appears:

$$A = \frac{1}{\sqrt{6}} \begin{pmatrix} -1 & i & -i & 1 & -1 & 1 \\ -i & 1 & i & -1 & -1 & 1 \\ i & -i & -1 & i & -i & 1 \\ 1 & -1 & -i & -1 & i & 1 \\ -1 & -1 & i & -i & 1 & 1 \\ 1 & 1 & 1 & 1 & 1 & 1 \end{pmatrix}.$$

It is clear that $(\sqrt{6}a_{ij})^4 = 1$ for all i, j. So, in accordance with (1.18), for any $X \in N(\mathcal{H}_0, \mathcal{H})$, where $\mathcal{H} = A\mathcal{H}_0A^{-1}$ and $\mathcal{A} = A\mathcal{A}_0A^{-1}$ as above, we have $(g_r/g_s)^4 = 1$ for all r, s. Standard arguments using (1.19) show that the exponents of $I(\mathcal{H}_0, \mathcal{H})$ and $I^*(\mathcal{A}_0, \mathcal{A})$ divide 4. Hence, the conjectures are false if $n = 6$.

3) Finally, assume that $n = 15$. Then the matrix $A = (\lambda - 1)E_{15} + J$, where $\lambda = (-13 - \sqrt{161})/2$, is a GH-matrix. Set $\mathcal{H} = A\mathcal{H}_0A^{-1}$. Then, by (1.18), for $X \in N(\mathcal{H}_0, \mathcal{H})$ we have that

$$\frac{g_{\pi(k)}}{g_k} = \frac{a(k, \pi(k)).a(\pi(k), k)}{a(k, k).a(\pi(k), \pi(k))}.$$

In particular, when $\pi(k) \neq k$ we would have $g_k/g_{\pi(k)} = \lambda^2 > 1$, and this contradicts (1.19): $(g_r/g_s)^l = 1$ for $l = | < \pi > |$. So $\pi(k) = k$ for every k; in other words, the group $I(\mathcal{H}_0, \mathcal{H})$ is trivial. Hence Conjecture 1.5.3 is false if $n = 15$. □

The problem of the validity of Conjecture 1.5.2 for $n = 15$ remains open.

We see the proper relationship between the uniqueness problem for orthogonal pairs and the existence and uniqueness problem for ODs of the Lie algebra $\mathcal{L}$ as the following. A complete classification of orthogonal pairs (up to conjugacy) would allow us to parametrize the set of GH-matrices (and the set $\mathcal{C}$, see §1.4), which would lead in the end to a classification of orthogonal decompositions.

1.6 A connection with Hecke algebras

In this section we indicate a further approach to the problem of describing orthogonal pairs and orthogonal decompositions. It is based on a consideration of Hecke algebras.

Definition 1.6.1. Let $\Gamma = (V, E)$ be a non-directed graph with vertex set V and edge set E, having no loops and no multiple edges. Let q be a complex number. The *Hecke algebra associated with the graph* Γ is the associative algebra $H(q, \Gamma)$ with unit 1 generated by elements x_v, $v \in V$ with the following defining relations:

$$\begin{cases} (x_v - 1)(x_v - q) = 0; & \\ x_u x_v x_u = x_v x_u x_v & \text{if} \quad (u, v) \in E, \\ x_u x_v = x_v x_u & \text{if} \quad (u, v) \notin E. \end{cases} \tag{1.21}$$

With every orthogonal pair $(\mathcal{H}_1, \mathcal{H}_2)$ of Cartan subalgebras of the Lie algebra $\mathcal{L}$ of type A_{n-1}, we shall associate a certain Hecke algebra, as follows. For $\mathcal{H}_1$, take the standard Cartan subalgebra $\mathcal{H}_0$. Suppose that $\mathcal{H}_2 = A\mathcal{H}_0 A^{-1}$ for some GH-matrix A. Set

$$P_i = E_{ii} = \operatorname{diag}(0, \ldots, 0, \underbrace{1}_{i}, 0, \ldots, 0), \; Q_i = AP_iA^{-1}, \; 1 \le i \le n.$$

The elements P_i are called *minimal projectors*, or *primitive idempotents* of the Cartan subalgebra $\mathcal{H}_0$. Clearly, $(P_i)^2 = P_i$, $(Q_i)^2 = Q_i$ and $P_iP_j = 0$, $Q_iQ_j = 0$ if $i \neq j$. In addition, the following relations hold:

$$P_iQ_jP_i = \frac{1}{n}P_i, \; Q_jP_iQ_j = \frac{1}{n}Q_j \tag{1.22}$$

for all i, j. Indeed, the (k, l)-coefficient of the matrix $P_iQ_jP_i$ is $\delta_{ik} a_{ij} \hat{a}_{ji} \delta_{il} = (1/n)\delta_{ik}\delta_{il}$. Similarly, the (k, l)-coefficient of $Q_jP_iQ_j$ is $a_{kj}\hat{a}_{ji}a_{ij}\hat{a}_{jl} = (1/n)a_{kj}\hat{a}_{jl}$. Here we have used the defining identity for a GH-matrix: $A^{-1} = \hat{A} = (\hat{a}_{ij})$. In fact, (1.22) is equivalent to the requirement that A be a GH-matrix, in other words, that $\mathcal{H}_1$ and $\mathcal{H}_2$ be orthogonal Cartan subalgebras.

Now we choose complex numbers a, b such that the matrices $\hat{P}_i = aP_i + bE_n$ and $\hat{Q}_j = aQ_j + bE_n$ satisfy relations like (1.21). Direct calculations show that triplet (a, b, q) takes one of the following values:

$$a = \frac{-n \pm \sqrt{n^2 - 4n}}{2}, \; b = 1, q = a + b = \frac{2 - n \pm \sqrt{n^2 - 4n}}{2};$$

$$a = \frac{n \pm \sqrt{n^2 - 4n}}{2}, \; b = q = \frac{2 - n \mp \sqrt{n^2 - 4n}}{2}.$$

The calculations carried out above lead us to introduce the following graphs:

$$\Gamma_1 = (V_1, E_1); \begin{cases} V_1 = \{u_1, \dots, u_n, v_1, \dots, v_n\}, \\ E_1 = \{(u_i, v_j) \mid 1 \leq i, j \leq n\}, \end{cases}$$

$$\Gamma_2 = (V_2, E_2); \begin{cases} V_2 = \{v_{ij} \mid 1 \leq i \leq n+1, 1 \leq j \leq n\}, \\ E_2 = \{(v_{ij}, v_{kl}) \mid 1 \leq i \neq k \leq n+1, 1 \leq j, l \leq n\}. \end{cases}$$

Proposition 1.6.2. *The following statements hold.*

(i) *A classification of orthogonal pairs of Cartan subalgebras of the Lie algebra $\mathcal{L}$ of type A_{n-1} (up to* Inn($\mathcal{L}$)*-conjugacy) is equivalent to a classification of representations*

$$\Phi : H(q, \Gamma_1) \to M_n(\mathbb{C})$$

of the Hecke algebra $H(q, \Gamma_1)$, $q = \frac{2-n\pm\sqrt{n^2-4n}}{2}$, such that

$$\Phi(x_u)\Phi(x_v) = \Phi(x_u) + \Phi(x_v) - E_n$$

for all $u, v \in V_1$, $(u, v) \notin E_1$.

(ii) *A classification of the orthogonal decompositions of the Lie algebra $\mathcal{L}$ of type A_{n-1} (up to* Inn($\mathcal{L}$)*-conjugacy) is equivalent to a classification of representations*

$$\Phi : H(q, \Gamma_2) \to M_n(\mathbb{C})$$

of the Hecke algebra $H(q, \Gamma_2)$, $q = \frac{2-n\pm\sqrt{n^2-4n}}{2}$, such that

$$\Phi(x_u)\Phi(x_v) = \Phi(x_u) + \Phi(x_v) - E_n$$

for all $u, v \in V_2$, $(u, v) \notin E_2$.

Proof. If we consider an orthogonal pair or an orthogonal decomposition of the Lie algebra $\mathcal{L}$, the calculations carried out above show that we are in fact dealing with a certain representation of the form indicated of the Hecke algebra $H(q, \Gamma_i)$, $i = 1, 2$. Conjugacy of pairs (of ODs, respectively) under Inn($\mathcal{L}$) means equivalence of corresponding representations of $H(q, \Gamma_1)$ (of $H(q, \Gamma_2)$, respectively). □

To conclude this section, we indicate the corresponding graphs for the Lie algebra of type A_1.

Graph Γ_1 and its supplementary graph

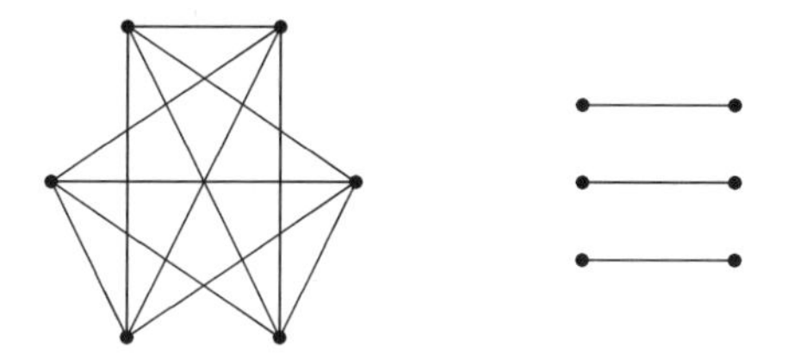

Graph Γ_2 and its supplementary graph

Commentary

1) The standard construction of ODs for the Lie algebra of type A_{q-1}, $q = p^m$, is explained in [KKU 1], [KKU 2]. In addition, the authors formulated a general concept of ODs of complex simple Lie algebras following a pioneering work of Thompson [Tho 2]. In fact, they had to reject a rather restrictive multiplicativity condition (in this connection, see Chapter 3), because, for type A_n, the condition can hold only if $n = 1$. Later, it will be explained (see Chapter 5) that the standard construction does in fact give all the most interesting ODs.

The central problem stated at the beginning of this chapter concerning the primality of $n + 1$ as a necessary (and sufficient) condition for the existence of ODs for Lie algebras of type A_n, did not arise immediately. Experimental data for $n \leq 8$, $n \neq 5$ and persistent but unsuccessful attempts to construct an OD in the case $n = 5$, recall the humming of the all-seeing Winnie-the-Pooh in the Russian performance of [KKU 2] and B. Zahoder, the talented translator into Russian of A. A. Milne's "Winnie-the-Pooh":

> Возьмем это самое слово «А-пять»,
> Зачем мы его произносим,
> Хотя мы свободно могли бы сказать,
> «А-шесть», и «А-семь», и «А-восемь» !?

The delicate play on words, which is incomprehensible for English-language readers, has been explained to a certain extent by Queen, who has translated [KKU 2] into English:

> Let's take again this word we said AFORE
> and on it let us ruminate.
> Why not go on a little more,
> with A-FIVE, and A-SIX, A-SEVEN, A-EIGHT?

She added in a note: "The verse given in Russian here bears no relation to anything I have been able to find in the Pooh books, and so I have rather freely translated it back into English so as to preserve at least some of its aptness. In the Russian, the word for "again" puns with the pronunciation of "A_5", and is contrasted with "A_6", "A_7", "A_8". "

2) The term "J-decompositions" and also the connection between J-decompositions and spreads, the latter of which have had a wide circulation in combinatorics and finite geometry for some long time [Demb], [Lun], was introduced by Ivanov [Iva 1]. The question as to whether every orthogonal decomposition of the Lie algebra of type A_n is a J-decomposition is still open, even in the transitive case (in this connection, see Commentary to Chapter 5). There exists a voluminous literature on spreads and finite geometries (see for example [Kal] and [Lun]). A classification of all spreads (in other words, all translation planes) has been obtained for small orders q. One could restate these results in terms of J-decompositions and thus get a corresponding classification. As an example, we state the following result. Every J-decomposition of the Lie algebra of type A_7 is standard (this follows from the fact that every spread of order 8 is desarguesian).

According to Theorem 1.2.9, there exists a transitive symplectic spread of order 2^{4n+2} which is not irreducible and which is strongly related to the Lüneburg-Tits translation plane (see [Lun]). Many other examples, also of even order, have been constructed by Kantor [Kan 4, 5]. It would be very interesting to make similar examples for odd orders. Another open question is how to recognize the symplecticness of a given spread by means of its inner geometric structure.

The exceptional spread of order 27, which is described in Theorem 1.2.7, corresponds to the Hering translation plane [Her 1]. In our view, the most interesting general result concerning transitive spreads is the classification of flag-transitive spreads with insoluble automorphism groups by Hering [Her 3] and Buekenhout, Delandtsheer, Doyen, Kleidman, Liebeck and Saxl [BDDKLS]. We have mentioned that transitive spreads of order q^2 (with kernel of order q) have been studied by Baker and Ebert [BaE 1, 2].

3) The question concerning the splitting of the group $\mathrm{Inn}(\mathcal{D}) = \mathbb{Z}_p^{2m}.\Sigma L_2(p^m)$, where $\mathcal{D}$ is a standard OD is quite simple for odd primes p and seemed to require only technical refinements in the case $p = 2$. Abdukhalikov [Abd 2] was the first to observe that it is in fact non-split if $p = 2$ and $m > 2$. Later we shall see that

non-split extensions $\mathbb{Z}_2^3 \circ SL_3(2)$ and $\mathbb{Z}_2^5 \circ SL_5(2)$ (the Dempwolff group [Demp 1]) occur as automorphism groups of orthogonal decompositions for Lie algebras of types G_2 and E_8.

4) The types A_1, A_2, A_3 and A_4 stand alone in the series A_n, for which the question of describing all ODs up to conjugacy has been solved completely. It is not difficult to see that there exists just one J-decomposition of the Lie algebra of type A_{p-1}, where p is a prime. It could be that it is, in general, the unique OD for A_{p-1}. It would be interesting to compute the number of orbits of J-decompositions for Lie algebras of type A_{q-1}, $q = p^m$. One finds asymptotic estimates for the number of translation planes of specific orders in [Ebe] and [Kan 4]. As far as we know, a classification of translation planes of order p^m, $m \geq 2$, (without the transitivity condition) has been achieved only for $p^m \leq 25$ (in this connection, see [Lun] and [Ost]). The case $p^m = 25$ is treated in [Cze], [Czo], and the case $p^m = 16$ in [She].

Another approach to this problem is based on considering the graph $\Gamma = (V, E)$ to be described now. The vertex set V is the set of maximal totally isotropic planes in four-dimensional symplectic space over $\mathbb{F}_p$. Furthermore, two different planes are joined by an edge in E if and only if their intersection has dimension 1. The graph Γ has rank 3 (in this connection, see [BaI]), and the above-mentioned problem about describing J-decompositions for A_{p^2-1} is equivalent to the classification of maximal cocliques, that is, maximal empty-subgraphs (graphs without edges), in Γ. In general, the Klein correspondence (see e.g. [MaS]) turns out to be very useful for the investigation of spreads of orders q^2 (with kernel of order q).

5) The connection between ODs and GH-matrices is mentioned in [KKU 2], where these matrices are called *suitable*. The prefix GH (generalized Hadamard) seems to us to be more "suitable", since in some sense these matrices are true generalizations of classical Hadamard matrices (if A is a Hadamard matrix of degree n, then $\frac{1}{\sqrt{n}}A$ is a GH-matrix).

A description of all GH-matrices of given degree n would allow us to solve the Winnie-the-Pooh problem (see 1)). Unfortunately, as yet, all we can do is dispose of many types of GH-matrices. Ufnarovskii [Ufn 2] has found all GH-matrices A of degree 6 such that $A^2 = \lambda E_6$. Up to conjugation by a monomial matrix and multiplication by a scalar, every matrix of this kind has the following form:

$$\begin{pmatrix} -1 & a & -a & -c & c & 1 \\ 1/a & 1 & x & -cx/a & -c & 1 \\ -1/a & 1/x & -1 & 1/a & -1/x & 1 \\ -1/c & -a/cx & a & 1 & 1/x & 1 \\ 1/c & -1/c & -x & x & -1 & 1 \\ 1 & 1 & 1 & 1 & 1 & 1 \end{pmatrix},$$

where $a, c, x \in \mathbb{C}$, $x(a-c) = ac - 2a - 1$, $x^{-1}(a-c) = ac + 2c - 1$. Tchanyšhev has described the following non-standard GH-matrix of arbitrary degree n: $A =$

$(\lambda - 1)E_n + J$, where E_n is the unit matrix of degree n as usual, all coefficients of J are equal to 1, and λ satisfies the equation $t^2 + (n-2)t + 1 = 0$. He has also found all GH-matrices of degree at most 5.

Kryliuk has made the following observation: if $n \equiv 1 \pmod 2$ and n is not a square, then there does not exist an OD of the Lie algebra $\mathcal{L} = sl_n(\mathbb{C})$ of the form $\mathcal{L} = \sum_{i=0}^{n} X^i \mathcal{H}_0 X^{-i}$, where $X \in M_n(\mathbb{C})$, $X^{n+1} = E_n$. Indeed, suppose that we have some OD of this kind. Then $A = (a_{ij}) = X^{(n+1)/2}$ is a GH-matrix and $A^2 = E$, that is, $A = A^{-1} = \hat{A}$. Hence $a_{ii} = \pm\frac{1}{\sqrt{n}}$ and $\mathrm{Tr}A = \frac{m}{\sqrt{n}}$ for some $m \in \mathbb{Z}$ on the one hand. On the other hand, since A is an involutory matrix of odd degree, it follows that $\mathrm{Tr}A = l$ is a nonzero integer. Thus $\sqrt{n} = m/l$ is rational, a contradiction.

In fact, we have proved the following assertion. *Suppose as above that $n \equiv 1 \pmod 2$, n is not a square, and $\mathcal{D} : \mathcal{L} = \sum_{i=0}^{n} \mathcal{H}_i$ is an OD of the Lie algebra of type A_{n-1}. Then every involution in* $\mathrm{Aut}(\mathcal{D})$ *fixes every component $\mathcal{H}_i$. In particular, $\mathcal{D}$ is not transitive.*

The construction of GH-matrices using the Paley graph is due to de la Harpe and Jones [HaJ]. Later, from a letter of de la Harpe (of 7 January 1992), the authors learnt that Munemasa and Watatani obtained a result slightly weaker than Theorem 1.5.16, in [MuW], based on studying association schemes. In the same letter, the question of a classification of all GH-matrices of small prime degrees $p = 7, 11, 13, \ldots$ is raised.

6) The idea of introducing Hecke algebras into the problem of classifying ODs is due to Bondal [Bon 2]. To get a perspective of what this idea means will probably depend on the development of the representation theory of Hecke algebras.

It is worth mentioning one more approach to the Winnie-the-Pooh problem, using the Hasse principle to some extent. It would replace the Lie algebra $\mathcal{L} = sl_{n+1}(\mathbb{C})$ by the Lie algebra $\mathcal{L}_r$ of type A_n over an algebraically closed field $\mathbb{F}$ of arbitrary characteristic $r \nmid (n+1)$. Then the Killing form remains non-degenerate on $\mathcal{L}_r$, and one can raise the question about ODs for $\mathcal{L}_r$. If $\mathcal{L}$ possesses an OD, then the same property is evidently satisfied by $\mathcal{L}_r$ for all $r \nmid (n+1)$. *Is the converse true?*

7) Returning to the field $\mathbb{C}$, it is reasonable to introduce a "Winnie-the-Pooh number" $wp(\mathcal{L})$ for any simple complex Lie algebra $\mathcal{L}$ – the maximal possible number of Cartan subalgebras of $\mathcal{L}$ that are orthogonal to the standard Cartan subalgebra and to each other. As will be shown, for Lie algebras of all types other than types A_n with $n+1$ not primary and C_n with $n \neq 2^m$, the Winnie-the-Pooh number is just the Coxeter number $h = h(\mathcal{L})$. Because of this, Problem 0.1 stated in the Introduction takes the following form:

Generalized Winnie-the-Pooh problem. *Prove that every simple complex Lie algebra $\mathcal{L}$ of type A_n with $n+1$ not primary, and type C_n with $n \neq 2^m$, satisfies the inequality $wp(\mathcal{L}) < h(\mathcal{L})$. Determine the exact value (or give upper bounds) of $wp(\mathcal{L})$ for these types.*

It is not difficult to show that

$$h(A_{n-1}) = n \geq wp(A_{n-1}) \geq \phi(n) + 1,$$

where $\phi(n)$ is the Euler function.

To engage the reader's interest, we suggest a quite elementary formulation of the Winnie-the-Pooh problem. Let $\mathcal{L} = sl_n(\mathbb{C})$ be the Lie algebra of type A_{n-1} and

$$\mathcal{L} = \mathcal{H}_0 \oplus \mathcal{H}_1 \oplus \ldots \oplus \mathcal{H}_n$$

an orthogonal decomposition. We know that if

$$\mathcal{H}_0 = \{\mathrm{diag}(\lambda_1, \lambda_2, \ldots, \lambda_n) \mid \sum_i \lambda_i = 0\}$$

is the standard Cartan subalgebra and $\mathcal{H}_s = X_s \mathcal{H}_0 X_s^{-1}$, the orthogonality condition for the pair $(\mathcal{H}_0, \mathcal{H}_s)$ is equivalent to the condition that X_s be a GH-matrix. Now if $s, t > 0$ and $s \neq t$, then by definition of OD we have

$$0 = K(\mathcal{H}_s, \mathcal{H}_t) = K(X_s \mathcal{H}_0 X_s^{-1}, X_t \mathcal{H}_0 X_t^{-1}) = K(\mathcal{H}_0, X_s^{-1} X_t \mathcal{H}_0 X_t^{-1} X_s).$$

So the matrices $X_s^{-1} X_t$, and X_s and X_t are GH-matrices. However, if $X = (x_{ij})$, $Y = (y_{ij})$ and $X^{-1}Y$ are GH-matrices, then in addition to the relations

$$\sum_{k=1}^{n} \frac{x_{ki}}{x_{kj}} = n\delta_{ij} = \sum_{k=1}^{n} \frac{x_{ik}}{x_{jk}}, \quad \sum_{k=1}^{n} \frac{y_{ki}}{y_{kj}} = n\delta_{ij} = \sum_{k=1}^{n} \frac{y_{ik}}{y_{jk}},$$

which were obtained in the proof of Lemma 1.5.5, the following equations must hold:

$$\left(\sum_{k=1}^{n} \frac{x_{ki}}{y_{kj}}\right) \cdot \left(\sum_{k=1}^{n} \frac{y_{kj}}{x_{ki}}\right) = n;$$

$$\sum_{s=1}^{n} \left(\sum_{k=1}^{n} \frac{y_{ks}}{x_{ki}}\right) \left(\sum_{k=1}^{n} \frac{x_{kj}}{y_{ks}}\right) = n^2 \delta_{ij}, \; 1 \leq i, j \leq n.$$

Using the non-degeneracy of the restriction of the Killing form to an arbitrary Cartan subalgebra, it is easy to show that our arguments are reversible; that is, the following statement holds:

The Lie algebra of type A_{n-1} possesses an OD if and only if there exist n GH-matrices $X_1, \ldots, X_n$ of degree n such that $X_i^{-1} X_j$ are GH-matrices for all i, j with $1 \leq i, j \leq n$, $i \neq j$.

The system of equations introduced above defines a certain affine algebraic variety. *What is this variety, in fact?*

8) The Garrick Club of London was founded in 1831, and would have gone out of existence long ago for reasons of bankruptcy, had not its late member A. A. Milne bequeathed it a fixed portion of the royalties from reprints of his book "Winnie-the-Pooh".

The authors promise to transfer \$1 to this honourable club.

Chapter 2

The types B_n, C_n and D_n

In this chapter, orthogonal decompositions for other classical Lie algebras are constructed. The results concerning orthogonal Lie algebras are characterized by their great variety. In this context, information about orthogonal decompositions of Lie algebras of type C_n is rather meagre in the sense that constructions available at present of ODs for this type are very limited.

2.1 Type C_n

As yet, ODs for Lie algebras of type C_n are known only in the case $n = 2^m$. It is very probable that the following conjecture is true:

Conjecture 2.1.1. *The Lie algebra of type C_n possesses an OD if and only if $n = 2^m$.*

This conjecture is open even in the simplest case $n = 3$, where we confront practically the same difficulties as in the case of the Winnie-the-Pooh problem for the Lie algebra of type A_5.

We realize the Lie algebra $\mathcal{L}$ of type C_{2^m} in the form

$$\{X \in M_{2^{m+1}}(\mathbb{C}) \mid X.J + J.{}^tX = 0\},$$

where $J = \begin{pmatrix} 0 & E_n \\ -E_n & 0 \end{pmatrix}$, and $n = 2^m$. Set $D = \begin{pmatrix} 1 & 0 \\ 0 & -1 \end{pmatrix}$, $P = \begin{pmatrix} 0 & 1 \\ 1 & 0 \end{pmatrix}$; then $D_8 =< D, P >= 2_+^{1+2}$ is the dihedral group of order 8 and $Q_8 =< iD, iP >= 2_-^{1+2}$ is the quaternion group of order 8. Recall (see §1.1) that we have fixed some special basis $\{e_1, \ldots, e_{m+1}, f_1, \ldots, f_{m+1}\}$ in the space $W = \mathbb{F}_2^{2(m+1)}$, and set

$$J_u = J_{(a_1,b_1)} \otimes \ldots \otimes J_{(a_{m+1},b_{m+1})}$$

for $u = (a_1, \dots, a_{m+1}, b_1, \dots, b_{m+1}) \in W$, where $J_{(a,b)} = D^a P^b$. We write further:

$$q(u) = \sum_{i=1}^{m+1} a_i b_i + (a_1 + b_1), \ < u \mid v >= q(u) + q(v) + q(u+v),$$

$$Q = \{u \in W \mid q(u) = 1\}.$$

Then W is a non-degenerate quadratic space of type $(-)$ (that is, it has Witt index $m = (\dim W)/2 - 1$). The operations on the matrices J_u are performed in accordance with formulas (1.3) and (1.4). In particular, the matrices of the shape $i^s J_u$ form the group $\mathbb{Z}_4 * \hat{H}$ under the usual matrix multiplication, where $\hat{H} = 2_-^{1+2(m+1)} = Q_8 * D_8 * \dots * D_8$.

Lemma 2.1.2. *The Lie algebra $\mathcal{L}$ of type C_{2^m} has $\{J_u \mid u \in Q\}$ as basis; it is an eigen-basis for the group $\hat{H} = 2_-^{1+2(m+1)}$, which acts on $\mathcal{L}$ with kernel $\mathbb{Z}_2$. Moreover, all the linear characters of $\hat{H}$ in this action are different.*

Proof. Set $J_u = J_{(a_1,b_1)} \otimes J_v$, where $v = (a_2, \dots, a_{m+1}, b_2, \dots, b_{m+1})$. Note that $J = DP \otimes E_n$. Then

$$J.^t(J_u) = (-1)^{B(u,u)} J.J_u = (-1)^{B(u,u)} (DP \otimes E_n)(J_{(a_1,b_1)} \otimes J_v) =$$

$$= (-1)^{B(u,u)} (DP.J_{(a_1,b_1)}) \otimes J_v = (-1)^{B(u,u)+a_1+b_1} (J_{(a_1,b_1)}.DP) \otimes J_v =$$

$$= (-1)^{q(u)} J_u.J.$$

In particular, $J_u \in \mathcal{L}$ if and only if $u \in Q$. All the remaining assertions of the lemma are easy to check. □

By analogy with Definition 1.2.1 we introduce:

Definition 2.1.3. An OD $\mathcal{L} = \oplus_i \mathcal{H}_i$ of the Lie algebra $\mathcal{L}$ is said to be a *J-decomposition* if each of its components $\mathcal{H}_i$ possesses a basis of the form $\{J_u \mid u \in Q_i\}$ for some $Q_i \subseteq Q$.

For convenience we set $W_i =< Q_i >_{\mathbb{F}_2}$.

Lemma 2.1.4. *The collection $\pi = \{W_i \mid 1 \le i \le 2^{m+1}\}$ forms a symplectic spread of the space W. In particular, every J-decomposition of the Lie algebra $\mathcal{L}$ of type C_{2^m} is obtained by restricting some J-decomposition of the Lie algebra $\hat{\mathcal{L}}$ of type $A_{2^{m+1}-1}$ to $\mathcal{L}$:*

$$\hat{\mathcal{L}} = \oplus_i \hat{\mathcal{H}}_i, \ \mathcal{L} = \oplus_i \mathcal{H}_i, \ \mathcal{H}_i = \hat{\mathcal{H}}_i \cap \mathcal{L}.$$

In addition, transitivity of an OD for $\mathcal{L}$ implies transitivity of an OD for $\hat{\mathcal{L}}$.

Proof. We know that $[J_u, J_v] = 0$ if and only if $< u \mid v >= 0$. So $< u \mid v >= 0$ for all $u, v \in Q_i$, and W_i is a totally isotropic subspace. Furthermore, $|W_i| \geq 1 + |Q_i| = 2^m + 1$, so that $\dim W_i \geq m + 1$. Thus W_i is a maximal totally isotropic subspace. By definition, we have $Q_i \subseteq Q \cap W_i$; however, q is a quadratic form of type $(-)$, so the converse is also true, that is, $Q_i = Q \cap W_i$. It remains to check that $W = \cup W_i$. Suppose that some vector a belongs to $W \setminus \cup W_i$. Then it is clear that $q(a) = 0$. Set $a^\perp = \{u \in W \mid < a \mid u >= 0\}$ and $Q^+ = Q \cap a^\perp$. Because a is singular, the subspace $a^\perp / < a >$ is a non-degenerate quadratic space of type $(-)$. Hence the number of non-singular vectors belonging to $a^\perp / < a >$ is equal to $2^{m-1}(2^m + 1)$, and $|Q^+| = 2^m(2^m + 1)$ since $q(u) = q(u + a)$ for every $u \in a^\perp$. Now we compute $|Q^+|$ in another way. Note that $\dim(W_i \cap a^\perp) \leq m$; otherwise we would have $W_i \subseteq a^\perp$, $a \in (W_i)^\perp = W_i$, which contradicts our assumptions. Furthermore, W_i is totally isotropic, so the restriction $q\mid_{(W_i \cap a^\perp)}$ is a linear functional and the set $\{b \in W_i \cap a^\perp \mid q(b) = 1\}$ has cardinality at most 2^{m-1}. Hence

$$2^m(2^m + 1) = |Q^+| = |Q \cap a^\perp| = |(\cup_i Q_i) \cap a^\perp| = |\cup (Q_i \cap a^\perp)| \leq$$

$$\leq (2^{m+1} + 1)2^{m-1},$$

a contradiction. □

Setting $r = 2n = 2^{m+1}$, we note that $\mathbb{F}_{r^2} = \mathbb{F}_r(\vartheta)$, where $\vartheta^{r+1} = 1$. Identifying $W = \mathbb{F}_r \oplus \mathbb{F}_r = \mathbb{F}_2^{2(m+1)}$ with $\mathbb{F}_{r^2} = \mathbb{F}_r \oplus \vartheta\mathbb{F}_r$, we consider the quadratic form

$$\hat{q}(u) = \mathrm{Tr}_{\mathbb{F}_r/\mathbb{F}_2}(\varepsilon u^{r+1})$$

on it, where $\varepsilon^{-1} = \vartheta + \vartheta^r \in \mathbb{F}_r^*$. Then for $u = \alpha + \vartheta\beta$, $v = \gamma + \vartheta\delta$, $\alpha, \beta, \gamma, \delta \in \mathbb{F}_r$ we have (note that the index $\mathbb{F}_r/\mathbb{F}_2$ in the notation for the trace is omitted)

$$\hat{q}(u) = \mathrm{Tr}(\varepsilon(\alpha + \vartheta\beta)(\alpha + \vartheta\beta)^r) = \mathrm{Tr}(\varepsilon(\alpha + \vartheta\beta)(\alpha + \vartheta^r\beta)) =$$

$$= \mathrm{Tr}(\varepsilon(\alpha^2 + \beta^2 + (\vartheta + \vartheta^r)\alpha\beta)) = \mathrm{Tr}(\alpha\beta + \varepsilon(\alpha^2 + \beta^2)).$$

Similarly,

$$\hat{q}(v) = \mathrm{Tr}(\gamma\delta + \varepsilon(\gamma^2 + \delta^2)),\ \hat{q}(u + v) = \mathrm{Tr}((\alpha + \gamma)(\beta + \delta) + \varepsilon(\alpha^2 + \beta^2 + \gamma^2 + \delta^2)).$$

In particular,

$$\hat{q}(u) + \hat{q}(v) + \hat{q}(u + v) = \mathrm{Tr}_{\mathbb{F}_r/\mathbb{F}_2}(\alpha\delta + \beta\gamma) = \mathrm{Tr}_{\mathbb{F}_r/\mathbb{F}_2}(u \circ v) =< u \mid v >$$

(see §1.1). Furthermore, we set $R = \{\alpha \in \mathbb{F}_r \mid \mathrm{Tr}_{\mathbb{F}_r/\mathbb{F}_2}(\alpha) = 1\}$. It is clear that $|R| = r/2$. For $u \in \mathbb{F}_{r^2}$, we have

$$\hat{q}(u) = 1 \Leftrightarrow \mathrm{Tr}_{\mathbb{F}_r/\mathbb{F}_2}(\varepsilon u^{r+1}) = 1 \Leftrightarrow \varepsilon u^{r+1} \in R.$$

Choose a generating element τ of the group $\mathbb{F}_{r^2}^*$ such that $\vartheta = \tau^{r-1}$. Every element from the set $\varepsilon^{-1}R$, like any nonzero element from $\mathbb{F}_r$, can be written in the form $\tau^{i(r+1)}$, $0 \le i < r-1$. So, setting $u = \tau^j$, $0 \le j < r^2 - 1$, we have $\hat{q}(u) = 1$ if and only if $\tau^{j(r+1)} = \tau^{i(r+1)}$ for some i, $0 \le i < r-1$; that is, $j = i + k(r-1)$ for some integer k. In other words, $\hat{q}(u) = 1$ if and only if $u = \vartheta^k \tau^i$. Therefore, the equation $\hat{q}(u) = 1$ has precisely $r(r+1)/2 = 2^m(2^{m+1}+1)$ roots in $\mathbb{F}_{r^2} = W$. So $\hat{q}$ is a quadratic form of type $(-)$, associated with the symplectic form $< * \mid * >$. By choosing the basis $\{e_1, \dots, f_{m+1}\}$ appropriately, we can assume that $q = \hat{q}$. Hence,

$$q(u) = \mathrm{Tr}_{\mathbb{F}_r/\mathbb{F}_2}(\varepsilon u^{r+1}) = \mathrm{Tr}_{\mathbb{F}_r/\mathbb{F}_2}(\alpha\beta + \varepsilon(\alpha^2 + \beta^2))$$

for $u = (\alpha, \beta) \in W$.

Now we consider the standard OD

$$\hat{\mathcal{D}} : \hat{\mathcal{L}} = \bigoplus_{\alpha \in \mathbb{F}_r} \hat{\mathcal{H}}_\alpha \bigoplus \hat{\mathcal{H}}_\infty$$

for the Lie algebra $\hat{\mathcal{L}}$ of type $A_{2^{m+1}-1}$ described in §1.1. Under our identification $W = \mathbb{F}_r \oplus \vartheta\mathbb{F}_r$, we have $\hat{\mathcal{H}}_\infty = < J_{\lambda\vartheta} \mid \lambda \in \mathbb{F}_r^* >_{\mathbb{C}}$ and $\hat{\mathcal{H}}_\alpha = < J_{\lambda(1+\alpha\vartheta)} \mid \lambda \in \mathbb{F}_r^* >_{\mathbb{C}}$ for $\alpha \in \mathbb{F}_r$. Let

$$\mathcal{D} : \mathcal{L} = \bigoplus_{\alpha \in \mathbb{F}_r} \mathcal{H}_\alpha \bigoplus \mathcal{H}_\infty$$

be the corresponding OD for $\mathcal{L}$, where $\mathcal{H}_\beta = \hat{\mathcal{H}}_\beta \cap \mathcal{L}$ for $\beta \in \mathbb{F}_r \cup \{\infty\}$.

Proposition 2.1.5. *The OD so constructed is a transitive OD of the Lie algebra $\mathcal{L}$ of type C_{2^m}. In addition,*

$$\mathrm{Aut}(\mathcal{D}) = \mathbb{Z}_2^{2(m+1)}.(D_{2(2^{m+1}+1)}.\mathbb{Z}_{m+1}).$$

Proof. 1) Suppose that $\Phi \in \mathrm{Aut}(\mathcal{D})$. Like every automorphism of $\mathcal{L}$, Φ can be written in the form $X \mapsto FXF^{-1}$ for some symplectic matrix F of degree $r = 2^{m+1}$. Thus, Φ can be extended to an automorphism of the associative algebra $M_r(\mathbb{C})$. In the notation of the proof of Lemma 2.1.4, the equality $< Q_i >_{\mathbb{F}_2} = W_i$ implies immediately that the Lie algebra $\mathcal{H}_i$ generates the associative subalgebra $< \hat{\mathcal{H}}_i, E_r >_{\mathbb{C}}$. Hence, Φ can be extended to an inner automorphism of the Lie algebra $\hat{\mathcal{L}}$ that fixes the decomposition $\hat{\mathcal{D}}$: $\Phi \in \mathrm{Inn}(\hat{\mathcal{D}})$. It is shown in §1.3 that

$$\mathrm{Inn}(\hat{\mathcal{D}}) = \mathbb{Z}_2^{2(m+1)}.A,$$

where $\mathbb{Z}_2^{2(m+1)} = \hat{H}/Z(\hat{H})$ and $A = \Sigma L_2(r)$ consists of transformations of the $\mathbb{F}_2$-space $W = \mathbb{F}_r \oplus \mathbb{F}_r$ of the form

$$f : (\alpha, \beta) \mapsto (x\alpha^{2^k} + y\beta^{2^k}, z\alpha^{2^k} + t\beta^{2^k})$$

for some $0 \le k \le m$, $\begin{pmatrix} x & y \\ z & t \end{pmatrix} \in SL_2(r)$.

Conversely, every automorphism $\Phi \in \mathrm{Inn}(\hat{\mathcal{D}})$ which maps the set of lines $< J_u >_{\mathbb{C}}$, $u \in Q$, into itself, is obviously contained in $\mathrm{Aut}(\mathcal{D})$. Thus we have shown that $\mathrm{Aut}(\mathcal{D}) = \mathbb{Z}_2^{2(m+1)}.B$, where

$$B = \{f \in A \mid q(f(u)) = q(u), \forall u \in W\}.$$

2) Next we determine the group B. Note that ε, like every element of the field $\mathbb{F}_r$, $r = 2^{m+1}$, has a unique square root $\nu = \sqrt{\varepsilon}$. Now consider the map

$$f_1 : (\alpha, \beta) \mapsto (\nu^{-1}\alpha^2, \nu(\alpha^2 + \beta^2))$$

lying in A. Then, for any $u = (\alpha, \beta) \in W$, we have

$$q(f_1(u)) = \mathrm{Tr}_{\mathbb{F}_r/\mathbb{F}_2}(\alpha^4 + \alpha^2\beta^2 + \varepsilon\nu^{-2}\alpha^4 + \varepsilon\nu^2\alpha^4 + \varepsilon\nu^2\beta^4) =$$

$$= \mathrm{Tr}_{\mathbb{F}_r/\mathbb{F}_2}(\alpha^2\beta^2 + \varepsilon^2(\alpha^4 + \beta^4)) = \mathrm{Tr}_{\mathbb{F}_r/\mathbb{F}_2}((\alpha\beta + \varepsilon(\alpha^2 + \beta^2))^2) =$$

$$= \mathrm{Tr}_{\mathbb{F}_r/\mathbb{F}_2}(\alpha\beta + \varepsilon(\alpha^2 + \beta^2)) = q(u),$$

that is, $f_1 \in B$. Here, f_1 corresponds to the Frobenius automorphism $\alpha \mapsto \alpha^2$ of $\mathbb{F}_r$.

Next we consider an arbitrary element $g \in B$. Changing g by using f_1, one can suppose that $g \in SL_2(r)$, that is,

$$g : (\alpha, \beta) \mapsto (x\alpha + y\beta, z\alpha + t\beta) \tag{2.1}$$

for some $x, y, z, t \in \mathbb{F}_r$, $xt + yz = 1$. The condition $q(g(u)) = q(u)$ for all $u \in W$ is equivalent to the following two equations:

$$x^2 + z^2 + \varepsilon^{-1}xz = 1, \tag{2.2}$$

$$y^2 + t^2 + \varepsilon^{-1}yt = 1. \tag{2.3}$$

When $x = 0$ it follows from (2.1) – (2.3) that $y = z = 1$ and $t \in \{0, \varepsilon^{-1}\}$. Now assume that $x \neq 0$. Setting $t = x^{-1}(1 + yz)$ in (2.3), we see that $y^2(x^2 + z^2 + \varepsilon^{-1}xz) + x^2 + \varepsilon^{-1}xy = 1$. This equation, together with (2.2), implies that $y = z$ or $y = z + \varepsilon^{-1}x$. By definition, $\varepsilon^{-1} = \vartheta + \vartheta^{-1}$, so (2.2) is equivalent to the equality $(\vartheta x + z)(\vartheta^{-1}x + z) = 1$. In other words, $\vartheta x + z = a$ and $\vartheta^{-1}x + z = a^{-1}$ for some $a \in \mathbb{F}_{r^2}^*$. Hence,

$$x = \varepsilon(a + a^{-1}),\ z = \varepsilon(a\vartheta^{-1} + \vartheta a^{-1}).$$

Note that for $b \in \mathbb{F}_{r^2}^*$, the element $b + b^{-1}$ is contained in $\mathbb{F}_r$ if and only if $b \in \mathbb{F}_r^* \cup \{\vartheta, \vartheta^2, \ldots, \vartheta^r\}$. (In fact,

$$b + b^{-1} \in \mathbb{F}_r \Leftrightarrow b + b^{-1} = (b + b^{-1})^r = b^r + b^{-r} \Leftrightarrow$$

$$\Leftrightarrow 0 = b^{2r} + b^{r+1} + b^{r-1} + 1 = (b^{r+1} - 1)(b^{r-1} - 1).)$$

In our case $a + a^{-1} \in \mathbb{F}_r$ (since $x, \varepsilon \in \mathbb{F}_r$), whereas $a \notin \mathbb{F}_r$ (since $\vartheta \notin \mathbb{F}_r$). So $a \in \{\vartheta, \vartheta^2, \dots, \vartheta^r\}$.

We have thus shown that the system of equations (2.1) – (2.3) has exactly $2r + 2 = 2(r + 1)$ solutions. In other words,

$$|B| = 2(r + 1)(m + 1).$$

3) Next, we determine the structure of B. To this end, we consider two explicit elements of B:

$$f_2 : (\alpha, \beta) \mapsto (\beta, \alpha),\ f_3 : (\alpha, \beta) \mapsto (\beta, \alpha + \varepsilon^{-1}\beta).$$

It is clear that $f_2 \in B$, $|f_2| = 2$ and $f_2 f_3 f_2^{-1} = f_3^{-1}$. Furthermore, f_3 is nothing other than multiplication by ϑ: $u = (\alpha, \beta) = \alpha + \vartheta\beta \mapsto \vartheta u$. So $|f_3| = |\vartheta| = r + 1$ and $f_3 \in B$:

$$q(f_3(u)) = \mathrm{Tr}_{\mathbb{F}_r/\mathbb{F}_2}(\varepsilon(\vartheta u)^{r+1}) = \mathrm{Tr}_{\mathbb{F}_r/\mathbb{F}_2}(\varepsilon u^{r+1}) = q(u).$$

Therefore $< f_2, f_3 > \cong D_{2(r+1)}$ and $B =< f_1, f_2, f_3 > \cong D_{2(r+1)}.\mathbb{Z}_{m+1}$, as stated.

4) Finally, the transitivity of the given OD follows from the fact that f_3 permutes the $r + 1$ components of $\mathcal{D}$ cyclically. □

On replacing the form $q(u)$ by the quadratic form $u = (\alpha, \beta) \mapsto \mathrm{Tr}_{\mathbb{F}_r/\mathbb{F}_2}(\alpha\beta)$ and the group $\hat{H} = 2_-^{1+2(m+1)}$ by the group $2_+^{1+2(m+1)} = D_8 * \dots * D_8$ in the construction described above, we get the following result:

Proposition 2.1.6. *The Lie algebra of type D_{2^m} ($m \geq 3$) has a transitive OD $\mathcal{D}$ with automorphism group*

$$\mathrm{Aut}(\mathcal{D}) = \mathbb{Z}_2^{2(m+1)}.(D_{2(2^{m+1}-1)}.\mathbb{Z}_{m+1}).$$ □

Recall that in §1.4 we gave a scheme for proving the uniqueness of ODs for Lie algebras of small rank. This scheme, which is given in detail in [KKU 4], yields the following result:

Theorem 2.1.7. *The complex simple Lie algebra $\mathcal{L}$ of type C_2 admits an OD which is unique up to* $\mathrm{Aut}(\mathcal{L})$*-conjugacy.*

We shall omit the details of the proof, and simply give the list of all Cartan subalgebras $\mathcal{H} =< A, B >_{\mathbb{C}}$ which are orthogonal to the standard Cartan subalgebra

$$\mathcal{H}_0 =< \mathrm{diag}(1, 0, -1, 0), \mathrm{diag}(0, 1, 0, -1) >_{\mathbb{C}} .$$

(Here the standard matrix model of the Lie algebra $\mathcal{L}$ of type C_2 is used:

$$A = \begin{pmatrix} A_1 & A_2 \\ A_3 & A_4 \end{pmatrix} \in \mathcal{L} \Leftrightarrow A_4 = -{}^tA_1,\ {}^tA_i = A_i, i = 2, 3.)$$

$$1)\ A = \begin{pmatrix} 0 & 0 & a & 0 \\ 1 & 0 & 0 & c \\ d & 0 & 0 & -1 \\ 0 & f & 0 & 0 \end{pmatrix},\ B = \begin{pmatrix} 0 & af & acf & 0 \\ 0 & 0 & 0 & a(1+cd) \\ f(1+cd) & 0 & 0 & 0 \\ 0 & adf & -af & 0 \end{pmatrix}.$$

Here $af \neq 0$, $cd \neq -1$.

$$2)\ A = \begin{pmatrix} 0 & 0 & a & 0 \\ 1 & 0 & 0 & c \\ -1/c & 0 & 0 & -1 \\ 0 & f & 0 & 0 \end{pmatrix},\ B = \begin{pmatrix} 0 & 1 & c & 0 \\ 0 & 0 & 0 & z \\ t & 0 & 0 & 0 \\ 0 & -1/c & -1 & 0 \end{pmatrix}.$$

Here $c \neq 0$ and either $a = 0$, $z = 0$, $ft \neq 0$, or $az \neq 0$, $f = 0$, $t = 0$.

$$3)\ A = \begin{pmatrix} 0 & 0 & a & 0 \\ 0 & 0 & 0 & c \\ d & 1 & 0 & 0 \\ 1 & f & 0 & 0 \end{pmatrix},\ B = \begin{pmatrix} 0 & 0 & -acf & ac \\ 0 & 0 & ac & -acd \\ c(1-df) & 0 & 0 & 0 \\ 0 & a(1-df) & 0 & 0 \end{pmatrix},$$

$ac \neq 0$, $df \neq 1$.

$$4)\ A = \begin{pmatrix} 0 & 0 & a & 0 \\ 0 & 0 & 0 & c \\ d & 1 & 0 & 0 \\ 1 & 1/d & 0 & 0 \end{pmatrix},\ B = \begin{pmatrix} 0 & 0 & 1/d & -1 \\ 0 & 0 & -1 & d \\ t & 0 & 0 & 0 \\ 0 & v & 0 & 0 \end{pmatrix}.$$

Here $d \neq 0$ and either $a = 0$, $v = 0$, $ct \neq 0$, or $av \neq 0$, $c = 0$, $t = 0$.

$$5)\ A = \begin{pmatrix} 0 & 0 & 0 & 0 \\ 0 & 0 & 0 & 1 \\ 0 & 0 & 0 & 0 \\ 0 & f & 0 & 0 \end{pmatrix},\ B = \begin{pmatrix} 0 & 0 & 1 & 0 \\ 0 & 0 & 0 & 0 \\ t & 0 & 0 & 0 \\ 0 & 0 & 0 & 0 \end{pmatrix},\ tf \neq 0.$$

$$6)\ A = \begin{pmatrix} 0 & 0 & 1 & 0 \\ 0 & 0 & 0 & c \\ cf & 0 & 0 & 0 \\ 0 & f & 0 & 0 \end{pmatrix},\ B = \begin{pmatrix} 0 & 0 & 0 & 1 \\ 0 & 0 & 1 & 0 \\ 0 & f & 0 & 0 \\ f & 0 & 0 & 0 \end{pmatrix},\ cf \neq 0.$$

$$7)\ A = \begin{pmatrix} 0 & 0 & 1 & 0 \\ 0 & 0 & 0 & c \\ cf & 0 & 0 & 0 \\ 0 & f & 0 & 0 \end{pmatrix},\ B = \begin{pmatrix} 0 & 0 & 0 & y \\ 0 & 0 & y & 0 \\ 0 & fy & 0 & 0 \\ fy & 0 & 0 & 0 \end{pmatrix},\ cfy \neq 0.$$

$$8)\ A = \begin{pmatrix} 0 & 0 & 1 & 0 \\ 0 & 0 & 0 & c \\ cf & 0 & 0 & 0 \\ 0 & f & 0 & 0 \end{pmatrix},\ B = \begin{pmatrix} 0 & 1 & 0 & y \\ -c & 0 & y & z \\ 0 & fy & 0 & c \\ fy & fz/c & -1 & 0 \end{pmatrix},\ cf \neq 0.$$

To conclude this section we prove:

Proposition 2.1.8. *The Lie algebra $\mathcal{L}$ of type C_n admits an irreducible J-decomposition if and only if $n = 2$.*

Proof. Suppose that $\mathcal{L}$ admits an irreducible J-decomposition $\mathcal{D}$ and, as ever, set $G = \mathrm{Aut}(\mathcal{D})$, and let $\hat{G}$ be the complete inverse image of G in $Sp_{2n}(\mathbb{C})$. Later (see Chapter 4) we shall see that $n = 2^m$, $\hat{G} = \hat{H}.G_0$, $\hat{H} = 2_-^{1+2(m+1)}$ and $G_0 \subseteq O^-_{2m+2}(2)$. According to Lemma 2.1.4, this J-decomposition corresponds to some symplectic spread π of the space $W = \mathbb{F}_2^{2(m+1)}$. Furthermore, $G_0 \subseteq \mathrm{Aut}(\pi)$ acts transitively on the components of π and the set Q of non-singular points of W; we have $|Q| = 2^m(2^{m+1} + 1)$. According to Lemma 1.2.11, π corresponds to some flag-transitive finite space $\mathcal{V}$. By Proposition 1.2.12, we have to consider the three following cases.

1) π is desarguesian.

Since in our case the number of points belonging to one line is $N = 2^m$, then, as is shown in [BDDKLS], G_0 is transitive on $W \setminus \{0\}$. But Q is a G_0-orbit, a contradiction.

2) $m = 4e + 1$, $q = 2^{2e+1}$ and $Sz(q) \lhd G_0 \subseteq \mathrm{Aut}(Sz(q))$.

Setting $S = Sz(q)$, we have $G_0 = S\mathbb{Z}_a$, where $a|(2e + 1)$. Because G_0 is transitive on the set $\{W_1, \dots, W_{q^2+1}\}$ of components of π, S is $(1/2)$-transitive on the same set. Thus $(S : S_1) = d$, $d|(q^2 + 1)$ and $d \geq (q^2 + 1)/a > 1$, where $S_1 = St_S(W_1)$. Here, S_1 is a proper subgroup of odd index in S. By the Borel-Tits Theorem, S_1 is embedded in some parabolic subgroup P of S. But $(S : P) = q^2+1$, so in fact $S_1 = P$ and $d = q^2 + 1$. Setting $G_1 = St_{G_0}(W_1)$, we have $S_1 \lhd G_1$ and $G_1/S_1 = G_0/S = \mathbb{Z}_a$. Recall that G_1 acts transitively on the $2^m = q^2/2$ points of the set $Q_1 = Q \cap W_1$, and a is odd. Therefore S_1 is also transitive on Q_1, and so S_1 possesses a subgroup S_2 of index $q^2/2$. It is well known that $S_1 = U.H$, where U has order q^2 and contains $q - 1$ involutions, all of which are central in U, and $H = \mathbb{Z}_{q-1}$ permutes these involutions transitively. Now $|S_2| = 2(q - 1)$, so $US_2 = S_1$ and $\mathbb{Z}_2 = U \cap S_2 \lhd S_2$. In other words, H is isomorphic to $S_2/(U \cap S_2)$ and fixes some involution from $U \cap S_2$, a contradiction.

3) $G_0 \subseteq \Gamma L_1(2^{2(m+1)})$.

Since $|Q| = 2^m(2^{m+1} + 1)$ divides

$$|\Gamma L_1(2^{2(m+1)})| = 2(m + 1)(2^{2(m+1)} - 1)$$

by the transitivity of G_0 on Q, we must have that 2^{m-1} divides $m + 1$. In other words, $m = 1$ or $m = 3$. In the case $m = 1$, $\mathcal{L}$ has type C_2 and by Theorem 2.1.7, all ODs of $\mathcal{L}$ are irreducible and conjugate. Suppose that $m = 3$. Then $G_0 \subseteq \Gamma L_1(2^8) = (\mathbb{Z}_{17} \times \mathbb{Z}_{15}).\mathbb{Z}_8$. Setting $G_1 = St_{G_0}(u)$ for $u \in Q$, we have $|G_1| = r$, where $r|15$. So G_1 is a Hall subgroup of the soluble group $\Gamma L_1(2^8)$. As a consequence of this last assertion, we see that G_1 is just the normal subgroup $\mathbb{Z}_r$

of $\mathbb{Z}_{15}$ and $\Gamma L_1(2^8)$. If $\mathbb{F}_{256}^* =< \varepsilon >$, then $\mathbb{Z}_{15}$ is generated by the map $x \mapsto \varepsilon^{17} x$. But G_1 fixes the point $u \in \mathbb{F}_{256}^*$, so $r = 1$ and

$$G_0 = \mathbb{Z}_{17}.\mathbb{Z}_8 =< f > . < g >,$$

where $f : x \mapsto \varepsilon^{15} x$ and $g : x \mapsto \varepsilon^k x^2$ for some k. The group $< f >$ decomposes $\mathbb{F}_{256}^*$ into 15 orbits Ω_j, where $\Omega_j = \{\varepsilon^i \mid i \equiv j (\text{mod } 15)\}$ for $j \in \mathbb{Z}_{15}$, and $g(\Omega_j) = \Omega_{k+2j}$. In particular, $g^4(\Omega_j) = \Omega_j$ for all j. Hence the length of an arbitrary G_0-orbit is not larger than $4|\Omega_j| = 4.17$, and so G_0 cannot be transitive on the set Q of cardinality 8.17, a contradiction. □

2.2 Partitions of complete graphs and E-decompositions of Lie algebras of types B_n and D_n: automorphism groups

We realize the Lie algebra $\mathcal{L}$ of type D_n as the set of skew-symmetric $2n \times 2n$-matrices: $\mathcal{L} =< E(i,j) \mid 1 \leq i \neq j \leq 2n >_{\mathbb{C}}$, where $E(i,j) = E_{ij} - E_{ji}$ and $E_{ij} = (\delta_{ik}\delta_{jl})_{1 \leq k,l \leq 2n}$ are the standard matrix units.

Definition 2.2.1. An orthogonal decomposition $\mathcal{L} = \oplus_{k=1}^m \mathcal{H}_k$ is said to be an *E-decomposition* if every Cartan subalgebra $\mathcal{H}_k$ has a basis consisting of some elements $E(i,j)$.

The matrices $E(i,j)$ satisfy the following obvious relations:

$$E(i,j) = -E(j,i),\ (E(i,j))^3 = -E(i,j), \tag{2.4}$$

$$\text{Tr}(E(i,j).E(k,l)) = \begin{cases} 0 & \text{if } \{i,j\} \neq \{k,l\}, \\ \pm 2 & \text{if } \{i,j\} = \{k,l\}, \end{cases} \tag{2.5}$$

$$[E(i,j), E(k,l)] = \begin{cases} E(i,l) & \text{if } j = k, \\ 0 & \text{if } \{i,j\} \cap \{k,l\} = \emptyset. \end{cases} \tag{2.6}$$

These relations show that construction of an E-decomposition $\mathcal{D}$ of $\mathcal{L}$ is equivalent to constructing a partition $\mathcal{P}$ of the set $X(2)$ of all unordered pairs $\{i,j\}$, $1 \leq i \neq j \leq 2n$, into disjoint subsets M_k, $1 \leq k \leq 2n-1$ such that

$$|M_k| = n \text{ and } \alpha \cap \beta = \emptyset \text{ for all } \alpha, \beta \in M_k,\ \alpha \neq \beta. \tag{2.7}$$

Of course, we can view $\mathcal{P}$ as a partition of the complete graph with vertex set $X = \{1, 2, \ldots, 2n\}$ and edge set $X(2)$. Such partitions we shall call *admissible*. The equivalence relation is given by

$$\mathcal{H}_k =< E(i,j) \mid \{i,j\} \in M_k >_{\mathbb{C}} . \tag{2.8}$$

The commutativity of $\mathcal{H}_k$, the non-degeneracy of the restriction of the Killing form K to $\mathcal{H}_k$ and the pairwise orthogonality of the $\mathcal{H}_k$, $1 \leq k \leq 2n-1$, follow from relations (2.4) – (2.6).

Note that a linear combination

$$\sum_{\{i,j\}\in M_k, i<j} \mu_{ij} E(i,j)$$

has rank $2s$, where s is the number of nonzero coefficients μ_{ij}. So

$$\Phi(E(i,j)) = \lambda_{ij} E(k,l) \tag{2.9}$$

for every $\Phi \in \mathrm{Aut}(\mathcal{D})$. Here, it is necessary to remark that $\Phi(Z) = AZA^{-1}$ for every $\Phi \in \mathrm{Aut}(\mathcal{D})$, $Z \in \mathcal{L}$ and for a suitable orthogonal matrix A. So,

$$\mathrm{rank}(\Phi(Z)) = \mathrm{rank}(Z),\ \Phi((E(i,j))^3) = (\Phi(E(i,j)))^3. \tag{2.10}$$

This fact, which will be needed in what follows, holds good for any $n > 4$. In the case $n = 4$ there is an outer automorphism T of order 3, and the question as to whether T fixes a given orthogonal decomposition or not must be investigated individually for every concrete OD.

Relation (2.9) shows also that every automorphism $\Phi \in \mathrm{Aut}(\mathcal{D})$ is defined by the set of coefficients λ_{ij} together with some bijection $\varphi : X(2) \to X(2)$ satisfying the following conditions:

$$\varphi(M_k) = M_l, \tag{2.11}$$

$$|\varphi(\{i,j\}) \cap \varphi(\{k,l\})| = |\{i,j\} \cap \{k,l\}|. \tag{2.12}$$

The last condition follows from (2.6).

Lemma 2.2.2. *Every bijection $X(2) \to X(2)$ satisfying condition* (2.12) *is induced by some permutation π of the set X:*

$$\varphi(\{i,j\}) = \{\pi(i), \pi(j)\}.$$

Proof. From (2.12) it follows that

$$\varphi : \{1,2\} \mapsto \{a_1, a_2\},\ \{1,3\} \mapsto \{a_1, a_3\}$$

for some a_1, a_2, a_3. In this case, the image of $\{1,i\}$ is $\{a_1, a_i\}$ for some a_i, or $\{a_2, a_3\}$. However, the latter contradicts the fact that φ is a bijection, since $\varphi(\{2,3\}) = \{a_2, a_3\}$ by (2.12). So

$$\varphi : \{1,i\} \mapsto \{a_1, a_i\},\ \{1,j\} \mapsto \{a_1, a_j\},$$

and by (2.12) we have

$$\varphi : \{i, j\} \mapsto \{a_i, a_j\} = \{\pi(i), \pi(j)\},$$

where $\pi(i) = a_i$. □

As we have shown, every automorphism $\Phi \in \mathrm{Aut}(\mathcal{D})$ has the form

$$\Phi(E(i, j)) = \lambda_{ij} E(\pi(i), \pi(j)).$$

Here, the permutation π must preserve the corresponding admissible partition $\mathcal{P}$. Applying Φ to relations (2.4) and (2.6), we find that

$$\lambda_{ij} = \lambda_{ji},\ \lambda_{ij}^2 = 1,\ \lambda_{ij}\lambda_{jk} = \lambda_{ik}. \tag{2.13}$$

It is convenient to set $\lambda_{ii} = 1$. From these relations it follows that $\lambda_{2n,i} = (-1)^{f(i)+f(2n)}$ for some map $f : X \to \mathbb{Z}_2$. Moreover, $\lambda_{ij} = \lambda_{2n,i}\lambda_{2n,j} = (-1)^{f(i)+f(j)}$. The coefficients λ_{ij} defined in such a way satisfy (2.13) for any f. Note that the functions f and f', where $f'(i) = f(i) + 1$, define the same collections $\{\lambda_{ij}\}$. The functions $f : X \to \mathbb{Z}_2$ form an elementary abelian group under the operation of pointwise addition. The element of its factor-group by the subgroup $\{\mathbf{0}, \mathbf{1}\}$, where $\mathbf{0}(i) = 0$ and $\mathbf{1}(i) = 1$ for all i, corresponding to the functions f and the f', will be denoted by $\bar{f}$. Finally we have $\Phi = \Phi_{\pi,\bar{f}}$:

$$\Phi_{\pi,\bar{f}}(E(i, j)) = (-1)^{f(i)+f(j)} E(\pi(i), \pi(j)). \tag{2.14}$$

The group $\mathrm{Aut}(\mathcal{D})$ has the two following naturally distinguished subgroups: the normal elementary abelian subgroup $\bar{X} = \{\Phi_{e,\bar{f}}\}$ of order 2^{2n-1} (where $e(i) = i$ for all i) and the subgroup $\{\Phi_{\pi,\bar{\mathbf{0}}}\}$ isomorphic to the group $\mathrm{Aut}(\mathcal{P})$ of automorphisms of the given admissible partition

$$\mathcal{P} : X(2) = \bigcup_{k=1}^{2n-1} M_k.$$

Moreover, $\mathrm{Aut}(\mathcal{D}) \cong \bar{X}.\mathrm{Aut}(\mathcal{P})$ is a semidirect product:

$$\Phi_{\pi,\bar{f}} = \Phi_{e,\bar{f}}.\Phi_{\pi,\bar{\mathbf{0}}},\ (\Phi_{\pi,\bar{\mathbf{0}}})^{-1}\Phi_{e,\bar{f}}\Phi_{\pi,\bar{\mathbf{0}}} = \Phi_{\pi,\overline{f\circ\pi}}.$$

Thus we have shown:

Proposition 2.2.3. *Suppose that $\mathcal{D}$ is an E-decomposition of the Lie algebra $\mathcal{L}$ of type D_n corresponding to a partition $\mathcal{P} : X(2) = \cup M_k$ of the complete graph $X(2)$. Then* $\mathrm{Aut}(\mathcal{D}) = \bar{X}.\mathrm{Aut}(\mathcal{P})$, *where $\bar{X}$ is an elementary abelian group of order* 2^{2n-1} *and*

$$\mathrm{Aut}(\mathcal{P}) = \{\pi \in \mathrm{Sym}(X) \mid \forall k, \exists l, \pi(M_k) = M_l\}.$$ □

If we go over to the Lie algebra $\mathcal{L}'$ of type B_{n-1}, then we can realize $\mathcal{L}'$ as the set of skew-symmetric matrices of order $2n-1$:

$$\mathcal{L}' =< E(i,j) \mid 1 \le i \neq j \le 2n-1 >_{\mathbb{C}} .$$

Similarly, set $X' = \{1, 2, \dots, 2n-1\}$, and let $X'(2)$ be the set of unordered pairs of elements of X', in other words, the complete graph with $2n-1$ vertices. Then construction of an E-decomposition of $\mathcal{L}'$ is equivalent to constructing a partition of the set $X'(2)$ into subsets M'_k satisfying condition (2.7) with n replaced by $n-1$. Without loss of generality, we can assume that M_k contains the pair $\{k, 2n\}$ and that M'_k has the form

$$M'_k = \big\{\{i_1, j_1\}, \dots, \{i_{n-1}, j_{n-1}\} \mid i_s, j_s \in \{1, 2, \dots, 2n-1\} \backslash \{k\}\big\}.$$

Corresponding the partition $\mathcal{P} : X(2) = \bigcup_{k=1}^{2n-1} M_k$, we form the partition $\mathcal{P}' : X'(2) = \bigcup_{k=1}^{2n-1} M'_k$, where $M'_k = M_k \backslash \{k, 2n\}$. Conversely, every such partition $\mathcal{P}' : X'(2) = \bigcup_{k=1}^{2n-1} M'_k$ corresponds to the partition $\mathcal{P} : X(2) = \bigcup_{k=1}^{2n-1} M_k$, where $M_k = M'_k \bigcup \{\{k, 2n\}\}$. We have thus established a bijective correspondence between E-decompositions of $\mathcal{L}$ and of $\mathcal{L}'$.

For every automorphism $\Phi \in \mathrm{Aut}(\mathcal{D}')$, where the E-decomposition $\mathcal{D}'$ corresponds to the partition $\mathcal{P}'$ of $X'(2)$, the same arguments as in the case of $\mathcal{L}$ allow us to show that $\Phi = \Phi_{\pi', f'}$. Here

$$\pi' \in \mathrm{Aut}(\mathcal{P}') = \left\{\varphi \in \mathrm{Sym}(X') \mid \forall k, \exists l, \varphi(M'_k) = M'_l\right\},$$

and

$$f' \in \overline{X'} = \left\{ g : X' \to \mathbb{Z}_2 \mid \sum_{i \in X'} g(i) = 0 \right\}.$$

We have identified the factor-group of the group $\{h : X' \to \mathbb{Z}_2\}$ by the subgroup $\{\mathbf{0}', \mathbf{1}'\}$, where $\mathbf{0}'(i) = 0$ and $\mathbf{1}'(i) = 1$, with $\overline{X'}$. Thus we have proved:

Proposition 2.2.4. *Suppose that $\mathcal{D}$ is an E-decomposition of the Lie algebra $\mathcal{L}$ of type B_n corresponding to a partition $\mathcal{P} : X(2) = \cup M_k$ of the complete graph $X(2)$, where $X = \{1, 2, \dots, 2n+1\}$. Then* $\mathrm{Aut}(\mathcal{D}) = \bar{X}.\mathrm{Aut}(\mathcal{P})$, *where*

$$\bar{X} = \left\{ f : X \to \mathbb{Z}_2 \mid \sum_{i \in X} f(i) = 0 \right\} \cong \mathbb{Z}_2^{2n}$$

and

$$\mathrm{Aut}(\mathcal{P}) = \{\pi \in \mathrm{Sym}(X) \mid \forall k, \exists l, \pi(M_k) = M_l\}. \qquad \square$$

Coming back to an embedding of the Lie algebra $\mathcal{L}'$ of type B_{n-1} in the Lie algebra $\mathcal{L}$ of type D_n, we note that there is also a natural embedding $\mathrm{Aut}(\mathcal{D}') \hookrightarrow \mathrm{Aut}(\mathcal{D})$

for the E-decompositions $\mathcal{D}'$ and $\mathcal{D}$, which are constructed from corresponding admissible partitions $\mathcal{P}'$ and $\mathcal{P}$ of $X'(2)$ and $X(2)$. Namely, we view elements of $\mathcal{L}'$ as skew-symmetric $2n \times 2n$-matrices with zero last column and zero last row, that is, we use the standard embedding $\mathcal{L}' \hookrightarrow \mathcal{L}$. In this case, $\mathcal{H}_k = \mathcal{H}'_k \oplus < E(k, 2n) >_{\mathbb{C}}$. Extend each automorphism $\Phi_{\pi', f'}$ to an automorphism $\Phi_{\pi, \bar{f}}$ by setting

$$\pi(i) = \pi'(i) \text{ for } i < 2n \text{ and } \pi(2n) = 2n,$$

$$f(i) = f'(i) \text{ for } i < 2n \text{ and } f(2n) = 0.$$

It is easy to describe the converse procedure of determining the subgroup fixing $\mathcal{D}'$ from $\mathrm{Aut}(\mathcal{D})$. For this it is enough to consider the elements $\Phi_{\pi, \bar{f}}$ with $\pi(2n) = 2n$ and $\sum_{i=1}^{2n} f(i) = 0$.

Propositions 2.2.3 and 2.2.4 admit the following slight modification, without change in the arguments:

Proposition 2.2.5. *Suppose that the E-decompositions $\mathcal{D}_1 : \mathcal{L} = \oplus_i \mathcal{H}_i^{(1)}$ and $\mathcal{D}_2 : \mathcal{L} = \oplus_i \mathcal{H}_i^{(2)}$ (of the Lie algebra $\mathcal{L}$ of type D_n or B_n) correspond to partitions $\mathcal{P}_1$ and $\mathcal{P}_2$ of the complete graph $X(2)$. Then $\mathcal{D}_1$ and $\mathcal{D}_2$ are* $\mathrm{Aut}(\mathcal{L})$*-conjugate if and only if $\mathcal{P}_1$ and $\mathcal{P}_2$ are* $\mathrm{Sym}(X)$*-conjugate. Every such conjugation is realized by a map*

$$E(i, j) \mapsto (-1)^{f(i)+f(j)} E(\pi(i), \pi(j))$$

for some $f \in \bar{X}$ and some $\pi \in \mathrm{Sym}(X)$ such that $\pi(\mathcal{P}_1) = \mathcal{P}_2$. □

Warning. In the case of the Lie algebra $\mathcal{L}$ of type D_4, Propositions 2.2.3 and 2.2.5 are in fact true only when $\mathrm{Aut}(\mathcal{L})$ is replaced by $\mathrm{Inn}(\mathcal{L})$.

2.3 Partitions of complete graphs and E-decompositions of Lie algebras of types B_n and D_n: admissible partitions

A partition $\mathcal{P} : X(2) = \bigcup_{k=1}^{2n-1} M_k$ of the complete graph $X(2)$ with $2n$ vertices satisfying condition (2.7) is called in graph theory an *1-factorization*, or a *parallelism*. Furthermore, the components M_k are usually called *1-factors* or *parallel classes*. However, as always, we shall use the term "*admissible partition*".

Note at once that the number $N(2n)$ of all admissible partitions (up to $\mathrm{Sym}(X)$-conjugacy) grows very rapidly: $\ln(N(2n)) \approx 2n^2 \ln(2n)$ by [Cam 1]. On the other hand (see [SeS]), for small n we have the following results:

$$N(2) = N(4) = N(6) = 1,\ N(8) = 6,\ N(10) = 396,\ N(12) \geq 56391.$$

Clearly, it is very difficult to classify all the admissible partitions. There is extensive activity on describing the so-called *perfect* 1-factorizations (see [SeS]). In our terms, this notion means that we are dealing with E-decompositions $\mathcal{L} = \oplus\mathcal{H}_i$ of the Lie algebra $\mathcal{L}$ of type D_n such that $\mathcal{L}$ is generated by any two different subalgebras $\mathcal{H}_i$ and $\mathcal{H}_j$. If $N_p(2n)$ is the number of all perfect 1-factorizations, then

$$N_p(2) = N_p(4) = N_p(6) = N_p(8) = N_p(10) = 1,$$
$$N_p(12) = 5,\ N_p(14) \geq 21;$$

this is proved in [SeS]. We shall pay no attention to perfect 1-factorizations, although they are met with implicitly in our constructions. Our goal is a more modest one: construct admissible partitions with automorphism groups acting transitively on their components.

Let A be an arbitrary group of order $2n-1$ with identity element e. Denote by $X = A \cup \{\infty\}$ the set of cardinality $2n$ that contains all elements of A and the additional element ∞ with the property that $z.\infty = \infty.z = \infty$ for all $z \in A$. Enumerate the elements of X by integers $1, 2, \dots, 2n$ so that the element ∞ gets the number $2n$. Then we can index columns and rows of skew-symmetric $2n \times 2n$ matrices by the elements of X; moreover, the last column and the last row are of numeral ∞. Thus, to construct an admissible partition, we must indicate a partition $\mathcal{P}$ of the set $X(2) = \{\{u, v\} \mid u, v \in X, u \neq v\}$ of unordered pairs into disjoint subsets M_k such that

$$|M_k| = n,\ X(2) = \cup_k M_k,\ \alpha \cap \beta = \emptyset \text{ for any } \alpha, \beta \in M_k \text{ with } \alpha \neq \beta. \tag{2.15}$$

At this point we explain a method of constructing partitions that is based on the natural action of A on the set $X(2)$:

$$\{u, v\} \mapsto \{zu, zv\},\ z \in A.$$

Under this action, the set $X(2)$ is partitioned into n orbits M^j of length $2n-1$: $X(2) = \cup_{j=1}^{n} M^j$. Indeed, one orbit consists of the elements $\{\infty, z\}$, $z \in A$. The length of any other orbit with representative $\{u, v\}$ can be less than $2n-1$ only in the case that the stabilizer $St_A(\{u, v\})$ is not trivial, that is, $\{zu, zv\} = \{u, v\}$ for some $z \in A \backslash \{e\}$. From the identity $\{zu, zv\} = \{u, v\}$ it follows that $zu = v$, $zv = u$, which implies that $z^2 = (vu^{-1})(uv^{-1}) = e$, that is, $z = e$ (since $|A|$ is odd), a contradiction.

From now on we shall index components M_k by the elements of A. It is enough to form the set M_e consisting of an element from each orbit M^j, so that condition (2.15) is satisfied. After choosing M_e, the required admissible partition is constructed as follows:

$$X(2) = \bigcup_{z \in A} M_z,\ M_z = \big\{\{zu, zv\} \mid \{u, v\} \in M_e\big\}.$$

As was just shown, each pair occurs only once, and condition (2.15) is satisfied by M_z because it is satisfied by M_e.

Now we study in detail the admissible partition $\mathcal{P}(A)$ corresponding to the following choice of the set M_e:

$$M_e = \left\{ \{\infty, e\}, \{u, u^{-1}\} \mid u \in A \backslash \{e\} \right\}.$$

It is convenient to use the following property:

$$\{u, v\} \in M_z \Leftrightarrow \{u, v\} = \{\infty, z\} \text{ or } u = zv^{-1}z. \tag{2.16}$$

Only the last equation requires explanation. Assume that $u, v \neq \infty$ and $\{u, v\} = \{zz^{-1}u, zz^{-1}v\} \in M_z$. Then $\{z^{-1}u, z^{-1}v\} \in M_e$ and so $z^{-1}u = (z^{-1}v)^{-1}$, that is, $u = zv^{-1}z$.

We shall also frequently use the following property of groups of odd order:

$$\text{if } u, v \in A \text{ then } u^2 = v^2 \Leftrightarrow u = v. \tag{2.17}$$

In particular, this property means that every element $u \in A$ can be written in the form $u = v^2 = w^4$ and so on. It is easy to deduce from this that the choice of the set M_e is the correct one. Indeed, suppose that the pairs $\{u, u^{-1}\}$ and $\{v, v^{-1}\}$ are contained in the same orbit M^j. Then $\{zu, zu^{-1}\} = \{v, v^{-1}\}$ for some $z \in A$. Hence $zu = (zu^{-1})^{-1}$, $(zu)^2 = (zu^{-1})^{-1}zu = u^2$, $zu = u$ by (2.17), so that $z = e$ and $\{u, u^{-1}\} = \{v, v^{-1}\}$.

Note that every element $u \in A$ corresponds to an automorphism $\hat{u} \in \mathrm{Aut}(\mathcal{P}(A))$, which is defined via the natural action of A on $X(2)$.The subgroup $\mathcal{A} = \{\hat{u} \mid u \in A\}$ of $\mathrm{Aut}(\mathcal{P}(A))$ is obviously isomorphic to A, and acts transitively and regularly on the set $\{M_z\}$, as desired.

Lemma 2.3.1. *Suppose that $\varphi \in \mathrm{Aut}(\mathcal{P}(A))$. Then the permutation π corresponding to φ (see Lemma 2.2.2) always fixes the symbol ∞.*

Proof. The permutation $\pi_u \in \mathrm{Sym}(X)$ corresponding to the automorphism $\hat{u}$ fixes ∞ for any $u \in A$. So, using the transitivity of $\mathcal{A}$ on the set $\{M_z\}$, we may suppose that $\varphi(M_e) = M_e$. In this case, $\pi(u^{-1}) = \pi(u)^{-1}$ for every $u \in A\backslash\{e\}$ such that $\pi(u) \neq \infty, e$. We shall write $(u, v) \Rightarrow (x, y)$ whenever $\pi(u) = x$, $\pi(v) = y$. There are only two possibilities that contradict the statement of the lemma: $(\infty, e) \Rightarrow (x, x^{-1})$ for some $x \in A\backslash\{e\}$, or $(\infty, e) \Rightarrow (e, \infty)$.

a) Suppose that $(\infty, e) \Rightarrow (x, x^{-1})$. Since $n \geq 4$ and thus $|X| \geq 8$, we can choose an element $u \in A$ such that

$$(u, u^{-1}) \Rightarrow (y, y^{-1}) \text{ and } (u^2, u^{-2}) \Rightarrow (z, z^{-1})$$

for some $y, z \in A$, that is, the images of these pairs do not contain ∞. The pairs $\{\infty, u\}$ and $\{u^2, e\}$ belong to the same set M_u, and therefore their images $\{x, y\}$ and $\{z, x^{-1}\}$ are also contained in some M_a, which means that

$$(x, y) = (ap, ap^{-1}) \text{ and } (z, x^{-1}) = (aq, aq^{-1})$$

for some $p, q \in A\backslash\{e\}$. Similarly,

$$(\infty, u^{-1}) \Rightarrow (x, y^{-1}) = (bs, bs^{-1}) \text{ and } (e, u^{-2}) \Rightarrow (x^{-1}, z^{-1}) = (bt, bt^{-1})$$

for $b, s, t \in A$. From these relations it follows that $ap = (aq^{-1})^{-1} = bs = (bt)^{-1}$, $ap^{-1} = (bs^{-1})^{-1}$ and $aq = (bt^{-1})^{-1}$. The first chain of relations implies that $ap(aq^{-1})^{-1} = (bt)^{-1}bs$, that is, $apqa^{-1} = t^{-1}s$. From the second and third chains we obtain $(ap^{-1})^{-1}aq = bs^{-1}(bt^{-1})^{-1}$, that is, $pq = bs^{-1}tb^{-1}$. Comparing the two identities thus deduced, we have $abs^{-1}tb^{-1}a^{-1} = t^{-1}s$, which implies that $abs^{-1}t = t^{-1}sab$. Multiplying both sides of the latter equation on the left by $abs^{-1}t$, we get $(abs^{-1}t)^2 = (ab)^2$, so that $abs^{-1}t = ab$ by (2.17), that is, $s^{-1}t = e$ and $s = t$. This means in turn that $x = bs = bt = x^{-1}$ and $x = e$, a contradiction.

b) Assume that $(\infty, e) \Rightarrow (e, \infty)$. Then, for any $u \in A\backslash\{e\}$, we can write $(u, u^{-1}) \Rightarrow (x^4, x^{-4})$ for some $x \in A\backslash\{e\}$. This statement holds, because by (2.17) the map $x \mapsto x^4$ is a bijection from A onto itself.

Since $\{\infty, u\}, \{u^2, e\} \in M_u$ and $(\infty, u) \Rightarrow (e, x^4) \in M_{x^2}$, we have $(u^2, e) \Rightarrow (\pi(u^2), \infty) \in M_{x^2}$, which means that $\pi(u^2) = x^2$. Furthermore, either $u^3 = e$, or

$$(\infty, u^2) \Rightarrow (e, x^2) \in M_x \text{ and } (u^3, u) \Rightarrow (\pi(u^3), x^4) \in M_x.$$

In the second case we have $\pi(u^3) = x^{-2}$. Moreover, because $u^2 \neq e$ we have $\pi(u^3) = x^{-2} = \pi(u^2)^{-1} = \pi(u^{-2})$, which implies that $u^3 = u^{-2}$, that is, $u^5 = e$. Thus we have shown that $w^3 = e$ or $w^5 = e$ for all $w \in A$.

Since $|X| \geq 8$, there exists an element $v \in A$ which is not contained in the cyclic subgroup $< u >$. Suppose that $(v, v^{-1}) \Rightarrow (y^4, y^{-4})$ for some $y \in A\backslash\{e\}$. Clearly, the pair (v, y) satisfies the same relations as the pair (u, x); here u and v can have different orders.

Suppose that $(vu, u^{-1}v^{-1}) \Rightarrow (z^2, z^{-2})$. Then, from the conditions

$$(\infty, vu), (vu^2, v) \in M_{vu}, \ (\infty, vu) \Rightarrow (e, z^2) \in M_z,$$

$$(vu^2, v) \Rightarrow (\pi(vu^2), y^4)$$

it follows that $\pi(vu^2) = zy^{-4}z$. Similarly, the relations

$$(\infty, v), (vu^{-1}, vu) \in M_v, \ (\infty, v) \Rightarrow (e, y^4) \in M_{y^2},$$

$$(vu^{-1}, vu) \Rightarrow (\pi(vu^{-1}), z^2)$$

lead to the equality $\pi(vu^{-1}) = y^2z^{-2}y^2$.

We consider now the following two cases.

b1) There is an element $u \in A$ of order 3. Then $u^2 = u^{-1}$ and $vu^2 = vu^{-1}$. Because of these equations, the following sequence of equalities holds: $\pi(vu^2) = \pi(vu^{-1})$, $zy^{-4}z = y^2z^{-2}y^2$, $y^{-4} = z^{-1}y^2z^{-2}y^2z^{-1}$, $(y^{-2})^2 = (z^{-1}y^2z^{-1})^2$, $y^{-2} =$

$z^{-1}y^2z^{-1}$, $(y^2z^{-1})^2 = e$, $y^2z^{-1} = e$, $z = y^2$, $z^2 = y^4$, $\pi(vu) = \pi(v)$, $vu = v$ and $u = e$, a contradiction.

b2) A has no elements of order 3, that is, A is a 5-group. Then $\{\infty, vu^2\} \in M_{vu^2}$ and $\{vu^{-1}, v\} = \{vu^2u^{-3}, vu^2u^3\} \in M_{vu^2}$. Therefore, from the conditions

$$(vu^{-1}, v) \Rightarrow (y^2z^{-2}y^2, y^4) = (y^2z^{-1}y^2.y^{-2}z^{-1}y^2, y^2z^{-1}y^2.(y^{-2}z^{-1}y^2)^{-1}) \in M_t$$

where $t = y^2z^{-1}y^2$, it follows that $(\infty, vu^2) \Rightarrow (e, zy^{-4}z) \in M_t$. Applying (2.16), we obtain that $zy^{-4}z = te^{-1}t = y^2z^{-1}y^2y^2z^{-1}y^2$, which implies the following sequence of equalities:

$$y^4z^{-1}(zy^{-4}z)z^{-1} = y^4z^{-1}(y^2z^{-1}y^4z^{-1}y^2)z^{-1},\ (y^4z^{-1}y^2z^{-1})^2 = e,$$

$$y^4z^{-1}y^2z^{-1} = e,\ y^2z^{-1}y^2z^{-1} = y^{-2},\ y^2z^{-1} = y^{-1},\ z = y^3 \text{ and } z^2 = y^6.$$

But $y^5 = e$, so $z^2 = y^{-4}$. Thus we have $\pi(vu) = \pi(v^4)$, $vu = v^4$, that is, $u = v^3$. From the last identity it follows that $v = v^6 = u^2$, which contradicts the choice of v: $v \notin < u >$. □

Remark 2.3.2. The Lie algebra of type A_3 can be viewed as the Lie algebra of type D_3 realized as the set of skew-symmetric 6×6 matrices. So we can rediscover its (unique) OD via admissible partitions of the set $X(2)$. In this case $|A| = 5$, and the permutation group corresponding to the automorphisms of the decomposition acts transitively on the set $\{\infty, 0, 1, 2, 3, 4\}$. Hence, *the restriction $n \geq 4$ is necessary for Lemma* 2.3.1.

Lemma 2.3.3. *A permutation $\pi \in \mathrm{Sym}(X)$ defines an automorphism φ of the partition $\mathcal{P}(A)$ such that $\varphi(M_e) = M_e$ if and only if*

$$\pi(\infty) = \infty \text{ and } \pi(uvu) = \pi(u)\pi(v)\pi(u) \text{ for all } u, v \in A. \tag{2.18}$$

Proof. If $\varphi(M_e) = M_e$, then $\pi(\infty) = \infty$ by Lemma 2.3.1. Since $\{\infty, u\}$, $\{uvu, v^{-1}\} \in M_u$ and $(\infty, u) \Rightarrow (\infty, \pi(u)) \in M_{\pi(u)}$, we have $(uvu, v^{-1}) \Rightarrow (\pi(uvu), \pi(v^{-1})) \in M_{\pi(u)}$. According to (2.16), the latter implies that $\pi(uvu) = \pi(u)\pi(v^{-1})^{-1}\pi(u)$. But, since $\varphi(M_e) = M_e$ and $\pi(\infty) = \infty$, it follows that $\pi(v^{-1})^{-1} = \pi(v)$, which completes our arguments.

Conversely, suppose that $\pi(\infty) = \infty$ and $\pi(uvu) = \pi(u)\pi(v)\pi(u)$. Setting $u = v = e$ we have $\pi(e) = \pi(e)^3$, that is, $\pi(e) = e$. Furthermore, setting $v = u^{-1}$, we obtain $\pi(u) = \pi(u)\pi(u^{-1})\pi(u)$, that is, $\pi(u^{-1}) = (\pi(u))^{-1}$. By (2.16), any element from M_z is either $\{\infty, z\}$ or $\{u, zu^{-1}z\}$ for some $u \in A\backslash\{z\}$. Now the bijection φ of the set $X(2)$ corresponding to π acts as follows:

$$\varphi : \{\infty, z\} \mapsto \{\infty, \pi(z)\} \in M_{\pi(z)},$$

$$\varphi : \{u, zu^{-1}z\} \mapsto \{\pi(u), \pi(z)\pi(u)^{-1}\pi(z)\} \in M_{\pi(z)}.$$

This means that $\varphi(M_z) = M_{\pi(z)}$, that is, $\varphi \in \mathrm{Aut}(\mathcal{P}(A))$; moreover, $\varphi(M_e) = M_e$. □

Set

$$\mathcal{B} = \big\{\hat{\pi} : \{u, v\} \mapsto \{\pi(u), \pi(v)\} \mid$$

$$\pi \in \mathrm{Sym}(X), \pi(\infty) = \infty, \pi(aba) = \pi(a)\pi(b)\pi(a)\big\}.$$

Theorem 2.3.4. *The group* $\mathrm{Aut}(\mathcal{P}(A))$ *admits the factorization* $\mathrm{Aut}(\mathcal{P}(A)) = \mathcal{A}\mathcal{B}$ *and contains a subgroup isomorphic to the holomorph* $\Gamma(A)$ *of* A.

Proof. Consider an arbitrary automorphism $\varphi \in \mathrm{Aut}(\mathcal{P}(A))$, and suppose that $\varphi(M_e) = M_u$. Then the automorphism $(\hat{u})^{-1}\varphi$ fixes M_e, and so $(\hat{u})^{-1}\varphi = \hat{\pi} \in \mathcal{B}$ by Lemmas 2.3.1 and 2.3.3. Hence $\varphi = \hat{u}\hat{\pi} \in \mathcal{A}\mathcal{B}$. Furthermore, every automorphism σ of A satisfies the equation $\sigma(uvu) = \sigma(u)\sigma(v)\sigma(u)$, and defines an automorphism $\hat{\sigma}$ (if in addition we set $\sigma(\infty) = \infty$) of $\mathcal{P}(A)$. Finally, for every $u \in A$ we have $\hat{\sigma}\hat{u}(\hat{\sigma})^{-1} = \hat{v}$, where $v = \sigma(u)$. □

For an abelian group A it is easy to deduce the following refinement of Theorem 2.3.4:

Corollary 2.3.5. *If* A *is abelian, then* $\mathrm{Aut}(\mathcal{P}(A)) \cong \Gamma(A)$.

Proof. It is enough to check that, in this case, the restriction to A of any permutation $\pi \in \mathrm{Sym}(X)$ satisfying (2.18) is an automorphism of A. Indeed, setting $v = e$ in (2.18), we have $\pi(u^2) = \pi(u)^2$. Hence,

$$\pi(vu^2) = \pi(uvu) = \pi(u)\pi(v)\pi(u) = \pi(v)\pi(u)^2 = \pi(v)\pi(u^2).$$

According to (2.17), every element $w \in A$ can be written in the form $w = u^2$ for some $u \in A$, so that $\pi(vw) = \pi(v)\pi(w)$ (for all $v, w \in A$), and this completes the proof. □

If A is nonabelian, then in addition to automorphisms of $\mathcal{P}(A)$ that are induced by automorphisms of A, we have also the bijection $\hat{\omega}$ lying in the centre $Z(\mathcal{B})$ of $\mathcal{B}$, where $\omega(u) = u^{-1}$. It can be verified immediately that when A is the (nonabelian) metacyclic group

$$A = \left\langle a, b \mid a^p = e = b^q, bab^{-1} = a^r, r^q \equiv 1 (\mathrm{mod}\, p)\right\rangle$$

of order pq, where p and q are primes and $p \equiv 1 (\mathrm{mod}\, q)$, the group $\mathcal{B}$ is isomorphic to the direct product $\mathrm{Aut}(A) \times < \omega >$. On the other hand, $\mathcal{B}$ can be very different from this direct product.

Definition 2.3.6. A bijective map $\pi : A \mapsto A$ from a group A onto itself is said to be a *generalized automorphism*, if

$$\forall u, v \in A, \pi(uvu) = \pi(u)\pi(v)\pi(u). \tag{2.19}$$

A is called *generalized homogeneous* if the group $\mathrm{Aut}_g(A)$ of all its generalized automorphisms acts transitively on $A\backslash\{e\}$.

Theorem 2.3.7. *A group A of odd order N is generalized homogeneous if and only if*

$$N = p^m,\ \exp(A) = p \text{ and } cl(A) \le 2$$

for some prime p. Moreover, in this case $\mathrm{Aut}_g(A) = GL_m(p)$.

Proof. Let A be a generalized homogeneous group.

1) Firstly we remark that A has prime exponent: $\exp(A) = p$ for some prime p. For, setting $v = u^k$ in (2.19), when $\pi \in \mathrm{Aut}_g(A)$ we have $\pi(u^{k+2}) = \pi(u)\pi(u^k)\pi(u)$; so an easy induction on k leads to the relation $\pi(u^k) = \pi(u)^k$. In particular, π preserves the order of any element from A, since $\pi(e) = e$. This proves that $\exp(A) = p > 2$ and $N = p^m$.

2) Consider the *generalized centre* of A:

$$Z_g(A) = \{u \in A \mid \forall a \in A, ua^2u = au^2a\}.$$

It is clear that $Z(A) \subseteq Z_g(A)$; but A is a p-group and so $Z(A) \neq \{e\}$, therefore $Z_g(A) \neq \{e\}$. Moreover, $Z_g(A)$ is invariant under $\mathrm{Aut}_g(A)$; for any $u \in Z_g(A)$ we have:

$$\pi(a)\pi(u)^2\pi(a) = \pi(au^2a) = \pi(ua^2u) = \pi(u)\pi(a)^2\pi(a).$$

Hence, from the generalized homogeneity of A it follows that $Z_g(A) = A$.

Recall that in any group A of odd order, extraction of the square root $\sqrt{a}$ of each $a \in A$ is uniquely defined. So, the assertion just proved means that the binary operation

$$(u, v) \mapsto u \circ v = \sqrt{u}.v.\sqrt{u}$$

is correctly defined in A and is commutative. Furthermore, e is also the identity element for $(A, \circ)$, and the inverse element for u in $(A, \circ)$ is simply the inverse u^{-1} of u in A.

3) Next, we show that $\circ$ is associative. To this end we introduce the *generalized associative centre*:

$$A_g(A) = \{a \in A \mid \forall x, y \in A, x \circ (a \circ y) = (x \circ a) \circ y\}.$$

Then, for any $a \in A_g(A)$ we have

$$xyayx = x^2 \circ (y^2 \circ a) = x^2 \circ (a \circ y^2) = (x^2 \circ a) \circ y^2 = y^2 \circ (x^2 \circ a) = yxaxy,$$

which is equivalent to the following commutation relation:

$$a = [y^{-1}, x^{-1}]a[x, y]. \tag{2.20}$$

Note that the commutativity of $(A, \circ)$ is equivalent to the identity

$$[x, y] = [x^{-1}, y^{-1}]. \tag{2.21}$$

(Indeed, $(A, \circ)$ is commutative $\Leftrightarrow \forall x, y \in A,\ xy^2x = yx^2y \Leftrightarrow \Leftrightarrow yx.xyx^{-1}y^{-1} = xy \Leftrightarrow xyx^{-1}y^{-1} = x^{-1}y^{-1}xy$.) So (2.20) is equivalent to the identity

$$[[x, y], a] = e. \tag{2.22}$$

In particular, $Z(A) \subseteq A_g(A)$, on the one hand. On the other hand, the set $A_g(A)$ is invariant under $\mathrm{Aut}_g(A)$, because $\mathrm{Aut}_g(A)$ consists precisely of (the usual) automorphisms of the system $(A, \circ)$. Hence, from the generalized homogeneity of A it follows that $A_g(A) = A$.

Thus, we have proved the associativity of $\circ$ and the fact that $(A, \circ)$ is an elementary abelian p-group of order p^m. In particular, $\mathrm{Aut}_g(A) = \mathrm{Aut}((A, \circ)) = GL_m(p)$. Note also that the group A satisfies the identity $[[x, y], z] = e$ (see (2.22)), so A is a nilpotent group of class ≤ 2.

4) Conversely, let A be an arbitrary p-group of exponent p and of nilpotency class ≤ 2. Then the following identity holds in A:

$$[x, y^{-1}] = [x, y]^{-1}.$$

Hence, $[x^{-1}, y^{-1}] = [y^{-1}, x^{-1}]^{-1} = [y^{-1}, x] = [x, y^{-1}]^{-1} = [x, y]$, that is, A satisfies identity (2.21), which is equivalent to the commutativity of $(A, \circ)$. Furthermore, assuming that (2.21) holds, the condition $cl(A) \leq 2$ implies the associativity of $(A, \circ)$. Consequently, $(A, \circ)$ is an elementary abelian p-group and $\mathrm{Aut}_g(A) = \mathrm{Aut}((A, \circ)) = GL_m(p)$ acts transitively on the set $A \backslash \{e\}$. □

Example 2.3.8. Setting $A = p_+^{1+2m}$, $p > 2$, we have $\mathrm{Aut}_g(A) = GL_{2m+1}(p)$ and $\mathrm{Aut}(\mathcal{P}(A)) = AGL_{2m+1}(p)$.

It should be noted that, although there is a great number of groups A satisfying the conditions of Theorem 2.3.7, all partitions $\mathcal{P}(A)$ corresponding to them are in fact $\mathrm{Sym}(X)$-conjugate. Indeed, let A_0 be an elementary abelian p-group of order $|A|$. Then we can identify the group A_0 with $(A, \circ)$ (see the proof of Theorem 2.3.7). Clearly, the first component

$$M_e = \Big\{\{\infty, e\}, \{a, a^{-1}\} \mid a \in A \backslash \{e\}\Big\}$$

is the same for $\mathcal{P}(A)$ and $\mathcal{P}(A_0)$. Next we show that the set

$$M_z = \Big\{\{\infty, z\}, \{za, za^{-1}\} \mid a \in A \backslash \{e\}\Big\}$$

coincides with

$$M_z^0 = \left\{ \{\infty, z\}, \{z \circ a, z \circ a^{-1}\} \mid a \in A \backslash \{e\} \right\}.$$

For, the pair (za, za^{-1}) is the same as the pair $(z \circ b, z \circ b^{-1})$ provided that $b = \sqrt{z} a \sqrt{z}^{-1}$:

$$z \circ b = \sqrt{z}.\sqrt{z} a \sqrt{z}^{-1}.\sqrt{z} = za, \; z \circ b^{-1} = \sqrt{z}.\sqrt{z} a^{-1} \sqrt{z}^{-1}.\sqrt{z} = za^{-1}.$$

In accordance with the construction described in §2.2, the admissible partition $\mathcal{P}(A)$, $|A| = 2n - 1$, corresponds to some OD $\mathcal{D}$ of the Lie algebra $\mathcal{L}$ of type D_n. The description of Aut($\mathcal{D}$) is given by Proposition 2.2.3 and Theorem 2.3.4. Note here that an outer automorphism T of order 3 of the Lie algebra $\mathcal{L}$ of type D_4 does not fix the decomposition $\mathcal{D}$ corresponding to the partition $\mathcal{P}(A)$.

According to the same construction, the OD $\mathcal{D}$ of the Lie algebra $\mathcal{L}$ of type D_n obtained via the admissible partition $\mathcal{P}(A)$ corresponds to some OD $\mathcal{D}'$ of the Lie algebra $\mathcal{L}'$ of type B_{n-1}, which can be obtained via the admissible partition $\mathcal{P}'(A)$. Here $M_z' = M_z \backslash \{\{\infty, z\}\}$. Since $\hat{u}\hat{\pi}(\infty, z) = (\infty, u\pi(z))$ for all $\hat{u} \in \mathcal{A}$, $\hat{\pi} \in \mathcal{B}$, we have

$$\mathrm{Aut}(\mathcal{P}'(A)) \cong \mathrm{Aut}(\mathcal{P}(A)).$$

The description of Aut($\mathcal{D}'$) is contained in Proposition 2.2.4 and Theorem 2.3.4.

To conclude this section we remark that the E-decomposition corresponding to the partition $\mathcal{P}'(A)$, where $A = \mathbb{Z}_p^m$ and p an odd prime, is an irreducible E-decomposition of the Lie algebra of type $B_{(p^m-1)/2}$. A classification of all IODs, in particular, all irreducible E-decompositions, of the Lie algebra of type B_n will be carried out in Chapter 6. A classification of the irreducible E-decompositions of the Lie algebra of type D_n is given in the following section.

2.4 Classification of the irreducible E-decompositions of the Lie algebra of type D_n

Let $\mathcal{D}$ be an E-decomposition of the Lie algebra $\mathcal{L}$ of type D_n corresponding to an admissible partition $\mathcal{P} : X(2) = \bigcup_{k=1}^{2n-1} M_k$ of the complete graph $X(2)$, where $X = \{1, 2, \dots, 2n\}$. From Proposition 2.2.3, it follows in particular that $\mathcal{D}$ is irreducible (that is, the group Aut($\mathcal{D}$) $= \bar{X}$.Aut($\mathcal{P}$) acts absolutely irreducibly on $\mathcal{L}$) if and only if the permutation group

$$\mathrm{Aut}(\mathcal{P}) = \{\pi \in \mathrm{Sym}(X) \mid \forall k, \exists l, \pi(M_k) = M_l\}$$

acts transitively on (the edge set of the graph) $X(2)$; in other words, if and only if Aut($\mathcal{P}$) is doubly homogeneous on X. As always, we assume that $|X| = 2n > 2$.

Lemma 2.4.1. *A group $G \subseteq \mathrm{Sym}(X)$ acts transitively on $X(2)$ if and only if G is doubly transitive on X.*

Proof. Clearly, double transitivity of G on X implies transitivity of G on $X(2)$. Conversely, suppose that G is doubly homogeneous on X. Then G is transitive on X. For, if $X = X' \cup X''$ is a partition of X into two disjoint non-empty G-invariant components, then one can suppose that $|X'| \geq 2$ and $|X''| \geq 1$. Choose three different points $x, y \in X'$ and $z \in X''$. Obviously, no element of G maps the edge $\{x, y\}$ onto the edge $\{x, z\}$, contrary to our assumption.

Thus G is transitive on X. But $|X|$ is even, so G is of even order. In this case, some involution $g \in G$ permutes the ends x, y of an edge $\{x, y\}$. From this and the transitivity of G on the set of edges, it follows that G is doubly transitive. □

As was shown by Kantor [Kan 1], Lemma 2.4.1 is false for X of odd cardinality.

Lemma 2.4.2. *Suppose that $G \subseteq \mathrm{Sym}(X)$ is doubly transitive. Then the following conditions are equivalent:*

(i) *there exists an admissible partition $\mathcal{P}$ of the set $X(2)$ such that $G \subseteq \mathrm{Aut}(\mathcal{P})$.*

(ii) *G possesses a subgroup H of index $2n - 1$ which contains some edge stabilizer $G_{\{x,y\}}$, $\{x, y\} \in X(2)$, and which is transitive on X.*

Proof. (i) $\Rightarrow$ (ii). Let $\mathcal{P}$ be an admissible G-invariant partition, and suppose that $\{x, y\} \in M_k$ for some k. Set $H = St_G(M_k)$. Then $(G : H) = 2n - 1$, because G acts transitively on the set $\{M_k\}$. Clearly, every element of $G_{\{x,y\}}$ fixes M_k and so is contained in H. Further, let G_{xy} be the pointwise stabilizer of the pair (x, y) and H_x the stabilizer of x in H. Then any element of H_x fixes $\{x, y\}$ as the unique edge from M_k with end x. Thus $G_{xy} = H_x$. Hence,

$$(H : H_x) = (H : G_{xy}) = (G : G_{xy})/(G : H) = 2n(2n - 1)/(2n - 1) = |X|,$$

that is, H is transitive on X.

(ii) $\Rightarrow$ (i). Suppose that H contains some edge stabilizer $G_{\{x,y\}}$, H is transitive on X and $(G : H) = 2n-1$. Since H_x contains G_{xy} and $(H : H_x) = 2n = (H : G_{xy})$, we have $H_x = G_{xy}$. Then the partition $G = \cup_{k=1}^{2n-1} g_k H$ $(g_1 = e)$ leads in a unique way to the required admissible partition $\mathcal{P}$: $M_1 = \{x, y\}^H$ and $M_k = g_k(M_1)$. □

Lemma 2.4.3. *Let $\mathcal{P}$ be an admissible partition, and H a subgroup of $\mathrm{Aut}(\mathcal{P})$. Then H fixes either a single point or an even number of points of X.*

Proof. Suppose that $X' = \{x_1 = x, x_2 = y, x_3, \ldots, x_m\}$ is the set of H-fixed points and that $m \geq 2$. Then $\{x, y\} \in M_k$ for some component M_k of $\mathcal{P}$. Clearly, H fixes M_k. By definition of $\mathcal{P}$, for any $i \leq m$ one can find a unique point $z \in X$ such that $\{x_i, z\} \in M_k$. But H fixes x_i and M_k, so H fixes z and therefore $z = x_j$ for some j. Thus we obtain a partition of X' into disjoint pairs. In particular, m is even. □

Now we go over to a classification of the irreducible E-decompositions of the Lie algebra $\mathcal{L}$ of type D_n. As was shown above, assuming irreducibility of an E-decomposition $\mathcal{D}$ corresponding to a partition $\mathcal{P}$, the group $G = \mathrm{Aut}(\mathcal{P})$ acts doubly transitively on the set X, $|X| = 2n$. Set $S = \mathrm{soc}(G)$. By the O'Nan-Scott Theorem (see [LPS]), S is either an elementary abelian p-group (and then $|S| = |X| = 2n$, that is, $p = 2$), or a nonabelian simple group.

Proposition 2.4.4. *Suppose that $S = \mathrm{soc}(\mathrm{Aut}(\mathcal{P}))$ is an elementary abelian 2-group and $n = 2^m$. Then an admissible partition $\mathcal{P}$ exists and is conjugate to the following standard partition:*

$$X(2) = \bigcup_{u \in W \setminus \{0\}} M_u,\ M_u = \big\{\{x, u + x\} \mid x \in W\big\}, \tag{2.23}$$

where X is identified with the space $W = \mathbb{F}_2^{m+1}$. Moreover,

$$\mathrm{Aut}(\mathcal{P}) = AGL_{m+1}(2).$$

Proof. Suppose that $\mathcal{P}$ is an admissible partition with $G = \mathrm{Aut}(\mathcal{P})$ and $S = \mathrm{soc}(G) = \mathbb{Z}_2^{m+1}$. Then, we can identify X with the space $W = \mathbb{F}_2^{m+1}$ and define a doubly transitive permutation action of G on X as follows. The normal subgroup S acts by translations $x \mapsto x + v$. Furthermore, $G = S.G_0$, where $G_0 = St_G(0)$ acts naturally on X as a subgroup of $GL(W)$.

Set $G_1 = St_G(M_1)$. Then, as a subgroup of odd index in G, G_1 contains some Sylow 2-subgroup of G; but S is a normal 2-subgroup of G, so $S \subseteq G_1$. In particular, if $\{0, u\} \in M_1$, then

$$M_1 = \big\{\{x, u + x\} \mid x \in W\big\}.$$

Thus M_1 coincides with the set M_u from (2.23). Making the elements of G_0 act on M_1, we come to the conclusion that the partition $\mathcal{P}$ coincides with the partition (2.23).

Regarding the group $\mathrm{Aut}(\mathcal{P})$, we start from the obvious inclusion $AGL_{m+1}(2) \subseteq \mathrm{Aut}(\mathcal{P})$. Changing any given element $g \in \mathrm{Aut}(\mathcal{P})$ by using $AGL_{m+1}(2)$, one can suppose that $g(0) = 0$. Then, for any $x, y \in W$ with $x \neq y$, we have $\{x, y\}$, $\{0, x + y\} \in M_{x+y}$. So the pairs $\{g(x), g(y)\}$ and $\{0, g(x + y)\}$ belong to the same component M_z, which implies that $g(x+y) = 0+g(x+y) = z = g(x)+g(y)$. Thus, the equality $g(x + y) = g(x) + g(y)$ holds for any $x, y \in W$, that is, $g \in GL(W)$. □

Proposition 2.4.5. *Suppose that $S = \mathrm{soc}(\mathrm{Aut}(\mathcal{P}))$ is a nonabelian simple group. Then $(S, 2n) = (L_2(5), 6)$, $(L_2(11), 12)$ or $(L_2(8), 28)$. In each case, an admissible partition $\mathcal{P}$ exists and, up to conjugacy, is unique. Furthermore, $\mathrm{Aut}(\mathcal{P}) = PGL_2(5)$, $L_2(11)$ or $P\Gamma L_2(8)$, respectively.*

Proof. According to [Cam 3], all possibilities for the pair $(S, 2n)$ are listed in Table A2. In accordance with this Table, we break up the proof into a small number of steps.

1) Suppose that $G = \text{Aut}(\mathcal{P})$ is a 3-transitive permutation group on X. Then by [Cam 1], $\mathcal{P}$ is conjugate to the standard partition (2.23); but in that case, S is abelian by Proposition 2.4.4, a contradiction. This means that

$$(S, 2n) \neq (\mathbb{A}_{2n}, 2n),\ (M_{11}, 12),\ (M_{12}, 12),\ (M_{22}, 22) \text{ or } (M_{24}, 24).$$

2) Denote by $P(H)$ the minimal index of proper subgroups of a group H. For most finite simple groups of Lie type, the values of this parameter are found in [Coo 1] and [KlL 2]; they are reproduced in Table A6. From this Table and Table A2, it is clear that we have $P(G) \geq P(S) = 2n$ in most cases. On the other hand, according to Lemma 2.4.2 we have $P(S) \leq 2n - 1$. Thus, it remains to consider the cases

$$(S, 2n) = (L_2(q), q+1),\ q = 5, 7, 9, 11;\ (L_2(8), 28);\ (U_3(5), 126);$$

$$(Sp_{2d}(2), 2^{d-1}(2^d+1)),\ d \geq 3;\ (HS, 176);\ (Co_3, 276).$$

3) Note (see [ATLAS]) that $U_3(5)$ has no subgroups of index dividing $2n-1 = 125$, so the case $(U_3(5), 126)$ is eliminated. The same argument can be applied to the cases $(HS, 176)$ and $(Co_3, 276)$. Now consider the case $(Sp_{2d}(2), 2^{d-1}(2^d+1))$, $d \geq 3$. According to Lemma 2.4.2, $S = Sp_{2d}(2)$ has a proper subgroup S_1 of index dividing $2^{d-1}(2^d+1) - 1$. In particular, $(S : S_1)$ is odd, so by the Borel-Tits Theorem, S_1 is contained in some maximal parabolic subgroup P of S. But in that case $(S : S_1) \geq (S : P) \geq 2^{2d} - 1$, a contradiction.

4) Next, suppose that $(S, 2n) = (L_2(5), 6)$.

In the beginning of §2.3, we mentioned that the number $N(6)$ of all admissible partitions of the complete graph with 6 vertices is equal to 1. It is easy to show that this unique partition $\mathcal{P}$ has automorphism group $\text{Aut}(\mathcal{P}) = PGL_2(5)$, and that it corresponds to an irreducible E-decomposition $\mathcal{D}$ of the Lie algebra of type D_3 with automorphism group $\text{Aut}(\mathcal{D}) = \mathbb{Z}_2^5.PGL_2(5)$.

5) Now suppose that $(S, 2n) = (L_2(7), 8)$.

In this case one can identify X with the projective line $\mathbb{PF}_7^2$, on which the groups $S = L_2(7)$ and $G \subseteq PGL_2(7)$ act naturally: $X = \{\bar{0}, \bar{1}, \ldots, \bar{6}, \infty\}$, where $\bar{i} = <(1, i)>_{\mathbb{F}_7}$ for $0 \leq i \leq 6$ and $\infty = <(0, 1)>_{\mathbb{F}_7}$. Without loss of generality we can suppose that $\alpha = \{\bar{0}, \infty\} \in M_1$. Then it is clear that the group

$$St_S(\alpha) = \left\langle \begin{pmatrix} 2 & 0 \\ 0 & 4 \end{pmatrix}, \begin{pmatrix} 0 & 1 \\ -1 & 0 \end{pmatrix} \right\rangle$$

fixes M_1. This leads to the following three possibilities:

$$M_1 = \left\{ \{\bar{0}, \infty\}, \{\bar{1}, \bar{3}\}, \{\bar{2}, \bar{6}\}, \{\bar{4}, \bar{5}\} \right\}, \tag{2.24}$$

$$M_1 = \left\{\{\bar{0}, \infty\}, \{\bar{1}, \bar{5}\}, \{\bar{2}, \bar{3}\}, \{\bar{4}, \bar{6}\}\right\}, \tag{2.25}$$

$$M_1 = \left\{\{\bar{0}, \infty\}, \{\bar{1}, \bar{6}\}, \{\bar{2}, \bar{5}\}, \{\bar{3}, \bar{4}\}\right\}. \tag{2.26}$$

Direct calculations show that if (2.24) or (2.25) holds, the resulting partition $\mathcal{P}$ has automorphism group $\mathrm{Aut}(\mathcal{P}) = \mathbb{Z}_2^3.SL_3(2)$ (and it is conjugate to the standard partition (2.23) with $m = 2$). In particular, $S = \mathbb{Z}_2^3$ is abelian, a contradiction. In the case of (2.26), the resulting partition is not invariant under S, which is again a contradiction.

The fact that condition (2.24) leads to an S-invariant admissible partition $\mathcal{P}$ with $\mathrm{Aut}(\mathcal{P}) = \mathbb{Z}_2^3.SL_3(2)$ (together with the isomorphism $SL_3(2) \cong L_2(7)$) gives an explanation of the well-known equality

$$H^1(SL_3(2), \mathbb{F}_2^3) = \mathbb{F}_2.$$

6) Here we assume that $(S, 2n) = (L_2(11), 12)$.

Then $S \lhd G \subseteq S.\mathbb{Z}_2 = PGL_2(11)$. By Lemma 2.4.2, G has subgroups of index 11. But $PGL_2(11)$ has no such subgroups (see [ATLAS]), so $\mathrm{Aut}(\mathcal{P}) = G = L_2(11)$.

Conversely, suppose $G = S = L_2(11)$. Then G has a unique doubly transitive permutation representation of degree 12, which extends to a doubly transitive permutation representation of $G.\mathbb{Z}_2 = PGL_2(11) = \mathrm{Aut}(G)$. Furthermore, G has two conjugacy classes of subgroups H of index 11 isomorphic to $\mathbb{A}_5$, which are permuted by an outer automorphism $\varphi \in \mathrm{Aut}(G)\backslash G$. Consider some subgroup H from one of these classes. Let $G_{\{x,y\}}$ denote, as always, the setwise stabilizer of the pair $\{x, y\} \in X(2)$, where G acts doubly transitively on X. Then $G_{\{x,y\}} \cong D_{10}$. In fact $G_{\{x,y\}}$ coincides with the normalizer $N_G(F)$ of some Sylow 5-subgroup F of G. On the other hand, the normalizer (in H) of a Sylow 5-subgroup of H is also isomorphic to D_{10}. So one can suppose without loss of generality that $H \supseteq G_{\{x,y\}}$.

Next we note that H is transitive on X. In fact, the stabilizer G_z of $z \in X$ is isomorphic to $\mathbb{Z}_{11}.\mathbb{Z}_5$. Hence,

$$|H \cap G_z| = |\mathbb{A}_5 \cap (\mathbb{Z}_{11}.\mathbb{Z}_5)| \leq 5, \ |z^H| = (H : (H \cap G_z)) \geq 60/5 = 12 = |X|,$$

that is, $z^H = X$ as desired.

Thus, the subgroup H satisfies condition (ii) of Lemma 2.4.2. Consequently, there exists a (unique) admissible partition $\mathcal{P} : X(2) = \bigcup_{k=1}^{11} M_k$ such that $G \subseteq \mathrm{Aut}(\mathcal{P})$ and $H = St_G(M_1)$. As the list of doubly transitive permutation groups of degree 12 consists precisely of the groups $L_2(11)$, $PGL_2(11)$, M_{11}, M_{12}, $\mathbb{A}_{12}$ and $\mathbb{S}_{12}$ (see [Sim]), it follows from our arguments that $G = \mathrm{Aut}(\mathcal{P})$.

Replacing H by some subgroup H' from the other conjugacy class, we obtain a new partition $\mathcal{P}'$, which can be mapped onto $\mathcal{P}$ by an automorphism $\varphi \in \mathrm{Aut}(G)\backslash G$.

7) Finally, suppose that $(S, 2n) = (L_2(8), 28)$.

Then $S \lhd G \subseteq \mathrm{Aut}(S) = S.\mathbb{Z}_3 = P\Gamma L_2(8)$. However, the restriction of the (unique) doubly transitive permutation representation of degree 28 of $P\Gamma L_2(8)$ to S is not doubly transitive, so $G = P\Gamma L_2(8)$.

Conversely, suppose that $G = P\Gamma L_2(8)$. Note that any subgroup H of index 27 in G is contained in S and is isomorphic to $\mathbb{Z}_2^3.\mathbb{Z}_7$; moreover, S has only one conjugacy class of such subgroups. For, suppose that $(G : H) = 27$. Then $3 = (G : S)$ is divisible by

$$(G : HS) = (G : H) : (S : S \cap H);$$

hence $(S : S \cap H)$ divides 27 and is divisible by 9. From [ATLAS] it follows that $(S : S \cap H) = 9$, $HS = S$, and so $H \subseteq S$ as stated.

In what follows, we assume that $(G : H) = 27$. Since $(G : H)$ is odd, H contains some Sylow 2-subgroup T of G. However, the setwise stabilizer $G_{\{x,y\}}$ of $\{x, y\} \in X(2)$ has order 4, and therefore one can suppose that $H \supseteq T \supseteq G_{\{x,y\}}$. Furthermore, the point stabilizer S_z of $z \in X$ is isomorphic to D_{18}. Hence,

$$|S_z \cap H| = |(\mathbb{Z}_2^3.\mathbb{Z}_7) \cap D_{18}| \leq 2, \ |z^H| = |H : (S_z : H)| \geq 2^3.7/2 = 28,$$

that is, $z^H = X$ and H is transitive on X. Thus, the subgroup H satisfies condition (ii) of Lemma 2.4.2. Therefore, there exists a (unique) admissible partition $\mathcal{P}$ of the set $X(2)$ such that $G \subseteq \mathrm{Aut}(\mathcal{P})$. As the list of doubly transitive permutation groups of degree 28 consists precisely of the groups $P\Gamma L_2(8)$, $L_2(27)$, $L_2(27).\mathbb{Z}_2$, $L_2(27).\mathbb{Z}_3$, $L_2(27).\mathbb{Z}_6$, $U_3(3)$, $G_2(2)$, $Sp_6(2)$, $\mathbb{A}_{28}$ and $\mathbb{S}_{28}$ (see [DiM]), our arguments allow us to conclude that $G = \mathrm{Aut}(\mathcal{P})$. □

In fact we have proved:

Theorem 2.4.6. *The complex simple Lie algebra $\mathcal{L}$ of type D_n has an irreducible E-decomposition $\mathcal{D}$ if and only if $n = 3, 4, 6, 14$ or $n = 2^m$, $m \geq 3$. In every case, such a decomposition is unique up to conjugacy. The corresponding automorphism group* $\mathrm{Aut}(\mathcal{D})$ *is equal to* $\mathbb{Z}_2^5.\mathbb{S}_5$, $(\mathbb{Z}_2^7.ASL_3(2))\mathbb{Z}_3$, $\mathbb{Z}_2^{11}.L_2(11)$, $\mathbb{Z}_2^{27}.P\Gamma L_2(8)$ *or* $\mathbb{Z}_2^{2^{m+1}-1}.ASL_{m+1}(2)$. □

Commentary

1) By analogy with the case of Lie algebras of type A_{p^m-1}, it would be possible to try to classify all (or all transitive) J-decompositions of Lie algebras of type C_{2^m}. Here the methods of finite group theory are evidently insufficient for this purpose: we deal exclusively with soluble groups (see [BDDKLS]!).

2) The general construction of E-decompositions, which plays roughly the same role in the case of Lie algebras of types B_n and D_n as J-decompositions do in

the case of type A_n, was first explained in [KKU 6]. However, the term "E-decomposition" appears only later [Iva 3]. It is worth mentioning that Lie algebras of types B_n and D_n possess many other ODs, with automorphism groups that are not so rich. Following [KKU 1], we reproduce one of these constructions.

Consider the standard realization of the Lie algebra $\mathcal{L}$ of type D_n (see [Bour 2]):

$$\mathcal{L} = \left\langle F_k; A_{i,j}^{\epsilon_i,\epsilon_j} \mid 1 \le k \le n, 1 \le i < j \le n; \epsilon_i, \epsilon_j = \pm 1 \right\rangle_{\mathbb{C}},$$

where

$$F_k = E_{k,k} - E_{k+n,k+n};$$

$$A_{i,j}^{\epsilon_i,\epsilon_j} = \epsilon_j(E_{i,j} - E_{j+n,i+n}) + \epsilon_i(E_{j,i} - E_{i+n,j+n}) +$$
$$+ \epsilon_i\epsilon_j(E_{i+n,j} - E_{j+n,i}) + (E_{j,i+n} - E_{i,j+n}),$$

and $E_{i,j}$ are the standard matrix units. The indicated basis of $\mathcal{L}$ consists of matrices of rank 2, and is orthogonal with respect to the Killing form. Note that $A_{i,j}^{\epsilon_i,\epsilon_j} = -A_{j,i}^{-\epsilon_j,-\epsilon_i}$. We introduce sequences $[\epsilon_1, \epsilon_2] = \epsilon_1, \epsilon_2, \epsilon_3, \ldots$, where $\epsilon_1, \epsilon_2 \in \{\pm 1\}$ and $\epsilon_k = (\epsilon_1)^k \epsilon_2$ for $k \ge 3$. For any cycle $\sigma = (i_1, \ldots, i_k)$ and for any subsequence consisting of the first k terms of $[\epsilon_1, \epsilon_2]$, we have a corresponding k-dimensional commutative subalgebra

$$\mathcal{H}_{\sigma}^{\epsilon_1,\epsilon_2} = \langle A_{i_1,i_2}^{\epsilon_1,\epsilon_2}, A_{i_2,i_3}^{\epsilon_2,\epsilon_3}, \ldots, A_{i_{k-1},i_k}^{\epsilon_{k-1},\epsilon_k}, A_{i_k,i_1}^{\epsilon_k,\epsilon_1} \rangle_{\mathbb{C}}.$$

For definiteness, we shall assume that $i_1 = \min\{i_s\}$. Now let $\sigma = \sigma_1 \ldots \sigma_r$ be a permutation of degree n which has no fixed points, and suppose that $\sigma_1, \ldots, \sigma_r$ are independent cycles. Then

$$\mathcal{H}_{\sigma}^{\epsilon_1,\epsilon_2} = \mathcal{H}_{\sigma_1}^{\epsilon_1,\epsilon_2} \oplus \mathcal{H}_{\sigma_2}^{\epsilon_1,\epsilon_2} \oplus \ldots \oplus \mathcal{H}_{\sigma_r}^{\epsilon_1,\epsilon_2}$$

is a Cartan subalgebra of $\mathcal{L}$. The symbol $\mathcal{H} =< F_1, \ldots, F_n >$ will denote the standard Cartan subalgebra.

Proposition 2A. *Suppose that* $\sigma = (1, 2, \ldots, n)$*. Then the decompositions*

$$\mathcal{L} = \mathcal{H} \oplus \sum_{1 \le k \le m, \epsilon_i = \pm 1} \mathcal{H}_{\sigma^k}^{\epsilon_1,\epsilon_2}$$

when $n = 2m + 1$*; and*

$$\mathcal{L} = \mathcal{H} \oplus \mathcal{H}_{\sigma^m}^{1,1} \oplus \mathcal{H}_{\sigma^m}^{1,-1} \oplus \sum_{1 \le k < m, \epsilon_i = \pm 1} \mathcal{H}_{\sigma^k}^{\epsilon_1,\epsilon_2}$$

when $n = 2m$*, constitute ODs of the Lie algebra of type* D_n. □

Now assume that $\mathcal{L}'$ is a Lie algebra of type B_{n-1}. We shall let $B_{i,j}^{\epsilon_i,\epsilon_j}$ denote the matrix obtained from $A_{i,j}^{\epsilon_i,\epsilon_j} \in \mathcal{L}$ by striking out the $(n + 1)$th row and the

$(n+1)$th column. Because the latter are zero, this does not affect commutativity. We introduce the $(2n-1)\times(2n-1)$ matrices

$$B_{i,1}^{\epsilon_i,1}=0,\ B_{1,i}^{1,\epsilon_i}=\epsilon_i(E_{1,i+n-1}-E_{i,1})+(E_{1,i}-E_{i+n-1,1}).$$

Proposition 2B. *If* $\mathcal{H}_\sigma^{\epsilon_1,\epsilon_2}=\left\langle A_{i,j}^{\epsilon_i,\epsilon_j}\right\rangle_{\mathbb{C}}$ *is a Cartan subalgebra of* $\mathcal{L}$, *then* $\overline{\mathcal{H}_\sigma^{\epsilon_1,\epsilon_2}}=\left\langle B_{i,j}^{\epsilon_i,\epsilon_j}\right\rangle_{\mathbb{C}}$ *is a Cartan subalgebra of* $\mathcal{L}'$. *Furthermore, suppose that* $\mathcal{L}=\mathcal{H}\oplus\sum_i\mathcal{H}_i$ *is an OD of the Lie algebra* $\mathcal{L}$ *of type* D_n. *Then* $\mathcal{L}'=\overline{\mathcal{H}}\oplus\sum_i\overline{\mathcal{H}_i}$ *is an OD of the Lie algebra* $\mathcal{L}'$ *of type* B_{n-1} *(*$\overline{\mathcal{H}}$ *is the standard subalgebra of diagonal matrices).* □

The decompositions described in Propositions 2A and 2B (the proof is omitted) have very poor automorphism groups. For example, when $n>4$ and $\mathcal{L}$ is a Lie algebra of type D_n, any automorphism of the OD indicated acts by the rule

$$\mathcal{H}_{\sigma^k}^{\epsilon_1,\epsilon_2}\to\mathcal{H}_{\sigma^k}^{\epsilon_1',\epsilon_2'},$$

and the automorphism group has order $2^{2n-1+\delta}$, where $\delta=1$ if n is a prime and $\delta=0$ otherwise.

3) At first sight, the admissible partition (2.23) is not compatible with the construction in §2.3. However, if we consider W to be the additive group of the field $\mathbb{F}_q$, $q=2^{m+1}$, and take A to be the multiplicative group $\mathbb{F}_q^*$, and, finally, take the zero element 0 of $\mathbb{F}_q$ for ∞, then (2.23) can be rewritten in the form

$$M_e=\left\{\{u,u+e\}\mid u\in\mathbb{F}_q\right\},\ M_z=\left\{\{zu,z(u+e)\}\mid u\in\mathbb{F}_q\right\},\ z\in\mathbb{F}_q^*.$$

It was mentioned in §2.3 that the set M_e can be chosen in many different ways, although in this section only one choice is considered in detail. In the particular case of the Lie algebra $\mathcal{L}$ of type D_5 and the cyclic group $A=\{0,1,\ldots,8\}$ of order 9, one can make another choice of the set M_0:

$$M_0=\left\{\{\infty,0\},\{1,2\},\{3,7\},\{4,6\},\{5,8\}\right\},$$

corresponding to the cyclic group $\mathrm{Aut}(\mathcal{P})\cong\mathbb{Z}_9$. As can be seen, there are many possibilities for constructing different E-decompositions.

4) Theorem 2.4.6 was obtained by Ivanov [Iva 3] and, independently, Burichenko (unpublished). It was with the appearance of this paper [Iva 3] that irreducible orthogonal decompositions (IODs) began to be of such great importance. This topic will be considered in more detail in subsequent chapters of the book.

Chapter 3

Jordan subgroups and orthogonal decompositions

The constructions of transitive orthogonal decompositions (TODs) at the beginning of the eighties and the resulting list of complex simple Lie algebras admitting a TOD are an astonishing reminder of the earlier results (see [Ale 1]) on the classification of Jordan subgroups in complex Lie groups. In fact, one can give an internal explanation of this coincidence. It is that the presence of a Jordan subgroup in the Lie group $\mathcal{G} = \mathrm{Inn}(\mathcal{L})$ of all inner automorphisms of the given complex simple Lie algebra $\mathcal{L}$ implies the existence of a TOD for $\mathcal{L}$. Attempts to prove the converse implication have still not met with success (except for Lie algebras of small ranks). However, as is shown later, the existence of an irreducible orthogonal decomposition (IOD) for $\mathcal{L}$ is enough to guarantee the existence of a Jordan subgroup in $\mathrm{Inn}(\mathcal{L})$.

3.1 General construction of TODs

Definition 3.1.1. Let $\mathcal{G}$ be a complex simple Lie group of adjoint type. A finite abelian subgroup $J \subset \mathcal{G}$ is called a *Jordan subgroup* if the following conditions hold:

(i) $N = N_{\mathcal{G}}(J)$ is finite;

(ii) J is a minimal normal subgroup of N;

(iii) if J' is an abelian subgroup of $\mathcal{G}$ such that $J' \supseteq J$ and $N_{\mathcal{G}}(J') \supseteq N_{\mathcal{G}}(J)$, then $N_{\mathcal{G}}(J') = N$.

The classification theorem [Ale 1] shows that the possibilities for the pair $(\mathcal{G}, J)$, up to $\mathrm{Aut}(\mathcal{L})$-conjugacy, are those listed in Table A1. Note that in the case of classical groups $\mathcal{G}$, their Jordan subgroups were actually known before the introduction of the general concept of Jordan subgroup (see e.g. [Sup 1]).

Theorem 3.1.2. *Let $\mathcal{L}$ be a complex simple Lie algebra and set $\mathcal{G} = \mathrm{Inn}(\mathcal{L})$. Suppose that $\mathcal{G}$ admits a Jordan subgroup J. Then $\mathcal{L}$ admits a transitive orthogonal decomposition $\mathcal{D} : \mathcal{L} = \oplus_{k=1}^{h+1} \mathcal{H}_k$ (h is the Coxeter number) such that*

(i) *J fixes every component $\mathcal{H}_k$; and*

(ii) $\mathrm{Aut}(\mathcal{D}) \subseteq N_{\mathcal{A}}(J)$, *where* $\mathcal{A} = \mathrm{Aut}(\mathcal{L})$.

Proof. The proof is carried out by considering separate cases in accordance with Table A1.

1) *$\mathcal{L}$ is of types A_{p^m-1}, $m \geq 1$; $C_{2^{m-1}}$, $m \geq 2$; and $D_{2^{m-1}}$, $m \geq 3$.*

In this case, $\mathcal{G}$ is one of $PSL_{p^m}(\mathbb{C})$, $PSp_{2^m}(\mathbb{C})$ and $PSO_{2^m}(\mathbb{C})$, respectively. Let $\hat{\mathcal{G}}$ stand for the cover $SL_{p^m}(\mathbb{C})$, $Sp_{2^m}(\mathbb{C})$ or $SO_{2^m}(\mathbb{C})$, respectively. As is shown in [Ale 1], one can choose a subgroup $\hat{J}$ of $\hat{\mathcal{G}}$ such that $\hat{J}$ covers J in the natural homomorphism $\hat{\mathcal{G}} \to \mathcal{G}$; moreover, $\hat{J}$ is p_+^{1+2m}, 2_-^{1+2m} or 2_+^{1+2m}, respectively. The required TOD is then as indicated in §1.1 (for type A_{p^m-1}) and in §2.1 (for types $C_{2^{m-1}}$ and $D_{2^{m-1}}$; see Proposition 2.1.5 and Proposition 2.1.6). Here we note only that in these sections the group $\hat{J}$ is denoted by the symbol $\hat{H}$.

2) *$\mathcal{L}$ is of types B_n or D_n.*

In this case, J is nothing but the group consisting of diagonal ± 1-matrices with determinant 1 of degree $2n+1$ (for type B_n) or the factor group of the group consisting of diagonal ± 1-matrices with determinant 1 of degree $2n$ by the subgroup $\{\pm E_{2n}\}$ (for type D_n). The required TOD is constructed in §2.3.

3) *Exceptional cases: D_4, G_2, F_4, E_6 and E_8.*

In these cases, the required TOD can be given by the standard construction $\mathcal{D} : \mathcal{L} = \oplus_{k=1}^{h+1} \mathcal{H}_k$, where

$$\mathcal{H}_k = C_{\mathcal{L}}(A_k) = \{x \in \mathcal{L} \mid \forall f \in A_k, f(x) = x\} \tag{3.1}$$

and $\{A_1, \ldots, A_{h+1}\}$ is the set of all maximal subgroups of the corresponding Jordan subgroup J. To prove that formula (3.1) leads to a TOD for $\mathcal{L}$, we use the following information about the Jordan subgroup $J = \mathbb{Z}_p^m$, given in [Ale 1]. Let χ be the character of the adjoint representation of $\mathcal{G}$ on $\mathcal{L}$. Then

$$\chi_J = l \sum_{\rho \in \mathrm{Irr}(J) \setminus \{1_J\}} \rho. \tag{3.2}$$

In particular, $l = \dim \mathcal{L}/(p^m - 1)$ takes the values 4, 2, 2, 3, 8 or 2 when the pair $(\mathcal{G}, J)$ is equal to $(D_4, \mathbb{Z}_2^3)$, $(G_2, \mathbb{Z}_2^3)$, $(F_4, \mathbb{Z}_3^3)$, $(E_6, \mathbb{Z}_3^3)$, $(E_8, \mathbb{Z}_2^5)$ or $(E_8, \mathbb{Z}_5^3)$ respectively. In accordance with (3.2), the space $\mathcal{L}$ is decomposed into a direct sum of eigensubspaces

$$\mathcal{L} = \bigoplus_{\rho \in \mathrm{Irr}(J) \setminus \{1_J\}} \mathcal{L}_\rho,$$

where $\mathcal{L}_\rho = \{x \in \mathcal{L} \mid \forall f \in J, f(x) = \rho(f)x\}$ for $\rho \in \mathrm{Irr}(J)$. It is obvious that, in this case,

$$\mathcal{H}_k = C_{\mathcal{L}}(A_k) = \sum_{\rho \in \mathrm{Irr}(J)\setminus\{1_J\}, \rho|_{A_k} = 1_{A_k}} \mathcal{L}_\rho$$

and $\dim \mathcal{H}_k = l(p-1) = \dim \mathcal{L}.(p-1)/(p^m-1)$ is just rank $\mathcal{L}$. It can be checked immediately that in all cases, the number $h+1$ really is $(p^m-1)/(p-1)$. Now we have

$$[\mathcal{L}_\alpha, \mathcal{L}_\beta] \subseteq \mathcal{L}_{\alpha\beta}$$

for all $\alpha, \beta \in \mathrm{Irr}(J)$. Therefore, when $x \in \mathcal{L}_\alpha$ and $y \in \mathcal{L}_\beta$, the operator $\mathrm{ad}x.\mathrm{ad}y$ maps $\mathcal{L}_\gamma$ into $\mathcal{L}_{\alpha\beta\gamma}$ for every $\gamma \in \mathrm{Irr}(J)$. Thus

$$K(x, y) = \mathrm{Tr}(\mathrm{ad}x.\mathrm{ad}y) = 0$$

if $\alpha.\beta \neq 1_J$. In particular,

$$K(\mathcal{H}_i, \mathcal{H}_j) = 0 \tag{3.3}$$

for all i, j with $1 \leq i \neq j \leq h+1$. Furthermore, each subalgebra $\mathcal{H}_k$ is nilpotent. In fact, if $J = \langle A_k, \varphi \rangle$ then the automorphism φ of prime order p fixes $\mathcal{H}_k$. Moreover, the fixed-point subspace $C_{\mathcal{H}_k}(\varphi)$ of φ is trivial: $C_{\mathcal{H}_k}(\varphi) = C_{\mathcal{L}}(J) = \mathcal{L}_{1_J} = 0$ (see (3.2)). Thus, the automorphism φ of prime order acts fixed-point-freely on the Lie algebra $\mathcal{H}_k$. Hence, $\mathcal{H}_k$ is nilpotent by [SSL]. Finally, from (3.3) it follows that the restriction of the Killing form K to $\mathcal{H}_k$ is non-degenerate. So, by Lemma 1.1.1, $\mathcal{H}_k$ is a Cartan subalgebra. Correspondingly, $\mathcal{D}$ is an $N_{\mathcal{G}}(J)$-invariant OD. In particular, $\mathcal{D}$ is even an irreducible OD.

As is shown in §§1.3, 2.1 and 2.2, the TODs of classical Lie algebras indicated satisfy the inclusion $\mathrm{Aut}(\mathcal{D}) \subseteq N_{\mathcal{G}}(J)$. For the exceptional cases we have in fact equality $\mathrm{Aut}(\mathcal{D}) = N_{\mathcal{G}}(J)$. We postpone the proof until Chapter 4. □

Definition 3.1.1 of Jordan subgroups can be restated naturally for any simple algebraic group $\mathcal{G}$ of adjoint type over an algebraically closed field $\mathbb{F}$ of arbitrary characteristic. Similarly, the classification theorem of Alekseevskii can be restated without change (except for the natural restriction $\mathrm{char}\mathbb{F} \neq \exp(J)$) (see [Bor 1]). Moreover, when $r = \mathrm{char}\mathbb{F} > 0$, Jordan subgroups J of $\mathcal{G}$ are contained in the group $\mathcal{G}(\mathbb{F}_q)$ of $\mathbb{F}_q$-points of $\mathcal{G}$, for some finite field $\mathbb{F}_q$. The Jordan subgroups defined in this way play an essential role in the description of the different maximal subgroup classes of $\mathcal{G}$ and of $\mathcal{G}(\mathbb{F}_q)$. Namely, the following results hold.

Theorem 3.1.3 [Bor 2]. *Let $\mathcal{G}$ be a simple algebraic group of adjoint type over an algebraically closed field $\mathbb{F}$ of characteristic 0, and X a finite subgroup of $\mathcal{G}$. Then one of the following assertions holds:*

(i) *X normalizes some nontrivial proper connected subgroup $\mathcal{H}$ of $\mathcal{G}$,*

(ii) *X is an almost simple group (that is, $X_0 \lhd X \subseteq \mathrm{Aut}(X_0)$ for some non-abelian simple group X_0),*

(iii) *$X \subseteq N_{\mathcal{G}}(J)$ for some Jordan subgroup J of $\mathcal{G}$,*
(iv) *$\mathcal{G}$ is of type E_8 and $Y \lhd X \subseteq N_{\mathcal{G}}(Y)$, where Y is the Borovik group.* □

In this, the Borovik group Y is defined uniquely (up to $\mathcal{G}$-conjugacy) by the following properties: $Y = A \times B$, $A \cong \mathbb{A}_5$, $B \cong \mathbb{A}_6$, $N_{\mathcal{G}}(Y)/Y \cong \mathbb{Z}_2^2$, $C_{\mathcal{G}}(B) = A$ and $C_{\mathcal{G}}(A) \cong \mathbb{S}_6$ (see also [Bor 5]).

Furthermore, let $\mathcal{G}$ be a simple algebraic group of adjoint type defined over a finite field $\mathbb{F}_q$, $\mathcal{G}(\mathbb{F}_q)$ its point group over $\mathbb{F}_q$, and $G_0 = \mathcal{G}(\mathbb{F}_q)'$. Then, with some exceptions for small q, the group G_0 is a finite simple group [Ste 2]. Suppose that $G_0 \subseteq G_1 \subseteq \mathrm{Aut}(G_0)$ and $X \subseteq G_1$.

Theorem 3.1.4 [Bor 2], [CLSS]. *Under these assumptions, one of the following statements holds.*
(i) *X normalizes some nontrivial connected $\mathbb{F}_q$-defined subgroup $\mathcal{H} \subset \mathcal{G}$.*
(ii) *X is an almost simple group.*
(iii) *$X \cap G_0 \subseteq N_{\mathcal{G}}(J)$ for some Jordan subgroup J of $\mathcal{G}$.*
(iv) *$\mathcal{G}$ is of type E_8, $r = \mathrm{char}\mathbb{F}_q > 5$, $\mathbb{F}_q \ni \sqrt{5}$ and $Y \subset X \cap G_0 \subset N_{\mathcal{G}}(Y)$, where Y is the Borovik group.* □

Thus, the normalizers of Jordan subgroups in simple algebraic groups usually represent *maximal finite* subgroups.

In the case when G_0 is a classical finite simple group, $q = r$ (and $\mathcal{G}$ possesses a Jordan subgroup J), Aschbacher's Theorem [Asch 1] states that, with several exceptions, the normalizers $N_{G_0}(J)$ of Jordan subgroups are *local maximal* subgroups (of class $\mathcal{C}_6$ for types A_{p^m-1}, C_{2^m}, D_{2^m} and of class $\mathcal{C}_2$ for types B_n, D_n; see [KlL 2]). Recall that a subgroup X of G is said to be *local*, if $E \subseteq X \subseteq N_G(E)$ for some elementary abelian subgroup E. From Theorem 3.1.4 it follows, in particular, that a similar statement holds for finite simple groups G_0 of exceptional types. The cases when $N_{G_0}(J)$ are local maximal subgroups of G_0 are listed in the following Table 3.1, which is taken from [CLSS].

Table 3.1
The normalizers of Jordan subgroups
which are local maximal subgroups of exceptional finite simple groups

G_0	J	$N_{G_0}(J)$	Conditions
$G_2(r)$	$\mathbb{Z}_2^3$	$\mathbb{Z}_2^3.SL_3(2)$	$r > 2$
$F_4(r)$	$\mathbb{Z}_3^3$	$\mathbb{Z}_3^3.SL_3(3)$	$r \geq 5$
$E_6^\epsilon(r)$	$\mathbb{Z}_3^3$	$3^{3+3}.SL_3(3)$	$\epsilon = \pm 1, 3 \mid (r - \epsilon), r \geq 5$
$E_8(r)$	$\mathbb{Z}_2^5$	$2^{5+10} \circ SL_5(2)$	
$E_8(q)$	$\mathbb{Z}_5^3$	$\mathbb{Z}_5^3.SL_3(5)$	$q = r^a$; $r \neq 2, 5$; $a = \begin{cases} 1 & \text{if } 5 \mid (r^2 - 1) \\ 2 & \text{if } 5 \mid (r^2 + 1) \end{cases}$

Let us give some explanations. In Table 3.1 we use the notation $E_6^\epsilon(q)$ for $E_6(q)$ if $\epsilon = 1$ and ${}^2E_6(q)$ if $\epsilon = -1$. The condition $r \neq \exp(J)$ is necessary, by the Borel-Tits Theorem. The condition that $r \neq 2$ when J is $\mathbb{Z}_3^3$ or $\mathbb{Z}_5^3$ and G_0 is $F_4(r)$ or $E_8(r^a)$ occurs because of the embeddings

$$\mathbb{Z}_3^3.SL_3(3) \subset L_4(3) \subset F_4(2)$$

and

$$\mathbb{Z}_5^3.SL_3(5) \subset L_4(5) \subset E_8(4).$$

The embedding $L_4(3) \subset F_4(2)$ is known [Coo 2], [ATLAS], [NoW], and the fact that $L_4(5) \subset E_8(4)$ was first discovered in [CLSS]. Similarly, we have the following sequence of embeddings:

$$3^{3+3}.SL_3(3) \subset \Omega_7(3) \subset Fi_{22} \subset {}^2E_6(2) \subset E_6(4).$$

Quite recently [Bur 3], a new explanation of the embeddings $L_4(3) \subset F_4(2)$ and $\Omega_7(3) \subset {}^2E_6(2)$ via Moufang loops has been given. It is of great interest to note that all the (in our context) "exceptional" groups $L_4(3)$, $L_4(5)$, $\Omega_7(3)$ and Fi_{22} also appear as the full automorphism groups of invariant lattices of type F_4 [Bur 2], [BuT], E_8 [Tiep 16] and E_6 [BuT]. For the details, see Chapters 12 and 13.

3.2 Root orthogonal decompositions (RODs)

As yet, all our constructions of orthogonal decompositions use Jordan subgroups in varying degrees. At this point we consider a new construction, which is used in an essential way in the subsequent section, and which is of independent interest.

Definition 3.2.1. A decomposition $\mathcal{L} = \mathcal{H}_0 \oplus \mathcal{H}_1 \oplus \ldots \oplus \mathcal{H}_h$ of a simple Lie algebra $\mathcal{L}$ is called a *root decomposition* if $\mathcal{H}_0$ is the standard Cartan subalgebra and every subalgebra $\mathcal{H}_k$, $k \geq 1$, possesses a basis consisting of vectors of the type $aX_\alpha + bX_{-\alpha}$, where α is a positive root with respect to $\mathcal{H}_0$ and X_α, $X_{-\alpha}$ are corresponding root vectors. (Thus,

$$\mathcal{L} = \mathcal{H}_0 \oplus \sum_{\alpha \in R} < X_\alpha >_{\mathbb{C}},$$

where R the corresponding root system, is the Cartan decomposition with respect to $\mathcal{H}_0$).

For any basis vector $aX_\alpha + bX_{-\alpha}$, the coefficients a and b are not equal to 0, because the ad-nilpotent elements X_α and $X_{-\alpha}$ cannot be contained in a Cartan subalgebra. It is also clear that the elements $aX_\alpha + bX_{-\alpha}$ and $a'X_{\alpha'} + b'X_{-\alpha'}$ can belong to the same Cartan subalgebra only if $\alpha + \alpha'$ and $\alpha - \alpha'$ are not roots. Thus we can associate with each Cartan subalgebra $\mathcal{H}_k$, $k \geq 1$, some collection $\Sigma_k^+ = \{\sigma_1, \ldots, \sigma_n\}$, $n = \text{rank } \mathcal{L}$, consisting of positive roots and possessing the following property (cf. §2.2):

$$\text{if } i \neq j \text{ then } \sigma_i \neq \sigma_j, \text{ and } \sigma_i \pm \sigma_j \text{ are not roots.} \tag{3.4}$$

Note that for the given decomposition each positive root α enters exactly two such collections, say, Σ_k^+, Σ_l^+.

As always, we use the tables of root systems given in [Bour 2]. We shall denote by X^+ the set of positive roots of a root system of type X.

Proposition 3.2.2. *The Lie algebras of types A_n, $n \geq 2$; C_n, $n \geq 2$; D_{2m+1}, $m \geq 2$; B_{2m}, $m \geq 2$; F_4 and E_6 admit no root decompositions.*

Proof. 1) *Type A_n, $n \geq 2$.*

Here the positive roots are $\varepsilon_i - \varepsilon_j$, where $1 \leq i < j \leq n+1$. Suppose that $\varepsilon_1 - \varepsilon_2 \in \Sigma_k^+$ for some k. According to (3.4) a root $\varepsilon_i - \varepsilon_j \neq \varepsilon_1 - \varepsilon_2$ can belong to Σ_k^+ only in the case where $\{i, j\} \cap \{1, 2\} = \emptyset$. The set of such roots forms a root system of type A_{n-2}. It is impossible to choose in the last system more than $n-2$ roots that satisfy condition (3.4). This completes the proof.

2) *Type C_n, $n \geq 2$.*

Here the positive roots are $\varepsilon_i \pm \varepsilon_j$, $1 \leq i < j \leq n$; $2\varepsilon_i$, $1 \leq i \leq n$. Let $\varepsilon_1 - \varepsilon_2 \in \Sigma_k^+$. Then in addition to $\varepsilon_1 - \varepsilon_2$, the set Σ_k^+ can contain only the roots of type $\varepsilon_i \pm \varepsilon_j$, $3 \leq i < j \leq n$; $2\varepsilon_i$, $3 \leq i \leq n$, all of which form a positive root system of type C_{n-2}^+. However, in the last system, one can choose at most $n-2$ roots satisfying condition (3.4).

3) *Type D_{2m+1}, $m \geq 1$.*

Note that type D_3 is isomorphic to type A_3, and this is considered here as a special case, in order to facilitate our discussion of 4). In this case, $D_{2m+1}^+ = \{\varepsilon_i \pm \varepsilon_j \mid 1 \leq i < j \leq 2m+1\}$. If $\varepsilon_i + \varepsilon_j \in \Sigma_k^+$, then, apart from this root, Σ_k^+ can contain only the roots $\varepsilon_s \pm \varepsilon_t$ with $\{i, j\} \cap \{s, t\} = \emptyset$, and $\varepsilon_i - \varepsilon_j$. Hence, there exists at most $2m$ positive roots satisfying condition (3.4).

4) *Type B_{2m}, $m \geq 2$.*

Here the positive roots are $\varepsilon_i \pm \varepsilon_j$, $1 \leq i < j \leq 2m$; ε_i, $1 \leq i \leq 2m$. Suppose that $\varepsilon_{2m} \in \Sigma_k^+$. Then, in addition to this root, Σ_k^+ can contain only the roots $\varepsilon_i \pm \varepsilon_j$, $1 \leq i < j \leq 2m-1$, the totality of which form a positive root system of type D_{2m-1}^+. But in 3) it was shown that the latter system has at most $2m-2$ roots satisfying condition (3.4).

5) *Type F_4.*

Here, the positive roots are ε_i, $1 \le i \le 4$; $\varepsilon_i \pm \varepsilon_j$, $1 \le i < j \le 4$; $(\varepsilon_1 \pm \varepsilon_2 \pm \varepsilon_3 \pm \varepsilon_4)/2$. If $\varepsilon_1 \in \Sigma_k^+$, then, apart from this root, Σ_k^+ can contain only the roots $\varepsilon_i \pm \varepsilon_j$, $2 \le i < j \le 4$, among which no three satisfy condition (3.4).

6) *Type E_6.*

Here, the positive roots are the following: $\pm\varepsilon_i + \varepsilon_j$, $1 \le i < j \le 5$; $(\varepsilon + \sum_{i=1}^{5}(-1)^{\nu(i)}\varepsilon_i)/2$, where $\varepsilon = \varepsilon_8 - \varepsilon_7 - \varepsilon_6$ and $\sum_i \nu(i) \equiv 0 (\operatorname{mod} 2)$ (see [Bour 2]). If Σ_k^+ contains the roots $\varepsilon_s + \varepsilon_t$ and $-\varepsilon_s + \varepsilon_t$, then no root of the kind $(\varepsilon + \ldots)/2$ can belong to Σ_k^+. On the other hand, this collection does not contain more than two other roots like $\pm\varepsilon_i + \varepsilon_j$, a contradiction. So we can suppose that, in addition to $-\varepsilon_1 + \varepsilon_2$, the collection Σ_k^+ contains at most one other root like $\pm\varepsilon_s + \varepsilon_t$. In this case, Σ_k^+ must contain at least four roots of the type $(\varepsilon + \sum_{i=1}^{5}(-1)^{\nu(i)}\varepsilon_i)/2$, where moreover $\nu(1) = \nu(2)$. However, in the set

$$\left\{(\varepsilon + \sum_{i=1}^{5}(-1)^{\nu(i)}\varepsilon_i)/2 \,\middle|\, \sum_i \nu(i) \equiv \nu(1) + \nu(2) \equiv 0 (\operatorname{mod} 2)\right\}$$

one can find at most two roots satisfying condition (3.4), a contradiction. □

In all remaining cases, the Lie algebras admit RODs. Postponing to the next section the discussion of the cases A_1, G_2 and E_8, we establish now the existence of RODs for the Lie algebras of types B_{2m-1}, D_{2m} and E_7. We do not try to describe all possible RODs.

Suppose that the set Σ^+ of positive roots of the Lie algebra $\mathcal{L}$ is decomposed into components Σ_k^+ satisfying condition (3.4). Starting from $\Sigma_k^+ = \{\sigma_1, \ldots, \sigma_n\}$, we construct the two subalgebras $\mathcal{H}_k^\tau = < E_{\sigma_1}^\tau, \ldots, E_{\sigma_n}^\tau >_{\mathbb{C}}$, $\tau = 0, 1$. Here and below, $E_\sigma^\tau = X_\sigma + (-1)^\tau X_{-\sigma}$. It is easy to see that

$$K(E_\sigma^\tau, E_{\sigma'}^{\tau'}) = 0 \text{ if either } \sigma \ne \sigma', \text{ or } \sigma = \sigma' \text{ and } \tau \ne \tau'.$$

Moreover, $K(\mathcal{H}_0, \mathcal{H}_k^\tau) = 0$. Hence, the commutative subalgebras $\mathcal{H}_0$ and $\mathcal{H}_k^\tau$, where $\tau = 0, 1$ and $1 \le k \le |\Sigma^+|/n = h/2$ (h being the Coxeter number), are pairwise orthogonal with respect to the Killing form. By Lemma 1.1.1, the decomposition

$$\mathcal{L} = \mathcal{H}_0 \oplus \mathcal{H}_1^0 \oplus \mathcal{H}_1^1 \oplus \mathcal{H}_2^0 \oplus \mathcal{H}_2^1 \oplus \ldots$$

is an ROD.

At this point we indicate a partition of Σ^+ with property (3.4) for the Lie algebras mentioned above.

(i) *Type E_7.*

Here, the positive roots are $\pm\varepsilon_i + \varepsilon_j$, $1 \le i < j \le 6$; $\varepsilon = -\varepsilon_7 + \varepsilon_8$; $(\varepsilon + \sum_{i=1}^{6}(-1)^{\nu(i)}\varepsilon_i)/2$, where $\sum_i \nu(i) \equiv 1 (\operatorname{mod} 2)$. We shall indicate one admissible partition of Σ^+, using the following brief notation: we write $1+2$ instead of $\varepsilon_1+\varepsilon_2$, $e+1+2-3+4-5-6$ instead of $(\varepsilon+\varepsilon_1+\varepsilon_2-\varepsilon_3+\varepsilon_4-\varepsilon_5-\varepsilon_6)/2$, and so on.

$$\Sigma_1^+ : -7+8,\ 1+2,\ -1+2,\ 3+4,\ -3+4,\ 5+6,\ -5+6$$

$$\Sigma_2^+ : \begin{array}{l} 1+3,\ 2+5,\ 4+6, \\ e+1+2-3+4-5-6, \\ e-1+2+3-4-5+6, \\ e-1-2+3+4+5-6, \\ e+1-2-3-4+5+6 \end{array} \qquad \Sigma_3^+ : \begin{array}{l} 1+4,\ 2+6,\ 3+5, \\ e-1-2-3+4+5+6, \\ e+1-2+3-4-5+6, \\ e+1+2+3+4+5-6, \\ e-1+2-3-4+5+6 \end{array}$$

$$\Sigma_4^+ : \begin{array}{l} 1+5,\ -2+3,\ -4+6, \\ e-1-2-3-4+5-6, \\ e-1+2+3+4+5+6, \\ e+1-2-3+4-5+6, \\ e+1+2+3-4-5-6 \end{array} \qquad \Sigma_5^+ : \begin{array}{l} 1+6,\ -2+4,\ -3+5, \\ e+1+2+3+4+5-6, \\ e+1-2-3-4-5-6, \\ e-1+2-3+4-5+6, \\ e-1-2+3-4+5+6 \end{array}$$

$$\Sigma_6^+ : \begin{array}{l} -1+3,\ -2+6,\ 4+5, \\ e-1-2-3+4-5-6, \\ e+1-2+3-4+5-6, \\ e-1+2-3-4+5+6, \\ e+1+2+3+4-5+6 \end{array} \qquad \Sigma_7^+ : \begin{array}{l} -1+4,\ -2+5,\ 3+6, \\ e+1+2-3+4+5+6, \\ e+1-2+3+4-5-6, \\ e-1+2+3-4+5-6, \\ e-1-2-3-4-5+6 \end{array}$$

$$\Sigma_8^+ : \begin{array}{l} -1+5,\ 2+4,\ -3+6, \\ e-1+2-3-4-5-6, \\ e-1-2+3+4-5+6, \\ e+1-2-3+4+5-6, \\ e+1+2+3-4+5+6 \end{array} \qquad \Sigma_9^+ : \begin{array}{l} -1+6,\ 2+3,\ -4+5, \\ e-1+2-3+4+5-6, \\ e+1-2+3+4+5+6, \\ e+1+2-3-4-5+6, \\ e-1-2+3-4-5-6 \end{array}$$

(ii) *Type* D_{2m}.

Here the positive roots are $\varepsilon_i \pm \varepsilon_j$, $1 \le i < j \le 2m$. It is convenient to introduce the following notation:

$$(i,j) = \varepsilon_i + \varepsilon_j,\ [i,j] = \varepsilon_i - \varepsilon_j;$$

and we set $\Sigma_k^+ = \Omega_k \cup \Xi_k$, where

$$\Omega_k = \{(i_1,j_1),(i_2,j_2),\dots,(i_m,j_m)\},\ \Xi_k = \{[i_1,j_1],[i_2,j_2],\dots,[i_m,j_m]\}.$$

Then condition (3.4) is equivalent to the following:

$$\{i_s,j_s\} \cap \{i_t,j_t\} = \emptyset \text{ if } s \ne t.$$

Example 3.2.3. An ROD of the Lie algebra of type D_4 $(m = 2)$:

$$\mathcal{H}_0 \text{ - standard Cartan subalgebra; } \tau = 0, 1;$$

$$\mathcal{H}_1^\tau = \langle E^\tau_{\varepsilon_1+\varepsilon_2}, E^\tau_{\varepsilon_1-\varepsilon_2}, E^\tau_{\varepsilon_3+\varepsilon_4}, E^\tau_{\varepsilon_3-\varepsilon_4} \rangle_{\mathbb{C}},$$

$$\mathcal{H}_2^\tau = \langle E^\tau_{\varepsilon_1+\varepsilon_3}, E^\tau_{\varepsilon_1-\varepsilon_3}, E^\tau_{\varepsilon_2+\varepsilon_4}, E^\tau_{\varepsilon_2-\varepsilon_4} \rangle_{\mathbb{C}},$$

$$\mathcal{H}_3^\tau = \langle E^\tau_{\varepsilon_1+\varepsilon_4}, E^\tau_{\varepsilon_1-\varepsilon_4}, E^\tau_{\varepsilon_2+\varepsilon_3}, E^\tau_{\varepsilon_2-\varepsilon_3} \rangle_{\mathbb{C}}.$$

So, to construct an ROD for the Lie algebra of type D_{2m} it is enough to indicate a partition of the set $\Omega = \{(i, j) \mid 1 \le i < j \le 2m\}$ into $2m - 1$ subsets $\Omega_k = \{(i_1, j_1), (i_2, j_2), \ldots, (i_m, j_m)\}$ such that $\{i_s, j_s\} \cap \{i_t, j_t\} = \emptyset$ if $s \ne t$. The required partition is nothing but an admissible partition in the sense of §2.2 rewritten in different notation. One of the possible constructions of such partitions is explained in §2.3.

(iii) *Type* B_{2m-1}.

In this case $\Sigma^+ = \{\varepsilon_1, \ldots, \varepsilon_{2m-1}; \varepsilon_i \pm \varepsilon_j, 1 \le i < j \le 2m - 1\}$. Starting from the ROD of the Lie algebra of type D_{2m} constructed in (ii), we consider the corresponding admissible partition of the set

$$\Omega = \{(i, j) \mid 1 \le i < j \le 2m\} = \{\varepsilon_i + \varepsilon_j \mid 1 \le i < j \le 2m\}.$$

With every set $\Omega_k \cup \Xi_k = \{\ldots; \varepsilon_{i_k} + \varepsilon_{2m}, \varepsilon_{i_k} - \varepsilon_{2m}\}$ (where "..." denotes roots $\varepsilon_{i_s} \pm \varepsilon_{j_s}$ with indices $i_s < j_s < 2m$), we associate the set $\Omega_k' = \{\ldots; \varepsilon_{i_k}\}$. In the latter formula "..." denotes the same roots as in the case of $\Omega_k \cup \Xi_k$. The desired partition of the set Σ^+ defining an ROD for the Lie algebra of type B_{2m-1} may be given as follows:

$$\Sigma^+ = \Omega_1' \cup \Omega_2' \cup \ldots \cup \Omega_{2m-1}'.$$

Example 3.2.4. An ROD of the Lie algebra of type B_3 (see also Example 3.2.3):

$$\mathcal{H}_0 \text{ - standard Cartan subalgebra; } \tau = 0, 1;$$

$$\mathcal{H}_1^\tau = \langle E^\tau_{\varepsilon_1+\varepsilon_2}, E^\tau_{\varepsilon_1-\varepsilon_2}, E^\tau_{\varepsilon_3} \rangle_{\mathbb{C}},$$

$$\mathcal{H}_2^\tau = \langle E^\tau_{\varepsilon_1+\varepsilon_3}, E^\tau_{\varepsilon_1-\varepsilon_3}, E^\tau_{\varepsilon_2} \rangle_{\mathbb{C}},$$

$$\mathcal{H}_3^\tau = \langle E^\tau_{\varepsilon_2+\varepsilon_3}, E^\tau_{\varepsilon_2-\varepsilon_3}, E^\tau_{\varepsilon_1} \rangle_{\mathbb{C}}.$$

The most interesting result of this section is the construction of an OD for the Lie algebra of type E_7. As we know (see Table A1), the Lie group of type E_7 has no Jordan subgroups.

3.3 Multiplicative orthogonal decompositions (MODs)

This section is devoted to describing all the decompositions with the most rigid structure.

Definition 3.3.1. An OD $\mathcal{D} : \mathcal{L} = \oplus_{i=0}^{h} \mathcal{H}_i$ of a simple Lie algebra $\mathcal{L}$ is said to be *multiplicative* (abbreviated by MOD), if for all i, j there exists k such that $[\mathcal{H}_i, \mathcal{H}_j] \subseteq \mathcal{H}_k$.

In fact, as shown below, the orthogonality condition $K(\mathcal{H}_i, \mathcal{H}_j) = 0$, $i \neq j$, is a consequence of multiplicativity. The investigation of MODs is based on the following two lemmas. The first explains the relationship between MODs and RODs.

Lemma 3.3.2. *Every MOD of a Lie algebra $\mathcal{L}$ containing a standard Cartan subalgebra $\mathcal{H}_0$ is an ROD.*

Proof. For every two different positive roots σ_i and σ_j (with respect to $\mathcal{H}_0$), one can find an element $h_{ij} \in \mathcal{H}_0$ such that

$$(\operatorname{ad} h_{ij}) X_{\sigma_i} = X_{\sigma_i}, \ (\operatorname{ad} h_{ij}) X_{\sigma_j} = 0.$$

(Recall that $\mathcal{L} = \mathcal{H}_0 \oplus \sum_{\sigma \in R} < X_\sigma >_{\mathbb{C}}$ is the Cartan decomposition with respect to $\mathcal{H}_0$). Indeed, setting $h_{ij} = \sum_{k=1}^{n} x_k h_k$, where $\mathcal{H}_0 =< h_1, \ldots, h_n >_{\mathbb{C}}$, we come to the following system of equations:

$$\sum_{k=1}^{n} x_k \sigma_i(h_k) = 1, \ \sum_{k=1}^{n} x_k \sigma_j(h_k) = 0.$$

This system is soluble because the roots σ_i and σ_j are not proportional. Set

$$A = \sum_{i=1}^{m} (a_i X_{\sigma_i} + b_i X_{-\sigma_i}) \in \mathcal{H}_k,$$

where $\mathcal{H}_k$ is some Cartan subalgebra from the given MOD $\mathcal{D}$. Then the elements

$$A_s = \Big(\prod_{1 \leq j \leq m, j \neq s} (\operatorname{ad} h_{sj})^2 \Big) A = a_s X_{\sigma_s} + b_s X_{-\sigma_s}$$

belong to the subspace

$$[\mathcal{H}_0, [\ldots, [\mathcal{H}_0, [\mathcal{H}_0, \mathcal{H}_k]] \ldots]],$$

($\mathcal{H}_0$ enters $2(m-1)$ times); by the definition of MOD, this is contained in some Cartan subalgebra $\mathcal{H}_l$, where the index l does not depend on s. But in this case,

$A = \sum_s A_s \in \mathcal{H}_l$ and so $l = k$. This means that $\mathcal{H}_k$ is spanned by elements of the type $aX_\sigma + bX_{-\sigma}$. □

Lemma 3.3.3. *If the Lie algebra $\mathcal{L}$ admits an MOD $\mathcal{D}$, then*

$$\frac{\dim \mathcal{L}}{\operatorname{rank} \mathcal{L}} = 2^d - 1$$

for some natural number d, and $\mathcal{D}$ can be rewritten in the form

$$\mathcal{L} = \bigoplus_{v \in \mathbb{F}_2^d \setminus \{0\}} \mathcal{H}_v,$$

where $[\mathcal{H}_u, \mathcal{H}_v] = \mathcal{H}_{u+v}$ for $u \neq v$.

Proof. Let $\mathcal{D}$ be an MOD for $\mathcal{L}$, and suppose that $[\mathcal{H}_i, \mathcal{H}_j] \subseteq \mathcal{H}_k$, $i \neq j$. Consider an automorphism Φ of $\mathcal{L}$ such that $\Phi(\mathcal{H}_i) = \mathcal{H}_0$ is the standard Cartan subalgebra. Then $\Phi(\mathcal{D}) : \mathcal{L} = \oplus_s \Phi(\mathcal{H}_s)$ is also an MOD for $\mathcal{L}$, and by Lemma 3.3.2 we can write

$$\Phi(\mathcal{H}_j) = \langle a_1 X_{\sigma_1} + b_1 X_{-\sigma_1}, \ldots, a_n X_{\sigma_n} + b_n X_{-\sigma_n} \rangle_{\mathbb{C}}.$$

From this it is clear that

$$[\Phi(\mathcal{H}_i), \Phi(\mathcal{H}_j)] = \langle a_1 X_{\sigma_1} - b_1 X_{-\sigma_1}, \ldots, a_n X_{\sigma_n} - b_n X_{-\sigma_n} \rangle_{\mathbb{C}}$$

is a Cartan subalgebra, and so

$$\Phi(\mathcal{H}_k) = \langle a_1 X_{\sigma_1} - b_1 X_{-\sigma_1}, \ldots, a_n X_{\sigma_n} - b_n X_{-\sigma_n} \rangle_{\mathbb{C}} = [\Phi(\mathcal{H}_i), \Phi(\mathcal{H}_j)]. \quad (3.5)$$

Commutating $\Phi(\mathcal{H}_k)$ with the standard subalgebra $\Phi(\mathcal{H}_i)$ once again, we obtain

$$[\Phi(\mathcal{H}_i), \Phi(\mathcal{H}_k)] = \langle a_1 X_{\sigma_1} + b_1 X_{-\sigma_1}, \ldots, a_n X_{\sigma_n} + b_n X_{-\sigma_n} \rangle_{\mathbb{C}} = \Phi(\mathcal{H}_j). \quad (3.6)$$

Applying the automorphism Φ^{-1} to the relations (3.5) and (3.6), we come to the following useful relations:

$$[\mathcal{H}_i, \mathcal{H}_j] = \mathcal{H}_k, \; [\mathcal{H}_i, [\mathcal{H}_i, \mathcal{H}_j]] = \mathcal{H}_j. \quad (3.7)$$

The lemma can now be proved using the following inductive argument. Suppose that there is a multiplicatively closed system $\Pi_l = \{\mathcal{H}_{i_1}, \ldots, \mathcal{H}_{i_q}\}$ consisting of $q = 2^l - 1$ Cartan subalgebras, and that $\mathcal{H}_i \notin \Pi_l$. According to (3.7), $[\mathcal{H}_i, \mathcal{H}_{i_\mu}]$ is a Cartan subalgebra for each $\mu = 1, 2, \ldots, q$. All the subalgebras

$$\mathcal{H}_i; \mathcal{H}_{i_1}, \ldots, \mathcal{H}_{i_q}; [\mathcal{H}_i, \mathcal{H}_{i_1}], \ldots, [\mathcal{H}_i, \mathcal{H}_{i_q}]$$

are different. Indeed, by (3.7) the equality $\mathcal{H}_{i_\nu} = [\mathcal{H}_i, \mathcal{H}_{i_\mu}]$ would give

$$\mathcal{H}_i = [[\mathcal{H}_i, \mathcal{H}_{i_\mu}], \mathcal{H}_{i_\mu}] = [\mathcal{H}_{i_\nu}, \mathcal{H}_{i_\mu}] \in \Pi_l,$$

which contradicts the choice of $\mathcal{H}_i$. Furthermore, the possibility $\mathcal{H}_i = [\mathcal{H}_i, \mathcal{H}_{i_\mu}]$ is eliminated, since in the contrary case we would have

$$\mathcal{H}_{i_\mu} = [\mathcal{H}_i, [\mathcal{H}_i, \mathcal{H}_{i_\mu}]] = [\mathcal{H}_i, \mathcal{H}_i] = 0.$$

Finally, the equality $[\mathcal{H}_i, \mathcal{H}_{i_\nu}] = [\mathcal{H}_i, \mathcal{H}_{i_\mu}]$ implies that

$$\mathcal{H}_{i_\nu} = [\mathcal{H}_i, [\mathcal{H}_i, \mathcal{H}_{i_\nu}]] = [\mathcal{H}_i, [\mathcal{H}_i, \mathcal{H}_{i_\mu}]] = \mathcal{H}_{i_\mu}.$$

Hence we obtain a system Π_{l+1} consisting of $2q+1 = 2^{l+1}-1$ Cartan subalgebras. We show now that this system is also multiplicatively closed. The nontrivial cases are considered below.

From the Jacoby identity and the first relation of (3.7), it follows that when $\nu \neq \mu$,

$$[\mathcal{H}_i, [\mathcal{H}_{i_\nu}, \mathcal{H}_{i_\mu}]] = [[\mathcal{H}_i, \mathcal{H}_{i_\nu}], \mathcal{H}_{i_\mu}] + [[\mathcal{H}_i, \mathcal{H}_{i_\mu}], \mathcal{H}_{i_\nu}],$$

where every commutator subspace is a Cartan subalgebra from the given MOD $\mathcal{D}$. But the sum in the notation of $\mathcal{D}$ is direct, so in fact these three commutator subspaces coincide with each other. Thus

$$[[\mathcal{H}_i, \mathcal{H}_{i_\nu}], \mathcal{H}_{i_\mu}] = [\mathcal{H}_i, [\mathcal{H}_{i_\nu}, \mathcal{H}_{i_\mu}]] \in \Pi_{l+1},$$

because $[\mathcal{H}_{i_\nu}, \mathcal{H}_{i_\mu}] \in \Pi_l$.

Similarly, when $\nu \neq \mu$ it follows from the equalities

$$[\mathcal{H}_{i_\nu}, \mathcal{H}_{i_\mu}] = [\mathcal{H}_{i_\nu}, [\mathcal{H}_i, [\mathcal{H}_i, \mathcal{H}_{i_\mu}]]] =$$

$$= [[\mathcal{H}_i, \mathcal{H}_{i_\nu}], [\mathcal{H}_i, \mathcal{H}_{i_\mu}]] + [\mathcal{H}_i, [\mathcal{H}_{i_\nu}, [\mathcal{H}_i, \mathcal{H}_{i_\mu}]]]$$

that

$$[[\mathcal{H}_i, \mathcal{H}_{i_\nu}], [\mathcal{H}_i, \mathcal{H}_{i_\mu}]] = [\mathcal{H}_{i_\nu}, \mathcal{H}_{i_\mu}] \in \Pi_l \subset \Pi_{l+1}.$$

Thus, starting from the multiplicatively closed system Π_l of cardinality $2^l - 1$ we come to a multiplicatively closed system Π_{l+1} of cardinality $2^{l+1} - 1$. Repeating this procedure, we exhaust all Cartan subalgebras from $\mathcal{D}$ and obtain the first statement of Lemma.

Our inductive argument also prompts the following method of indexing subalgebras $\mathcal{H}_i$ by nonzero elements of the vector space $\mathbb{F}_2^d$. Namely, assume that subalgebras from the multiplicatively closed system Π_l considered above are numbered by nonzero elements of a subspace $\mathbb{F}_2^l \subset \mathbb{F}_2^d : \Pi_l = \{\mathcal{H}_u \mid u \in \mathbb{F}_2^l \backslash \{0\}\}$. Then we index a subalgebra $\mathcal{H}_i$ not belonging to Π_l by the symbol $\mathcal{H}_v$, where v is an arbitrary element from $\mathbb{F}_2^d \backslash \mathbb{F}_2^l$. In this case we can denote subalgebras $[\mathcal{H}_u, \mathcal{H}_v]$, $u \in \mathbb{F}_2^l \backslash \{0\}$, by the symbols $\mathcal{H}_{v+u}$. The agreement between this notation and the multiplication of subalgebras with the additive structure of $\mathbb{F}_2^d$ follows from (3.7) and the relations obtained checking that Π_{l+1} is multiplicatively closed. □

Remark 3.3.4. The definition of MOD includes the orthogonality condition. In fact, it is a consequence of the multiplicativity property. This can be seen from the following argument. In Lemma 3.3.2, it is shown that a multiplicative decomposition (not necessarily orthogonal!) containing a standard Cartan subalgebra is a root decomposition. The orthogonality can be violated only by pairs of elements of the type

$$aX_\sigma + bX_{-\sigma} \in \mathcal{H}_k,\ cX_\sigma + dX_{-\sigma} \in \mathcal{H}_l,\ k \neq l.$$

But in the proof of Lemma 3.3.3 (which does not use the orthogonality property), it is shown that $c = \lambda a$ and $d = -\lambda b$; and in this case the corresponding elements are orthogonal.

Theorem 3.3.5. *A complex simple Lie algebra $\mathcal{L}$ admits an MOD if and only if $\mathcal{L}$ is of types A_1; B_{2^m-1}, $m \geq 2$; D_{2^m}, $m \geq 2$; G_2 or E_8.*

Proof. For Lie algebras $\mathcal{L}$ which are not of the types indicated, the absence of MODs follows from Proposition 3.2.2 and Lemmas 3.3.2 and 3.3.3. An explicit construction of MODs for Lie algebras which are of the types indicated is described below.

(i) *Type* A_1 :

$$\mathcal{L} = sl_2(\mathbb{C}) = \left\langle \begin{pmatrix} 1 & 0 \\ 0 & -1 \end{pmatrix} \right\rangle_{\mathbb{C}} \oplus \left\langle \begin{pmatrix} 0 & 1 \\ 1 & 0 \end{pmatrix} \right\rangle_{\mathbb{C}} \oplus \left\langle \begin{pmatrix} 0 & 1 \\ -1 & 0 \end{pmatrix} \right\rangle_{\mathbb{C}}.$$

Here one can consider a compact form $\mathcal{L}^{(c)}$ of $\mathcal{L}$ which is actually defined by its MOD : $\mathcal{L}^{(c)} = < A > \oplus < B > \oplus < C >$, where $[A, B] = C$, $[B, C] = A$ and $[C, A] = B$.

(ii) *Type* G_2. In a Lie algebra $\mathcal{L}$ of type G_2, we consider a Chevalley basis $\{H_{\alpha_1}, H_{\alpha_2}; X_\alpha \mid \alpha \in \Sigma\}$ which is normed as in [Bour 2]. Here α_1 and α_2 are simple roots; so α_1, $\alpha_1+\alpha_2$ and $2\alpha_1+\alpha_2$ are short roots, and α_2, $3\alpha_1+\alpha_2$ and $3\alpha_1+2\alpha_2$ are long roots. It is clear that the Cartan subalgebras

$$\mathcal{H} = \langle H_{\alpha_1}, H_{\alpha_2} \rangle_{\mathbb{C}},\ \mathcal{H}_1^\tau = \langle E_{\alpha_1}^\tau, E_{3\alpha_1+2\alpha_2}^\tau \rangle_{\mathbb{C}},$$

$$\mathcal{H}_2^\tau = \langle E_{\alpha_1+\alpha_2}^\tau, E_{3\alpha_1+\alpha_2}^\tau \rangle_{\mathbb{C}},\ \mathcal{H}_3^\tau = \langle E_{2\alpha_1+\alpha_2}^\tau, E_{\alpha_2}^\tau \rangle_{\mathbb{C}};$$

$\tau = 0, 1$, form an MOD of $\mathcal{L}$; the notation is that of §3.2.

(iii) *Type* E_8. This case was investigated in full by Thompson [Tho 2], who also established the Aut($\mathcal{L}$)-conjugacy of MODs of $\mathcal{L}$. Here we shall indicate a partition of the system Σ^+ of positive roots, which corresponds to an MOD and is written in rather intuitive terms. The positive roots are $\pm\varepsilon_i + \varepsilon_j$, $1 \leq i < j \leq 8$ and

$(\varepsilon_8 + \sum_{i=1}^{7}(-1)^{\nu(i)}\varepsilon_i)/2$, with even sum $\sum_i \nu(i)$. As in the case of type E_7 (see §3.2), we shall denote them briefly as $\pm i + j$ and $8\pm1\pm2\pm3\pm4\pm5\pm6\pm7$. Subsets of the admissible partition of Σ^+ are indexed by nonzero elements of $\mathbb{F}_2^4$, and the corresponding Cartan subalgebras are indexed by nonzero elements of $\mathbb{F}_2^5$. Here, in accordance with the notation introduced in §3.2, the subset $\Sigma^+_{\alpha\beta\gamma\delta}$ corresponds to the two Cartan subalgebras $\mathcal{H}_{\tau\alpha\beta\gamma\delta} = \mathcal{H}^{\tau}_{\alpha\beta\gamma\delta}$, $\tau = 0, 1$. The standard subalgebra has the numeral 10000.

$$\Sigma^+_{0100} : 1+2,\ -1+2,\ 3+4,\ -3+4,\ 5+6,\ -5+6,\ 7+8,\ -7+8$$

$$\Sigma^+_{0010} : 1+3,\ -1+3,\ 2+4,\ -2+4,\ 5+7,\ -5+7,\ 6+8,\ -6+8$$

$$\Sigma^+_{0110} : 1+4,\ -1+4,\ 2+3,\ -2+3,\ 5+8,\ -5+8,\ 6+7,\ -6+7$$

$$\Sigma^+_{0001} : 1+5,\ -1+5,\ 2+6,\ -2+6,\ 3+7,\ -3+7,\ 4+8,\ -4+8$$

$$\Sigma^+_{0101} : 1+6,\ -1+6,\ 2+5,\ -2+5,\ 4+7,\ -4+7,\ 3+8,\ -3+8$$

$$\Sigma^+_{0011} : 1+7,\ -1+7,\ 3+5,\ -3+5,\ 4+6,\ -4+6,\ 2+8,\ -2+8$$

$$\Sigma^+_{0111} : 1+8,\ -1+8,\ 2+7,\ -2+7,\ 3+6,\ -3+6,\ 4+5,\ -4+5$$

Σ^+_{1000} :	Σ^+_{1010} :
$8+1+2+3+4+5+6+7$,	$8-1+2-3+4+5+6+7$,
$8-1-2-3-4+5+6+7$,	$8+1-2+3-4+5+6+7$,
$8-1-2+3+4-5-6+7$,	$8+1+2+3+4-5+6-7$,
$8+1+2-3-4-5-6+7$,	$8-1-2-3-4-5+6-7$,
$8-1+2-3+4-5+6-7$,	$8+1-2-3+4-5-6+7$,
$8+1-2+3-4-5+6-7$,	$8-1+2+3-4-5-6+7$,
$8+1-2-3+4+5-6-7$,	$8-1-2+3+4+5-6-7$,
$8-1+2+3-4+5-6-7$	$8+1+2-3-4+5-6-7$

Σ^+_{1100} :	Σ^+_{1110} :
$8-1-2+3+4+5+6+7$,	$8+1-2-3+4+5+6+7$,
$8+1+2-3-4+5+6+7$,	$8-1+2+3-4+5+6+7$,
$8+1+2+3+4-5-6+7$,	$8-1-2+3+4-5+6-7$,
$8-1-2-3-4-5-6+7$,	$8+1+2-3-4-5+6-7$,
$8+1-2-3+4-5+6-7$,	$8-1+2-3+4-5-6+7$,
$8-1+2+3-4-5+6-7$,	$8+1-2+3-4-5-6+7$,
$8-1+2-3+4+5-6-7$,	$8+1+2+3+4+5-6-7$,
$8+1-2+3-4+5-6-7$	$8-1-2-3-4+5-6-7$

$$\begin{array}{ll}
\Sigma^+_{1001}: & \Sigma^+_{1011}: \\
8-1+2+3+4-5+6+7, & 8-1+2+3+4+5+6-7, \\
8+1-2+3+4+5-6+7, & 8+1+2-3+4-5+6+7, \\
8+1+2-3+4+5+6-7, & 8+1+2+3-4+5-6+7, \\
8-1-2-3+4-5-6-7, & 8-1+2-3-4-5-6-7, \\
8+1-2-3-4-5+6+7, & 8+1-2-3-4+5+6-7, \\
8-1+2-3-4+5-6+7, & 8-1-2+3-4-5+6+7, \\
8-1-2+3-4+5+6-7, & 8-1-2-3+4+5-6+7, \\
8+1+2+3-4-5-6-7 & 8+1-2+3+4-5-6-7
\end{array}$$

$$\begin{array}{ll}
\Sigma^+_{1101}: & \Sigma^+_{1111}: \\
8-1+2+3+4+5-6+7, & 8+1-2-3-4-5-6-7, \\
8+1-2+3+4-5+6+7, & 8+1-2+3+4+5+6-7, \\
8+1+2+3-4+5+6-7, & 8+1+2-3+4+5-6+7, \\
8-1-2+3-4-5-6-7, & 8+1+2+3-4-5+6+7, \\
8+1-2-3-4+5-6+7, & 8-1+2+3+4-5-6-7, \\
8-1+2-3-4-5+6+7, & 8-1+2-3-4+5+6-7, \\
8-1-2-3+4+5+6-7, & 8-1-2+3-4+5-6+7, \\
8+1+2-3+4-5-6-7 & 8-1-2-3+4-5+6+7
\end{array}$$

(iv) MODs of the Lie algebra $\mathcal{L}$ of type D_{2^m} and of the Lie algebra $\mathcal{L}'$ of type B_{2^m-1} were actually constructed in Chapter 2. Namely, consider a skew-symmetric realization of $\mathcal{L}$ in the form

$$\mathcal{L} = \left\langle E(u, v) = E_{uv} - E_{vu} \mid u, v \in W = \mathbb{F}_2^{m+1} \right\rangle_{\mathbb{C}},$$

where we identify the set $\{1, 2, \dots, 2^{m+1}\}$ with the point set of the space W. Then formula (2.23), that is,

$$\mathcal{L} = \bigoplus_{u \in W \setminus \{0\}} \mathcal{H}_u, \; \mathcal{H}_u = < E(x, u + x) \mid x \in W >_{\mathbb{C}},$$

yields an MOD of $\mathcal{L}$ because of the identity $[\mathcal{H}_u, \mathcal{H}_v] = \mathcal{H}_{u+v}$ (if $u \neq v$). Similarly, we realize $\mathcal{L}'$ in the form

$$\mathcal{L}' = \langle E(u, v) \mid u, v \in W \setminus \{0\} \rangle_{\mathbb{C}} .$$

Then the decomposition

$$\mathcal{L}' = \bigoplus_{u \in W \setminus \{0\}} \mathcal{H}'_u, \; \mathcal{H}'_u = < E(x, u + x) \mid x \in W \setminus \{0, u\} >_{\mathbb{C}} \tag{3.8}$$

is an MOD of $\mathcal{L}'$. □

Remark 3.3.6. In the case of Lie algebras of types A_1, D_4, G_2 and E_8, the standard transitive OD constructed in §3.1 (see (3.1)) and related to a Jordan 2-subgroup is in fact an MOD. Indeed, here we have

$$\{\mathcal{H}_1, \dots, \mathcal{H}_{h+1}\} \equiv \{\mathcal{L}_{\rho_1}, \dots, \mathcal{L}_{\rho_{h+1}}\},$$

where $\{\rho_1, \dots, \rho_{h+1}\} = \mathrm{Irr}(J)\backslash\{1_J\}$. Hence, the multiplicativity of the decomposition follows from the relation

$$\forall\alpha, \beta \in \mathrm{Irr}(J),\ [\mathcal{L}_\alpha, \mathcal{L}_\beta] \subseteq \mathcal{L}_{\alpha\beta}$$

for eigensubspaces.

As always, the given MOD $\mathcal{D}$ of the Lie algebra $\mathcal{L}$ of type X corresponds to the automorphism group consisting of all automorphisms $\Phi \in \mathrm{Aut}(\mathcal{L})$ fixing $\mathcal{D}$; we denote it by $\mathrm{Aut}_{MOD}(X)$.

Theorem 3.3.7. *The multiplicative decompositions constructed in Theorem* 3.3.5 *correspond to the following automorphism groups:*

(i) $\mathrm{Aut}_{MOD}(A_1) = ASL_2(2) \cong \mathbb{S}_4$*;*

(ii) $\mathrm{Aut}_{MOD}(G_2) = \mathbb{Z}_2^3 \circ SL_3(2)$*;*

(iii) $\mathrm{Aut}_{MOD}(E_8) = 2^{5+10} \circ SL_5(2)$*;*

(iv) $\mathrm{Aut}_{MOD}(D_{2^m}) = \mathbb{Z}_2^{2^{m+1}-1}.ASL_{m+1}(2)$ *if* $m \geq 3$ *and* $\mathrm{Aut}_{MOD}(D_4) = (\mathbb{Z}_2^7.ASL_3(2))\mathbb{Z}_3$*;*

(v) $\mathrm{Aut}_{MOD}(B_{2^m-1}) = \mathbb{Z}_2^{2^{m+1}-2}.SL_{m+1}(2)$.

Proof. (i) See Corollary 1.3.4.

(ii) and (iii) will be deduced in Chapter 4. See also [Tho 2] and [Hes].

(iv) According to Proposition 2.2.3, when $m > 2$ we have

$$\mathrm{Aut}_{MOD}(D_{2^m}) = \mathbb{Z}_2^{2^{m+1}-1}.\mathrm{Aut}(\mathcal{P}),$$

where $\mathcal{P}$ is an admissible partition corresponding to $\mathcal{D}$. Furthermore, $\mathrm{Aut}(\mathcal{P}) = ASL_{m+1}(2)$ by Proposition 2.4.4. If $m = 2$ we need to be convinced that there is an outer automorphism T of order 3 of the Lie algebra $\mathcal{L}$ of type D_4 fixing $\mathcal{D}$. Of course, we can appeal to Remark 3.3.6 and (the last line of) Table A1; however, we shall check this fact directly. We know that T is induced by an automorphism s of order 3

$$s : \alpha_1 \mapsto \alpha_3 \mapsto \alpha_4 \mapsto \alpha_1,\ \alpha_2 \mapsto \alpha_2$$

of the simple root system $\{\alpha_1, \alpha_2, \alpha_3, \alpha_4\}$ of $\mathcal{L}$; here we use the notation of [Bour 2]. In a Chevalley basis $\{H_{\alpha_i}, X_\alpha \mid i = 1, \dots, 4; \alpha \in \Sigma\}$ of $\mathcal{L}$ we set $T(H_\sigma) = H_{s(\sigma)}$ and $T(X_\sigma) = X_{s(\sigma)}$. From the definition of the MOD of $\mathcal{L}$ given in Example 3.2.3, it follows that $T(\mathcal{H}_0) = \mathcal{H}_0$ and $T(\mathcal{H}_k^\tau) = \mathcal{H}_k^\tau$. For example, we indicate the action of T on the Cartan subalgebra $\mathcal{H}_1^\tau =< E^\tau_{\varepsilon_1+\varepsilon_2}, E^\tau_{\varepsilon_1-\varepsilon_2}, E^\tau_{\varepsilon_3+\varepsilon_4}, E^\tau_{\varepsilon_3-\varepsilon_4} >_{\mathbb{C}}$:

$$T(E^\tau_{\varepsilon_1+\varepsilon_2}) = T(E^\tau_{\alpha_1+2\alpha_2+\alpha_3+\alpha_4}) = E^\tau_{\alpha_3+2\alpha_2+\alpha_4+\alpha_1} = E^\tau_{\varepsilon_1+\varepsilon_2}.$$

Similarly,

$$T : E^\tau_{\varepsilon_1-\varepsilon_2} \mapsto E^\tau_{\varepsilon_3-\varepsilon_4} \mapsto E^\tau_{\varepsilon_3+\varepsilon_4} \mapsto E^\tau_{\varepsilon_1-\varepsilon_2}.$$

Note that T *does not normalize the subgroup* $\mathbb{Z}_2^7.ASL_3(2)$ *of* $\mathrm{Aut}_{MOD}(D_4)$.

(v) Using the embedding of the Lie algebra $\mathcal{L}'$ of type B_{2^m-1} in the Lie algebra $\mathcal{L}$ of type D_{2^m} (see §2.2), by Proposition 2.2.4 we have also an embedding $\mathrm{Aut}_{MOD}(B_{2^m-1}) \hookrightarrow \mathrm{Aut}_{MOD}(D_{2^m})$. Furthermore, $\mathrm{Aut}_{MOD}(B_{2^m-1})$ consists of all elements $\Phi_{\pi,\bar{f}} \in \mathrm{Aut}_{MOD}(D_{2^m})$ such that

$$\sum_{u\in W} f(u) = 0, \ \pi(0) = 0.$$

From the first condition it follows that $\mathrm{Aut}_{MOD}(B_{2^m-1})$ has a normal subgroup isomorphic to $\mathbb{Z}_2^{2^{m+1}-2}$, and the second condition implies that $\mathrm{Aut}(\mathcal{P}') \cong SL_{m+1}(2)$, where $\mathcal{P}'$ is a partition corresponding to the MOD $\mathcal{D}'$ of $\mathcal{L}'$.

Note that all extensions mentioned in the formulation of Theorem 3.3.7 are split, except those in cases (ii) and (iii). The non-splitness of the extension $2^{5+10} \circ SL_5(2)$ follows from the fact that $\mathrm{Aut}_{MOD}(E_8) \subseteq E_8(\mathbb{C})$, whereas $SL_5(2)$ cannot be embedded in the Lie group $E_8(\mathbb{C})$ (see [CoG]). Concerning the non-splitness of the extension $\mathrm{Aut}_{MOD}(G_2) = \mathbb{Z}_2^3 \circ SL_3(2)$, in [KKU 5] an element $A \in \mathrm{Aut}_{MOD}(G_2)$ is found with the following properties:

a) $A^4 \in J = \mathbb{Z}_2^3$, $A^2 \notin J$;

b) $(\Phi A)^4 \neq 1$ for all $\Phi \in J$.

The uniqueness of the non-split extensions $\mathbb{Z}_2^3 \circ SL_3(2)$ and $\mathbb{Z}_2^5 \circ SL_5(2)$ is established in [San] and [Demp 1]. □

Although Theorem 3.3.7 gives us the automorphism groups of concrete MODs, this result is of an absolute nature, since all MODs of the given Lie algebra $\mathcal{L}$ are $\mathrm{Aut}(\mathcal{L})$-conjugate.

Theorem 3.3.8. *All MODs of the given complex simple Lie algebra* $\mathcal{L}$ *are* $\mathrm{Aut}(\mathcal{L})$-*conjugate.*

Proof. 1) All Cartan subalgebras of $\mathcal{L}$ are conjugate under $\mathrm{Aut}(\mathcal{L})$, so any OD of $\mathcal{L}$ is conjugate to some OD containing a standard Cartan subalgebra. Since the multiplicativity condition is preserved by conjugation, we can restrict our consideration to the class of those MODs which are RODs. Following Lemma 3.3.3, we use the indexing by elements of $W = \mathbb{F}_2^d$, and then the given MOD is written as $\mathcal{L} = \oplus_{0\neq v\in W}\mathcal{H}_v$. Let $\mathcal{H}_{v_0}$ be the standard Cartan subalgebra and $U \subset W$ be a subspace of codimension 1 of W not containing v_0. For each positive root α of $\mathcal{L}$ (with respect to $\mathcal{H}_{v_0}$), one can find precisely two elements like $X_\alpha + aX_{-\alpha}$ and $X_\alpha - aX_{-\alpha}$ entering a basis of our ROD. If in addition $X_\alpha + aX_{-\alpha} \in \mathcal{H}_u$, $u \in U\backslash\{0\}$, then $X_\alpha - aX_{-\alpha} \in [\mathcal{H}_{v_0}, \mathcal{H}_u] = \mathcal{H}_{u+v_0}$. Consider a set $\{\alpha_1, \dots, \alpha_n\}$ of simple roots of $\mathcal{L}$ (with respect to $\mathcal{H}_{v_0}$) and the corresponding collection of basis elements $X_{\alpha_i} + (a_i)^2 X_{-\alpha_i}$, which are contained in $\mathcal{L}_U = \oplus_{0\neq u\in U}\mathcal{H}_u$. Here for $(a_i)^2$ we take either a_i or $-a_i$. Because of the multiplicativity property, all commutators of these elements are also contained in $\mathcal{L}_U$ and are basis elements of the same kind $X_\alpha + a^2 X_{-\alpha}$. From this, it is easy to see that each positive root $\alpha = s_1\alpha_1 + \dots + s_n\alpha_n$ corresponds to the basis element

$$X_\alpha + (a_1)^{2s_1} \dots (a_n)^{2s_n} X_{-\alpha} \in \mathcal{L}_U.$$

Define an automorphism $\Phi \in \mathrm{Aut}(\mathcal{L})$ as follows: $\Phi(h) = h$ for each $h \in \mathcal{H}_{v_0}$, and $\Phi(X_{\alpha_i} + (a_i)^2 X_{-\alpha_i}) = a_i(X_{\alpha_i} + X_{-\alpha_i})$. The automorphism Φ chosen in such a way satisfies the following condition: in the resulting MOD $\mathcal{L} = \oplus\Phi(\mathcal{H}_v)$, the Cartan subalgebra $\mathcal{H}_v$ admits a basis consisting of elements of type $X_\alpha + X_{-\alpha}$ if $v \in U\backslash\{0\}$, and of type $X_\alpha - X_{-\alpha}$ if $v \in U + v_0$. Such a decomposition will be called *reduced.*

2) Let Σ^+ be the set of positive roots of $\mathcal{L}$ with respect to $\mathcal{H}_{v_0}$. Now with every Cartan subalgebra $\mathcal{H}_v$ entering a reduced OD one can associate a collection $\Sigma_v^+ = \{\sigma_1, \dots, \sigma_n\}$ as follows: $\alpha \in \Sigma_v^+$ if and only if $X_\alpha + X_{-\alpha} \in \mathcal{H}_v$ or $X_\alpha - X_{-\alpha} \in \mathcal{H}_v$. Furthermore, $\Sigma_{v+v_0}^+ = \Sigma_v^+$. It is easy to see that there exists a bijective correspondence between the set of reduced MODs of $\mathcal{L}$ and the set of partitions $\Sigma^+ = \cup_{0\neq u\in U}\Sigma_u^+$ satisfying (3.4) and the following additional condition:

$$\text{If } \alpha \in \Sigma_u^+, \beta \in \Sigma_v^+ \text{ and } \gamma = \alpha \pm \beta \in \Sigma^+, \text{ then } \gamma \in \Sigma_{u+v}^+. \tag{3.9}$$

The condition (3.4) means that we are dealing with an ROD, and, together with (3.9), it means that we are in fact dealing with an MOD.

For the Lie algebra of type A_1, a partition of Σ^+ is trivial, because there exists only one positive root. For the Lie algebra of type G_2, a partition of Σ^+ with the properties indicated is also unique (see the proof of Theorem 3.3.5 (ii)). Hence, the conjugacy of MODs for Lie algebras of types A_1 and G_2 follows. The same property was established by Thompson in the case of the Lie algebra of type E_8, using similar arguments [Tho 2]; see also Chapter 4. It remains to consider the types B_{2^m-1} and D_{2^m}; moreover, from that follows it is clear that it is sufficient to consider the case D_{2^m}.

3) In the given reduced MOD of the Lie algebra $\mathcal{L}$ of type D_{2^m}, the elements $E^\tau_{\varepsilon_k-\varepsilon_1}$ and $E^\tau_{\varepsilon_k+\varepsilon_1}$ (see the notation and arguments of §3.2) belong to the same Cartan subalgebra. Consequently, for any Cartan subalgebra entering the MOD, other than standard ones, there exists a basis consisting of elements of the form

$$E^\tau_{\varepsilon_k-\varepsilon_1} \pm E^\tau_{\varepsilon_k+\varepsilon_1}. \tag{3.10}$$

If we go over to the skew-symmetric realization $\mathcal{L} = \langle E(i,j) \mid 1 \le i \ne j \le 2n \rangle_{\mathbb{C}}$ (using the standard isomorphism), then as images of elements like (3.10) we obtain elements of the form $E(i,j)$ (with some coefficients, which are of no interest for us). Hence, each MOD is conjugate to some E-decomposition (in the skew-symmetric realization). Now we show that one can index rows and columns of (skew-symmetric) matrices by elements of W in such a way that the given reduced MOD coincides with (2.23). In fact, as mentioned above, one can index the Cartan subalgebras entering the MOD by elements of W. With the index $2n$ we associate the zero vector. Furthermore, if $E(i,2n) \in \mathcal{H}_{v(i)}$, then with i we associate the vector $v(i) \in W$. Now every element $E(i,j)$ can be written as $E(v(i), v(j))$. We still have to check that this notation agrees with the structure of the MOD. Suppose that $E(v(i), v(j)) \in \mathcal{H}_v$. When $j = 2n$, we have $E(v(i), v(2n)) = E(i, 2n) \in \mathcal{H}_{v(i)}$, that is, $v = v(i) = v(i) + v(2n)$. If $i, j \ne 2n$, then

$$E(v(i), v(j)) = [E(v(i), 0), E(0, v(j))] \in [\mathcal{H}_{v(i)}, \mathcal{H}_{v(j)}] \subseteq \mathcal{H}_{v(i)+v(j)},$$

that is, $v = v(i) + v(j)$, and this completes our argument. □

Remark 3.3.9. Shpektorov (unpublished) notes that if

$$\mathcal{D} : \mathcal{L} = \mathcal{H}_0 \oplus \mathcal{H}_1^0 \oplus \mathcal{H}_1^1 \oplus \mathcal{H}_2^0 \oplus \mathcal{H}_2^1 \oplus \dots$$

is the ROD of the Lie algebra $\mathcal{L}$ of type E_7 constructed in §3.2, then $\mathrm{Aut}(\mathcal{D})$ fixes the standard Cartan subalgebra $\mathcal{H}_0$. Moreover, an ROD of the kind indicated can be a transitive OD if and only if it is an MOD. In fact, $[\mathcal{H}_0, \mathcal{H}_k^\tau] = \mathcal{H}_k^{\tau+1}$. Assuming the existence of $\Phi \in \mathrm{Aut}(\mathcal{D})$ such that $\Phi(\mathcal{H}_k^\tau) = \mathcal{H}_0$ for any k, τ, we have

$$\Phi([\mathcal{H}_k^\tau, \mathcal{H}_{k'}^{\tau'}]) = [\Phi(\mathcal{H}_k^\tau), \Phi(\mathcal{H}_{k'}^{\tau'})] = [\mathcal{H}_0, \mathcal{H}_{k''}^{\tau''}] = \mathcal{H}_{k''}^{\tau''+1}$$

for some k'', τ''. From this there follows the multiplicativity property:

$$[\mathcal{H}_k^\tau, \mathcal{H}_{k'}^{\tau'}] = \Phi^{-1}(\mathcal{H}_{k''}^{\tau''+1}) = \mathcal{H}_{k'''}^{\tau'''}.$$

However, this remark relates to only one special ROD; meanwhile RODs can be constructed in many different ways. Namely, assume that h subsets Σ_k^+ satisfying condition (3.4) are distinguished in the set Σ^+ (h the Coxeter number); in addition, suppose that each positive root enters precisely two of these subsets. Starting from a subset Σ_k^+ one can construct a Cartan subalgebra. Furthermore, if $\sigma \in \Sigma_k^+ \cap \Sigma_l^+$, then one should include an element $X_\sigma + a_\sigma X_{-\sigma}$ into $\mathcal{H}_k$ and an element $X_\sigma - a_\sigma X_{-\sigma}$ into $\mathcal{H}_l$, for some complex number $a_\sigma \ne 0$.

Supplementing and generalizing the arguments carried out above, which refer only to the case E_7, we can deduce the following statement.

If an ROD of a given complex simple Lie algebra is transitive, then it is an MOD.

The authors have no information concerning the existence of TODs for the Lie algebra of type E_7.

Commentary

1) The construction of a certain class of ODs of Lie algebras of types F_4 and E_6, which is expounded in [KKU 7], was first suggested in the School on "Lie algebras and their applications to mathematics and physics" (Zimenki, 1982), with the active contribution of Alekseevskii. However, the automorphism groups of these decompositions were not computed. Later, Burichenko [Bur 1] (see also [Kos 1]) arrived at an OD in the same class, from the viewpoint of group theory. In fact he used a Jordan subgroup $\mathbb{Z}_3^3$ of the complex Lie groups $F_4(\mathbb{C})$ and $E_6(\mathbb{C})$, and formula (3.1). Independently, Borovik [Bor 3] described a general construction enabling one to make a TOD, starting from a Jordan subgroup. This result was stated in the following form.

Theorem 3A. *Let $\mathcal{L}$ be a simple complex Lie algebra, set $\mathcal{G} = \mathrm{Inn}(\mathcal{L})$ and let J be a Jordan subgroup of $\mathcal{G}$. A subgroup $A \subseteq J$ is said to be toroidal if $C_{\mathcal{G}}(A)$ is a maximal torus in $\mathcal{G}$ and A is maximal among those subgroups in J with this property. Then the following assertions hold.*

(i) *J contains toroidal subgroups, and all toroidal subgroups are conjugate under $N_{\mathcal{G}}(J)$;*

(ii) *if $\{A_0, A_1, \dots, A_h\}$ is a collection of toroidal subgroups such that $A_iA_j = J$ whenever $i \neq j$ and having the largest possible length, then the subalgebras $\mathcal{H}_i = C_{\mathcal{L}}(A_i)$ form an OD of $\mathcal{L}$;*

(iii) *the collection $\{A_0, A_1, \dots, A_h\}$ can be chosen to be invariant under some subgroup X of $N_{\mathcal{G}}(J)$ acting transitively on $\{A_0, A_1, \dots, A_h\}$ and on the corresponding OD $\mathcal{L} = \oplus_{i=0}^{h}\mathcal{H}_i$.* □

This theorem is proved by analysing separate cases, using the TODs of classical algebras constructed earlier. Regarding the exceptional Lie algebras, this theorem yields a new transitive OD of the Lie algebra of type E_8 which we shall call *Borovik decomposition.* Furthermore, the task of constructing suitable collections of toroidal subgroups in J is equivalent to that of constructing symplectic spreads (in the case of types A_n and C_n) and admissible partitions (in the case of types B_n and D_n), which is studied in detail in Chapters 1 and 2. For this reason, we prefer to formulate the results of Borovik in the form of Theorem 3.1.2.

2) Our approach to MODs (see [KKU 3, 7]) is based on the fact that MODs are RODs and on the description of RODs. Moreover, the notion of RODs led us to the only known OD for the type E_7. In some sense, all RODs of Lie algebras of type A_1, B_{2m-1} and D_{2m} have actually been classified: for type A_1, the OD is unique, while for types B_{2m-1} and D_{2m} the classification of RODs reduces to the classification of admissible partitions of the complete graph with $2m-1$ and $2m$ vertices, respectively. It would be interesting to find all RODs for the Lie algebras of types E_7 and E_8. The construction of an OD of the Lie algebra of type

D_n expounded in Proposition 2A is a slight modification of the construction of an ROD: if the latter uses pairs of elements $X_\alpha \pm aX_{-\alpha}$, then the former deals with quadruple collections of roots.

3) The first OD discovered by Thompson [Tho 2] for the Lie algebra of type E_8 is in fact an MOD. After the beginning of the systematic investigation of ODs in [KKU 1, 2, 3], it became clear immediately that ODs with a rigid multiplicative structure form a rather narrow class among the general variety of ODs. The classification and conjugacy of MODs were established in [KKU 3] as well as in the works of Jackson [Jac] and Hesselink [Hes]. The last author considered a general problem about graduations of Lie algebras. The orthogonality property is not mentioned by them; however, as we have seen (see Remark 3.3.4), it is implied by the multiplicativity property. MODs of Lie algebras of types A_1, G_2, D_4 and E_8 are presented implicitly in Alekseevskii's paper [Ale 1]. Finally, note that the uniqueness of MODs of Lie algebras of types G_2, D_4 and E_8 follows immediately from the results of Chapter 4 classifying IODs of the corresponding Lie algebras.

4) The partition of the system Σ^+ of positive roots of type E_8 described in the proof of Theorem 3.3.5, is graphic, but difficult to memorise. Here we expound another partition based on a quite different realization of the root system Σ of type E_8. Namely, set $W = \mathbb{F}_2^3$, and let $\mathcal{V}_2(W)$ be the variety of all (vector) planes in W, and $\{e_u \mid u \in W\}$ an orthonormal basis of $\mathbb{C}^8$: $(e_u|e_v) = \delta_{uv}$. Then it is easy to see that the set

$$\Sigma = \{\pm\sqrt{2}e_a, (\pm e_a \pm e_b \pm e_c \pm e_d)/\sqrt{2}\},$$

where a, b, c and d are pairwise different vectors in W with $a + b + c + d = 0$, comprises a root system of type E_8. A partition of Σ of the required sort consists of the following components:

$$\Sigma_1 = \left\{\pm\sqrt{2}e_a \mid a \in W\right\},$$

$$\Sigma_{A,\delta} = \left\{(\sum_{b\in B} \varepsilon_b e_b)/\sqrt{2} \mid \varepsilon_b = \pm 1, \prod_{b\in B} \varepsilon_b = \delta,\ B = A \text{ or } B = W\backslash A\right\}.$$

Here A runs over $\mathcal{V}_2(W)$ and $\delta = \pm 1$. Corresponding to each such component Σ' and each coefficient $\varepsilon = \pm 1$, one can build a Cartan subalgebra

$$\mathcal{H}(\Sigma', \varepsilon) = \langle X_\alpha + \varepsilon X_{-\alpha} \mid \alpha \in \Sigma' \rangle_{\mathbb{C}}.$$

The resulting 30 subalgebras, together with the standard subalgebra $\mathcal{H}_0$, form an MOD as required.

5) Here we expound an intuitive construction of an OD of the Lie algebra $\mathcal{L}$ of type G_2. Realize $\mathcal{L}$ in the form of the differential algebra $\mathrm{Der}(\mathcal{C})$ of the Cayley algebra $\mathcal{C}$, the algebra of matrices of the form

$$\begin{pmatrix} a & u \\ u^* & b \end{pmatrix},\ a, b \in \mathbb{C},\ u \in V = \mathbb{C}^3,\ u^* \in V^* = \mathrm{Hom}_{\mathbb{C}}(V, \mathbb{C}),$$

with multiplication

$$\begin{pmatrix} a & u \\ u^* & b \end{pmatrix} \cdot \begin{pmatrix} c & v \\ v^* & d \end{pmatrix} = \begin{pmatrix} ac+ < u, v^* > & av + du - u^* \times v^* \\ cu^* + bv^* + u \times v & bd+ < v, u^* > \end{pmatrix},$$

involutory antiautomorphism

$$x = \begin{pmatrix} a & u \\ u^* & b \end{pmatrix} \mapsto \bar{x} = \begin{pmatrix} b & -u \\ -u^* & a \end{pmatrix},$$

and norm $N(x) = x\bar{x} = \bar{x}x = ab- < u, u^* >$. Here $< u, u^* >$ is the standard pairing between V and V^*. Furthermore, if $\{e_1, e_2, e_3\}$ and $\{f_1, f_2, f_3\}$ are mutually dual bases of V and V^*, then the vector multiplications $V \times V \to V^*$ and $V^* \times V^* \to V$ are defined by the formulae

$$e_i \times e_j = \delta f_k,\ f_i \times f_j = \delta e_k,$$

where $\delta = 0$ if $i = j$ and $\delta = 1$ if $i \neq j$, $k = \{1, 2, 3\} \backslash \{i, j\}$ and the 3-cycles $(1, 2, 3)$ and (i, j, k) coincide. $\mathcal{C}$ admits the following graduation modulo 3:

$$\mathcal{C} = \mathcal{C}_{-1} \oplus \mathcal{C}_0 \oplus \mathcal{C}_1,$$

where

$$\mathcal{C}_{-1} = \left\{ \begin{pmatrix} 0 & 0 \\ u^* & 0 \end{pmatrix} \right\} \cong V^*,\ \mathcal{C}_0 = \left\{ \begin{pmatrix} a & 0 \\ 0 & b \end{pmatrix} \right\} \cong \mathbb{C}^2,$$

$$\mathcal{C}_1 = \left\{ \begin{pmatrix} 0 & u \\ 0 & 0 \end{pmatrix} \right\} \cong V.$$

Correspondingly, $\mathcal{L} = \mathrm{Der}(\mathcal{C})$ has graduation

$$\mathcal{L} = \mathcal{L}_{-1} \oplus \mathcal{L}_0 \oplus \mathcal{L}_1 \cong V^* \oplus sl(V) \oplus V,$$

where the Lie multiplication is defined as follows:

$$[A, u] = Au,\ [A, u^*] = -{}^t\!Au^*,\ [u, v] = -2u \times v,\ [u^*, v^*] = 2u^* \times v^*,$$

$$[u, u^*] = 3u \otimes u^* - < u, u^* > E_3,\ [A, B] = AB - BA.$$

Here $A, B \in sl(V) = \{f \in \mathrm{End}(V) \mid \mathrm{Tr} f = 0\}$; $u, v \in V$; $u^*, v^* \in V^*$ and the operator ${}^t\!A$ is defined by the identity $< Au, u^* >=< u, {}^t\!Au^* >$ (see [VGO, Chapter 5]). It is easy to see that

$$K(A, u) = K(A, u^*) = K(u, v) = K(u^*, v^*) = 0,$$

$$K(A, B) = 8\mathrm{Tr}(AB),\ K(e_i, f_j) = 24\delta_{ij},$$

where, as always, K denotes the Killing form.

Now we set

$$\mathcal{H}_0 = \{\mathrm{diag}(a, b, -a-b) \mid a, b \in \mathbb{C}\},$$

and define, for a triple $\lambda = (\lambda_1, \lambda_2, \lambda_3)$ of nonzero complex numbers, the following commutative subalgebras:

$$\mathcal{H}_k^{\varepsilon}(\lambda) = \langle E_{ij} + \varepsilon E_{ji}, e_k + \varepsilon \lambda_k f_k \rangle_{\mathbb{C}},$$

where $\varepsilon = \pm 1$, $k = 1, 2, 3$ and $\{i, j\} = \{1, 2, 3\} \backslash \{k\}$. It can be checked immediately that these subalgebras are pairwise orthogonal with respect to the Killing form K; hence, by Lemma 1.1.1, they form an OD $\mathcal{D}(\lambda)$ of the Lie algebra $\mathcal{L}$:

$$\mathcal{D}(\lambda) : \mathcal{L} = \mathcal{H}_0 \oplus \sum_{\varepsilon=\pm 1, 1 \leq k \leq 3} \mathcal{H}_k^{\varepsilon}(\lambda).$$

When $\lambda = (1, 1, 1)$ the decomposition $\mathcal{D} = \mathcal{D}(\lambda)$ is an MOD:

$$[\mathcal{H}_0, \mathcal{H}_k^{\varepsilon}] = \mathcal{H}_k^{-\varepsilon},\ [\mathcal{H}_k^{\varepsilon}, \mathcal{H}_k^{-\varepsilon}] = \mathcal{H}_0,\ [\mathcal{H}_i^{\alpha}, \mathcal{H}_j^{\beta}] = \mathcal{H}_k^{-\alpha\beta},$$

(if $i \neq j$ and $k \in \{1, 2, 3\} \backslash \{i, j\}$).

However, in general $\mathcal{D}(\lambda)$ is not an MOD (nor a TOD). In fact, the construction explained here represents only a "linear" realization of the ROD suggested in §3.3.

A loop approach to the groups $\mathrm{Aut}_{MOD}(G_2) = \mathbb{Z}_2^3 \circ SL_3(2)$ and $\mathrm{Aut}_{MOD}(D_4) = (\mathbb{Z}_2^7 . ASL_3(2))\mathbb{Z}_3$ is indicated in [Gri 5] and in [Gri 6] respectively.

To conclude this commentary, we remark that some preliminary results concerning the classification of TODs of the Lie algebra of type G_2 have been obtained by Tchakerian (unpublished).

Chapter 4

Irreducible orthogonal decompositions of Lie algebras with special Coxeter number

As can be seen from the preceding chapters, in principal the variety of all ODs of a given Lie algebra $\mathcal{L}$ may be so large that the complete classification of $\mathrm{Aut}(\mathcal{L})$-orbits of these decompositions is out of reach. However, from the practical point of view, *irreducible ODs*, that is, ODs $\mathcal{D}$ where the automorphism group $\mathrm{Aut}(\mathcal{D})$ acts absolutely irreducibly on $\mathcal{L}$, are of the greatest interest. The investigation of this class of ODs, which is still quite an extensive one, is considerably more advanced.

4.1 The irreducibility condition and the finiteness theorem

Recall that our ultimate aim is to study invariant integral lattices associated with ODs of complex simple Lie algebras. It is natural to restrict oneself to the case where there exists only a finite number of $\mathbb{Z}$-similarity classes of such lattices.

In this connection, the following theorem plays a basic role; it is also of an independent interest.

Theorem 4.1.1. *Let G be a finite group and V a finite-dimensional $\mathbb{Q}G$-module. Then the following conditions are equivalent:*

(i) *V has only a finite number of $\mathbb{Z}$-similarity classes of full finitely generated $\mathbb{Z}G$-modules Λ (that is, $\Lambda \otimes_{\mathbb{Z}} \mathbb{Q} = V$);*

(ii) *V is absolutely irreducible.*

Proof. (i) $\Rightarrow$ (ii). Let V be a reducible $\mathbb{Q}G$-module, so that $V = V_1 \oplus V_2$, where V_1 and V_2 are nonzero $\mathbb{Q}G$-modules. Consider $\mathbb{Z}G$-modules $\Lambda_1 \subset V_1$ and $\Lambda_2 \subset V_2$ such that $\Lambda_1 + \Lambda_2$ is a full finitely generated $\mathbb{Z}G$-module in V. Then the $\mathbb{Z}G$-modules $\Lambda_{(s)} = \Lambda_1 + s\Lambda_2$, where $s \in \mathbb{Q}$ and $s > 0$, are pairwise non-similar, contrary to condition (i). So the $\mathbb{Q}G$-module V is irreducible. Now suppose that

V is not absolutely irreducible. Set $D = \mathrm{Hom}_{\mathbb{Q}G}(V, V)$. By Schur's Lemma, D is a division ring. By the well known absolute irreducibility criterion (see e.g. [CuR, Theorem 29.13]) the dimension $[D : \mathbb{Q}]$ is greater than 1. If K is a maximal subfield of D, then we also have $[K : \mathbb{Q}] > 1$. The following remark will be needed:

Let E be a finite field extension of $\mathbb{Q}$, $[E : \mathbb{Q}] > 1$, R an order of E and $R^1 = \{\alpha \in R | N_{E/\mathbb{Q}}(\alpha) = \pm 1\}$ the group of units of R. Then the index of the subgroup $R^1\mathbb{Q}^*$ in E^* is infinite.

For, without loss of generality one can suppose that R is maximal. Our remark then follows by theorems on the finiteness of the divisor class group and the infinity of simple divisors of degree one in an algebraic number field (see [BoS]).

Suppose now that Λ is a full finitely generated $\mathbb{Z}G$-module in V, and set $R = \{\alpha \in K | \alpha\Lambda \subseteq \Lambda\}$. It is easy to see that R is an order in K. By our remark, there exists an infinite set $\{\alpha_i | i \in \mathbb{N}\}$ of elements of K lying in different cosets of K^* modulo $R^1\mathbb{Q}^*$. Then the $\mathbb{Z}G$-modules $\alpha_i\Lambda$, $i \in \mathbb{N}$, are pairwise non-similar, again a contradiction. Consequently, V is absolutely irreducible.

(ii) $\Rightarrow$ (i). Let V be an absolutely irreducible $\mathbb{Q}G$-module. Fix a basis $\{e_1, \ldots, e_n\}$ of V, and consider the $\mathbb{Z}$-module $\Gamma = < e_1, \ldots, e_n >_{\mathbb{Z}}$. After choice of a basis, the $\mathbb{Q}$-algebra $\mathrm{End}_{\mathbb{Q}}(V)$ is identified with the matrix algebra $M_n(\mathbb{Q})$. Suppose that the $\mathbb{Q}G$-module structure of V is given by the homomorphism $\Phi : G \to GL_n(\mathbb{Q})$. Because of the absolute irreducibility of V, the ring $A = < \Phi(g) | g \in G >_{\mathbb{Z}}$ is an order in $M_n(\mathbb{Q})$. Hence, there exists $d \in \mathbb{N}$ such that $d.M_n(\mathbb{Q}) \subseteq A$.

Now for any full finitely generated $\mathbb{Z}G$-module Λ' in V, there exists a $\mathbb{Z}G$-module Λ, which is similar to Λ' and such that $\Lambda \subseteq \Gamma$ but $\Lambda \not\subseteq m\Gamma$ for all $m \in \mathbb{Z}$, $m \geq 2$. We show that every such $\mathbb{Z}G$-module Λ contains $2d\Gamma$. Since $2d\Gamma$ has finite index in Γ, the latter sentence implies assertion (i).

Choose an element $v = a_1e_1 + \ldots + a_ne_n \in \Lambda$ such that $\gcd(a_1, \ldots, a_n) = 1$. Set $D_1 = \mathrm{diag}(-1, 1, \ldots, 1)$, and let P_i be a permutation matrix such that $P_ie_i = e_1$. Then

$$d.P_i(v) - d.D_1P_i(v) = 2da_ie_1;$$

however, dP_i and dD_1P_i belong to A, so that $2da_ie_1 \in \Lambda$ for all i. Since $\gcd(a_1, \ldots, a_n) = 1$, we must have $2de_1 \in \Lambda$. Similarly, $2de_i \in \Lambda$ for all i, that is, $2d\Gamma \subseteq \Lambda$. □

Corollary 4.1.2. *Let $\mathcal{D}$ be an OD of a Lie algebra $\mathcal{L}$. Then the following conditions are equivalent:*

(i) *$\mathcal{D}$ admits only a finite number of $\mathbb{Z}$-similarity classes of invariant lattices;*

(ii) *$\mathcal{L}$ is an irreducible* $\mathrm{Aut}(\mathcal{D})$*-module.* □

The following statement is an analogue of Remark 3.3.4.

Proposition 4.1.3. *Let $\mathcal{D} : \mathcal{L} = \oplus_{i=0}^{h} \mathcal{H}_i$ be a decomposition of a complex simple Lie algebra $\mathcal{L}$ into a direct sum of Cartan subalgebras such that the automorphism group $G = \mathrm{Aut}(\mathcal{D})$ acts irreducibly on $\mathcal{L}$. Then $\mathcal{D}$ is orthogonal, and so it is an IOD.*

Proof. Recall the following standard fact [Bour 1]: If V is an irreducible $\mathbb{C}G$-module with a G-invariant bilinear form, the form is unique up to multiplication by a scalar.

By its definition, $G = \mathrm{Aut}(\mathcal{D})$ preserves the Killing form K of $\mathcal{L}$. We shall define a new form B on $\mathcal{L}$. To this end, set $G_0 = St_G(\mathcal{H}_0)$, and expand $G = \bigcup_{i=0}^{h} g_i G_0$, where $g_i(\mathcal{H}_0) = \mathcal{H}_i$. Note that G_0 acts on the dual space $\mathcal{H}_0^* =< R >_{\mathbb{C}}$ (R is the corresponding root system) as a subgroup of the extended Weyl group $\mathrm{Aut}(R)$. In particular, $\mathcal{H}_0^*$ admits some G_0-invariant non-degenerate bilinear form. The same holds for $\mathcal{H}_0$: G_0 preserves some non-degenerate bilinear form B_0 defined on $\mathcal{H}_0$. For $x \in \mathcal{H}_i$ and $y \in \mathcal{H}_j$ we set

$$B(x, y) = \begin{cases} B_0(g_i^{-1}(x), g_i^{-1}(y)) & \text{if} \quad i = j, \\ 0 & \text{if} \quad i \neq j. \end{cases}$$

This definition does not depend on the choice of representatives g_i: if $f(\mathcal{H}_0) = \mathcal{H}_i$, that is, if $f = g_i g$ for some $g \in G_0$, then

$$B_0(f^{-1}x, f^{-1}y) = B_0(g^{-1}g_i^{-1}x, g^{-1}g_i^{-1}y) = B_0(g_i^{-1}x, g_i^{-1}y),$$

as B_0 is a G_0-invariant form. Extend B by linearity to the whole of $\mathcal{L} = \oplus_i \mathcal{H}_i$. It is easy to see that B is G-invariant. For, suppose that $g \in G$, $x \in \mathcal{H}_i$, $y \in \mathcal{H}_j$, $g(\mathcal{H}_i) = \mathcal{H}_k$ and $g(\mathcal{H}_j) = \mathcal{H}_l$. If $i \neq j$, then $k \neq l$ and $B(gx, gy) = 0 = B(x, y)$. When $i = j$ we have

$$B(gx, gy) = B_0(g_k^{-1}gx, g_k^{-1}gy) = B_0(g_k^{-1}gg_i x', g_k^{-1}gg_i y'),$$

where $x' = g_i^{-1}x$ and $y' = g_i^{-1}y$ belong to $\mathcal{H}_0$. Of course, $g_k^{-1}gg_i \in G_0$; hence,

$$B_0(g_k^{-1}gg_i x', g_k^{-1}gg_i y') = B_0(x', y') = B_0(g_i^{-1}x, g_i^{-1}y) = B(x, y).$$

According to our remark made at the beginning of the proof, the forms B and K are proportional: $B = \lambda K$, $\lambda \in \mathbb{C}$. Furthermore, $\lambda \neq 0$ because $B|_{\mathcal{H}_0} = B_0$ is nonzero. By the definition of B, $\mathcal{D}$ is orthogonal with respect to B; hence, $\mathcal{D}$ is orthogonal with respect to K. □

The subsequent paragraphs of this chapter are devoted to the classification of IODs of simple complex Lie algebras with Coxeter number h such that $h + 1$ is prime.

4.2 General outline of the arguments

From here to the end of the chapter we use the following notation. Let $\mathcal{L}$ be a simple finite dimensional Lie algebra over $\mathbb{C}$, h its Coxeter number, R its root system, $A(R)$ its (full) automorphism group, $\mathcal{A} = \mathrm{Aut}(\mathcal{L})$ and $\mathcal{G} = \mathcal{A}^0 = \mathrm{Inn}(\mathcal{L})$. Suppose that $\mathcal{L}$ has an IOD, that is, an OD

$$\mathcal{D} : \mathcal{L} = \bigoplus_{i=1}^{h+1} \mathcal{H}_i$$

whose automorphism group $G = \mathrm{Aut}(\mathcal{D})$ acts on $\mathcal{L}$ absolutely irreducibly with character χ. Set

$$G_i = St_G(\mathcal{H}_i) = \{\varphi | \varphi \in G, \varphi(\mathcal{H}_i) = \mathcal{H}_i\}.$$

The following assertion is obvious:

Lemma 4.2.1. (i) *G is finite.*

(ii) *χ is integer-valued.*

Proof. (i) Clearly, $|G| \leq |\mathrm{Aut}(R)|^{h+1}.(h+1)!$.

(ii) Take $g \in G$. If g moves $\mathcal{H}_i$, then the contribution of $\mathcal{H}_i$ to $\chi(g)$ is 0. If $g(\mathcal{H}_i) = \mathcal{H}_i$, then this contribution is the trace $\mathrm{Tr}(g|_{\mathcal{H}_i^*})$, which is an integer since $g|_{\mathcal{H}_i^*} \in \mathrm{Aut}(R)$. □

Lemma 4.2.2. (i) *The representation ρ of G_1 on $\mathcal{H}_1$ is irreducible. Moreover, $\chi = \mathrm{Ind}_{G_1}^G(\rho)$.*

(ii) *If $\mathcal{L}$ is a Lie algebra of type A_n ($n > 1$), then G_1 admits a doubly transitive permutation representation τ of degree $n+1$ such that* $\mathrm{Ker}\,\tau \supseteq \mathrm{Ker}\,\rho$, $(\mathrm{Ker}\,\tau : \mathrm{Ker}\,\rho) \leq 2$, *and* $\mathrm{Ker}\,\tau$ *is soluble.*

(iii) *If $\mathcal{L}$ is a Lie algebra of type B_n, C_n or D_n, then G_1 admits a transitive permutation representation τ of degree n such that* $\mathrm{Ker}\,\tau$ *is soluble.*

Remark. Sometimes we shall denote a (complex) representation and its character by the same Greek letter.

Proof. (i) This is obvious.

(ii) We consider the representation ρ^* of the group G_1 on $\mathcal{H}_1^* = < R >_{\mathbb{C}}$, dual to ρ. If the root system R is realized in the form $\{e_i - e_j | 1 \leq i \neq j \leq n+1\}$, where $\{e_1, \ldots, e_{n+1}\}$ is an orthonormal basis of the space $\mathbb{R}^{n+1}$, then $\rho^*(G_1)$ is an irreducible subgroup of the group $\mathrm{Aut}(R) = \mathbb{Z}_2 \times \mathbb{S}_{n+1}$. We write each element $g \in \rho^*(G_1)$ in the form $\delta(g).\pi(g)$, where $\delta(g) \in \mathbb{Z}_2$ and $\pi(g) \in \mathbb{S}_{n+1}$, and set $P = \{\pi(g) | g \in \rho^*(G_1)\}$. Then P acts irreducibly on $< R >_{\mathbb{C}}$, and so P is a 2-transitive permutation group on the symbols $e_1, \ldots, e_{n+1}$. The composition of ρ^* with this action is exactly the desired τ. Obviously, $\operatorname{Ker}\rho = \operatorname{Ker}\rho^*$ and $(\operatorname{Ker}\tau : \operatorname{Ker}\rho) \leq 2$. Since $\operatorname{Ker}\rho$ lies in the maximal torus of $\mathcal{A}$ corresponding to $\mathcal{H}_1$, $\operatorname{Ker}\rho$ is abelian, so that $\operatorname{Ker}\tau$ is soluble.

(iii) Similar to (ii). □

In the case where $\mathcal{L}$ is a classical Lie algebra, the group $\mathcal{G}$ possesses a natural projective representation on a module V, on which $\mathcal{L}$ acts according to its natural realization. Namely, $\mathcal{G} = \hat{\mathcal{G}}/Z$, where $Z = Z(\hat{\mathcal{G}})$, and $V = \mathbb{C}^{n+1}$, $\hat{\mathcal{G}} = SL(V)$ if $\mathcal{L}$ is of type A_n; $V = \mathbb{C}^{2n}$, $\hat{\mathcal{G}} = Sp(V)$ if $\mathcal{L}$ is of type C_n; $V = \mathbb{C}^{2n+1}$, $\hat{\mathcal{G}} = \mathcal{G} = SO(V)$ if $\mathcal{L}$ is of type B_n, and $V = \mathbb{C}^{2n}$, $\hat{\mathcal{G}} = SO(V)$ if $\mathcal{L}$ is of type D_n. Of course, in cases $B-D$, V is equipped with a non-degenerate bilinear symplectic or orthogonal form $(*, *)$. The group $\hat{\mathcal{G}}$ is simply-connected in the cases A and C; in cases B and D, we have $\pi_1(\hat{\mathcal{G}}) = \mathbb{Z}_2$. Set $G^0 = G \cap \mathcal{G}$, $G_i^0 = G^0 \cap G_i$, let $\hat{G}^0$ be the full preimage of G^0 in $\hat{\mathcal{G}}$ and λ the $\hat{G}^0$-character afforded by V. One can extend the character χ of G^0 to some character of $\hat{G}^0$ (which we denote by the same symbol χ). The characters χ and λ are related to each other as follows:

Lemma 4.2.3. (i) *For $g \in \hat{G}^0$ we have $\chi(g) = |\lambda(g)|^2 - 1$ in case A, $\chi(g) = \{\lambda(g)^2 + \lambda(g^2)\}/2$ in case C, and $\chi(g) = \{\lambda(g)^2 - \lambda(g^2)\}/2$ in cases B and D.*

(ii) *The character λ is irreducible.*

Proof. (i) It is known that, in case A, the $\hat{G}^0$-module $\mathcal{L}$ can be identified with $(V \otimes_{\mathbb{C}} V^*)/ < E_{n+1} >_{\mathbb{C}}$, where $V^* = \operatorname{Hom}_{\mathbb{C}}(V, \mathbb{C})$ and E_{n+1} is the unit matrix of degree $n+1$. Hence, $\chi(g) = |\lambda(g)|^2 - 1$. In cases B and D (respectively, in case C), by using the form $(*, *)$ one can identify firstly V^* with V, and then $\mathcal{L}$ with the exterior square $\wedge^2(V)$ (respectively, the symmetric square $S^2(V)$). Hence $\chi(g) = \{\lambda(g)^2 \mp \lambda(g^2)\}/2$.

(ii) Suppose that λ is reducible, that is, $\lambda = \lambda_1 + \lambda_2$ for some $\hat{G}^0$-characters λ_1 and λ_2. In case A we have:

$$\chi = |\lambda_1 + \lambda_2|^2 - 1_{\hat{G}^0} = (|\lambda_1|^2 - 1_{\hat{G}^0}) + (|\lambda_2|^2 - 1_{\hat{G}^0}) + (\lambda_1\overline{\lambda_2} + \lambda_2\overline{\lambda_1}) + 1_{\hat{G}^0}.$$

Furthermore, $|\lambda_i|^2 - 1_{\hat{G}^0}$ is either 0 or a character of $\hat{G}^0$ since

$$(|\lambda_i|^2, 1_{\hat{G}^0})_{\hat{G}^0} = (\lambda_i, \lambda_i)_{\hat{G}^0} \geq 1.$$

Hence, $(\chi, 1_{\hat{G}^0})_{\hat{G}^0} > 0$, $(\chi_{G^0}, 1_{G^0})_{G^0} > 0$, that is,

$$\mathcal{M} = C_{\mathcal{L}}(G^0) = \{x \in \mathcal{L} | \forall g \in G^0, g(x) = x\}$$

is nonzero. But G^0 is a normal subgroup of G, so $\mathcal{M}$ is a G-module. Furthermore, it is easy to see that $\mathcal{M} \neq \mathcal{L}$. In other words, $\mathcal{L}$ is a reducible G-module, a contradiction.

In the remaining cases $B-D$ we have

$$\chi(g) = \{(\lambda_1+\lambda_2)^2(g) \pm (\lambda_1+\lambda_2)(g^2)\}/2 =$$

$$= \hat{\lambda}_1(g) + \hat{\lambda}_2(g) + (\lambda_1\lambda_2)(g),$$

where the maps $\hat{\lambda}_i : \hat{G}^0 \to \mathbb{C}$ given by the rule

$$\hat{\lambda}_i(g) = \{\lambda_i(g)^2 \pm \lambda_i(g^2)\}/2$$

are either 0 or $\hat{G}^0$-characters. Hence, if $d_i = \deg \lambda_i$, then $\mathcal{L}$ can be decomposed into a direct sum of G^0-submodules of dimensions $d_1(d_1 \pm 1)/2$, $d_2(d_2 \pm 1)/2$ and $d_1 d_2$. On the other hand, we have $G/G^0 \subseteq \mathrm{Out}(\mathcal{L}) = \mathcal{A}/\mathcal{G} = \mathbb{S}_3$ if $\mathcal{L}$ is of type D_4 and $\mathcal{A}/\mathcal{G} \subseteq \mathbb{Z}_2$ in the other cases. So, by Clifford's Theorem [Hup], we obtain the following decomposition: $\chi_{G^0} = \rho_1 + \ldots + \rho_s$, where $\rho_1, \ldots, \rho_s$ are irreducible characters of G^0 of the same degree $\dim \mathcal{L}/s$, and s divides $\gcd(\dim \mathcal{L}, |G/G^0|)$. Consequently, $s \leq 2$, contrary to the above-mentioned decomposition of the G^0-module $\mathcal{L}$. □

In passing we have proved the following statement:

Lemma 4.2.4. *The character χ_{G^0} is either irreducible or a sum of two irreducible characters of the same degree.* □

We now consider

$$K = \{\varphi \in G | \forall i, \varphi(\mathcal{H}_i) = \mathcal{H}_i\}.$$

In what follows, we shall refer to the situation with $K = 1$ as the *P-case*. We shall speak of the *E-case* for $K \neq 1$ when K contains a minimal normal subgroup H of G that is elementary abelian. All the remaining possibilities are referred to together as the *S-case*.

Concerning a minimal normal subgroup H of G lying in K, the following lemma holds:

Lemma 4.2.5. (i) *The subspace $C_{\mathcal{L}}(H) = \{x \in \mathcal{L} | \forall h \in H, h(x) = x\}$ is trivial.*
(ii) $H \subseteq \mathrm{Inn}(\mathcal{L}) = \mathcal{G}$.

Proof. (i) $C_{\mathcal{L}}(H)$ is a proper G-submodule of $\mathcal{L}$, so $C_{\mathcal{L}}(H) = 0$.

(ii) Assume that $H \not\subseteq \mathcal{G}$. Then $H \cap \mathcal{G} \subset H$, and $H \cap \mathcal{G}$ is a normal subgroup of G; but H is minimal, so $H \cap \mathcal{G} = 1$. Hence

$$H = H/(H \cap \mathcal{G}) \cong H\mathcal{G}/\mathcal{G} \subseteq \mathcal{A}/\mathcal{G}.$$

The last group is equal to 1, $\mathbb{Z}_2$ or $\mathbb{S}_3$ (see [SSL]), and H is characteristically simple, so that $H \cong \mathbb{Z}_2$ or $\mathbb{Z}_3$. But then $C_{\mathcal{L}}(H) \neq 0$ by [SSL], a contradiction. □

In what follows, in the E-case we shall sometimes consider instead of H some minimal normal subgroup H^0 of G^0 lying in H. The appropriateness of this replacement is explained by the following assertion (cf. Lemma 4.2.5):

Lemma 4.2.6. *$C_{\mathcal{L}}(H^0) = 0$ if $\mathcal{L}$ is not of type D_4.*

Proof. We have to consider only the case $H^0 \subset H$, $G^0 \subset G = < G^0, f >$. Assume that $\mathcal{L}_0 = C_{\mathcal{L}}(H^0) \neq 0$.

1) Set $H^1 = fH^0f^{-1}$. Then, like H^0, H^1 is also a normal subgroup of G^0. By the minimality of H^0 we have either $H^0 = H^1$ or $H^0 \cap H^1 = 1$. In the former case we would have that H^0 is a normal subgroup of G with $H^0 \subset H$, a contradiction. So $H^0 \cap H^1 = 1$. But then $H \supseteq H^0 \times H^1 \lhd G$, and hence $H = H^0 \times H^1$. Set $\mathcal{L}_1 = C_{\mathcal{L}}(H^1)$. Then $\mathcal{L}_1 = f(\mathcal{L}_0) \neq 0$. Since $H^0, H^1 \lhd G^0$, the subalgebras $\mathcal{L}_0$ and $\mathcal{L}_1$ are G^0-invariant. However, $0 = C_{\mathcal{L}}(H) = \mathcal{L}_0 \cap \mathcal{L}_1$. Hence, according to Lemma 4.2.4, we have $\mathcal{L} = \mathcal{L}_0 \oplus \mathcal{L}_1$ and $\dim \mathcal{L}_0 = \dim \mathcal{L}_1 = (\dim \mathcal{L})/2$.

2) Decompose $\mathcal{L}$ into a direct sum of eigensubspaces with respect to H^0:

$$\mathcal{L} = \bigoplus_{\alpha \in \mathrm{Irr}(H^0)} \mathcal{L}(\alpha).$$

Of course, $\mathcal{L}_0 = \mathcal{L}(1_{H^0})$ and $[\mathcal{L}(\alpha), \mathcal{L}(\beta)] \subseteq \mathcal{L}(\alpha\beta)$. So, setting

$$\mathcal{L}' = \sum_{\alpha \neq 1_{H^0}} \mathcal{L}(\alpha),$$

we have $[\mathcal{L}_0, \mathcal{L}'] \subseteq \mathcal{L}'$. Because of the H^0-invariance of $\mathcal{L}_1$, we may also decompose this space into a direct sum of eigenspaces (with respect to H^0), and it is easy to convince ourselves that $\mathcal{L}_1 \subseteq \mathcal{L}'$. But $\mathcal{L} = \mathcal{L}_0 \oplus \mathcal{L}_1 = \mathcal{L}_0 \oplus \mathcal{L}'$, so $\mathcal{L}_1 = \mathcal{L}'$ and $[\mathcal{L}_0, \mathcal{L}_1] \subseteq \mathcal{L}_1$. On the other hand, it is clear that $[\mathcal{L}_1, \mathcal{L}_1] \subseteq \mathcal{L}_1$ since $\mathcal{L}_1 = C_{\mathcal{L}}(H^1)$. Consequently, $[\mathcal{L}, \mathcal{L}_1] \subseteq \mathcal{L}_1$, that is, $\mathcal{L}_1$ is an ideal of the Lie algebra $\mathcal{L}$. The last sentence contradicts the simplicity of $\mathcal{L}$. □

Our first goal is to prove that the P- and S-cases are impossible, and that, in the E-case, H is a Jordan subgroup of $\mathcal{A}$. Next we claim that, when G contains a Jordan subgroup, an IOD can almost always be obtained by the construction described in Theorem 3.1.2.

4.3 Regular automorphisms of prime order and Jordan subgroups

Throughout §4.3, we let $\mathcal{L}$ be a simple Lie algebra for which $h + 1 = r$ is prime, and which has an IOD $\mathcal{D}$; we set $G = \mathrm{Aut}(\mathcal{D})$. Since an IOD is transitive, $|G|$ is divisible by r. It turns out that the following statement holds:

Proposition 4.3.1. *Let φ be an element of G with $|\varphi| = r$. Then φ is a regular automorphism (in the sense that the fixed-point subalgebra $C_{\mathcal{L}}(\varphi)$ is commutative). Moreover, $C_{\mathcal{A}}(\varphi)$ is soluble.*

Proof. 1) Let us first show that $\mathrm{Tr}\varphi = 0$. Indeed, if $\mathrm{Tr}\varphi \neq 0$, then φ leaves each subalgebra $\mathcal{H}_i$ fixed. But r does not divide $|A(R)|$, and so it follows from the inclusion $\varphi|_{\mathcal{H}_i^*} \in A(R)$ that $\varphi|_{\mathcal{H}_i} = 1_{\mathcal{H}_i}$ and $\varphi = 1_{\mathcal{L}}$, a contradiction.

2) According to [SAG], φ leaves some Cartan subalgebra $\mathcal{H}$ pointwise fixed. We set $\varepsilon = \exp(2\pi i/r)$ and $n_j = \dim \mathrm{Ker}\,(\varphi - \varepsilon^j.1_{\mathcal{L}})$, $0 \leq j \leq r-1$. Then $0 = \mathrm{Tr}\varphi = \sum_j n_j\varepsilon^j$, and so $n_0 = n_1 = \ldots = n_{r-1} = (\dim \mathcal{L})/r = \dim \mathcal{H}$. Consequently, $\mathrm{Ker}\,(\varphi - 1_{\mathcal{L}}) = \mathcal{H}$ is an abelian subalgebra, and φ is a regular automorphism.

3) According to [Kac], there is just one conjugacy class of regular automorphisms φ of order r. Thus, we can define φ as follows. Let

$$\mathcal{L} = \mathcal{H} \oplus \sum_{\alpha \in R} < X_\alpha >_{\mathbb{C}}$$

be the Cartan decomposition corresponding to a Cartan subalgebra $\mathcal{H}$. Then

$$\varphi(X_\alpha) = \varepsilon^{f(\alpha)} X_\alpha,$$

where $f \in \mathrm{Hom}_{\mathbb{Z}}(R, \mathbb{F}_r)$, and $f(\alpha_1) = \ldots = f(\alpha_n) = 1$ if $\{\alpha_1, \ldots, \alpha_n\}$ is a simple root system. If we set $\mathcal{L}_j = \mathrm{Ker}\,(\varphi - \varepsilon^j.1_{\mathcal{L}})$, then $\mathcal{L}_0 = \mathcal{H}$ and $\mathcal{L}_1 = \, < X_{\alpha_1}, \ldots, X_{\alpha_n} >_{\mathbb{C}}$. Now suppose that $\psi \in C_{\mathcal{A}}(\varphi)$. Then ψ fixes $\mathcal{H}$, that is, $\psi \in N_{\mathcal{A}}(\mathcal{H}) = T.A(R)$, where T is the maximal torus corresponding to $\mathcal{H}$. In particular, ψ induces some permutation ω on the root system R. But ψ fixes $\mathcal{L}_1$, so ω permutes the roots $\alpha_1, \ldots, \alpha_n$. Thus ψ is an automorphism of the Dynkin diagram corresponding to $\mathcal{L}$. We have shown that $D \subseteq \mathbb{S}_3$, where $D = C_{\mathcal{A}}(\varphi)T/T$. Since $C_{\mathcal{A}}(\varphi) \cap T$ is abelian, $C_{\mathcal{A}}(\varphi)$ is soluble. □

Let us go over to the S-case.

Lemma 4.3.2. *Suppose that the S-case occurs, so that $H = L^m = L \times \ldots \times L$, where L is a nonabelian simple group. Then r divides $(G : C_G(H))$ and $|\mathrm{Aut}(L)|$.*

Proof. Suppose that r does not divide $(G : C_G(H))$. Then r divides $|C_G(H)|$, and we can find an element $\varphi \in C_G(H)$ with $|\varphi| = r$. According to Lemma 4.3.1, the subgroup H of $C_G(\varphi) \subseteq C_{\mathcal{A}}(\varphi)$ is soluble, contrary to our assumption that $H = L^m$. Hence r divides $(G : C_G(H))$. But

$$G/C_G(H) \hookrightarrow \mathrm{Aut}(H) = (\mathrm{Aut}(L))^m.\mathbb{S}_m,$$

so all we need do is to show that r does not divide $|\mathbb{S}_m|$, that is, $m < r$. Suppose to the contrary that $m \geq r$. Choose an odd prime number q dividing $|L|$. Then clearly, q^r divides $|H|$. Note that the kernel of the representation of H on $\mathcal{H}_1$ is abelian. But $H = L^m$ has no nontrivial abelian composition factors, therefore H acts faithfully on $\mathcal{H}_1$ and $H \hookrightarrow A(R)$. In particular, q^r divides $|A(R)|$, a contradiction. □

As we have seen, L is a nonabelian simple group such that the prime r divides $|\mathrm{Aut}(L)|$ but not $|L|$ (the last because of the embedding $L \hookrightarrow A(R)$). Concerning finite simple groups of this sort, we have:

Lemma 4.3.3. *L is a finite group of Lie type defined over a field $\mathbb{F}_{p^e}$, where p a prime and $r|e$. In particular, p^r divides $|L|$.*

Proof. It is known [Gor 2] that the outer automorphism group $\mathrm{Out}(L)$ can be embedded in $\mathbb{Z}_2^2$ if L is an alternating or sporadic simple group. Therefore, L is a group of Lie type defined over some finite field $\mathbb{F}_{p^e}$. In this case, the order of $\mathrm{Out}(L)$ is one of those listed in Table A3, and it is easy to see that r must divide e. □

Now we are able to prove the following:

Proposition 4.3.4. *If $h + 1 = r$ is a prime, then the S-case is impossible.*

Proof. Assume the contrary, that is, $H = L^m$, where L is a nonabelian simple group. According to Lemmas 4.3.2 and 4.3.3, L is embedded in $A(R)$ and L is a finite group of Lie type defined over a field $\mathbb{F}_{p^e}$, where $r \mid e$. In particular, p^r divides $|A(R)|$. It is easy to check directly that the last condition cannot be satisfied if $\mathcal{L}$ is a Lie algebra of type A_n $(h = n + 1)$; B_n, C_n $(h = 2n)$; G_2, F_4, E_6, E_7 or E_8 $(h = 6, 12, 12, 18$ or 30, respectively). If $\mathcal{L}$ is of type D_n, then $n \geq 5$ since $A(D_4)$ is soluble. When $n \geq 5$ we have $A(D_n) = E.S$, where $E = \mathbb{Z}_2^n$ and $S = \mathbb{S}_n$; furthermore, $h = 2n - 2$. Note that L is simple, so that $L \cap E = 1$, and this implies the embedding $L \hookrightarrow S$. Thus we conclude that $n!$ is divisible by p^{2n-1}, a contradiction. □

Next, we consider the E-case. Here an essential role is played by the following assertion:

Lemma 4.3.5. *Suppose that $\varphi \in G$, $|\varphi| = r$ and $\psi \in C_G(\varphi)$. Then (the integer) $\mathrm{Tr}\psi$ is divisible by r. If in addition $\psi \in C_K(\varphi)$, then $\dim C_{\mathcal{L}}(\psi)$ is also divisible by r.*

Proof. We have mentioned that φ permutes the r subalgebras $\mathcal{H}_i$ cyclically. Therefore, if ψ fixes some subalgebra $\mathcal{H}_i$, then in fact ψ leaves all subalgebras $\mathcal{H}_j$ fixed and, as a consequence, r divides $\mathrm{Tr}\psi$ and $\dim C_{\mathcal{L}}(\psi)$. On the contrary, ψ moves every subalgebra $\mathcal{H}_j$, and so $\mathrm{Tr}\psi = 0$. □

Lemma 4.3.6. *Let $\mathcal{L}$ be a Lie algebra of type G_2, F_4, E_6, E_7 or E_8, and ψ an inner automorphism of $\mathcal{L}$ of prime order p, where p is one of the torsion primes (in the sense of* [SAG]*). Then either* $\mathrm{Tr}\psi$ *or* $\dim C_{\mathcal{L}}(\psi)$ *is not divisible by* $r = h + 1$.

Proof. A direct verification using the description of automorphisms of finite order given in [Kac] is what is needed here. In the case of E_8, one can use Table 4 from [CoG]. It is interesting to note that the Lie algebra of type E_7 has an automorphism ψ of order 7 such that $\dim \mathrm{Ker}\,(\psi - \varepsilon^i.1_{\mathcal{L}}) = 19$ for each $i = 0, 1, \dots, 6$, where $\varepsilon = \exp(2\pi i/7)$. □

Proposition 4.3.7. *Let $\mathcal{L}$ be an exceptional Lie algebra. Suppose that the E-case occurs for some IOD $\mathcal{D}$ of $\mathcal{L}$, that is, $H \cong \mathbb{Z}_p^m$. Then*

(i) $\mathcal{L}$ is not of type E_7;

(ii) H is a Jordan subgroup of order 2^5, 2^3, 3^3, 3^3 or 5^3 ($\mathcal{L}$ is of type E_8, G_2, F_4, E_6 or E_8, respectively). Moreover, in each case, the subgroup H is unique up to conjugacy.

Proof. 1) Note first that p is one of the torsion primes. Indeed, suppose the contrary. Set $\mathcal{G} = \mathrm{Inn}(\mathcal{L})$, and let $\hat{\mathcal{G}}$ be its simply-connected covering. Then $\mathcal{G} = \hat{\mathcal{G}}/Z(\hat{\mathcal{G}})$, where the centre $Z(\hat{\mathcal{G}})$ is 1, 1, $\mathbb{Z}_3$, $\mathbb{Z}_2$ or 1, if $\mathcal{L}$ is of type G_2, F_4, E_6, E_7 or E_8, respectively. Since by our assumption p does not divide $|Z(\hat{\mathcal{G}})|$, H can be lifted to an elementary abelian subgroup of $\hat{\mathcal{G}}$. By [SAG], the last subgroup is contained in a maximal torus; hence $C_{\mathcal{L}}(H) \neq 0$, and this contradicts Lemma 4.2.5.

2) Now we show that r divides $|H|-1$. Consider an element $\varphi \in G$ with $|\varphi| = r$. Then φ normalizes H. By Lemma 4.2.5, H consists of inner automorphisms of $\mathcal{L}$. So, according to Lemmas 4.3.5 and 4.3.6, φ acts (by conjugation) on $H\backslash\{1\}$ fixed-point-freely. In particular, $r = |\varphi|$ divides $|H\backslash\{1\}| = p^m - 1$.

3) Set $n = \mathrm{rank}\mathcal{L}$. By [Ada] and [CSe], $m \leq n+1$. The last inequality, together with the conditions that r divides $p^m - 1$ and p is one of the torsion primes, leads to the following possibilities for the pair $(\mathcal{L}, H)$:

a) $(G_2, \mathbb{Z}_2^3)$;

b) $(F_4, \mathbb{Z}_3^3)$;

c) $(E_6, \mathbb{Z}_3^3$ or $\mathbb{Z}_3^6)$;

d) $(E_8, \mathbb{Z}_2^5, \mathbb{Z}_5^3$ or $\mathbb{Z}_5^6)$.

Note that the case E_7 is absent. Furthermore, by [CSe], the subgroup $\mathbb{Z}_3^6$ in the Lie group of type E_6 lies in a maximal torus and, as a consequence, $C_{\mathcal{L}}(H) \neq 0$. Now assume that $\mathcal{L}$ is a Lie algebra of type E_8 and $H = \mathbb{Z}_2^5$. Then a regular automorphism φ of order 31 acts transitively on the set $H\backslash\{1\}$. This means that, if χ is the character of G afforded by $\mathcal{L}$, then χ takes the same value l on all elements $h \in H\backslash\{1\}$. By [CoG], $l = -8$ and the subgroup $H = \mathbb{Z}_2^5$ is unique. Next we consider the case $(\mathcal{L}, H) = (E_8, \mathbb{Z}_5^6)$. It is known [SAG] that H, like every nilpotent subgroup of $\mathcal{G}$, lies in the normalizer $N = N_{\mathcal{G}}(T)$ of some maximal torus T of $\mathcal{G}$.

Set $H_0 = H \cap T$, $T_5 = \{t \in T | t^5 = 1\}$ and $P = HT/T \subseteq N/T = W = W(R)$. Because H is not contained in T, P is a nontrivial 5-subgroup of W. By [SAG], the group W has two conjugacy classes (A_4 and $2A_4$ in the notation of Carter [SAG]) of elements of order 5, with representatives ω_1 and ω_2, say. It is not difficult to see that $|C_{T_5}(\omega_1)| = 5^4$ and $|C_{T_5}(\omega_2)| = 5^2$. Furthermore, the commutativity of H implies that $H_0 \subseteq C_{T_5}(P)$. Hence, if $|P| = 5$, then $|H| \le 5^5$. If $|P| = 5^2$, then $P \in \mathrm{Syl}_5(W)$ and P contains an element of class $2A_4$, and this implies that $|H_0| \le 5^2$ and $|H| \le 5^4$. We have therefore shown that the case $(E_8, \mathbb{Z}_5^6)$ is impossible.

4) It remains for us to prove that the subgroup $\mathbb{Z}_2^3$ ($\mathbb{Z}_3^3$, $\mathbb{Z}_3^3$ and $\mathbb{Z}_5^3$ respectively), when $\mathcal{L}$ is of type G_2 (F_4, E_6 and E_8 respectively) is a Jordan subgroup, and that this subgroup is unique. In all these cases a regular automorphism φ of order r acts transitively on the set of nontrivial cyclic subgroups of H. But the character χ of H on $\mathcal{L}$ is integer-valued (see Lemma 4.2.1), and so there exists an integer l such that $\chi(x) = -l$ for every $x \in H \backslash \{1\}$ (of course, $\chi(1) = \dim \mathcal{L}$). Since $C_{\mathcal{L}}(H) = 0$ we have

$$0 = (1_H, \chi_H)_H = \frac{\dim \mathcal{L} - l(p^3 - 1)}{p^3},$$

that is, $l = \dim \mathcal{L}/(p^3 - 1) = 2$ (2, 3 and 2 respectively). Hence, $\chi_H = l(\rho_H - 1_H)$, where ρ_H is the regular character of H. Now we establish that the subgroup $H = \mathbb{Z}_p^3$ with the character $l(\rho_H - 1_H)$ (on $\mathcal{L}$) is unique up to conjugacy in $\mathcal{G} = \mathrm{Inn}(\mathcal{L})$. Since a Jordan subgroup has the same character, we conclude at the same time that H is a Jordan subgroup.

5) Let H_0 be a maximal subgroup of H. It is easy to show that H_0 centralizes some Cartan subalgebra $\mathcal{H}$. Here we shall see that $C_{\mathcal{L}}(H_0) = \mathcal{H}$. Indeed, if

$$\mathcal{L} = \mathcal{H} \oplus \sum_{\alpha \in R} < X_\alpha >_{\mathbb{C}}$$

is the corresponding Cartan decomposition, then each element $x \in H_0$ can be written in the form $x : X_\alpha \mapsto \varepsilon^{\tilde{x}(\alpha)} X_\alpha$, where $\varepsilon = \exp(2\pi i/p)$ and $\tilde{x} \in \mathrm{Hom}_{\mathbb{Z}}(R, \mathbb{F}_p)$. Therefore

$$C_{\mathcal{L}}(H_0) = \mathcal{H} \oplus \sum_{\alpha \in R_0} \langle X_\alpha \rangle_{\mathbb{C}},$$

where $R_0 = \{\alpha \in R | \forall h \in H_0, \tilde{h}(\alpha) = 0\}$. In particular, if $C_{\mathcal{L}}(H_0) \neq \mathcal{H}$, then $C_{\mathcal{L}}(H_0)$ is an insoluble Lie algebra. But this Lie algebra has a fixed-point-free automorphism ψ of prime order p, where $H =< H_0, \psi >$, namely,

$$C_{C_{\mathcal{L}}(H_0)}(\psi) = C_{\mathcal{L}}(\psi) \cap C_{\mathcal{L}}(H_0) = C_{\mathcal{L}}(H) = 0.$$

According to [SSL], the Lie algebra $C_{\mathcal{L}}(H_0)$ is nilpotent. This contradiction means that $C_{\mathcal{L}}(H_0) = \mathcal{H}$.

6) As $[H_0, \psi] = 1$, ψ leaves $\mathcal{H}$ fixed. Suppose that ψ corresponds to an element ω of the Weyl group $W(R) = N_{\mathcal{G}}(T)/T$, where T is the maximal torus corresponding to $\mathcal{H}$. Since $C_{\mathcal{L}}(H) = 0$, ω acts fixed-point-freely on $\mathcal{H}^*$ and $|\omega| = p$. A direct check using the results of Carter [SAG], shows the following.

(α) There exists only one conjugacy class of elements ω of order p in $W(R)$ such that $C_{\mathcal{H}^*}(\omega) = 0$.

(β) If $T_p = \{t \in T | t^p = 1\}$, then $C_{T_p}(\omega)$ is equal to $\mathbb{Z}_p^2$ in the cases G_2, F_4 and E_8, and to $\mathbb{Z}_p^3$ in the case of E_6. In addition, in that case one can show that:

a) $W(E_6)$ acts transitively on the set $\mathcal{M}$ of pairs $(\tilde{a}, \tilde{b})$ such that $\tilde{a}, \tilde{b} \in \mathrm{Hom}_{\mathbb{Z}}(R, \mathbb{F}_3)$, and for any $\alpha \in R$, either $\tilde{a}(\alpha) \neq 0$ or $\tilde{b}(\alpha) \neq 0$;

b) set $C = \{\delta \in W(E_6) | \delta \circ \tilde{a} = \tilde{a}, \delta \circ \tilde{b} = \tilde{b}\}$ for $(\tilde{a}, \tilde{b}) \in \mathcal{M}$. (Recall that $(\delta \circ \tilde{c})(\alpha) = \tilde{c}(\delta^{-1}(\alpha))$ if $\delta \in W(R)$, $\tilde{c} \in \mathrm{Hom}_{\mathbb{Z}}(R, \mathbb{F}_3)$ and $\alpha \in R$). Then C is nonabelian. Furthermore, each element of order 3 of C that acts fixed-point-freely on $\mathcal{H}^*$ lies in the centre of C.

7) Before establishing the fact that H is a Jordan subgroup, we make the following observation. Let ψ and ϑ be elements of $N_{\mathcal{G}}(T)$ which correspond to the same element $\omega \in W(R)$, where $|\omega| = p$ and $C_{\mathcal{H}^*}(\omega) = 0$. Then there exists an element $t \in T$ such that $t^{-1}\psi t = \vartheta$. Indeed, consider the map

$$f : T \to T,\ f(t) = [\psi, t].$$

As T is abelian, for $s, t \in T$ we have

$$f(s)f(t) = (\psi^{-1}s^{-1}\psi)s(\psi^{-1}t^{-1}\psi t) = (\psi^{-1}s^{-1}\psi)(\psi^{-1}t^{-1}\psi t)s = f(st),$$

that is, f is a homomorphism. Furthermore, the condition $C_{\mathcal{H}^*}(\omega) = 0$ implies that $\mathrm{Ker} f$ is a finite group, and so $\dim f(T) = \dim T$. It is also clear that $f(T)$ is connected. Hence, $f(T) = T$. In particular, if $\vartheta = \psi s$, $s \in T$, then there exists an element $t \in T$ such that $f(t) = s$, that is, $t^{-1}\psi t = \psi s = \vartheta$ as desired.

8) Finally, we deduce that any two subgroups $H, H' \cong \mathbb{Z}_p^3$ of $\mathcal{G}$ which act on $\mathcal{L}$ with character $l(\rho_H - 1_H)$ are conjugate. To this end we fix maximal subgroups H_0 and H_0' of H and H', and set $H =< H_0, \psi >$, $H' =< H_0', \psi' >$. As $C_{\mathcal{L}}(H_0)$ and $C_{\mathcal{L}}(H_0')$ are Cartan subalgebras, we can suppose that $C_{\mathcal{L}}(H_0) = C_{\mathcal{L}}(H_0') = \mathcal{H}$. In step 6) we showed that one can find an element $g \in N_{\mathcal{G}}(T)$ such that $g^{-1}H_0 g = H_0'$ and $g^{-1}\psi g \psi'^{-1} \in T$. Furthermore, by 7), there exists an element $t \in T$ such that $t^{-1}g^{-1}\psi g t = \psi'$. Because T is abelian, we have also that $t^{-1}g^{-1}H_0 g t = H_0'$. In other words, $t^{-1}g^{-1}Hgt = H'$. This completes the proof of Proposition 4.3.7. □

Now we treat the E-case for Lie algebras of types A and C. Let $\hat{G}^0$ and $\hat{H}^0$ be the full preimages of G^0 and H^0 in $\hat{\mathcal{G}}$. Recall that H^0 is a minimal normal subgroup of G^0 that lies in H. The structure of $\hat{H}^0$ is explained in the following simple statement, whose proof is omitted.

Lemma 4.3.8. *$\hat{H}^0$ is a nilpotent group of class 2 with the centre $Z(\hat{H}^0) = Z = Z(\hat{\mathcal{G}})$. If H^0 is a p-group for some prime p, then p divides $|Z|$. The map $(x, y) \mapsto [\hat{x}, \hat{y}]$ defines a non-degenerate alternating bilinear form on $\hat{H}^0$ taking values in $\Omega_p(Z)$.* □

The following assertion is well-known [Hup].

Lemma 4.3.9. *Let P be a nilpotent group of class 2 and θ a faithful irreducible character of it. Then θ vanishes on $P \backslash Z(P)$.* □

We are now able to prove the following.

Proposition 4.3.10. *Suppose that the Lie algebra $\mathcal{L}$ of type A_n admits an IOD $\mathcal{D}$ and that the E-case occurs. Suppose in addition that the notations G, H and H^0 have the sense indicated above. Then*

(i) *$n = p^m - 1$ and $|H^0| = p^{2m}$ for some prime power p^m;*

(ii) *$H = H^0$;*

(iii) *H is a Jordan subgroup and is uniquely defined up to $\mathcal{G}$-conjugacy.*

Proof. 1) According to Lemma 4.2.3, the group $\hat{G}^0$ acts faithfully and irreducibly on $V = \mathbb{C}^{n+1}$ with character λ. By Clifford's Theorem [Hup],

$$\lambda_{\tilde{H}^0} = l \sum_{i=1}^{t} \theta_i,$$

where $\tilde{H}^0 = O_p(\hat{H}^0)$, $\theta_i \in \mathrm{Irr}(\tilde{H}^0)$ and $\theta_1, \ldots, \theta_t$ are all the distinct $\hat{G}^0$-conjugates of θ_1. Note that $\hat{H}^0 = \tilde{H}^0 \times O_{p'}(\hat{H}^0)$ and $Z(\tilde{H}^0) = O_p(Z)$. In addition, $O_{p'}(\hat{H}^0)$ acts trivially on $\mathcal{L}$. Suppose that the character θ_1 is not faithful, that is, $P = \mathrm{Ker}\ \theta_1 \neq 1$. Then $1 \neq P \cap Z(\tilde{H}^0)$, which implies that $1 \neq P \cap \Omega_p(Z(\tilde{H}^0))$. But $\Omega_p(Z(\tilde{H}^0)) = \Omega_p(Z) = \mathbb{Z}_p$, and so in fact we have $P \supseteq \Omega_p(Z)$. Since $\Omega_p(Z) = \Omega_p(Z(\hat{H}^0))$ is a normal subgroup of $\hat{G}^0$ and $\theta_1, \ldots, \theta_t$ are $\hat{G}^0$-conjugate, we come to the conclusion that $\mathrm{Ker}\ \theta_i \supseteq \Omega_p(Z)$ for every i, $1 \leq i \leq t$. Hence, $\mathrm{Ker}\ \lambda \supseteq \Omega_p(Z)$, contrary to the exactness of the character λ. Thus $\theta_1, \ldots, \theta_t$ are faithful. Now for $\hat{x} \in \tilde{H}^0 \backslash Z(\tilde{H}^0)$, by Lemma 4.3.9 we have $\theta_i(\hat{x}) = 0$, $i = 1, \ldots, t$. This means that $\lambda(\hat{x}) = 0$ and $\chi(\hat{x}) = -1$ (see Lemma 4.2.3). Furthermore, $\tilde{H}^0 / Z(\tilde{H}^0) \cong H^0$ and $Z(\tilde{H}^0)$ acts trivially on $\mathcal{L}$. Therefore we have shown that $\chi(x) = -1$ for $x \in H^0 \backslash \{1\}$ (and $\chi(1) = \dim \mathcal{L}$). Applying Lemma 4.2.6 and setting $|H^0| = p^l$, we get

$$0 = (\chi_{H^0}, 1_{H^0})_{H^0} = \frac{\dim \mathcal{L} - (p^l - 1)}{p^l} = \frac{(n+1)^2 - p^l}{p^l},$$

that is, $l = 2m$ is even, $n = p^m - 1$ and $\chi_{H^0} = \rho_{H^0} - 1_{H^0}$, where ρ_{H^0} is the regular character of H^0.

2) The rest of the proof is devoted to proving that a subgroup $H^0 = \mathbb{Z}_p^{2m}$ with the character $\rho_{H^0} - 1_{H^0}$ on $\mathcal{L}$ is unique in $\mathcal{G} = PSL_{p^m}(\mathbb{C})$.

By Lemma 4.3.8, considered as an $\mathbb{F}_p$-space, H^0 admits a non-degenerate alternating form, namely, the form $(x, y) \mapsto [\hat{x}, \hat{y}]$. Consider a subspace I that is maximal totally isotropic with respect to this form. Then $|I| = p^m$. Because of the isotropy of I, the full preimage $\hat{I}$ is abelian, and so by [SAG] the fixed-point subalgebra $C_{\mathcal{L}}(I)$ contains some Cartan subalgebra $\mathcal{H}$. On the other hand, as H^0 acts on $\mathcal{L}$ with character $\rho_{H^0} - 1_{H^0}$, we must have that $\dim C_{\mathcal{L}}(I) = p^m - 1 = \operatorname{rank}\mathcal{L}$. Hence, $C_{\mathcal{L}}(I) = \mathcal{H}$. Now H^0 centralizes I, therefore H^0 leaves $\mathcal{H}$ fixed, so that $H^0 \subseteq N_{\mathcal{G}}(T)$, where $T = C_{\mathcal{G}}(\mathcal{H})$ is a maximal torus. It is clear that the kernel $\operatorname{Ker}(H^0|_{\mathcal{H}})$ is an abelian subgroup containing I and lying in T. By the maximality of I, we get that $\operatorname{Ker}(H^0|_{\mathcal{H}}) = I$. This means that

$$H^0/I \hookrightarrow W(R) = \mathbb{S}_{p^m}.$$

Moreover, H^0/I acts on $\mathcal{H}$ with character

$$\sum_{\substack{\sigma \in \operatorname{Irr}(H^0) \\ \sigma \neq 1_{H^0}, \sigma_I = 1_I}} \sigma = \sum_{\substack{\sigma \in \operatorname{Irr}(H^0/I) \\ \sigma \neq 1_{H^0/I}}} \sigma = \rho_{H^0/I} - 1_{H^0/I}.$$

In other words, the elementary abelian permutation group $H^0/I = \mathbb{Z}_p^m$ acts regularly on a set of $q = p^m$ symbols $e_1, \dots, e_q$. Here $\{e_1, \dots, e_q\}$ is an orthonormal basis of $\mathbb{R}^q$, and $\{e_i - e_j | 1 \leq i \neq j \leq q\}$ is a root system of type A_{q-1}.

3) As $H^0/I = \mathbb{Z}_p^m$ acts regularly on $\{e_1, \dots, e_q\}$, we can identify this set with the additive group of the field $W = \mathbb{F}_q$, and H^0/I with the group consisting of translations $l_v : u \mapsto u + v$ of W. Set $T_p = \{t \in T | t^p = 1\}$ and $C = C_{T_p}(H^0/I)$. In order to avoid confusion, we rewrite the set $\{e_1, \dots, e_q\}$ in the form $\{e_u | u \in W\}$, and then translations l_u satisfy $l_u(e_v) = e_{u+v}$. Furthermore, each element $f \in T_p$ can be written in the form $f(e_u - e_0) = \varepsilon^{\bar{f}(u)}$, where $\varepsilon = \exp(2\pi i/p)$, $\bar{f} : W \to \mathbb{F}_p$ and $\bar{f}(0) = 0$. Then $f \in C$ if and only if the following sequence of equalities holds:

$$f(e_u - e_0) = f(l_v(e_u - e_0)) = f(e_{u+v} - e_v) = f(e_{u+v} - e_0).(f(e_v - e_0))^{-1} \Leftrightarrow$$

$$\Leftrightarrow \bar{f}(u + v) = \bar{f}(u) + \bar{f}(v),\ \forall u, v \in W.$$

Thus $C = \{f \in T_p | \bar{f} \in W^*\} \cong \mathbb{Z}_p^m$, where $W^* = \operatorname{Hom}_{\mathbb{F}_p}(W, \mathbb{F}_p)$. But $C \supseteq I$, and therefore $I = C = C_{T_p}(H^0/I)$.

4) In this step we show that $C_{\mathcal{G}}(H^0) = H^0$, from which the desired identity $H = H^0$ will follow. Denote $C = C_{\mathcal{G}}(H^0)$. Since $[C, H^0] = 1$, C leaves $\mathcal{H} = C_{\mathcal{L}}(I)$ fixed, that is, $C \subseteq N_{\mathcal{G}}(T)$. Set $D = CT/T$ and $L = H^0T/T \cong H^0/I$. Clearly, $[D, L] = 1$; moreover, since $C \supseteq H^0$, we have $D \supseteq L$. We have shown above that L is a regular abelian permutation group on the symbols $e_1, \dots, e_q$. Hence

it is easy to see that $D = L$. Furthermore, the intersection $C \cap T$ is contained in $C_T(L)$. As is shown in step 3), $C_{T_p}(L) = I$. Here we claim that $C_T(L) \subseteq T_p$. Take $f \in C_T(L)$ and consider a root $\alpha = \alpha_0 = e_v - e_0$ and a translation $l_v \in L$. Set $\alpha_i = (l_v)^i \alpha_0 = e_{(i+1)v} - e_{iv}$ for $0 \leq i \leq p-1$, so that $\alpha_0 + \alpha_1 + \ldots + \alpha_{p-1} = 0$. Hence,

$$1 = f(\alpha_0 + \alpha_1 + \ldots + \alpha_{p-1}) = f(\alpha_0) f(\alpha_1) \ldots f(\alpha_{p-1}) = (f(\alpha_0))^p.$$

This property holds for any root $\alpha = e_v - e_0$. Consequently $f^p = 1$, that is, $f \in T_p$. Thus we have shown that $C \supseteq H^0$, $C \cap T = H^0 \cap T$ and $CT/T = H^0T/T$. In other words, $C = H^0$ as asserted.

6) From now on we can write H instead of H^0. Suppose that $H_1, H_2 \cong \mathbb{Z}_p^{2m}$ are subgroups of $\mathcal{G}$ which act on $\mathcal{L}$ with characters $\rho_{H_i} - 1_{H_i}$, $i = 1, 2$. We have to show that they are conjugate. Choose two maximal totally isotropic subspaces I_1 and I_2 of H_1 and H_2. Since the $C_{\mathcal{L}}(I_i)$ are Cartan subalgebras, without loss of generality one can suppose that $\mathcal{H} = C_{\mathcal{L}}(I_1) = C_{\mathcal{L}}(I_2)$. But H_i/I_i, $i = 1, 2$, are regular elementary abelian permutation groups of degree $q = p^m$, and so they are conjugate (in $W(R) = N_{\mathcal{G}}(T)/T$). In what follows, we shall assume that $H_1/I_1 = H_2/I_2$. In this case, $I_1 = C_{T_p}(H_1/I_1) = C_{T_p}(H_2/I_2) = I_2$. Setting $I = I_1 = I_2$, we consider the subgroup $L = H_1/I = H_2/I$ of $W(R)$. Identify L with a complement L_1 of I in H_1: $H_1 = L_1 \oplus I$. Similarly, $H_2 = L_2 \oplus I$. Then $L_2 = \{\varphi(l)l | l \in L_1\}$, where φ maps L_1 into T. For $x, y \in L_1$ we have $\varphi(xy)xy = \varphi(x)x.\varphi(y)y$, that is,

$$\varphi(xy) = \varphi(x)x\varphi(y)x^{-1} = \varphi(x).(x \circ \varphi(y)),$$

where $\circ$ denotes the natural action of L on T. Hence φ is a 1-cocycle. It can be shown that, for an appropriate choice of L_2 (as a complement of I in H_2), φ is a 1-coboundary. But in that case, H_1 and H_2 are conjugate. Indeed, $\varphi(x) = s^{-1}(x \circ s)$ for some $s \in T$. Hence, $L_2 = \{s^{-1}xs | x \in L_1\}$ and $H_2 = s^{-1}H_1 s$. This completes the proof of Proposition 4.3.10. □

Assume now that $\mathcal{L}$ is a Lie algebra of type C_n admitting an IOD $\mathcal{D}$, and set $\mathcal{G} = \mathrm{Inn}(\mathcal{L}) = \mathrm{Aut}(\mathcal{L}) = PSp_{2n}(\mathbb{C})$, $\hat{\mathcal{G}} = Sp_{2n}(\mathbb{C})$ and $Z = Z(\hat{\mathcal{G}}) = \mathbb{Z}_2$. Assume that the notations G, H, $\hat{G}$ and $\hat{H}$ are of the usual ones. Following the proof of Proposition 4.3.10 and applying Lemma 4.2.3, we obtain:

Proposition 4.3.11. *Suppose that the Lie algebra $\mathcal{L}$ of type C_n admits an IOD $\mathcal{D}$ and that the E-case occurs. Then*

(i) $n = 2^m$ *for some integer m;*

(ii) $H = \mathbb{Z}_2^{2(m+1)}$ *is a Jordan subgroup;*

(iii) $\hat{H} = 2_-^{1+2(m+1)}$ *is an extraspecial 2-group of type $(-)$; moreover, $\hat{H}$ is defined uniquely up to $\mathcal{G}$-conjugacy.* □

4.4 $h+1=r$: the P-case

The analysis of the P-case is based on certain classification results on finite permutation groups.

Proposition 4.4.1. *If G is a (faithful) transitive permutation group of prime degree p, then one of the following assertions holds.*
(i) $\mathbb{Z}_p \subseteq G \subseteq \mathbb{Z}_p.\mathbb{Z}_{p-1}$,
(ii) $\mathbb{A}_p \subseteq G \subseteq \mathbb{S}_p$,
(iii) $p=11$ *and either* $G=L_2(11)$ *or* $G=M_{11}$,
(iv) $p=23$ *and* $G=M_{23}$,
(v) $p=(q^k-1)/(q-1)$, q *is a prime power and* $L_k(q) \subseteq G \subseteq \mathrm{Aut}(L_k(q))$.

Proof. If G is soluble, then $\mathrm{soc}(G)$ is abelian and (i) holds. If G is insoluble, then G is doubly transitive (see [Wie]), and one can appeal to Theorem 1.49 of [Gor 2]. □

Proposition 4.4.2. *If G is a (faithful) doubly transitive permutation group of degree $p+1$, p a prime, then one of the following assertions holds.*
(i) $\mathbb{A}_{p+1} \subseteq G \subseteq \mathbb{S}_{p+1}$,
(ii) $L_2(p) \subseteq G \subseteq PGL_2(p)$,
(iii) $p=11$ *and either* $G=M_{11}$ *or* $G=M_{12}$,
(iv) $p=23$ *and* $G=M_{24}$,
(v) $p=2^d-1$ *for some integer d and* $G=\mathbb{Z}_2^d.G_1$, *where either* $\mathbb{Z}_p \subseteq G_1 \subseteq \mathbb{Z}_p.\mathbb{Z}_d$ *or* $G_1=SL_d(2)$.

Proof. 1) By the O'Nan-Scott Theorem [LPS] G has a unique minimal normal subgroup $H=\mathrm{soc}(G)$. In addition, $C_G(H)=H$ if H is abelian and $C_G(H)=1$ if H is nonabelian. First assume that H is nonabelian. From Table A2, we obtain one of the possibilities (i) – (iv).

2) Now assume that $H=\mathbb{Z}_r^d$ is an elementary abelian r-group. Then $p+1=r^d$, so that $r=2$ and $p=2^d-1$ is a Mersenne prime. As the cases $p=2,3$ are trivial, we shall assume that $p \geq 7$. Let G_1 be a point stabilizer. Because H is a regular permutation group, we must have $G_1 \cap H=1$ and $|G_1|=(G:H)$; that is, $G=H.G_1$ and $G_1 \subseteq \mathrm{Aut}(H)=SL_d(2)$. Clearly, G_1 is a transitive permutation group of prime degree $p=2^d-1$ lying in $SL_d(2)$. Applying Proposition 4.4.1, we obtain three possibilities for G_1, namely, 4.4.1(i), 4.4.1(ii) and 4.4.1(v). As $p=2^d-1$, $\mathbb{A}_p$ cannot be embedded in $SL_d(2)$. Furthermore, let P be a subgroup of order p lying in $S=SL_d(2)$ and in $L=L_k(q)$. Then $N_S(P)=P.\mathbb{Z}_d$ and $N_L(P) \supseteq P.\mathbb{Z}_k$. So, in case 4.4.1(i) we must have $P \subseteq G_1 \subseteq P.\mathbb{Z}_d$. If case 4.4.1(v) occurs, then k divides d. But d is prime, hence in fact we have $k=d$, $q=2$ and $G_1=SL_d(2)$. □

Proposition 4.4.3. *Let $\mathcal{L}$ be a Lie algebra of one of the following types: A_{p-2}, $B_{(p-1)/2}$, $C_{(p-1)/2}$, $D_{(p+1)/2}$ (p a prime); G_2, F_4, E_6, E_7 and E_8. Then the P-case does not occur for any IOD of $\mathcal{L}$.*

Proof. Suppose the contrary: the P-case occurs for an IOD $\mathcal{D}$ of $\mathcal{L}$, that is, the group $G = \mathrm{Aut}(\mathcal{D})$, which lies in $\mathbb{S}_{h+1}$, acts irreducibly on $\mathcal{L}$ (h is the Coxeter number).

1) First we show that G cannot contain $\mathbb{A}_{h+1}$. Suppose that $\mathbb{A}_{h+1} \subseteq G \subseteq \mathbb{S}_{h+1}$. When $h \leq 5$, $\mathbb{A}_{h+1}$ and $\mathbb{S}_{h+1}$ have no irreducible characters of degree equal to $\dim \mathcal{L}$. Suppose that $h \geq 6$, and consider the subgroup $G_1 = St_G(\mathcal{H}_1)$ of index $h+1$ in G. There is just one conjugacy class of subgroups of this index in G, and therefore we can suppose that $\mathbb{A}_h \subseteq G_1 \subseteq \mathbb{S}_h$. As G_1 has no nontrivial abelian normal subgroups, G_1 acts faithfully on $\mathcal{H}_1$, that is, $G_1 \hookrightarrow A(R)$, where R is the root system of $\mathcal{L}$. Clearly, such embeddings are impossible for the Lie algebra $\mathcal{L}$ of type G_2 ($h = 6$); F_4, E_6 ($h = 12$); E_7 ($h = 18$); E_8 ($h = 30$); B_n, C_n ($h = 2n$) or D_n ($h = 2n - 2$). Consider the case where $\mathcal{L}$ is of type A_{h-1}. Let ϕ be the character of the representation Φ of G_1 on $\mathcal{H}_1^*$, and χ the G-character on $\mathcal{L}$. Then $\chi = \mathrm{Ind}_{G_1}^{G}(\bar{\phi})$, $\chi \in \mathrm{Irr}(G)$ and $\phi \in \mathrm{Irr}(G_1)$. We know that Φ embeds G_1 in $A(R) = \mathbb{Z}_2 \times \mathbb{S}_h$. Consider the subgroup $G' = \mathbb{A}_{h+1}$ of G and set $G'_1 = G' \cap G_1$. Clearly, G' acts transitively on the set $\{\mathcal{H}_i\}$, so that $\chi|_{G'} = \mathrm{Ind}_{G'_1}^{G'}(\bar{\phi}')$, where $\phi' = \phi|_{G'_1}$. Furthermore, Φ embeds G'_1 in $\mathbb{A}_h \subseteq W(R)$. Let $\Psi : G'_1 \to \mathbb{A}_h$ be the embedding corresponding to the natural 2-transitive permutation representation, with character $1_{G'_1} + \theta$, say, where $\theta \in \mathrm{Irr}(G'_1)$. If $h \neq 6$ then $\mathrm{Aut}(\mathbb{A}_h) = \mathbb{S}_h$ (see [Gor 2]), and therefore one can find an element $s \in \mathbb{S}_h$ such that $\Phi(g) = s\Psi(g)s^{-1}$ for $g \in G'_1$. Note that θ can be extended to a character of $\mathbb{S}_h$, and hence $\phi' = \theta$ and $\bar{\phi}' = \theta$. Moreover, $1_{G'_1} + \theta$ is the restriction to G'_1 of some irreducible character $\tilde{\theta}$ of G'. Consequently,

$$(\chi_{G'}, \tilde{\theta})_{G'} = (\mathrm{Ind}_{G'_1}^{G'}(\bar{\phi}'), \tilde{\theta})_{G'} = (\theta, 1_{G'_1} + \theta)_{G'_1} = 1,$$

and also $\deg \chi = h^2 - 1 > 2 \deg \tilde{\theta} = 2(h+1)$. Taking the condition $(G : G') \leq 2$ in account and applying Lemma 4.2.4, we obtain that $(\chi, \chi)_G > 1$, a contradiction.

If $h = 6$ one has to modify the arguments slightly, because $\mathrm{Out}(\mathbb{A}_6) = \mathbb{Z}_2^2$. In fact we have proved:

Lemma 4.4.4. *Suppose that the Lie algebra $\mathcal{L}$ admits an IOD $\mathcal{D}$ and that the P-case occurs. Then $G = \mathrm{Aut}(\mathcal{D})$ does not contain $\mathbb{A}_{h+1}$.* □

2) Let $\mathcal{L}$ be of type E_8. Then $h = 30$, and either $G \subseteq \mathbb{Z}_{31}.\mathbb{Z}_{30}$ or $G \in \{SL_3(5), SL_5(2)\}$. None of these groups has irreducible characters of degree 248. The cases G_2, F_4, E_6 and E_7 are similar.

3) Let $\mathcal{L}$ be of type A_{p-2} (p a prime, $p \geq 5$). By Proposition 4.4.1 and Lemma 4.4.4, we have to examine two cases:

a) $p = 11$, $G \in \{L_2(11), M_{11}\}$; $p = 23$, $G = M_{23}$,
b) $p = (q^k - 1)/(q - 1)$, $L_k(q) \subseteq G \subseteq \mathrm{Aut}(L_k(q))$.

Case a) is immediately eliminated, as the groups $L_2(11)$, M_{11} and M_{23} have no irreducible characters of degree $p(p-2)$. Consider case b), and note first of all that the following assertion holds.

Lemma 4.4.5. *Suppose that q is a power of a prime r and that $p = (q^k - 1)/(q - 1)$ is prime for some $k \geq 2$. Then*
(i) *k is prime and k does not divide $q - 1$;*
(ii) *$q = r^{k^m}$ for some integer $m \geq 0$, and $p \equiv 1 \pmod{k}$.* □

Concerning these integers k, q we have

$$SL_k(q) \cong L_k(q) = PGL_k(q), \ \mathrm{Aut}(L_k(q)) = (L_k(q).\mathbb{Z}_{k^m}).\mathbb{Z}_2.$$

As $\dim \mathcal{L} = p(p-2) \equiv -1 \pmod{2k}$, by Clifford's Theorem the normal subgroup $SL_k(q)$ of G acts irreducibly on $\mathcal{L}$, and we can suppose without loss of generality that $G = SL_k(q)$. Moreover, $k > 2$ because when $k = 2$ we have $p = q + 1$ and $|SL_2(q)| = q(q^2 - 1) < \{p(p-2)\}^2$. For similar reasons, we must have $(k, q) \neq (3, 2), (3, 3)$. Set $G_1 = St_G(\mathcal{H}_1)$. Since the index $(G : G_1) = (q^k - 1)/(q - 1)$ is coprime to q, by the Borel-Tits Theorem we can assume that

$$G_1 = \left\{ \begin{pmatrix} \det a^{-1} & * \\ 0 & a \end{pmatrix}, \ a \in GL_{k-1}(q) \right\} = E.(F.S).\mathbb{Z}_{q-1},$$

where $E = \mathbb{F}_q^{k-1}$, $S = L_{k-1}(q)$ and $F \subseteq \mathbb{Z}_{q-1}$. By Lemma 4.2.2, G_1 has a 2-transitive permutation representation τ of degree $p - 1$ with soluble kernel $\mathrm{Ker}\,\tau$. First we show that $E \subseteq \mathrm{Ker}\,\tau$. Indeed, if $E \not\subseteq \mathrm{Ker}\,\tau$, then

$$E/(E \cap \mathrm{Ker}\,\tau) \cong E.\mathrm{Ker}\,\tau/\mathrm{Ker}\,\tau \lhd G_1/\mathrm{Ker}\,\tau,$$

and so $E/(E \cap \mathrm{Ker}\,\tau)$ is a transitive abelian (and therefore regular) permutation group of degree $p - 1$. This implies that the order $|E| = q^{k-1}$ is divisible by $p - 1 = q + q^2 + \ldots + q^{k-1}$, a contradiction. Similarly, $F \subseteq \mathrm{Ker}\,\tau$. Thus $G_2 = S.\mathbb{Z}_{q-1} = L_{k-1}(q).\mathbb{Z}_{q-1}$ has a 2-transitive permutation representation τ of degree $t = p - 1$ with soluble kernel. Of course, S is simple, and so is a minimal normal subgroup of $G_3 = G_2/\mathrm{Ker}\,\tau$, which means that $\mathrm{soc}(G_3) = S$. According to Table A2, one of the following cases occurs.
a) $t = p - 1 = (q^{k-1} - 1)/(q - 1)$;
b) $t = 5, 6$, $S = L_2(5) \cong L_2(4)$;
c) $t = 6$, $S = L_2(9)$;
d) $t = 11$, $S = L_2(11)$;
e) $t = 28$, $S = L_2(8)$;
f) $t = 7, 8$, $S = L_2(7) \cong L_3(2)$;

g) $t = 8$, $S = L_4(2)$.

All these possibilities are eliminated, because $t+1 = (q^k - 1)/(q-1)$ must be prime.

4) The cases $B_{(p-1)/2}$, $C_{(p-1)/2}$ and $D_{(p+1)/2}$ are treated similarly. □

4.5 $h+1 = r$: classification of IODs

We are now able to prove the main result of this chapter:

Theorem 4.5.1. *The Lie algebras of types A_{p-2} (p a prime, $p \neq 2^d + 1$), $C_{(p-1)/2}$ (p a prime, $p \geq 7$), and E_7 have no IODs. Furthermore, the number of* $\mathrm{Aut}(\mathcal{L})$*-conjugacy classes of IODs for the Lie algebras $\mathcal{L}$ of type G_2, F_4, E_6 or E_8 is equal to* 1, 1, 1 *or* 2*, respectively.*

The remaining part of the paragraph is devoted to the proof of this theorem. Let $\mathcal{L}$ be a Lie algebra of one of types mentioned in Theorem 4.5.1, admitting an IOD

$$\mathcal{D} : \mathcal{L} = \bigoplus_{i=1}^{h+1} \mathcal{H}_i,$$

and set $G = \mathrm{Aut}(\mathcal{D})$. Then, by Propositions 4.3.4 and 4.4.3, the S- and the P-cases are impossible. Furthermore, by Propositions 4.3.7, 4.3.10 and 4.3.11, the E-case can occur only if $\mathcal{L}$ is of types A_{p^n-1} (p a prime), C_{2^n}, G_2, F_4, E_6, E_8 (and, possibly, B_n, D_n). Consequently, we have just shown the absence of IODs for types A_{p-2} ($p \neq 2^d + 1$) and E_7. The case of $C_{(p-1)/2}$ is treated in Proposition 2.1.8. We claim now that in the cases G_2, F_4, E_6 and E_8, IODs must coincide with the standard decompositions constructed from the corresponding Jordan subgroups (see §3.1). The arguments given will follow a uniform scheme. Therefore, we carry out the proof here only in cases when additional difficulties appear.

A. The case where $\mathcal{L}$ is of type E_8, and $H = \mathbb{Z}_5^3$.

Here $G = \mathrm{Aut}(\mathcal{D})$ is contained in $N_{\mathcal{G}}(H) = H.SL_3(5)$ (see [Ale 1]); furthermore, this last extension is split (see [CoG]). As G acts irreducibly on $\mathcal{L}$, 248 divides the order $|G|$. Using the description of maximal subgroups in $SL_3(5)$ given in [ATLAS], we can see that in fact G must be $H.SL_3(5)$. Consider $G_1 = St_G(\mathcal{H}_1)$ and $P_1 = \mathbb{Z}_5^3.5_+^{1+2} \in \mathrm{Syl}_5(G_1)$. Set $Q_1 = P_1 \cap T_1$, where T_1 is the maximal torus corresponding to $\mathcal{H}_1$. We would like to show that $|Q_1 \cap H| = 5^2$. Since $C_{\mathcal{L}}(H) = 0$, we have $H \not\subseteq Q_1$ and so $|Q_1 \cap H| < 5^3$. Furthermore, 5^3 does not divide $|W(E_8)|$, which implies that $Q_1 \cap H \neq 1$. Suppose that $|Q_1 \cap H| = 5$. Then

$P_1/Q_1 \cong P_1T_1/T_1 \hookrightarrow W(E_8)$ and so $|P_1/Q_1| \le 5^2$, that is, $|Q_1| \ge 5^4$. Hence,

$$|Q_1H| = \frac{|Q_1|.|H|}{|Q_1 \cap H|} \ge \frac{5^4.5^3}{5} = 5^6.$$

In other words, $Q_1H = P_1$. In particular, the abelian group $Q_1/(Q_1 \cap H)$ is isomorphic to the extraspecial group $P_1/H = 5_+^{1+2}$, a contradiction.

Thus $Q_1 \cap H = \mathbb{Z}_5^2$, that is, $H \cap T_1 = \mathbb{Z}_5^2$. Since H acts on $\mathcal{L}$ with character $2(\rho_H - 1_H)$ (see the proof of Proposition 4.3.7), we must have $\dim C_{\mathcal{L}}(H \cap T_1) = 8$. But $C_{\mathcal{L}}(H \cap T_1) \supseteq \mathcal{H}_1$, so in fact $C_{\mathcal{L}}(H \cap T_1) = \mathcal{H}_1$. Consequently, the IOD $\mathcal{D}$ is of the form

$$\mathcal{L} = \bigoplus_{(H:H')=5} C_{\mathcal{L}}(H'),$$

that is, $\mathcal{D}$ is just the standard IOD constructed from the Jordan subgroup $\mathbb{Z}_5^3$ (see formula (3.1)).

B. The case where $\mathcal{L}$ is of type E_8 and $H = \mathbb{Z}_2^5$.

Here we know [Ale 1] that $C = C_{\mathcal{G}}(H) = 2^{5+10}$ and $N = N_{\mathcal{G}}(H) = C \circ S$, where $S = SL_5(2)$; moreover, the extension $N = C \circ S$ is non-split [CoG] and the standard IOD is an MOD. It is shown in [Gri 2] that C is a special 2-group of order 2^{15} with the following properties:

(i) $C' = Z(C) = \Phi(C) = H$;

(ii) $C = C_N(H)$, $N/C \cong S$;

(iii) N has a factorization $N = CD$, $C \cap D = H$, where $D = H \circ S$ is the Dempwolff group [Demp 1];

(iv) $C/H = \mathbb{Z}_2^{10}$; and H and C/H are irreducible $\mathbb{F}_2S$-modules.

Firstly we establish an auxiliary statement which is needed in Chapter 13.

Lemma 4.5.2. *A subgroup $G \subseteq N = N_{\mathcal{G}}(H)$ acts absolutely irreducibly on $\mathcal{L}$ if and only if one of the following assertions holds:*

(i) $G \supseteq D = H \circ S = \mathbb{Z}_2^5 \circ SL_5(2)$.

(ii) $I = C.\mathbb{Z}_{31} \subseteq G \subseteq J = N_N(I) = I.\mathbb{Z}_5$.

Moreover, all subgroups of type D or I in N are conjugate.

Proof. 1) The irreducibility of D and I on $\mathcal{L}$.

The irreducibility of D on $\mathcal{L}$ was noted by Thompson [Tho 2]. Consider now the standard OD,

$$\mathcal{D}' : \mathcal{L} = \bigoplus_{i=1}^{31} \mathcal{H}_i'$$

constructed from the Jordan subgroup H in §3.1. Then C leaves each subalgebra $\mathcal{H}_i'$ fixed. Denote by K_i and ρ_i the kernel and the character respectively of $N_i =$

$St_N(\mathcal{H}'_i)$ on $\mathcal{H}'_i$, and set $C_i = C \cap K_i$. By Clifford's Theorem, C acts on $\mathcal{L}$ with character $\chi_C = e\sum_{j=1}^{t}\vartheta_j$, where $e \in \mathbb{N}$ and $\vartheta_j \in \mathrm{Irr}(C)$. Clearly,

$$[C/C_i, C/C_i] = C'C_i/C_i = HC_i/C_i \cong H/(C_i \cap H) = \mathbb{Z}_2.$$

In particular, C/C_i is nonabelian and $|C/C_i| \geq 2^3$. Furthermore, $A_i = N_i/K_i$ is embedded in $W(E_8)$ and $\mathbb{Z}_2 = H/(H \cap K_i) \cong HK_i/K_i \lhd A_i$. Hence, the group $B_i = A_i/\mathbb{Z}_2$ is embedded in $W(E_8)/\mathbb{Z}_2 = O_8^+(2)$; moreover, B_i has the nontrivial normal 2-subgroup $(C/C_i)/\mathbb{Z}_2$. Note that $N_i = C.\mathbb{Z}_2^4.\mathbb{A}_8$, and so B_i has a composition factor $\mathbb{A}_8$. Looking over the list of maximal 2-local subgroups of $O_8^+(2)$, we obtain that $\mathbb{Z}_2^6.\mathbb{A}_8 \subseteq B_i \subseteq \mathbb{Z}_2^6.\mathbb{S}_8$ and, in addition, $\mathbb{Z}_2^6$ is a minimal normal subgroup of B_i. Therefore, $(C/C_i)/\mathbb{Z}_2 = \mathbb{Z}_2^6$, $|C/C_i| = 2^7$ and $Z(C/C_i) = \mathbb{Z}_2$. Thus C/C_i is an extraspecial 2-group of order 2^7: $C/C_i = 2_\pm^{1+6}$. But the faithful representation of C/C_i of degree 8 on $\mathcal{H}'_i$ is real, and so $C/C_i = 2_+^{1+6}$. In particular, C/C_i is irreducible on $\mathcal{H}'_i$. We conclude that the character $\rho_i|_C$ is irreducible, the kernel $\mathrm{Ker}\,\rho_i = C_i$ is of order 2^8, and $H \cap C_i = \mathbb{Z}_2^4$. Consequently $e = 1$, $t = 31$ and

$$\{\vartheta_1, \ldots, \vartheta_t\} = \{\rho_1|_C, \ldots, \rho_{31}|_C\},$$

which means finally that $I = C.\mathbb{Z}_{31}$ is irreducible on $\mathcal{L}$. Using Thompson's construction of the group N, one can see that in fact $C_i = \mathbb{Z}_2^8$.

2) The conjugacy of subgroups of type I in N follows by Sylow's Theorem. To convince oneself of the conjugacy of subgroups of type D, it is sufficient to prove the conjugacy of any two subgroups $S_1 \cong S_2 \cong SL_5(2)$ in $N/H = \mathbb{Z}_2^{10}.S$. Since $\mathrm{Syl}_{31}(S_i) \subseteq \mathrm{Syl}_{31}(N/H)$, we can suppose that $S_1 \cap S_2 \supseteq P \cong \mathbb{Z}_{31}$. Then

$$2^{10}|S_1| = |N/H| \geq |S_1S_2| = \frac{|S_1|.|S_2|}{|S_1 \cap S_2|},$$

that is, $31m = |S_1 \cap S_2| \geq 2^{-10}|S_2| = 3^2.5.7.31$. Looking over the list of maximal subgroups of $SL_5(2)$, we have $S_1 \cap S_2 = S_2$; in other words, $S_1 = S_2$ as stated.

3) Here we show that any element φ of order 31 of N acts fixed-point-freely on H and on C/H. Note that C/H is isomorphic to $\wedge^2(H^*)$ as an $SL_5(2)$-module. Namely,

$$(\wedge^2(H^*)) \wedge (\wedge^2(H^*)) \cong \wedge^4(H^*) \cong H.$$

Thus, if ζ is a 31-st primitive root of unity in the algebraic closure $\bar{\mathbb{F}}_2$ of $\mathbb{F}_2$, then φ has spectrum

$$\{\zeta, \zeta^2, \zeta^4, \zeta^8, \zeta^{16}\}$$

and

$$\{\zeta^{-3}, \zeta^{-5}, \zeta^{-9}, \zeta^{-17}, \zeta^{-6}, \zeta^{-10}, \zeta^{-18}, \zeta^{-12}, \zeta^{-20}, \zeta^{-24}\},$$

on H and on C/H, respectively.

In the rest of the proof, we assume that G is a subgroup of N that is irreducible on $\mathcal{L}$.

4) Firstly suppose that $G \supseteq C$. Then either $G = N$ or $I \subseteq G \subseteq N_N(I) = J = (C.\mathbb{Z}_{31}).\mathbb{Z}_5$. Indeed, $248 = \dim \mathcal{L}$ divides $|G|$, and so 31 divides $|G|$. It remains to inspect the list of subgroups of $SL_5(2) = N/C$ whose orders are divisible by 31.

5) Secondly, we prove that if $G \supseteq H$ but $G \not\supseteq C$ then $G \cong D$. From 3) it follows that $(G \cap C)/H$ is 1, $\mathbb{Z}_2^5$ or $\mathbb{Z}_2^{10}$. Furthermore, in step 4) we have shown that $CG = N$ or $I \subseteq CG \subseteq J$. Suppose that $CG \subseteq J$. Then

$$155.|C| = |J| \geq |CG| = \frac{|C|.|G|}{|C \cap G|},$$

that is, $|C \cap G| \geq |G|/155 > 248^2/155 > 2^8$. Hence, $C \cap G = 2^{5+5}$. On the other hand, $C \cap G$ is irreducible on $\mathcal{H}_i'$. Indeed, $G \cap I \lhd G$, $G/(G \cap I) \cong IG/I \subseteq J/I = \mathbb{Z}_5$, and so by Clifford's Theorem, $G \cap I$ is irreducible on $\mathcal{L}$. Furthermore, $(G \cap I)C = I$, $\mathbb{Z}_{31} = I/C = C(G \cap I)/C \cong (G \cap I)/(C \cap G)$, that is, $G \cap I = (C \cap G).\mathbb{Z}_{31}$. The subgroup $\mathbb{Z}_{31}$ permutes the set $\{\mathcal{H}_i'\}_{i=1}^{31}$ regularly, so that $C \cap G$ is irreducible on each $\mathcal{H}_i'$. Thus, the group $A = C \cap G = 2^{5+5}$ has an irreducible character of degree 8. According to [Ser 3], $8^2 = 2^6$ divides $(A : Z(A))$, that is, $|Z(A)| \leq 2^4$. Meanwhile $Z(A) \supseteq H = \mathbb{Z}_2^5$, a contradiction. Hence $CG = N$. Set $\bar{C} = C/H$, $\bar{G} = G/H$ and $\bar{N} = N/H$. Then $\bar{C}\bar{G} = \bar{N}$, and in particular, $\bar{C} \cap \bar{G} \lhd \bar{N}$. But $\bar{C}$ is a minimal normal subgroup of $\bar{N} = \mathbb{Z}_2^{10}.S$ and $\bar{G} \not\supseteq \bar{C}$, so that $\bar{C} \cap \bar{G} = 1$, and $\bar{G} \cong \bar{C}\bar{G}/\bar{C} = \bar{N}/\bar{C} = S$. In other words, $G = H.S \cong D$, as required.

6) Finally, we show that G always contains H. Suppose the contrary. As $G \cap H \lhd G$, it follows from the results of 3) that $G \cap H = 1$ and $HG = H.G$. If $HG \not\supseteq C$, then by 4), $HG = D$, that is, D splits over $H = O_2(D)$, a contradiction. So $HG \supseteq C$. In particular, C splits over $H = Z(C)$, again a contradiction. □

Now we note that if $G = \mathrm{Aut}(\mathcal{D}) \supseteq C$, where $\mathcal{D}$ is an OD of the Lie algebra $\mathcal{L}$ of type E_8, then $\mathcal{D}$ is just the standard MOD. Indeed, as a normal 2-subgroup of G, C leaves each Cartan subalgebra $\mathcal{H}_i$ fixed. But, as is shown in the proof of Lemma 4.5.2, $\chi_C = \sum_{i=1}^{31} \vartheta_i$, where $\vartheta_i \in \mathrm{Irr}(C)$ and ϑ_i is the C-character afforded by $\mathcal{H}_i'$; moreover, $\vartheta_i \neq \vartheta_j$ whenever $i \neq j$. Hence, $\{\mathcal{H}_i\}_{i=1}^{31} \equiv \{\mathcal{H}_i'\}_{i=1}^{31}$. Bearing Lemma 4.5.2 in mind, we only need to consider the case $G = D$. This case is treated in the following statement, which is also needed in Chapter 13.

Lemma 4.5.3. *Let $\mathcal{D} : \mathcal{L} = \oplus_{i=1}^m V_i$ be a decomposition of the $\mathbb{C}$-space $\mathcal{L}$ into a direct sum of its (proper) subspaces that is invariant under the Dempwolff subgroup $D = \mathbb{Z}_2^5 \circ SL_5(2)$. Then $m = 31$ and $\mathcal{D}$ is an MOD.*

Proof. Set $D = H \circ S$, $H = O_2(D)$, $D_1 = St_D(V_1)$ and $S_1 = HD_1/H$. Then $(S : S_1) = (D : HD_1) = (D : D_1)/(H : H \cap D_1)$, and so $(S : S_1)$ divides both m and 248. (Of course, the irreducibility of D on $\mathcal{L}$ implies that D permutes the m components V_i of $\mathcal{D}$ transitively). The group $S = SL_5(2)$ has the following maximal subgroups: $\mathbb{Z}_2^4.SL_4(2)$, $\mathbb{Z}_2^6.(\mathbb{S}_3 \times SL_3(2))$, and $\mathbb{Z}_{31}.\mathbb{Z}_5$ of index 31, 155 and 64512 respectively. Hence, $S_1 = S$ or $S_1 \subseteq P = \mathbb{Z}_2^4.SL_4(2)$.

1) Note that $S_1 \subseteq P$. In fact, if $S_1 = S$, then $HD_1 = D$. Here we have $H \cap D_1 \lhd D_1$ and $[H, H \cap D_1] = 1$, so that $H \cap D_1 \lhd HD_1 = D$. By the minimality of H, we must have either $H \cap D_1 = 1$ or $H \subseteq D_1$. In the former case $D = H.D_1$ splits over H, a contradiction. In the latter case $D_1 = HD_1 = D$, $m = 1$ and $V_1 = \mathcal{L}$, again a contradiction.

2) In this step we show that either $S_1 = P$, or $S_1 = \mathbb{Z}_2^4.\mathbb{A}_7$, $D_1 \supseteq H$, and $m = 248$. Suppose that $S_1 \neq P$. Then the index $(P : S_1) = (S : S_1)/31$ divides $248/31 = 8$, on the one hand. On the other hand, any permutation representation of P of degree less than 16 is trivial on $\mathbb{Z}_2^4 = O_2(P)$, and so $S_1 \supseteq O_2(P)$. Furthermore, $P/O_2(P) \cong \mathbb{A}_8$ has no proper subgroups of index less than 8. Hence, $S_1 = O_2(P).\mathbb{A}_7$, $m = 248$ and $H \subseteq D_1$.

3) The group P, as a subgroup of index 31 in S, is either a centralizer $C_S(h)$ for some $h \in H \backslash \{1\}$, or a normalizer $N_S(W)$ for some subgroup W of index 2 in H. On this step we establish that $P = N_S(W)$. Assume the contrary, so that $P = C_S(h)$. Note that $D_1 \supseteq H$. This is certainly true if $S_1 \neq P$ (see 2)). If $S_1 = P$, then $(H : H \cap D_1) = m/(S : S_1)$ divides 8. But $S_1 = P$ acts on $H \backslash \{1\}$ with orbits $\{h\}$ and $H \backslash < h >$, so $H \cap D_1 = H$, and thus $H \lhd D_1$.

It is well known that the embedding of $\mathbb{A}_7$ in $SL_4(2)$ acts transitively on nonzero vectors of the natural $\mathbb{F}_2$-module for $SL_4(2)$. Therefore, S_1 and D_1 have the orbits $\{h\}$ and $H \backslash < h >$ on $H \backslash \{1\}$. But in that case, D_1 acts on $\mathrm{Irr}(H) \backslash \{1_H\}$ with two orbits, namely,

$$\{\sigma \in \mathrm{Irr}(H) \backslash \{1_H\} | \sigma(h) = 1\}, \text{ of length 15,}$$

$$\{\sigma \in \mathrm{Irr}(H) | \sigma(h) \neq 1\}, \text{ of length 16.}$$

On the other hand, applying Clifford's Theorem to the action of D_1 on V_1, we see that D_1 fixes a certain set of nontrivial H-characters of cardinality not greater than $\dim V_1 \leq 8$, a contradiction.

4) We have shown that $S_1 \subseteq P = N_S(W)$. Set $\mathcal{H}_1 = C_{\mathcal{L}}(W)$. As above, it follows from the results of 2) that D_1 has two orbits, of length 15 and 16, on $H \backslash \{1\}$, namely $W \backslash \{1\}$ and $H \backslash W$; furthermore, D_1 has two orbits on $\mathrm{Irr}(H) \backslash \{1_H\}$, namely:

$$\{\sigma \in \mathrm{Irr}(H) \backslash \{1_H\} | \sigma_W = 1_W\}, \text{ of length 1,}$$

$$\{\sigma \in \mathrm{Irr}(H) | \sigma_W \neq 1_W\}, \text{ of length 30.}$$

Note that $D_1 \supseteq H$. This is certainly true if $S_1 \neq P$ (see 2)). If $S_1 = P$, then $(H : H \cap D_1) = m/(S : S_1)$ divides 8 and so $|H \cap D_1| \geq 4$. Hence, either $H \cap D_1 = W$ or $H \subseteq D_1$. In the former case, $W \lhd D_1$; moreover, we have shown that D_1 is transitive on the set $\mathrm{Irr}(W) \backslash \{1_W\}$, and $\dim V_1 < 15$. Hence W is trivial on V_1; in other words, $V_1 \subseteq \mathcal{H}_1$, which implies that $H \subseteq D_1$, a contradiction. Thus $H \lhd D_1$. Applying Clifford's Theorem to the representation of D_1 on V_1, we conclude that H acts by scalars on V_1, $V_1 \subseteq \mathcal{H}_1$ and $D_1 \subseteq St_D(\mathcal{H}_1)$.

5) Set $N = N_{\mathcal{G}}(H)$, $N_1 = St_N(\mathcal{H}_1)$, $Q_1 = St_D(\mathcal{H}_1)$, and let K_1 be the kernel of N_1 on $\mathcal{H}_1$. At this point we show that the group $\tilde{Q}_1 = Q_1/(Q_1 \cap K_1)$ is in fact a double cover of $\mathbb{A}_8$. Clearly, $|Q_1| = 2^9|\mathbb{A}_8|$. Furthermore, $W = H \cap K_1 \lhd Q_1$, and so $\mathbb{Z}_2 = H/(H \cap K_1) \lhd \tilde{Q}_1$, $\mathbb{Z}_2 \subseteq Z(\tilde{Q}_1)$. In addition to $\mathbb{Z}_2$, $\tilde{Q}_1$ has another composition factor, namely $\mathbb{A}_8 = Q_1/O_2(Q_1)$. In item 1) of the proof of Lemma 4.5.2, we proved that the group $\tilde{C}_1 = C/(C \cap K_1)$ is isomorphic to 2_+^{1+6}. In addition $C \lhd N$. Hence we get a natural homomorphism

$$\pi : \tilde{Q}_1 \to \mathrm{Aut}(\tilde{C}_1)' = \mathbb{Z}_2^6 \cdot \mathbb{A}_8,$$

and moreover, this last group does split over its minimal normal subgroup $\mathbb{Z}_2^6$, as is shown in [Gri 2]. From the irreducibility of C on $\mathcal{H}_1$ it follows that $\mathrm{Ker}\,\pi = \mathbb{Z}_2$. Thus, the group $\mathrm{Im}\pi = \tilde{Q}_1/\mathbb{Z}_2$ has $\mathbb{A}_8$ as composition factor. Moreover, $|\mathrm{Im}\pi| \leq (Q_1 : H) = 2^4|\mathbb{A}_8|$. Hence $\mathrm{Im}\pi = \mathbb{A}_8$ and $\tilde{Q}_1 = \mathbb{Z}_2.\mathbb{A}_8$. But $\mathbb{Z}_2 \times \mathbb{A}_8$ has no irreducible characters of degree 8, and so $\tilde{Q}_1 = \mathbb{Z}_2 \circ \mathbb{A}_8$ as asserted. In fact, $Q_1 \cap K_1 = \mathbb{Z}_4^4$.

6) Consequently, the representation of Q_1 on $\mathcal{H}_1$ is in fact a faithful irreducible representation of degree 8 of $\mathbb{Z}_2 \circ \mathbb{A}_8$. As is shown in [Gow 3], this representation is primitive, that is, it cannot be induced from proper subgroups. Hence, $V_1 = \mathcal{H}_1$, $D_1 = Q_1$, $m = 31$ and $\mathcal{D}$ is just the standard MOD. □

This completes the analysis of case B.

C. The case where $\mathcal{L}$ is of type E_6 and $H = \mathbb{Z}_3^3$.

1) Embed a Lie algebra $\mathcal{L}_0$ of type F_4 in $\mathcal{L}$. It is known [Ale 1] that every automorphism of $\mathcal{L}_0$ extends to an automorphism of $\mathcal{L}$, and in addition that a Jordan subgroup $H = \mathbb{Z}_3^3$ of $\mathrm{Aut}(\mathcal{L}_0) = \mathcal{G}_0$ becomes a Jordan subgroup of $\mathcal{G} = \mathrm{Aut}(\mathcal{L})$. Furthermore, $C = C_{\mathcal{G}}(H) = 3^{3+3}$ and $N = N_{\mathcal{G}}(H) = C.S$, where $S = SL_3(3)$. Since $H^2(SL_3(3), \mathbb{F}_3^3) = 0$ (see [Sah]), the last extension splits. Concerning the group C, we have the following properties:

a) $C' = Z(C) = H$;

b) $\exp(C) = 3$;

c) H and C/H are irreducible $\mathbb{F}_3 S$-modules.

Let V and V^* be irreducible 3-dimensional $\mathbb{F}_3 S$-modules. Then C can be identified with $\{(u, x) | u \in V, x \in V^*\}$, where $(u, x).(v, y) = (u + v, x + y + u \wedge v)$. Here we identify $V \wedge V = \wedge^2(V)$ with V^*. Moreover, $H = \{(0, x) | x \in V^*\}$.

As in case B, one can show that $CG = N$, and either $G = N$ or $G = H.S$; moreover, any two subgroups of type $H.S$ in N are conjugate. One of these subgroups, namely $N_{\mathcal{G}_0}(H)$, leaves $\mathcal{L}_0$ fixed. Hence, any subgroup of type $H.S$ acts reducibly on $\mathcal{L}$. This proves that $G = N$.

2) As always, set $G_1 = St_G(\mathcal{H}_1)$, so that $G_1 \supseteq C$. Indeed, G acts primitively on $\{\mathcal{H}_1, \ldots, \mathcal{H}_{13}\}$, and C is a normal 3-subgroup of G. We have to show that $|H \cap T_1| = 3^2$, where T_1 is the maximal torus corresponding to $\mathcal{H}_1$, and this

implies immediately that $\mathcal{D}$ coincides with the standard OD. Since $C_{\mathcal{L}}(H) = 0$, it follows that $H \not\subseteq T_1$. Suppose that $|H \cap T_1| \leq 3$. Taking $P_1 \in \mathrm{Syl}_3(G_1)$, one can suppose that $P_1 = C.3_+^{1+2}$. Note that the subgroup $Q_1 = P_1 \cap T_1$ is abelian. Furthermore, $P_1/Q_1 \cong P_1T_1/T_1 \hookrightarrow W(E_6)$, and so $|P_1 \cap Q_1| \leq 3^4$, $|Q_1| \geq 3^5$. Next, consider the subgroup $C \cap Q_1$. Clearly, $Q_1/(C \cap Q_1) \cong Q_1C/C$ is an abelian subgroup of $P_1/C = 3_+^{1+2}$, so that $|Q_1/(C \cap Q_1)| \leq 3^2$ and $|C \cap Q_1| \geq 3^3$. Using the explicit construction of the group C, one can show that $(A : A \cap H) \leq 3$ for every abelian subgroup A of C. Applying this remark to the subgroup $C \cap Q_1$, we obtain

$$|C \cap Q_1| \leq 3|C \cap Q_1 \cap H| = 3|H \cap T_1| \leq 3^2,$$

a contradiction.

This completes the analysis of case C and the proof of Theorem 4.5.1. □

4.6 A characterization of the multiplicative orthogonal decompositions of the Lie algebra of type D_4

The aim of this section is to prove the following statement.

Theorem 4.6.1. *Every irreducible orthogonal decomposition of the Lie algebra of type D_4 is multiplicative.*

Let $\mathcal{L}$ be a simple complex finite-dimensional Lie algebra of type D_4. We constructed an IOD of $\mathcal{L}$ in §2.4, and it turned out be an MOD. Furthermore, the $\mathrm{Aut}(\mathcal{L})$-conjugacy of MODs is proved in §3.3.

We introduce the following notation, which differs slightly from that used before. Let $\mathcal{L}$ be the Lie algebra of skew-symmetric complex matrices of degree 8, considered together with the natural action on the orthogonal space $V = \mathbb{C}^8$. Furthermore set $\mathcal{A} = \mathrm{Aut}(\mathcal{L})$, $\mathcal{G} = PO_8(\mathbb{C})$ and $\mathcal{G}^0 = \mathrm{Inn}(\mathcal{L}) = PSO_8(\mathbb{C})$. Suppose that $\mathcal{L}$ admits an IOD $\mathcal{D} : \mathcal{L} = \oplus_{i=1}^{7} \mathcal{H}_i$ with automorphism group $\tilde{G} = \mathrm{Aut}(\mathcal{D})$. Set $G = \tilde{G} \cap \mathcal{G}$, $G^0 = \tilde{G} \cap \mathcal{G}^0$, $\hat{\mathcal{G}} = O_8(\mathbb{C})$, and let $\hat{G}$ be the full preimage of G in $\hat{\mathcal{G}}$, χ the $\tilde{G}$-character afforded by $\mathcal{L}$, and λ be the $\hat{G}$-character afforded by V. Finally, set $\tilde{G}_1 = St_{\tilde{G}}(\mathcal{H}_1)$, $G_1 = G \cap \tilde{G}_1$, and let ρ be the $\tilde{G}_1$-character afforded by $\mathcal{H}_1$. The following statement holds by analogy with Lemma 4.2.4.

Lemma 4.6.2. *Under the above assumptions, one of the following assertions holds.*

(i) *G is irreducible on $\mathcal{L}$,*

(ii) *$\tilde{G}/G^0 = \mathbb{S}_3$, $G/G^0 = \mathbb{Z}_2$ and $\chi|_{G^0} = 2\zeta$ for some $\zeta \in \mathrm{Irr}(G^0)$,*

(iii) *$G = G^0$, $\tilde{G}/G = \mathbb{Z}_2$ and $\chi|_G = \zeta_1 + \zeta_2$ for some $\zeta_1, \zeta_2 \in \mathrm{Irr}(G)$, $\zeta_1 \neq \zeta_2$.*

Proof. By Clifford's Theorem, we have $\chi|_{G^0} = e\sum_{i=1}^{t} \zeta_i$ for some $e, t \in \mathbb{N}$ and $\zeta_i \in \mathrm{Irr}(G^0)$, $i = 1, \dots, t$; here $\zeta_1, \dots, \zeta_t$ are pairwise different. Suppose that assertion (i) does not hold. Then, in particular, $et > 1$. On the other hand, et divides $\gcd(|\tilde{G}/G^0|, \dim \mathcal{L})$, which is not greater than 2. Hence, in fact we have $et = 2$. If $e = 2$, then $t = 1$ and

$$1 = (\chi, \chi)_{\tilde{G}} \geq \frac{1}{|\tilde{G}|} \sum_{x \in G^0} |\chi(x)|^2 = \frac{|G^0|}{|\tilde{G}|} . (\chi|_{G^0}, \chi|_{G^0})_{G^0} = \frac{4|G^0|}{|\tilde{G}|},$$

that is, $(\tilde{G} : G^0) \geq 4$. But $\tilde{G}/G^0 \hookrightarrow \mathbb{S}_3$, and so $\tilde{G}/G^0 = \mathbb{S}_3$; hence (ii) holds. Finally, suppose that $e = 1$ and $t = 2$, so that $\chi|_{G^0} = \zeta_1 + \zeta_2$, where $\zeta_1 \neq \zeta_2$. Set $S = St_{\tilde{G}}(\zeta_1)$. Because $\tilde{G}$ permutes the two characters ζ_1 and ζ_2 transitively, we have $\tilde{G}/S = \mathbb{Z}_2$. But $\tilde{G}/G^0 \hookrightarrow \mathbb{S}_3$ and $S \supseteq G^0$, therefore in fact we have either $\tilde{G}/G^0 = \mathbb{Z}_2$, or $\tilde{G}/G^0 = \mathbb{S}_3$ and $S/G^0 = \mathbb{Z}_3$. In the former case, it follows from the reducibility of G on $\mathcal{L}$ that $G = G^0$, that is, (iii) holds. In the latter case we have $G/G^0 = \mathbb{Z}_2$. In particular, $G \not\subseteq S$, so that G permutes the characters ζ_1 and ζ_2 transitively and thus G is irreducible on $\mathcal{L}$, a contradiction. □

Denote by K the kernel of the permutation representation of $\tilde{G}$ on $\{\mathcal{H}_1, \dots, \mathcal{H}_7\}$.

Lemma 4.6.3. *The group K is soluble and nontrivial.*

Proof. First of all, note that the group $\tilde{G}_1$ is soluble. Indeed, if K_1 denotes the kernel of $\tilde{G}_1$ on $\mathcal{H}_1$, then K_1 is contained in some maximal torus T_1, that is, K_1 is abelian. Furthermore, $\tilde{G}_1/K_1$ is embedded in the group $A(D_4)$ of order $2^7.3^2$; hence, $\tilde{G}_1$ is soluble. But $K \subseteq \tilde{G}_1$, and so K is soluble. If $K = 1$, then $\tilde{G}$ can be embedded in $\mathbb{S}_7$; in addition, $|\tilde{G}| > (\dim \mathcal{L})^2 = 784$. Hence $\mathbb{A}_7 \subseteq \tilde{G} \subseteq \mathbb{S}_7$ and $\mathbb{A}_6 \subseteq \tilde{G}_1$, that is, $\tilde{G}_1$ is insoluble, a contradiction. □

Let $\tilde{H}$ be a minimal normal subgroup of $\tilde{G}$ lying in K. It is not difficult to see that $\tilde{H} \subseteq G^0$. In particular, $\tilde{H}$ is a nontrivial normal subgroup of G. Thus we can choose a minimal normal subgroup H of G lying in $\tilde{H}$. Certainly, H is elementary abelian. The role of H is explained in the following statement (cf. Lemma 4.2.6).

Lemma 4.6.4. *The fixed point subspace $C_{\mathcal{L}}(H) = \{x \in \mathcal{L} \mid \forall f \in H, f(x) = x\}$ is trivial.*

Proof. Apply Lemma 4.6.2. If one of the conditions 4.6.2(i) and 4.6.2(iii) holds, then the statement of Lemma 4.6.4 follows by Lemma 4.2.6. If condition 4.6.2(ii) holds and $\mathcal{L}_0 = C_{\mathcal{L}}(H) \neq 0$, then G^0 acts on $\mathcal{L}_0$ with character ζ and, in addition, $\zeta|_H = \dim \mathcal{L}_0.1_H$. But in that case, $\chi|_H = 2\zeta|_H = \dim \mathcal{L}.1_H$ and $H \subseteq \operatorname{Ker} \chi$, a contradiction. □

Denote by $\hat{H}$ the full preimage of H in $\hat{\mathcal{G}}$.

Lemma 4.6.5. *One of the following assertions holds.*

(i) *$\hat{H}$ is an elementary abelian 2-group,*

(ii) *$\hat{H} \cong 2_+^{1+6}$ is an extraspecial 2-group. Moreover, $\hat{H}$ is conjugate in $\hat{\mathcal{G}}$ to the subgroup $D_8 \otimes D_8 \otimes D_8$, where $D_8 =< D, P >$ and*

$$D = \begin{pmatrix} 1 & 0 \\ 0 & -1 \end{pmatrix}, \; P = \begin{pmatrix} 0 & 1 \\ 1 & 0 \end{pmatrix}.$$

Proof. First of all, we see that $\hat{H}$ is a 2-group. Indeed, if $H \cong \mathbb{Z}_p^m$ and $p \neq 2$, then H is embedded in a maximal torus of $\hat{\mathcal{G}}$ and $C_{\mathcal{L}}(H) \neq 0$, contrary to Lemma 4.6.4. We now distinguish two cases.

1) The group $\hat{H}$ is abelian. Setting $Z = Z(\hat{\mathcal{G}}) = \mathbb{Z}_2$, we then have $Z \lhd \Omega_2(\hat{H})$ char $\hat{H} \lhd \hat{G}$ and $\Omega_2(\hat{H})/Z \lhd G$. By the minimality of H, we have either $\Omega_2(\hat{H}) = Z$, or $\Omega_2(\hat{H}) = \hat{H}$. In the former case $H = \mathbb{Z}_2$ and so $C_{\mathcal{L}}(H) \neq 0$, a contradiction. In the latter case, $\hat{H}$ is elementary abelian.

2) The group $\hat{H}$ is nonabelian. As $\hat{H}/Z = H$ is elementary abelian, we must have $[\hat{H}, \hat{H}] = \Phi(\hat{H}) = Z$. But $Z \lhd Z(\hat{H})$ char $\hat{H} \lhd \hat{G}$ and so $Z(\hat{H})/Z$ is a normal subgroup of G; moreover, $Z(\hat{H})/Z \subset H$. Hence, $Z(\hat{H}) = Z$ and $\hat{H}$ is an extraspecial 2-group: $\hat{H} = 2_\pm^{1+2m}$ for some integer m. Let ϑ be the unique faithful irreducible character of $\hat{H}$. As $\hat{H}$ is faithful on V and $\hat{G}$ is irreducible on V, by Clifford's Theorem we have

$$\lambda|_{\hat{H}} = e\vartheta \tag{4.1}$$

for some $e \in \mathbb{N}$. Furthermore, by Lemma 4.6.4, $(\chi_H, 1_H)_H = 0$. Calculating $(\chi_H, 1_H)_H$ using Lemma 4.2.3 and identity (4.1), we obtain $m = 3$ and $\hat{H} = 2_+^{1+6}$. □

Repeating the proof of this lemma word for word, we obtain the following analogue of Propositions 4.3.10 and 4.3.11.

Proposition 4.6.6. *Suppose that the Lie algebra $\mathcal{L}$ of type D_n admits an IOD and that the E-case occurs. Then one of the following assertions holds:*

(i) *$\hat{H}$ is an elementary abelian 2-group,*

(ii) *$n = 2^m$ and $\hat{H} = 2_+^{1+2(m+1)}$. Moreover, $\hat{H}$ is uniquely defined up to conjugacy in $\hat{\mathcal{G}} = O_{2^{m+1}}(\mathbb{C})$.* □

Note that the abundance of different ODs for Lie algebras of type D_n, in comparison with the rigidity of the cases A_n and C_n, is explained by the possibility 4.6.6(i).

In accordance with Lemma 4.6.5, suppose first that $\hat{H}$ is an extraspecial 2-group: $\hat{H} = D_8 \otimes D_8 \otimes D_8$. Following Chapter 1, we set

$$J_u = D^{a_1}P^{b_1} \otimes D^{a_2}P^{b_2} \otimes D^{a_3}P^{b_3}$$

for $u = (a_1, a_2, a_3, b_1, b_2, b_3) \in W = \mathbb{F}_2^6$. We define the following quadratic and alternating forms on W:

$$q(u) = a_1b_1 + a_2b_2 + a_3b_3, \; < u|v >= q(u) + q(v) + q(u + v).$$

The following lemma is trivial (cf. Lemma 2.1.2).

Lemma 4.6.7. *The basis $\{J_u | u \in W, q(u) = 1\}$ is an eigenbasis for H with respect to its action on $\mathcal{L}$. Furthermore, $\chi|_H = \sum_{u \in W, q(u)=1} \tau_u$, where $\tau_u(J_v) = (-1)^{<u|v>}$.* □

Apart from the basis $\{J_u | q(u) = 1\}$, the $\mathbb{C}$-space $\mathcal{L}$ admits also the so-called *E-basis* $\{E(i, j) | 1 \leq i < j \leq 8\}$, where $E(i, j) = E_{ij} - E_{ji}$ and $E_{ij} = (\delta_{ik}\delta_{jl})$ (see Chapter 2). It is not difficult to see that E-basis is an eigenbasis for the group I in its action on $\mathcal{L}$, where

$$I = \hat{I}/Z, \; \hat{I} = \{\text{diag}(a_1, \ldots, a_8) | a_i = \pm 1, \prod_i a_i = 1\}. \tag{4.2}$$

Lemma 4.6.8. *In the case where $\hat{H}$ is an extraspecial 2-group, the decomposition $\mathcal{D}$ is conjugate to some E-decomposition.*

Proof. When $\hat{H} = 2_+^{1+6}$, by Lemma 4.6.7 each subalgebra $\mathcal{H}_i$ has a basis consisting of some elements J_u. The lemma now follows from the fact that the subgroups $H = \hat{H}/Z$ and $I = \hat{I}/Z$, both of which are Jordan subgroups of order 2^6 in $\mathcal{G}^0$, are $\mathcal{A}$-conjugate. □

Lemma 4.6.8 reduces Theorem 4.6.1 to the following lemma in the case $\hat{H} = 2_+^{1+6}$.

Lemma 4.6.9. *Every irreducible E-decomposition of the Lie algebra of type D_4 is multiplicative.*

Proof. This is a special case of Theorem 2.4.6. □

Assume now that $\hat{H}$ is abelian. In this case, one can suppose that $H \subseteq T$ and $G \subseteq T.\mathbb{S}_8$, where

$$T = \hat{T}/Z, \; \hat{T} = \{\text{diag}(a_1, \ldots, a_8) | a_i = \pm 1\},$$

and that $P = GT/T$ and $P^0 = G^0T/T$ are doubly transitive permutation groups of degree 8.

Considering the action of an element of order 7 of P^0 on $G^0 \cap T$ by conjugation, we see that $|G^0 \cap T|$ is either 2^3 or 2^6. The same holds for H. If $|G^0 \cap T| = 2^6$, then $G^0 \cap T = I$ (see (4.2)). In this case, I leaves each subalgebra $\mathcal{H}_i$ fixed, and so $\mathcal{D}$ is an E-decomposition; hence, by Lemma 4.6.9, $\mathcal{D}$ is an MOD. Therefore, in what follows we shall suppose that

$$G^0 \cap T = H \cong \mathbb{Z}_2^3. \tag{4.3}$$

The proof of the following statement will be omitted.

Lemma 4.6.10 [Ale 1]. *When* (4.3) *holds, the group H is a Jordan subgroup of $\mathcal{G}^0.\mathbb{Z}_3$ and $\mathcal{A}$. Furthermore, $C_{\mathcal{A}}(H) = 2^{3+6}.\mathbb{S}_3$, $C_{\mathcal{G}}(H) = \mathbb{Z}_2^7.\mathbb{Z}_2^3$, $C_{\mathcal{G}^0}(H) = H.\mathbb{Z}_2^6$, $N_{\mathcal{G}^0}(H)/C_{\mathcal{G}^0}(H) = SL_3(2)$. Finally, if the symbol U denotes the natural $\mathbb{F}_2SL_3(2)$-module, then H and $C_{\mathcal{G}^0}(H)/H$ are $SL_3(2)$-modules of type U^* and $U \oplus U$, respectively.* □

Here and below $C = C_{G^0}(H)$ and $Q = G^0/C$.

Lemma 4.6.11. *Suppose that* (4.3) *holds and* 3 *divides* $|G^0|$. *Then $\mathcal{D}$ is an MOD.*

Proof. The group H acts on $\mathcal{L}$ with character $\chi|_H = 4\sum_{i=1}^7 \lambda_i$, where $\mathrm{Irr}(H) = \{1_H, \lambda_1, \ldots, \lambda_7\}$.

1) Since 3 divides $|G^0|$ and C is a 2-group, $|Q|$ is divisible by 3. Moreover, $G_1^0 = St_{G^0}(\mathcal{H}_1) = C.Q_1$, where $(Q : Q_1) = 7$. Every element φ of order 3 of Q acts on $\{\lambda_1, \ldots, \lambda_7\}$ with one fixed point and two orbits of length 3. We shall suppose that $\varphi \in Q_1$ has the orbits $\{\lambda_1\}$, $\{\lambda_2, \lambda_3, \lambda_4\}$ and $\{\lambda_5, \lambda_6, \lambda_7\}$. By Clifford's Theorem, either G_1^0 is irreducible on $\mathcal{H}_1$, or G_1^0 has two irreducible submodules of dimension 2 on $\mathcal{H}_1$. Hence, either $\rho|_H = 4\lambda_1$, or

$$\rho|_H = \lambda_1 + \lambda_2 + \lambda_3 + \lambda_4 \quad \text{and} \quad Q_1 \quad \text{has} \quad \{\lambda_1, \ldots, \lambda_4\} \quad \text{as orbit on} \quad \mathrm{Irr}(H). \tag{4.4}$$

Under the former hypothesis, $\mathcal{H}_1$ is the λ_1-eigensubspace for H, which implies immediately the multiplicativity of $\mathcal{D}$. So we shall suppose that (4.4) holds. Clearly, in that case $|Q_1|$ is divisible by $3.4 = 12$ and $|Q|$ is divisible by $3.4.7 = 84$. But $Q \subseteq SL_3(2)$, and so in fact $Q = SL_3(2)$.

2) Next we assume that $H \neq C$. By Lemma 4.6.10, C/H is a Q-module of type $U \oplus U$ or U. But by (4.3), C does not contain T, and so $C/H \cong U$. It can be shown that $C \cong \mathbb{Z}_2^6$. Consider the full preimage $\hat{C}$ of C in $\hat{\mathcal{G}}^0 = SO_8(\mathbb{C})$. If $\hat{C}$ is abelian, then $\hat{C} \subseteq C_{\hat{\mathcal{G}}^0}(\hat{H}) = \hat{I}$, and so $C = \hat{I}/Z = I$, $G^0 \cap T = I$, contrary to (4.3). Hence $\hat{C}$ is nonabelian. If $Z(\hat{C}) = Z$, then $\hat{C}$ is extraspecial and we can appeal to Lemmas 4.6.8 and 4.6.9. Finally, suppose that $\hat{C} \supset Z(\hat{C}) \supset Z$. By Lemma 4.6.10, $Z(\hat{C})/Z$ is a Q-module of type U^* or U; in particular, $|Z(\hat{C})| = 2^4$. As $\hat{C}$ is faithful on V, by Clifford's Theorem we have

$$\lambda|_{\hat{C}} = e\sum_{i=1}^{t} \tau_i$$

for some $e, t \in \mathbb{N}$ and some $\tau_i \in \mathrm{Irr}(\hat{C})$ with $\deg \tau_i > 1$. However $(\deg \tau_i)^2$ divides $(\hat{C} : Z(\hat{C})) = 2^3$, so that $\deg \tau_i = 2$. The group Q has no proper subgroups of index at most 4, hence $t = 1$, $e = 4$ and $\lambda|_{\hat{C}} = 4\tau_1$. The last equality contradicts the fact that $\lambda|_{\hat{H}}$ is the sum of 8 different characters.

3) Finally, suppose that $C = H$. Then G_1^0 acts faithfully on $\mathcal{H}_1$, and this means that $G_1^0 \cong W(D_4) = \mathbb{Z}_2^3.\mathbb{A}_4$. As $H \lhd G_1^0$ and $H \cong \mathbb{Z}_2^3$, one can deduce that the extension G_1^0 of H by G_1^0/H splits. By Gaschütz's Theorem (see [Hup]), the extension G^0 of H also splits, and thus $G^0 = H.Q \cong ASL_3(2)$. In particular, G^0 has no irreducible characters of degree 28, and it has only one irreducible character ζ of degree 14. Hence, $\chi|_{G^0} = 2\zeta$. By Lemma 4.6.2, we must have that $G/G^0 = \mathbb{Z}_2$, which implies that $C_G(H)/H = \mathbb{Z}_2$. As is easy to see, the last equality contradicts Lemma 4.6.10. □

To prove the theorem, we still need to consider the case where 3 does not divide $|G^0|$. Then 3 does not divide $|P|$, and P is a doubly transitive subgroup of $\mathbb{S}_8$, so that $P = \mathbb{Z}_2^3.\mathbb{Z}_7$. In this case one can show that H is a normal subgroup of $\tilde{G}$ and that the group $\tilde{C} = C_{\tilde{G}}(H)$ is irreducible on $\mathcal{H}_1$. By Schur's Lemma, the group $H \subseteq Z(\tilde{C})$ is scalar on $\mathcal{H}_1$, so that $\rho|_H = 4\lambda_1$ and $\mathcal{D}$ is an MOD. This completes the proof of Theorem 4.6.1. □

4.7 The non-existence of IODs for Lie algebras of types C_p and D_p

The aim of this section is to prove the following result:

Theorem 4.7.1. *The complex simple Lie algebras of types C_p and D_p (p a prime, $p \geq 5$) have no IODs.*

Note that Theorem 4.5.1 says that the Lie algebra of type C_3 also has no IODs. Furthermore, by Theorems 1.4.1 and 2.1.7, Lie algebras of types C_2 and $D_3 \cong A_3$ have exactly one conjugacy class of ODs.

We fix the following notation. Let $\mathcal{L}$ be a simple Lie algebra of type C_p or D_p (p a prime, $p \geq 5$) that admits an IOD

$$\mathcal{D} : \mathcal{L} = \bigoplus_{i=1}^{n} \mathcal{H}_i$$

($n - 1 = h$ is the Coxeter number). Set $G = \mathrm{Aut}(\mathcal{D})$, $G_1 = St_G(\mathcal{H}_1)$, and let K be the kernel of the permutation representation of G on $\{\mathcal{H}_1, \ldots, \mathcal{H}_n\}$. Denote by χ and ρ the character of G on $\mathcal{L}$ and of G_1 on $\mathcal{H}_1$ respectively.

A. Firstly we show that the P-case (that is, $K = 1$) is impossible.

Lemma 4.7.2. *Suppose that $K = 1$. Then the triple (G, n, p) can take only the following values: $(\mathbb{A}_n, n, (n\mp 1)/2)$, $(\mathbb{S}_n, n, (n\mp 1)/2)$, $(M_{23}, 23, 11)$, $(M_{11}, 11, 5)$, $(L_2(11), 11, 5)$, $(\mathbb{A}_7, 15, 7)$, and $(SL_d(2), 2^d - 1, 2^{d-1} - 1)$, $d \geq 4$.*

Proof. 1) As $K = 1$, we can embed G in $\mathbb{S}_n$. Furthermore, the group G_1 is represented on $\mathcal{H}_1^*$ as an irreducible monomial subgroup of $\mathbb{Z}_2 \wr \mathbb{S}_p$. Therefore p divides $|G|$. Suppose that G is imprimitive (as a subgroup of $\mathbb{S}_n$). Then there exist integers k, l with $k, l > 1$, $k.l = n$ such that $G \subseteq \mathbb{S}_k \wr \mathbb{S}_l$. But $n = 2p \pm 1$, and so $k, l < p$ and p does not divide $|G|$, a contradiction. Hence G is a faithful primitive permutation of degree $n = 2p\pm 1$, whose order is divisible by p. Primitive permutation groups containing an element of large prime order are classified by Liebeck and Saxl in [LiS].

2) Suppose that $\mathrm{soc}(G)$ is nonabelian. Applying [LiS] we obtain for (G, n, p) the possibilities indicated in the statement of the lemma. Now suppose that $\mathrm{soc}(G)$ is abelian. According to [LiS] we have $\mathrm{soc}(G) = V = \mathbb{Z}_q^{kr}$ (q a prime), $n = q^{kr}$, $G = V.G_0$ and $SL_k(q^r) \subseteq G_0 \subseteq \Gamma L_k(q^r)$. If $k = 1$, then $|G| \leq rq^r(q^r - 1) < (\dim \mathcal{L})^2$ and G cannot be irreducible on $\mathcal{L}$. If $k \geq 2$, then $G_0 \supseteq SL_k(q^r)$ acts transitively on the set $\mathrm{Irr}(V)\backslash\{1_V\}$. Hence, by Clifford's Theorem, either V acts trivially on $\mathcal{L}$, or the integer $n.(n\mp 1)/2 = \dim \mathcal{L}$ is divisible by $|\mathrm{Irr}(V)| - 1 = n - 1$. Both cases are impossible. □

Lemma 4.7.3. $K \neq 1$.

Proof. Assuming the contrary, we apply Lemma 4.7.2.

1) First we note that the irreducible representation of G_1 on $\mathcal{H}_1$ induces a transitive permutation representation of G_1 of degree p. Furthermore, by Lemma 4.4.4 $G \not\supseteq \mathbb{A}_n$. If $G = M_{23}$, then $G_1 = M_{22}$ and $p = 11$. If $G = M_{11}$, then $G_1 = \mathbb{A}_6.\mathbb{Z}_2$ and $p = 5$. In any of these cases the group G_1 has no transitive permutation representations of degree p.

2) If $G = \mathbb{A}_7$ or $G = \mathbb{A}_8 \cong SL_4(2)$ then $\dim \mathcal{L} = 105$. If $G = L_2(11)$ then $\dim \mathcal{L} = 55$. In any of these cases G cannot have irreducible representations of degree equal to $\dim \mathcal{L}$.

3) Finally, suppose that $G = SL_d(2)$, $d \geq 5$ and $p = 2^{d-1} - 1$. Setting $W = \mathbb{F}_2^d$, one can suppose that $G = SL(W)$ and $G_1 = St_G(a)$ for some $a \in W\backslash\{0\}$. We realize the root system of $\mathcal{L}$ in the form $\{\pm e_i \pm e_j, \pm 2e_i\}$, where $\{e_1, \ldots, e_p\}$ is an orthonormal basis of the space $\mathcal{H}_1^*$. Acting on $\mathcal{H}_1^*$, the group G_1 permutes transitively the p pairs $\pm e_i$, $1 \leq i \leq p$. Set $G_2 = St_{G_1}(\{\pm e_1\})$, and let δ be the character of G_2 afforded by $< e_1 >_{\mathbb{C}}$. Then by Lemma 4.2.2 we have

$$\chi = \mathrm{Ind}_{G_1}^{G}(\rho) = \mathrm{Ind}_{G_1}^{G}(\mathrm{Ind}_{G_2}^{G_1}(\delta)) = \mathrm{Ind}_{G_2}^{G}(\delta).$$

Furthermore, $\delta \in \mathrm{Hom}(G_2, \mathbb{Z}_2)$, and the irreducibility of χ implies that δ is surjective. Thus G_2 is a subgroup of index $2^{d-1} - 1$ in G_1 having the subgroup $G_3 = \mathrm{Ker}\,\delta$ of index 2. It can be shown that there exists an element $b \in W \setminus < a >$ such that $G_2 = St_G(\{b, a+b\})$, and then $G_3 = St_G(a, b)$ is the unique subgroup of index 2 in G_2. Take $g \in G$ such that $g(a) = b$ and $g(b) = a$; set $G_2' = gG_2g^{-1} = St_G(\{a, a+b\})$ and $\delta'(x) = \delta(g^{-1}xg)$ for $x \in G_2'$. Then

$$G_3' = \mathrm{Ker}\,\delta' = gG_3g^{-1} = St_G(a, b) = G_3 = G_2 \cap G_2'.$$

Thus $\delta = \delta' = 1$ on $G_2 \cap G_2'$. By Mackey's criterion [Ser 3], the character $\chi = \mathrm{Ind}_{G_2}^G(\delta)$ cannot be irreducible, a contradiction. □

B. The S-case. Here we assume that all minimal normal subgroups of G lying in K are nonabelian. (Recall that $K \neq 1$ by Lemma 4.7.3). In particular, K and G_1 are insoluble. As mentioned above, the group G_1 has a transitive permutation representation of prime degree p (on the pairs $\pm e_i$) with soluble kernel $\tilde{K}_1$. By Wielandt's Theorem [Wie], $L = \mathrm{soc}(G_1/\tilde{K}_1)$ is a nonabelian simple group. Concerning L, the following statement holds:

Lemma 4.7.4. $|\mathrm{Out}(L)| > \sqrt{n}/2$.

Proof. 1) One can suppose that L is embedded in $K \subseteq G_1$ as a minimal normal subgroup of G. As $L \lhd G_1$, L is simple and $\mathcal{H}_1$ is an irreducible $\mathbb{C}G_1$-module of prime dimension, Clifford's Theorem tells us that L is irreducible on $\mathcal{H}_1$. The same holds for every component $\mathcal{H}_i$.

2) Set $C = C_G(L)$ and $C_1 = C \cap G_1$. Since $C \lhd G$ and G is transitive on the set $\{\mathcal{H}_i\}$, C is (1/2)-transitive on it, that is,

$$(C : C_1) = k, \text{ where } k \text{ divides } n. \tag{4.5}$$

Let T be the maximal torus corresponding to $\mathcal{H}_1$ and set $C_2 = C_1 \cap T$. By Schur's Lemma, the group $C_1/C_2 \cong C_1T/T$ acts faithfully by scalars on $\mathcal{H}_1$, so that $C_1/C_2 \subseteq \mathbb{Z}_2$. On the other hand, one can show that $C_2 \subseteq C_T(L) = \mathbb{Z}_2$. We have therefore shown that

$$|C| = 2^m.k, \text{ where } m \leq 2. \tag{4.6}$$

3) Setting $M = CL$, we have $C \cap L = Z(L) = 1$ and so $M = C \times L$ is a normal subgroup of G. By Clifford's Theorem, $\chi|_M = e\sum_{i=1}^t \xi_i$ for some $e, t \in \mathbb{N}$ and $\xi_i \in \mathrm{Irr}(M)$. Furthermore, for every i there exist $\alpha_i \in \mathrm{Irr}(C)$ and $\beta_i \in \mathrm{Irr}(L)$ such that $\xi_i(xy) = \alpha_i(x).\beta_i(y)$ for all $x \in C$ and $y \in L$. In particular, $\chi|_L = e\sum_{i=1}^t(\deg \alpha_i).\beta_i$. By 1), we must have $\deg \beta_i = \dim \mathcal{H}_1 = p$. Hence, $\deg \alpha_i = \deg \chi/tep = n/te$, on the one hand. On the other hand, by (4.5) and

(4.6) we have $\deg \alpha_i < |C|^{1/2} \leq 2\sqrt{n}$. Moreover, te divides the index $(G : M)$ and $G/M \hookrightarrow \mathrm{Out}(L)$. Consequently, we arrive at the inequality:

$$|\mathrm{Out}(L)| \geq te = n/\deg \alpha_i > \sqrt{n}/2. \qquad \square$$

The list of nonabelian socles L of transitive permutation groups of prime degree p is presented in Proposition 4.4.1. A direct verification shows that none of these groups L satisfies the inequality $|\mathrm{Out}(L)| > \sqrt{2p-1}/2$. This contradiction, together with Lemma 4.7.4, means that the S-case is impossible.

C. The E-case. The results of items A and B show that K contains some elementary abelian subgroup $H \neq 1$ which is a minimal normal subgroup of G. It is shown in Proposition 4.3.11 that for Lie algebras of type C_m the E-case can occur only if $m = 2^l$. Hence $\mathcal{L}$ must be of type D_p. Set $\mathcal{A} = \mathrm{Aut}(\mathcal{L}) = PO_{2p}(\mathbb{C})$, $\hat{\mathcal{A}} = O_{2p}(\mathbb{C})$, and let $\hat{G}$, $\hat{H}$ be the full preimages of G and H in $\hat{\mathcal{A}}$. By Proposition 4.6.6, either $\hat{H}$ is an extraspecial 2-group and $\mathrm{rank}\mathcal{L} = 2^l$, or $\hat{H}$ is an elementary abelian 2-group and $\hat{G}$ acts doubly transitively on the set of pairs $\pm v_i$, $1 \leq i \leq 2p$, where $\{v_1, \ldots, v_{2p}\}$ is an orthonormal basis of the space $V = \mathbb{C}^{2p}$. In our case $\mathrm{rank}\mathcal{L} = p$, so the latter possibility occurs. Let $\hat{E}$ and P be the kernel and image of $\hat{G}$ in its action on $\{\pm v_i\}$. Because $2p$ is not a prime-power, the socle $R = \mathrm{soc}(P)$ is a nonabelian simple group. Using Table A2, we find the following possibilities for R:

$$R = \mathbb{A}_{2p}, \quad \text{or } R = M_{22} \text{ (and } p = 11\text{)}, \quad \text{or } R = L_2(q) \text{ (and } q = 2p-1\text{)}. \qquad (4.7)$$

In particular, $|\mathrm{Out}(L)|$ is a 2-group. Recall that G acts transitively on the set of $(2p-1)$ Cartan subalgebras $\mathcal{H}_i$. But $\hat{E}$ and $\mathrm{Out}(R)$ are 2-groups, and so R is also transitive on the set $\{\mathcal{H}_i\}$, that is, R has a subgroup of index $2p-1$. This condition, considered together with (4.7), leads to the unique possibility $R = L_2(5)$ and $p = 3$, which is realized when $\mathcal{L}$ is of type $D_3 \cong A_3$.

This completes the proof of Theorem 4.7.1. $\square$

Commentary

1) The general Theorem 4.1.1 was first proved by Ivanov [Iva 3]. Although this result is close to the classical Jordan-Zassenhaus Theorem (see [CuR]), in fact the latter theorem yields only the implication (ii) $\Rightarrow$ (i) of Ivanov's Theorem. For, on the assumption of the absolute irreducibility of the $\mathbb{Q}$-module V and the isomorphism of full lattices $\Lambda_1, \Lambda_2 \subset V$, it follows from Schur's Lemma that Λ_1 and Λ_2 are similar. Furthermore, by the Jordan-Zassenhaus Theorem, there exists only a finite number of *isomorphism classes* of full lattices; hence, in our case, we have only a finite number of *similarity classes* of full lattices.

2) In the same paper [Iva 3], Ivanov studied for the first time a special class of IODs (namely, the class of irreducible E-decompositions) for Lie algebras of type D_n. Later, the general problem of describing all IODs is systematically investigated in the works of the second author [Tiep 7 – 11], see also [KoT 1, 2]. He also suggested the division of this problem into the E-, the S- and the P-cases. It seems very plausible that for any OD $\mathcal{D} : \mathcal{L} = \oplus_i \mathcal{H}_i$, the kernel

$$K = \{\varphi \in \mathrm{Aut}(\mathcal{L}) | \forall i, \varphi(\mathcal{H}_i) = \mathcal{H}_i\}$$

is soluble. A step toward the proof of this fact has been made in [Tiep 26].

3) Theorem 4.5.1 on the non-existence of IODs for Lie algebras of types A_{p-2} (p a prime, $p \neq 2^d + 1$) and $C_{(p-1)/2}$ (p a prime, $p > 5$) actually gives a partial solution of the following problem:

Weak Winnie-the-Pooh problem. *Prove that a finite-dimensional complex simple Lie algebra $\mathcal{L}$ has an IOD if and only if $\mathcal{L}$ is of one of the following types: A_n, $n = p^m - 1$ for some prime power p^m; B_n, $n = [(p^m - 1)/2]$ for some prime power p^m; C_2; D_n, $n = 6, 14$ or $n = 2^m$; G_2, F_4, E_6 and E_8.*

When $p = 7$, Theorem 4.5.1 guarantees the absence of IODs for Lie algebras of types A_5 and C_3. Recall (see the Commentary to Chapter 1) that the Winnie-the-Pooh problem proper for these Lie algebras is still open. Up to now, the weak Winnie-the-Pooh problem has been solved completely for the Lie algebras of types A_n (see Chapter 5), B_n (see Chapter 6); C_p, $C_{(p-1)/2}$, D_p, $D_{(p+1)/2}$ (p a prime); G_2, F_4, E_6, E_7 and E_8. It can be solved by methods similar to those of §4.7 for the Lie algebras of types C_n and D_n when n or $2n \pm 1$ has a "large" prime divisor $p > \sqrt{n}$. However, the cases C_n and D_n are still open for general n.

4) The main conjecture about the class of transitive ODs is the following:

Conjecture 4A. *A finite-dimensional complex simple Lie algebra $\mathcal{L}$ has a TOD if and only if $\mathcal{L}$ is of one of the following types: A_n, where $n + 1$ is a prime-power; B_n, C_{2^m}, D_n, G_2, F_4, E_6 and E_8.*

The existence of TODs for Lie algebras of the types mentioned is resumed by Borovik's Theorem 3.1.2. However, the converse implication is much more complicated, and indeed is still an unsolved problem.

5) The scheme for studying IODs suggested in this chapter is based on powerful methods and results of finite group theory. In principle, it would be interesting to relax the dependence on group theory, perhaps using stronger geometric and combinatorial methods.

Chapter 5

Classification of irreducible orthogonal decompositions of complex simple Lie algebras of type A_n

---

The aim of the chapter is to prove Conjecture 0.2 for Lie algebras of type A_n:

Theorem 5.0.1. *The complex simple Lie algebra $\mathcal{L}$ of type A_n admits an IOD if and only if $n+1$ is a prime power: $n = p^m - 1$. The total number of* $\mathrm{Aut}(\mathcal{L})$*-conjugacy classes of IODs is* 1 *if $p^m \neq 27$ and* 2 *if $p^m = 27$.*

The proof requires the classification of finite simple groups; namely, we use the classification of doubly transitive permutation groups with nonabelian socle (see Table A2) and the validity of Schreier's conjecture (see [Gor 2]). The notation and the general outline of the arguments are taken from §4.2: $\mathcal{L} = sl_{n+1}(\mathbb{C})$ is the simple Lie algebra of type A_n, $h = n+1$ its Coxeter number, R its root system, $\mathcal{A} = \mathrm{Aut}(\mathcal{L})$, $\mathcal{G} = \mathrm{Inn}(\mathcal{L}) = PSL_{n+1}(\mathbb{C})$, and $V = \mathbb{C}^{n+1}$ is the space on which $\mathcal{L}$ is represented. Suppose that $\mathcal{L}$ has an IOD

$$\mathcal{D} : \mathcal{L} = \bigoplus_{i=1}^{h+1} \mathcal{H}_i$$

with automorphism group $G = \mathrm{Aut}(\mathcal{D})$ acting on $\mathcal{L}$ with character $\chi \in \mathrm{Irr}(G)$. Set $G^0 = G \cap \mathcal{G}$, $\hat{\mathcal{G}} = SL_{n+1}(\mathbb{C})$, $Z = Z(\hat{\mathcal{G}})$, let $\hat{G}^0$ be the full preimage of G^0 in $\hat{\mathcal{G}}$, and λ the $\hat{G}^0$-character afforded by V. Finally, let K be the kernel of the permutation representation of G on $\{\mathcal{H}_1, \ldots, \mathcal{H}_{n+2}\}$, set $G_i = \{g \in G \mid g(\mathcal{H}_i) = \mathcal{H}_i\}$, $G_i^0 = G^0 \cap G_i$, and let R_i be the kernel of G_i on $\mathcal{H}_i$.

Because the OD is unique and irreducible when $n = 1$, from now on we shall suppose that $n > 1$. Recall (see §4.2) that our proof is divided into the E-, the S- and the P-cases. The E-case is treated fully in §1.2. (By Proposition 4.3.10 in the E-case we have $n = p^m - 1$ for some prime power p^m; moreover, $G = \mathrm{Aut}(\mathcal{D})$ admits a normal subgroup $H = \mathbb{Z}_p^{2m}$, lying in K and having an inverse image $\hat{H} = p_+^{1+2m}$ in $\hat{\mathcal{G}}$. Thus $\hat{H}$ leaves each Cartan subalgebra $\mathcal{H}_i$ fixed. In other words, $\mathcal{D}$ is a J-decomposition in the sense of §1.2). The S-case is considered in §5.1, and §§5.2 – 5.4 are devoted to the P-case.

5.1 The S-case

Let H be a minimal normal subgroup of G contained in K. Recall that all such subgroups are nonabelian in the S-case. Moreover, the following assertion holds; in it the subgroup $K_1 = \mathrm{Ker}\tau$ is that introduced in Lemma 4.2.2.

Lemma 5.1.1. *H is a nonabelian simple group. Furthermore,* $\mathrm{soc}(G_1/K_1) = H$.

Proof. From the minimality and non-commutativity of H, it follows that $H = L_1 \times \ldots \times L_m = L^m$, where $L_i \cong L$ is a nonabelian simple group. By Lemma 4.2.2, K_1 is soluble and G_1/K_1 is a faithful doubly transitive permutation group of degree $n+1$. Since $H \subseteq K \subseteq G_1$ and $H \lhd G$, H is a normal subgroup of G_1. Furthermore, $K_1 \cap H$ is a soluble normal subgroup of $H = L^m$, and so $K_1 \cap H = 1$. Hence, $H \cong HK_1/K_1 \lhd G_1/K_1$, that is, we can suppose that H is embedded in G_1/K_1 as a normal subgroup. Set $T = \mathrm{soc}(G_1/K_1)$. Note that, for any $i \geq 1$ and $x \in G_1/K_1$, the subgroup xL_ix^{-1} is also a normal subgroup of order $|L|$ in H. Therefore, one can find an index j such that $xL_ix^{-1} = L_j$. In other words, the group G_1/K_1 permutes the m subgroups $L_1, \ldots, L_m$. Let $\mathcal{O}_1, \ldots, \mathcal{O}_r$ be the G_1/K_1-orbits on the set $\{L_1, \ldots, L_m\}$, and set

$$H_k = \prod_{L_i \in \mathcal{O}_k} L_i$$

for $k = 1, \ldots, r$. Clearly, H_k is a minimal normal subgroup of G_1/K_1 and $H = H_1 \times \ldots \times H_r$. Hence, $L^m = H \lhd T$. In particular, T is nonabelian. By the O'Nan-Scott Theorem, T is a nonabelian simple group, so that $m = 1$ and $T = H = L$. □

By Lemma 4.2.5, $H \subseteq \mathcal{G}$. Let $\hat{H}$ be the full preimage of H in $\hat{\mathcal{G}}$.

Lemma 5.1.2. *The group $\hat{H}$ is absolutely irreducible on V.*

Proof. Assume the contrary, so that $\lambda|_{\hat{H}} = \lambda_1 + \lambda_2$ for some $\hat{H}$-characters λ_1 and λ_2. Then by Lemma 4.2.3 we have

$$\chi_{\hat{H}} = |\lambda_{\hat{H}}|^2 - 1_{\hat{H}} = |\lambda_1 + \lambda_2|^2 - 1_{\hat{H}} =$$

$$= (|\lambda_1|^2 - 1_{\hat{H}}) + (|\lambda_2|^2 - 1_{\hat{H}}) + (\overline{\lambda_1}\lambda_2 + \lambda_1\overline{\lambda_2}) + 1_{\hat{H}}.$$

Note that $(|\lambda_i|^2, 1_{\hat{H}})_{\hat{H}} = (\lambda_i, \lambda_i)_{\hat{H}} \geq 1$, and so $|\lambda_i|^2 - 1_{\hat{H}}$ is either 0 or an $\hat{H}$-character. Hence, $(\chi_H, 1_H)_H > 0$, that is, $C_{\mathcal{L}}(H) \neq 0$, contrary to Lemma 4.2.5. □

An essential role in the analysis of the S-case is played by the following statement:

Proposition 5.1.3. *Under our assumptions, $C_G(H) = 1$.*

Proof. Set $C^0 = C_{G^0}(H)$, and let $\hat{C}^0$ be the full preimage of C^0 in $\hat{\mathcal{G}}$. With every element $c \in \hat{C}^0$ we associate the function

$$f_c : h \mapsto f_c(h) = chc^{-1}h^{-1} \in Z,\ h \in \hat{H}.$$

Then

$$f_c(h_1)f_c(h_2) = ch_1c^{-1}h_1^{-1}ch_2c^{-1}h_2^{-1} =$$
$$= ch_1c^{-1}.ch_2c^{-1}h_2^{-1}.h_1^{-1} = ch_1h_2c^{-1}h_2^{-1}h_1^{-1} = f_c(h_1h_2),$$

that is, $f_c \in \mathrm{Hom}(\hat{H}, Z)$. Clearly, $\mathrm{Ker} f_c \supseteq Z$, but Z is abelian and $H = \hat{H}/Z$ is simple; therefore $\mathrm{Ker} f_c = \hat{H}$ and $f_c = 1$. Hence, $\hat{C}^0$ centralizes $\hat{H}$, on the one hand. On the other hand, $\hat{H}$ is absolutely irreducible on V by Lemma 5.1.2, and so, by Schur's Lemma, $\hat{C}^0$ is scalar, that is $\hat{C}^0 = Z$ and $C^0 = 1$. Now set $C = C_G(H)$. Then $C \cap G^0 = C^0 = 1$, hence $C \cong CG^0/G^0 \subseteq G/G^0 \subseteq \mathbb{Z}_2$. If in addition $C \neq 1$, then $C = \mathbb{Z}_2$ and this implies that $C_{\mathcal{L}}(C) = \{x \in \mathcal{L} \mid \forall c \in C, c(x) = x\}$ is a nontrivial G-submodule of $\mathcal{L}$ (see [SSL]), contrary to the irreducibility of G on $\mathcal{L}$. Thus $C = 1$. □

From the preceding proposition, it follows that $H \lhd G \subseteq \mathrm{Aut}(H)$. But G acts transitively on the set $\{\mathcal{H}_1, \dots, \mathcal{H}_{n+2}\}$, and the subgroup H of K leaves each subalgebra $\mathcal{H}_i$ fixed. Hence,

$$|\mathrm{Out}(H)| \text{ is divisible by } n+2. \tag{5.1}$$

On the other hand, by Lemma 5.1.1,

$$H \text{ is the nonabelian socle of a 2-transitive group of degree } n+1. \tag{5.2}$$

Lemma 5.1.4. *Conditions* (5.1) *and* (5.2) *are incompatible.*

Proof. Assume the contrary. If H is a sporadic or alternating group and $H \neq \mathbb{A}_6$, then $|\mathrm{Out}(H)| \leq 2$ and $|\mathrm{Out}(H)|$ is not divisible by $n+2$, which is at least 3. If $H = \mathbb{A}_6$, then $|\mathrm{Out}(H)| = 4$, which implies that $n = 2$. But $\mathbb{A}_6$ has no subgroups of index 3, and so (5.2) does not hold. Now we can suppose that H is a finite group of Lie type and appeal to Tables A2 and A3.

1) $H \neq L_2(11)$, $L_2(8)$ or $Sp_{2d}(2)$, because the outer automorphism group of these groups has order not greater than 3.

2) If $H = {}^2A_2(q)$, ${}^2B_2(q)$ or ${}^2G_2(q)$, then $n = q^3$, q^2 or q^3, respectively (see Table A2). Furthermore, $|\mathrm{Out}(H)| = \gcd(3, q+1) \cdot f$, f or f, respectively, where $q^2 = p^f$ in the first case and $q = p^f$ in the remaining cases, p prime. Clearly, in any case $|\mathrm{Out}(H)| < n+2$.

3) Finally, suppose that $H = L_d(q)$ and $n+1 = (q^d - 1)/(q-1)$ $(d \geq 2)$. Here, in accordance with Table A3, $|\mathrm{Out}(H)|$ divides $\gcd(d, q-1).2f$ if $q = p^f$ for some prime p. Clearly, $|\mathrm{Out}(H)| < n+2$.

These arguments show that (5.1) and (5.2) are incompatible. □

From Lemma 5.1.4 it follows immediately that the S-case is impossible.

5.2 The P-case. I. Generic position

Recall that in the P-case the automorphism group $G = \mathrm{Aut}(\mathcal{D})$ acts faithfully and transitively on $\{\mathcal{H}_1, \ldots, \mathcal{H}_{n+2}\}$. The goal of this and the subsequent sections is to prove that the following property holds in the P-case:

$(2T)$: the group G acts 2-transitively on the set $\{\mathcal{H}_1, \ldots, \mathcal{H}_{n+2}\}$.

Set $L = \mathrm{soc}(G_1/K_1)$. We recall that G_1/K_1 is a faithful 2-transitive permutation group of degree $n+1$. By the O'Nan-Scott Theorem, L is either an elementary abelian p-group: $L \cong \mathbb{Z}_p^m$ and then $n = p^m - 1$, or a nonabelian simple group. The words "generic position" in the title of the section mean that throughout this section the group L is supposed to be a nonabelian simple group.

We start with the following statement:

Lemma 5.2.1. *Let φ be an automorphism of the Lie algebra $\mathcal{L}$ leaving two Cartan subalgebras $\mathcal{H}$ and $\mathcal{H}'$ pointwise fixed, and suppose in addition that $\mathcal{H} \cap \mathcal{H}' = 0$. Then φ is trivial.*

Proof. Set $\mathcal{C} = C_{\mathcal{L}}(\varphi) = \{x \in \mathcal{L} \mid \varphi(x) = x\}$ and $\mathcal{Z} = Z(\mathcal{C})$. Then certainly $\mathcal{C}$ contains $\mathcal{H}$ and $\mathcal{H}'$. Furthermore,

$$[\mathcal{H} + \mathcal{Z}, \mathcal{H} + \mathcal{Z}] = [\mathcal{H}, \mathcal{H}],$$

$$[[\mathcal{H} + \mathcal{Z}, \mathcal{H} + \mathcal{Z}], \mathcal{H} + \mathcal{Z}] = [[\mathcal{H}, \mathcal{H}], \mathcal{H}], \ldots,$$

that is, $\mathcal{H} + \mathcal{Z}$ is a nilpotent subalgebra. But $\mathcal{H}$ is a maximal nilpotent subalgebra of $\mathcal{L}$, so that $\mathcal{H} + \mathcal{Z} = \mathcal{H}$ and $\mathcal{Z} \subseteq \mathcal{H}$. Similarly, $\mathcal{Z} \subseteq \mathcal{H}'$. Hence, $\mathcal{Z} \subseteq \mathcal{H} \cap \mathcal{H}' = 0$. Thus, the centralizer $\mathcal{C}$ of φ has trivial centre. It is known [Jac] that the inclusion $\mathcal{C} \supseteq \mathcal{H}$ implies that φ is an inner automorphism. Using the standard description of inner automorphisms given by Kac [Kac], one can see that $Z(\mathcal{C})$ can be trivial only in the case when all but one of the labels on the Dynkin diagram $A_n^{(1)}$ corresponding to φ are zero: $(0, 0, \ldots, a, 0, \ldots, 0)$. But these labels are relatively prime, so $a = 1$. Hence, $\varphi = 1$. □

Set, as always, $G^0 = G \cap \mathcal{G}$, $G_i^0 = G_i \cap G^0$ and $G_{12}^0 = G_1^0 \cap G_2^0$. Consider the following two conditions:

$(2T^0)$: G^0 acts 2-transitively on $\{\mathcal{H}_1, \dots, \mathcal{H}_{n+2}\}$,

and

$S(k)$: L has no proper subgroups of index $\leq k$.

Proposition 5.2.2. *Condition $S(n)$ implies condition $(2T^0)$.*

Proof. Assume that $S(n)$ holds but that $(2T^0)$ does not.

1) Let R_1 be the kernel of G_1 on $\mathcal{H}_1$. Here we show that G_1^0/R_1 is a 2-transitive permutation group with socle L.

Clearly, R_1 is a normal subgroup of G_1^0. Now we have the embeddings:

$$\begin{array}{ccc} A = G_1^0/R_1 & \hookrightarrow & W(R) = \mathbb{S}_{n+1} \\ \downarrow & & \downarrow \\ B = G_1/R_1 & \hookrightarrow & A(R) = \mathbb{Z}_2 \times \mathbb{S}_{n+1} \end{array}$$

Using these embeddings, one can write each element $\varphi \in G_1/R_1$ in the form $\eta^i[\varphi]$, where $\mathbb{Z}_2 = Z(A(R)) =< \eta >$ and $[\varphi] \in \mathbb{S}_{n+1}$. Set $C = \{[\varphi] \mid \varphi \in G_1/R_1\}$. Then $A \subseteq C$. Furthermore,

$$B/A = G_1/G_1^0 = G_1/(G_1 \cap \mathcal{G}) \cong G_1\mathcal{G}/\mathcal{G} \subseteq \mathcal{A}/\mathcal{G} = \mathbb{Z}_2,$$

that is, $B/A \subseteq \mathbb{Z}_2$. The map $\varphi \mapsto [\varphi]$ defines an epimorphism $B \to C$ with kernel contained in $\mathbb{Z}_2$. Hence, $|B|/|C| = 1$ or 2, and so $C/A = 1$ or $\mathbb{Z}_2$. In particular, A is a normal subgroup of C. Recall that C is a 2-transitive permutation group of degree $n+1$ with socle L. We shall check that $A \supseteq L$. Indeed, $A \cap L \lhd L$; but L is simple, and so either $A \supseteq L$ or $A \cap L = 1$. In the latter case $[A, L] = 1$ and $A \subseteq C_C(L) = 1$, a contradiction. It is recorded in Table A2 that the (nonabelian) socle L of the 2-transitive permutation group C is always 2-transitive, except the case $L = L_2(8)$ and $C = P\Gamma L_2(8) = L.\mathbb{Z}_3$. But in this exceptional case, A is a normal subgroup of index not greater than 2 in C and so coincides with C. We have shown that $A \supseteq L$ and A is 2-transitive.

2) In this step we show that

$$(G_1^0 : G_{12}^0) \leq n. \tag{5.3}$$

Indeed, suppose that $(G_1^0 : G_{12}^0) > n$. Then the group G_1^0 permutes the $n+1$ subalgebras $\mathcal{H}_2, \dots, \mathcal{H}_{n+2}$ transitively. Furthermore, $(G : G^0) \leq 2 < n+2$ and $(G : G_1) = n+2$, therefore G^0 cannot leave $\mathcal{H}_1$ fixed. In other words, G^0 permutes the $n+2$ subalgebras $\mathcal{H}_1, \dots, \mathcal{H}_{n+2}$ transitively, and we arrive at condition $(2T^0)$, contrary to our assumption.

3) Next we show that G_{12}^0 acts faithfully on the spaces $(\mathcal{H}_1)^*$ and $(\mathcal{H}_2)^*$ as a transitive group of degree $n+1$ with socle L.

It was shown in 1) that $A = G_1^0/R_1$ has socle L. Set $D = R_1G_{12}^0/R_1$. Then by (5.3),

$$(A : D) = (G_1^0 : R_1G_{12}^0) \leq (G_1^0 : G_{12}^0) \leq n.$$

But $L \lhd A$ and $D \subseteq A$, and so LD is a subgroup of A, which implies that $|A| \geq |LD| \geq |L| \cdot |D|/|L \cap D|$. Hence, $(L : L \cap D) \leq (A : D) \leq n$. Now we recall that L satisfies condition $S(n)$. Thus we must have $L \cap D = L$ and $L \subseteq D$. This shows that $L \lhd R_1 G_{12}^0 / R_1$.

Now let R_2 denote the kernel of G_2 on $\mathcal{H}_2$, and set $M_1 = G_{12}^0 \cap R_1$, $M_2 = G_{12}^0 \cap R_2$. Then M_1 and M_2 are normal subgroups of G_{12}^0; moreover, $M_1 \cap M_2 \subseteq R_1 \cap R_2 = 1$ by Lemma 5.2.1. Hence,

$$M_2 = M_2/(M_1 \cap M_2) \cong M_1 M_2 / M_1 \lhd G_{12}^0 / M_1 =$$

$$= G_{12}^0/(G_{12}^0 \cap R_1) \cong R_1 G_{12}^0 / R_1 = D.$$

Thus D has an abelian normal subgroup M isomorphic to M_2, on the one hand. On the other hand $L \lhd D$, which implies that $M \cap L \lhd L$; but L is simple and M is abelian. Hence $M \cap L = 1$ and $[M, L] = 1$, that is, M centralizes L. But in that case, M is contained in $C_A(L)$, which is trivial because A is a 2-transitive group with socle L. Consequently, $M = 1$ and $M_2 = 1$. As the roles of M_1 and M_2 are symmetric, we can conclude that $M_1 = M_2 = 1$ and G_{12}^0 acts faithfully on $(\mathcal{H}_1)^*$ and $(\mathcal{H}_2)^*$. Furthermore,

$$G_{12}^0 = G_{12}^0/M_1 = G_{12}^0/(G_{12}^0 \cap R_1) \cong D.$$

It is easy to see that $\mathrm{soc}(G_{12}^0) \cong \mathrm{soc}(D) = L$.

4) As a main result of 1) – 3) we obtain that L is embedded in G_{12}^0 and therefore in G_1^0 as well. So we can now assume that $L \subseteq G_1^0$. As a subgroup of G_1^0, L fixes $\mathcal{H}_1$ and acts on $\{\mathcal{H}_2, \ldots, \mathcal{H}_{n+2}\}$. Let $L_i = St_L(\mathcal{H}_i)$ for $i > 1$. Then $(L : L_i) \leq n+1$. But L satisfies $S(n)$, and so either $(L : L_i) = n + 1$, or $L = L_i$. In the former case L permutes $\{\mathcal{H}_2, \ldots, \mathcal{H}_{n+2}\}$ transitively, and by analogy with 1), we arrive at property $(2T^0)$. In the latter case L leaves all subalgebras $\mathcal{H}_i$ fixed; in other words, $1 \neq L \subseteq K$, which contradicts the main P-case assumption. This proves Proposition 5.2.2. □

An analogue of Proposition 5.2.2 is the following statement:

Proposition 5.2.3. *Suppose that*

(i) *L satisfies condition $S((n+2)/2)$;*

(ii) *$L = G_1^0/R_1$, where R_1 is the kernel of G_1^0 on $\mathcal{H}_1$.*

Then property $(2T^0)$ holds.

Proof. 1) Assume the contrary: G^0 is not 2-transitive on $\{\mathcal{H}_1, \ldots, \mathcal{H}_{n+2}\}$. Then, as we have shown in item 1) of the proof of Proposition 5.2.2, G_1^0 (here we are using the notation of that proof) is intransitive on $\{\mathcal{H}_2, \ldots, \mathcal{H}_{n+2}\}$. Hence, at least one G_1^0-orbit on this set, say that containing $\mathcal{H}_2$, has length $a \leq (n+1)/2$. Clearly, in that case $(G_1^0 : G_{12}^0) = a$. Note that we have also that $(G_2^0 : G_{12}^0) = a$. Indeed,

G is transitive on $\{\mathcal{H}_1, \ldots, \mathcal{H}_{n+2}\}$, and $G^0 \lhd G$, therefore G^0 is $(1/2)$-transitive on the same set. The last sentence implies that $|G_2^0| = |G_1^0| = a|G_{12}^0|$.

2) In this step we establish the isomorphism $G_{12}^0 \cong L$ and also that $G_1^0 = R_1 \times G_{12}^0$.

Clearly,

$$(L : R_1 G_{12}^0/R_1) = (G_1^0 : R_1 G_{12}^0) \leq (G_1^0 : G_{12}^0) = a \leq (n+1)/2.$$

But L satisfies condition $S((n+1)/2)$, and so $R_1 G_{12}^0/R_1 = G_1^0/R_1$ and $R_1 G_{12}^0 = G_1^0$. Again denote by R_2 the kernel of G_2 on $\mathcal{H}_2$ and set $M_1 = R_1 \cap G_{12}^0$, $M_2 = R_2 \cap G_{12}^0$. Then M_1, $M_2 \lhd G_{12}^0$ and $M_1 \cap M_2 \subseteq R_1 \cap R_2 = 1$. Hence,

$$M_2 \cong M_1 M_2/M_1 \lhd G_{12}^0/M_1 = G_{12}^0/(R_1 \cap G_{12}^0) \cong R_1 G_{12}^0/R_1 = L.$$

But L is simple and M_2 is abelian, and therefore $M_2 = 1$. As the roles of M_1 and M_2 are symmetric, we can state that $M_1 = 1$. Consequently,

$$G_{12}^0 = G_{12}^0/M_1 \cong R_1 G_{12}^0/R_1 = L,$$

and $G_1^0 = R_1.G_{12}^0$.

Recall that R_1 is an abelian group of order $(G_1^0 : G_{12}^0) = a \leq (n+1)/2$. In particular, $|R_1 \backslash \{1\}| \leq (n+1)/2$. Consider the action of $G_{12}^0 \cong L$ by conjugation on $R_1 \backslash \{1\}$. The resulting orbits have length not greater than $(n+1)/2$, whereas L satisfies condition $S((n+1)/2)$. So all orbits are of length 1. In other words, G_{12}^0 centralizes R_1 and $G_1^0 = R_1 \times G_{12}^0$, as claimed.

3) Next we show that G^0 acts primitively on $\{\mathcal{H}_1, \ldots, \mathcal{H}_{n+2}\}$. Firstly suppose that G^0 is intransitive. Because of the $(1/2)$-transitivity of G^0 on the same set, G^0 has orbits of the same length $b \leq (n+2)/2$ on $\{\mathcal{H}_1, \ldots, \mathcal{H}_{n+2}\}$. Consider the action of $G_{12}^0 \cong L$ on these orbits. As L satisfies $S((n+2)/2)$, L must leave every subalgebra $\mathcal{H}_i$ fixed, which is impossible in the P-case. So G^0 is transitive. Now suppose that G^0 is imprimitive, say G^0 permutes b blocks Ω_j, each of which contains c subalgebras $\mathcal{H}_i$, where $b \cdot c = n+2$ and $b, c > 1$. Of course, $b, c \leq (n+2)/2$. Consider the action of L on the set $\{\Omega_j\}$. Because L has no proper subgroups of index at most $(n+2)/2$, L must fix every block Ω_j. Furthermore, the length of Ω_j is equal to $c \leq (n+2)/2$, therefore in its action on Ω_j, L leaves every subalgebra $\mathcal{H}_i \in \Omega_j$ fixed. In other words, $1 \neq L \subseteq K$, a contradiction.

4) Finally, set $N = N_{G^0}(G_{12}^0)$. Then, by 2), we have $G_1^0, G_2^0 \subseteq N$. In 3) we showed that G_1^0 and G_2^0 are maximal in G^0, and therefore $N = G^0$ or $N = G_1^0 = G_2^0$. In the former case, we obtain $L \cong G_{12}^0 \lhd G^0$. Thus $L \lhd G^0$, $L \subseteq G_1^0$ and G^0 is transitive on $\{\mathcal{H}_1, \ldots, \mathcal{H}_{n+2}\}$. This implies that $L \subseteq G_i^0$ for each i, that is $K \neq 1$, a contradiction. In the latter case consider an element $\varphi \in G^0$ such that $\varphi(\mathcal{H}_1) = \mathcal{H}_2$. Then

$$G_1^0 = G_2^0 = St_{G^0}(\mathcal{H}_2) = \varphi.St_{G^0}(\mathcal{H}_1).\varphi^{-1} = \varphi.G_1^0.\varphi^{-1},$$

that is, $\varphi \in N_{G^0}(G_1^0)\backslash G_1^0$. This means that $G_1^0 \subset N_{G^0}(G_1^0) \subseteq G$. Because of the maximality of G_1^0, we must have that $N_{G^0}(G_1^0) = G^0$ and $G_1^0 \lhd G^0$. But G^0 is transitive on $\{\mathcal{H}_1, \ldots, \mathcal{H}_{n+2}\}$, therefore from the normality of G_1^0 in G^0 it follows that $G_1^0 = G_i^0$ for each i, that is, K is not trivial, a contradiction. This proves Proposition 5.2.3. □

Now let $\mathcal{C}$ denote the list of pairs $(L, N = n+1)$, where L is the nonabelian socle of a 2-transitive permutation group of degree $N = n+1$ (see Table A2), and $\mathcal{C}^0$ the following part of $\mathcal{C}$: $(L, N) = (\mathbb{A}_5, 6)$, $(\mathbb{A}_6, 10)$, $(\mathbb{A}_7, 15)$, $(\mathbb{A}_8, 15)$, $(L_2(7), 8)$, $(L_2(11), 12)$, $(L_2(8), 28)$, $(U_3(5), 126)$, $(M_{11}, 12)$, $(HS, 176)$, $(Sp_{2d}(2), 2^{d-1}(2^d+1))$ $(d > 2)$.

Lemma 5.2.4. *For any pair $(L, n+1)$ belonging to the list $\mathcal{C}\backslash\mathcal{C}^0$, condition $S(n)$ holds.*

Proof. Let $P(X)$ denote, as usual, the minimal index of proper subgroups of a finite group X. Then we have $P(\mathbb{A}_5) = 5$, $P(\mathbb{A}_6) = 6$, $P(\mathbb{A}_7) = 7$, $P(\mathbb{A}_8) = 8$, $P(L_2(7)) = 7$, $P(L_2(11)) = 11$, $P(L_2(8)) = 9$, $P(U_3(5)) = 50$, $P(M_{11}) = 11$, $P(HS) = 100$, and $P(Sp_{2d}(2)) = 2^{d-1}(2^d - 1)$ (see Table A6). Thus, $S(n)$ does not hold for any pair $(L, n+1) \in \mathcal{C}^0$. As it is clear from Table A6, $S(n)$ holds for all remaining pairs, as for them we have $P(L) = n+1$. □

Corollary 5.2.5. *Condition $(2T^0)$ holds for $(L, n+1) \in \mathcal{C}\backslash\mathcal{C}^0$.* □

Lemma 5.2.6. *Condition $(2T^0)$ holds if the pair $(L, n+1)$ takes one of the following values:* $(\mathbb{A}_5, 6)$, $(\mathbb{A}_6, 10)$, $(L_2(11), 12)$, $(L_2(8), 28)$, $(U_3(5), 126)$, $(M_{11}, 12)$, $(Sp_{2d}(2), 2^{d-1}(2^d+1))$.

Proof. For the first six values, the integer $n+2$ is prime. As $(G : G^0) \leq 2$ and G is transitive on $\{\mathcal{H}_1, \ldots, \mathcal{H}_{n+2}\}$, G^0 is also transitive on the same set. Furthermore, in the generic position of the P-case, G^0 acts faithfully on $\{\mathcal{H}_1, \ldots, \mathcal{H}_{n+2}\}$ and G^0 is insoluble, since G^0 has L as section. So, by Wielandt's Theorem [Wie], G^0 is 2-transitive. Finally, set $(L, n+1) = (Sp_{2d}(2), 2^{d-1}(2^d+1))$. Then L satisfies the following conditions:

(i) $S((n+2)/2)$ (as $P(L) = 2^{d-1}(2^d - 1)$);

(ii) $\mathrm{Aut}(L) = L$ (see Table A3).

But $L \lhd G_1^0/R_1 \subseteq \mathrm{Aut}(L) = L$, and so $G_1^0/R_1 = L$. Hence, $(2T^0)$ follows from Proposition 5.2.3. □

Lemma 5.2.7. *Condition $(2T^0)$ holds for $(L, n+1) = (L_2(7), 8)$ and $(HS, 176)$.*

Proof. When $(L, n+1) = (L_2(7), 8)$, G^0 is a faithful transitive permutation group of degree 9. If G^0 is imprimitive, then $G^0 \subseteq \mathbb{S}_3 \wr \mathbb{S}_3$, that is, G^0 is soluble, a contradiction. So G^0 is primitive; moreover, 7 divides $|G^0|$. From [DiM] it follows that G^0 is 2-transitive. Now suppose that $(L, n+1) = (HS, 176)$. Here HS is the Higman-Sims simple group of order $2^9.3^2.5^3.7.11$ and G^0 is a faithful transitive permutation group of degree $177 = 3.59$. If G^0 is primitive, then $G^0 \rhd \mathbb{A}_{177}$, according to [DiM], and we arrive at property $(2T^0)$. Assuming that G^0 is imprimitive, we distinguish the following two cases.

1) G^0 permutes three blocks, each of which contains 59 subalgebras. Let G_B be a block stabilizer. Then G_B is a transitive permutation group of degree 59 and, in addition, 11 divides $|G_B|$. Hence, G_B has $\mathbb{A}_{59}$ as composition factor by [DiM], and G_1^0 has $\mathbb{A}_{58}$ as composition factor.

2) G^0 permutes 59 blocks, each of which contains 3 subalgebras. Let S consist of all the elements $g \in G^0$ leaving every block fixed. Then $S \subseteq (\mathbb{S}_3)^{59}$, G^0/S is a faithful transitive permutation group of degree 59 and, in addition, 11 divides $|G^0/S|$. By [DiM] again, we must have $G^0/S \rhd \mathbb{A}_{59}$. If G_B is a block stabilizer, then G_B, as a subgroup of index 59 in G^0 containing S, is of the form $S.A$, where $\mathbb{A}_{58} \subseteq A \subseteq \mathbb{S}_{58}$. But G_1^0 is a subgroup of index 3 in G_B, and so G_1^0 has $\mathbb{A}_{58}$ as composition factor.

Thus we have shown that G_1^0 has a composition factor isomorphic to $\mathbb{A}_{58}$. On the other hand, $HS \lhd G_1^0/R_1 \subseteq \mathrm{Aut}(HS)$, that is, HS is a unique nonabelian composition factor of G_1^0, a contradiction. □

Lemma 5.2.8. *Condition* $(2T^0)$ *holds for* $(L, n+1) = (\mathbb{A}_8, 15)$.

Proof. Recall that G is a faithful transitive permutation group of degree 16.

1) Firstly suppose that G is primitive. By [DiM], either $\mathbb{A}_{16} \lhd G \subseteq \mathbb{S}_{16}$ or $\mathbb{Z}_2^4 \lhd G = \mathbb{Z}_2^4.G_1 \subseteq \mathbb{Z}_2^4.SL_4(2)$. In the former case we have also that $\mathbb{A}_{16} \lhd G^0 \subseteq \mathbb{S}_{16}$, that is, condition $(2T^0)$ follows. In the latter case, $G_1^0 \subseteq G_1 \subseteq SL_4(2) = \mathbb{A}_8$, but $G_1^0/R_1 \rhd \mathbb{A}_8$; therefore in fact we have $G_1^0 = G_1 = \mathbb{A}_8$, $G^0 = G = \mathbb{Z}_2^4.\mathbb{A}_8$, and we arrive again at property $(2T^0)$.

2) Secondly, suppose that G is imprimitive. Since 7 divides $|G|$, G cannot be embedded in $\mathbb{S}_4 \wr \mathbb{S}_4$, and it remains to consider the following two cases.

2a) G permutes 8 blocks, each of which contains 2 subalgebras, that is, $G \hookrightarrow \mathbb{S}_2 \wr \mathbb{S}_8$. Let S consist of all the elements $g \in G$ leaving every block fixed, and let G_B denote a block stabilizer. Then $G = S.H$ and $G_B = S.H_1$, where $S \subseteq \mathbb{Z}_2^8$, $H \subseteq \mathbb{S}_8$ and $(H : H_1) = 8$. In particular, G_B cannot have $\mathbb{A}_8$ as composition factor. Meanwhile $G_B \supset G_1 \supseteq G_1^0$, $G_B/G_1 = \mathbb{Z}_2$, $G_1/G_1^0 \subseteq \mathbb{Z}_2$ and G_1^0 has $L = \mathbb{A}_8$ as composition factor, a contradiction.

2b) G permutes two blocks, each of which contains 8 subalgebras, that is, $G \hookrightarrow \mathbb{S}_8 \wr \mathbb{S}_2$. Let $S = G_B$ be a block stabilizer, so that $G/S = \mathbb{Z}_2$. Set

$$E_1 = \{g \in S \mid g \text{ acts trivially on the first block}\},$$

$$E_2 = \{g \in S \mid g \text{ acts trivially on the second block}\}.$$

Then $E_1 \cap E_2 \subseteq K = 1$ and $E_1, E_2 \lhd S$. Writing S in the form $E_1.H$, where $H \subseteq \mathbb{S}_8$, we have $G_1 = E_1.H_1$, where $(H : H_1) = 8$. Clearly, H_1 has no composition factors isomorphic to $\mathbb{A}_8$, on the one hand. On the other hand, $G_1/G_1^0 \subseteq \mathbb{Z}_2$ and G_1^0 has $L = \mathbb{A}_8$ as composition factor; therefore E_1 has $\mathbb{A}_8$ as composition factor. Furthermore,

$$E_1 \cong E_1E_2/E_2 \lhd S/E_2 \subseteq \mathbb{S}_8,$$

so in fact we have $\mathbb{A}_8 \lhd E_1 \subseteq \mathbb{S}_8$. Similarly, $\mathbb{A}_8 \lhd E_2 \subseteq \mathbb{S}_8$. Hence, $S \rhd E_1 \times E_2 \rhd \mathbb{A}_8 \times \mathbb{A}_8$. In particular, $|S|$ is divisible by $(8!/2)^2$ and $|H| = |S|/|E_1|$ is divisible by $(8!/2)^2/8! = 8!/4$. But $H \subseteq \mathbb{S}_8$, and so the last condition implies that $\mathbb{A}_8 \lhd H \subseteq \mathbb{S}_8$ and $\mathbb{A}_7 \lhd H_1 \subseteq \mathbb{S}_7$. Thus G_1 has $\mathbb{A}_7$ and $\mathbb{A}_8$ as composition factors; meanwhile $L = \mathbb{A}_8$ is the unique nonabelian composition factor of G_1, a contradiction. □

Before studying the last possibility $(L, n+1) = (\mathbb{A}_7, 15)$, we establish a general statement.

Lemma 5.2.9. *Let $\mathcal{H}$ be a Cartan subalgebra of $\mathcal{L}$ and T the corresponding maximal torus. Suppose in addition that H is a doubly transitive subgroup of $W(R) = N_{\mathcal{G}}(T)/T$. Then $C_T(H) = 1$.*

Proof. Assuming the contrary, that is, $C_T(H) \neq 1$, we consider an element $f \in C_T(H)\setminus\{1\}$. One can suppose that $|f| = p$ for some prime number p. Then f can be identified with a functional belonging to $P = \mathrm{Hom}_{\mathbb{Z}}(R, \mathbb{F}_p)$. Express the root system R in the form $\{e_i - e_j \mid 1 \leq i \neq j \leq n+1\}$. Clearly, the Weyl group $W(R)$ acts on R via permutations of the basis vectors e_i. Furthermore, H acts on P by the rule $(h \circ g)(\delta) = g(h^{-1}(\delta))$, where $h \in H$, $g \in P$ and $\delta \in R$. As $f \in C_T(H)$, we have $f(h^{-1}(\delta)) = f(\delta)$ for all $h \in H$ and $\delta \in R$. Take $\delta = e_1 - e_2$ and $h \in H$ such that $h(e_i) = e_1$ and $h(e_j) = e_2$. Then $f(e_i - e_j) = f(e_1 - e_2) = \alpha \in \mathbb{F}_p$ whenever $i \neq j$. In particular, $f(e_1 - e_2) = f(e_1 - e_3) = f(e_2 - e_3) = \alpha$. However, in this case we have

$$\alpha = f(e_1 - e_3) = f((e_1 - e_2) + (e_2 - e_3)) = f(e_1 - e_2) + f(e_2 - e_3) = \alpha + \alpha.$$

Hence, $\alpha = 0$ and $f = 1$, a contradiction. □

Lemma 5.2.10. *Condition $(2T^0)$ holds for $(L, n+1) = (\mathbb{A}_7, 15)$.*

Proof. First of all we make the following observations:

(i) $P(\mathbb{A}_7) = 7$;

(ii) if P is a faithful 2-transitive permutation group of degree 15 with socle isomorphic to $\mathbb{A}_7$, then $P \cong \mathbb{A}_7$ (although $\mathrm{Out}(\mathbb{A}_7) = \mathbb{Z}_2$). In particular, in the notation of Proposition 5.2.3 we have $G_1^0/R_1 = \mathbb{A}_7$.

1) Suppose that

$$\text{some } G_1^0\text{-orbit on } \{\mathcal{H}_2, \ldots, \mathcal{H}_{16}\}, \text{ say that containing } \mathcal{H}_2, \text{ has length } a \leq 6. \tag{5.4}$$

The same arguments as in items 1) and 2) of the proof of Proposition 5.2.3, show that $G_1^0 = R_1 \times G_{12}^0$ and $G_{12}^0 \cong L = \mathbb{A}_7$ under this assumption. But in that case, $R_1 \subseteq C_{T_1}(L)$, where T_1 is the maximal torus corresponding to $\mathcal{H}_1$. By Lemma 5.2.9, $R_1 = 1$. Thus $G_1^0 = G_{12}^0 = G_2^0 = L = \mathbb{A}_7$.

2) Recall (see [ATLAS]) that $\mathbb{A}_7$ has two complex irreducible characters χ_5 and χ_6 (in the notation of [ATLAS]) of degree 14; moreover, if δ_{15} denotes the permutation character corresponding to a 2-transitive permutation representation of degree 15 of $L = \mathbb{A}_7$, then $\delta_{15} = 1_L + \chi_6$. Furthermore, $\mathbb{A}_7$ has a unique irreducible character χ_2 of degree 6, and so the permutation character δ_7 corresponding to a 2-transitive permutation representation of degree 7 of L is simply $1_L + \chi_2$. Here, we show that if L leaves a Cartan subalgebra $\mathcal{H}$ fixed, then the L-character τ_L afforded by $\mathcal{H}$ is χ_6, $2\chi_2 + 2.1_L$ or $\chi_2 + 8.1_L$. Indeed, if T is the maximal torus corresponding to $\mathcal{H}$, then $L \cap T \lhd L$, but L is simple and T is abelian, so $L \cap T = 1$, that is, L acts faithfully on $\mathcal{H}$ (and on $\mathcal{H}^*$). From the simplicity of L, it follows also that $L \subseteq W(R) = \mathbb{S}_{15}$. Consider the permutation representation of degree 15 of L arising here, and denote a point stabilizer by L_1. If $(L : L_1) = 15$, then $L_1 = L_2(7)$, L acts 2-transitively and the character τ_L of L on $\mathcal{H}^*$ and on $\mathcal{H}$ is $\delta_{15} - 1_L = \chi_6$. If $(L : L_1) < 15$, then either $L_1 = L$ or $L_1 \subseteq \mathbb{A}_6 \subset L$. In the latter case, if $L_1 \subset \mathbb{A}_6$, then $(\mathbb{A}_6 : L_1) \geq 6$ and $(L : L_1) \geq 42 > 15$, a contradiction. Hence $L_1 = \mathbb{A}_6$. Thus we have either $L_1 = L$, or $(L : L_1) = 7$ and $L_1 = \mathbb{A}_6$. In other words, L acts on 15 symbols with orbits of lengths $(7, 7, 1)$ or $(7, 1, 1, \ldots, 1)$. Correspondingly, τ_L is $2\chi_2 + 2.1_L$ or $\chi_2 + 8.1_L$ as required.

3) Next we show that under hypothesis (5.4), the group $G_{12} = G_1 \cap G_2$ acts on $\mathcal{H}_1$ and on $\mathcal{H}_2$ with the same character. Recall that $(G : G^0) \leq 2$, so that $(G_1 : G_1^0) \leq 2$. Moreover, if ρ is the G_1-character afforded by $\mathcal{H}_1$, then $\chi = \mathrm{Ind}_{G_1}^G(\rho)$ and $\rho \in \mathrm{Irr}(G_1)$. However, $(G_1 : L) \leq 2$, therefore ρ_L is a sum of at most two irreducible characters of L. Taking account of the results of 2), we see that $\rho_L = \chi_6$. Similarly, if ρ' is the character of G_2 on $\mathcal{H}_2$, then $\rho'|_L = \chi_6$. Hence, the desired statement follows immediately if $G_{12} = L$. If $G_{12} \supset L$, then we must have $G_1 = G_2 = G_{12} = L.\mathbb{Z}_2$. It is not difficult to see that $G_{12} = \mathrm{Aut}(L) = \mathbb{S}_7$ or $G_{12} = \mathbb{Z}_2 \times L$ in this case. With the former assumption, $\mathbb{S}_7$ would have a subgroup of index 15, a contradiction. Under the latter assumption, set $\mathbb{Z}_2 =< \varphi >$. As L is irreducible on $\mathcal{H}_1$, by Schur's Lemma φ is scalar on $\mathcal{H}_1$, that is, $\varphi|_{\mathcal{H}_1} = \pm 1_{\mathcal{H}_1}$. If in addition $\varphi|_{\mathcal{H}_1} = 1_{\mathcal{H}_1}$, then φ belongs to the maximal torus T_1 corresponding to $\mathcal{H}_1$. In other words, $\varphi \in C_{T_1}(L)$ and $C_{T_1}(L) \neq 1$, contrary to Lemma 5.2.9. Hence, $\varphi|_{\mathcal{H}_1} = -1_{\mathcal{H}_1}$. Similarly, $\varphi|_{\mathcal{H}_2} = -1_{\mathcal{H}_2}$. Consequently, $\rho|_{G_{12}} = \rho'|_{G_{12}}$ as stated.

4) Now we claim that condition (5.4) cannot hold. Indeed, by Mackey's Theorem [Ser 3], we have

$$1 = (\chi, \chi)_G \geq 1 + (\rho|_{G_{12}}, \rho'|_{G_{12}})_{G_{12}}.$$

In particular, $\rho|_{G_{12}} \neq \rho'|_{G_{12}}$. On the other hand, we have shown above that condition (5.4) implies that $\rho|_{G_{12}} = \rho'|_{G_{12}}$.

5) As we have seen, the length of any G_1^0-orbit on $\{\mathcal{H}_2, \dots, \mathcal{H}_{16}\}$ is always greater than 6. If in addition property $(2T^0)$ does not hold, then we necessarily get some G_1^0-orbit of length 8. Thus we can suppose that G_{12}^0 is a subgroup of index 8 in G_1^0. Recall that $G_1^0 = R_1.L$. Hence the index $(L : G_{12}^0 R_1/R_1) = (G_1^0 : G_{12}^0 R_1)$ divides $(G_1^0 : G_{12}^0) = 8$. But L has no subgroups of index 2, 4 or 8, and so $R_1 G_{12}^0 = G_1^0$. Setting $M_1 = R_1 \cap G_{12}^0$, we have

$$(R_1 : M_1) = (G_1^0 : G_{12}^0) = 8. \tag{5.5}$$

Let R_2 be the kernel of G_2 on $\mathcal{H}_2$ and set $M_2 = R_2 \cap G_{12}^0$. Then $M_1, M_2 \lhd G_{12}^0$ and $M_1 \cap M_2 = 1$. This implies that

$$M_2 \cong M_1 M_2/M_1 \lhd G_{12}^0/M_1 \cong G_1^0/R_1 = L \cong \mathbb{A}_7;$$

moreover, M_2 is abelian. Thus $M_2 = 1$. Similarly, $M_1 = 1$. Now equality (5.5) means that R_1 is an abelian group of order 8. If we consider $W = \Omega_2(R_1)$ as an $\mathbb{F}_2$-space acted on by the group $L = \mathbb{A}_7$, then $1 \leq \dim_{\mathbb{F}_2}(W) \leq 3$. However, $\mathrm{Hom}(\mathbb{A}_7, SL_3(2)) = 1$, so L acts trivially on W. In other words, $W \subseteq C_{T_1}(L)$ and $C_{T_1}(L) \neq 1$, contrary to Lemma 5.2.9. This final contradiction shows that condition $(2T^0)$ holds. This proves Lemma 5.2.10. □

The following proposition summarizes this section:

Proposition 5.2.11. *Suppose that, in the notation introduced above, the group $L = \mathrm{soc}(G_1/K_1)$ is nonabelian and simple. Then the group $G^0 = G \cap \mathrm{Inn}(\mathcal{L})$ acts 2-transitively on the set $\{\mathcal{H}_i \mid i = 1, \dots, n+2\}$.* □

5.3 The *P*-case. II. Affine obstruction

Throughout the section we shall suppose that $W = \mathrm{soc}(G_1/K_1)$ is an elementary abelian p-group: $W \cong \mathbb{Z}_p^m$ for some prime p. Then G_1/K_1 is an affine doubly transitive permutation group of degree $n + 1 = p^m$, which explains the word "affine" in the title of the section. Whereas in the analysis of the generic position the main role is played by pushing up the socle $\mathrm{soc}(G_1/K_1)$ to G_1, here, in the solution of the affine obstruction, the principal role is assigned to a proof of the fact that either W or its dual group W^* (together with its normalizer in G_1) is embedded in G_1.

We start with the following auxiliary statement:

Lemma 5.3.1. *Let H be a finite group with a normal subgroup $E \cong \mathbb{Z}_p^m$. Then H acts on $E\backslash\{1\}$ and on $\mathrm{Irr}(E)\backslash\{1_E\}$ by conjugation with the same permutation character; in particular, with the same total number of orbits.*

Proof. 1) Identify $\mathrm{Irr}(E)$ with the dual group $E^* = \mathrm{Hom}(E, \mathbb{F}_p)$. We have to prove that the permutation characters α and β of H on $E\backslash\{0\}$ and on $E^*\backslash\{0\}$ are the same. Here we use additive notation for the groups E and E^*. Suppose that the action of an element $h \in H$ on E corresponds to linear operator $\varphi \in GL(E)$. Then

$$\alpha(h) = \{ \text{ the number of } \varphi\text{-fixed nonzero vectors } v \in E\} = p^s - 1,$$

where s is the number of Jordan cells of φ with eigenvalue 1. Note that α and β are real-valued characters, so that $\beta(h) = \beta(h^{-1})$. Moreover, if Φ is the matrix of φ in a basis $\{e_i\}$, then in the dual basis $\{f_j\}$ (that is, $f_j(e_i) = \delta_{ij}$) the operator φ acting on E^* has matrix ${}^t\Phi^{-1}$. Thus, we have to show that Φ and ${}^t\Phi$ have the same numbers of Jordan cells with eigenvalue 1. Extending $\mathbb{F}_p$ to its algebraic closure $\bar{\mathbb{F}}_p$, one can suppose that $\{e_1, \dots, e_m\}$ is a Jordan basis for φ, namely, φ has matrix

$$\Phi = \mathrm{diag}(J_{a_1}(1), \dots, J_{a_s}(1), A),$$

in this basis, where

$$J_l(1) = \underbrace{\begin{pmatrix} 1 & 1 & 0 & \dots & 0 & 0 \\ 0 & 1 & 1 & \dots & 0 & 0 \\ \vdots & \vdots & \vdots & \ddots & \vdots & \vdots \\ 0 & 0 & 0 & \dots & 1 & 1 \\ 0 & 0 & 0 & \dots & 0 & 1 \end{pmatrix}}_{l}$$

is the Jordan cell of size $l \times l$, and the $r \times r$ matrix $A - E_r$ is non-singular. Then we have

$${}^t\Phi = \mathrm{diag}({}^tJ_{a_1}(1), \dots, {}^tJ_{a_s}(1), {}^tA).$$

When the basis $\{e_i\}$ is replaced by the basis

$$\{e_{a_1+\dots+a_s}, e_{a_1+\dots+a_s-1}, \dots, e_1, e_{a_1+\dots+a_s+1}, e_{a_1+\dots+a_s+2}, \dots, e_m\},$$

${}^t\Phi$ is replaced by the matrix

$$\mathrm{diag}(J_{a_s}(1), \dots, J_{a_1}(1), {}^tA),$$

that is, ${}^t\Phi$ has exactly s Jordan cells with eigenvalue 1.

2) The number of H-orbits on $E\backslash\{1\}$ and on $E^*\backslash\{1\}$ is equal to $(\alpha, 1_H)_H$ and $(\beta, 1_H)_H$, respectively. □

Corollary 5.3.2. *Suppose that H is a finite group having a normal subgroup $E \cong \mathbb{Z}_p^m$, and that H acts transitively on $E\backslash\{1\}$. Then*

(i) *every irreducible character $\rho \in \mathrm{Irr}(H)$ is trivial on E or has degree divisible by $p^m - 1$; and*

(ii) *every permutation representation of H of degree less than p^m is trivial on E.*

Proof. (i) By Clifford's Theorem, $\rho_E = e\sum_{i=1}^{t} \vartheta_i$, where $\{\vartheta_1, \dots, \vartheta_t\}$ is some H-orbit on $\mathrm{Irr}(E)$. If $\vartheta_i = 1_E$ for some i, then $t = 1$, $\rho_E = \rho(1).1_E$ and ρ is trivial on E. Suppose that $\vartheta_i \neq 1_E$. Since H is transitive on $E\backslash\{1\}$, by Lemma 5.3.1 it is also transitive on $\mathrm{Irr}(E)\backslash\{1_E\}$. Hence, $t = p^m - 1$ and $\deg\rho = e(p^m - 1)$.

(ii) Let ρ be a permutation character of H such that $\deg\rho < p^m$. Decompose ρ as a sum $e.1_H + \rho_1 + \ldots + \rho_k$, where $e \geq 1$ and $\rho_i \in \mathrm{Irr}(H)\backslash\{1_H\}$. Then $\deg\rho_i < p^m - 1$, and so, by (i), ρ_i is trivial on E. Hence, $\rho_E = \rho(1).1_E$ and $\mathrm{Ker}\rho \supseteq E$. □

Remark 5.3.3. In some sense, assertion 5.3.2(ii) plays the role of the property $S(p^m - 1)$ in §5.2.

The aim of §5.3 is to prove the following statement:

Proposition 5.3.4. *Suppose that, in the P-case, the group $W = \mathrm{soc}(G_1/K_1)$ is elementary abelian: $W \cong \mathbb{Z}_p^m$. Then the condition*

(2*T*)*: G acts 2-transitively on $\{\mathcal{H}_1, \dots, \mathcal{H}_{n+2}\}$*
is satisfied.

We shall distinguish two cases: $O_p(K_1) = 1$ and $O_p(K_1) \neq 1$.

Lemma 5.3.5. *Proposition* 5.3.4 *holds if $O_p(K_1) = 1$.*

Proof. Let R_1 be the kernel of G_1 on $\mathcal{H}_1$. Recall that $R_1 = K_1 \cap \mathcal{G}$ is abelian and $(K_1 : R_1) = 1$ or 2.

1) We can assume that p does not divide $|K_1|$.

In fact, if $p > 2$ and p divides $|K_1|$, then p divides $|R_1|$. But R_1 is abelian, and so $1 \neq O_p(R_1)$ char $R_1 \lhd K_1$, that is, $O_p(K_1) \neq 1$. If $p = 2$, then $\dim\mathcal{L} = 2^{2m} - 1$ is odd, and so by Clifford's Theorem, the group $G^0 = G \cap \mathcal{G}$ is irreducible on $\mathcal{L}$. In this case we can replace G by G^0, K_1 by R_1, and then, since $O_2(R_1) = 1$ and R_1 is abelian, it follows that $|R_1|$ is odd.

2) Set $L = K_1.W$. Let E be a Sylow p-subgroup of L. As $p \nmid |K_1|$, we have $E \cong W$ and $L = K_1.E$ splits over K_1. Set $N = N_{G_1}(E)$. In this part of the proof we show that N acts transitively on $E \backslash \{1\}$.

First, we claim that $NL = G_1$. Indeed, let $g \in G_1$. Then $gEg^{-1} \subseteq gLg^{-1} = L$ (as $L \lhd G_1$); hence gEg^{-1} is a subgroup of order p^m of L. But in this case $gEg^{-1} \in \mathrm{Syl}_p(L)$, and so there is an element $l \in L$ such that $lEl^{-1} = gEg^{-1}$. Thus $l^{-1}gEg^{-1}l = E$, $l^{-1}g = n \in N$ and $g = ln \in NL$ as required.

Recall that G_1/K_1 is a faithful 2-transitive permutation group with abelian socle W, so G_1/K_1 acts transitively on $W \backslash \{1\}$. Furthermore, $W = L/K_1 = (K_1.E)/K_1$. Hence, for any $e, f \in E \backslash \{1\}$ one can find $g \in G_1$ and $k_1 \in K_1$ such that $geg^{-1} = fk_1$. But $G_1 = NL = NK_1E$, so $g = nk_2e_1$ for some $n \in N$, $k_2 \in K_1$ and $e_1 \in E$. We have

$$fk_1 = geg^{-1} = nk_2e_1ee_1^{-1}k_2^{-1}n^{-1} = nk_2ek_2^{-1}n^{-1} =$$

$$= ne.(e^{-1}k_2e).k_2^{-1}n^{-1} =$$

$$= ne.k_3k_2^{-1}n^{-1} \text{ (where } k_3 = e^{-1}k_2e \in K_1 \lhd L)$$

$$= nen^{-1}.nk_3k_2^{-1}n^{-1} =$$

$$= nen^{-1}.k_4 \text{ (where } k_4 = nk_3k_2^{-1}n^{-1} \in K_1 \lhd G_1),$$

that is, $k_1k_4^{-1} = f^{-1}.nen^{-1} \in E \cap K_1 = 1$. Hence, $f = nen^{-1}$ for $n \in N$ as required.

3) Finally, we consider the action of N on $\{\mathcal{H}_2, \ldots, \mathcal{H}_{p^m+1}\}$. If this action is transitive, then it is clear that property $(2T)$ holds. In the contrary case, every N-orbit Ω_j on this set has length less than p^m. By Corollary 5.3.2(ii), E acts trivially on every orbit Ω_j, and so E leaves every subalgebra $\mathcal{H}_i$ fixed; this is impossible in the P-case. □

In the analysis of the case where $O_p(K_1) \neq 1$, we shall need the following slight modification of Lemma 5.3.1 and Corollary 5.3.2:

Lemma 5.3.6. *Let H be a finite group with a normal subgroup $E \cong \mathbb{Z}_p^m$, and p an odd prime. Then H acts on the factor-sets $(E \backslash \{1\})/\{\pm 1\}$ and $(E^* \backslash \{1\})/\{\pm 1\}$ with the same permutation character. If, in addition, H acts transitively on the factor-set $(E \backslash \{1\})/\{\pm 1\}$, then every permutation representation of H of degree less than p^m is trivial on E.*

Proof. 1) Let α denote the permutation character of H on the factor-set $(E \backslash \{1\})/\{\pm 1\}$. This set is nothing but the set of pairs $\{v, v^{-1}\}$, where $v \in E \backslash \{1\}$. Now we go over to additive notation for E and E^*. Suppose that an element $h \in H$

acts on E as the linear operator $\varphi \in GL(E)$. Then

$$\alpha(h) = \{\text{the number of vectors } v \in E\backslash\{0\} \text{ such that } \varphi(v) = \pm v\}/2 =$$

$$= (p^{s_+} + p^{s_-})/2 - 1,$$

where s_+ (or s_-) denotes the number of Jordan cells with eigenvalue 1 (or -1) of φ. Using the same arguments as in the proof of Lemma 5.3.1, we obtain that $\alpha = \beta$, where β is the permutation character of H on the factor-set $(E^*\backslash\{0\})/\{\pm 1\}$.

2) Now suppose that H is transitive on the factor-set $(E\backslash\{0\})/\{\pm 1\}$, and consider a permutation character ρ of H of degree less than p^m. Decompose ρ as a sum $e.1_H + \rho_1 + \ldots + \rho_k$, where $e \geq 1$ and $\rho_i \in \mathrm{Irr}(H)\backslash\{1_H\}$. By Clifford's Theorem,

$$\rho_i|_E = e_i \sum_{j=1}^{t_i} \vartheta_{ij},$$

where $e_i \geq 1$ and $\{\vartheta_{i1}, \ldots, \vartheta_{it_i}\}$ is some H-orbit on $\mathrm{Irr}(E)$. By the results of 1), H is transitive on the set of pairs $\{\alpha, \bar{\alpha}\}$, where $\alpha \in \mathrm{Irr}(E)\backslash\{1_E\}$, and $\bar{\alpha}$ denotes the complex conjugate of α. Hence we have to consider the following two possibilities:

(i) H is transitive on $\mathrm{Irr}(E)\backslash\{1_E\}$,

(ii) H has two orbits, Ω_1 and Ω_2, on $\mathrm{Irr}(E)\backslash\{1_E\}$; moreover,

$$|\Omega_1| = |\Omega_2| = (p^m - 1)/2,$$

and for all $\alpha \in \mathrm{Irr}(E)\backslash\{1_E\}$ we have

$$\alpha \in \Omega_1 \Leftrightarrow \bar{\alpha} \in \Omega_2.$$

Note that $\deg \rho_i < p^m - 1$, so under assumption (i), ρ_i is trivial on E for each i, and ρ is trivial on E. Consider possibility (ii), and suppose that ρ is not trivial on E. Then the set $I = \{i | E \not\subseteq \mathrm{Ker}\rho_i\}$ is not empty. Furthermore, for each $i \in I$ we have $\deg \rho_i = e_i.(p^m - 1)/2$. Hence, $|I| \leq 2(\deg \rho - 1)/(p^m - 1) < 2$, that is, $|I| = 1$. Thus we can suppose that

$$\rho_E = a.1_E + e_1 \sum_{\alpha \in \Omega_1} \alpha$$

for some $a \in \mathbb{N}$. As $\deg \rho < p^m$, we have $e_1 = 1$. Recall that ρ is a permutation character, so that $\rho = \bar{\rho}$. In particular,

$$a.1_E + \sum_{\alpha \in \Omega_1} \alpha = \rho_E = \overline{\rho_E} = a.1_E + \sum_{\alpha \in \Omega_1} \bar{\alpha} = a.1_E + \sum_{\alpha \in \Omega_2} \alpha,$$

that is,

$$\sum_{\alpha \in \Omega_1} \alpha = \sum_{\alpha \in \Omega_2} \alpha,$$

a contradiction. □

Lemma 5.3.7. *Proposition* 5.3.4 *holds, if* $O_p(K_1) \neq 1$.

Proof. Set $Q = O_p(K_1)$, and recall that $W = \mathrm{soc}(G_1/K_1) \cong \mathbb{Z}_p^m$.

1) First, suppose that $p = 2$. Then one can assume that $Q \subseteq G^0$ and that G_1^0/R_1 is a 2-transitive permutation group of degree p^m with socle W. Indeed, in this case $\dim \mathcal{L} = 2^{2m} - 1$ is odd, and so by Clifford's Theorem G^0 is irreducible on $\mathcal{L}$, as a subgroup of index not greater than 2 in G. Hence, we can replace G by G^0, G_1 by G_1^0 and K_1 by R_1.

2) Now suppose that $p > 2$. Then we claim that $Q \subseteq G^0$, and G_1^0/R_1 contains a subgroup $U \cong \mathbb{Z}_p^m$ such that U is normalized by G_1/R_1, and G_1/R_1 acts transitively on $U \backslash \{1\}$. For, every automorphism of $\mathcal{L}$ of odd order is inner, and so $Q \subseteq G^0$. Furthermore, set $B = G_1/R_1$ and $A = G_1^0/R_1$, and we use the notation of step 1) of the proof of Proposition 5.2.2. Recall that the group

$$G_1/K_1 = C = \{[\varphi] \mid \varphi \in B\}$$

is a 2-transitive permutation group with socle W; moreover, the map $\varphi \mapsto [\varphi]$ defines an epimorphism $B \to C$ with kernel K_1/R_1. Let $\tilde{W}$ be the full preimage of W under this epimorphism. Then $\tilde{W} \lhd B$ and $|\tilde{W}| = p^m$ or $2p^m$. Consider the latter possibility, in which case $\tilde{W} = Z.W$, where $Z = \mathbb{Z}_2 =< \eta >$. Taking $U \in \mathrm{Syl}_p(\tilde{W})$, we have $U \cong W = \mathbb{Z}_p^m$. Furthermore, $\tilde{W} = Z.U$ splits over Z. But Z centralizes U, and so in fact we have $\tilde{W} = Z \times U$ and $U = O_p(\tilde{W}) \lhd B$. In the former case, that where $|\tilde{W}| = p^m$, we set $U = \tilde{W}$. Thus we have constructed a normal subgroup $U \cong \mathbb{Z}_p^m$ of B. Now again write every element $\varphi \in U$ in the form $\eta^i[\varphi]$. Then $1 = \varphi^p = \eta^{ip}[\varphi]^p = \eta^i[\varphi]^p$, which implies that $\eta^i = [\varphi]^{-p} \in \mathbb{Z}_2 \cap \mathbb{S}_{n+1} = 1$, that is, $\varphi = [\varphi] \in \mathbb{S}_{n+1} = W(R)$. This means that φ is an inner automorphism. Thus we have shown that U is a normal subgroup of $A = G_1^0/R_1$ and of B. Now we recall that $W = U\bar{K}_1/\bar{K}_1$, where $\bar{K}_1 = K_1/R_1$, and $C = G_1/K_1$ acts transitively on $W\backslash\{1\}$. Hence, for any $\varphi, \psi \in U\backslash\{1\}$ one can find $g \in G_1/R_1$ and $i = 0$ or 1 such that $g\varphi g^{-1} = \psi.\eta^i$. But in this case $\eta^i = \psi^{-1}.g\varphi g^{-1} \in \mathbb{Z}_2 \cap \mathbb{S}_{n+1} = 1$, that is, $g\varphi g^{-1} = \psi$. In other words, G_1/R_1 acts transitively on $U\backslash\{1\}$ as stated.

As a result of steps 1) and 2), we can suppose that $W \cong \mathbb{Z}_p^m$ is a normal subgroup of G_1/R_1 lying in G_1^0/R_1, and that G_1/R_1 acts transitively on $W\backslash\{1\}$.

3) Recall that $Q = O_p(K_1) \neq 1$. Hence, $1 \neq \Omega_p(Z(Q))$ char $K_1 \lhd G_1$. Choose a minimal normal subgroup E of G_1 lying in $\Omega_p(Z(Q))$. Of course, $E \cong \mathbb{Z}_p^r$ for some natural r, and $E \subseteq R_1$. Here we shall show that E is centralized by W. Indeed, $E \lhd G_1$, but $[E, R_1] = 1$, so in fact G_1/R_1 acts on E. As W is an elementary abelian p-group acting on the $\mathbb{F}_p$-space E, we must have $D = C_E(W) \neq 1$. Clearly, $D \lhd G_1$, so by the minimality of E we obtain that $D = E$, and W centralizes E.

4) Recall that $W = \mathbb{Z}_p^m$ is a subgroup of the Weyl group $W(R)$ which acts transitively on p^m symbols. We shall identify these p^m symbols with vectors in the $\mathbb{F}_p$-space W (so that W acts on them by translations), and the root system R with

the set $\{e_u - e_v \mid u, v \in W, u \neq v\}$. Let T be the maximal torus corresponding to $\mathcal{H}_1$. In the proof of Proposition 4.3.10 we showed that

$$C_T(W) = \{f : X_{e_u - e_0} \mapsto \zeta^{\bar{f}(u)} X_{e_u - e_0} \mid \bar{f} \in W^*\},$$

where $\zeta = \exp(2\pi i/p)$ and $\{H_1, \dots, H_n, X_\alpha \mid \alpha \in R\}$ is a Chevalley basis. Now we explain the action of G_1/R_1 on $C_T(W)$. Every element $\varphi \in G_1/R_1 \subseteq A(R)$ can be written in the form $\eta^i[\varphi]$, where $< \eta >= \mathbb{Z}_2 = Z(A(R))$ and $[\varphi] \in \mathbb{S}_{p^m} = W(R)$. Clearly, $[\varphi] \in C = W.C_0$, where C_0 is the stabilizer of the symbol 0. We shall suppose that $[\varphi] \in C_0 \subseteq GL(W)$. Then for $f \in C_T(W)$, we have

$$\varphi f \varphi^{-1} : X_{e_u - e_0} \mapsto \zeta^{(-1)^i \bar{f}([\varphi]^{-1}(u))} X_{e_u - e_0}.$$

Hence, if we identify $C_T(W)$ with W^*, then an element $\varphi = \eta^i[\varphi] \in G_1/R_1$, where $[\varphi] \in C_0$, acts on $C_T(W) \cong W^*$ by the rule

$$\bar{f} \mapsto (-1)^i [\varphi] \circ \bar{f}. \tag{5.6}$$

5) We claim that E coincides with $C_T(W) = \mathbb{Z}_p^m$, and that G_1 acts transitively on the factor-set $(E\backslash\{1\})/\{\pm 1\}$. For, C_0 acts transitively on $W\backslash\{0\}$, because of the double transitivity of the group $C = W.C_0$. By Lemma 5.3.1, C_0 is transitive on $W^*\backslash\{0\}$. From formula (5.6), it now follows immediately that G_1/R_1 acts transitively on the factor-set $(C_T(W)\backslash\{1\})/\{\pm 1\}$. (When $p = 2$ one can suppose in (5.6) that $i = 0$ and so G_1/R_1 is transitive on $C_T(W)\backslash\{1\}$). Recall that E is a nontrivial G_1/R_1-invariant subgroup of $C_T(W)$. Hence when $p = 2$, E obviously coincides with $C_T(W)$. If $p > 2$, then

$$|C_T(W)| = p^m \geq p^r = |E| \geq 1 + \frac{p^m - 1}{2} > p^{m-1},$$

so $r = m$ and again we have $E = C_T(W)$.

6) Finally, consider the permutation representation of degree p^m of the group $G_1 \rhd E = \mathbb{Z}_p^m$ on $\{\mathcal{H}_2, \dots, \mathcal{H}_{p^m+1}\}$. If this representation is intransitive, then it is trivial on E by Lemma 5.3.6, that is, $1 \neq E \subseteq K$, which is impossible in the P-case. Hence, G_1 is transitive on $\{\mathcal{H}_2, \dots, \mathcal{H}_{p^m+1}\}$, and G is 2-transitive on $\{\mathcal{H}_1, \dots, \mathcal{H}_{p^m+1}\}$. This completes the proof of Lemma 5.3.7. □

5.4 The *P*-case. III. Completion of the proof

From Lemma 4.2.2 and Propositions 5.2.11 and 5.3.4, we have the following conclusions in the P-case:

(i) $G = \mathrm{Aut}(\mathcal{D})$ is a faithful 2-transitive permutation group of degree $n + 2$;

(ii) the point stabilizer G_1 admits a 2-transitive permutation representation of degree $n + 1$ with soluble kernel K_1.

Lemma 5.4.1. *In the notation introduced above,* $\mathrm{soc}(G)$ *is nonabelian.*

Proof. Assume the contrary, so that $W = \mathrm{soc}(G)$ is abelian. Then $W = \mathbb{Z}_q^l$ for some prime power q^l; furthermore, $n = q^l - 2$, $G = W.G_1$ and G_1 acts transitively on $W \backslash \{1\}$. By Corollary 5.3.2(i), $n(n+2) = \dim \mathcal{L} = \deg \chi$ is divisible by $|W| - 1 = n + 1$, a contradiction. □

By Lemma 4.4.4, we have $\mathrm{soc}(G) \neq \mathbb{A}_{n+2}$ in all cases.

Lemma 5.4.2. *The P-case cannot occur if* $n = p^m - 1$ *for some prime power* p^m.

Proof. By Lemma 5.4.1 and Table A2, we can suppose that the pair $(L = \mathrm{soc}(G), n+2 = p^m + 1)$ takes one of the following values: $(L_2(q), q+1)$ $(q = p^m \geq 5)$; $({}^2A_2(q), q^3+1)$; $({}^2B_2(q), q^2+1)$ $(q = 2^{2e+1} > 2)$; $({}^2G_2(q), q^3+1)$ $(q = 3^{2e+1} > 3)$; $(L_2(8), 28)$; $(M_{11}, 12)$; $(Sp_6(2), 28)$; $(M_{12}, 12)$ and $(M_{24}, 24)$. The first six cases are eliminated because $|\mathrm{Aut}(L)| < (\dim \mathcal{L})^2 = (n(n+2))^2$, and so G cannot have an irreducible character of degree $\dim \mathcal{L}$. If $L = Sp_6(2)$, then $\dim \mathcal{L} = 728$, but $G = L$ has no irreducible characters of degree 728 (see [ATLAS]). Similarly, $G = L = M_{24}$ has no irreducible characters of degree 528. Finally, suppose that $G = L = M_{12}$. As G is simple, G is contained in $\mathrm{Inn}(\mathcal{L}) = PSL_{11}(\mathbb{C})$. But the Schur multiplier of G is $\mathbb{Z}_2$ (see [ATLAS]), so we can suppose that $G \subseteq SL_{11}(\mathbb{C})$. Let χ and λ be G-characters afforded by $\mathcal{L}$ and $\mathbb{C}^{11}$. By Lemma 4.2.3, $\lambda \in \mathrm{Irr}(G)$ and $\chi = |\lambda|^2 - 1_G$. The group M_{12} has exactly two irreducible characters λ_1 and λ_2 of degree 11 [ATLAS], and both of $|\lambda_i|^2 - 1_G$, $i = 1, 2$, are reducible. This completes the proof of Lemma 5.4.2. □

Lemma 5.4.3. *The P-case cannot occur if* $n+1$ *is not primary.*

Proof. Assume that the P-case occurs. Because $n+1$ is not primary, $S = \mathrm{soc}(G_1/K_1)$ is a nonabelian simple group and $S \lhd G_1/K_1 \subseteq \mathrm{Aut}(S)$. Furthermore, by Lemma 5.4.1, $L = \mathrm{soc}(G)$ is also a nonabelian simple group. According to Table A2, we have to consider the following seven cases.

1) $(L, n+2) = (L_d(q), (q^d - 1)/(q-1))$, $d \geq 3$.

First, suppose that $d = 3$ and $q \leq 4$. When $q = 2$, $G = L = SL_3(2)$ and $G_1 = \mathbb{S}_4$ is soluble. If $q = 3$, then $L = SL_3(3)$, $L_1 = \mathbb{Z}_3^2.(\mathbb{Z}_2.\mathbb{S}_4)$ is soluble, where $L_1 = G_1 \cap L$, and so G_1 is also soluble. (Here and in what follows we shall use the fact that $G_1/L_1 \hookrightarrow \mathrm{Out}(L)$ is soluble, so that S is a composition factor for L_1 as well). If $q = 4$, then $L = L_3(4)$, $L_1 = \mathbb{Z}_2^4.\mathbb{A}_5$, and so $S = \mathbb{A}_5$. But in that case, S cannot be the socle of a 2-transitive permutation group of degree 20.

Secondly suppose that $d \geq 4$. Then L_1 has a unique nonabelian composition factor $S = L_{d-1}(q)$, and S can be the socle of a 2-transitive permutation group of only one degree, namely $(q^{d-1} - 1)/(q-1)$. Here there are two exceptions. The first is $(d, q) = (4, 2)$, where $n + 1 = 14$ and $S = SL_3(2)$ can be the socle of a 2-transitive permutation group of degree 7 or 8. The second is $(d, q) = (5, 2)$, where $n + 1 = 30$ and $S = SL_4(2)$ can be the socle of a 2-transitive permutation group of degree 8 or 15. In any case, there do not exist 2-transitive permutation groups with socle S and of degree $n + 1 = (q^d - 1)/(q - 1) - 1 = q(q^{d-1} - 1)/(q - 1)$.

Finally, suppose that $d = 3$ and $q \geq 5$. Then L_1 has a unique nonabelian composition factor, namely $S = L_2(q)$, and S cannot be the socle of a 2-transitive permutation group of degree $n + 1 = q(q + 1)$.

2) $(L, n + 2) = (Sp_{2d}(2), 2^{d-1}(2^d \pm 1))$, $d \geq 3$.

Here $G = L$, and $G_1 = \Omega^+_{2d}(2).\mathbb{Z}_2$ or $G_1 = \Omega^-_{2d}(2).\mathbb{Z}_2$. When $d \geq 4$ there are no 2-transitive permutation groups with socle $S = \Omega^+_{2d}(2)$ or $S = \Omega^-_{2d}(2)$. If $d = 3$, then $G_1 = \mathbb{A}_8.\mathbb{Z}_2$ or $G_1 = U_4(2).\mathbb{Z}_2$, that is, $S = \mathbb{A}_8$ or $S = U_4(2)$. Again, there do not exist 2-transitive permutation groups of degree $n+1 = 35$ (27 respectively) with socle $\mathbb{A}_8$ ($U_4(2)$ respectively).

3) $(L, n + 2) = (L_2(11), 11)$.

Here $G = L$, $S = G_1 = \mathbb{A}_5$ and $n + 1 = 10$. There are no 2-transitive permutation groups of degree 10 with socle $\mathbb{A}_5$.

4) $(L, n + 2) = (\mathbb{A}_7, 15)$.

There are no 2-transitive permutation groups of degree $n + 1 = 14$ with socle $S = G_1 = L_2(7)$.

5) $(L, n + 2) = (M_{11}, 11)$, $(M_{23}, 23)$.

Here $G = L$ has no irreducible characters of degree $99 = 9.11$ (degree $483 = 21.23$ respectively).

6) $(L, n + 2) = (M_{22}, 22)$.

Here $L \lhd G \subseteq \mathrm{Aut}(L) = L.\mathbb{Z}_2$ and G has no irreducible characters of degree $440 = 20.22$.

7) $(L, n + 2) = (HS, 176)$, $(Co_3, 276)$.

Here $G = L$, $S = U_3(5)$ or $S = McL$. There do not exist 2-transitive permutation groups with socle S of degree 175 (degree 275 respectively).

This completes the proof of Lemma 5.4.3. □

The results of §§5.2 – 5.4 show that the P-case cannot occur for any IOD of the Lie algebra $\mathcal{L}$ of type A_n. This completes the proof of Theorem 5.0.1.

Commentary

It seems that one can, on the one hand, relax the dependence of Theorem 5.0.1 on the classification of finite simple groups; and, on the other hand, strengthen this theorem. Namely, it looks very plausible that, under the weaker assumption that the group $G = \mathrm{Aut}(\mathcal{D})$ acts transitively on the set of $(n+1)(n+2)$ minimal projectors of the Lie algebra $\mathcal{L}$ of type A_n (see §1.6) (but is not necessarily irreducible on $\mathcal{L}$), a given orthogonal decomposition $\mathcal{D}$ must be a J-decomposition. Quite recently Ivanov [Iva 5] has proved that, under this assumption, the S-case cannot occur, and that the E-case means simply that $\mathcal{D}$ is a J-decomposition. His proof is rather elementary, and does not require the classification of finite simple groups. Combining this nice result with the Kryliuk observation (see Commentary 5 to Chapter 1), we get the following statement:

Let n be an odd integer, which is not a square. Suppose that the automorphism group Aut($\mathcal{D}$) *of a given orthogonal decomposition* $\mathcal{D}$ *of the simple Lie algebra of type* A_{n-1} *acts transitively on the set of* $n(n+1)$ *minimal projectors. Then n is a prime power, and* $\mathcal{D}$ *is a* J*-decomposition.*

Chapter 6

Classification of irreducible orthogonal decompositions of complex simple Lie algebras of type B_n

In this chapter we prove the following result, which gives a solution of the weak Winnie-the-Pooh problem for Lie algebras of type B_n.

Theorem 6.0.1. *The complex simple Lie algebra $\mathcal{L}$ of type B_n admits an IOD if and only if its rank n is of the form $[(p^m - 1)/2]$ for some prime power p^m.*

We shall also compute the number $d(\mathcal{L})$ of $\mathrm{Aut}(\mathcal{L})$-conjugacy classes of IODs of $\mathcal{L}$.

Theorem 6.0.2. *Suppose that $\mathcal{L}$ is of rank n. Then*

$$d(\mathcal{L}) = \begin{cases} 1 & \text{if } n = 1, 2; \\ 2 & \text{if } n = 3; \\ 1 & \text{if } n = 2^m - 1 \text{ but } 2n+1 \text{ is not prime}; \\ (n+3)/2 & \text{if } n = 2^m - 1 > 3 \text{ and } 2n+1 \text{ is a prime}; \\ 1 & \text{if } n = \frac{p^m-1}{2} \equiv 0 \pmod 2 \text{ for some prime } p; \\ (n+1)/2 & \text{if } n = (p-1)/2 \equiv 1 \pmod 2 \text{ for some} \\ & \text{prime } p \text{ other than a Mersenne prime}; \\ 1 + N(p,m) & \text{if } n = \frac{p^m-1}{2} \equiv 1 \pmod 2 \text{ for some prime } p \\ & \text{and } m > 1; \\ 0 & \text{otherwise.} \end{cases}$$

Here

$$\sum_{\substack{r \mid \gcd(m,p^m-1) \\ rs \mid m}} \frac{rs(p^{m/s}-3)}{4m} \geq N(p,m) \geq$$

$$\geq \sum_{s \mid m, s>1} \frac{p^s - p^{(s-1)/2}}{4s} + \frac{p-3}{4}.$$

For the remainder of the chapter we fix the following notation. Let V be a $(2n+1)$-dimensional complex space, $(*, *)$ a non-degenerate symmetric bilinear form on V with fixed standard basis $\{e_i, 1 \leq i \leq 2n+1\}$ where $(e_i, e_j) = \delta_{ij}$. Let

$$\mathcal{L} = so_{2n+1}(\mathbb{C}) = \{X \in \mathrm{End}(V) | {}^t X = -X\}$$

be the simple Lie algebra of type B_n with fixed basis

$$\{E(i, j) = E_{ij} - E_{ji}, 1 \leq i < j \leq 2n+1\},$$

where $E_{ij} = (\delta_{ik}\delta_{jl})_{1 \leq k,l \leq 2n+1}$, and set $\mathcal{G} = \mathrm{Aut}(\mathcal{L}) = SO_{2n+1}(\mathbb{C})$. Suppose that $\mathcal{L}$ admits an IOD $\mathcal{D}$ with automorphism group $G = \mathrm{Aut}(\mathcal{D})$ which acts (naturally) on $\mathcal{L}$ and on V with characters χ and ρ, respectively. By our assumptions, $\chi \in \mathrm{Irr}(G)$. Let G_1 denote the stabilizer $St_G(\mathcal{H}_1)$ of Cartan subalgebra $\mathcal{H}_1$, a component of $\mathcal{D}$.

Since the orthogonal decomposition is unique and irreducible for $n = 1$ and 2, we shall assume in what follows that $n \geq 3$.

Our strategy is as follows. In §6.1 we deduce the existence of an orthonormal basis $\{e_1, \ldots, e_{2n+1}\}$ of V in which the group G is monomial; that is, elements of G permute the $(2n+1)$ pairs $\pm e_i$ of basis vectors transitively. Moreover, if H is the image of G in this permutation representation, H has rank 2 or 3. As a consequence of §6.2, we obtain that each IOD is a decomposition of a special form, namely, an E-decomposition in the sense of §2.2. The final section §6.3 is devoted to the study of E-decompositions.

6.1 The monomiality of $G = \mathrm{Aut}(\mathcal{D})$

We start with a simple statement.

Lemma 6.1.1. *Let $\mathcal{H}$ be a Cartan subalgebra of $\mathcal{L}$. Then*

(i) *the subspace $V_0(\mathcal{H}) = \{v \in V | \forall h \in \mathcal{H}, h(v) = 0\}$ is one-dimensional and anisotropic;*

(ii) *$\mathcal{H}$ contains an element h such that the linear operator h^2 acts on V with spectrum $\{0, -1, -1, \ldots, -1\}$.*

Proof. By the conjugacy of Cartan subalgebras we can suppose that

$$\mathcal{H} = < E(1, 2), E(3, 4), \ldots, E(2n-1, 2n) >_{\mathbb{C}} .$$

Then $V_0(\mathcal{H}) = < e_{2n+1} >_{\mathbb{C}}$, and we may take

$$h = E(1, 2) + \ldots + E(2n-1, 2n).$$

□

We shall need the following two lemmas, in which $U^\perp$ denotes the orthogonal complement to a subspace U of V.

Lemma 6.1.2. *Let U, W be subspaces of V such that*
(i) $U \cap W = 0$, $U \oplus W \neq V$;
(ii) *the subspace*

$$\mathcal{M} = \{f \in \mathcal{L} | f(U) \subseteq W, f(W) \subseteq U, f((U \oplus W)^\perp) = 0\}$$

contains a Cartan subalgebra $\mathcal{H}$ of $\mathcal{L}$.
Then $\dim U = \dim W = n$.

Proof. Consider the element $h \in \mathcal{H}$ described in Lemma 6.1.1. Then $(U \oplus W)^\perp \subseteq \mathrm{Ker}(h^2)$. But $(U \oplus W)^\perp \neq 0$ by our assumptions, and $\dim \mathrm{Ker}(h^2) = 1$ by Lemma 6.1.1. Hence $\mathrm{Ker}(h^2) = (U \oplus W)^\perp$, $\dim U + \dim W = 2n$. Now assume that $\dim U \neq \dim W (\neq n)$. Then we may suppose $\dim U < \dim W$ and we have $h^2(W) \subseteq W$, whereas

$$\dim h^2(W) = \dim h(h(W)) \leq \dim h(U) \leq \dim U < \dim W.$$

This implies that $\mathrm{Ker}(h^2) \cap W \neq 0$. Since $\mathrm{Ker}(h^2) = (U \oplus W)^\perp$, we obtain that $W \cap (U \oplus W)^\perp \neq 0$, and so $((U \oplus W)^\perp, (*, *))$ is isotropic, a contradiction. □

Lemma 6.1.3. *Let U be a subspace of V such that the subspace $\mathcal{N} = \{f \in L|$ $f(U) \subseteq U, f(U^\perp) = 0\}$ contains some Cartan subalgebra $\mathcal{H}$ of $\mathcal{L}$. Then* $\dim U \geq$ $2n$.

Proof. Again we consider the element $h \in \mathcal{H}$ as described in Lemma 6.1.1. Then $h^2(U^\perp) = 0$, $U^\perp \subseteq \mathrm{Ker}(h^2)$ and $\dim U^\perp \leq \dim \mathrm{Ker}(h^2) = 1$. □

We may state the following about the behaviour of the character ρ:

Lemma 6.1.4. *ρ is realized over $\mathbb{R}$. Moreover, there exists a non-principal linear character $\delta : G_1 \to \{\pm 1\}$ of the subgroup $G_1 = St_G(\mathcal{H}_1)$ such that $\rho = \mathrm{Ind}_{G_1}^G(\delta)$. In particular, ρ is a monomial character and G_1 has a normal subgroup of index* 2.

Proof. As a compact subgroup of $O_{2n+1}(\mathbb{C}) = O(V)$, G is embedded in the maximal compact subgroup $O_{2n+1}(\mathbb{R})$. Hence, ρ is realized over $\mathbb{R}$. By Lemma 6.1.1, the subspace $V_0(\mathcal{H}_1)$ is generated over $\mathbb{C}$ by some anisotropic vector v. Because $V_0(\mathcal{H}_1)$ is obviously G_1-invariant, we can consider the character δ of G_1 afforded by $V_0(\mathcal{H}_1)$. For $g \in G_1$ we have

$$(v, v) = (gv, gv) = (\delta(g)v, \delta(g)v) = \delta^2(g)(v, v);$$

hence $\delta^2(g) = 1$, and $\delta : G_1 \to \{\pm 1\}$. Note that $(\rho_{G_1}, \delta)_{G_1} \geq 1$. Hence, by Frobenius reciprocity, we have

$$1 \leq (\rho_{G_1}, \delta)_{G_1} = (\rho, \mathrm{Ind}_{G_1}^G(\delta))_G.$$

But $\deg \rho = \deg(\mathrm{Ind}_{G_1}^G(\delta)) = 2n+1$, and ρ is irreducible, hence ρ and $\mathrm{Ind}_{G_1}^G(\delta)$ are the same. Furthermore, if δ is principal then

$$(\rho, 1_G)_G = (\mathrm{Ind}_{G_1}^G(1_{G_1}), 1_G)_G = (1_{G_1}, 1_{G_1})_{G_1} = 1,$$

which contradicts the irreducibility of ρ. So, the homomorphism δ of G_1 into $\{\pm 1\} = \mathbb{Z}_2$ is surjective, and its kernel is the required normal subgroup of G. □

Next we shall list some properties of the character ρ_{G_1}.

Proposition 6.1.5. *ρ_{G_1} is multiplicity-free. Furthermore, the degree of every irreducible constituent of $\rho_{G_1} - \delta$ is n or $2n$. In particular, ρ_{G_1} is a sum of at most 3 irreducible constituents.*

Proof. 1) Decompose the character ρ_{G_1} into a sum of homogeneous constituents: $\rho_{G_1} = \sum_{i=1}^t m_i\rho_i$, where $\rho_1, \ldots, \rho_t$ are pairwise different irreducible characters of G_1. By Lemma 4.2.3, for $g \in G_1$ we have

$$\chi_{G_1}(g) = \frac{1}{2}\left\{\sum_i m_i\rho_i(g)\right\}^2 - \frac{1}{2}\sum_i m_i\rho_i(g^2) =$$

$$= \frac{1}{2}\sum_i m_i\left\{\rho_i(g)^2 - \rho_i(g^2)\right\} + \frac{1}{2}\sum_i m_i(m_i-1)\rho_i(g)^2 +$$

$$+ \sum_{1\leq i<j\leq t} m_im_j\rho_i(g)\rho_j(g),$$

that is,

$$\chi_{G_1} = \sum_{i=1}^t \left(m_i\hat{\rho}_i + \frac{m_i(m_i-1)}{2}(\rho_i)^2\right) + \sum_{1\leq i<j\leq t} m_im_j\rho_i\rho_j,$$

where the $\hat{\rho}_i(g) = (\rho_i(g)^2 - \rho_i(g^2))/2$ are also characters of G_1 (or 0). (If ρ_i is afforded by a G_1-module W_i, then $\hat{\rho}_i$ is afforded by the exterior square $\wedge^2(W_i)$). Suppose, for example, that $m_1 > 1$. If ρ_1 is real-valued, then

$$(\chi_{G_1}, 1_{G_1})_{G_1} \geq \frac{m_1(m_1-1)}{2}((\rho_1)^2, 1_{G_1})_{G_1} = \frac{m_1(m_1-1)}{2} \geq 1.$$

If $\rho_1 \neq \bar{\rho}_1$, then by the reality of ρ_{G_1} (see Lemma 6.1.4), $\bar{\rho}_1$ must occur among the ρ_i, $1 \leq i \leq t$; say, $\rho_2 = \bar{\rho}_1$. Therefore

$$(\chi_{G_1}, 1_{G_1})_{G_1} \geq m_1m_2(\rho_1\rho_2, 1_{G_1})_{G_1} = m_1m_2(\rho_1\bar{\rho}_1, 1_{G_1})_{G_1} = m_1m_2 \geq 1.$$

Thus in any case we have $(\chi_{G_1}, 1_{G_1})_{G_1} \geq 1$, and hence

$$1 \leq (\chi_{G_1}, 1_{G_1})_{G_1} = (\chi, \mathrm{Ind}_{G_1}^G(1_{G_1}))_G,$$

which is a contradiction because χ is irreducible and $\deg(\mathrm{Ind}_{G_1}^G(1_{G_1})) < \deg \chi$. Consequently, all multiplicities m_i are 1, and all characters ρ_i are real-valued.

2) Now we decompose the multiplicity-free character ρ_{G_1} as follows:

$$\rho_{G_1} = \delta + \sum_{i=1}^{t} \rho_i,$$

where $\delta, \rho_1, \ldots, \rho_t$ belong to $\mathrm{Irr}(G_1)$ and are pairwise different. By the results of 1), we have a decomposition $V = U_0 \oplus U_1 \oplus \ldots \oplus U_t$ where $U_0 = V_0(\mathcal{H}_1) =< v >_{\mathbb{C}}$, and $U_1, \ldots, U_t$ are pairwise orthogonal non-degenerate subspaces, on which G_1 has characters $\delta, \rho_1, \ldots, \rho_t$, respectively. Then $\mathcal{L}$ is decomposed into a direct sum of G_1-submodules,

$$\mathcal{L} = \left(\sum_{0 \leq i \leq t} \mathcal{C}_i \right) \oplus \left(\sum_{0 \leq i < j \leq t} \mathcal{D}_{ij} \right),$$

where

$$\mathcal{C}_i = \{f \in \mathcal{L} | f(U_i) \subseteq U_i, f((U_i)^{\perp}) = 0\}$$

and

$$\mathcal{D}_{ij} = \{f \in \mathcal{L} | f(U_i) \subseteq U_j, f(U_j) \subseteq U_i, f((U_i \oplus U_j)^{\perp}) = 0\}$$

for $0 \leq i < j \leq t$.

If $t = 1$, then $\rho_{G_1} = \delta + \rho_1$ and $\deg \rho_1 = 2n$. Therefore we suppose that $t \geq 2$. Let τ denote the character of G_1 afforded by $\mathcal{H}_1$. Obviously we have $\mathrm{Ind}_{G_1}^G(\tau) = \chi \in \mathrm{Irr}(G)$, hence $\tau \in \mathrm{Irr}(G_1)$. Furthermore, $1 = (\chi, \chi)_G = (\mathrm{Ind}_{G_1}^G(\tau), \chi)_G = (\tau, \chi_{G_1})_{G_1}$, that is, τ enters χ_{G_1} with multiplicity 1. Consequently, as the unique G_1-submodule of $\mathcal{L}$ with character τ, $\mathcal{H}_1$ must be contained in one of submodules $\mathcal{C}_i$, $\mathcal{D}_{ij}$. First suppose that $\mathcal{H}_1 \subseteq \mathcal{C}_i$. As $\mathcal{C}_0 = 0$, we have $i > 0$ and $\dim U_i \geq 2n$ by Lemma 6.1.3; hence $t = 1$, contrary to our assumption. Thus we have $\mathcal{H}_1 \subseteq \mathcal{D}_{ij}$. By Lemma 6.1.2 it follows that $\dim U_i = \dim U_j = n$; hence $t = 2$, $V =< v >_{\mathbb{C}} \oplus U_1 \oplus U_2$ and $\rho_{G_1} = \delta + \rho_1 + \rho_2$, $\deg \rho_1 = \deg \rho_2 = n$. □

In fact the monomiality of the character ρ was proved in Lemma 6.1.4. But we can state much more about ρ:

Proposition 6.1.6. *The space V admits an orthonormal basis $\{e_1, \ldots, e_{2n+1}\}$ in which G acts monomially; namely, G permutes the $(2n+1)$ pairs $\pm e_i$ with point stabilizer $G_1 = St_G(\{\pm e_1\})$. Moreover, the permutation action of G on these pairs is transitive and 2-homogeneous.*

Proof. 1) By Lemma 6.1.4 we have $\rho = \mathrm{Ind}_{G_1}^{G}(\delta)$, where δ is the character of G_1 on $V_0(\mathcal{H}_1) =< v >_{\mathbb{C}}$ and v is anisotropic. Choose a vector e_1 with norm 1 such that $< e_1 >_{\mathbb{C}} =< v >_{\mathbb{C}}$. Partition G into left cosets, $G = \bigcup_{i=1}^{2n+1} g_i G_1$ (where $g_1 = 1$), and set $e_i = g_i(e_1)$, $1 \leq i \leq 2n+1$. Obviously $V =< e_1, \ldots, e_{2n+1} >_{\mathbb{C}}$, and for $g \in G$ and $i = 1, \ldots, 2n+1$ we have $g(e_i) = \pm e_j$, where $gg_i \in g_j G_1$. Moreover, $(e_i, e_i) = (g_i(e_1), g_i(e_1)) = (e_1, e_1) = 1$. It remains to establish the pairwise orthogonality of $e_1, \ldots, e_{2n+1}$. Consider the subspace $U =< e_2, \ldots, e_{2n+1} >_{\mathbb{C}}$, which is, like $U' = (< e_1 >_{\mathbb{C}})^{\perp}$, a G_1-submodule. Suppose that $U \neq U'$. Then the G_1-module V has at least two one-dimensional composition factors, namely $< e_1 >_{\mathbb{C}}$ and $U/(U \cap U')$, contrary to Proposition 6.1.5. Hence $U = U'$, that is, $(e_1, e_i) = 0$ for $i > 1$. It is then easy to see that $(e_i, e_j) = 0$ for $1 \leq i < j \leq 2n+1$ as required.

2) We have mentioned that G permutes $(2n+1)$ pairs $\pm e_i$ of basis vectors. Therefore, the irreducibility of G on V implies that this permutation action is transitive. Furthermore, $G_1(e_1) = \pm e_1$ and $(G : G_1) = 2n+1$, so that $G_1 = St_G(\{\pm e_1\})$. Now consider the special basis $\{E(i,j) \mid 1 \leq i < j \leq 2n+1\}$ of $\mathcal{L}$. Note that G acts on the set $\{E(i,j) \mid 1 \leq i \neq j \leq 2n+1\}$; namely, if $g(e_i) = \varepsilon_i e_{\pi(i)}$ for $\varepsilon_i = \pm 1$ and $\pi \in \mathbb{S}_{2n+1}$, then $gE(i,j)g^{-1} = \varepsilon_i \varepsilon_j E(\pi(i), \pi(j))$. Hence from the irreducibility of G on $\mathcal{L}$, it follows immediately that the action of G on $\{\pm e_1, \ldots, \pm e_{2n+1}\}$ is 2-homogeneous. □

We refer to Proposition 6.1.6, and let E and H be the kernel and image of G in its action on $\{\pm e_1, \ldots, \pm e_{2n+1}\}$. Then $E \subseteq \mathbb{Z}_2^{2n+1}$ and $H = G/E \subseteq \mathrm{Sym}(X) = \mathbb{S}_{2n+1}$, where we have identified the set $\{\pm e_1, \ldots, \pm e_{2n+1}\}$ with $X = \{1, \ldots, 2n+1\}$. We fix the factorization

$$G = E.H. \tag{6.1}$$

Corollary 6.1.7. *One of the following assertions holds:*

1) *the permutation group H is 2-transitive;*

2) $2n+1 = p^l \equiv 3 \pmod 4$ *for some prime power p^l, H has rank 3, odd order and subdegrees 1, n, n; and $ASL_1(p^l) \subseteq H \subseteq A\Sigma L_1(p^l)$.*

Proof. This is a consequence of Kantor's Theorem [Kan 1]. □

To illustrate our method, we consider at this point the model case rank$H = 3$. Recall once more that the notion of E-decompositions was introduced in Chapter 2 (see Definition 2.2.1).

Proposition 6.1.8. *Suppose that H has rank 3. Then the IOD $\mathcal{D}$ is an E-decomposition.*

Proof. Note that a sufficient condition for $\mathcal{D}$ to be an E-decomposition is the following: the group E acts on lines $< E(i,j) >_{\mathbb{C}}$ with pairwise different characters

χ_{ij}. This is the case, because E leaves each subalgebra $\mathcal{H}_i$ fixed. Now assume that $\chi_{ij} = \chi_{kl}$ for some $\{i,j\} \neq \{k,l\}$. Set $G_{ij} = St_G(< E(i,j) >_{\mathbb{C}})$, $G_{kl} = St_G(< E(k,l) >_{\mathbb{C}})$ and $S = G_{ij} \cap G_{kl}$; then $S = E.R$ where R, as a subgroup of H, has odd order. Let α and β be the characters of S on the lines $< E(i,j) >_{\mathbb{C}}$ and $< E(k,l) >_{\mathbb{C}}$, respectively. Because S preserves the Killing form, which is nondegenerate on these lines, it follows that $\alpha, \beta \in \mathrm{Hom}(S, \mathbb{Z}_2)$. Setting $\gamma = \alpha.\beta^{-1}$ we have $\gamma \in \mathrm{Hom}(S, \mathbb{Z}_2)$, whereas $\alpha|_E = \chi_{ij} = \chi_{kl} = \beta|_E$, so that $E \subseteq \mathrm{Ker}\gamma$ and $\gamma \in \mathrm{Hom}(R, \mathbb{Z}_2)$. Since R is of odd order, we must have $\gamma = 1_S$ and $\alpha = \beta$. It is then easy to see that $\chi = \mathrm{Ind}_{G_{ij}}^{G}(\alpha)$. By Mackey's Theorem [Ser 3] it now follows that χ is reducible since $\alpha = \beta$; a contradiction. □

6.2 Every IOD is an E-decomposition

The purpose of this section is to prove that every irreducible orthogonal decomposition of $\mathcal{L}$ is necessarily an E-decomposition. Recall that we have fixed the factorization $G = E.H$ (see (6.1)). Moreover, throughout the section, H is assumed to be doubly transitive.

Proposition 6.2.1. *Suppose that H is 2-transitive. Then E is nontrivial.*

(In the terminology of Chapter 4, this means that we are dealing with the E-case).

Proof. Assume the contrary. Then we deduce the reducibility of the group $G = H$ on $\mathcal{L}$, which is impossible by our principal hypothesis: $\mathcal{D}$ is an IOD. By the O'Nan-Scott Theorem, we must distinguish two cases, those where $\mathrm{soc}(H)$ is abelian and nonabelian.

Case I. $\mathrm{soc}(H) = W = \mathbb{Z}_p^m$ *is abelian.*

In this case we have $p^m = 2n + 1$ and $H = W.H_0$, where $H_0 \subseteq GL(W)$ acts transitively on the nonzero points of W considered as a linear space over $\mathbb{F}_p$. By Clifford's Theorem, we have $\chi_w = e\sum_{i=1}^{t} \rho_i$, where $\{\rho_1, \ldots, \rho_t\}$ is some H_0-orbit on $\mathrm{Irr}(W)$. If $\rho_1 = 1_W$ then $t = 1$, $\chi_w = e.1_W$ and W acts trivially on $\mathcal{L}$, a contradiction. If $\rho_1 \neq 1_W$ then $t = p^m - 1 = 2n$ by Lemma 5.3.1 and $n(2n+1) = \dim \mathcal{L} = \deg \chi = et = 2ne$, again a contradiction.

Case II. $\mathrm{soc}(H) = T$ *is a nonabelian simple group.*

In this case $T \lhd H \subseteq \mathrm{Aut}(T)$. According to Table A2, we have to consider the following possibilities for the pair $(T, 2n+1)$.

1) $(T, 2n+1) = (\mathbb{A}_{2n+1}, 2n+1)$.

Here obviously we have $\mathbb{A}_{2n+1} \subseteq G \subseteq \mathbb{S}_{2n+1}$, and $\mathbb{A}_{2n} \subseteq G_1 \subseteq \mathbb{S}_{2n}$. Recall that $n \geq 3$, so G_1 has no non-trivial soluble normal subgroups. Hence G_1 acts faithfully on $\mathcal{H}_1$, and we obtain an embedding of G_1 in the Weyl group $W(\mathcal{L}) = \mathbb{Z}_2^n.\mathbb{S}_n$, a contradiction.

2) $(T, 2n+1) = (L_2(q), q+1)$, $(PSU_3(q), q^3+1)$, $({}^2B_2(q), q^2+1)$, or $(L_2(11), 11)$.

Here we have $(\dim \mathcal{L})^2 > |\mathrm{Aut}(T)|$, so that G cannot be irreducible on $\mathcal{L}$. The only exception is the case when $T = L_2(8)$ and $G = L_2(8).\mathbb{Z}_3$, but then G has no irreducible characters of degree equal to $\dim \mathcal{L} = 36$.

3) $(T, 2n+1) = (\mathbb{A}_7, 15)$, $(M_{11}, 11)$, or $(M_{23}, 23)$.

If $(T, 2n+1) = (\mathbb{A}_7, 15)$ then $G = \mathbb{A}_7$ and G has no irreducible characters of degree $\dim \mathcal{L} = 105$. If $(T, 2n+1) = (M_{23}, 23)$, then $G = M_{23}$, $G_1 = M_{22}$, and we get an embedding of M_{22} in $W(\mathcal{L}) = \mathbb{Z}_2^{11}.\mathbb{S}_{11}$, a contradiction. If $(T, 2n+1) = (M_{11}, 11)$ then $G = M_{11}$, $G_1 = M_{10}$ and we get a homomorphism of M_{10} into $W(\mathcal{L}) = \mathbb{Z}_2^5.\mathbb{S}_5$ with soluble kernel, which is a contradiction because M_{10} has $\mathbb{A}_6$ as composition factor.

4) $(T, 2n+1) = (PSL_d(q), (q^d-1)/(q-1))$, $d \geq 3$.

Here we have $T_1 = T \cap G_1 = U.S$, where U is an elementary abelian group of order q^{d-1}, $S \cong GL_{d-1}(q)/\mathbb{Z}_{\gcd(d,q-1)}$ and S acts (by conjugation) transitively on $U \backslash \{1\}$. First we remark that T acts irreducibly on V. In fact T, like G, permutes the $(2n+1)$ pairs $\pm e_i$ 2-transitively. Let ε be the character of $T_1 = St_T(< e_1 >_{\mathbb{C}})$ on the line $< e_1 >_{\mathbb{C}}$. Then $\rho_T = \mathrm{Ind}_{T_1}^T(\varepsilon)$, and, if $T = T_1 \cup T_1 x T_1$, we have $(\rho_T, \rho_T)_T = (\varepsilon, \varepsilon)_{T_1} + (\alpha, \beta)_{S_1}$ where $S_1 = T_1 \cap x T_1 x^{-1}$, and α and β are the characters of S_1 on $< e_1 >_{\mathbb{C}}$ and $x < e_1 >_{\mathbb{C}}$, respectively. Since α, β and ε are linear, we must have $(\rho_T, \rho_T)_T \leq 2$. On the other hand, T is a normal subgroup of G and $\dim V$ is odd; hence, by Clifford's Theorem, ρ_T is a sum of an odd number of irreducible constituents of the same degree. Consequently, $(\rho_T, \rho_T)_T = 1$ as required.

Assume now that q is even. As we have remarked, $\rho_T = \mathrm{Ind}_{T_1}^T(\varepsilon)$ is irreducible, so $\varepsilon \neq 1_{T_1}$, that is, $\varepsilon(T_1) = \{\pm 1\}$. Set $R = \mathrm{Ker}\varepsilon$, so that $(T_1 : R) = 2$. It is easy to see that $R \supseteq U$, $(S : R/U) = 2$, and S has $\mathbb{Z}_2$ as composition factor. This last assertion is impossible if $(d, q) \neq (3, 2)$, because $S = GL_{d-1}(q)/\mathbb{Z}_{\gcd(d,q-1)}$. If $(d, q) = (3, 2)$, then $T = L_3(2)$ and $(\dim \mathcal{L})^2 = 441 > |\mathrm{Aut}(T)|$, a contradiction.

Finally, suppose that q is odd. By Proposition 6.1.5, the degree of every irreducible constituent of ρ_{G_1} divides $2n = q(q^{d-1}-1)/(q-1)$. The same thing holds for ρ_{T_1}, since $T_1 \lhd G_1$. On the other hand, if $\alpha \in \mathrm{Irr}(T_1)$ and α_U is non-trivial, then Lemma 5.3.1 tells us that $\deg \alpha$ is divisible by $q^{d-1}-1$. Therefore, each irreducible constituent α of ρ_{T_1} is trivial on U. In other words, $\rho_{T_1}|_U = \rho_U$ is trivial, a contradiction. □

Unlike the case of Lie algebras of type A_n, where the existence of a non-trivial abelian normal subgroup E of G acting trivially on $\{\mathcal{H}_i\}$ has almost meant the success of our classification (see Chapter 5), we are still far from the completion of our proof in the case of type B_n even though Proposition 6.2.1 has been established.

In the remainder of the proof the study of the induced permutation representation of H on the set $X(2)$ of unordered pairs $\{i, j\}$, $1 \le i \ne j \le 2n+1$, plays an essential role.

Firstly we establish a simple statement:

Lemma 6.2.2. *The character ρ_E of the group E on V is a sum of $(2n+1)$ different non-principal linear characters: $\rho_E = \sum_{i=1}^{2n+1} \rho_i$. The character χ_E of E on $\mathcal{L}$ is $\sum_{i<j} \rho_i\rho_j$. Moreover, $\prod_{i=1}^{2n+1} \rho_i = 1_E$.*

Proof. Let ρ_i be the character of E on the line $< e_i >_{\mathbb{C}}$, for $1 \le i \le 2n+1$. Then $\rho_i \in \mathrm{Hom}(E, \mathbb{Z}_2)$, $\rho_E = \sum_i \rho_i$, and the character of E on $< E(i, j) >_{\mathbb{C}}$ is $\rho_i\rho_j$; that is, $\chi_E = \prod_{i<j} \rho_i\rho_j$. Since $E \subseteq SO_{2n+1}(\mathbb{C})$, we must have $\prod_i \rho_i = 1_E$. Assume now that $\rho_i = \rho_j$ for some i, j, $1 \le i \ne j \le 2n+1$. Then $\rho_i\rho_j = 1_E$ and $(\chi_E, 1_E)_E > 0$, that is, the subspace $C_{\mathcal{L}}(E) = \{x \in \mathcal{L} \mid \forall e \in E, e(x) = x\}$ is a nonzero G-submodule of $\mathcal{L}$. By the simplicity of the G-module $\mathcal{L}$ we have $C_{\mathcal{L}}(E) = \mathcal{L}$, that is, E acts trivially on $\mathcal{L}$, contrary to Proposition 6.2.1. Similarly we can see that $\rho_i \ne 1_E$, $i = 1, \ldots, 2n+1$. □

For the rest of the chapter we fix the decomposition

$$\rho_E = \sum_{i=1}^{2n+1} \rho_i \tag{6.2}$$

described in Lemma 6.2.2.

Lemma 6.2.3. *The IOD $\mathcal{D}$ is an E-decomposition if one of the following conditions holds:*

1) *the characters $\rho_i\rho_j$ and $\rho_k\rho_l$ are different whenever $\{i, j\} \ne \{k, l\}$;*

2) *the permutation representation of H on the set $X(2)$ of unordered pairs $\{i, j\}$, $1 \le i \ne j \le 2n+1$, is primitive.*

Proof. In the proof of Proposition 6.1.8, we noted that condition 1) implies that $\mathcal{D}$ is an E-decomposition. Now assume that 2) holds. We define an equivalence relation $\sim$ on $X(2)$, which we use in what follows:

$$\{i, j\} \sim \{k, l\} \Leftrightarrow \rho_i\rho_j = \rho_k\rho_l. \tag{6.3}$$

Since $E \lhd G$, this relation is G-invariant; that is, it partitions $X(2)$ into a disjoint union of imprimitivity sets for G. So 2) implies 1). □

Proposition 6.2.4. *Suppose as above that H is 2-transitive, and that $\mathrm{soc}(H) = T$ is nonabelian. Then the IOD $\mathcal{D}$ is an E-decomposition.*

Proof. According to Table A2, we have to consider the following possibilities for the pair $(T, 2n+1)$.

1) $(T, 2n+1) = (\mathbb{A}_{2n+1}, 2n+1)$, $(M_{11}, 11)$, $(M_{23}, 23)$, or $(L_2(q), q+1)$.

Here T acts 3-transitively on X, so by Cameron's Theorem [Cam 2] T and H act primitively on $X(2)$, and we can apply Lemma 6.2.3.

2) $(T, 2n+1) = (L_2(11), 11)$.

It is easy to see from Lemma 6.2.2 that $C_G(E) = E$, and therefore T is embedded in $\mathrm{Aut}(E)$. Assuming that $E = \mathbb{Z}_2^m$, $m \geq 1$, we have in particular that 11 divides $|SL_m(2)|$, so that $m \geq 10$. But in that case, E is conjugate to the diagonal subgroup $\{\mathrm{diag}(x_1, \ldots, x_{11}) \mid x_i = \pm 1, \prod_i x_i = 1\}$ of $SO_{11}(\mathbb{C})$. In particular, the characters $\rho_i\rho_j$, $1 \leq i \neq j \leq 11$, are pairwise distinct and we can apply Lemma 6.2.3.

3) $(T, 2n+1) = ({}^2B_2(q), q^2+1)$, $q = 2^{2e+1} > 2$.

Here we claim that the action of T on $X(2)$ induced by its 2-transitive permutation action on X, $|X| = q^2+1$, is primitive; in other words, some setwise stabilizer $T_2 = St_T(\{a, b\})$, $a, b \in X$, $a \neq b$, is maximal in T. Assume that this is not so. We have $|T_2| = |T|/|X(2)| = 2(q-1)$. By the description of maximal subgroups of T [Suz 1], we have an embedding of T_2 in a Frobenius group $B = U.A$, where $U = \mathbb{Z}_2^{2e+1}.\mathbb{Z}_2^{2e+1}$ and $A = \mathbb{Z}_{q-1}$. It is well-known that B has exactly $(q-1)$ involutions; they are contained in U and permuted transitively by A. Now since $T_2 \subset B$ it follows that $T_2 \cap U = \mathbb{Z}_2 =< u >$, where u is the unique (central) involution of T_2 and $C_B(u) \supseteq T_2$, a contradiction. Hence T acts primitively on $X(2)$, and we can apply Lemma 6.2.3.

For the rest of the proof of Proposition 6.2.4, we are interested in blocks Ω_i, into which the equivalence relation $\sim$ described in (6.3) partitions $X(2)$:

$$X(2) = \bigcup_{i=1}^{N} \Omega_i, \quad |\Omega_i| = t = \frac{n(2n+1)}{N}. \tag{6.4}$$

Note that for any two different points $\alpha = \{a, b\}$, $\beta = \{c, d\}$ of Ω_i, we always have $\{a, b\} \cap \{c, d\} = \emptyset$ (the "*non-intersection property*", abbreviated to NIP). In particular,

$$t \leq [\mid X \mid /2] = n, \; N \geq 2n+1. \tag{6.5}$$

We argue by proving that, in the remaining cases, any partition (6.4) satisfying NIP is trivial in the sense that $t = 1$. This means that condition 1) of Lemma 6.2.3 is satisfied.

We start with the following case:

4) $(T, 2n+1) = (\mathbb{A}_7, 15)$.

Here we can identify X with the set $W \setminus \{0\}$ of nonzero points of the space $W = \mathbb{F}_2^4$ on which $T = H = \mathbb{A}_7$ acts as a subgroup of $GL(W) \cong \mathbb{A}_8$. We shall suppose that $t > 1$; then from (6.4) and (6.5) it follows that $t = 3$, 5 or 7. First we remark that if $\alpha = \{a, b\} \in \Omega_1$ then $a + b \notin \beta$ for every $\beta \in \Omega_1$. Indeed, suppose that $\beta = \{a+b, c\} \in \Omega_1$. Set $Q = St_H(\alpha)$, then $\mid Q \mid = \mid H \mid / \mid X(2) \mid = 24$. Obviously

$c \neq a, b, a+b$, that is, a, b, c are linearly independent and $St_{\mathbb{A}_8}(a, b, c) = \mathbb{Z}_2^3$. Take an element $\varphi \in Q$ of order 3. Then $\varphi(a) = a$, $\varphi(b) = b$, but $\varphi(c) = d \neq c$, because $St_H(a, b, c) \subseteq St_{\mathbb{A}_8}(a, b, c) = \mathbb{Z}_2^3$ has no elements of order 3. Hence $\varphi(\alpha) = \alpha$, $\varphi \in St_H(\Omega_1)$ and $\gamma = \varphi(\beta) = \{a+b, d\} \in \Omega_1$, whereas $\beta \cap \gamma = \{a+b\}$, contrary to NIP.

Now if $t = 3$, $\Omega_1 = \{\{a, b\}, \{c, d\}, \{e, f\}\}$, then obviously Q fixes the set $Y = \{c, d, e, f\} \subseteq W\backslash < a, b >_{\mathbb{F}_2}$ with $|Y| = 4$. If $t = 5$, $\Omega_1 = \{\{a, b\}, \{c_1, d_1\}, \ldots, \{c_4, d_4\}\}$, then Q fixes the set

$$Y = (W\backslash < a, b >_{\mathbb{F}_2})\backslash\{c_1, d_1, \ldots, c_4, d_4\}$$

with $| Y |= 4$. In any case, considering the action of an element φ of order 3 on the set Y of cardinality 4, we find a new fixed point $g \in W\backslash < a, b >_{\mathbb{F}_2}$ for φ, and meanwhile $\varphi \notin St_{\mathbb{A}_8}(a, b, g) = \mathbb{Z}_2^3$. This contradiction means that $t = 7$ and we can suppose that $\Omega_1 = \{\{a_1, b_1\}, \ldots, \{a_7, b_7\}\}$ where $\{a_1, b_1, \ldots, a_7, b_7\} = W\backslash < a >_{\mathbb{F}_2}$. As we have remarked above, $a_1 + b_1 \neq a_i, b_i, 0$, and therefore $a_1 + b_1 = \ldots = a_7 + b_7 = a$. Consequently, $\Omega_i = \{\{x, x + a_i\} \mid x \neq 0, a_i\}$ for some vector $a_i \in W\backslash\{0\}$. This means that $\rho_x \rho_{x+a} = \rho_y \rho_{y+a}$ for all $x, y \in W\backslash < a >_{\mathbb{F}_2}$. From the identity $\prod_{x \in W\backslash\{0\}} \rho_x = 1_E$ it follows immediately that $\rho_a \rho_b = \rho_{a+b}$ for any two different vectors $a, b \in W\backslash\{0\}$. So by Lemma 6.2.2 we have $\chi_E = 7 \sum_{x \in W\backslash\{0\}} \rho_x$. Suppose that $\mathcal{H}_1$ is stabilized by $H_1 = St_H(a)$ and E has character $\rho_{a_1} + \ldots + \rho_{a_7}$ on $\mathcal{H}_1$. Then H_1 leaves $\{a_1, \ldots, a_7\}$ fixed, while H_1 has orbits $\{a\}$ and $W\backslash < a >_{\mathbb{F}_2}$. Hence $a_1 = \ldots = a_7 = a$, and

$$\mathcal{H}_1 = \langle E(x, x+a) \mid x \in W\backslash < a >_{\mathbb{F}_2} \rangle_{\mathbb{C}}.$$

In other words, $\mathcal{D}$ is an E-decomposition.

5) $(T, 2n+1) = (PSU_3(q), q^3 + 1)$, $q = 2^e > 2$.

a) Suppose that $t > 1$, $\Omega_1 = \{\{a_1, b_1\}, \ldots, \{a_t, b_t\}\}$, and let $T_2 = St_T(\{a_1, b_1\})$ be a point stabilizer and $T_b = St_T(\Omega_1)$ a block stabilizer. Note (see Table A2) that T, as socle of H, is 2-transitive. Then $t = (T_b : T_2)$, $\{a_1, b_1, \ldots, a_t, b_t\} = (a_1)^{T_b}$, and hence, by NIP, $|(a_1)^{T_b}| = 2t$ is even. Thus, since $t > 1$ we deduce that $| (a_1)^{T_b} |= q + 1$ is odd, and get the required contradiction.

b) We describe some 2-transitive representation of T at this point. Set $\hat{T} = SU_3(q)$, and let $\hat{A}$ be the full preimage of $A \subseteq T$ in $\hat{T}$. We define a non-degenerate Hermitian form $< *, * >$ on the 3-dimensional $\mathbb{F}_{q^2}$-space $W =< u, v, w >$ by setting $< u, u >=< v, v >=< u, w >=< v, w >= 0$, $< u, v >=< w, w >= 1$. Here, $\bar{\alpha} = \alpha^q$ for $\alpha \in \mathbb{F}_{q^2}$. Then we can identify X with the set of (q^3+1) isotropic 1-spaces of W and suppose that $a_1 =< u >$, $b_1 =< v >$. It is easy to see that $\hat{T}_2 = \mathbb{Z}_{q^2-1}.\mathbb{Z}_2$, where

$$\mathbb{Z}_{q^2-1} = \left\{\varphi_\lambda : u \mapsto \lambda u,\ v \mapsto \lambda^{-q} v,\ w \mapsto \lambda^{q-1} w \mid \lambda \in \mathbb{F}_{q^2}^*\right\},$$

$$\mathbb{Z}_2 =< \vartheta : u \mapsto v,\ v \mapsto u,\ w \mapsto w > .$$

In particular, $\mathcal{O}_1 = \{< w >\}$, $\mathcal{O}_2 = \{< u >, < v >\}$, and $\mathcal{O}_3 = \{< u + \lambda v > | \lambda \in \mathbb{F}_q^*\}$ are T_2-orbits on $\mathbb{P}W$. Set $N = St_T(< w >)$. Below (see c)) we shall prove that N is the unique maximal subgroup of T containing T_2. Since $T_2 \subset T_b \subset T$, we must have that $T_2 \subset T_b \subseteq N$. Now note that $T_2 = St_N(\mathcal{O}_2)$ and that N fixes the set $\mathcal{O}_2 \cup \mathcal{O}_3$ of isotropic 1-spaces which are orthogonal to w. Hence, $(a_1)^{T_b} = (< u >)^{T_b} = \mathcal{O}_2 \cup \mathcal{O}_3$ has length $q + 1$, as stated.

c) Suppose now that M is a maximal subgroup of T containing T_2. It is well-known [Kle 2] that the order of a maximal subgroup R of T is one of the following: $(q+1)^2q(q-1)/d$ (if R is the stabilizer of a non-isotropic 1-space), $q^3(q^2-1)/d$ (if R is the stabilizer of an isotropic 1-space), $6(q+1)^2/d$, $3(q^2-q+1)/d$, $\nu(\rho^3+1)\rho^3(\rho^2-1)/d$ (where $\nu = 1$ or 3, $q = \rho^m$ and $m \geq 3$ is odd), 36, 72, 168, 216, 360, 720 or 2520. Here $d = \gcd(3, q-1)$. Since $| T_2 | = 2(q^2-1)/d$, we have $| M | \neq 6(q+1)^2/d$ (if $q \neq 4$), $3(q^2-q+1)/d$, $\nu(\rho^3+1)\rho^3(\rho^2-1)/d$. If $q = 4$ and $| M | = 6(q+1)^2/d$, then $M = \mathbb{Z}_5^2.\mathbb{S}_3$, $T_2 = \mathbb{Z}_{15}.\mathbb{Z}_2$ and T_2 cannot be contained in M. It is easy to see that $| M | \notin \{36, 72, 168, 216, 360, 720, 2520\}$. Finally, note that T_2 has a unique fixed point $< w >$ in $\mathbb{P}W$. Hence $M = St_T(< w >) = N$, as stated.

6) $(T, 2n+1) = (PSL_d(q), (q^d-1)/(q-1))$, $q > 2$, $d \geq 3$.

We shall identify X with the projective space $\mathbb{P}W = \{\bar{u} =< u >_{\mathbb{F}_q} | u \in W \backslash \{0\}\}$, where $W = \mathbb{F}_q^d$. Suppose that $t > 1$, and set $T_b = St_T(\Omega_1)$.

a) First, we show that q is necessarily odd in this case, and that Ω_1 has the form

$$\left\{\{\bar{a}, \bar{b}\}, \{\overline{a+\lambda b}, \overline{a-\lambda b}\} \mid \lambda \in \mathbb{F}_q^*\right\}.$$

For, suppose that $\alpha = \{\bar{a}, \bar{b}\} \neq \beta = \{\bar{c}, \bar{d}\}$, $\alpha, \beta \in \Omega_1$. If $d \notin < a, b, c >_{\mathbb{F}_q}$, then taking the automorphism

$$\varphi \in SL(W) : a \mapsto a,\ b \mapsto b,\ c \mapsto c,\ d \mapsto d + a,$$

we have $\varphi(\alpha) = \alpha$, $\varphi(\beta) = \gamma = \{\bar{c}, \overline{d+a}\} \neq \beta$, while $\beta, \gamma \in \Omega_1$; that is, Ω_1 does not satisfy NIP. Thus $d \in < a, b, c >_{\mathbb{F}_q}$, and similarly $c \in < a, b, d >_{\mathbb{F}_q}$. Suppose that $c \notin < a, b >_{\mathbb{F}_q}$. Then we can assume that $d = xa + yb + c$, $x, y \in \mathbb{F}_q$, $x \neq 0$. Since $q \geq 3$, we can choose $\lambda \in \mathbb{F}_q \backslash \{0, 1\}$ and consider the automorphism

$$\varphi \in SL(W) : a \mapsto a,\ b \mapsto \lambda^{-1}b,\ c \mapsto \lambda c.$$

Then

$$\varphi(\alpha) = \alpha,\ \varphi(\beta) = \gamma = \{\bar{c}, \overline{x\lambda^{-1}a + y\lambda^{-2}b + c}\} \neq \beta,$$

and NIP does not hold. Hence $c, d \in < a, b >_{\mathbb{F}_q}$, and we can assume that $c = a + xb$, $d = a + yb$, $x, y \in \mathbb{F}_q^*$. Taking

$$\varphi \in SL(W) : a \mapsto xb,\ b \mapsto x^{-1}a,\ e \mapsto -e$$

for some $e \in W\backslash < a, b >$, we have

$$\varphi(\beta) = \gamma = \{\bar{c}, \overline{a + x^2y^{-1}b}\} \in \Omega_1, \ \varphi(\alpha) = \alpha.$$

By NIP, we must have $x^2y^{-1} = y$, $y = -x$. In particular, q is odd. Considering

$$\varphi \in SL(W) : a \mapsto a, \ b \mapsto \lambda b, \ e \mapsto \lambda^{-1}e$$

for some $e \in W\backslash < a, b >$, we have

$$\varphi(\alpha) = \alpha, \ \varphi(\beta) = \gamma = \{\overline{a + x\lambda b}, \overline{a - x\lambda b}\} \in \Omega_1$$

for any $\lambda \in \mathbb{F}_q^*$. In other words, Ω_1 is of the form mentioned above. In particular, $t =| \Omega_1 |= (q + 1)/2$. (As a matter of fact, we can show that $q = 3$ or 5).

b) Now we shall convince ourselves that $(L : L \cap T_b) \geq 2n$ for the stabilizer $L = St_T(\bar{c})$ of any line $\bar{c} \in \mathbb{P}W$. Indeed, we have shown that

$$\Omega_1 = \left\{\{\bar{a}, \bar{b}\}, \ \{\overline{a + \lambda b}, \overline{a - \lambda b}\} \mid \lambda \in \mathbb{F}_q^*\right\}$$

for some $a, b \in W$, $a \neq b$. If $c \in < a, b >_{\mathbb{F}_q}$, then it is easy to see that $(L : L \cap T_b) = 2n$. Assuming that $c \notin < a, b >_{\mathbb{F}_q}$, we set $T_2 = St_T(\{\bar{a}, \bar{b}\})$ and $S = St_T(\bar{a}, \bar{b})$. Then

$$(L : L \cap S) = (St_T(\bar{c}) : St_T(\bar{a}, \bar{b}, \bar{c})) = (q^d - q)(q^d - q^2)/(q - 1)^2.$$

Next we have $(T_b : S) = (T_b : T_2)(T_2 : S) = 2 \mid \Omega_1 \mid= q + 1$. Now S and $L \cap T_b$ are subgroups of T_b, and so $|T_b| \geq |S(L \cap T_b)| = |S|.|L \cap T_b|/|L \cap S|$, and this implies that $(L \cap T_b : L \cap S) \leq (T_b : S) = q + 1$. Hence

$$(L : L \cap T_b) = (L : L \cap S)/(L \cap T_b : L \cap S) \geq \frac{(q^d - q)(q^d - q^2)}{(q + 1)(q - 1)^2} =$$

$$= 2n.\frac{(q^d - q^2)}{(q^2 - 1)} > 2n.$$

c) Finally, we deduce a contradiction from the assumption that $t > 1$. Set $G_1 = St_T(\mathcal{H}_1)$, so that $G_1 = E.H_1$, where $H_1 = St_H(\bar{c})$ for some line $\bar{c} \in \mathbb{P}W$. If $T_1 = T \cap H_1$, then the inequality proved in b) means that every T_1-orbit on the set $\{\rho_{\bar{a}}\rho_{\bar{b}} \mid \bar{a}, \bar{b} \in \mathbb{P}W, \bar{a} \neq \bar{b}\}$ has length not less than $2n$. On the other hand, considering the action of $E.T_1$ on the space $\mathcal{H}_1$ of dimension n, we get a T_1-invariant set of cardinality not greater than n consisting of some characters $\rho_{\bar{a}}\rho_{\bar{b}}$, which is a contradiction.

7) $(T, 2n + 1) = (SL_d(2), 2^d - 1)$, $d \geq 3$.

Suppose again that $t > 1$. We can identify X with $W\backslash\{0\}$, where $W = \mathbb{F}_2^d$. In [Tiep 4] (see also Chapter 11) it is shown that if the partition (6.4) is T-invariant, then one of the following assertions holds:

(i) $t = 3$ and $\Omega_i = t(A) = \{\{x, y\} \mid x, y \in A\backslash\{0\}, x \neq y\}$ for some plane A of W;

(ii) $t = 2^{d-1} - 1 (= n)$ and $\Omega_i = l(a) = \{\{x, x + a\} \mid x \in W\backslash < a >\}$ for some vector $a \in W\backslash\{0\}$.

By NIP, we have $\Omega_i = l(a)$. Then arguments similar to those used in case 4) allow us to state that $\mathcal{H}_i =< E(x, y) \mid \{x, y\} \in l(a) >_{\mathbb{C}}$ for some vector $a \in W\backslash\{0\}$. In other words, $\mathcal{D}$ is an E-decomposition. In fact, $\mathcal{D}$ is just the MOD described in (3.8). This proves Proposition 6.2.4. □

The remainder of the section is devoted to the case where $\mathrm{soc}(H) = U$ is elementary abelian. In this situation, we view U as the $\mathbb{F}_p$-space $\mathbb{F}_p^m$ for some prime number $p > 2$, and identify X with U, with U acting on itself by translations, and $H_0 = St_H(0)$ acting transitively on $U\backslash\{0\}$ as a subgroup of $GL(U)$.

Proposition 6.2.5. *Suppose as above that H is 2-transitive and that* $\mathrm{soc}(H) = U$ *is abelian. Then the IOD $\mathcal{D}$ is an E-decomposition. Moreover, $\mathcal{D}$ is conjugate to the E-decomposition*

$$\mathcal{L} = \bigoplus_{a \in U} \mathcal{H}_a, \ \mathcal{H}_a =< E(x, 2a - x) \mid x \in U\backslash\{a\} >_{\mathbb{C}} . \tag{6.6}$$

Proof. 1) By Proposition 6.2.1, we have $E \neq 1$. We shall show that

$$C_E(U) = 1. \tag{6.7}$$

Clearly, we may view U as a Sylow p-subgroup of $E.U$, so that the extension $E.U$ splits. Suppose that $f \in C_E(U)$. Since U acts regularly on the set of 1-spaces $< e_u >_{\mathbb{C}}$, $u \in U$, and E has character ρ_u on the line $< e_u >_{\mathbb{C}}$, we have $\rho_u(f) = a$ for all $u \in U$ and some $a = \pm 1$. But $1 = \prod_{u \in U} \rho_u(f) = a^{p^m}$ (see Lemma 6.2.2), hence $a = 1$ and $f = 1$, as required.

2) Recall that $E \lhd G$, so we may consider E as an $\mathbb{F}_2 H$-module. By Maschke's Theorem, we have a decomposition of $\bar{E} = E \otimes_{\mathbb{F}_2} \bar{\mathbb{F}}_2$ (where $\bar{\mathbb{F}}_2$ is the algebraic closure of $\mathbb{F}_2$) into a direct sum of irreducible $\bar{\mathbb{F}}_2 U$-submodules: $\bar{E} = F_1 \oplus \ldots \oplus F_t$. Because U is abelian, $\dim F_i = 1$. Moreover, if ζ denotes a primitive p-th root of unity in $\bar{\mathbb{F}}_2$, then U acts on $F_i =< f_i >$ by the rule $u(f_i) = \zeta^{\lambda_i(u)} f_i$ for some linear function $\lambda_i \in U^* = \mathrm{Hom}(U, \mathbb{F}_p)$. Setting $\Lambda = \{\lambda_1, \ldots, \lambda_t\}$, we note that Λ is H_0-invariant. Indeed, if $h \in H_0$ then U acts on $hF_i =< hf_i >$ as follows:

$$u(hf_i) = h(h^{-1}uh)(f_i) = \zeta^{\lambda_i(h^{-1}uh)} hf_i = \zeta^{(h \circ \lambda_i)(u)} hf_i.$$

Because the $\bar{\mathbb{F}}_2 U$-module $\bar{E}$ is completely reducible, the function $h \circ \lambda_i$ must belong to Λ. By Lemma 5.3.1, the transitivity of H_0 on $U\backslash\{0\}$ implies the transitivity of H_0 on $U^*\backslash\{0\}$, and hence either

$$\dim_{\mathbb{F}_2} E = \dim_{\bar{\mathbb{F}}_2} \bar{E} = t = |\Lambda| \geq p^m - 1,$$

or $\lambda_1 = \ldots = \lambda_t = 0$. But in the latter case U acts trivially on $\bar{E}$ and on E, so that $C_E(U) = E$, contrary to (6.7). Therefore, $\dim_{\mathbb{F}_2} E \geq p^m - 1$.

3) It is now easy to see that E is just the diagonal subgroup

$$\{\mathrm{diag}(a_1, \ldots, a_{p^m}) \mid a_i = \pm 1, \prod_i a_i = 1\}.$$

In such a case, condition 1) of Lemma 6.2.3 is satisfied, and we can apply the lemma. Hence $\mathcal{D}$ is an E-decomposition. Suppose that $\mathcal{H}_0 = \ < E(a_1, b_1), \ldots, E(a_n, b_n) >_{\mathbb{C}}$, where $n = (p^m - 1)/2$. Then, the group $H_0 = St_H(0) \subseteq GL(U)$ leaves the set $\{a_1 + b_1, \ldots, a_n + b_n\}$ fixed. But H_0 has two orbits, $\{0\}$ and $U \backslash \{0\}$ of cardinality 1 and $2n$ on U. Hence $a_1 + b_1 = \ldots = a_n + b_n = 0$; that is,

$$\mathcal{H}_0 = < E(x, -x) \mid x \in U \backslash \{0\} >_{\mathbb{C}} .$$

If $\mathcal{H}_a = t_a(\mathcal{H}_0)$ for a translation $t_a : u \mapsto u + a$, then

$$\mathcal{H}_a = < E(y, 2a - y) \mid y \in U \backslash \{a\} >_{\mathbb{C}} .$$

This proves Proposition 6.2.5. □

6.3 Study of E-decompositions

Recall that the notion of E-decomposition was introduced in §2.2. There is a one-to-one correspondence between E-decompositions $\mathcal{D}$ of the Lie algebra $\mathcal{L}$ of type B_n and admissible partitions

$$\mathcal{P} : X(2) = \bigcup_{k=1}^{2n+1} M_k$$

of the set $X(2) = \big\{\{i, j\} \mid 1 \leq i \neq j \leq 2n + 1\big\}$, that is, partitions satisfying the non-intersection property (NIP):

$$\forall k,\ \forall \alpha, \beta \in M_k,\ \mid \alpha \cap \beta \mid = 0 \text{ or } 2.$$

Namely,

$$\mathcal{H}_k = < E(i, j) \mid \{i, j\} \in M_k >_{\mathbb{C}} .$$

Here $X = \{1, 2, \ldots, 2n + 1\}$, as above. In what follows, we shall assume that

$$\bigcup_{\alpha \in M_k} \alpha = X \backslash \{k\},\ \forall k = 1, 2, \ldots, 2n + 1.$$

The study of E-decompositions will be based on Propositions 2.2.4 and 2.2.5 and the following simple statement (cf. Lemma 2.4.3):

Lemma 6.3.1. *Let L be any automorphism group of an admissible partition $\mathcal{P}$. Then the number of fixed points of L on X is either zero or odd.*

Proof. Suppose that L fixes exactly l points $1, 2, \ldots, l$, where $l \geq 2$. Since $\bigcup_{\alpha \in M_1} \alpha = X \backslash \{1\}$, L fixes M_1. Suppose now that $\alpha_i = \{i, a_i\} \in M_1$ for $2 \leq i \leq l$. Then for $\varphi \in L$, we have $M_1 \ni \varphi(\alpha_i) = \{i, \varphi(a_i)\}$, hence (by NIP), $\varphi(a_i) = a_i$ for $2 \leq i \leq l$. Thus the set $\{2, 3, \ldots, l\}$ is split into pairwise non-intersecting pairs $\{i, a_i\}$. Consequently, $l - 1$ is even. □

Proposition 6.3.2. *Suppose that the Lie algebra $\mathcal{L}$ of type B_n has an E-decomposition $\mathcal{D}$ with irreducible automorphism group $G = E.H$ (see Proposition 2.2.4). Let H be a 2-transitive subgroup of $\mathrm{Sym}(X)$, and suppose that $T = \mathrm{soc}(H)$ is nonabelian. Then $n = 2^m - 1$ and $H = T = SL_{m+1}(2)$. Moreover, if we identify X with the set $W \backslash \{0\}$ for $W = \mathbb{F}_2^{m+1}$, then $\mathcal{D}$ is conjugate to the E-decomposition*

$$\mathcal{L} = \bigoplus_{a \in W \backslash \{0\}} \mathcal{H}_a, \quad \mathcal{H}_a = \langle E(x, x+a) \mid x \in W \backslash < a >_{\mathbb{F}_2} \rangle_{\mathbb{C}}$$

introduced in (3.8), and which is in fact an MOD.

Proof. Suppose that $\mathcal{D}$ corresponds to an admissible partition $\mathcal{P}$ of $X(2)$. According to Table A2, we have to consider the following possibilities for the pair $(T, 2n+1)$:

1) $(T, 2n+1) = (\mathbb{A}_{2n+1}, 2n+1)$.

Applying Lemma 6.3.1 to the group $L = St_T(1, 2)$ fixing exactly two points, we come to a contradiction.

2) $(T, 2n+1) = (L_2(q), q+1)$, $({}^2B_2(q), q^2+1)$.

Here T acts on X as a Zassenhaus permutation group. Then $L = St_T(1, 2)$ fixes exactly two points, and we can appeal to Lemma 6.3.1.

3) $(T, 2n+1) = (M_{11}, 11)$, $(M_{23}, 23)$.

Here T is 4-transitive, then $L = St_T(1, 2)$ fixes exactly two points, and we can appeal to Lemma 6.3.1.

4) $(T, 2n+1) = (PSL_d(q), (q^d-1)/(q-1))$, $d \geq 3$, $q \geq 3$.

Identifying X with the projective space $\mathbb{P}U$, where $U = \mathbb{F}_q^d =< e_1, \ldots, e_d >$ and setting $L = St_T(< e_1 >, < e_2 >)$, we see that L fixes exactly two points.

5) $(T, 2n+1) = (SL_d(2), 2^d-1)$, $d \geq 3$, or $(\mathbb{A}_7, 15)$.

These cases were considered in the proof of Proposition 6.2.4.

6) $(T, 2n+1) = (PSU_3(q), q^3+1)$.

Consider again the space $U = (\mathbb{F}_{q^2})^3 =< e_1, e_2, e_3 >$ with the Hermitian form $< *, * >$, and identify X with the set of (q^3+1) isotropic 1-spaces of U (see the proof of Proposition 6.2.4). Then the stabilizer L of two points in T fixes exactly two points.

7) $(T, 2n+1) = (L_2(11), 11)$.
In this case $T = H$ and $H_1 = St_H(1) = \mathbb{A}_5$. If

$$M_1 = \big\{\{2,3\}, \{4,5\}, \dots, \{10,11\}\big\},$$

for example, then $\mathbb{A}_5$ acts transitively but imprimitively on $\{2, 3, \dots, 11\}$. In particular, the group $St_H(\{2,3\}) = \mathbb{A}_4 = \mathbb{Z}_2^2.\mathbb{Z}_3$ has the subgroup $St_H(2,3)$ of index 2, a contradiction. □

Theorem 6.0.1 now follows from Propositions 6.1.6, 6.2.4, 6.2.5 and 6.3.2, and Corollary 6.1.7.

The rank 2 analysis leads to the following two orthogonal decompositions described in (3.8) for $\mathcal{L} = B_{2^m-1}$:

$$\mathcal{L} = \bigoplus_{a \in W\setminus\{0\}} \mathcal{H}_a, \quad \mathcal{H}_a = \big\langle E(x, x+a) \mid x \in W\setminus <a>_{\mathbb{F}_2} \big\rangle_{\mathbb{C}}, \; W = \mathbb{F}_2^{m+1},$$

and in (6.6) for $\mathcal{L} = B_{(p^m-1)/2}$:

$$\mathcal{L} = \bigoplus_{a \in U} \mathcal{H}_a, \quad \mathcal{H}_a = \langle E(x, 2a - x) \mid x \in U\setminus\{a\}\rangle_{\mathbb{C}}, \; U = \mathbb{F}_p^m.$$

Proposition 6.3.3. *The decompositions indicated have full automorphism group* $\mathbb{Z}_2^{2\mathrm{rank}(\mathcal{L})}.H$*, where* $H = SL_{m+1}(2)$ *for* (3.8) *and* $H = AGL_m(p)$ *for* (6.6).

Proof. For (3.8), this statement was proved in Theorem 3.3.7. Consider the group $H = \mathrm{Aut}(\mathcal{P})$ for (6.6). Obviously, $A = AGL_m(p) \subseteq H \subseteq \mathrm{Sym}(U)$. We make the following observations.

1) $H = AH_0$ where $H_0 = St_H(0)$;

2) $H_0 = GL(U)H_{0,a}$, where $H_{0,a} = St_H(0, a)$;

3) if $\varphi \in H_0$ fixes $a \in U\setminus\{0\}$, then φ fixes all points λa, $\lambda \in \mathbb{F}_p^*$. (Indeed, φ fixes 0, a, M_0 and M_a (here and below $M_u = \big\{\{x, 2u - x\} \mid x \in U\setminus\{u\}\big\}$). However, the points $\{0, 2a\}$ and $\varphi(\{0, 2a\}) = \{0, \varphi(2a)\}$ belong to M_a, so $\varphi(2a) = 2a$ by NIP, and so on);

4) if $\varphi \in H_0$ fixes two non-collinear points $a, b \in U\setminus\{0\}$, then φ fixes all points of the plane $< a, b >_{\mathbb{F}_p}$. (Indeed, for $x \in \mathbb{F}_p$ φ fixes b, $xa/2$ and $M_{xa/2}$. But the points $\{b, xa - b\}$ and $\varphi(\{b, xa - b\}) = \{b, \varphi(xa - b)\}$ belong to $M_{xa/2}$, so $\varphi(xa - b) = xa - b$).

Using these remarks, we can show that $H = AGL_m(p)$. Hence, by Proposition 2.2.4, $\mathrm{Aut}(\mathcal{D}) = \mathbb{Z}_2^{2\mathrm{rank}(\mathcal{L})}.AGL_m(p)$. □

Now we continue the rank 3 analysis. Recall that, by Corollary 6.1.7, in this case we have $\mathrm{Aut}(\mathcal{D}) = G = \bar{U}.H$, where $U = \mathbb{Z}_p^m$, $H = U.H_0$ and $H_0 \subseteq \Sigma L_1(p^m)$,

$p^m \equiv -1 \pmod 4$, H is a transitive permutation group of odd order, degree p^m and subdegrees 1, $(p^m - 1)/2$ and $(p^m - 1)/2$. Furthermore, the IOD $\mathcal{D}$ is an E-decomposition by Proposition 6.1.8. Here

$$\Sigma L_1(p^m) = \left\{ f : x \mapsto \lambda x^{p^k} \mid \lambda \in \mathbb{F}_q^*, \lambda \text{ is a quadratic residue}, 0 \le k < m \right\},$$

where $q = p^m$. In particular, H_0 acts on $X = U$, which in what follows is identified with $\mathbb{F}_q$, with the three orbits $\{0\}$, U^+ and U^-, where

$$U^+ = \{u \in \mathbb{F}_q^* \mid u \text{ is a quadratic residue}\},$$

$$U^- = \{u \in \mathbb{F}_q^* \mid u \text{ is a quadratic non-residue}\}.$$

Suppose that the admissible partition $\mathcal{P} : U(2) = \bigcup_{v \in \mathbb{F}_q} M_v$ (where $\bigcup_{\alpha \in M_v} = \mathbb{F}_q \setminus \{v\}$) corresponds to the E-decomposition $\mathcal{D}$ with the given automorphism group $G = \mathrm{Aut}(\mathcal{D})$.

Lemma 6.3.4. *Suppose that* $\{1, \nu\} \in M_0$. *Then*

(i) H_0 *has orbits* $\{0\}$, U^+ *and* U^- *on* U;

(ii) ν *is a quadratic non-residue;*

(iii) $St_{H_0}(1) \subseteq St_{H_0}(\nu)$.

Furthermore, $\mathcal{P}$ *is uniquely defined by the pair* (H_0, ν). *Conversely, if a pair* (H_0, ν) *satisfies conditions* (i), (ii) *and* (iii), *then there exists a (unique)* $(U.H_0)$-*invariant partition* $\mathcal{P}$.

Proof. 1) Suppose that the partition $\mathcal{P}$ is given. Property (i) has been mentioned above. Furthermore, suppose that $\nu \in U^+$. By the transitivity of H_0 on U^+, we can find an element $\varphi \in H_0$ such that $\varphi(1) = \nu$. Then $\varphi(M_0) = M_0$ and $\varphi(\{1, \nu\}) = \{\nu, \varphi(\nu)\} \in M_0$, so $\varphi(\nu) = 1$ by NIP. Thus φ has even order, whereas $|H|$ is odd, a contradiction. Hence $\nu \in U^-$. Finally, if $\varphi \in St_{H_0}(1)$, then $\varphi(M_0) = M_0$ and $\varphi(\{1, \nu\}) = \{1, \varphi(\nu)\}$, so that $\varphi(\nu) = \nu$ as required.

2) Conversely, suppose that (H_0, ν) is a pair satisfying (i), (ii) and (iii). Of course, the required partition $\mathcal{P}$ must be of the form

$$M_0 = \{1, \nu\}^{H_0}, \; M_v = M_0 + v := \left\{ \{x + v, y + v\} \mid \{x, y\} \in M_0 \right\}.$$

We have to check that M_v satisfies NIP, and it is sufficient to check this for M_0. Thus, assume that $\alpha \cap \beta \neq \emptyset$, where $\alpha = \{g_1(1), g_1(\nu)\}$, $\beta = \{g_2(1), g_2(\nu)\}$, and $g_1, g_2 \in H_0$. Since $g_1(1), g_2(1) \in U^+$ and $g_1(\nu), g_2(\nu) \in U^-$, we can assume that $g_1(\nu) = g_2(\nu)$, that is, $g_1^{-1} g_2 \in St_{H_0}(\nu)$. Note that $|St_{H_0}(1)| = |St_{H_0}(\nu)| = 2|H_0|/(p^m - 1)$; therefore, since $St_{H_0}(1) \subseteq St_{H_0}(\nu)$, we have $St_{H_0}(1) = St_{H_0}(\nu)$. Hence $g_1^{-1} g_2 \in St_{H_0}(1)$, $g_1(1) = g_2(1)$ and $\alpha = \beta$. □

From now on, the partition $\mathcal{P}$ constructed from the pair (H_0, ν) will be denoted by $\mathcal{P}(H_0, \nu)$.

Lemma 6.3.5. *Suppose that* (H_0, ν) *is a pair satisfying conditions* (i), (ii) *and* (iii) *of Lemma* 6.3.4, *and set* $\mathcal{P} = \mathcal{P}(H_0, \nu)$. *Then*

1) *if* $\nu = -1$, *then* $\mathrm{Aut}(\mathcal{P}) = AGL_m(p)$ *and the corresponding E-decomposition coincides with decomposition* (6.6)*;*

2) *if* $p^m = 7$ *and* $\nu \neq -1$, *then* $\mathrm{Aut}(\mathcal{P}) = SL_3(2)$, *and the corresponding E-decomposition is conjugate to decomposition* (3.8) *(of the Lie algebra of type* B_3*);*

3) *if* $p^m \neq 7$ *and* $\nu \neq -1$, *then* $\mathrm{Aut}(\mathcal{P})$ *is a transitive permutation group of rank* 3 *contained in* $U.\Sigma L_1(p^m)$.

Proof. 1) This is obvious.

2) Direct calculations show that the corresponding OD is an MOD in such a case. Furthermore, decomposition (3.8) is also multiplicative, and by Theorem 3.3.8, any two MODs of the given Lie algebra are $\mathrm{Aut}(\mathcal{L})$-conjugate.

3) Set $P = \mathrm{Aut}(\mathcal{P})$, so that P contains $U.H_0$ and so it is 2-homogeneous permutation group on U. If P is not 2-transitive, then by Kantor's Theorem [Kan 1], $\mathrm{rank}P = 3$ and P is contained in $U.\Sigma L_1(p^m)$. Suppose that P is 2-transitive. Note that $\mathcal{P}$ corresponds to an irreducible E-decomposition $\mathcal{D}$ with automorphism group $\mathrm{Aut}(\mathcal{D}) = \mathbb{Z}_2^{2\mathrm{rank}(\mathcal{L})}.P$. Firstly we assume that $T = \mathrm{soc}(P)$ is nonabelian. Then, by Proposition 6.3.2, $T = SL_d(2)$ and $7 < p^m = 2^d - 1$. This implies that $m = 1$ and $p = 2^d - 1$ is a Mersenne prime. We have the inclusion

$$S = SL_d(2) = P \supseteq H = U.H_0 = \mathbb{Z}_p.\mathbb{Z}_{(p-1)/2}.$$

Set $Q = \mathbb{Z}_p$; then $Q \in \mathrm{Syl}_p(SL_d(2))$ and $N_{SL_d(2)}(Q) = Q.\mathbb{Z}_d$. In particular, $d \geq (p-1)/2 = 2^{d-1} - 1 > 3$, a contradiction. Thus we must have $\mathrm{soc}(P) = T = \mathbb{Z}_p^m$. In such a case, U acts regularly on p^m points, hence $C_P(U) = U$. On the other hand, consideration of the p-modular representation of the abelian p-group U on T gives that $C_P(U) \cap T \neq 1$. Then $U \cap T \neq 1$, while $U \cap T \lhd U.H_0$ and U is a minimal normal subgroup of $U.H_0$. Consequently, $U = T$. By Proposition 6.2.5, our E-decomposition is the same as that in (6.6). In particular, $E(1, -1) \in \mathcal{H}_0$, $\{1, -1\} \in M_0$ and $\nu = -1$, contrary to our assumption. □

Corollary 6.3.6. *The Lie algebra of type* B_3 *has exactly two IODs, and their full automorphism groups are* $\mathbb{Z}_2^6.SL_3(2)$ *and* $\mathbb{Z}_2^6.(\mathbb{Z}_7.\mathbb{Z}_6)$. □

Now we shall be interested in the possible groups H_0. Set

$$\mathcal{S} = \{L \mid L \subseteq \Sigma L_1(p^m),\, (L : St_L(1)) = (p^m - 1)/2\}.$$

Clearly, $H_0 \in \mathcal{S}$.

Proposition 6.3.7. 1) *Each element* $L \in \mathcal{S}$ *is uniquely defined by the three parameters* $r, s \in \mathbb{N}$, $\lambda \in \mathbb{F}_q^*$, *for which* $|L| = s(p^m - 1)/2$, $L \cap GL_1(p^m) = \mathbb{Z}_{(p^m-1)/2r}$

and L contains the element $f : x \mapsto \lambda x^{p^{m/rs}}$, $q = p^m$. Here $r \mid \gcd(m, (p^m - 1)/2)$, $s \mid (m/r)$, $\lambda \in \{\varepsilon^2, \varepsilon^4, ..., \varepsilon^{2r}\}$, where $\mathbb{F}_q^* =< \varepsilon >$. For r, s, λ with these properties, a group $L \in \mathcal{S}$ with parameters r, s, λ exists if and only if the following condition is satisfied: either $r = 1$, or else $r > 1$ and the following statements hold:

$$\frac{(\gcd(r, p^{m/rs} - 1)).(\gcd(r, p^{m/s} - 1))}{p^{m/rs} - 1} > r, \tag{6.8}$$

$$\gcd(r, \frac{p^{m/s} - 1}{p^{m/rs} - 1}) \nmid \frac{p^{im/rs} - 1}{p^{m/rs} - 1}, \quad 1 \leq i < r. \tag{6.9}$$

There exist at most $\gcd(r, p^{m/rs} - 1)$ *classes of* $\Gamma L_1(q)$*-conjugate groups in* $\mathcal{S}$ *with given parameters r, s.*

2) *If $H_0 \in \mathcal{S}$, then there exists an element $\nu \in \mathbb{F}_q^*$ such that (H_0, ν) satisfies conditions* (i), (ii) *and* (iii) *of Lemma* 6.3.4.

Proof. 1) It is obvious that we can find such $r, s \in \mathbb{N}$, and furthermore that $L =< L \cap GL_1(q), f >$. Moreover, we can choose λ from $\mathbb{Z}_{(q-1)/2} =< \varepsilon^2 >$ modulo $\mathbb{Z}_{(q-1)/2r} =< \varepsilon^{2r} >$. Note that, if $\lambda = \varepsilon^{2t}$, then

$$f^i : x \mapsto \lambda^{1+p^{m/rs}+p^{2m/rs}+...+p^{(i-1)m/rs}} x^{p^{im/rs}}.$$

Moreover, $St_L(1) = \mathbb{Z}_s =< x \mapsto x^{p^{m/s}} >$, so we must have:

$$r \text{ divides } t\frac{p^{m/s} - 1}{p^{m/rs} - 1}, \tag{6.10}$$

$$r \nmid t\frac{p^{im/rs} - 1}{p^{m/rs} - 1} \text{ for every } i = 1, \ldots, r - 1. \tag{6.11}$$

In what follows, we set $d = \gcd(r, \frac{p^{m/s}-1}{p^{m/rs}-1})$. Then, it follows from (6.10) and (6.11) that $(r/d) \mid t$ but $d \nmid \frac{p^{im/rs}-1}{p^{m/rs}-1}$. That is, (6.9) is a necessary condition for the existence of a group $L \in \mathcal{S}$. Let $\mathcal{C}$ denote the $\Gamma L_1(q)$-conjugacy class of subgroups which contains L. Then an arbitrary $L' \in \mathcal{C}$ also has parameters r, s. Furthermore, set

$$\Lambda = \left\{\alpha \in \mathbb{F}_q^* \mid \exists L' \in \mathcal{C}, L' \ni g : x \mapsto \alpha x^{p^{m/rs}}\right\}.$$

It is easy to see that

a) $\Lambda \subseteq A =< \varepsilon^2 >$,

b) if $\alpha \in \Lambda$, then $\alpha.\varepsilon^{2r}$ and $\alpha.\varepsilon^{p^{m/rs}-1}$ belong to Λ.

Hence Λ contains (some) cosets of A modulo the subgroup

$$B = \left\langle \varepsilon^{2r}, \varepsilon^{p^{m/rs}-1} \right\rangle = \left\langle \varepsilon^{2\gcd(r, p^{m/rs}-1)} \right\rangle.$$

In particular, for given r, s there exist at most $\gcd(r, p^{m/rs} - 1)$ $\Gamma L_1(q)$-conjugacy classes of groups in $\mathcal{S}$ with parameters r, s.

Of course, for any s with $s \mid m$, there exists a (unique) group $L \in \mathcal{S}$ with parameters $r = 1$, s, namely $L = \mathbb{Z}_{(q-1)/2}.\mathbb{Z}_s$. Suppose that $r > 1$. As we have shown above, there exists a subgroup $L \in \mathcal{S}$ with parameters r, s, $\lambda = \varepsilon^{2t}$, where $1 \leq t < \gcd(r, p^{m/rs} - 1)$. Then from (6.10) it follows that $(r/d) \mid t$, and hence that $d.(\gcd(r, p^{m/rs} - 1)) > r$; that is, (6.8) is a necessary condition for the existence of a group $L \in \mathcal{S}$. Conversely, suppose that r, s satisfy (6.8) and (6.9). Then we can set $\lambda = \varepsilon^{2t}$, where $t = r/d$ (recall that $d = \gcd(r, \frac{p^{m/s}-1}{p^{m/rs}-1})$). Obviously, we have $1 \leq t < \gcd(r, p^{m/rs} - 1)$ from (6.8), and so (6.10) and (6.11) follow immediately. Note that the group $L =< \mathbb{Z}_{(q-1)/2r}, f >$, where $\mathbb{Z}_{(q-1)/2r} \subset GL_1(q)$ and $f : x \mapsto \lambda x^{p^{m/rs}}$, belongs to $\mathcal{S}$.

2) Suppose that $H_0 \in \mathcal{S}$. Because $St_{H_0}(1) = St_{H_0}(-1)$ has index $(q-1)/2$ in H_0, and H_0 fixes the sets $U^+ \ni 1$ and $U^- \ni -1$ of cardinality $(q-1)/2$, it follows that H_0 has orbits $\{0\}$, U^+ and U^- on $U = \mathbb{F}_q$. Taking $\nu \in \mathbb{F}_{p^{m/s}} \cap U^-$, we have $St_{H_0}(1) \subseteq St_{H_0}(\nu)$. □

In what follows, we shall need the following statement:

Lemma 6.3.8. *Suppose that* $2, p_1, \ldots, p_t$ *are all the distinct prime divisors of* $(p^m - 1)$*, that is,* $p^m - 1 = 2\prod_{i=1}^{t} p_i^{\alpha_i}$. *Set* $e = 2\prod_{p_i \mid m} p_i^{\alpha_i}$. *Then*

(i) *the group* $\Gamma L_1(q)$ *has a unique subgroup* Φ *of order* $(p^m - 1)/e$ *(recall that* $q = p^m \equiv -1 \pmod 4$*);*

(ii) $\Phi \subseteq GL_1(q)$ *and so* Φ *is cyclic;*

(iii) *every* $L \in \mathcal{S}$ *contains* Φ*;*

(iv) $N_{GL_m(p)}(\Phi) = \Gamma L_1(q)$.

Proof. 1) As a representative for Φ, take the cyclic subgroup $\mathbb{Z}_{(q-1)/e}$ of the group $F = GL_1(q) \cong \mathbb{Z}_{q-1}$. Next, suppose that $\Phi \subseteq \Gamma = \Gamma L_1(q)$ and $\mid \Phi \mid = (q-1)/e$. Then $F\Phi \subseteq \Gamma$, $\mathbb{Z}_m = \Gamma/F \supseteq F\Phi/F \cong \Phi/(\Phi \cap F)$, whereas $\gcd(\mid \Phi \mid, m) = 1$, so that $\Phi = \Phi \cap F$ and $\Phi \subseteq F$. Obviously, $\Phi \lhd \Gamma L_1(q)$. Furthermore, if $L \in \mathcal{S}$, then $L \cap F = \mathbb{Z}_{(q-1)/2r}$, where $r \mid \gcd(m, p^m - 1)$. But $2r \mid e$, and hence $L \supseteq \Phi$.

2) We show next that $N_{GL_m(p)}(\Phi) \subseteq \Gamma L_1(q)$. Consider the maps

$$A : x \mapsto \varepsilon x, \; B : x \mapsto x^p$$

as linear operators on the $\mathbb{F}_p$-space $U = \mathbb{F}_q$ (recall that $\mathbb{F}_q^* =< \varepsilon >$). Then A has simple spectrum $\{\varepsilon, \varepsilon^p, \ldots, \varepsilon^{p^{m-1}}\}$. Choose a basis $\{e_0, \ldots, e_{m-1}\}$ of the $\mathbb{F}_q$-space $U \otimes_{\mathbb{F}_p} \mathbb{F}_q$ in which we have $A(e_i) = \varepsilon^{p^i} e_i$. Because $BAB^{-1} = A^p$, the map B operates by the rule:

$$B :< e_{m-1} > \to < e_{m-2} > \to \ldots \to < e_1 > \to < e_0 > \to < e_{m-1} > .$$

Note that the relator $\varphi = A^e$ of the cyclic group Φ also has simple spectrum. In fact, if this does not hold, then we have $(p^m - 1) \mid e(p^k - 1)$ for some k, $1 \leq k < m$. As always, we can choose some prime divisor t of $(p^m - 1)$ not dividing $\prod_{1 \leq i < m}(p^i - 1)$. Then we have $t \mid e$, so that $t \mid m$, say $m = tl$. But in this case

$$0 \equiv p^m - 1 = p^{tl} - 1 = (p^l)^t - 1 \equiv p^l - 1 \pmod t,$$

and $t \mid (p^l - 1)$, which contradicts the choice of t.

Now suppose that $n \in N_{GL_m(p)}(\Phi)$, so that $n^{-1}\varphi n = \varphi^k$ for some k. Note that $\varphi = n\varphi^k n^{-1}$, and hence

$$\varphi(n(e_0)) = n\varphi^k n^{-1}(n(e_0)) = n\varphi^k(e_0) = \varepsilon^{ek} n(e_0),$$

that is, $\varepsilon^{ek} \in \mathrm{Spec}(\varphi)$. Thus $\varepsilon^{ek} = \varepsilon^{ep^i}$ for some i, $0 \leq i \leq m-1$. Then $p^m - 1 = \mid \varepsilon \mid$ divides $e(k - p^i)$, $\mid \varphi \mid = (p^m - 1)/e$ divides $(k - p^i)$, $n^{-1}\varphi n = \varphi^k = \varphi^{p^i} = B^i \varphi B^{-i}$, and $\psi = nB^i$ centralizes φ. By the simplicity of $\mathrm{Spec}(\varphi)$, we have $\psi(e_j) = \lambda_j e_j$ for some $\lambda_j \in \mathbb{F}_q^*$. Without loss of generality, we may suppose that

$$B : e_{m-1} \mapsto e_{m-2} \mapsto \ldots \mapsto e_2 \mapsto e_1 \mapsto e_0 \mapsto e_{m-1}.$$

In particular, $\mathrm{Ker}(B - 1) = < e_0 + e_1 + \ldots + e_{m-1} >_{\mathbb{F}_q}$. But $B(1) = 1^p = 1$, so we may suppose that $e_0 + \ldots + e_{m-1} = 1 \in U$. Then since $\psi \in GL(U)$, it follows that

$$\sum_{j=0}^{m-1} \lambda_j e_j = \psi(\sum_j e_j) = \psi(1) \in U \backslash \{0\}.$$

But $< A > = \mathbb{Z}_{q-1}$ acts regularly on $U \backslash \{0\}$, so we have $\psi(1) = A^l(1)$ for some integer l. Now

$$\sum_j \lambda_j e_j = \psi(1) = A^l(\sum_j e_j) = \sum_j \varepsilon^{lp^j} e_j,$$

hence $\lambda_j = \varepsilon^{lp^j}$, $\psi(e_j) = A^l(e_j)$ for all j; in other words, $\psi = A^l$. Consequently, $n = \psi B^{-i} = A^l B^{-i} \in \Gamma L_1(q)$. □

Remark 6.3.9. In some sense, the generator φ of the cyclic group $\Phi = \mathbb{Z}_{(q-1)/e}$ plays the role of a Singer cycle (see [Kan 2]).

Lemma 6.3.10. *Suppose that $p^m > 7$ and $\mathcal{D}_1$ and $\mathcal{D}_2$ are IODs of the Lie algebra $\mathcal{L} = B_{(p^m-1)/2}$ corresponding to the partitions $\mathcal{P}_1 = \mathcal{P}(H_1, \nu_1)$ and $\mathcal{P}_2 = \mathcal{P}(H_2, \nu_2)$, where $\nu_1, \nu_2 \neq -1$ and $H_1, H_2 \in \mathcal{S}$. Assume further that φ is an automorphism of $\mathcal{L}$ that maps $\mathcal{D}_1$ onto $\mathcal{D}_2$. Then $\varphi \in \mathbb{Z}_2^{p^m-1}.(U.\Gamma L_1(p^m))$.*

Proof. By Proposition 2.2.5, φ operates as follows:

$$\varphi : E(u, v) \mapsto (-)^{f(u)+f(v)} E(\pi(u), \pi(v)),$$

where $f \in \bar{U} = \mathbb{Z}_2^{p^m-1}$, $\pi \in \mathrm{Sym}(U)$ and $\pi(\mathcal{P}_1) = \mathcal{P}_2$. By Lemma 6.3.5, $\mathrm{Aut}(\mathcal{P}_i) = U.H_i'$, where $H_i \subseteq H_i' \subseteq \Sigma L_1(p^m)$ for $i = 1, 2$. Obviously, $\mathrm{Aut}(\mathcal{P}_2) = \pi.\mathrm{Aut}(\mathcal{P}_1).\pi^{-1}$. But $U = \mathrm{soc}(\mathrm{Aut}(\mathcal{P}_i))$ for $i = 1, 2$, so $U = \pi U \pi^{-1}$ and $\pi \in N_{\mathrm{Sym}(U)}(U) = U.GL(U)$. Without loss of generality, we can suppose that $\pi \in GL(U)$, and then $\pi H_1' \pi^{-1} = H_2'$. Now we recall that, by Lemma 6.3.8, all the groups $\Gamma L_1(p^m)$, H_1', H_2', H_1 and H_2 contain the same group Φ of order $(p^m - 1)/e$; moreover, Φ is the unique subgroup of order $(p^m - 1)/e$ of $\Gamma L_1(p^m)$, and $N_{GL(U)}(\Phi) = \Gamma L_1(p^m)$. Hence $\pi \Phi \pi^{-1} = \Phi$ and $\pi \in \Gamma L_1(p^m)$. □

Corollary 6.3.11. *Suppose that a group $H_0 \in \mathcal{S}$ has the parameters r, s. Then for all $\nu \neq -1$ in the set*

$$U^- \bigcap \left\{ \mathbb{F}_{p^{m/s}} \setminus \bigcup_{i \mid \frac{m}{s}, i < \frac{m}{s}} \mathbb{F}_{p^i} \right\},$$

the IOD corresponding to the partition $\mathcal{P} = \mathcal{P}(H_0, \nu)$ has (full) automorphism group $\mathbb{Z}_2^{p^m-1}.(U.H_0)$.

Proof. It is sufficient to prove that each φ in $\mathrm{Aut}(\mathcal{P})$ belongs to $U.H_0$. By Lemma 6.3.10, we have $\varphi \in U.\Sigma L_1(p^m)$. We can suppose that $\varphi \in \Sigma L_1(p^m)$, because $U \subseteq \mathrm{Aut}(\mathcal{P})$ and U acts regularly on U. Since $H_0 \subseteq \mathrm{Aut}(\mathcal{P})$, and H_0 and $\Sigma L_1(p^m)$ are transitive on U^+, we can suppose that $\varphi(1) = 1$. But $\varphi(M_0) = M_0$ and $\{1, \nu\} \in M_0$, so we must have that $\varphi(\nu) = \nu$. By the choice of ν, we get that $\varphi \in \mathbb{Z}_s = < f : x \mapsto x^{p^{m/s}} > = St_{H_0}(1)$, so that $\varphi \in H_0$ as stated. □

Definition 6.3.12. For $r \mid \gcd(m, p^m - 1)$ and $s \mid \frac{m}{r}$, let $S(p, m, r, s)$ denote the total number of groups $L \in \mathcal{S}$ with parameters r, s. For a prime number $p \equiv 3 \pmod 4$ and $l \in \mathbb{N}$, let $Q(p, l)$ denote the total number of all quadratic non-residues $\nu \in \mathbb{F}_{p^l} \setminus \{-1\}$ not contained in any proper subfield of $\mathbb{F}_{p^l}$.

Lemma 6.3.13. 1) $S(p, m, r, s) \leq r$.
2) $Q(p, 1) = (p - 3)/2$. *Furthermore, if $l > 1$ is odd, then*

$$(p^l - 3)/2 \geq Q(p, l) \geq (p^l - p^{(l-1)/2})/2.$$

Proof. 1) This follows from Proposition 6.3.7.
2) All the statements are obvious except the last estimate $Q(p, l) \geq (p^l - p^{(l-1)/2})/2$. Assume that l has exactly t distinct prime divisors $p_1 < \ldots < p_t$. Then

$$2Q(p, l) = \left| \mathbb{F}_{p^l} \setminus \bigcup_{d \mid l, d < l} \mathbb{F}_{p^d} \right| \geq p^l - \sum_{i=1}^{t} p^{l/p_i} \geq p^l - tp^{l/p_1}.$$

Since $p \geq 3$ and $p_1 \geq 3$, we have $l \geq 3^t$, so that $tp^{l/p_1} \leq p^{(l-1)/2}$. □

Proposition 6.3.14. *The Lie algebra $\mathcal{L}$ of type B_n ($n \geq 3$) admits an IOD $\mathcal{D}$ with full automorphism group* $\mathrm{Aut}(\mathcal{D}) = E.H$*, where $E \subseteq \mathbb{Z}_2^{2n}$ and H is a transitive subgroup of rank 3 of $\mathbb{S}_{2n+1}$, if and only if $n = (p^m - 1)/2$ for some prime power $p^m \equiv -1 \pmod 4$ with $p^m > 7$. Each such IOD has full automorphism group* $\mathrm{Aut}(\mathcal{D}) = \mathbb{Z}_2^{p^m-1}.(\mathbb{Z}_p^m.H_0)$*, where $H_0 \subseteq \Sigma L_1(p^m)$. If $N(p,m)$ denotes the total number of* $\mathrm{Aut}(\mathcal{L})$*-conjugacy classes of IODs with the indicated property of having rank 3, then*

$$\sum_{r|\gcd(m,p^m-1),s|\frac{m}{r}} S(p,m,r,s).\frac{s(p^{m/s}-3)}{4m} \geq N(p,m) \geq$$

$$\geq \sum_{r|\gcd(m,p^m-1),s|\frac{m}{r}} S(p,m,r,s).Q(p,\frac{m}{s}).\frac{s}{2m}.$$

Proof. All the statements except the last estimate were proved above. From Lemma 6.3.10, it follows that $N(p,m)$ is the total number of $\Gamma L_1(p^m)$-conjugacy classes of partitions $\mathcal{P}$ of the set $U(2) = \{\{a,b\} \mid a,b \in U = \mathbb{F}_{p^m}, a \neq b\}$ with $\mathrm{Aut}(\mathcal{P}) = U.H_0$. By Lemma 6.3.4, $\mathcal{P} = \mathcal{P}(H_0,\nu)$. The group H_0 has parameters r,s, where $r \mid \gcd(m,p^m-1)$, $s \mid \frac{m}{r}$, and there exist exactly $S(p,m,r,s)$ groups with the same parameters. Next, there exist at most $(p^{m/s}-3)/2$ (because $St_{H_0}(1) \subseteq St_{H_0}(\nu)$) and at least $Q(p,\frac{m}{s})$ (see Corollary 6.3.11) possibilities for the quadratic non-residue $\nu \neq -1$. Finally, the length of the $\Gamma L_1(p^m)$-orbit of $\mathcal{P}$ equals $(\Gamma L_1(p^m) : H_0) = 2m/s$. □

Corollary 6.3.15. *In the notation of Proposition 6.3.14, we have*

$$\sum_{r|\gcd(m,p^m-1),s|\frac{m}{r}} \frac{rs(p^{m/s}-3)}{4m} \geq N(p,m) \geq \sum_{s|m,s>1} \frac{p^s - p^{(s-1)/2}}{4s} + \frac{p-3}{4}.$$

Proof. This follows from Proposition 6.3.14 and Lemma 6.3.13, because $S(p,m,1,s) = 1$. □

Theorem 6.0.2 now follows from Propositions 6.2.4, 6.3.2, 6.2.5 and 6.3.14, and Corollaries 6.3.6 and 6.3.15.

Question 6.3.16. *Describe all the members $L \in \mathcal{S}$, for which the group $U.L$ is the full automorphism group* $\mathrm{Aut}(\mathcal{P})$ *of some admissible partition $\mathcal{P}$.*

It is possible to show that if $\mathcal{P} = \mathcal{P}(H_0,\nu)$ and $\mathrm{Aut}(\mathcal{P}) = U.H_0'$, then H_0' is a maximal member with respect to inclusion of the set

$$\mathcal{S}' = \{L \in \mathcal{S} \mid L \supseteq H_0, St_L(1) \subseteq St_L(\nu)\}.$$

Commentary

Burichenko (unpublished) was the first to study E-decompositions of Lie algebras of type B_n. In particular, he established Lemma 6.3.1 and Proposition 6.3.2. The general case of the weak Winnie-the-Pooh problem for type B_n is treated in [Tiep 11]. Unfortunately, the scheme used in this chapter works only for the type B_n. This scheme is not applicable to types C_n and D_n. As in case A_n (see Commentary to Chapter 5), it would be also interesting to eliminate the dependence of our proof of Theorems 6.0.1 and 6.0.2 on the classification of doubly transitive permutation groups.

Chapter 7

Orthogonal decompositions of semisimple associative algebras

Let us consider an orthogonal decomposition of a complex simple Lie algebra $\mathcal{L}$ of type A_n, say, $n = p^m - 1$ for some prime power p^m:

$$\mathcal{L} = \bigoplus_{i=0}^{n+1} \mathcal{H}_i.$$

We shall assume that $\mathcal{L}$ is realized by complex matrices of degree $n+1$ with zero trace. Then the associative commutative subalgebras

$$\tilde{\mathcal{H}}_i = < E_{n+1} >_{\mathbb{C}} \oplus \mathcal{H}_i$$

of type $(n+1)M_1(\mathbb{C})$, where $0 \leq i \leq n+1$, are pairwise orthogonal in the sense of §1.5. The system $\{\tilde{\mathcal{H}}_i\}_{i=0}^{n+1}$ constitutes what we shall call below an *orthogonal decomposition of the semisimple associative algebra* $M_{n+1}(\mathbb{C})$. In fact, the notion of ODs of associative algebras is much wider; it includes, in addition to the orthogonal decompositions of simple Lie algebras of type A_n, topics such as finite affine planes, classical Hadamard matrices and finite split groups.

7.1 Definitions and examples

Throughout this chapter, we shall consider finite-dimensional semisimple associative algebras over an algebraically closed field $\mathbb{F}$ of characteristic 0. By the Molin-Wedderburn theorem, any such algebra $\mathcal{A}$ is isomorphic to a direct sum of matrix algebras over $\mathbb{F}$. If

$$\mathcal{A} \cong \oplus_n m_n M_n(\mathbb{F}),$$

we shall say that $\mathcal{A}$ has type $\oplus m_n M_n$. Throughout this chapter, the term "algebra" means a finite direct sum of matrix algebras. Note that the left and the right regular representations of $\mathcal{A}$ yield the same trace, which we shall denote by the symbol $\mathrm{tr}_{\mathcal{A}}$. For example, for any element $x \in \mathcal{A} = M_n(\mathbb{F})$ we have $\mathrm{tr}_{\mathcal{A}} x = n\mathrm{Tr} x$.

It is obvious that the map

$$(x, y) \mapsto (x|y) = \mathrm{tr}_{\mathcal{A}} xy$$

defines a non-degenerate symmetric invariant (associative) form on $\mathcal{A}$. Let $1_{\mathcal{A}}$ denote the unit element of $\mathcal{A}$.

Definition 7.1.1. A subalgebra $\mathcal{B}$ of $\mathcal{A}$ is said to be *tame* if $\mathcal{B}$ is semisimple, $\dim \mathcal{B} > 1$ and $\mathcal{B}$ contains the unit element of $\mathcal{A}$, that is, $1_{\mathcal{B}} = 1_{\mathcal{A}}$.

Definition 7.1.2. The system $\mathcal{D} = \{\mathcal{B}_i\}$ constitutes an *orthogonal decomposition* of $\mathcal{A}$, if

(i) each subalgebra $\mathcal{B}_i \in \mathcal{D}$ is a proper tame subalgebra;
(ii) $\mathcal{A}$ is linearly generated by the subalgebras in $\mathcal{D}$;
(iii) $\mathcal{B}_i \cap \mathcal{B}_j = < 1_{\mathcal{A}} >_{\mathbb{F}}$ for $i \neq j$;
(iv) for any $i, j, i \neq j$, and for any elements $x \in \mathcal{B}_i$, $y \in \mathcal{B}_j$ with

$$(x|1_{\mathcal{A}}) = 0 = (1_{\mathcal{A}}|y),$$

we have $(x|y) = 0$.

An OD of this sort is said to be *homogeneous* if all the subalgebras $\mathcal{B}_i \in \mathcal{D}$ are mutually isomorphic.

In what follows, the following properties of orthogonal decompositions will be required. Let $\mathcal{D}$ be an OD of $\mathcal{A}$ and take $\mathcal{B} \in \mathcal{D}$. Then the symbol $x_{\mathcal{B}}$ denotes the orthogonal projection of an element x of $\mathcal{A}$ onto $\mathcal{B}$. Similarly, $x_{<1>}$ denotes the orthogonal projection of x onto $< 1_{\mathcal{A}} >_{\mathbb{F}}$, so we have

$$x_{<1>} = (\mathrm{tr}_{\mathcal{A}} x / \dim \mathcal{A}).1_{\mathcal{A}}.$$

Proposition 7.1.3. *Let $\mathcal{D} = \{\mathcal{B}_i\}_{i=1}^r$ be an OD of an algebra $\mathcal{A}$ of dimension n. Then:*

(i) $\mathrm{tr}_{\mathcal{A}} xy = (1/n).\mathrm{tr}_{\mathcal{A}} x.\mathrm{tr}_{\mathcal{A}} y$ *for all* $x \in \mathcal{B}_i$, $y \in \mathcal{B}_j$, $i \neq j$;
(ii) $x = \sum_{i=1}^r x_{\mathcal{B}_i} - (r-1)x_{<1>}$ *for all* $x \in \mathcal{A}$;
(iii) $(x|x) = \sum_{i=1}^r (x_{\mathcal{B}_i}|x_{\mathcal{B}_i}) - (r-1)(x_{<1>}|x_{<1>})$ *for all* $x \in \mathcal{A}$.

Proof. (i) Writing $x = x_{<1>} + x'$, $y = y_{<1>} + y'$, we have

$$\mathrm{tr}_{\mathcal{A}} x' = \mathrm{tr}_{\mathcal{A}} y' = \mathrm{tr}_{\mathcal{A}} x' y_{<1>} = \mathrm{tr}_{\mathcal{A}} x_{<1>} y' = 0.$$

In addition, condition (iv) from Definition 7.1.2 implies that $\mathrm{tr}_{\mathcal{A}} x'y' = 0$. Hence,

$$\mathrm{tr}_{\mathcal{A}} xy = \mathrm{tr}_{\mathcal{A}} x_{<1>} y_{<1>} = (1/n)(\mathrm{tr}_{\mathcal{A}} x)(\mathrm{tr}_{\mathcal{A}} y).$$

(ii), (iii). Let $\mathcal{B}_i^0$ be the orthogonal complement to $< 1_{\mathcal{A}} >_{\mathbb{F}}$ in $\mathcal{B}_i$. Then we have the following decomposition of $\mathcal{A}$ into a direct sum of subspaces: $\mathcal{A} = < 1_{\mathcal{A}} >_{\mathbb{F}} \oplus(\oplus_{i=1}^r \mathcal{B}_i^0)$. Assertions (ii) and (iii) now follow by the properties of orthogonal projection. □

Next, we exhibit some examples of orthogonal decompositions.

Example 7.1.4. As mentioned at the beginning of the chapter, each OD of the Lie algebra $\mathcal{L}$ of type A_n corresponds to a homogeneous OD of type $(n+1)M_1$ of the associative algebra $\mathcal{A} = M_{n+1}(\mathbb{F})$ (that is, every component of the orthogonal decomposition is of type $(n+1)M_1$). Conversely, let $\mathcal{D} = \{\mathcal{B}_1, \dots, \mathcal{B}_{n+2}\}$ be a homogeneous OD of type $(n+1)M_1$ of $\mathcal{A}$. Then the system of Lie subalgebras $\mathcal{H}_i = \mathcal{B}_i \cap \mathcal{L}$, $1 \leq i \leq n+2$, constitutes an OD of $\mathcal{L}$.

Example 7.1.5. Let $A = \{P, L\}$ be a finite affine plane of order n (see [Hall 1]), where P is the point-set and L is the line-set. Denote by $\mathcal{A}$ the algebra of all $\mathbb{F}$-valued functions on P, endowed with pointwise addition and pointwise multiplication operations. Then $\mathcal{A}$ is a commutative associative algebra of type n^2M_1. Let χ_x denote the characteristic function of a subset X of P. With every collection of parallel lines $\sigma = \{l_1, \dots, l_n\}$, we associate the subalgebra $\mathcal{B}_\sigma \subseteq \mathcal{A}$ generated by the functions χ_{l_i}, $1 \leq i \leq n$. Suppose now that σ and σ' are two distinct collections of parallel lines, and consider lines $l \in \sigma$, $l' \in \sigma'$. Then

$$\mathrm{tr}_{\mathcal{A}}\chi_l\chi_{l'} = \mathrm{tr}_{\mathcal{A}}\chi_{l\cap l'} = 1 = \frac{1}{n^2}\mathrm{tr}_{\mathcal{A}}\chi_l \cdot \mathrm{tr}_{\mathcal{A}}\chi_{l'}.$$

In other words, $\mathcal{B}_\sigma$ and $\mathcal{B}_{\sigma'}$ *intersect orthogonally in the unit subalgebra*, that is, they satisfy conditions (iii) and (iv) of Definition 7.1.2. On calculating the dimensions of the algebras $\mathcal{B}_\sigma$ and $\mathcal{A}$, it is easy to see that the system $\mathcal{D} = \{\mathcal{B}_\sigma\}$, with σ running over the set of all the collections of parallel lines, generates $\mathcal{A}$ linearly. So, $\mathcal{D}$ is a homogeneous OD of type nM_1 of $\mathcal{A}$.

Conversely, every homogeneous orthogonal decomposition $\mathcal{D}$ of type nM_1 of $\mathcal{A} = n^2M_1(\mathbb{F})$ corresponds to some affine plane $A = \{P, L\}$ of order n. Namely, P is the set of all the primitive idempotents of $\mathcal{A}$; L is the set of all primitive idempotents of all the subalgebras from $\mathcal{D}$. Furthermore, a point $p \in P$ is incident to a line $l \in L$ if and only if $pl \neq 0$. (If, say, $l = \sum_{i=1}^n p_i$, then $pl \neq 0$ exactly when $p = p_i$, $i = 1, \dots, n$).

Example 7.1.6. Let $H = \{P, B\}$ be an extended Hadamard $3-(4(n+1), 2(n+1), n)$ block design (see [CaL] and [Hall 1]), where P is the point set of cardinality $4(n+1)$, and B is the block set. The last set splits into a union of pairs of mutually supplementary blocks:

$$B = \bigcup_i \{b_i, b_i'\},$$

where $b'_i = P\backslash b_i$ and $|b_i \cap b_j| = n+1$ provided that $i \neq j$. Let $\mathcal{A}$ stand once more for the algebra of all $\mathbb{F}$-valued functions on P. Then for $i \neq j$ we have:

$$\mathrm{tr}_{\mathcal{A}} \chi_{b_i} \chi_{b_j} = \mathrm{tr}_{\mathcal{A}} \chi_{b_i \cap b_j} = n+1 = \frac{1}{4(n+1)} \mathrm{tr}_{\mathcal{A}} \chi_{b_i} \cdot \mathrm{tr}_{\mathcal{A}} \chi_{b_j}.$$

In other words, the subalgebras

$$\mathcal{B}_i =< \chi_{b_i}, \chi_{b'_i} >_{\mathbb{F}}, \ \mathcal{B}_j =< \chi_{b_j}, \chi_{b'_j} >_{\mathbb{F}}$$

intersect orthogonally in the unit subalgebra $< 1_{\mathcal{A}} >_{\mathbb{F}}$; together, they constitute a homogeneous OD of type $2M_1$ of $\mathcal{A} = 4(n+1)M_1(\mathbb{F})$. Conversely, every homogeneous OD of type $2M_1$ of $\mathcal{A}$ corresponds to some extended Hadamard $3-(4(n+1), 2(n+1), n)$ block design, or, in other words, a Hadamard matrix of order $4(n+1)$.

The three examples considered above are constructions of homogeneous ODs. We exhibit next an example of a different kind.

Example 7.1.7. Let G be a finite split group (for the definition, see [Suz 2] and [BuG]) and let $G = \bigcup_{i=1}^m H_i$ be a splitting by subgroups H_i. This means that the pairwise intersections of the H_i, $i = 1, \ldots, m$, are trivial and their union is G. Then the system of group algebras $\mathbb{F}H_i$ forms an OD of the group algebra $\mathcal{A} = \mathbb{F}G$. To verify this assertion, it is enough to show that $\mathbb{F}H_i$, $\mathbb{F}H_j$ intersect orthogonally in the unit subalgebra $< 1 >_{\mathbb{F}}$. Indeed, as before let B_k^0 denote the orthogonal complement to $< 1 >_{\mathbb{F}}$ in $\mathbb{F}H_k$. Then

$$B_k^0 = \langle h | h \in H_k \backslash \{1\} \rangle_{\mathbb{F}},$$

and for $x \in H_i\backslash\{1\}$, $y \in H_j\backslash\{1\}$, we have $\mathrm{tr}_{\mathcal{A}} xy = 0$ since $xy \neq 1$. Hence, $(B_i^0 | B_j^0) = 0$ as stated.

For example, if $G = \mathbb{S}_3 =< (123) > \cup < (12) > \cup < (23) > \cup < (31) >$, then we obtain in the way indicated above an OD of $\mathcal{A} = 2M_1(\mathbb{F}) \oplus M_2(\mathbb{F})$ into a sum of a subalgebra of type $3M_1$ and three subalgebras of type $2M_1$.

7.2 The divisibility conjecture

We start this section with the following definition.

Definition 7.2.1. Let $\mathcal{B}$ be a tame subalgebra of $\mathcal{A}$. We shall say that $\mathcal{B}$ *divides* $\mathcal{A}$ if $\mathcal{A}$ is a free right $\mathcal{B}$-module.

To be precise, we would have to talk about *right divisibility*. However, as is clear from the divisibility criterion explained below, the notions of right divisibility and left divisibility are the same, that is, $\mathcal{A}$ is a free left $\mathcal{B}$-module if and only if $\mathcal{A}$ is a free right $\mathcal{B}$-module.

Proposition 7.2.2 (The Divisibility Criterion). *A tame subalgebra $\mathcal{B}$ divides $\mathcal{A}$ if and only if*

$$\mathrm{tr}_{\mathcal{B}} x = \frac{\dim \mathcal{B}}{\dim \mathcal{A}} \mathrm{tr}_{\mathcal{A}} x \tag{7.1}$$

for every $x \in \mathcal{B}$.

Proof. The necessity is obvious, and we establish sufficiency. Let $1_{\mathcal{A}} = e_1 + \cdots + e_s$ be a representation of $1_{\mathcal{A}}$ as a sum of primitive idempotents of $\mathcal{B}$. Every right $\mathcal{B}$-module $(e_i\mathcal{A})_{\mathcal{B}}$ decomposes as a direct sum of simple $\mathcal{B}$-modules: $(e_i\mathcal{A})_{\mathcal{B}} = V_1 \oplus \cdots \oplus V_k$. Then $e_i = v_1 + \cdots + v_k$, where $v_j \in V_j$ for $j = 1, \dots, k$. Multiplying the last equality on the right by e_i and using the uniqueness of such representations for $e_i \in \mathcal{B}$, we get that $v_j e_i = v_j$, $j = 1, \dots, k$. Hence, every $\mathcal{B}$-module $V_j = v_j\mathcal{B} = v_j e_i \mathcal{B}$ is a homomorphic image of the module $e_i\mathcal{B}$, and so, by the simplicity of this module, $V_j \cong e_i\mathcal{B}$. Thus

$$(e_i\mathcal{A})_{\mathcal{B}} \cong \underbrace{e_i\mathcal{B} \oplus \cdots \oplus e_i\mathcal{B}}_{n_i},$$

where $n_i = \mathrm{tr}_{\mathcal{A}} e_i / \mathrm{tr}_{\mathcal{B}} e_i$. By our hypothesis, the number $\mathrm{tr}_{\mathcal{A}} e_i / \mathrm{tr}_{\mathcal{B}} e_i$ is equal to $\dim \mathcal{A} / \dim \mathcal{B}$, and so does not depend on the index i. Denoting this number by n, we have

$$\mathcal{A}_{\mathcal{B}} = \sum_{i=1}^{s} (e_i\mathcal{A})_{\mathcal{B}} \cong \bigoplus_{i=1}^{s} n_i e_i \mathcal{B} = n \bigoplus_{i=1}^{s} e_i \mathcal{B} \cong \underbrace{\mathcal{B} \oplus \cdots \oplus \mathcal{B}}_{n}. \qquad \square$$

Remark 7.2.3. It is clear from this proof that $\mathcal{B}$ divides $\mathcal{A}$ if identity (7.1) holds for all primitive idempotents x of $\mathcal{B}$.

The notion of divisibility is important in the light of the following conjecture.

Conjecture 7.2.4 (The Divisibility Conjecture). *If $\mathcal{D} = \{\mathcal{B}_1, \dots, \mathcal{B}_r\}$ is an OD of an algebra $\mathcal{A}$, then each subalgebra $\mathcal{B}_i$ divides $\mathcal{A}$.*

The proof of this very natural conjecture would lead to a series of non-existence criteria for ODs and many desirable results. Below, the label "C" indicates that the given statement holds under the assumption that Conjecture 7.2.4 is true.

Proposition 7.2.5C. *If n is a square-free natural number, then the algebra $\mathcal{A} = nM_1(\mathbb{F})$ has no ODs.*

Proof. Suppose the contrary: $\mathcal{D} = \{\mathcal{B}_1, \ldots, \mathcal{B}_r\}$ is an OD of $\mathcal{A}$. Note that provided $i \neq j$, the product $n_i n_j$ divides $n = \dim \mathcal{A}$, where $n_k = \dim \mathcal{B}_k$. In fact, let e (f respectively) be a primitive idempotent of $\mathcal{B}_i$ ($\mathcal{B}_j$ respectively). Then, by Proposition 7.1.3(i), we have

$$\mathrm{tr}_{\mathcal{A}} ef = \frac{1}{n} \mathrm{tr}_{\mathcal{A}} e . \mathrm{tr}_{\mathcal{A}} f.$$

By the commutativity of $\mathcal{A}$, ef is an idempotent and so $\mathrm{tr}_{\mathcal{A}} ef$ is a non-negative integer (it is enough to remark that any idempotent in the algebra $M_n(\mathbb{F})$ is conjugate to an element of type

$$\mathrm{diag}(\underbrace{1, \ldots, 1}_{m}, 0, \ldots, 0)$$

for some integer m, $1 \leq m \leq n$). Furthermore, if $1_{\mathcal{A}} = f_1 + \cdots + f_m$ is a decomposition of $1_{\mathcal{A}}$ into a sum of primitive idempotents of $\mathcal{B}_j$, then by Conjecture 7.2.4 and Proposition 7.2.2, we have

$$\mathrm{tr}_{\mathcal{A}} f_k = \mathrm{tr}_{\mathcal{A}} f, \ 1 \leq k \leq m.$$

From the decomposition $e = \sum_{k=1}^{m} ef_k$, it now follows that

$$\mathrm{tr}_{\mathcal{A}} e = \sum_{k=1}^{m} \mathrm{tr}_{\mathcal{A}} ef_k = n_j \mathrm{tr}_{\mathcal{A}} ef.$$

Again by Conjecture 7.2.4, we have $n = n_i \mathrm{tr}_{\mathcal{A}} e$. Hence, $n = n_i n_j \mathrm{tr}_{\mathcal{A}} ef$.

As n is square-free, the product $n_1 \ldots n_r$ divides n. But in that case, $n_1 + \cdots + n_r < n$, since $n_k > 1$ for each k (see Definition 7.1.1). This means that the subalgebras $\mathcal{B}_1, \ldots, \mathcal{B}_r$ cannot generate $\mathcal{A}$ linearly, and this is a contradiction. □

As another consequence of Conjecture 7.2.4, we establish the equivalence of the notions of homogeneous ODs of an algebra of type vM_1 and of *quasi-symmetric soluble Euclidean block designs on v points.* By an Euclidean soluble block design, we mean a block design in which non-parallel blocks have non-empty intersections (see [CaL]).

Let $\mathcal{A} = vM_1(\mathbb{F})$, and let $J = \{e_1, \ldots, e_v\}$ be the set of all primitive idempotents of $\mathcal{A}$. Every partition $J = \bigcup_{i=1}^{d} J_i$ of J into blocks J_i corresponds to a tame subalgebra $\mathcal{B}$ with primitive idempotents $e_{J_i} = \sum_{j \in J_i} e_j$, $i = 1, \ldots, d$. Conversely, every tame subalgebra $\mathcal{B}$ leads to a partition of J into blocks corresponding to primitive idempotents of $\mathcal{B}$.

Proposition 7.2.6C. *Let $\mathcal{D} = \{\mathcal{B}_1, \ldots, \mathcal{B}_r\}$ be a homogeneous OD of type dM_1 of the algebra $\mathcal{A} = vM_1(\mathbb{F})$. Let B denote the set of all blocks corresponding to* *subalgebras from $\mathcal{D}$. Then the pair (J, B) is a quasi-symmetric soluble Euclidean $2-(v, k, \lambda)$ block design, where the parameters v, k and λ satisfy the relations*

$$v = d^2(sd+1-s),\ k = d(sd+1-s),\ \lambda = sd+1. \tag{7.2}$$

Here $s = (r-d-1)/d^2 \in \mathbb{N}$ and $d \geq 2$.

Conversely, every quasi-symmetric soluble Euclidean 2-(v, k, λ) block design corresponds to some homogeneous OD of $\mathcal{A}$.

Proof. By Conjecture 7.2.4, all blocks in B contain the same number $k = v/d$ of elements. We claim that every two idempotents $e_i, e_j \in J$ are contained in exactly λ blocks. Indeed, suppose that e_i belongs to the block corresponding to the primitive idempotent e_{ti} of $\mathcal{B}_t$, $t = 1, \dots, r$. Then by Proposition 7.1.3(ii) we have:

$$e_i = \sum_{t=1}^{r} \frac{1}{k} e_{ti} - \frac{r-1}{v} 1_{\mathcal{A}}.$$

On scalar multiplication of both parts of this equality by e_j, we obtain:

$$0 = (e_i|e_j) = \sum_{t=1}^{r} \frac{1}{k}(e_{ti}|e_j) - \frac{r-1}{v}.$$

Thus, the number

$$\lambda = \sum_{t=1}^{r} (e_{ti}|e_j) = \frac{k(r-1)}{v} \tag{7.3}$$

does not depend on i nor on j. Hence, the pair (J, B) is a 2-design. In this case, $r(k-1) = (v-1)\lambda$; moreover, for the dimensions of the components of $\mathcal{D}$, the additional relation $v-1 = r(d-1)$ holds. Therefore,

$$\lambda = \frac{r(k-1)}{v-1} = \frac{k-1}{d-1}.$$

We show that $\lambda - 1$ is divisible by d. During the proof of Proposition 7.2.5C, we established that $\dim \mathcal{A}$ (which is v in our case) is divisible by $\dim \mathcal{B}_i \cdot \dim \mathcal{B}_j$ (d^2 in our case). Hence, $k = v/d = dm$ for some natural m. Thus,

$$\lambda = \frac{k-1}{d-1} = \frac{dm-1}{d-1} = m + \frac{m-1}{d-1}.$$

Therefore $s = (m-1)/(d-1)$ is an integer.

Thus, we have shown that $\lambda = sd+1$, $k = dm = d(sd+1-s)$ and $v = dk = d^2(sd+1-s)$. Moreover, it follows from relation (7.3) that $\lambda = (r-1)/d$, that is, $s = (\lambda-1)/d = (r-d-1)/d^2$.

To prove the reverse conclusion of the Proposition, we consider a primitive idempotent f of $\mathcal{B}_i$ and the decomposition $1_{\mathcal{A}} = g_1 + \cdots + g_d$ of the unit element into a sum of primitive idempotents of $\mathcal{B}_j$. By the quasi-symmetry of the block design we have

$$\mathrm{tr}_{\mathcal{A}} f g_1 = \ldots = \mathrm{tr}_{\mathcal{A}} f g_d = \frac{\mathrm{tr}_{\mathcal{A}} f}{d} = \frac{\mathrm{tr}_{\mathcal{A}} f . \mathrm{tr}_{\mathcal{A}} g_t}{v}, \ 1 \leq t \leq d,$$

and so, by Proposition 7.1.3, the subalgebras $\mathcal{B}_i$, $\mathcal{B}_j$ intersect orthogonally in the unit subalgebra. □

In connection with the last statement, we formulate here another conjecture, the proof of which would allow us to state the equivalence of the notions of ODs of commutative algebras and of (arbitrary) quasi-symmetric block designs.

Definition 7.2.7. An orthogonal decomposition $\mathcal{D}$ of an algebra $\mathcal{A}$ is said to be *reduced* if no subalgebra from $\mathcal{D}$ admits an OD.

Conjecture 7.2.8. *Every reduced OD of a commutative algebra is homogeneous.*

Clearly, if $\mathcal{A}$ is a non-commutative algebra with an OD

$$\mathcal{D} = \{\mathcal{B}_1, \ldots, \mathcal{B}_r\},$$

then either all the components $\mathcal{B}_i$ are commutative, or at least one of them is non-commutative. In the first case, we speak of a *commutative OD*, in the second case of a *non-commutative OD*.

Example 7.2.9. The algebra $\mathcal{A} = nM_1(\mathbb{F}) \oplus M_2(\mathbb{F})$ does not admit non-commutative ODs.

For, let $\mathcal{B}$ be a non-commutative component of an OD $\mathcal{D}$ of the given algebra $\mathcal{A}$. Let $\pi : \mathcal{A} \to J$ be the projection of $\mathcal{A}$ onto the non-commutative ideal $J = M_2(\mathbb{F})$. Consider another component $\mathcal{B}'$ from $\mathcal{D}$, and its primitive idempotent e with projection $f = \pi(e)$. From the non-commutativity of $\mathcal{B}$ and the simplicity of J, it follows that $J \subseteq \mathcal{B}$. Then by Proposition 7.1.3 we have for any element $x \in J$:

$$\mathrm{tr}_{\mathcal{A}} e x = \frac{\mathrm{tr}_{\mathcal{A}} e . \mathrm{tr}_{\mathcal{A}} x}{\dim \mathcal{A}} = \mathrm{tr}_{A} \left(\left(\frac{\mathrm{tr}_{\mathcal{A}} e}{\dim \mathcal{A}} 1_A \right) x \right).$$

Hence,

$$0 = \mathrm{tr}_{\mathcal{A}} \left(e - \frac{\mathrm{tr}_{\mathcal{A}} e}{\dim \mathcal{A}} 1_A \right) x = \mathrm{tr}_{\mathcal{A}} \left(f - \frac{\mathrm{tr}_{\mathcal{A}} e}{\dim \mathcal{A}} \pi(1_A) \right) x.$$

By the non-degeneracy of the trace form restricted to J, we obtain

$$f = \frac{\mathrm{tr}_{\mathcal{A}} e}{\dim \mathcal{A}} \pi(1_{\mathcal{A}}).$$

But $f^2 = f$, so that

$$\mathrm{tr}_{\mathcal{A}} e = \dim \mathcal{A} = \mathrm{tr}_{\mathcal{A}} 1_{\mathcal{A}}.$$

Consequently, the idempotents e and $1_{\mathcal{A}}$ have the same projection into J and the same trace. This means that they coincide. Because e is chosen as an arbitrary primitive idempotent of $\mathcal{B}'$, the last assertion implies that $\mathcal{B}' =< 1_{\mathcal{A}} >_{\mathbb{F}}$, a contradiction (every component of $\mathcal{D}$ has dimension greater than 1!)

As we mentioned above, the matrix algebra $M_{p^m}(\mathbb{F})$ has a commutative homogeneous OD of type $p^m M_1$. Examples of non-commutative ODs of many algebras will be described in the next section.

7.3 A construction of ODs

We start with the following concept, which is of independent interest.

Definition 7.3.1. Let G be a finite group of order n. Two ordered rows $(s_1, \dots, s_n)$, $(t_1, \dots, t_n)$ consisting of elements of G are said to be *left-orthogonal* if $G = \{s_1 t_1^{-1}, \dots, s_n t_n^{-1}\}$. A system consisting of n pairwise left-orthogonal rows is said to be *complete*. A group G admitting a complete system of left-orthogonal rows is called an *S-group*.

The notion of *right-orthogonal* rows is similar: $(s_1, \dots, s_n)$ and $(t_1, \dots, t_n)$ are right-orthogonal if $G = \{s_1^{-1} t_1, \dots, s_n^{-1} t_n\}$.

Clearly, left-orthogonality and right-orthogonality are symmetrical relations. Furthermore, in general left-orthogonal rows may not be right-orthogonal. For example, consider the group

$$G = \mathbb{S}_3 = \left\langle a, b | a^3 = b^2 = 1,\ bab = a^{-1} \right\rangle.$$

Then the rows $(1, a, a^2, b, b, ba^2)$ and $(1, 1, 1, 1, a^2, 1)$ are left-orthogonal but not right-orthogonal. As regards complete systems of left-orthogonal and right-orthogonal rows, there is a bijective correspondence between them. In fact, let

$$X_1 = (x_{11}, \dots, x_{1n}), \ \dots \ , X_n = (x_{n1}, \dots, x_{nn})$$

be a complete system of left-orthogonal rows. Consider the system of rows "transpose" to it:

$$Y_1 = (x_{11}, x_{21}, \dots, x_{n1}), \ \dots \ , Y_n = (x_{1n}, x_{2n}, \dots, x_{nn}),$$

and the matrix $X = (x_{ij})$. Let $I =< \sum_{g \in G} g >_{\mathbb{C}}$ be the obvious 1-dimensional ideal of the group algebra $\mathbb{C}G$. Set $\hat{X} = (\hat{x}_{ij})$, where $\hat{x}_{ij} = \frac{1}{n} x_{ji}^{-1}$. Then left-orthogonality of the rows X_i is equivalent to the relation

$$X\hat{X} \equiv E_n (\mathrm{mod}\, I).$$

In turn, the last relation means that the map

$$f : A \mapsto AX$$

acting on the factor-algebra $\mathcal{G} = \mathbb{C}G/I$ has a right inverse, namely

$$g : A \mapsto A\hat{X}.$$

But $\mathcal{G}$ is finite-dimensional, and so g is also a left inverse for f. This means that

$$\hat{X}X \equiv E_n (\mathrm{mod}\, I),$$

in other words, $\{Y_1, \dots, Y_n\}$ is a complete system of right-orthogonal rows. Thus, the map

$$\{X_1, \dots, X_n\} \mapsto \{Y_1, \dots, Y_n\}$$

gives the required bijective correspondence.

Based on the right-orthogonality of the system $\{Y_1, \dots, Y_n\}$ described above, we might also prove that for a given group G of order n there are at most n left- (right-) orthogonal rows, in explanation of the term "complete system". Namely, let $X_i = (x_{i1}, \dots, x_{in})$, $1 \le i \le n+1$, be a system of $n+1$ pairwise left-orthogonal rows. Then by our hypothesis

$$G = \{x_{11}^{-1}x_{12}, x_{21}^{-1}x_{22}, \dots, x_{i-1,1}^{-1}x_{i-1,2}, x_{i+1,1}^{-1}x_{i+1,2}, \dots, x_{n+1,1}^{-1}x_{n+1,2}\}$$

for each $i = 1, 2, \dots, n+1$. This means that all the elements $x_{i1}^{-1}x_{i2}$ are equal to one other, a contradiction.

All known S-groups are elementary abelian p-groups, which admit the following construction of complete row systems (of course, for abelian groups the notions of left-orthogonality and right-orthogonality coincide). Let $G = \{x_1, \dots, x_q\}$, $q = p^m$, be an elementary abelian p-group realized as the additive group of the field $\mathbb{F}_q$. Let θ be a generator of the multiplicative group $\mathbb{F}_q^*$. Then the map $\Theta : x \mapsto \theta x$ generates the group of fixed-point-free automorphisms of G, of order $q-1$. It is easy to see that the row system

$$\{(0, \dots, 0), (\Theta^k(x_1), \dots, \Theta^k(x_q)) | k = 0, 1, \dots, q-2\}$$

is complete.

Note in passing that if a finite group G admits a group of fixed-point-free automorphisms of order $|G| - 1$, then G is an elementary abelian group. For this reason, the following conjecture is very plausible.

Conjecture 7.3.2. *Every S-group is an elementary abelian group.*

Theorem 7.3.3. *Suppose that G is an S-group of order n with an m-dimensional irreducible projective representation over $\mathbb{F}$. Then $nM_m(\mathbb{F})$ admits an OD consisting of n subalgebras of type M_m and one subalgebra of type nM_1.*

Proof. Let $\{(g_{i1}, \ldots, g_{in}) | i = 1, \ldots, n\}$ be a complete row system for G. Furthermore, let $\varphi : G \to GL_m(\mathbb{F})$ be a map defining the given projective representation of G. For every row we define a subalgebra

$$\mathcal{B}_i = \left\{ \operatorname{diag}\left(\varphi(g_{i1})^{-1} x \varphi(g_{i1}), \ldots, \varphi(g_{in})^{-1} x \varphi(g_{in}) \right) | x \in M_m(\mathbb{F}) \right\}$$

of the algebra $\mathcal{A} = nM_m(\mathbb{F})$. Set

$$\mathcal{E} = \{\operatorname{diag}(\lambda_1 E_m, \ldots, \lambda_n E_m) | \lambda_i \in \mathbb{F}\}.$$

We claim that $\mathcal{D} = \{\mathcal{B}_1, \ldots, \mathcal{B}_n, \mathcal{E}\}$ is an OD of $\mathcal{A}$. To see this, we consider elements

$$X_k = \operatorname{diag}\left(\varphi(g_{k1})^{-1} x_k \varphi(g_{k1}), \ldots, \varphi(g_{kn})^{-1} x_k \varphi(g_{kn}) \right) \in \mathcal{B}_k$$

where $k = i$ or $k = j$, $i \neq j$. Then

$$\operatorname{tr}_{\mathcal{A}} X_i X_j = m \sum_{s=1}^{n} \operatorname{Tr} \left\{ \varphi(g_{is})^{-1} x_i \varphi(g_{is}) . \varphi(g_{js})^{-1} x_j \varphi(g_{js}) \right\} =$$

$$= m \sum_{s=1}^{n} \operatorname{Tr} \left\{ \varphi(g_{js} g_{is}^{-1}) x_i \left(\varphi(g_{js} g_{is}^{-1}) \right)^{-1} x_j \right\} =$$

$$= m \sum_{g \in G} \operatorname{Tr}\{\varphi(g) x_i \varphi(g)^{-1} x_j\} = m \operatorname{Tr} \left\{ \left\{ \sum_{g \in G} \varphi(g) x_i \varphi(g)^{-1} \right\} x_j \right\}.$$

Here, $\operatorname{Tr} x$ denotes the trace of a matrix $x \in M_m(\mathbb{F})$, as before. It should be mentioned that in transformations

$$\varphi(x)\varphi(y) = \varphi(xy) C_{x,y}, \quad \varphi(x^{-1}) = \varphi(x)^{-1} C_x$$

of the kind used above, the scalars $C_{x,y}$, C_x defining the given projective representation, are deleted. Furthermore, because φ is irreducible, for every $x \in M_m(\mathbb{F})$ we have

$$\sum_{g \in G} \varphi(g) x \varphi(g)^{-1} = \frac{n \operatorname{Tr} x}{m} E_m.$$

Hence,

$$\operatorname{tr}_{\mathcal{A}} X_i X_j = n \operatorname{Tr} x_i . \operatorname{Tr} x_j.$$

It remains to note that $\mathrm{tr}_{\mathcal{A}} X_k = nm\mathrm{Tr} x_k$, $k = i, j$. Therefore,

$$\mathrm{tr}_{\mathcal{A}} X_i X_j = \frac{\mathrm{tr}_{\mathcal{A}} X_i . \mathrm{tr}_{\mathcal{A}} X_j}{nm^2} = \frac{\mathrm{tr}_{\mathcal{A}} X_i . \mathrm{tr}_{\mathcal{A}} X_j}{\dim \mathcal{A}}.$$

In the same way, one verifies that

$$\mathrm{tr}_{\mathcal{A}} E X_k = \frac{\mathrm{tr}_{\mathcal{A}} E . \mathrm{tr}_{\mathcal{A}} X_k}{\dim \mathcal{A}}$$

for every element $E \in \mathcal{E}$. By Proposition 7.1.3, all the components of $\mathcal{D}$ intersect orthogonally in the unit subalgebra. □

As an elementary abelian p-group of rank $2m$ has an irreducible projective representation of degree p^m, we obtain the following:

Corollary 7.3.4. *The algebra $q^2 M_q(\mathbb{F})$, q a prime power, admits an OD consisting of q^2 subalgebras of type M_q and one subalgebra of type $q^2 M_1$.* □

Commentary

All the material in this small chapter is written following the papers of Ivanov [Iva 2, 4]. The Ivanov approach establishes relationships between different combinatorial objects. Moreover, Conjectures 7.2.4 and 7.3.2 are of independent interest. Unfortunately, the authors are not able to correct the errors contained in the original text [Iva 2, 4], so some statements in the chapter (namely, Propositions 7.2.5C and 7.2.6C), are of a convential character.

If we investigate the existence problem for ODs for algebras of small dimension assuming the validity of Conjecture 7.2.4, the first difficulties appear in dimension 12. It is a question whether the algebra $3M_2(\mathbb{F})$ admits a homogeneous OD of type $2M_1$ or not. It is also unknown whether there is an algebra $\mathcal{A}$ with an OD such that the type of every component does not contain the term M_1.

Part II

Integral lattices and their automorphism groups

By an *integral lattice* we understand a pair (Λ, B), where Λ is a finitely generated torsion free $\mathbb{Z}$-module of rank n and B is a non-degenerate symmetric bilinear form (the *inner product*) taking only integer values on Λ. When it is clear from the context which form B is involved, the lattice (Λ, B) is denoted simply by Λ.

A lattice Λ is said to be *even* if $B(x, x) \in 2\mathbb{Z}$ for all $x \in \Lambda$. Λ is said to be *unimodular* if it coincides with its *dual* lattice

$$\Lambda^* = \{x \in \Lambda \otimes_{\mathbb{Z}} \mathbb{C} | B(x, \Lambda) \subseteq \mathbb{Z}\},$$

in other words, if the determinant of the Gram matrix of the form B is ± 1.

The *automorphism group* $\mathrm{Aut}(\Lambda) = \mathrm{Aut}_{\mathbb{Z}}(\Lambda)$ (or equivalently, the *isometry group*) of a lattice Λ consists of all linear transformations φ of the space $\Lambda \otimes_{\mathbb{Z}} \mathbb{C}$ that preserves Λ and B:

$$\varphi(\Lambda) = \Lambda, \ B(\varphi(x), \varphi(y)) = B(x, y).$$

Λ is said to be *Euclidean* (or equivalently, *positive definite*) if $B(x, x) > 0$ for all $x \in \Lambda \backslash \{0\}$. In this book we shall consider Euclidean lattices only. The positive definiteness of Λ implies that $\mathrm{Aut}(\Lambda)$ is finite, as it is a compact subgroup of $GL_n(\mathbb{Z})$. Generally speaking, the investigation of automorphism groups of Euclidean lattices is closely related to the study of integral representations of abstract finite groups.

A vector $x \in \Lambda \backslash \{0\}$ is said to be *minimal* if

$$B(x, x) = \min \{B(y, y) | y \in \Lambda \in \{0\}\},$$

and a *root* if $B(x, x) = 1$ or 2. There may be no such vectors in Λ; however, if they do exist, then they constitute a *root system* R in the classical sence (see [Bour 2]). If $\Lambda = < R >_{\mathbb{Z}}$, then Λ is called a *root lattice*.

A lattice Λ is said to be *decomposable* if

$$\Lambda = \Lambda' \oplus \Lambda''$$

for some nonzero sublattices Λ', Λ'' with $B(\Lambda', \Lambda'') = 0$. Otherwise, Λ is said to be *indecomposable*. Decomposition of a Euclidean lattice as a sum of indecomposable sublattices is unique (see [Cas]).

The class of Euclidean lattices contains the important subclass consisting of the unimodular lattices. It is known that the rank n of an even unimodular lattice is divisible by 8 (see [Ser 2]). Such lattices have been studied in depth for small n. For example, when $n = 8$ there exists just one even unimodular lattice E_8 (the Korkine-Zolotarev-Witt lattice), and it realizes the densest sphere-packing in Euclidean space $\mathbb{R}^8$. When $n = 16$, there are two such lattices. Furthermore, there exist precisely 24 even unimodular 24-dimensional Euclidean lattices (see [Nie], [Ven 1]). Among them there is just one lattice, denoted by Λ_{24}, that has no roots.

This lattice was first discovered by Leech (see [Lee]) in 1965, in connection with sphere-packings in $\mathbb{R}^{24}$. The group $\mathrm{Aut}(\Lambda_{24})$ was studied by Conway in 1969 (see [Con 1, 2]), and it led to the discovery of three sporadic simple groups. These elegant articles of Conway were a powerful stimulus to investigations in the area of Euclidean lattices.

Unfortunately, the number of even unimodular lattices in $\mathbb{R}^{32}$, which is given by the Minkowski-Siegel formula, is unimaginably great ($\approx 8.10^7$), and their complete classification is still out of the question. However, some interesting results have been obtained in the papers of Venkov and Koch (see [KoV 1, 2], [KoN], [Ven 3, 6, 8]). Any construction covering a reasonably wide class of even unimodular lattices is of great interest in this connection. At present, one can distinguish the following principal constructions in general lattice theory:

1. The classification of lattices in small dimensions. This concerns

– the lattices in $\mathbb{R}^n$, $n \leq 6$ (Brown, Bülow, Neubüser, Wondratschek and Zassenhaus; Ryshkov and Lomakina; Plesken and Hanrath);

– even unimodular lattices in $\mathbb{R}^{24}$ (Niemeier, Venkov);

– odd unimodular lattices in $\mathbb{R}^n$, $n \leq 16$ (Kneser), $n \leq 23$ (Conway and Sloane), $n = 24, 25$ (Borcherds), odd extremal unimodular lattices in $\mathbb{R}^{32}$ (Conway and Sloane);

– the classification of lattices with small determinants $(2, 3, \ldots)$ (Conway and Sloane);

– the classification of maximal finite irreducible subgroups of $GL_n(\mathbb{Z})$, $n \leq 24$ or n prime (Plesken, Pohst, Nebe).

2. "Rebuilding" constructions: "neighbour" lattices (Kneser), laminated lattices (Conway and Sloane).

3. "Code" constructions: constructions using the connection between lattices and linear codes (Barnes and Wall; Broue and Enguehard; Sloane, Venkov, Koch, Tasaka et al).

4. Algebraic constructions: constructions using orders in algebraic number fields (Conway and Sloane, Thompson, Feit, Tits, McKay, Bayer-Fluckiger, Craig, Quebbemann, Litsin, Rosenbloom, Tsfasman et al). Constructions invoking the arithmetics of algebraic curves (Elkies, Gross, Shioda).

5. Constructions attracting a connection with indefinite (Lorentzian) lattices (Conway and Sloane).

6. Group-theoretic constructions: Constructions using rational representations of various finite groups (Gow, Gross, Tiep).

These constructions give us examples of indecomposable even unimodular (root-free) Euclidean lattices in dimensions 24, 32, 40, 48, 64, 2^{2d+1} $(d \geq 2)$, $p+1$ ($p \equiv -1 (\mathrm{mod}\, 8)$ a prime, $p > 7$), $p^{2n} - 1$ (p an odd prime), $2p^2(p^2-1)$ (p a prime) and $2(p^n - 1)$($p \equiv 1(\mathrm{mod}\, 4)$ a prime, $p^n > 5$).

One of the main problems in Part II is to study integral lattices naturally related to orthogonal decompositions of complex simple Lie algebras. Suppose that $\mathcal{L}$ is a Lie algebra admitting an IOD $\mathcal{D}$:

$$\mathcal{L} = \bigoplus_{i=1}^{h+1} \mathcal{H}_i,$$

with automorphism group $G = \mathrm{Aut}(\mathcal{D})$. It is easy to see that G acts rationally on $\mathcal{L}$. Indeed, set $G_1 = St_G(\mathcal{H}_1)$. Then G_1 acts on $\mathcal{H}_1^*$ as a subgroup of the extended Weyl group (that is, the automorphism group of the root system). In particular, the action of G_1 on $\mathcal{H}_1$ is rational. Since the representation of G on $\mathcal{L}$ is induced by the representation of G_1 on $\mathcal{H}_1$ (this is a consequence of the transitivity of the decomposition $\mathcal{D}$), G acts rationally on $\mathcal{L}$.

In view of this remark, one can consider G'-invariant lattices in $\mathcal{L}$, where G' is any subgroup of G that acts irreducibly on $\mathcal{L}$. More precisely, by an *invariant lattice of type* $\mathcal{L}$ we mean any full $\mathbb{Z}$-module $\Lambda \subset \mathcal{L}$ with the following properties:

(i) the Killing form K of $\mathcal{L}$ is integral and positive definite on Λ;

(ii) Λ is fixed by some subgroup G' of G that is irreducible on $\mathcal{L}$.

It is natural to consider invariant lattices up to similarity: a lattice Γ is *similar* to a lattice Γ' (in symbols $\Gamma \sim \Gamma'$), if Γ is isometric to the lattice $\lambda\Gamma'$ for some $\lambda \in \mathbb{C}^*$. In view of the Finiteness Theorem 4.1.1, the number of similarity classes of invariant lattices is finite. Hence, in principle, they can be classified. Unfortunately, the solution of this problem in practice is beset by great difficulties, which are as yet insurmountable; for instance, in the cases A_{p^m-1} and E_8. On the other hand, there is much more known about the automorphism groups $\mathbb{G} = \mathrm{Aut}(\Lambda)$. For example, by using the classification of finite simple groups, one can show that any such group $\mathbb{G}$, viewed as a complex linear group, satisfies the following alternative $(\mathcal{A})$:

Either $\mathbb{G}$ *is imprimitive on* $\mathcal{L}$, *and then* $\mathbb{G}$ *has well-described composition structure;* or $\mathbb{G}$ *is primitive, and then the factor group* $\mathbb{G}/Z(\mathbb{G})$ *is an almost simple group, that is,*

$$L \lhd \mathbb{G}/Z(\mathbb{G}) \subseteq \mathrm{Aut}(L)$$

for some finite non-abelian simple group L. Moreover, the list of such "exceptional" groups L is well-defined in each case.

In the final Chapter 14, we consider other lattice constructions, inspired by the works of Elkies, Gow, Gross and Shioda.

Chapter 8

Invariant lattices of type G_2 and the finite simple group $G_2(3)$

It is known that a complex simple Lie algebra $\mathcal{L}$ of type G_2 has a unique IOD

$$\mathcal{D} : \mathcal{L} = \oplus_{i=1}^{7} \mathcal{H}_i,$$

which is also an MOD. In this chapter it will be shown that the group $G = \mathrm{Aut}(\mathcal{D})$ is a non-split extension of $\mathbb{Z}_2^3$ by $SL_3(2)$. In particular, this group has a unique irreducible complex representation of degree $14 = \dim \mathcal{L}$. Taking in $\mathcal{H}_1$ the lattice Λ^1 dual to the root lattice in $\mathcal{H}_1^*$ and "scattering" it by means of G, we obtain a G-invariant lattice $\Lambda_0 = \oplus_{i=1}^{7} \Lambda^i$ endowed with the inner product induced by the Killing form of $\mathcal{L}$. According to the Deuring-Noether Theorem [CuR], every faithful G-invariant lattice can be isometrically embedded in Λ_0. This means that we can restrict ourselves to a consideration of invariant sublattices of Λ_0. The main result of the chapter is the following:

Theorem 8.0.1. *Up to similarity and duality, there exist six G-invariant sublattices Λ_i, $0 \le i \le 5$, in Λ_0. Just one of them, namely Λ_5, does not contain $2\Lambda_0$ but contains $4\Lambda_0$. Its automorphism group* $\mathrm{Aut}(\Lambda_5)$ *is isomorphic to $\mathbb{Z}_2 \times G_2(3)$, where $G_2(3)$ is the Dickson-Chevalley simple group of order* 4245696. *The automorphism groups of the remaining sublattices Λ_i, $1 \le i \le 4$, are embedded in* $\mathrm{Aut}(\Lambda_0) = D_{12} \wr \mathbb{S}_7$ *as subgroups of index* 16.

8.1 Preliminaries

First of all, let us recall the notion of the *Cayley octonion algebra* $\mathcal{C}$ over a field $\mathbb{F}$. This is an 8-dimensional alternative algebra

$$\mathcal{C} = < 1, e_i | 1 \le i \le 7 >_{\mathbb{F}}$$

with unit element 1 and the following multiplication:

$$(e_i)^2 = -1,\ e_ie_j = -e_je_i \ \text{if}\ i \neq j,$$

$$e_{i+1}e_{i+2} = e_{i+4},$$

where the subscripts are taken modulo 7. The fact that $\mathcal{C}$ is alternative yields further relations. For example,

$$e_{i+1}e_{i+4} = e_{i+1}(e_{i+1}.e_{i+2}) = (e_{i+1})^2 e_{i+2} = -e_{i+2}.$$

Here we shall deal with the Cayley algebra $\mathcal{C}$ defined over the complex field $\mathbb{C}$. In this case, it is well known that $\mathcal{G} = \mathrm{Aut}(\mathcal{C})$ is a complex simple Lie group of type G_2. Correspondingly, the derivation algebra $\mathcal{L} = \mathrm{Der}(\mathcal{C})$ is a Lie algebra of type G_2. Clearly, $\mathcal{G}$ preserves the space V consisting of *octaves with zero trace*:

$$V = < e_1, \ldots, e_7 >_{\mathbb{C}},$$

and V is an irreducible $\mathcal{G}$-module. V can be endowed with a $\mathcal{G}$-invariant multiplication

$$* : V \otimes V \to V,$$

and an inner product

$$p : V \otimes V \to \mathbb{C},\ p(e_i, e_j) = \delta_{i,j}$$

such that

$$uv = u * v - p(u, v).1,$$

where uv is the product of the elements u, v in $\mathcal{C}$. Clearly, $*$ is skew-symmetric.

If $D \in \mathcal{L}$, then $D(1) = 0$ and $D(V) \subseteq V$. Conversely, a linear map $D \in \mathrm{End}_{\mathbb{C}}(V)$ belongs to $\mathcal{L}$ if and only if

$$D \in \mathrm{Der}(V, *),\ p(Dx, y) + p(x, Dy) = 0, \forall x, y \in V.$$

Identifying $\mathrm{End}_{\mathbb{C}}(V)$ with $V \otimes V$ in such a way that

$$(u \otimes v)w = p(u, w)v,$$

we obtain that $\mathcal{L} \subseteq V \wedge V$.

Clearly, $* : V \wedge V \to V$ is a $\mathcal{G}$-epimorphism. Since $\mathcal{G}$ acts irreducibly on V and on $\mathcal{L}$, and $\dim V + \dim \mathcal{L} = \dim(V \wedge V)$, we come to the conclusion that $\mathcal{L}$ coincides with the kernel of $*$ on $V \wedge V$.

Next we describe the so-called *Cayley loop*. Recall that a *loop* is a quasigroup with unit element, an operation $x \mapsto x^{-1}$ of taking inverses, in which the following identities are satisfied:

$$(x^{-1})^{-1} = x,\ x.x^{-1} = 1,\ x^{-1}(xy) = (yx)x^{-1} = y.$$

We shall consider the loops L possessing a loop epimorphism $L \to \mathbb{Z}_2^3$ with kernel $Z = \{1, e\} \cong \mathbb{Z}_2$. Identify $\mathbb{Z}_2^3$ with $\bar{L} = L/Z$ and denote by $\bar{x}$ the image of $x \in L$ in $\bar{L}$.

Clearly, e is contained in the *centre* and the *associative centre* of L, that is,

$$ex = xe;\ e(xy) = (ex)y,\ (xe)y = x(ey),\ (xy)e = x(ye), \forall x, y \in L.$$

Thus we may assert that L is an "extraspecial 2-loop of type 2^{1+3}".

Note that the subset $\mathcal{O} \subset \mathcal{C}$ consisting of the elements ± 1, $\pm e_i$, $1 \le i \le 7$, is such a loop, where the role of $e \in Z$ is played by -1. Moreover, $\mathcal{O}$ satisfies the following conditions:

(i) $x^2 = e$ for all $x \notin Z$;

(ii) $xy = eyx$ for all $x, y \notin Z$ with $\bar{x} \neq \bar{y}$;

(iii) $(xy)z = ex(yz)$ if x, y, z generate $L = \mathcal{O}$ modulo Z; and $(xy)z = x(yz)$ otherwise.

Conversely, properties (i) – (iii) determine such a loop L uniquely. To show this, we choose three elements e_1, e_2, e_3 such that $\bar{e}_1$, $\bar{e}_2$ and $\bar{e}_3$ generate $\bar{L}$. Then any element $x \in L$ has a unique expression of the form

$$x = \{a, b, c, d\} = \left((e_1)^a (e_2)^b\right)(e_3)^c e^d,\ a, b, c, d \in \{0, 1\}.$$

Moreover, the loop product can be rewritten as

$$\{a, b, c, d\}\{a', b', c', d'\} = \{a + a', b + b', c + c', d + d' + \kappa\},$$

where

$$\kappa = aa' + bb' + cc' + a'b + a'c + b'c + a'b'c + a'bc' + ab'c'.$$

Compute for example $x = \{0, 1, 0, 1\}.\{1, 1, 1, 0\} = (e_2 e)((e_1 e_2)e_3)$. Firstly, $x = (e_2.((e_1 e_2).e_3))e$. Secondly, in view of (iii) we have $e_2.((e_1 e_2).e_3) = e(e_2.(e_1 e_2)).e_3$. By (ii) and (i) we have

$$e_2.(e_1 e_2) = e_2.(ee_2.e_1) = e_2.(e_2 e_1)e = (e_2 e_2).e_1.e = ee_1 e = e_1.$$

Hence $x = e_1 e_3 = \{1, 0, 1, 0\}$ as required.

In passing, our arguments show that for any three elements $x, y, z \in L$ that are linearly independent in $\bar{L}$ there exists just one automorphism $\varphi \in G = \mathrm{Aut}(L)$ such that

$$\varphi : e_1 \mapsto x, e_2 \mapsto y, e_3 \mapsto z.$$

Denote by $\bar{\varphi}$ the map induced by φ acting on $\bar{L}$. As we have shown, the map

$$G \to \mathrm{Aut}(\bar{L}),\ \varphi \mapsto \bar{\varphi}$$

is surjective. Furthermore, $\bar{\varphi} = 1$ if and only if φ is of the form $\varphi(x) = x\tau(x)$, where $\tau : \bar{L} \to Z$ is a homomorphism. The subgroup E consisting of automorphisms of this kind is dual to $\bar{L}$ as $\mathrm{Aut}(\bar{L})$-module, and it is easy to see that G is non-split over E. Thus we obtain

$$G = \mathrm{Aut}(L) = \mathbb{Z}_2^3 \circ SL_3(2). \tag{8.1}$$

The uniqueness of the non-split extension $\mathbb{Z}_2^3 \circ SL_3(2)$ is established in [San].

Now we define a homomorphism $\tilde{\varphi}$ of the monoid algebra $\mathbb{C}L$ into $\mathcal{C}$ that extends the isomorphism $\varphi : L \cong \mathcal{O}$. Clearly, $\tilde{\varphi}$ is an epimorphism and its kernel is the principal ideal $(1+e)$ generated by $1+e$. Hence, $\mathcal{C}$ can be identified with $\mathbb{C}L/(1+e)$. Of course, φ fixes the element e and the ideal $(1+e)$, so $\tilde{\varphi}$ induces an isomorphism $\mathbb{C}L/(1+e) \cong \mathcal{C}$. From now on we identify L with $\mathcal{O}$.

We shall frequently view elements of $\bar{L}$ as characters of E. Set

$$L^{\#} = \bar{L}\backslash\{0\}.$$

In each coset $\alpha \in L/Z$, $\alpha \neq Z$, we choose a representative e_α, and then we have

$$V = \oplus_\alpha \mathbb{C}e_\alpha. \tag{8.2}$$

Here, $\oplus_\alpha$ and $\sum_\alpha$ run over $\alpha \in L^{\#}$.

Next we construct a multiplicative orthogonal decomposition for $\mathcal{L}$. Note that the components $\mathbb{C}e_\alpha$ in decomposition (8.2) are eigensubspaces corresponding to the nontrivial characters of the group $E = \mathbb{Z}_2^3$. For $\alpha \in \bar{L}$ set

$$\mathcal{H}_\alpha = \{x \in \mathcal{L} | \forall f \in E, f(x) = \alpha(f)x\}.$$

Since E acts fixed-point-freely on $V \wedge V$ and, all the more, on $\mathcal{L}$, we have $\mathcal{H}_0 = 0$. It is also clear that G fixes the decomposition

$$\mathcal{D} : \mathcal{L} = \oplus_\alpha \mathcal{H}_\alpha.$$

From the transitivity of $\bar{G}$ on $L^{\#}$ it follows that

$$\dim \mathcal{H}_\alpha = \dim \mathcal{L}/|L^{\#}| = 14/7 = 2$$

for all $\alpha \in L^{\#}$. Furthermore, for $x \in \mathcal{H}_\alpha$, $y \in \mathcal{H}_\beta$, and $f \in E$ we have

$$f([x, y]) = [f(x), f(y)] = \alpha(f)\beta(f)[x, y],$$

that is,

$$[\mathcal{H}_\alpha, \mathcal{H}_\beta] \subseteq \mathcal{H}_{\alpha+\beta}$$

($\bar{L}$ is viewed as an additive group here). In particular, $[\mathcal{H}_\alpha, \mathcal{H}_\alpha] = 0$, that is, $\mathcal{H}_\alpha$ is abelian.

Finally, as always we denote the Killing form of $\mathcal{L}$ by K. Then for the same x, y, f with $\alpha \neq \beta$ we have

$$K(x, y) = K(f(x), f(y)) = \alpha(f)\beta(f)K(x, y) = (\alpha + \beta)(f)K(x, y).$$

Choosing f such that $(\alpha+\beta)(f) \neq 1$, we obtain $K(x, y)=0$. Hence, $K(\mathcal{H}_\alpha, \mathcal{H}_\beta)=0$ whenever $\alpha \neq \beta$, and K is non-degenerate when restricted to $\mathcal{H}_\alpha$. In accordance with §1.1, $\mathcal{H}_\alpha$ is a Cartan subalgebra and $\mathcal{D}$ is an MOD.

Now we claim that

$$\mathrm{Aut}_{MOD}(\mathcal{L}) = \mathrm{Aut}(\mathcal{D}) = G. \tag{8.3}$$

To this end we need the following result.

Lemma 8.1.1. *$N_{\mathbb{G}}(E) = G$ and $C_{\mathbb{G}}(E) = E$.*

Proof. It is sufficient to show that $C_{\mathbb{G}}(E) = E$. If $\varphi \in C_{\mathbb{G}}(E)$, then φ leaves every E-eigensubspace in V fixed, that is, $\varphi(e_\alpha) = \lambda_\alpha e_\alpha$ for each $\alpha \in L^{\#}$ and some $\lambda_\alpha \in \mathbb{C}^*$. Because $e_\alpha * e_\beta = \pm e_{\alpha+\beta}$ for $\alpha \neq \beta$, we must have $\lambda_\alpha \lambda_\beta = \lambda_{\alpha+\beta}$. It is easy to see that any map $\lambda : L^{\#} \to \mathbb{C}^*$ with this property is obtained by restricting some homomorphism $\bar{L} \to \{\pm 1\}$ to $L^{\#}$, that is, λ is induced by some element of E. □

Now suppose that $\varphi \in \mathrm{Aut}(\mathcal{D})$. For every $f \in E$, the element $f' = \varphi f \varphi^{-1}$ leaves every subspace $\mathcal{H}_\alpha$ fixed, and so f' commutes with E. By Lemma 8.1.1, $f' \in E$, and this means that φ normalizes E. Again by Lemma 8.1.1, $\varphi \in G$, and this proves equality (8.3).

Finally, we claim that G acts irreducibly on $\mathcal{L}$. For this we choose a_i and b_i, $1 \leq i \leq 3$, such that $a_i b_i = e_1$ and

$$\{\bar{e}_1, \bar{a}_i, \bar{b}_i | 1 \leq i \leq 3\} = L^{\#}.$$

Set $g_i = a_i \wedge b_i \in V \wedge V$. It is easy to see that if $\varphi \in G$ and $\varphi(e_1) = e_1$, then $\varphi(g_i) = g_{\pi(i)}$ for some permutation $\pi \in \mathbb{S}_3$. If $\varphi(e_1) = -e_1$, then $\varphi(g_i) = -g_{\pi(i)}$. It is also obvious that every permutation $\pi \in \mathbb{S}_3$ can be realized in this way. Furthermore, the subalgebra $\mathcal{H} = \mathcal{H}_{\bar{e}_1}$ is just the set

$$\{x_1 g_1 + x_2 g_2 + x_3 g_3 | x_i \in \mathbb{C}, x_1 + x_2 + x_3 = 0\}.$$

Set $P = St_G(\pm e_1)$. From the above description, it follows that the image of P in $GL(\mathcal{H})$ is isomorphic to $\mathbb{Z}_2 \times \mathbb{S}_3$ and is an irreducible complex linear group. But $\mathcal{L} = \mathrm{Ind}_P^G(\mathcal{H})$, so by Clifford's Theorem it follows immediately that $\mathcal{L}$ is an irreducible G-module.

8.2 Invariant lattices in $\mathcal{L}$

We start with the following construction. Choose a system of representatives t_α, $\alpha \in L^\#$, for G/P such that $G = \cup_\alpha t_\alpha P$ and $\mathcal{H}_\alpha = t_\alpha(\mathcal{H})$ (P and $\mathcal{H}$ are as introduced at the end of §8.1). Let Γ be a P-invariant lattice in $\mathcal{H}$. Then

$$G\Gamma = \oplus_\alpha t_\alpha \Gamma = \mathrm{Ind}_P^G(\Gamma)$$

is a G-invariant lattice in $\mathcal{L}$. Below we shall set

$$\Gamma^{(\alpha)} = t_\alpha \Gamma.$$

Let pr_α denote the projection onto $\mathcal{H}_\alpha$, and $\mathrm{pr}_1 = \mathrm{pr}_{\bar{e}_1}$.

Lemma 8.2.1. *Let Λ be a G-invariant lattice in $\mathcal{L}$. Then $\Gamma = \mathrm{pr}_1\Lambda$ is a P-invariant lattice in $\mathcal{H}$, and*

$$4\mathrm{Ind}_P^G(\Gamma) \subseteq \Lambda \subseteq \mathrm{Ind}_P^G(\Gamma).$$

Proof. Note that $4\mathrm{pr}_1$ is an integral linear combination of operators from G. Indeed, set $\sigma = \sum_{f \in E_1} f$ for $E_1 = E \cap P$. Then $\sigma(x) = 4x$ for $x \in \mathcal{H}$ and $\sigma(\mathcal{H}_\alpha) = 0$ for $\alpha \neq \bar{e}_1$, that is, $\sigma = 4\mathrm{pr}_1$.

Now set $\Gamma = \mathrm{pr}_1\Lambda$. Then Γ is a finitely generated torsion-free abelian group. If $\mathrm{rank}(\Gamma) \leq 1$, then Λ cannot be a lattice in $\mathcal{L}$, a contradiction. Hence, $\mathrm{rank}(\Gamma) = 2$. Furthermore, $4\Gamma = 4\mathrm{pr}_1\Lambda \subset \Lambda$; therefore Γ is a P-invariant lattice in $\mathcal{H}$ and

$$4\mathrm{Ind}_P^G(\Gamma) = G(4\Gamma) \subseteq \Lambda.$$

Finally,

$$\Lambda \subseteq \oplus_\alpha \mathrm{pr}_\alpha \Lambda = \oplus_\alpha t_\alpha \Gamma = \mathrm{Ind}_P^G(\Gamma).$$

□

We know that P acts on $\mathcal{H}$ as $\mathbb{Z}_2 \times \mathbb{S}_3$. The following result is obvious.

Lemma 8.2.2. *Suppose that $\mathbb{S}_3$ acts faithfully on $\mathbb{C}^2$. Then $\mathbb{C}^2$ contains just two $\mathbb{S}_3$-invariant lattices up to similarity.* □

In our case, it is easy to exhibit P-invariant lattices in $\mathcal{H}$:

$$\Gamma_1 = \{n_1g_1 + n_2g_2 + n_3g_3 | n_i \in \mathbb{Z}, \sum_i n_i = 0\},$$

$$\Gamma_2 = \{n_1g_1 + n_2g_2 + n_3g_3 | n_i \in \mathbb{Z}, \sum_i n_i = 0, n_1 \equiv n_2 \equiv n_3 \pmod 3\};$$

moreover, $[\Gamma_1 : \Gamma_2] = 3$.

Thus we have to study G-submodules of $\mathrm{Ind}_P^G(\Gamma_i)/4\mathrm{Ind}_P^G(\Gamma_i)$. Note that the inclusion $\Gamma_2 \subset \Gamma_1$ induces an isomorphism

$$\mathrm{Ind}_P^G(\Gamma_2)/4\mathrm{Ind}_P^G(\Gamma_2) \cong \mathrm{Ind}_P^G(\Gamma_1)/4\mathrm{Ind}_P^G(\Gamma_1).$$

Moreover, $(\Gamma_1)^* = \frac{1}{3}\Gamma_2$. Therefore, up to duality one can suppose that

$$4\mathrm{Ind}_P^G(\Gamma_1) \subseteq \Lambda \subseteq \mathrm{Ind}_P^G(\Gamma_1). \tag{8.4}$$

Note also that $\mathcal{L} \subset V \wedge V \subset V \otimes V$ is endowed with the G-invariant form $(*|*)$ induced by the form

$$(a \otimes b|c \otimes d) = p(a, c)p(b, d)$$

defined on $V \otimes V$. For instance, $(g_i|g_j) = \delta_{i,j}$. Clearly, $(*|*)$ is proportional to the Killing form K.

The restriction $(*|*)|_{\Gamma_1}$ is an integral bilinear form with determinant 3, and so it induces a non-degenerate bilinear form on $U = \Gamma_1/2\Gamma_1$ (respectively, on $T = \Gamma_1/4\Gamma_1$) with values in $\mathbb{Z}_2$ (respectively, in $\mathbb{Z}_4$). Correspondingly, the modules

$$U^G = \mathrm{Ind}_P^G(U) = \mathrm{Ind}_P^G(\Gamma_1)/2\mathrm{Ind}_P^G(\Gamma_1),$$

$$T^G = \mathrm{Ind}_P^G(T) = \mathrm{Ind}_P^G(\Gamma_1)/4\mathrm{Ind}_P^G(\Gamma_1)$$

are endowed with non-degenerate invariant forms.

From now on a lattice Λ with (8.4) is called *shallow* if $\Lambda \supseteq 2\mathrm{Ind}_P^G(\Gamma_1)$, and *deep* otherwise.

Clearly, classification of shallow lattices is equivalent to classification of G-submodules of

$$U^G = \mathrm{Ind}_P^G(U) = \mathrm{Ind}_P^G(\Gamma_1)/2\mathrm{Ind}_P^G(\Gamma_1).$$

Because E acts trivially on U^G, we are in fact dealing with $\bar{G}$-modules, where $\bar{G} = G/E$. Set $\bar{P} = P/E$.

The following statement follows from the properties of induced modules.

Lemma 8.2.3. *Let A be a finite group, B a subgroup, and let V_1 and V_2 be an A- and a B-module respectively. Then V_1 is a factor-module for $\mathrm{Ind}_B^A(V_2)$ if and only if there exists a homomorphism $\psi : V_2 \to V_1|_B$ of B-modules such that $A\psi(V_2) = V_1$.* □

Corollary 8.2.4. *A $\bar{G}$-module M is a factor-module for U^G if and only if there exists a two-dimensional irreducible $\bar{P}$-submodule U' of M such that $\bar{G}U' = M$.*

Proof. It is sufficient to see that every two-dimensional irreducible $\bar{P}$-module is isomorphic to U. □

To avoid confusion, we denote the $\bar{G}$-module dual to E by R. Then it is clear that R^* is a factor-module for U^G. Furthermore, consider $M = R \otimes R^*$, the space of 3×3 matrices over $\mathbb{F}_2$ with the natural action of $SL_3(2)$. Then $M = M_8 \oplus I$, where M_8 consists of matrices with zero trace and I is generated by the unit matrix. It is obvious that M_8 is irreducible. Moreover, the subspace

$$U' = \left\{ \begin{pmatrix} 0 & * & * \\ 0 & 0 & 0 \\ 0 & 0 & 0 \end{pmatrix} \right\}$$

is an irreducible $\bar{P}$-module. So, M_8 is also a factor-module for U^G.

As we have mentioned above, U^G is endowed with a non-degenerate invariant form. Consequently, U^G is self-dual: $U^G \cong (U^G)^*$. In particular, U^G contains $(M_8)^*$ as a submodule. However, $(M_8)^* \cong M_8$ and $\dim U^G = 14 < 2\dim M_8 = 16$. We arrive at the conclusion that M_8 is a direct summand of U^G: $U^G = M_8 \oplus D$.

Regarding D, we know that $\dim D = 6$, $D \cong D^*$ and R^* is a factor module for D. Hence, R is a submodule of D. If in addition $D = R \oplus R^*$, then R is a factor-module for U^G and so R contains a two dimensional $\bar{P}$-invariant subspace, a contradiction. Hence, D is a non-split extension of R by R^*. We write briefly:

$$U^G = M_8 \oplus (R|R^*).$$

As a consequence, we see that there are just five nontrivial submodules of U^G, namely R, D, M_8, $R \oplus M_8$ and, of course, U^G.

Corollary 8.2.5. *There are precisely five shallow sublattices* Λ_i, $0 \leq i \leq 4$ *in* $\Lambda_0 = \mathrm{Ind}_P^G(\Gamma_1)$, *and* $\dim_{\mathbb{F}_2} \Lambda_i/2\Lambda_0$ *is* 3, 6, 8 *and* 11 *for* $i = 1, 2, 3$ *and* 4 *respectively.* □

The remainder of the section is devoted to the classification of deep lattices. Such lattices correspond to the G-submodules M of

$$T^G = \mathrm{Ind}_P^G(T) = \mathrm{Ind}_P^G(\Gamma_1)/4\mathrm{Ind}_P^G(\Gamma_1)$$

such that $M \not\supseteq 2T^G$.

There is a natural isomorphism of $\bar{G}$-modules:

$$2T^G \cong T^G/2T^G \cong U^G. \tag{8.5}$$

For $M \subseteq T^G$ we set

$$M_{(1)} = (M + 2T^G)/2T^G, \; M_{(2)} = M \cap 2T^G.$$

Clearly, $M_{(1)}$ is contained in $M_{(2)}$ (as submodules of U^G). Furthermore, if $N = M^\perp$ denotes the annihilator of M in T^G, then

$$N_{(1)} = M_{(2)}^\perp, \; N_{(2)} = M_{(1)}^\perp,$$

(of course, we mean here annihilators in U^G).

Next, we define a homomorphism

$$\nabla : (T^G/2T^G) \otimes E \to 2T^G$$

of $\bar{G}$-modules, such that $\nabla((t+2T^G)\otimes f) = f(t) - t$ for $t \in T^G$ and $f \in E$. Since E acts trivially on $T^G/2T^G$ and on $2T^G$, our definition is correct, and it is easy to verify that ∇ is in fact a $\bar{G}$-homomorphism. In view of the identification (8.5), we have a homomorphism

$$\nabla : U^G \otimes E \to U^G.$$

We write

$$U^G = \oplus_\alpha U_\alpha,$$

where $U_\alpha = t_\alpha U$. Suppose that $f \in E$. Then it is not difficult to see for $u \in U_\alpha$ that

$$\nabla(u \otimes f) = \begin{cases} 0 & \text{if} \quad f|_{\mathcal{H}_\alpha} = 1, \\ u & \text{if} \quad f|_{\mathcal{H}_\alpha} = -1. \end{cases}$$

Clearly, if M is a submodule of T^G, then $\nabla(M_{(1)} \otimes E) \subseteq M_{(2)}$.

Now we consider a deep lattice Λ and set $M = M(\Lambda) = \Lambda/4\mathrm{Ind}_P^G(\Gamma_1)$. Suppose first that $M_{(1)} \supseteq R$. Then $M_{(2)} \supseteq \nabla(R\otimes E)$. Note that $R\otimes E = R\otimes R^* \cong M_8 \oplus I$, and $\nabla(R\otimes E) \neq 0$. Hence, $\nabla(R\otimes E) = M_8$ and $M_{(2)} \supseteq M_8$, $M_{(2)} \supseteq M_{(1)} + M_8 \supseteq R + M_8$. In particular, when $M_{(1)} \supseteq D$ we have $M_{(2)} \supseteq D + M_8 = U^G$, a contradiction.

Next, suppose that $M_{(1)} \supseteq M_8$, so that $M_{(2)} \supseteq M_8$. Going over to $N = M^\perp$, we obtain that

$$N_{(1)}, N_{(2)} \subseteq M_8^\perp = D.$$

However, if $N_{(1)} \neq 0$, then $N_{(1)} \supseteq R$ and so $N_{(2)} \supseteq M_8$, a contradiction. Hence, $N_{(1)} = 0$ and $M_{(2)} = U^G$, again a contradiction.

Thus we have shown that the inclusions $M_{(1)} \supseteq D$ and $M_{(1)} \supseteq M_8$ are impossible, and so

$$M_{(1)} = R, \; M_{(2)} = M_8 \oplus R. \tag{8.6}$$

Furthermore, we claim that if a submodule M satisfying (8.6) exists, then it is unique. Indeed, the modules $A = M + 2T^G$ and $B = M \cap 2T^G$ are uniquely determined by the properties $A \supseteq 2T^G$, $A/2T^G \cong R$, $B \subseteq 2T^G$ and $2T^G/B \cong R^*$. Furthermore, A/B has the composition factors $A/2T^G \cong R$ and $2T^G/B \cong R^*$. Hence, in A/B there exist at most one G-submodule S such that $S \cong R$. Meanwhile M/B must be a submodule of such a kind.

Finally, we construct a deep lattice using our loop approach. For this we consider a lattice

$$\Delta_0 = < e_\alpha | \alpha \in L^\# >_{\mathbb{Z}}$$

in V endowed with the inner product p. Note that Δ_0 is closed under the multiplication $*$. Furthermore, we consider a lattice

$$\hat{\Delta}_0 = \Delta_0 \wedge \Delta_0 = < u \wedge v | u, v \in \Delta_0 >_{\mathbb{Z}}$$

in $V \wedge V$. Introduce a permutation $\bar{G}$-module

$$\tilde{V} = \oplus_\alpha \mathbb{F}_2 v_\alpha,$$

and map Δ_0 into $\tilde{V}$ by the rule

$$x = \sum_\alpha x_\alpha e_\alpha \mapsto \bar{x} = \sum_\alpha \bar{x}_\alpha v_\alpha,$$

where $\bar{x}_\alpha = x_\alpha \pmod 2$. Similarly, $\hat{\Delta}_0$ is mapped into $\tilde{V} \wedge \tilde{V}$:

$$x = \sum_{\alpha,\beta} x_{\alpha\beta} e_\alpha \wedge e_\beta \mapsto \bar{x} = \sum_{\alpha,\beta} \bar{x}_{\alpha\beta} v_\alpha \wedge v_\beta.$$

Clearly, $\{x \in \Delta_0 | \bar{x} = 0\} = 2\Delta_0$, and the same is true for $\hat{\Delta}_0$.
Next we define $* : \tilde{V} \wedge \tilde{V} \to \tilde{V}$ as follows:

$$*(v_\alpha \wedge v_\beta) = v_{\alpha+\beta}.$$

Then for every $v \in \hat{\Delta}_0$ we have

$$\overline{*v} = *\bar{v}.$$

We will now introduce some $\bar{G}$-submodules of $\tilde{V}$. To begin with, $\tilde{I} = < \sigma = \sum_\alpha v_\alpha >$ is a one-dimensional submodule. Furthermore, we set

$$\pi_i = \left\{ \sum_{j=1}^{3} t_j \bar{e}_j | t_j \in \mathbb{F}_2, t_i = 1 \right\} \subset L^\#,$$

$$\sigma_i = \sum_{\alpha \in \pi_i} v_\alpha, \; i = 1, 2, 3, \quad S = < \sigma_1, \sigma_2, \sigma_3 >_{\mathbb{F}_2} .$$

The subspace S can be described in the following way. Every element of $\tilde{V}$ can be viewed as a functional $L^\# \to \mathbb{F}_2$. Then S corresponds to the subspace consisting of the *linear* functionals $\bar{L} \to \mathbb{F}_2$. From this it follows that S is a $\bar{G}$-submodule. Set

$$\tilde{S} = S \oplus \tilde{I}.$$

It is convenient to rename the generators of L as $x = e_1$, $y = e_2$, $z = e_3$. Set $\Delta = \{u \in \Delta_0 | \bar{u} \in \tilde{S}\}$. Then $\Delta = < s, s_1, s_2, s_3; 2\Delta_0 >$, where

$$s = \sum_\alpha e_\alpha, \; s_i = \sum_{\alpha \in \pi_i} e_\alpha.$$

One can show that Δ is dual to the root lattice of type E_7. Finally, we set

$$\Lambda = (\Delta \wedge \Delta) \cap \mathcal{L}.$$

Proposition 8.2.6. Λ *is a deep lattice.*

The proof splits into several steps.

Lemma 8.2.7. $\hat{\Delta}_0 \cap \mathcal{L} = \mathrm{Ind}_P^G(\Gamma_1)$.

Proof. Clearly, $g_1 - g_2, g_1 - g_3 \in \hat{\Delta}_0 \cap \mathcal{L}$, that is, $\Gamma_1 \subseteq \hat{\Delta}_0 \cap \mathcal{L}$. Hence, $\Gamma_1^{(\alpha)} \subseteq \hat{\Delta}_0 \cap \mathcal{L}$ for every $\alpha \in L^{\#}$. □

Note that $\hat{\Delta}_0 \supseteq \Delta \wedge \Delta \supseteq 2\Delta_0 \wedge 2\Delta_0 = 4\hat{\Delta}_0$, so that by Lemma 8.2.7 the lattice Λ satisfies condition (8.4). We now determine $M_{(1)}$ and $M_{(2)}$ for $M = \Lambda / 4\mathrm{Ind}_P^G(\Gamma_1)$. We can consider $M_{(1)}$ and $M_{(2)}$ as subspaces of $\tilde{V} \wedge \tilde{V}$, using the embedding $U^G \hookrightarrow \hat{\Delta}_0 / 2\hat{\Delta}_0$.

Lemma 8.2.8. (i) $M_{(1)} = < \sigma_i \wedge \sigma_j | 1 \leq i, j \leq 3 >_{\mathbb{F}_2}$.
(ii) $\Lambda \cap 2\mathrm{Ind}_P^G(\Gamma_1) = (\Delta \wedge 2\Delta_0) \cap \mathcal{L}$.

Proof. Each element of $\Delta \wedge \Delta$ can be written in the form

$$v = \sum_{i=1}^{3} a_i s_i \wedge s + \sum_{1 \leq i < j \leq 3} b_{ij} s_i \wedge s_j + u,$$

where $u \in \Delta \wedge 2\Delta_0$ and $a_i, b_{ij} \in \mathbb{F}_2$. Then

$$\bar{v} = \sum_i a_i \bar{s} \wedge \bar{s}_i + \sum_{ij} b_{ij} \bar{s}_i \wedge \bar{s}_j.$$

Clearly, $\bar{s} = \sigma$, $\bar{s}_i = \sigma_i$; moreover, σ and σ_i are linearly independent in $\tilde{V}$, hence $\sigma \wedge \sigma_i$ and $\sigma_i \wedge \sigma_j$ are linearly independent in $\tilde{V} \wedge \tilde{V}$. This means that $\bar{v} = 0$ if and only if $a_i = b_{ij} = 0$ for all i and j, that is, $v \in \Delta \wedge 2\Delta_0$. This proves identity (ii).

Now suppose that $v \in \Lambda$, that is, $*v = 0$. Then $*(\bar{v}) = 0$, but $*(\sigma_i \wedge \sigma_j) = 0$ and $*(\sigma \wedge \sigma_i) = \sigma_i$, therefore we have $0 = \sum_i a_i \sigma_i$. In other words, $a_i = 0$ for all i and this proves that $M_{(1)} \subseteq < \sigma_i \wedge \sigma_j | 1 \leq i, j \leq 3 >$. On the other hand, it is easy to find $v \in \Lambda$ with, for example, $\bar{v} = \sigma_1 \wedge \sigma_2$. It is sufficient to set

$$v = s_1 \wedge s_2 - 2(x + y - xy) \wedge z,$$

where $s_1 = x + xy + xz + xy.z$, $s_2 = y + xy + yz + xy.z$ (recall that x, y, z are generators for L). □

Let J denote the kernel of the homomorphism $* : \tilde{V} \wedge \tilde{V} \to \tilde{V}$. Then it is clear that $M_{(1)}, M_{(2)} \subseteq J$. Note that the image of the lattice $\Delta \wedge \Delta_0$ in $\tilde{V} \wedge \tilde{V}$ is precisely $\tilde{V} \wedge \tilde{S}$. Thus we have

$$M_{(2)} \subseteq (\tilde{V} \wedge \tilde{S}) \cap J.$$

On the other hand, suppose that $a \in (\tilde{V} \wedge \tilde{S}) \cap J$, and consider $u_1 \in \Delta \wedge \Delta_0$ such that $\bar{u}_1 = a$. Since $*a = 0$, we have $*u_1 \in 2\Delta_0$. It is easy to see that, for every $m \in \Delta_0$, there exists $l \in \Delta_0 \wedge \Delta_0$ such that $*l = m$. In particular, there exists $u_2 \in \Delta_0 \wedge \Delta_0$ such that $*u_1 = 2(*u_2)$. Setting $u = u_1 - 2u_2$, we have $*u = 0$, $\bar{u} = a$ and $u \in \Delta \wedge \Delta_0$, because $2u_2 \in 2\Delta_0 \wedge \Delta_0 \subseteq \Delta \wedge \Delta_0$. This means that $a \in M_{(2)}$, and we obtain

Lemma 8.2.9. $M_{(2)} = (\tilde{V} \wedge \tilde{S}) \cap J$. □

Lemma 8.2.10. $\dim_{\mathbb{F}_2} M_{(2)} = 11$.

Proof. First we show that $Y = *(\tilde{V} \wedge \tilde{S})$ is equal to $\tilde{V}$. For $\alpha \in L^{\#}$, by definition of $*$ we have

$$*(\sigma \wedge v_\alpha) = \sigma + v_\alpha.$$

Hence $Y \supseteq Y_0$, where

$$Y_0 = < \sigma + v_\alpha | \alpha \in L^{\#} > = < \sum_\alpha a_\alpha v_\alpha | a_\alpha \in \mathbb{F}_2, \sum_\alpha a_\alpha = 0 > .$$

Furthermore, take $\alpha \in \pi_1$. Clearly, $*(\sigma_1 \wedge v_\alpha) = \sigma_1 + \sigma$ does not belong to Y_0, so that $Y = \tilde{V}$. Now we have

$$\dim M_{(2)} = \dim((\tilde{V} \wedge \tilde{S}) \cap J) =$$

$$= \dim(\tilde{V} \wedge \tilde{S}) - \dim(*(\tilde{V} \wedge \tilde{S})) = 18 - 7 = 11.$$ □

Lemma 8.2.11. (i) *$M_{(2)}$ is generated by elements of the form $(v_\alpha + v_\beta) \wedge (v_\gamma + v_\delta)$, where α, β, γ and δ are distinct elements of $L^{\#}$ with $\alpha + \beta + \gamma + \delta = 0$.*

(ii) *Define a map $\rho : \tilde{V} \wedge \tilde{V} \to \tilde{V}$ as follows:*

$$\rho(v_\alpha \wedge v_\beta) = v_\alpha + v_\beta.$$

Then an element x of $\tilde{V} \wedge \tilde{V}$ belongs to $M_{(2)}$ if and only if $*x = \rho(x) = 0$.

Proof. (i) Regarding α, β, γ, δ with the given properties, we note that $\pi = \{\alpha, \beta, \gamma, \delta\}$ is an affine plane in $L^{\#}$. Therefore,

$$m = (v_\alpha + v_\beta) \wedge (v_\gamma + v_\delta) =$$

$$= (v_\alpha + v_\beta) \wedge (v_\alpha + v_\beta + v_\gamma + v_\delta) \in \tilde{V} \wedge \tilde{S}.$$

Moreover, $*m = 0$, and thus

$$m \in (\tilde{V} \wedge \tilde{S}) \cap J = M_{(2)}.$$

Finally, elements m of this kind generate an 11-dimensional subspace, and so this subspace is $M_{(2)}$.

(ii) Because $\rho((v_\alpha + v_\beta) \wedge (v_\gamma + v_\delta)) = 0$, we must have that $\rho(M_{(2)}) = 0$. It remains to show that $\dim \rho(J) = 3 = \mathrm{codim}_J M_{(2)}$. For this we remark that J is generated by elements like $l = v_\alpha \wedge v_\beta + v_\gamma \wedge v_\delta$, where α, β, γ and δ are as in (i). Clearly, $\rho(l) = v_\alpha + v_\beta + v_\gamma + v_\delta \in S$, and so $\rho(J) = S$. □

This completes the proof of Proposition 8.2.6. □

One can define elements of Λ similar to the elements of $M_{(2)}$ considered above. Namely, let $a, b, c, d \in L\backslash Z$ be such that $\alpha = \bar{a}$, $\beta = \bar{b}$, $\gamma = \bar{c}$, $\delta = \bar{d}$ are distinct and $ab = cd$. Consider a vector

$$u = u(a, b, c, d) = a \wedge b - c \wedge d + a \wedge c \pm b \wedge d.$$

Then

$$\bar{u} = v_\alpha \wedge v_\beta + v_\gamma \wedge v_\delta + v_\alpha \wedge v_\gamma + v_\beta \wedge v_\delta \in M_{(2)},$$

and $*u = 0$. Hence, $2u \in \Lambda \cap 2\mathrm{Ind}_P^G(\Gamma_1)$. It is easy to see that $\Lambda \cap 2\mathrm{Ind}_P^G(\Gamma_1)$ is generated by elements of this sort. Furthermore, the elements $4(a \wedge b - c \wedge d)$ with a, b, c, d as above generate $\Lambda \cap 4\mathrm{Ind}_P^G(\Gamma_1) = 4\mathrm{Ind}_P^G(\Gamma_1)$.

Now we describe some metric properties of the deep lattice Λ constructed above.

Proposition 8.2.12. *If $x, y \in \Lambda$, then $4|(x|y)$ and $8|(x|x)$.*

Proof. Consider a bilinear form q on $\tilde{V} \wedge \tilde{V}$ which accepts $\{v_\alpha \wedge v_\beta\}$ as an orthonormal basis. Then q restricted to J is non-degenerate. Moreover, $q(M_{(1)}, M_{(2)}) = 0$. In particular, $4|(x|y)$ for all $x \in \Lambda \cap 2\mathrm{Ind}_P^G(\Gamma_1)$ and $y \in \Lambda$. This allows us to consider on

$$\Lambda/(\Lambda \cap 2\mathrm{Ind}_P^G(\Gamma_1)) = M_{(1)}$$

the following bilinear form $< *, * >$ with values in $\mathbb{F}_2$:

$$< a + \Lambda \cap 2\mathrm{Ind}_P^G(\Gamma_1), b + \Lambda \cap 2\mathrm{Ind}_P^G(\Gamma_1) > = \frac{1}{2}(a|b) \ (\mathrm{mod}\, 2).$$

This form is $\bar{G}$-invariant. But any such form on $M_{(1)}$ is trivial, therefore $< *, * >= 0$ and $4|(a|b)$ for all $a, b \in \Lambda$.

The fact that $8|(x|x)$ for $x \in \Lambda$ now follows since Λ possesses a generating system consisting of vectors with this property, namely, $2u(a, b, c, d)$ and the vector $v \in \Lambda\backslash 2\mathrm{Ind}_P^G(\Gamma_1)$ defined in the proof of Lemma 8.2.8. □

The proof of the following statement is omitted.

Proposition 8.2.13. *Minimal vectors in Λ are of norm 16. They form three G-orbits of lengths 84, 336 and 336.* □

8.3 Automorphism groups

We start with the following general statement.

Lemma 8.3.1. *Suppose that G is a finite group and*
(i) *V is a $\mathbb{C}G$-module with $\mathbb{Q}$-valued character χ;*
(ii) *the restriction of V to some subgroup H of G is irreducible and rational.*
Then the representation of G on V is also rational.

Proof. Suppose that H is represented by rational matrices in some basis $\{e_1, \ldots, e_n\}$ of V, and that the representation of G on V corresponds to the homomorphism

$$\Psi : G \to GL_n(\mathbb{C}),\ n = \dim V.$$

Set $\mathcal{H} =< \Psi(h) | h \in H >_{\mathbb{Q}}$. By Burnside's Theorem [CuR], it follows from the irreducibility of H on V that

$$\mathcal{H} \otimes_{\mathbb{Q}} \mathbb{C} = \mathrm{End}(V) = M_n(\mathbb{C}).$$

This means that $\dim_{\mathbb{Q}} \mathcal{H} = \dim_{\mathbb{C}} \mathcal{H} \otimes \mathbb{C} = n^2$, and $\mathcal{H} = M_n(\mathbb{Q})$. In particular, all the elementary matrices $E_{ij} = (\delta_{ik}\delta_{jl})_{1 \le k,l \le n}$ belong to $\mathcal{H}$, that is,

$$E_{ij} = \sum_{h \in H} a_h^{ij} \Psi(h)$$

for some $a_h^{ij} \in \mathbb{Q}$. Now for any $g \in G$ the coefficient $\Psi(g)_{ij}$ of the matrix $\Psi(g)$ is equal to

$$\Psi(g)_{ij} = \mathrm{Tr}\left(\Psi(g)E_{ji}\right) = \mathrm{Tr}\left\{\Psi(g)\sum_{h \in H} a_h^{ji}\Psi(h)\right\} = \sum_{h \in H} a_h^{ji}\chi(gh) \in \mathbb{Q},$$

that is, $\Psi(g) \in M_n(\mathbb{Q})$. □

We now proceed to the proof of Theorem 8.0.1.

The equality $\mathrm{Aut}(\Lambda_0) = D_{12} \wr \mathbb{S}_7$ follows from the fact that Λ_0 is simply a root lattice of type $7G_2$ (or equivalently, $7A_2$), and $\mathrm{Aut}(G_2) = D_{12} = \mathbb{Z}_2 \times \mathbb{S}_3$. Furthermore, the assertions concerning $\mathrm{Aut}(\Lambda_i)$, $1 \le i \le 4$ can be verified by direct calculation.

It remains to compute the group $\mathbb{G} = \mathrm{Aut}(\Lambda)$, where $\Lambda = \Lambda_5$ is a deep lattice. It is not difficult to see that $\mathbb{G}$ must contain a subgroup isomorphic to $G_2(3)$. Indeed, from [ATLAS] we have that

a) $G_2(3)$ has a (unique) integer-valued irreducible complex character χ of degree 14;

b) G can be embedded in $G_2(3)$ as a maximal subgroup;

c) the restriction $\chi|_G$ is the (unique) irreducible complex character of degree 14 of G.

Therefore, without loss of generality one can suppose that $\chi|_G$ is afforded by the G-module $\mathcal{L}$, and so can be realized over $\mathbb{Q}$. In view of Lemma 8.3.1, χ is afforded also by some rational representation. In particular, $G_2(3)$ stabilizes some G-invariant lattice Λ' in $\mathcal{L}$. Clearly, Λ' cannot be isometric or dual to any of the lattices Λ_i, $0 \leq i \leq 4$. Thus, we arrive at the conclusion that $\mathbb{G} \supseteq G_2(3)$.

However, by constructing an additional automorphism Φ we give another direct proof of this result.

Theorem 8.3.2. $\mathbb{G} = \mathrm{Aut}(\Lambda) = \mathbb{Z}_2\times < G, \Phi >$, *where*

$$< G, \Phi > \cong G_2(3).$$

The remainder of this section is devoted to the proof of Theorem 8.3.2.

We shall need the following auxiliary construction. Consider a space $\mathbb{R}^4$ with an orthonormal basis $\{w_i | 1 \leq i \leq 4\}$ and a root system

$$D = \{\pm w_i \pm w_j | i \neq j\}$$

of type D_4 in it. For

$$\mu = (1,2)(3,4), \ \nu = (1,3)(2,4), \ \eta = (1,4)(2,3)$$

we have $D = D_\mu \cup D_\nu \cup D_\eta$, where $D_\mu = \{\pm w_1 \pm w_2, \pm w_3 \pm w_4\}$, and so on. Note that $(x|y) = \pm 1$ if $x \in D_\mu, y \in D_\nu$; meanwhile, for $x, y \in D_\mu$ we have $(x|y) = 0, \pm 2$. This means that the decomposition $D = D_\mu \cup D_\nu \cup D_\eta$ is $\mathrm{Aut}(D)$-invariant.

Now set $A = \mathrm{Aut}(D) \cap SO_4(\mathbb{R})$, and let A_0 be the subgroup of A consisting of the monomial matrices with entries $0, \pm 1$ and determinant 1. Furthermore, A contains the element

$$\varphi = \frac{1}{2}\begin{pmatrix} 1 & 1 & 1 & -1 \\ 1 & 1 & -1 & 1 \\ 1 & -1 & 1 & 1 \\ 1 & -1 & -1 & -1 \end{pmatrix}.$$

The following assertion is obvious.

Lemma 8.3.3. $A = A_0. < \varphi >$ *and* $| < \varphi > | = 3$. □

Consider the decompositions

$$V = V_0 \oplus V_1, \tag{8.7}$$

$$V \wedge V = V_0 \wedge V_0 \oplus V_1 \wedge V_1 \oplus T_x \oplus T_y \oplus T_{xy}, \tag{8.8}$$

where $V_0 =< x, y, xy >_{\mathbb{C}}$, $V_1 =< z, xz, yz, xy.z >_{\mathbb{C}}$, $T_t = t \wedge V_1$ for $t = x, y, xy$. Recall that the generators e_1, e_2, e_3 of $\mathcal{C}$ have been renamed as x, y, z.

We make the group A act on the basis $\{z, xz, yz, xy.z\}$ of V_1 instead of on the basis $\{w_i | 1 \leq i \leq 4\}$. Furthermore, we put $\bar{x}, \bar{y}, \overline{xy} \in L^{\#}$ in correspondence with the permutations μ, ν, η, so that $D = D_{\bar{x}} \cup D_{\bar{y}} \cup D_{\overline{xy}}$, and A acts on $\{\bar{x}, \bar{y}, \overline{xy}\}$.

Suppose that $s \in A$ fixes the kernel X of the map

$$*|_{V_1 \wedge V_1} : V_1 \wedge V_1 \to V_0,$$

so that one can define the action of s on V_0. Furthermore, because the map

$$*|_{V_0 \wedge V_0} : V_0 \wedge V_0 \to V_0$$

is bijective, one can extend the action of s defined on V_0 to $V_0 \wedge V_0$. Next, we define an action of s on

$$V_0 \wedge V_1 = T_x \oplus T_y \oplus T_{xy}$$

by requiring that

(a) s acts on $\{T_x, T_y, T_{xy}\}$ as it does on $\{\bar{x}, \bar{y}, \overline{xy}\}$; and

(b) the diagram

$$\begin{array}{ccc} T_x \oplus T_y \oplus T_{xy} & \stackrel{*}{\longrightarrow} & V_1 \\ s \downarrow & & s \downarrow \\ T_x \oplus T_y \oplus T_{xy} & \stackrel{*}{\longrightarrow} & V_1 \end{array}$$

is commutative.

Since $*$ realizes isomorphisms between each of the spaces T_x, T_y, T_{xy} and V_1, our conditions determine s uniquely. From our description it is clear that the diagram

$$\begin{array}{ccc} V \wedge V & \stackrel{*}{\longrightarrow} & V \\ s \downarrow & & s \downarrow \\ V \wedge V & \stackrel{*}{\longrightarrow} & V \end{array}$$

is also commutative. In particular, s preserves $\mathcal{L} = \mathrm{Ker}(*)$.

By definition, Φ is obtained by the construction described as applied to the element $s = \varphi$. This is possible in view of the following statement.

Lemma 8.3.4. (i) *φ fixes X.*

(ii) *The induced action of φ on V_0 is as follows:*

$$\varphi : x \mapsto -y,\ y \mapsto -xy,\ xy \mapsto x.$$

Proof. (i) Clearly,

$$X = (V_1 \wedge V_1) \cap \mathrm{Ker}(*) = < h_x, h_y, h_{xy} >_{\mathbb{C}},$$

where

$$h_x = z \wedge xz + yz \wedge xy.z,$$
$$h_y = z \wedge yz - xz \wedge xy.z,$$
$$h_z = z \wedge xy.z + xz \wedge yz.$$

We have

$$\varphi(h_x) = \frac{1}{4}(z + xz + yz + xy.z) \wedge (z + xz - yz - xy.z) +$$
$$+ \frac{1}{4}(z - xz + yz - xy.z) \wedge (-z + xz + yz - xy.z) =$$
$$= \frac{1}{2}(z + xz) \wedge (-yz - xy.z) + \frac{1}{2}(z - xz) \wedge (yz - xy.z) =$$
$$= -z \wedge xy.z - xz \wedge yz = -h_{xy}.$$

Similarly, $\varphi(h_y) = -h_x$, $\varphi(h_{xy}) = h_y$.

(ii) Take $m \in V \wedge V$ such that $*m = x$, for example, $m = z \wedge xz$. Then

$$\varphi(x) = \varphi(*m) = *\varphi(m) =$$
$$= * \left(\frac{1}{4}(z + xz + yz + xy.z) \wedge (z + xz - yz - xy.z) \right) =$$
$$= * \left(\frac{1}{2}(z + xz) \wedge (-yz - xy.z) \right) =$$
$$= -\frac{1}{2}\,(z.yz + z.(xy)z + xz.yz + xz.(xy)z) =$$
$$= -\frac{1}{2}(y + xy - xy + y) = -y.$$

The values $\varphi(y)$, $\varphi(xy)$ are computed similarly. □

Now suppose that $s' \in A_0$. Then it is easy to see that s' has the form $s' = g|_{V_1}$ for some $g \in Q$, where

$$Q = St_G(< x, y, xy >_{\mathbb{C}}). \tag{8.9}$$

Hence, s' preserves X, and so one can define an action of it on $V \wedge V$. Of course, this action coincides with that of g (one of the reasons for this tautology is that the actions of g on $\{\bar{x}, \bar{y}, \overline{xy}\}$, $\{T_x, T_y, T_{xy}\}$ and $\{D_{\bar{x}}, D_{\bar{y}}, D_{\overline{xy}}\}$ are compatible). Thus s acts on $V \wedge V$, for each $s \in A$.

Our next goal is to show that $\Phi \in \mathrm{Aut}(\Lambda)$. One of the ways to check this is the following. Suppose that we can find vectors $u_1, \ldots, u_n$ such that

$$\mathbb{Z}Q < u_1, \ldots, u_n >= \Lambda$$

(the subgroup Q is defined in (8.9)). Then it is sufficient to verify that

$$\Phi^{\pm 1} u_i \in \Lambda.$$

Indeed, let $g \in Q$. Then by Lemma 8.3.3 we have $\Phi g = h\Phi^{\kappa}$ for some $h \in Q$ and $\kappa = \pm 1$. Hence, $\Phi g(u_i) = h\Phi^{\kappa}(u_i) \in \Lambda$.

In fact we can take $n = 2$, and

$$u_1 = 2(y \wedge xy + yz \wedge xy.z) + 2(y \wedge yz - xy \wedge xy.z),$$

$$u_2 = v = xy \wedge y + xy.z \wedge yz + x \wedge xy + xz \wedge xy.z +$$

$$+x \wedge y + xz \wedge yz + y \wedge xy.z + xy \wedge yz - xy \wedge xz -$$

$$-x \wedge xy.z + x \wedge yz - y \wedge xz + 2xy \wedge z.$$

Proposition 8.3.5. $\mathbb{Z}Q < u_1, v >= \Lambda$.

Proof. First we show that $\mathbb{Z}Qu_1 = \Lambda \cap 2\mathrm{Ind}_P^G(\Gamma_1)$. Consider $\psi \in Q$ such that $\psi(x) = x$, $\psi(y) = y$, $\psi(z) = -z$. Then

$$(1 - \psi)(u_1) = 4(y \wedge yz - xy \wedge xy.z) = 4a,$$

$$(1 + \psi)(u_1) = 4(y \wedge xy + yz \wedge xy.z) = 4b.$$

It is not difficult to see that

$$\mathbb{Z}Qa = \Gamma_1^{(\bar{z})} \oplus \Gamma_1^{(\overline{xz})} \oplus \Gamma_1^{(\overline{yz})} \oplus \Gamma_1^{(\overline{xyz})},$$

$$\mathbb{Z}Qb = \Gamma_1^{(\bar{x})} \oplus \Gamma_1^{(\bar{y})} \oplus \Gamma_1^{(\overline{xy})}.$$

Hence, $\mathbb{Z}Qu_1 \supseteq 4\mathrm{Ind}_P^G(\Gamma_1)$. Furthermore, the image of u_1 in $M_{(2)}$ is

$$v_{\bar{y}} \wedge v_{\overline{yz}} + v_{\overline{xy}} \wedge v_{\overline{xyz}} + v_{\bar{y}} \wedge v_{\overline{xy}} + v_{\overline{yz}} \wedge v_{\overline{xyz}},$$

and the $\mathbb{F}_2Q$-span of this element is $M_{(2)}$.

Finally, let $\bar{v} = \sigma_1 \wedge \sigma_2$ be the image of v in $M_{(1)}$. Because Q realizes every linear map belonging to $GL(< \sigma_1, \sigma_2, \sigma_3 >_{\mathbb{F}_2})$ and stabilizing σ_3, the span $\mathbb{F}_2Q\bar{v}$ contains the elements $(\sigma_1 + \sigma_3) \wedge \sigma_2$ and $\sigma_1 \wedge (\sigma_2 + \sigma_3)$, that is, it is just $M_{(1)}$ (see Lemma 8.2.8). □

Now we compute $\Phi(u_1)$. For this we write

$$u_1 = a - b + c + d,$$

where $a = 2y \wedge yz$, $b = 2xy \wedge xy.z$, $c = 2yz \wedge xy.z$ and $d = 2y \wedge xy$. Note that $\Phi(a)$ is the preimage of $\varphi(*a)$ in T_y under $*$. We have $*a = -2z$, $\varphi(-2z) = -z - xz - yz - xy.z$, which implies that

$$\Phi(a) = y \wedge yz - y \wedge z + y \wedge xy.z - y \wedge xz.$$

Similarly,

$$\Phi(b) = xy \wedge xy.z - xy \wedge z + xy \wedge xz - xy \wedge yz.$$

Furthermore,

$$\Phi(c) = 2 \cdot \frac{1}{4}(z - xz + yz - xy.z) \wedge (-z + xz + yz - xy.z) =$$

$$= (z - xz) \wedge (yz - xy.z) = z \wedge yz - z \wedge xy.z - xz \wedge yz + xz \wedge xy.z.$$

Finally, $\Phi(d)$ is determined by the conditions

$$\Phi(d) \in V_0 \wedge V_0,\ *(\Phi(c) + \Phi(d)) = 0,$$

that is, $\Phi(d) = 2x \wedge xy$. Combining these calculations, we obtain

$$\Phi(u_1) = y \wedge yz - y \wedge z + y \wedge xy.z-$$

$$-y \wedge xz - xy \wedge xy.z + xy \wedge z - xy \wedge xz + xy \wedge yz + z \wedge yz-$$

$$-z \wedge xy.z - xz \wedge yz + xz \wedge xy.z + 2x \wedge xy.$$

We claim that $\Phi(u_1) \in \Lambda$. Taking $g \in G$ such that

$$g : x \mapsto y,\ y \mapsto z,\ z \mapsto x,$$

we have

$$g(v) = -z \wedge yz + xz \wedge xy.z + y \wedge yz - xy \wedge xy.z + y \wedge z+$$

$$+xy \wedge yz + z \wedge xy.z + xz \wedge yz - xy \wedge yz - y \wedge xy.z-$$

$$-y \wedge xz - xy \wedge z - 2x \wedge yz.$$

Hence,

$$\Phi(u_1) + g(v) = 2\{y \wedge yz - xy \wedge xy.z+$$

$$+ xz \wedge xy.z - y \wedge xz + x \wedge xy - x \wedge yz\}.$$

The image of the last vector in $\tilde{V} \wedge \tilde{V}$ is

$$v_{\bar{y}} \wedge v_{\overline{yz}} + v_{\overline{xy}} \wedge v_{\overline{xyz}} + v_{\overline{xz}} \wedge v_{\overline{xyz}} + v_{\bar{y}} \wedge v_{\overline{xz}} + v_{\bar{x}} \wedge v_{\overline{xy}} + v_{\bar{x}} \wedge v_{\overline{yz}}.$$

Using Lemma 8.2.11(ii), one can see that the last element is contained in $M_{(2)}$. Hence, $\Phi(u_1) + g(v) \in \Lambda$ and $\Phi(u_1) \in \Lambda$.

Similarly, we have

$$\Phi(v) = -x \wedge xy.z - x \wedge yz + y \wedge xy.z + y \wedge xz + xy \wedge yz-$$

$$-xy \wedge xz - 2xy \wedge z - xz \wedge xy.z + xz \wedge yz + yz \wedge xy.z-$$

$$-xy \wedge y - x \wedge xy + x \wedge y.$$

This element coincides with $g'(v)$, where $g' \in G$ is such that

$$g' : x \mapsto -x,\ y \mapsto -y,\ z \mapsto -z.$$

Now we show that $\Phi^{-1}(u_1), \Phi^{-1}(v) \in \Lambda$. Here

$$\Phi^{-1}(v) = \Phi g'(v) = h\Phi(v),$$

where $h = \Phi g' \Phi^{-1}$. Clearly, h is obtained by the construction described as applied to the element

$$\varphi.\mathrm{diag}(-1, 1, 1, -1).\varphi^{-1} \in A_0,$$

and so $h \in Q$, $h\Phi(v) \in \Lambda$.

The calculation with $\Phi^{-1}(u_1)$ is similar to that with $\Phi(u_1)$, and so is omitted.

Thus we have shown that $\Phi(\Lambda) = \Lambda$. We claim that Φ is an isometry. To this end, we remark that the subspaces that occur in decomposition (8.8) are orthogonal to each other. Since the map

$$*|_{T_x} : T_x \to V_1$$

is a similarity, the restriction $\Phi|_{T_x}$ is an isometry. The same is true for T_y and T_{xy}. Furthermore, φ is an isometry of V_1, and so the transformation induced by φ on $V_1 \wedge V_1$ is also an isometry. Finally, it is easy to show that

$$\Phi : x \wedge y \mapsto y \wedge xy,\ x \wedge xy \mapsto x \wedge y,\ y \wedge xy \mapsto x \wedge xy,$$

that is, $\Phi|_{V_0 \wedge V_0}$ is an isometry. This completes the proof of the following result.

Proposition 8.3.6. *The map Φ constructed above is an automorphism of order 3 of the lattice Λ that is not contained in G.* □

We shall need some bilinear forms. The lattice Γ_1 has the basis $d_1 = g_1 - g_2$, $d_2 = g_2 - g_3$. In it the inner product has Gram matrix $\begin{pmatrix} 2 & -1 \\ -1 & 2 \end{pmatrix}$. Introduce a bilinear form $b' : \Gamma_1 \times \Gamma_1 \to \mathbb{F}_3$ as follows:

$$b'(u, v) = (u|v) \pmod 3.$$

Then $\mathrm{Ker}(b') =< 3d_1, d_1 - d_2 >_{\mathbb{Z}} = \Gamma_2$. We introduce the bilinear form b given by

$$b : (u, v) \mapsto \frac{1}{3}(u|v) \pmod 3,$$

on Γ_2, and find that $\mathrm{Ker}(b) = 3\Gamma_1$.

Correspondingly, one can define on $\mathrm{Ind}_P^G(\Gamma_1)$ a form

$$B' : (u, v) \mapsto (u|v) \pmod 3,$$

and on $\mathrm{Ind}_P^G(\Gamma_2) = \mathrm{Ker}(B')$ a form

$$B : (u, v) \mapsto \frac{1}{3}(u|v) \pmod 3$$

with kernel $\mathrm{Ker}(B) = 3\mathrm{Ind}_P^G(\Gamma_1)$. We shall denote the induced form on the space

$$W = \mathrm{Ind}_P^G(\Gamma_2)/3\mathrm{Ind}_P^G(\Gamma_1) = \mathrm{Ker}(B')/\mathrm{Ker}(B)$$

by the same symbol B.

Consider now the restriction B'_Λ. Since the factor-group $\mathrm{Ind}_P^G(\Gamma_1)/\Lambda$ is a 2-group, the embedding $\Lambda \subset \mathrm{Ind}_P^G(\Gamma_1)$ induces a natural isomorphism

$$\Lambda/\mathrm{Ker}(B'_\Lambda) \cong \mathrm{Ind}_P^G(\Gamma_1)/\mathrm{Ker}(B').$$

Set $\Lambda' = \mathrm{Ker}(B'_\Lambda)$ and $B_\Lambda = B_{\Lambda'}$. Then $\mathrm{Ker}(B_\Lambda) = 3\Lambda$, and $W_\Lambda = \Lambda'/3\Lambda$ is isomorphic to W. In particular, $\dim_{\mathbb{F}_3} W_\Lambda = 7$.

Lemma 8.3.7. *The lattice $\Lambda_0 = \mathrm{Ind}_P^G(\Gamma_1)$ is closed under the Lie product in $\mathcal{L}$.*

Proof. For any $u, v, w, t \in V$ we have

$$[u \wedge v, w \wedge t] = (v|w)(u \wedge t) - (v|t)(u \wedge w) +$$

$$+ (u|t)(v \wedge w) - (u|w)(v \wedge t).$$

If $u, v, w, t \in \Delta_0$, then the right-hand expression belongs to $\Delta_0 \wedge \Delta_0$, that is,

$$[\Delta_0 \wedge \Delta_0, \Delta_0 \wedge \Delta_0] \subseteq \Delta_0 \wedge \Delta_0.$$

Taking the intersection of this relation with the obvious relation $[\mathcal{L}, \mathcal{L}] \subseteq \mathcal{L}$, we arrive at the required conclusion. $\square$

Note that $[\mathrm{Ker}(B'), \mathrm{Ind}_P^G(\Gamma_1)] \subseteq \mathrm{Ker}(B')$. In fact, for $u \in \mathrm{Ker}(B')$ and $v, w \in \mathrm{Ind}_P^G(\Gamma_1)$, by the basic property of the Killing form and Lemma 8.3.7 we have

$$([u, v]|w) = (u|[v, w]) \equiv 0 \pmod 3,$$

as stated. In particular,

$$[\mathrm{Ind}_P^G(\Gamma_2), \mathrm{Ind}_P^G(\Gamma_2)] \subseteq \mathrm{Ind}_P^G(\Gamma_2).$$

Furthermore, since $\mathrm{Ker}(B) = 3\mathrm{Ind}_P^G(\Gamma_1)$ we have

$$[\mathrm{Ind}_P^G(\Gamma_2), \mathrm{Ker}(B)] \subseteq 3\mathrm{Ind}_P^G(\Gamma_2) \subseteq \mathrm{Ker}(B).$$

Thus, the Lie product on $\mathrm{Ind}_P^G(\Gamma_2)$ factors through $\mathrm{Ker}(B)$, that is, one can define a map

$$[\,,\,]_W : W \wedge W \to W.$$

Similarly, one can define a map

$$[\,,\,]_\Lambda : W_\Lambda \wedge W_\Lambda \to W_\Lambda.$$

We define a trilinear form ϑ on W and W_Λ as follows:

$$\vartheta(a, b, c) = B(a, [b, c]).$$

Lemma 8.3.8. *The form ϑ is skew-symmetric.*

Proof. Take $a, b, c \in W_\Lambda$, $a = u+3\Lambda$, $b = v+3\Lambda$, $c = w+3\Lambda$. Then by definition we have $[b, c] = [v, w] + 3\Lambda$, and so

$$\vartheta(a, b, c) = B(a, [b, c]) = \frac{1}{3}(u|[v, w]) \pmod 3.$$

Similarly, $B([a, b], c) = \frac{1}{3}([u, v]|w) \pmod 3$. But $(u|[v, w]) = ([u, v]|w)$, and therefore

$$-\vartheta(a, c, b) = \vartheta(a, b, c) = B(a, [b, c]) = B([a, b], c) =$$

$$= -B(c, [b, a]) = -\vartheta(c, b, a).$$ $\square$

Next we claim that the action of $< G, \Phi >$ on W_Λ preserves ϑ. This is obvious for G. Before proving it for Φ, we formulate a simple statement.

Lemma 8.3.9. *Let Q_1 be a subgroup of Q that induces even permutations on the set $\{\bar{z}, \overline{xz}, \overline{yz}, \overline{xyz}\}$. Then Φ normalizes Q_1.* □

Proposition 8.3.10. *Φ preserves the form ϑ.*

Proof. Since $\Phi \in \mathrm{Aut}(\Lambda)$, Φ fixes B. Hence, it is sufficient to show that Φ preserves $[\,,\,]_\Lambda$, that is,

$$[\Phi(u), \Phi(v)] \equiv \Phi([u, v]) \pmod{3\Lambda}, \ \forall u, v \in \Lambda'.$$

The action of the normal subgroup E of G induces a graduation

$$W_\Lambda = \oplus_\alpha W_\alpha.$$

All the subspaces W_α have the same dimension, indeed, they are of dimension 1: $W_\alpha =< d_\alpha >_{\mathbb{F}_3}$. We have to show that

$$[\Phi(d_\alpha), \Phi(d_\beta)] = \Phi([d_\alpha, d_\beta]). \tag{8.10}$$

It is easy to see that Q_1 acts on the set of pairs $\{\alpha, \beta\}$ with $\alpha \neq \beta$; the action has three orbits, with representatives

$$\{\bar{x}, \bar{y}\}, \ \{\bar{x}, \bar{z}\}, \ \{\bar{z}, \overline{xz}\}.$$

Note that if relation (8.10) holds for the pair $\{\alpha, \beta\}$, then it holds for any pair $\{\alpha', \beta'\} = \{g\alpha, g\beta\}$ with $g \in Q_1$. Indeed, we have $g(d_\alpha) = \varepsilon d_{\alpha'}$, $g(d_\beta) = \eta d_{\beta'}$ for some $\varepsilon, \eta = \pm 1$, and so

$$[\Phi(d_{\alpha'}), \Phi(d_{\beta'})] = \varepsilon\eta[\Phi g(d_\alpha), \Phi g(d_\beta)] = \varepsilon\eta[h\Phi(d_\alpha), h\Phi(d_\beta)] =$$

(where $h = \Phi g \Phi^{-1} \in Q_1$ by Lemma 8.3.9)

$$= \varepsilon\eta h([\Phi(d_\alpha), \Phi(d_\beta)]) = \varepsilon\eta h\Phi([d_\alpha, d_\beta]) = \Phi g(\varepsilon\eta[d_\alpha, d_\beta]) =$$

$$= \Phi([g(\varepsilon d_\alpha), g(\eta d_\beta)]) = \Phi([d_{\alpha'}, d_{\beta'}]),$$

as stated. Hence, it is sufficient to verify (8.10) for the pairs

$$\{\bar{x}, \bar{y}\}, \ \{\bar{x}, \bar{z}\}, \ \{\bar{z}, \overline{xz}\}.$$

We carry out this verification for the pair $\{\bar{x}, \bar{z}\}$. Take

$$d_{\bar{x}} = 4m + 3\Lambda, \ d_{\bar{z}} = 2l + 3\Lambda,$$

where

$$m = 2y \wedge xy - z \wedge xz + yz \wedge xy.z,$$

$$l = 2(x \wedge xz + y \wedge yz - 2xy \wedge xy.z).$$

Then

$$\Phi(l) = x \wedge xz - x \wedge z - x \wedge xy.z + x \wedge yz + y \wedge yz +$$

$$+ y \wedge xy.z - y \wedge z - y \wedge xz - 2xy \wedge xy.z + 2xy \wedge yz -$$

$$-2xy \wedge xz + 2xy \wedge z;$$

$$\Phi(m) = z \wedge yz + xz \wedge xy.z + 2x \wedge xy;$$

$$[\Phi(m), \Phi(l)] = y \wedge z + y \wedge xz + y \wedge yz + y \wedge xy.z +$$

$$+4xy \wedge z - 4xy \wedge xz - 4xy \wedge yz + 4xy \wedge xy.z +$$

$$+5x \wedge z - 5x \wedge xz + 5x \wedge yz - 5x \wedge xy.z;$$

$$[m, l] = 2(-x \wedge z - 5y \wedge xy.z - 4xy \wedge yz);$$

$$\Phi([m, l]) = -5y \wedge z - 5y \wedge xz - 5y \wedge yz - 5y \wedge xy.z +$$

$$+4xy \wedge z - 4xy \wedge xz - 4xy \wedge yz + 4xy \wedge xy.z -$$

$$-x \wedge z + x \wedge xz - x \wedge yz + x \wedge xy.z;$$

and so

$$[\Phi(m), \Phi(l)] - \Phi([m, l]) = 6(y \wedge z + y \wedge xz + y \wedge yz +$$

$$+y \wedge xy.z + x \wedge z - x \wedge xz + x \wedge yz - x \wedge xy.z) \in 3\Lambda.$$ □

We need the following statement, which is proved in [Asch 3].

Proposition 8.3.11. (i) *Let V be a 7-dimensional space over a field $\mathbb{F}$. There exist a skew-symmetric trilinear form*

$$\varsigma : \wedge^3(V) \to \mathbb{F}$$

(the Dickson form) and a quadratic form

$$F : V \to \mathbb{F}$$

with the following properties:

(a) *$O(V, \varsigma) = H\times <\upsilon E_7>$, where $H \cong G_2(\mathbb{F})$, E_7 denotes the identity operator on V, υ is a cube root of* 1 *in $\mathbb{F}$, and $\upsilon = 1$ if such a root does not exist.*

(b) *$O(V, \varsigma, F) = H$.*

(ii) *Suppose that $\mathbb{F}$ is either finite or algebraically closed, that K is an irreducible subgroup of $GL(V)$, and, finally, that ς_1 is a nonzero K-invariant skew-symmetric trilinear form on V. Then ς_1 is proportional to ς.*

(iii) *Suppose that $\mathbb{F}$ is as in* (ii) *and that $p = \mathrm{char}\mathbb{F} \neq 2$. Then H contains a subgroup G isomorphic to $\mathbb{Z}_2^3 \circ SL_3(2)$ that is unique up to conjugacy. If $\mathbb{F} = \mathbb{F}_p$, then G is a maximal subgroup of H.* □

Let $\overline{< G, \Phi >}$ denote the image of $< G, \Phi >$ in $GL(W_\Lambda)$.

Proposition 8.3.12. $\overline{< G, \Phi >} \cong G_2(3)$.

Proof. We have shown (Proposition 8.3.10) that $\overline{< G, \Phi >}$ preserves ϑ. Since G acts irreducibly on W_Λ, by Proposition 8.3.11(ii) we know that ϑ is proportional to the Dickson form ς. Furthermore, $\mathbb{F}_3$ does not contain a cube root of 1, therefore $O(W_\Lambda, \vartheta) \cong G_2(3)$ by Proposition 8.3.11(i). Finally, by Proposition 8.3.11(iii), G is maximal in $O(W_\Lambda, \vartheta)$. Hence,

$$\overline{< G, \Phi >} = O(W_\Lambda, \vartheta) \cong G_2(3).$$ □

Completion of the Proof of Theorem 8.3.2. At the beginning of the section we mentioned that $\mathbb{G} = \mathrm{Aut}(\Lambda)$ contains a subgroup H isomorphic to $G_2(3)$. In particular, $\mathbb{G}$ acts primitively on $\mathcal{L}$. Using this fact and following the arguments of §12.4, one can show that $Z = Z(\mathbb{G}) =< \mathbf{t} >\cong \mathbb{Z}_2$ is the maximal normal soluble subgroup of $\mathbb{G}$. Furthermore, using the obvious relation

$$14 = 2 \cdot 7$$

and the following observation:

Suppose that M is a finite simple group such that $G_2(3) \hookrightarrow \mathrm{Aut}(M)$. Then M cannot be embedded in $PGL_2(\mathbb{C})$,

one can show that $\mathbb{G}/Z$ is an almost simple group, that is,

$$M \lhd \mathbb{G}/Z \subseteq \mathrm{Aut}(M)$$

for some finite nonabelian simple group M. Consider the natural homomorphism

$$\tau : \mathbb{G} \to O(W_\Lambda, B_\Lambda) \cong O_7(3),$$

and note that $\tau|_Z$ is faithful. By Proposition 8.3.12, we have

$$\mathbb{Z}_2 \times G_2(3) \subseteq \mathrm{Im}\tau.$$

Using the description of maximal subgroups of

$$O_7(3) = (\Omega_7(3).\mathbb{Z}_2) \times \mathbb{Z}_2$$

given in [ATLAS], we come to one of the following possibilities:

$$\mathbb{Z}_2 \times \Omega_7(3) \subseteq \mathrm{Im}\tau \subseteq O_7(3), \tag{8.11}$$

$$\mathrm{Im}\tau = \mathbb{Z}_2 \times G_2(3). \tag{8.12}$$

Now we examine the kernel $K = \mathrm{Ker}\tau$ and denote by $\hat{M}$ the full preimage of M in $\mathbb{G}$: $\hat{M} = Z.M$. Suppose that

$$K' = K \cap \hat{M} \neq 1.$$

Then, since $K \cap Z = 1$ and M is simple, it follows that $K' \cong M$, and $\hat{M} = Z \times K'$. We identify M with K'. Note that the subgroup K possesses a natural homomorphism

$$\tau' : K \to O(3\Lambda/3\Lambda') \cong O_7(3)$$

with soluble kernel. Again by the simplicity of M, the restriction $\tau'|_M$ is injective, and this means that $M \hookrightarrow O_7(3)$. In particular, the outer automorphism group $\mathrm{Out}(M) = \mathrm{Aut}(M)/M$ is soluble. Hence, M is the unique nonabelian composition factor of $\mathbb{G}$ (and has multiplicity 1). On the other hand, the group $\mathrm{Im}\tau$ is insoluble (see (8.11) and (8.12)), which is a contradiction.

Thus, we can suppose that

$$\hat{M} \hookrightarrow \mathrm{Im}\tau, \ |K| \leq |\mathrm{Out}(M)|.$$

By (8.11) and (8.12), we have $M = G_2(3)$ or $M = \Omega_7(3)$. In particular, $\mathrm{Out}(M) = \mathbb{Z}_2$. Thus, K is a normal subgroup of order at most 2 in $\mathbb{G}$. In other words, $K \subseteq Z(\mathbb{G}) = Z$. But $K \cap Z = 1$, so in fact we have $K = 1$ and $\mathbb{G} = \mathrm{Im}\tau$.

Finally, the simple group $\Omega_7(3)$ cannot be embedded in $GL_{14}(\mathbb{C})$. We arrive at the conclusion that case (8.12) occurs. Furthermore, $\overline{< G, \Phi >} = < G, \Phi >$ fixes ϑ, and $\mathbf{t}$ does not fix ϑ. Hence, $\mathbb{G} = Z \times < G, \Phi >$, and this completes the proof of Theorem 8.3.2. □

Theorem 8.0.1 has now been completely proved. □

Commentary

1) The loop approach to $G_2(3)$ that allows us to construct the required additional automorphism Φ by hand was suggested quite recently by Burichenko. Our arguments depend neither on computer calculations nor on the classification of finite simple groups, and the price we pay is that the arguments are not elementary.

The original proof of Theorem 8.0.1 given in [KKU 5] is more intuitive and quite elementary, but it uses a computer! Of course, the difference between the two approachs concerns only the deep lattice $\Lambda = \Lambda_5$. Let us reproduce the main highlights of [KKU 5].

First, by using the Chevalley basis, an MOD is constructed, and generators of the group $G = \mathrm{Aut}_{MOD}(\mathcal{L}) = \mathbb{Z}_2^3 \circ SL_3(2)$ are calculated. In particular, one finds an element $A \in G$ such that $A^4 \in E = O_2(G)$ but $(XA)^4 \neq 1$ for all $X \in E$. (We are using the notation of [KKU 5]). This means that G is not split over E. This explicit realization of G supplements the information about this group contained in [Ale 1], [CoW] and [San]. Incidentally, the loop L related to G, as well as other loops, is studied in [Gri 5]. In this connection, see also [Gri 4, 6, 7].

Next, the classification of invariant lattices is given. In particular, a deep lattice Λ is constructed. The set $\mathcal{M}$ consisting of 756 minimal vectors of Λ is found; and a basis for Λ consisting of 14 vectors from $\mathcal{M}$ is determined.

Furthermore, a computer is used to find an additional automorphism D of order 13 lying in the stabilizer of a minimal vector. It then becomes possible to show that $\mathbb{G}$ acts transitively on $\mathcal{M}$ with point stabilizer isomorphic to $\mathrm{Aut}(SL_3(3))$. The group $\mathbb{G}$ also acts transitively and faithfully on the set of 52×756 ordered pairs (u, v), where $u, v \in \mathcal{M}$ and $(u|v) = 8 = \frac{1}{2}(u|u)$. The stabilizer of such a pair has order 216. Hence, $|\mathbb{G}| = 2^7.3^6.7.13 = 2|G_2(3)|$. The centre $Z = <\mathbf{t}> = \mathbb{Z}_2$ is a direct summand of $\mathbb{G}$, since a Sylow 2-subgroup of $\mathbb{G}$ can be written in the form $Z \times G_2$, where G_2 is a Sylow 2-subgroup of G.

The identification of a (unique) subgroup Γ of $\mathbb{G}$ of order $2^6.3^6.7.13$ with $G_2(3)$ can be achieved in two ways. First, the transitive permutation representations mentioned above and the explicit description of the corresponding point stabilizers enable us to establish the simplicity of Γ. Hence, by the equality of the orders $|\Gamma|$ and $|G_2(3)|$, it follows that $\Gamma \cong G_2(3)$.

The other method of finding the isomorphism $\Gamma \cong G_2(3)$ is based on the description of the centralizer of an involution in Γ and the result of [Jan].

The details of computer calculations can be found in [Ufn 1].

2) As mentioned, one of our main goals is to compute the automorphism groups of invariant lattices Λ. Usually, the automorphism group $G = \mathrm{Aut}(\mathcal{D})$ of a given orthogonal decomposition $\mathcal{D}$ acts on the set $\mathcal{M}$ of minimal vectors of Λ intransitively, splitting $\mathcal{M}$ into several orbits $\mathcal{M}_1, \ldots, \mathcal{M}_k$. If it is possible to glue these orbits together by means of some additional automorphism $\varphi \in \mathbb{G} = \mathrm{Aut}(\Lambda)$, then $\mathbb{G}$ is usually rich in structure and close to being simple. This is demonstrated by

the example of invariant lattices of type G_2 (in this chapter), and E_8 (see [Tho 3] and [Smi 1, 2]). However, is there such a gluing automorphism φ? A priori this is not clear, and sometimes the determination of such an automorphism makes heavy use of computers, as happened in [Smi 1] or [KKU 5]. The same may also be the case in the construction of "exceptional" lattices (in our case, deep lattices) as part of the classification of invariant lattices.

3) The results of this chapter stated in Theorem 8.0.1 led us to the alternative $(\mathcal{A})$ mentioned in the preamble to Part II. In our case, the role of the "exceptional" finite simple group L is played by the simple Dickson-Chevalley group $G_2(3)$. The remaining Chapters 9-13 are devoted, in particular, to establishing this alternative for the cases A_{p^n-1}, B_{2^n-1}, D_{2^n}, F_4, E_6 and E_8.

Chapter 9

Invariant lattices, the Leech lattice and even unimodular analogues of it in Lie algebras of type A_{p-1}

Although the remarkable Leech lattice has been realized in several ways by now (see [CoS 7], for example), it appears to us that it deserves at least one new construction. This is all the more so, as this chapter deals with a whole series of Euclidean even unimodular integral lattices of dimension $p^2 - 1$ (p runs through all odd primes; to each p there corresponds a finite family of even unimodular lattices). For $p = 5$, one of these lattices is isometric to the Leech lattice Λ_{24} — a fact reflected in the title of the chapter, which refers to the Lie algebras $\mathcal{L}$ of type A_{p-1} ($p = 3, 5, \ldots$) of complex $p \times p$ matrices with zero trace. This is not fortuitous, and cannot be explained merely by the arithmetic coincidences $\dim \Lambda_{24} = 24 = \dim \mathcal{L}$ and $8|(p^2 - 1)$. In fact, the structure of the Lie algebra on the space $\mathcal{L}$ is important at the initial stage, in forming the array of those integral lattices that we propose to study. There are too many different Euclidean even unimodular lattices of dimension $p^2 - 1$ with $p > 5$ (see [Ser 2]) for us to list them.

We follow the programme sketched in Chapter 8. The following are our main results:

1) *For every prime $p > 2$, we have classified (up to similarity) all invariant integral lattices that correspond to the standard orthogonal decomposition of the Lie algebra $\mathcal{L}$.*

2) *We have studied duality in the class of invariant lattices, and among the even unimodular lattices we have distinguished the lattices that do not contain roots.*

3) *We have obtained lower bounds for the minimal length of vectors of root-free lattices and have found certain automorphism groups.* A complete description of them will be given in Chapter 10.

We have not attempted to carry out here the beautiful calculations of Conway [Con 1], [Con 2]. Note that our arguments are absolute in the sense that the standard OD used here is the unique IOD of $\mathcal{L}$ (see Chapter 5).

To avoid confusion, let us fix some notation. Everywhere $< *| ** >_{\mathbb{Z}}$ is the $\mathbb{Z}$-linear span of the vectors $*$ subject to the condition $**$; $< *; * >$ is a skew-symmetric form; $(*|*)$ and $(*; *)$ are different inner products.

9.1 Preliminary results

Let us recall (see Chapter 1) how to construct an irreducible orthogonal decomposition (IOD) of the Lie algebra $\mathcal{L}$ of type A_{p-1} of all $p \times p$ matrices with zero trace. Throughout the chapter p is an odd prime. Set

$$D = \begin{pmatrix} 1 & & & \\ & \varepsilon & & \\ & & \ddots & \\ & & & \varepsilon^{p-1} \end{pmatrix}, \quad P = \begin{pmatrix} 0 & \dots & 0 & 1 \\ 1 & \dots & 0 & 0 \\ & \vdots & & \\ 0 & \dots & 1 & 0 \end{pmatrix},$$

and $J_u = D^a P^b$, where $u = (a, b) \in \mathbb{F}_p^2 \backslash (0, 0)$. As a rule, we denote elements of the prime field $\mathbb{F}_p$ by lower-case Latin letters; we identify the exponents a, b of powers X^a of a non-identity matrix $X \in M_p(\mathbb{C})$ such that $X^p = E_p$, and powers ε^b of a pth root $\varepsilon \neq 1$ of 1, with representatives $0, 1, \dots, p-1$ of the residue classes mod p. For convenience we put $V = \mathbb{F}_p^2$ and understand by $\mathbb{P}V$ the projective line with points $\sigma, \sigma_0, \dots$, which we identify with vector lines without the origin. The notation $(a, b) \in \sigma$ means that σ is determined by the vector $(a, b) \neq (0, 0)$.

The standard OD of $\mathcal{L}$ is of the form

$$\mathcal{D} : \mathcal{L} = \bigoplus_{\sigma \in \mathbb{P}V} \mathcal{H}_\sigma,$$

where $\mathcal{H}_\sigma =< J_{(a,b)} | (a, b) \in \sigma >_{\mathbb{C}}$. It was shown in Chapter 1 that the group of inner automorphisms of the given OD (that is, the automorphisms induced by maps $X \mapsto A^{-1} X A$ of $\mathcal{L}$) takes the form

$$G = \mathrm{Inn}(\mathcal{D}) = \{\hat{u}\hat{\varphi} | u \in V, \varphi \in SL(V)\},$$

where $\hat{u}(J_v) = \varepsilon^{u \circ v} J_v$, $u, v \in V$, $u \circ v = ac + bd$ for $u = (a, b)$ and $v = (c, d)$, and if $B(u, v) = -bc$, then

$$\hat{\varphi}(J_v) = \varepsilon^{\{B(\varphi(v), \varphi(v)) - B(v, v)\}/2} J_{\varphi(v)}.$$

We observe that G acts doubly transitively on the set of Cartan subalgebras that occur in the OD.

In future we fix the following notation: $(1, 0) \in \sigma_0$, $\mathcal{H}_0 = \mathcal{H}_{\sigma_0}$, and $G_0 = St_G(\mathcal{H}_0)$. More generally, the subscript σ_0 is replaced by 0 everywhere.

In every Cartan subalgebra $\mathcal{H}_\sigma$ we introduce the finitely generated $\mathbb{Z}$-module Λ_σ spanned by the basis dual to the system of simple roots. It is convenient to use the basis obtained from the one already mentioned by a triangular unimodular transformation. More precisely,

$$\Lambda_0 = < \frac{1}{p}\sum_{a=1}^{p-1} \varepsilon^{ka} J_{(a,0)} | 0 \leq k \leq p-2 >_{\mathbb{Z}}, \ \Lambda_\sigma = \hat{\varphi}(\Lambda_0),$$

where $\varphi(\sigma_0) = \sigma$. Our original integral lattice is the pair $(\Lambda, (X|Y))$, where

$$\Lambda = \bigoplus_{\sigma \in \mathbb{P}V} \Lambda_\sigma$$

and $(X|Y) = \frac{1}{2}K(X, Y) = p\mathrm{Tr}XY$.

It is easy to see that

$$(\frac{1}{p}\sum_a \varepsilon^{ka} J_{(a,0)} | \frac{1}{p}\sum_a \varepsilon^{la} J_{(a,0)}) = p\delta_{k,l} - 1,$$

that is, the form $(X|Y)$ is positive definite on Λ. Since

$$G_0|_{\mathcal{H}_0} = \{g_{s,t} : J_{(a,0)} \mapsto \varepsilon^{sa} J_{(ta,0)} | 0 \leq s \leq p-1, 1 \leq t \leq p-1\}, \tag{9.1}$$

the sublattice Λ_0 is invariant under G_0. Consequently, Λ is a G-invariant lattice; in other words, G is a subgroup of the group of isometries of Λ.

We say that a sublattice $\Lambda' \subseteq \Lambda$ is *indivisible* if there is no $n \in \mathbb{Z}$, $n > 1$, such that $\Lambda' \subseteq n\Lambda$. The first of our problems is to describe (up to similarity) all G-invariant sublattices of Λ.

We call the reader's attention to the following two circumstances. Firstly, when $p > 11$, G is the unique minimal irreducible subgroup of $\mathrm{Aut}(\mathcal{D})$. Secondly, when $p > 3$, G has just one real absolutely irreducible representation of degree $p^2 - 1$. Therefore, by the Deuring-Noether Theorem [CuR], every *invariant lattice of type* A_{p-1} can be isometrically embedded in the lattice Λ.

At this point we are interested in projections of Λ' onto the subalgebras $\mathcal{H}_\sigma$ that occur in the OD; however, since $\mathcal{H}_\sigma$ is the image of $\mathcal{H}_0$ under an isometry, the properties of the projections are "identical", and it is sufficient to restrict attention to $\mathcal{H}_0$.

Lemma 9.1.1. *Let Λ' be an indivisible G-invariant sublattice of Λ. Then its projection $\mathrm{pr}_0\Lambda'$ onto $\mathcal{H}_0$ is also an indivisible G_0-invariant lattice.*

Proof. The assertion follows immediately from the relation $\mathrm{pr}_0 . g = g . \mathrm{pr}_0$, $g \in G_0$ and the remark that $\mathrm{pr}_\sigma\Lambda' \subseteq n\Lambda_\sigma$ if $\mathrm{pr}_0\Lambda' \subseteq n\Lambda_0$, so that $\Lambda' \subseteq n\Lambda$. □

Lemma 9.1.2. *Let Λ' be an indivisible G_0-invariant sublattice. Then $\Lambda' \supseteq p(\mathrm{pr}_0\Lambda')$.*

Proof. Set $f_k = (0, k) \in V$, $0 \le k \le p-1$, so that $f_k \circ (a, b) = kb$. Then $\hat{f}_k \in G_0$ and

$$\sum_{k=0}^{p-1} \hat{f}_k(J_{(a,b)}) = \sum_{k=0}^{p-1} \varepsilon^{kb} J_{(a,b)} = \begin{cases} pJ_{(a,b)} & \text{if} \quad b = 0; \\ 0 & \text{if} \quad b \neq 0. \end{cases}$$

Therefore, for elements $X = \sum_{\sigma \in \mathbb{P}V} X_\sigma \in \Lambda'$ we have

$$p(\mathrm{pr}_0 X) = pX_0 = \sum_{k,\sigma} \hat{f}_k X_\sigma \in \Lambda'.$$

□

Theorem 9.1.3. *The lattice Λ_0 contains exactly $p-1$ indivisible G_0-invariant sublattices.*

Proof. 1) Recall that

$$\Lambda_0 = \left\{ \mathrm{diag}(n_0 - \frac{n}{p}, \ldots, n_{p-1} - \frac{n}{p}) | n_i \in \mathbb{Z}, n = \sum_{i=0}^{p-1} n_i \right\}.$$

We identify Λ_0 with the $\mathbb{Z}$-module

$$\mathbb{Z} < x >= \mathbb{Z}[x]/(\Phi(x)),$$

where $\Phi(x) = x^{p-1} + \cdots + x^2 + x + 1$ is a cyclotomic polynomial. Each element of $\mathbb{Z} < x >$ can be written as a polynomial $f(x) = \sum_{k=0}^{p-1} a_k x^k$, where $pa_k \in \mathbb{Z}$ and $\sum_{k=0}^{p-1} a_k = 0$. The required isomorphism between $\mathbb{Z} < x >$ and Λ_0 is given by the map

$$\Psi : f(x^{-1}) \mapsto \mathrm{diag}(a_0, a_1, \ldots, a_{p-1}); \tag{9.2}$$

that is,

$$\Psi(\varepsilon^{-na} x^a) = \frac{1}{p} \sum_{k=1}^{p-1} \varepsilon^{n(k-a)} J_{(k,0)}, \quad \Psi(\Phi(\varepsilon^{-a} x)) = J_{(a,0)}.$$

We observe that $x^{-1} = x^{p-1}$. Precisely such an isomorphism will be used in what follows. For reasons that we reveal below, we also write $\mathbb{Z} < x >$ instead of the more natural $\mathbb{Z}[\varepsilon]$, the ring of integral elements of the cyclotomic field $\mathbb{Q}(\varepsilon)$.

We endow $\mathbb{Z} < x >$ with the structure of a G_0-module, using the explicit form (9.1) for elements of G_0. Under the isomorphism Ψ, the polynomial x^n, $0 \le n \le p-1$ corresponds to the matrix

$$\frac{1}{p}\mathrm{diag}(-1, \ldots, -1, p-1, -1, \ldots, -1) = \frac{1}{p} \sum_{a=1}^{p-1} \varepsilon^{-(p-n)a} J_{(a,0)}.$$

Knowing that the actions of Ψ and $g_{s,t}$ commute, we introduce on $\mathbb{Z} < x >$ the structure of a G_0-module necessary to ensure that $g_{s,t}(x^n) = x^{(s+n)t^{-1}}$. Clearly, $\mathbb{Z} < x >$ can now be regarded as a ring. Consequently, we identify every G_0-invariant sublattice of Λ_0 with an ideal of $\mathbb{Z} < x >$ invariant under the Galois group of the cyclotomic extension $\mathbb{Q}(x) = \mathbb{Q}[x]/(\Phi(x))$.

2) The enumeration of the $Gal(\mathbb{Q}(x))$-invariant (simply, invariant) ideals of $\mathbb{Z} < x >$ is derived from the well-known theorem on ramified ideals (see [Weil]). We now see that *the indivisible invariant ideals in $\mathbb{Z} < x >$ are just the principal ideals generated by elements of the form* $(1-x)^k$, $0 \leq k \leq p-2$.

As we know, $\mathbb{Z} < x >$ is the integral closure of $\mathbb{Z}$ in the extension $\mathbb{Q} \subset \mathbb{Q}(x)$, and is therefore a Dedekind ring. In particular, we have uniqueness of decomposition into prime divisors for ideals in $\mathbb{Z} < x >$. Let $\mathcal{P}$ be a prime element of the semigroup of invariant ideals, and decompose it into prime divisors in the semigroup of all ideals: $\mathcal{P} = \mathcal{P}_1 \cdots \mathcal{P}_k$. Then $\mathcal{P} = \mathcal{P}^\gamma = \mathcal{P}_1^\gamma \cdots \mathcal{P}_k^\gamma$, where γ is any Galois automorphism, and from the uniqueness of the decomposition it follows that the set $\{\mathcal{P}_i\}$ splits into orbits under the action of the Galois group of the extension. The product of the ideals $\mathcal{P}_i$ from a single orbit is obviously also an invariant ideal, and $\mathcal{P}$ is the product of ideals of this form. Since $\mathcal{P}$ is a prime element among the invariant ideals, the number of orbits is equal to 1 and $\mathcal{P} = \mathcal{P}_1^{\gamma_1} \cdots \mathcal{P}_1^{\gamma_k}$, where $\gamma_1, \ldots, \gamma_k$ are Galois automorphisms and $\mathcal{P}_1$ is a prime ideal. Moreover, we have $N(\mathcal{P}) = N(\mathcal{P}_1)^k$ (N stands for the norm of an ideal), and since the norm of the prime ideal $\mathcal{P}_1$ is the principal ideal generated by a power of some prime p_1, we have $N(\mathcal{P}) = p_1^m$. But $\mathcal{P}$ is an invariant ideal, so $N(\mathcal{P}) = \mathcal{P}^{p-1}$. From the equality $p_1^m = \mathcal{P}^{p-1}$ and the uniqueness of the decomposition, it follows that $\gcd((p_1), \mathcal{P}) \neq (1)$. But $\gcd((p_1), \mathcal{P})$ is an invariant ideal, since $\mathcal{P}$ and (p_1) are invariant. Consequently, $\gcd((p_1), \mathcal{P}) = \mathcal{P}$ and $(p_1) = \mathcal{P}^l$ for some $l \neq 1$ (we are only interested in indivisible ideals $\mathcal{P}$). Since the ramification index of the ideal (p_1) is greater than 1, the theorem mentioned above asserts that $p_1 = p$. The decomposition of p into prime divisors in $\mathbb{Z} < x >$ is also known: $p = (1-x)^{p-1}$, and the ideal generated by $(1-x)^k$ is invariant under the action of the Galois group. From this we get assertion 2), which completes the proof of the theorem. $\square$

Before going over to invariant lattices, we consider the algebraic variety

$$M = \left\{(\varepsilon^a, \varepsilon^b) | (a, b) \in V \backslash (0, 0)\}\right\}.$$

The ring of functions on M has the form $\mathcal{A} = \mathbb{C}[x, y]/I$, where I is the ideal generated by the elements $x^p - 1$, $y^p - 1$, and $\Phi(x)\Phi(y)$.

We now set up a linear correspondence χ between the spaces $\mathcal{L}$ and $\mathcal{A}$, using the basis $(J_{(a,b)})$:

$$\chi(J_{(a,b)}) = f_{(a,b)}, \tag{9.3}$$

where

$$f_{(a,b)}(\varepsilon^c, \varepsilon^d) = \begin{cases} p^2 \varepsilon^{-B((a,b),(a,b))/2} & \text{if} \quad (c, d) = (a, b), \\ 0 & \text{if} \quad (c, d) \neq (a, b). \end{cases}$$

We recall that the bilinear form B on V was introduced in the beginning of the section.

We endow the ring $\mathcal{A}$ with the structure of a G-module, putting for any element $F(x, y) \in \mathcal{A}$:

a) $\hat{u}(F(x, y)) = x^a y^b F(x, y)$, where $u = (a, b) \in V$;

b) $\hat{\varphi}(F(x, y)) = F(x^a y^b, x^c y^d)$, where $\begin{pmatrix} a & b \\ c & d \end{pmatrix}$ is the matrix of the element $\varphi^{-1} \in SL_2(p)$. The compatibility is almost obvious:

$$\hat{u}(\chi(J_v)) = \hat{u}(f_v) = \varepsilon^{u \circ v} f_v = \chi(\varepsilon^{u \circ v} J_v) = \chi(\hat{u}(J_v));$$

$$\hat{\varphi}(\chi(J_v)) = \hat{\varphi}(f_v) = \varepsilon^{\{B(\varphi(v),\varphi(v))-B(v,v)\}/2} f_{\varphi(v)} =$$
$$= \chi(\varepsilon^{\{B(\varphi(v),\varphi(v))-B(v,v)\}/2} J_{\varphi(v)}) = \chi(\hat{\varphi}(J_v)).$$

Lemma 9.1.4. *The image of Λ_0 under the map χ is $\mathbb{Z} < x > \Phi(y)$.*

Proof. We first verify that $f_{(a,0)} = \Psi^{-1}(J_{(a,0)})\Phi(y)$ (for the definition of Ψ see (9.2)). In fact,

$$f_{(a,0)} = \Phi(\varepsilon^{-a} x)\Phi(y) = \{1 + \varepsilon^{-a} x + \varepsilon^{-2a} x^2 + \cdots + \varepsilon^{-(p-1)a} x^{p-1}\}\Phi(y) =$$
$$= \Psi^{-1}(J_{(a,0)})\Phi(y).$$

Since $\mathcal{H}_0 =< J_{(a,0)} | 1 \le a \le p-1 >$, for every $X \in \mathcal{H}_0$ we have $\chi(X) = \Psi^{-1}(X)\Phi(y)$, that is, $\chi(\Lambda_0) = \Psi^{-1}(\Lambda_0)\Phi(y) = \mathbb{Z} < x > \Phi(y)$. □

Proposition 9.1.5. *The following assertions hold.*

a) *For any point $\sigma \in \mathbb{P}V$ we have*

$$\chi(\Lambda_\sigma) = \mathbb{Z} < x^a y^b > \Phi(x^c y^d),$$

where $(d, -c) \in \sigma$, $(b, -a) \notin \sigma$.

b) *The value of $\Phi(x^c y^d)$ does not depend on the choice of $(d, -c)$ in σ, and the value of $x^a y^b \Phi(x^c y^d)$ does not depend on the choice of representative $(b, -a)$ from V/σ.*

c) *The image of any of the $p-1$ indivisible G_σ-invariant sublattices of Λ_σ has the form*

$$\Gamma_\sigma^k = (1 - x^{a_1} y^{b_1}) \cdots (1 - x^{a_k} y^{b_k}) \Phi(x^c y^d) \mathbb{Z} < x, y >,$$

where $G_\sigma = St_G(\mathcal{H}_\sigma)$, $\mathbb{Z} < x, y >= \mathbb{Z}[x, y]/(\Phi(x)\Phi(y), x^p - 1, y^p - 1)$, $(d, -c) \in \sigma$, $(b_i, -a_i) \notin \sigma$, and $0 \le k \le p-2$. The index $[\Gamma_\sigma^k : \Gamma_\sigma^{k+1}]$ is equal to p.

Proof. a) In fact,

$$\chi(\Lambda_\sigma) = \chi(\hat{\varphi}(\Lambda_0)) = \hat{\varphi}(\chi(\Lambda_0)),$$

where φ is an element of $SL_2(p)$ for which $\varphi(\sigma_0) = \sigma$.

b) The assertion follows from the obvious equality $\Phi(z) = \Phi(z^k) = z^l\Phi(z)$ valid for any z with $z^p = 1$, $k, l \in \mathbb{Z}$, p not dividing k.

c) In view of a), it is sufficient to consider the case $\sigma = \sigma_0$; that is, to prove that

$$\Gamma_0^k = \{\prod_{i=1}^{k}(1 - x^{a_i}y^{b_i})\}\Phi(x^c y^d)\mathbb{Z} < x, y >,$$

where $(d, -c) \in \sigma_0 =< (1, 0) >_{\mathbb{F}_p}$ and $(b_i, -a_i) \notin \sigma_0$. By b), we may assume that $(c, d) = (0, 1)$ and $(a_i, b_i) = (a_i, 0)$, $a_i \neq 0$. Hence we need to check that

$$\Gamma_0^k = \{\prod_{i=1}^{k}(1 - x^{a_i})\}\Phi(y)\mathbb{Z} < x, y > .$$

For $k = 0$ this follows from assertion a) and the obvious equality

$$\Phi(y)\mathbb{Z} < x, y >= \Phi(y)\mathbb{Z} < x > .$$

From the fact that the elements $1 - x^a$ and $1 - x$ are associated in $\mathbb{Z} < x >$ (see [Weil]) and also from the proof of Theorem 9.1.3 (see assertion 2)), we obtain the form stated for Γ_0^k with $k > 0$. Observe that the limitation $k \leq p - 2$ follows from the fact that $(1 - x)^{p-1}$ and p are associated (for $k \geq p - 1$ we obtain divisible lattices). This remark gives the assertion about the index:

$$[\Gamma_0^k : \Gamma_0^{k+1}] = [(1 - x)^k\Phi(y)\mathbb{Z} < x >: (1 - x)^{k+1}\Phi(y)\mathbb{Z} < x >] = p. \qquad \square$$

Remark. Proposition 9.1.5 enables us to go over from the lattice Λ to its image $\Gamma = \Gamma^0 = \chi(\Lambda)$. As we now know, to each indivisible G-invariant sublattice of Λ there corresponds an indivisible ideal in Γ that is invariant under $SL_2(p)$ in its representation

$$SL_2(p) \cong \{\varphi | \varphi^{-1} : x \mapsto x^a y^b, y \mapsto x^c y^d; ad - bc = 1\}.$$

We also note that *every indivisible $SL_2(p)$-invariant ideal $\Gamma' \lhd \Gamma$ contains $p^2\Gamma$.* In fact, $\Gamma' = \chi(\Lambda') \supset p\chi(\mathrm{pr}_\sigma\Lambda')$ by Lemma 9.1.2; $p\chi(\mathrm{pr}_\sigma\Lambda') = p\Gamma_\sigma^k$ for some k, $0 \leq k \leq p - 2$ by Proposition 9.1.5. Finally, the arguments used at the end of the proof of Proposition 9.1.5 give $p\Gamma_\sigma^k \supseteq p\Gamma_\sigma^{p-1} = p^2\Gamma_\sigma^0$. Combining the inclusions for different σ, we obtain that $\Gamma' \supset p^2\Gamma$.

Corollary 9.1.6. *The lattice $\Gamma^k = \bigoplus_{\sigma\in\mathbb{P}V}\Gamma_\sigma^k$, $0 \leq k \leq p - 2$, is indivisible and G-invariant, and it contains $p\Gamma$.*

Proof. The fact that Γ^k is indivisible and that $\Gamma^k \supset p\Gamma$ are obvious. Moreover, every element $\varphi \in SL_2(p)$ takes Γ^k_σ to $\Gamma^k_{\varphi(\sigma)}$ (direct verification using Proposition 9.1.5), and so $\varphi(\Gamma^k) = \Gamma^k$. □

In what follows, we shall also need the lattices $\Gamma^k = \bigoplus_{\sigma \in \mathbb{P}V} \Gamma^k_\sigma$ for $k > p-2$, which are not indivisible: $\Gamma^k \subseteq p\Gamma$.

We remark at this point that the factor-group $\Gamma/p\Gamma$ is isomorphic to the (p^2-1)-dimensional vector space over $\mathbb{F}_p$, which is an $SL_2(p)$-module.

Theorem 9.1.7. *The representation of $SL_2(p)$ on $\Gamma/p\Gamma$ is isomorphic to its natural representation in the space*

$$\mathcal{F}(V) = \{f : V \to \mathbb{F}_p | f((0,0)) = 0\}$$

of functions.

We preface the proof of the theorem with some assertions, one of which has independent significance.

Lemma 9.1.8. *The following inclusion holds:*

$$(1 - x^a y^b)\Gamma^k \subseteq \Gamma^{k+1} \textit{ for } (a,b) \neq (0,0).$$

Proof. It is sufficient to verify that $(1 - x^a y^b)\Gamma^k_0 \subseteq \Gamma^{k+1}_0$. More precisely,

$$(1 - x^a y^b)\Gamma^k_0 = \begin{cases} \Gamma^{k+1}_0 & \text{for } a \neq 0, \\ 0 & \text{for } a = 0. \end{cases}$$

In fact, by Proposition 9.1.5 we have

$$\Gamma^k_0 = \{\prod_{i=1}^{k} (1 - x^{a_i} y^{b_i})\} \Phi(y) \mathbb{Z} < x, y >,$$

where all the a_i are nonzero. If $a \neq 0$, then multiplication by $1 - x^a y^b$ gives Γ^{k+1}_0. If $a = 0$, we use the relation $(1 - y^b)\Phi(y) = 0$. □

We now introduce the *truncated logarithmic series*

$$L(x) = (x-1) - \frac{(x-1)^2}{2} + \cdots + (-1)^{p-2} \frac{(x-1)^{p-1}}{p-1}. \tag{9.4}$$

Its coefficients are interpreted as elements of the field $\mathbb{F}_p$. We note that, as an element of $\mathbb{Z} < x >$, $L(x)$ is divisible by $x - 1$ but not by $(x-1)^2$. The decisive role of $L(x)$ in the study of the structure of the $SL_2(p)$-module $\Gamma^0/p\Gamma^0$ is explained by the following property:

Proposition 9.1.9. *The following equality holds in Γ^0/Γ^p, and hence in $\Gamma^0/p\Gamma^0$:*

$$L(x^a y^b)\Phi(x^c y^d) = \{aL(x) + bL(y)\}\Phi(x^c y^d).$$

Proof. We observe that if $\log(1+x) = \sum_{k=1}^{\infty}(-1)^{k-1}x^k/k$ is the usual logarithmic series, then the difference $\log(1+x) - L(1+x)$ contains only terms x^k with $k \geq p$. From this and the equality $\log(1+u+v+uv) = \log(1+u) + \log(1+v)$ it follows that

$$L(1+u+v+uv) = L(1+u) + L(1+v) + G(u, v), \tag{9.5}$$

where $G(u, v)$ is a polynomial containing only terms of degree at least p in u and v combined. Applying (9.5) to $u = x^a - 1$ and $v = y^b - 1$, we obtain that

$$L(x^a y^b) = L(x^a) + L(y^b) + G(x^a - 1, y^b - 1).$$

From the definition of Γ^p (see Proposition 9.1.5 and the remark just after this proposition), $G(x^a - 1, y^b - 1)\Phi(x^c y^d) \in \Gamma^p$. Similarly, putting $u = x^s - 1$ and $v = x^t - 1$, $s + t = a$, we have

$$L(x^a)\Phi(x^c y^d) \equiv (L(x^s) + L(x^t))\Phi(x^c y^d) \pmod{\Gamma^p}.$$

Hence, $L(x^a)\Phi(x^c y^d) \equiv aL(x)\Phi(x^c y^d)$. Applying the same arguments to $L(y^b)$, we arrive at the required formula. □

Lemma 9.1.10. *Set $\Delta = \mathbb{Z} < x >$. Then the following is a basis for $\Delta/p\Delta$:*

$$\{1, L(x), \ldots, L^{p-2}(x)\}.$$

The proof is almost obvious, but we formalize it, observing that $\Delta/p\Delta = < 1, x, \ldots, x^{p-2} >_{\mathbb{F}_p}$. By analogy with (9.4) we introduce the *truncated exponential series* $E(x) = \sum_{k=0}^{p-2} x^k/k!$. From the two relations $\exp(\log(x)) = x$ and $(x-1)^{p-1}\Delta \subset p\Delta$, and following the scheme of the proof of Proposition 9.1.9, we obtain

$$L(x^n) \equiv nL(x) \pmod{p\Delta};\ E(L(x^n)) \equiv x^n \pmod{p\Delta}.$$

This implies, in particular, that $\Delta/p\Delta =< 1, L(x), \ldots, L^{p-2}(x) >$. □

Proof of Theorem 9.1.7. The required isomorphism Θ is constructed as follows (recall that the field $\mathbb{F}_p$ is identified with the set $\{0, 1, \ldots, p-1\}$). To an element

$L^a(x)L^b(y)\Phi(x^c y^d)$ of the space $\Gamma^0/p\Gamma^0$ with $a+b \le p-2$, $(d, -c) \in \sigma$, $\sigma \in \mathbb{P}V$, there corresponds the following function on $V = \mathbb{F}_p^2$:

$$f(u, v) = u^a v^b f_\sigma(u, v) = \begin{cases} u^a v^b & \text{if} \quad (u, v) \in \sigma; \\ 0 & \text{if} \quad (u, v) \notin \sigma, \end{cases} \tag{9.6}$$

where u and v are linear coordinates on V, and f_σ is the characteristic function of the set $\{(u, v) | (u, v) \in \sigma\} \subset V$.

On the basis of Lemma 9.1.10, we conclude that Θ is defined on the whole space $\Gamma^0/p\Gamma^0$. It is trivial to verify the fact that Θ is an isomorphism commuting with the action of $SL_2(p)$ $(x \mapsto x^{a'} y^{b'}, y \mapsto x^{c'} y^{d'};\ u \mapsto a'u + b'v, v \mapsto c'u + d'v)$:

$$\begin{array}{ccc} L^a(x)L^b(y)\Phi(x^c y^d) & \to & (a'L(x) + b'L(y))^a (c'L(x) + d'L(y))^b \Phi(x^{c''} y^{d''}) \\ \downarrow \Theta & & \downarrow \Theta \\ u^a v^b f_\sigma & \to & (a'u + b'v)^a (c'u + d'v)^b f_{\sigma'} \end{array}$$

(here, $c'' = a'c + c'd$, $d'' = b'c + d'd$). Note that $(d'', -c'') \in \sigma'$, since it is sufficient to restrict ourselves to the case $\sigma = \sigma_0$, and it is true for σ_0. □

We put

$$\mathcal{F}^k(V) = \{f \in \mathcal{F}(V) | f(\lambda u, \lambda v) = \lambda^k f(u, v), \forall \lambda \in \mathbb{F}_p\}. \tag{9.7}$$

Clearly, $\mathcal{F}^k(V)$ is a subspace invariant under $SL_2(p)$, and in addition the representation of $SL_2(p)$ in Γ^k/Γ^{k+1}, $k \le p-2$, is isomorphic to the representation of $SL_2(p)$ in $\mathcal{F}^k(V)$. In fact,

$$\Gamma^k/\Gamma^{k+1} = < L^a(x)L^b(y)\Phi(x^c y^d) | a+b = k, (c, d) \ne (0, 0) >_{\mathbb{F}_p},$$

and $\Theta(L^a(x)L^b(y)\Phi(x^c y^d)) = u^a v^b f_\sigma$, where $(d, -c) \in \sigma$. The elements $u^a v^b f_\sigma$ with $a + b = k$ generate $\mathcal{F}^k(V)$.

The finer structure of the $SL_2(p)$-module $\mathcal{F}^k(V)$ is revealed by the next result.

Proposition 9.1.11. *The following assertions hold:*

a) *For $1 \le k \le p-2$, the $SL_2(p)$-module $\mathcal{F}^k(V)$ can be embedded in the non-split exact sequence*

$$0 \to S^k(V^*) \to \mathcal{F}^k(V) \to S^{p-1-k}(V) \to 0.$$

For $k = 0$ the analogous sequence is split.

b) *There is an isomorphism of $SL_2(p)$-modules: $S^k(V) \cong S^k(V^*)$. Here, $S^k(V)$ and $S^k(V^*)$ are the spaces of polynomials of degree k on V and on V^*, respectively.*

Proof. Obviously, for $1 \le k \le p-2$ we can distinguish the unique submodule $S^k(V^*)$ of $\mathcal{F}^k(V)$. The fact that it is irreducible is well-known. For $k = 0$ there is an additional submodule W such that $\mathcal{F}^0(V) = S^0(V^*) \oplus W$, where W consists of the functions $f(u, v) = \sum_{\sigma \in \mathbb{P}V} c_\sigma f_\sigma$ with $\sum_\sigma c_\sigma = 0$. Assertion a) now follows from the fact that there is an $SL_2(p)$-invariant non-degenerate pairing between the modules $\mathcal{F}^k(V)$ and $\mathcal{F}^{p-1-k}(V)$.

In fact, let π be the projection of $\mathcal{F}^0(V)$ on $S^0(V^*)$ along W. Then the pairing is specified as follows: $(f, g) \mapsto \pi(fg)$, where $f \in \mathcal{F}^k(V)$ and $g \in \mathcal{F}^{p-1-k}(V)$ (we observe that $fg \in \mathcal{F}^0(V)$). For the invariance of the pairing, suppose that $\varphi \in SL_2(p)$. Then by definition $\varphi \circ f(u, v) = f(\varphi^{-1}(u, v))$, and so $(\varphi \circ f)(\varphi \circ g) = \varphi \circ (fg)$. Consequently,

$$\pi((\varphi \circ f)(\varphi \circ g)) = \pi(\varphi \circ (fg)) = \varphi(\pi(fg)) = \pi(fg),$$

since the action of φ on $S^0(V^*)$ is trivial. To prove the non-degeneracy of the pairing, we write $f \in \mathcal{F}^k(V)$ and $g \in \mathcal{F}^{p-1-k}(V)$ in the form $f = \sum_{\sigma \in \mathbb{P}V} c_\sigma w^k f_\sigma$ and $g = \sum_{\sigma \in \mathbb{P}V} d_\sigma w^{p-1-k} f_\sigma$, where

$$w^k f_\sigma = \begin{cases} v^k f_\sigma & \text{if } (0, 1) \in \sigma; \\ u^k f_\sigma & \text{otherwise.} \end{cases}$$

Since $f_\sigma f_{\sigma'} = 0$ for $\sigma \neq \sigma'$ and $w^{p-1} f_\sigma = f_\sigma$, we have $\pi(fg) = \pi(\sum_\sigma c_\sigma d_\sigma f_\sigma) = \sum_\sigma c_\sigma d_\sigma$. Therefore, the equality $\pi(fg) = 0$ valid for all $g \in \mathcal{F}^{p-1-k}(V)$ implies that $f = 0$.

The modules $S^k(V^*)$ and $S^{p-1-k}(V^*)$, which belong to $\mathcal{F}^k(V)$ and $\mathcal{F}^{p-1-k}(V)$ respectively, are annihilators of each other with respect to the pairing introduced above, since W consists of polynomials of degree $p-1$. But because the pairing is non-degenerate and $S^k(V^*)$ is the only proper submodule, it follows that the quotients $\mathcal{F}^{p-1-k}(V)/S^{p-1-k}(V^*)$ and $\mathcal{F}^k(V)/S^k(V^*)$ are dual to $S^k(V^*)$ and $S^{p-1-k}(V^*)$ respectively.

Finally, the $SL_2(p)$-modules V and V^* are isomorphic: the isomorphism is given by means of the skew-symmetric form $<(a, b); (c, d)> = ad - bc$. In particular, $S^k(V)$ and $S^k(V^*)$ are isomorphic. □

9.2 Classification of indivisible invariant lattices

We start with the case of invariant lattices containing $p\Lambda$ (or equivalently, $p\Gamma$).

Lemma 9.2.1. *Under the isomorphism Θ introduced at the very beginning of the proof of Theorem 9.1.7, the polynomials of degree at most $p-2$ form together with $p\Gamma$ an invariant sublattice $\tilde{\Gamma}$ (thus $\tilde{\Gamma}/p\Gamma \subset \Gamma/p\Gamma$, and $\tilde{\Gamma}$ is an invariant ideal).*

Proof. It is easy to see that the polynomials of degree at most $p-2$ correspond to the submodule generated by the polynomials $pL^a(x)L^b(y)$, $0 \le a+b \le p-2$. This submodule is $SL_2(p)$-invariant, and its preimage is an $SL_2(p)$-invariant ideal. □

Definition 9.2.2. The lattices

$$\Gamma^{k,l} = (\tilde{\Gamma} \cap \Gamma^k) + \Gamma^l,\ 0 \le k \le l \le p-1,\ k < p-1, \tag{9.8}$$

are called *lattices of type* I.

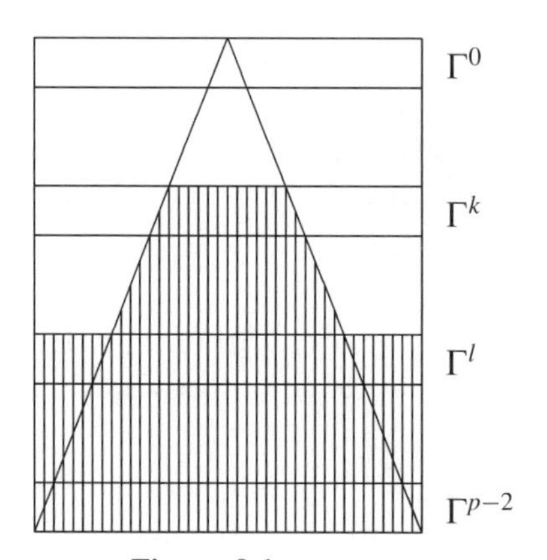

Figure 9.1

By Corollary 9.1.6 and Proposition 9.2.1, all these lattices are invariant. Since $\Gamma^{k,l} \supset \Gamma^{p-1} = p\Gamma$, lattices of type I are indivisible and contain $p\Gamma$. We represent the embedded sequence $\Gamma = \Gamma^0 \supset \Gamma^1 \supset \cdots \supset \Gamma^{p-2} \supset \Gamma^{p-1} = p\Gamma$ as a tower (see Figure 9.1). In this tower the lattice $\tilde{\Gamma}$ is represented as a pyramid that cuts the kth floor of the tower (that is, $\Gamma^k/\Gamma^{k+1} \cong \mathcal{F}^k(V)$) in a module $S^k(V^*)$ of dimension $k+1$. The shaded figure is a representation of $\Gamma^{k,l}$. According to Proposition 9.1.11, the quotient Γ^0/Γ^1 splits into a sum of two $SL_2(p)$-modules: $\Gamma^0/\Gamma^1 \cong S^0(V^*) \oplus W$. Let $\pi_1 : \Gamma^0 \to \Gamma^0/\Gamma^1$ be the natural projection. We introduce the lattice

$$\Gamma^w = \pi_1^{-1}(W). \tag{9.9}$$

Clearly, Γ^w is an invariant lattice containing Γ^1, since multiplication by $L(x)$ and $L(y)$ takes Γ^k into Γ^{k+1}, by Lemma 9.1.8.

We now introduce one more series of lattices by suitably "adjusting" each lattice $\Gamma^{k,p-k}$, $1 \le k < (p-1)/2$. We have

$$\Gamma^{k,p-k} = \Gamma^{k+1,p-k} \oplus S^k(V^*),$$

where, as we have already mentioned, $S^k(V^*)$ is the intersection of $\tilde{\Gamma}$ with the kth floor of the tower (see Figure 9.1). By Proposition 9.1.11, $\Gamma^k/\Gamma^{k+1,p-k}$ contains one further module of type $S^k(V^*)$, namely

$$\Gamma^{p-1-k}/(\Gamma^{k+1,p-k} \cap \Gamma^{p-1-k}) \cong S^k(V^*).$$

In what follows, any module of type $S^k(V^*)$ is denoted for brevity by S^k. Let τ be an isomorphism between the module $S^k(V^*)$ and the module S^k. Then for every $r \in \mathbb{F}_p^*$, $\Gamma^k/\Gamma^{k+1,p-k}$ contains the $SL_2(p)$-module

$$W_r = \{f + r\tau(f) | f \in S^k(V^*)\}.$$

Let $\pi_2 : \Gamma^k \to \Gamma^k/\Gamma^{k+1,p-k}$ be the natural epimorphism.

Definition 9.2.3. We call the module

$$\Delta^{k,r} = \pi_2^{-1}(W_r),\ 1 \le k < (p-1)/2,\ r \in \mathbb{F}_p^*, \tag{9.10}$$

a *lattice of type* II.

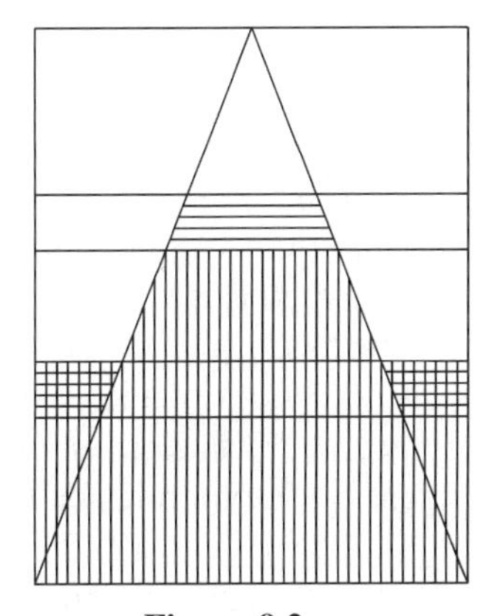

Figure 9.2

A graphical representation of these modules is shown in Figure 9.2: the isosceles trapezoid shaded by horizontal lines corresponds to the module $S^k(V^*)$, and the cross-hatched region corresponds to S^k. Obviously, $\Delta^{k,r}$ is invariant under the action of $SL_2(p)$ and contains $\Gamma^{k+1,p-k}$. Therefore $\Delta^{k,r}$ contains $p\Gamma$ and is indivisible. It remains to verify that $\Delta^{k,r}$ is invariant under multiplication by x and y, or equivalently by $L(x)$ and $L(y)$. But by Lemma 9.1.8, multiplication by $L(x)$ and $L(y)$ lowers the floor of the tower (see Figure 9.1) and therefore preserves $\Delta^{k,r}$.

One of the main results of the section is the following.

Theorem 9.2.4. *Lattices of types* I *and* II, *together with* Γ^w, *account for all the indivisible invariant sublattices of* Γ *containing* $p\Gamma$.

The proof splits into a number of steps.

1) By Lemma 9.1.1 and Theorem 9.1.3, the image of any indivisible invariant sublattice Γ' under projection onto the Cartan subalgebra $\mathcal{H}_\sigma$ is a sublattice of the form Γ_σ^k for some k with $0 \le k \le p-2$. This means that $\Gamma' \subseteq \Gamma^k$ and Γ' is not contained in Γ^{k+1}. Let l be the smallest non-negative integer for which $\Gamma' \supseteq \Gamma^l$. Obviously, $l \le p-1$. Also, in the case $k = l$ we have $\Gamma' = \Gamma^k = \Gamma^{k,k}$. Therefore in what follows we shall assume that $l > k$.

2) We claim that, *for* $k \le m < l$,

$$(\Gamma' \cap \Gamma^m + \Gamma^{m+1})/\Gamma^{m+1} = S^m(V^*).$$

The case $\Gamma' = \Gamma^w$ *is an exception.*

In fact, note first that, $(\Gamma' \cap \Gamma^m + \Gamma^{m+1})/\Gamma^{m+1}$ is an $SL_2(p)$-module in $\mathcal{F}^m(V)$. If it is zero, then $\Gamma' \cap \Gamma^m \subseteq \Gamma^{m+1}$. Hence, since Γ' is invariant, we have $\Gamma' \subseteq \Gamma^{m+1}$; but this contradicts the choice of k. If, on the contrary, it is complete (that is, it coincides with $\mathcal{F}^m(V)$), then, descending the tower through multiplying by $L(x)$ and $L(y)$, we get that

$$(\Gamma' \cap \Gamma^{l-1} + \Gamma^l)/\Gamma^l = \mathcal{F}^{l-1}(V) = \Gamma^{l-1}/\Gamma^l,$$

and so $\Gamma' + \Gamma^l \supseteq \Gamma^{l-1}$. But $\Gamma' \supseteq \Gamma^l$, and so $\Gamma' \supseteq \Gamma^{l-1}$, which contradicts the choice of l. Thus, $(\Gamma' \cap \Gamma^m + \Gamma^{m+1})/\Gamma^{m+1}$ is a proper submodule of $\mathcal{F}^m(V)$, and so by Proposition 9.1.11 it coincides with $S^m(V^*)$ (for $m > 0$), and with $S^0(V^*)$ or W (for $m = 0$). We need to show that the latter possibility (coincidence with W) is realized only for $\Gamma' = \Gamma^w$. By our hypothesis we have $(\Gamma' + \Gamma^1)/\Gamma^1 = W$. If $l = 1$, then $\Gamma' \supseteq \Gamma^1$ and $\Gamma'/\Gamma^1 = W$, that is, $\Gamma' = \pi_1^{-1}(W) = \Gamma^w$. Therefore, we can assume that $l > 1$. Multiplying both sides of the proposed equality by $L(x)$ and $L(y)$, we obtain, from the definition of W, an element of $(\Gamma' \cap \Gamma^1 + \Gamma^2)/\Gamma^2$ not lying in $S^1(V^*)$. Acting on this element by automorphisms from $SL_2(p)$, we obtain that $\mathcal{F}^1(V) = \Gamma^1/\Gamma^2$. Hence, $(\Gamma' \cap \Gamma^1 + \Gamma^2)/\Gamma^2 = \Gamma^1/\Gamma^2$, that is, $\Gamma' \cap \Gamma^1 + \Gamma^2 = \Gamma^1$. Descending the tower in the same way, we obtain $\Gamma' \cap \Gamma^{l-1} + \Gamma^l = \Gamma^{l-1}$, and because $\Gamma' \supseteq \Gamma^l$, we come to the inclusion $\Gamma' \supseteq \Gamma^{l-1}$, which contradicts the choice of l.

In what follows we assume that Γ' is an invariant lattice containing $p\Gamma$ and distinct from Γ^w.

3) *If* $\Gamma' \cap \Gamma^m = \Delta^{m,r}$ *for some* m, $1 \le m < (p-1)/2$, *then* $\Gamma' = \Delta^{m,r}$.

In fact, from the inclusion $\Gamma' \subseteq \Gamma^m$ it follows immediately that $\Gamma' = \Delta^{m,r}$. We now suppose that Γ' is not contained in Γ^m. As we showed in 2),

$$(\Gamma' \cap \Gamma^{m-1})/(\Gamma' \cap \Gamma^m) \cong (\Gamma' \cap \Gamma^{m-1} + \Gamma^m)/\Gamma^m = S^{m-1}(V^*).$$

On the other hand, the $SL_2(p)$-module $\Gamma^{m-1}/(\Gamma' \cap \Gamma^m) = \Gamma^{m-1}/\Delta^{m,r}$ is the direct sum

$$S^{m-1}(V^*) \oplus S^m \oplus \{S^{p-m} \oplus S^{p-1-m} \oplus \cdots \oplus S^{m+1}\},$$

where in the curly brackets we have the modules $S^{p-1-t} = \mathcal{F}^t(V)/S^t(V^*)$, $m-1 \leq t \leq p-2-m$. Clearly, $m-1 \notin \{m, m+1, \dots, p-m\}$. Consequently, the submodule $S^{m-1} = (\Gamma' \cap \Gamma^{m-1})/(\Gamma' \cap \Gamma^m)$ of $\Gamma^{m-1}/(\Gamma' \cap \Gamma^m)$ is just $S^{m-1}(V^*)$. Multilplication by $L(x)$ and $L(y)$ again enables us to deduce that $\Gamma' \cap \Gamma^m$ contains $S^m(V^*)$. However, this contradicts the construction of $\Delta^{m,r} = \Gamma' \cap \Gamma^m$.

4) Suppose that for some m, $1 \leq m \leq l$, we have $\Gamma' \cap \Gamma^m = \Gamma^{m,l}$. Then two cases are possible:

a) if $\Gamma' \subseteq \Gamma^m$, then $\Gamma' = \Gamma^{m,l}$;

b) if Γ' is not contained in Γ^m, then $\Gamma' \cap \Gamma^{m-1}$ is equal to $\Gamma^{m-1,l}$ or $\Delta^{m-1,r}$ for some $r \in \mathbb{F}_p^*$.

The case a) is obvious. For the proof of b) we consider the module

$$S^{m-1} = (\Gamma' \cap \Gamma^{m-1})/(\Gamma' \cap \Gamma^m).$$

As above (see 3)),

$$\Gamma^{m-1}/(\Gamma' \cap \Gamma^m) = S^{m-1}(V^*) \oplus \{S^{p-m} \oplus S^{p-1-m} \oplus \dots \oplus S^{p-l}\}.$$

For $m-1 \notin \{p-m, \dots, p-l\}$ the assertion is obvious: $(\Gamma' \cap \Gamma^{m-1})/(\Gamma' \cap \Gamma^m)$ is $S^{m-1}(V^*)$, and therefore $\Gamma' \cap \Gamma^{m-1} = \Gamma^{m-1,l}$. Now suppose that $m-1 = p-1-s$, $m-1 \leq s \leq l-1$. Then, denoting the isomorphism $S^{m-1}(V^*) \cong S^{p-1-s}$ by τ, we come to the equality

$$(\Gamma' \cap \Gamma^{m-1})/(\Gamma' \cap \Gamma^m) = W_r = \{f + r\tau(f) | f \in S^{m-1}(V^*)\}$$

for some $r \in \mathbb{F}_p^*$. Observe that the case $s = m-1$ is impossible: $\mathcal{F}^{m-1}(V)$ has only one proper submodule, namely $S^{m-1}(V^*)$. In the case where $m-1 < s < l-1$, multiplication by powers of $L(x)$ and $L(y)$ enables us to descend to the module $(\Gamma' \cap \Gamma^{l-1} + \Gamma^l)/\Gamma^l$, which is certainly different from $S^{l-1}(V^*)$; this contradicts the assumption that $\Gamma' \cap \Gamma^m = \Gamma^{m,l}$. There remains the last possibility $s = l-1$, which gives $\Gamma' \cap \Gamma^{m-1} = \Delta^{m-1,r}$, as required.

5) Finally, we consider an arbitrary invariant lattice $\Gamma' \supseteq p\Gamma$, $\Gamma' \neq \Gamma^w$, $\Gamma^k \supseteq \Gamma' \supseteq \Gamma^l$ (k and l are as defined in 1)). Starting from the equality $\Gamma' \cap \Gamma^l = \Gamma^{l,l}$ and ascending the tower by means of 4), we shall have $\Gamma' \cap \Gamma^m = \Gamma^{m,l}$ or $\Delta^{m,r}$ for any m with $k \leq m < l$. By 3), the equality $\Gamma' \cap \Gamma^m = \Delta^{m,r}$ gives us $\Gamma' = \Delta^{m,r}$. If $\Gamma' \cap \Gamma^m = \Gamma^{m,l}$ for all m, then $\Gamma' = \Gamma^{k,l}$. □

Now we consider invariant lattices that do not contain $p\Lambda$ ($p\Gamma$ respectively). To this end we introduce the lattice

$$\hat{\Gamma} = \left\langle pL^a(x)L^b(y) | a+b \geq 1 \right\rangle_{\mathbb{Z}} + p^2\Gamma,$$

which plays the same role for lattices not containing $p\Gamma$ as does the lattice

$$\tilde{\Gamma} = \left\langle pL^a(x)L^b(y) | a+b \geq 0 \right\rangle_{\mathbb{Z}} + p\Gamma$$

in the classification of lattices that do contain $p\Gamma$.

Lemma 9.2.5. *The lattice $\hat{\Gamma}$ is indivisible and invariant, and contains Γ^p but not $p\Gamma$: $(\hat{\Gamma} \cap \Gamma^{p-1})/\Gamma^p = pW$.*

Proof. a) It is easy to see that the factor-lattice $Q = (\hat{\Gamma} \cap \Gamma^{p-1} + \Gamma^p)/\Gamma^p$ is generated by the elements $pL^a(x)L^b(y)$ with $a+b = p-1$. On the other hand, W is generated by the elements $L^a(x)L^b(y)$ with $a+b = p-1$, so $Q = pW$. Let us consider a nonzero element of Q. Multiplying it by $L(x)$ and $L(y)$ we obtain an element of $p\mathcal{F}^1(V)$ not lying in $pS^1(V^*)$. Acting on this last element by automorphisms of $SL_2(p)$, we obtain the whole of $p\mathcal{F}^1(V)$. Hence, $(\hat{\Gamma} \cap \Gamma^p + \Gamma^{p+1})/\Gamma^{p+1} = \Gamma^p/\Gamma^{p+1}$ and, more generally,

$$(\hat{\Gamma} \cap \Gamma^l + \Gamma^{l+1})/\Gamma^{l+1} = \Gamma^l/\Gamma^{l+1}$$

for all $l \geq p$. Since $\hat{\Gamma} \supset p^2\Gamma = \Gamma^{2p-2}$ and $(\hat{\Gamma} \cap \Gamma^{2p-3} + \Gamma^{2p-2})/\Gamma^{2p-2} = \Gamma^{2p-3}/\Gamma^{2p-2}$, we have $\hat{\Gamma} \supseteq \Gamma^{2p-3}$. Continuing to ascend the tower in this way, we come to the required inclusion $\hat{\Gamma} \supseteq \Gamma^p$.

b) Using the property

$$L(x^a y^b)\Phi(x^c y^d) \equiv (aL(x) + bL(y))\Phi(x^c y^d) (\mathrm{mod}\, \Gamma^p)$$

established in Proposition 9.1.9, we get that $\hat{\Gamma}$ is invariant. Finally, $\hat{\Gamma} \supset p^2\Gamma$, so $\hat{\Gamma}$ is indivisible. □

Definition 9.2.6. By analogy with (9.8) we put

$$\Gamma^{k,p} = \hat{\Gamma} \cap \Gamma^k, \; 1 \leq k \leq p-2.$$

We observe that all the lattices $\Gamma^{k,p}$ are indivisible and invariant and do not contain $p\Gamma$.

Theorem 9.2.7. *The lattices $\Gamma^{k,p}$ with $1 \leq k \leq p-2$ account for all the indivisible invariant sublattices of Γ that do not contain $p\Gamma$.*

Proof. 1) Let Γ' be an arbitrary indivisible invariant sublattice of Γ not containing $p\Gamma$. Then $\Gamma^1 \supset \Gamma' \supset \hat{\Gamma} \cap \Gamma^{p-1}$. In fact, by Lemma 9.1.1 the projection of Γ' in $\mathcal{H}_\sigma$ is Γ^k_σ for some k with $0 \leq k \leq p-2$. Hence, $\Gamma' \subset \Gamma^k$; but Γ' is not contained in Γ^{k+1}, so by Lemma 9.1.2 we have $\Gamma' \supset p\Gamma^k = \Gamma^{k+p-1}$. Since Γ' does not contain $p\Gamma = \Gamma^{p-1}$ by hypothesis, we have $k > 0$ and $\Gamma' \subset \Gamma^1$. For $k \geq 1$ the arguments that we used in the course of the proof of Theorem 9.2.4 (see assertion 2)) show that

$$(\Gamma' \cap \Gamma^k + \Gamma^{k+1})/\Gamma^{k+1} = S^k(V^*) = (\hat{\Gamma} \cap \Gamma^k + \Gamma^{k+1})/\Gamma^{k+1}.$$

Hence, $\Gamma' \cap \Gamma^k + \Gamma^{k+1} = \hat{\Gamma} \cap \Gamma^k + \Gamma^{k+1}$. Descending the tower leads us to the inclusion $\Gamma' \cap \Gamma^l + \Gamma^{l+1} \supseteq \hat{\Gamma} \cap \Gamma^l + \Gamma^{l+1}$ for each $l \geq k$. By Lemma 9.2.5, for $l \geq p$ we have $\hat{\Gamma} \cap \Gamma^l + \Gamma^{l+1} = \Gamma^l$, and so $\Gamma' \cap \Gamma^l + \Gamma^{l+1} = \Gamma^l$. Since $\Gamma' \supset \Gamma^{k+p-1}$, ascending the floors of the tower leads us to the inclusion $\Gamma' \supset \Gamma^p$. Finally, from the equalities $\Gamma' \cap \Gamma^{p-1} + \Gamma^p = \hat{\Gamma} \cap \Gamma^{p-1} + \Gamma^p = \hat{\Gamma} \cap \Gamma^{p-1}$, on taking account of the last inclusion, we obtain that $\Gamma' \supset \hat{\Gamma} \cap \Gamma^{p-1}$.

2) As in 1), we choose k such that $\Gamma' \subset \Gamma^k$ and $\Gamma' \not\subseteq \Gamma^{k+1}$. We shall prove that $\Gamma' \cap \Gamma^l = \hat{\Gamma} \cap \Gamma^l$ by induction on l, $k \leq l \leq p-1$. For $l = p-1$ we know that $\Gamma' \supset \hat{\Gamma} \cap \Gamma^{p-1}$, and so we have the inclusion

$$p\mathcal{F}^0(V) \supset (\Gamma' \cap \Gamma^{p-1})/\Gamma^p \supseteq (\hat{\Gamma} \cap \Gamma^{p-1})/\Gamma^p = pW,$$

and since pW is a maximal submodule of $p\mathcal{F}^0(V)$ we have $\Gamma' \cap \Gamma^{p-1} = \hat{\Gamma} \cap \Gamma^{p-1}$.

Suppose that we already have $\Gamma' \cap \Gamma^l = \hat{\Gamma} \cap \Gamma^l$, $l > k$. We consider the submodule $S^{l-1} = (\Gamma' \cap \Gamma^{l-1})/(\Gamma' \cap \Gamma^l)$, which lies in the module

$$\Gamma^{l-1}/(\Gamma' \cap \Gamma^l) = S^{l-1}(V^*) \oplus \{S^{p-l} \oplus S^{p-1-l} \oplus \cdots \oplus S^1 \oplus pS^0\}.$$

The standard argument shows that $(\Gamma' \cap \Gamma^{l-1})/(\Gamma' \cap \Gamma^l) = S^{l-1}(V^*)$, that is, $\Gamma' \cap \Gamma^{l-1} = \hat{\Gamma} \cap \Gamma^{l-1}$. This completes the inductive step.

For $l = k$ the relation we have just proved gives us the chain of equalities $\Gamma' = \Gamma' \cap \Gamma^k = \hat{\Gamma} \cap \Gamma^k = \Gamma^{k,p}$. □

Theorems 9.2.4 and 9.2.7 can be summarized in Table 9.3. Thus the total number of indivisible invariant lattices is $(p-1)(2p+1)/2$.

Table 9.3

	$\Gamma' \supset p\Gamma$		$\Gamma' \not\supseteq p\Gamma$
Γ^w	Type I: $\Gamma^{k,l}$, $0 \leq k \leq l \leq p-1$, $k \neq p-1$	Type II: $\Delta^{k,r}$, $0 < k < \frac{p-1}{2}$, $r \in \mathbb{F}_p^*$	$\Gamma^{k,p}$, $1 \leq k \leq p-2$
		Number	
1	$\frac{p(p+1)}{2} - 1$	$\frac{(p-1)(p-3)}{2}$	$p-2$

It is worth mentioning that the lattices Γ^w and $\Gamma^{k,l}$ are invariant under a larger group of isometries. Namely, $\mathrm{Aut}(\Gamma^w)$ and $\mathrm{Aut}(\Gamma^{k,l})$ contain a subgroup consisting of ring automorphisms $\hat{\varphi}$ for $\varphi \in GL_2(p)$ (we have in mind the ring $\mathcal{A} = \mathbb{C}[x, y]/I$). As for the lattices $\Delta^{k,r}$, we have $\hat{\varphi}(\Delta^{k,r}) = \Delta^{k,r(\det\varphi)^k}$. Hence, the set $\{\Delta^{k,r}\}$ splits into several orbits. In particular, for $p = 5$ the orbit is unique, but for $p = 7$ there are three orbits with representatives $\Delta^{1,1}$, $\Delta^{2,1}$ and $\Delta^{2,3}$.

9.3 Metric properties of invariant lattices

Using the isomorphism χ (see (9.3)), we carry over the inner product $(X|Y)$ defined earlier on Λ to Γ. We first introduce an inner product on the ring of functions $\mathcal{A} = \mathbb{C}[x, y]/I$:

$$(f_u; f_v) = (\chi^{-1}(f_u)|\chi^{-1}(f_v)) = p\mathrm{Tr}J_u J_v =$$

$$= p\varepsilon^{B(u,v)}\mathrm{Tr}J_{u+v} = \begin{cases} 0 & \text{if } \; u+v \neq 0, \\ p^2\varepsilon^{-B(u,u)} & \text{if } \; u+v = 0. \end{cases}$$

This inner product is obviously G-invariant. Its restriction to the basic lattice Γ has the form given in the proposition below.

Proposition 9.3.1. *Set* $F_i(x, y) = x^{a_i}y^{b_i}\Phi(x^{c_i}y^{d_i})$, *where* $(d_i, -c_i) \in \sigma_i$, $i = 1, 2$. *Then*

$$(F_1; F_2) = \begin{cases} 0 & \textit{if} \;\; \sigma_1 \neq \sigma_2, \\ -1 & \textit{if} \;\; \sigma_1 = \sigma_2, \quad F_1 \neq F_2, \\ p-1 & \textit{if} \;\; \sigma_1 = \sigma_2, \quad F_1 = F_2. \end{cases}$$

Proof. For convenience of calculation, we introduce a new basis of $\mathcal{A}$ consisting of the "δ-functions" $g_u = p^{-2}\varepsilon^{B(u,u)/2}f_u$:

$$g_u(\varepsilon^c, \varepsilon^d) = \begin{cases} 0 & \text{if } \; (c, d) \neq u, \\ 1 & \text{if } \; (c, d) = u. \end{cases}$$

Then

$$(g_u; g_v) = \frac{1}{p^4}\varepsilon^{\{B(u,u)+B(v,v)\}/2}(f_u; f_v) = \begin{cases} 0 & \text{if } \; u+v \neq 0, \\ \frac{1}{p^2} & \text{if } \; u+v = 0. \end{cases}$$

We decompose F_i with respect to the basis (g_u):

$$F_i(x, y) = p \sum_{(k,l)\in\sigma_i} \varepsilon^{ka_i+lb_i} g_{(k,l)}, \; i = 1, 2.$$

It is immediately obvious that $(F_1; F_2) = 0$ if $\sigma_1 \neq \sigma_2$. If $\sigma_1 = \sigma_2 = \sigma$, then

$$(F_1; F_2) = p^2 \sum_{(k,l)\in\sigma} \varepsilon^{k(a_1-a_2)+l(b_1-b_2)}(g_{(k,l)}; g_{(-k,-l)}) =$$

$$= \sum_{(k,l)\in\sigma} \varepsilon^{k(a_1-a_2)+l(b_1-b_2)} = \sum_{i=1}^{p-1} \varepsilon^{i(k_0(a_1-a_2)+l_0(b_1-b_2))} =$$

$$= \begin{cases} -1 & \text{if} \quad k_0(a_1 - a_2) + l_0(b_1 - b_2) \neq 0, \quad \text{that is,} \quad F \neq F'; \\ p-1 & \text{if} \quad k_0(a_1 - a_2) + l_0(b_1 - b_2) = 0, \quad \text{that is,} \quad F = F'. \end{cases}$$

(We mean above that $(k_0, l_0) \in \sigma$). □

Corollary 9.3.2. a) *Set* $F_i = (\sum_{k=0}^{p-1} a_{ik} x^k)\Phi(y)$, *where* $i = 1, 2$, $a_{ik} \in \mathbb{Z}$ *and* $\sum_{k=0}^{p-1} a_{1k} = 0$. *Then*

$$(F_1; F_2) = p \sum_{k=0}^{p-1} a_{1k} a_{2k}. \tag{9.11}$$

In particular, for $0 \leq k, l \leq p-1$, $k > 0$ *we have*

$$((1-x)^k \Phi(y); (1-x)^l \Phi(y)) = p \binom{k+l}{k}. \tag{9.12}$$

b) *If* $0 \leq k, l, k', l' \leq p-1$, *then*

$$(px^k y^l; px^{k'} y^{l'}) = \begin{cases} p^2 - 1 & \text{if} \quad (k,l) = (k', l'), \\ -1 & \text{if} \quad (k,l) \neq (k', l'). \end{cases} \tag{9.13}$$

The proof is an immediate application of Proposition 9.3.1. □

Note that the functions occurring in Corollary 9.3.2 are elements of Λ. From now on, by the *norm* $||F||^2$ of a vector $F = F(x, y) \in \Gamma$ we mean $(F; F)$.

Proposition 9.3.3. *Suppose that* $f(x)$ *is a nonzero element of* $\mathbb{Z} < x >$ *such that* $(x-1)^k | f(x)$. *The following inequalities hold.*

a) *If* $1 \leq k \leq (p-1)/2$, *then*

$$||f(x)\Phi(y)||^2 \geq 2pk. \tag{9.14}$$

For $k | (p-1)$ *or* $k = 3$ $(p \geq 7)$ *or* $k = 4$ $(p \geq 13)$ *this estimate is best possible. For* $(p-1)/3 < k < (p-1)/2$, *it is not exact.*

b) *If* $(p+1)/2 \leq k \leq p-1$, *then*

$$||f(x)\Phi(y)||^2 \geq 2p^2(k - (p-1)/2). \tag{9.15}$$

The exactness of (9.15) *is equivalent to that of* (9.14).

Note, by the way, that inequalities (9.14) and (9.15) have also been established by Craig (see [CoS 7]).

Proof. a) We obtain (9.14) by induction on k. If $k = 1$, then the condition $(x-1) | f(x)$ can be rewritten as $f(x) = \sum_{i=0}^{p-1} a_i x^i$, $\sum_{i=0}^{p-1} a_i = 0$. Therefore, by

(9.11), we have

$$||f(x)\Phi(y)||^2 = p\sum_{i=0}^{p-1} a_i^2 = -2p \sum_{0\le i<j\le p-1} a_i a_j,$$

an integer divisible by $2p$. Hence either $f = 0$ or $||f(x)\Phi(y)||^2 \ge 2p$. In passing we note that

$$||f(x)\Phi(y)||^2 = 2p \Leftrightarrow f(x) = x^s - x^t,\ s \ne t. \tag{9.16}$$

Let us carry out the inductive step $k \to k+1$. Suppose that

$$||f(x)\Phi(y)||^2 < 2p(k+1),\ (x-1)^{k+1}|f(x)$$

for some function $f(x) \ne 0$. By the induction hypothesis, $||f(x)\Phi(y)||^2 \ge 2pk$, and the basis of the induction, already verified, ensures that $||f(x)\Phi(y)||^2$ is divisible by $2p$. Hence, $||f(x)\Phi(y)||^2 = 2pk$. Suppose that $f(x) = \sum_{i\in I} a_i x^{n_i} - \sum_{j\in J} b_j x^{m_j}$, $I \cup J \subseteq \{0, 1, \ldots, p-1\}$, $a_i > 0$ and $b_j > 0$. Since $f(1) = 0$, we have $\sum a_i = \sum b_j = c \in \mathbb{Z}$. Moreover,

$$2pk = ||f(x)\Phi(y)||^2 = p(\sum a_i^2 + \sum b_j^2) \ge p(\sum a_i + \sum b_j) = 2pc,$$

so $c \le k$. We rewrite $f(x)$ in the form $f(x) = \sum_{i=1}^{c}(x^{s_i} - x^{t_i})$, where $s_i \ne t_i$, but there may be coincidences among the exponents s_i (respectively t_i). Then, recalling that $f(x) \in \mathbb{Z} < x >$ and $(x-1)^{k+1}|f(x)$, we have $f^{(\nu)}(1) \equiv 0 \pmod p$ for $1 \le \nu \le k$; that is, $\sum_i s_i^\nu \equiv \sum_i t_i^\nu \pmod p$. On the basis of Newton's formulae, which express sums of powers in terms of the elementary symmetric functions, we arrive at the conclusion that $s_1, \ldots, s_c$ are the roots of the polynomial $\prod_{i=1}^{c}(x-t_i) \in \mathbb{F}_p[x]$. Hence, $\{s_1, \ldots, s_c\} = \{t_1, \ldots, t_c\}$; that is, $f = 0$, a contradiction.

We have actually proved that, if $f(x) = \sum_{i=0}^{p-1} a_i x^i \ne 0$, $a_i \in \mathbb{Z}$ and $\sum a_i = 0$, then

$$(x-1)^k|f(x) \Rightarrow \sum_{i=0}^{p-1} |a_i| \ge 2k. \tag{9.17}$$

Turning to the question of the exactness of (9.14), for small k we state explicitly the polynomials for which equality is attained in (9.14):

$$\begin{aligned}
k = 1 : f(x) &= x - 1;\\
k = 2 : f(x) &= (x-1)(x^2-1) = x^3 - x^2 - x + 1;\\
k = 3 : f(x) &= (x-1)(x^2-1)(x^3-1) = x^6 - x^5 - x^4 + x^2 + x - 1;\\
k = 4 : f(x) &= (x-1)(x^2-1)(x^3-1)(x^5-1) =\\
&= (x^{11} + x^7 + x^4 + 1) - (x^{10} + x^9 + x^2 + x).
\end{aligned}$$

From (9.17) it follows that if equality is attained in (9.14) for a function $f(x) = \sum_{i=0}^{p-1} a_i x^i \ne 0$ with $(x-1)^k|f(x)$, then $2k = \sum_i a_i^2 \ge \sum_i |a_i| \ge 2k$; that is,

$a_i = 0, \pm 1$ for all i. Therefore,

$$f(x) = \sum_{i=1}^{k} x^{s_i} - \sum_{i=1}^{k} x^{t_i}, \ |\{s_1, \dots, s_k, t_1, \dots, t_k\}| = 2k.$$

The above arguments show that the exactness of (9.14) is equivalent to the existence of $h(x) \in \mathbb{F}_p[x]$ and $a \in \mathbb{F}_p^*$ such that

$$h(x) = \prod_{i=1}^{k} (x - s_i), \ h(x) + a = \prod_{i=1}^{k} (x - t_i). \tag{9.18}$$

When $k|(p-1)$ we take $h(x) = x^k - 1$ and $a = 1 - b^k$, where $b^k \neq 0, 1$. We have $k < p - 1$, so such a $b \in \mathbb{F}_p^*$ exists. The nonexistence of such $h(x)$ and a for $(p-1)/3 < k < (p-1)/2$ (that is, the nonexactness of (9.14) for given k) follows from this more general assertion: *Suppose that $x^p - x = u(x)h(x)(h(x) + 2v(x))$, where $u, v, h \in \mathbb{F}_p[x]$, u and h are monic, and $\deg(h) = \deg(v) + k$. Then $p - 2k - 1 \geq \deg(h) \geq k$, or $u(x) = x + b$, $h(x) = (x + b)^{(p-1)/2} - v(x)$ and $v(x) = \pm 1$.* In fact, we have $x^p - x = u(h + v)^2 - uv^2$. Differentiating, we obtain $(h + v)|\{(uv^2)' - 1\}$. If $(uv^2)' - 1 = 0$, then $uv^2 = x + b$ for some $b \in \mathbb{F}_p$ (because $\deg(uv^2) < p$). Since $u(x)$ is monic, we have $u = x + b$, $v = \pm 1$, and then $h(x) = (x + b)^{(p-1)/2} - v$.

Suppose now that $(uv^2)' - 1$ is a non-zero polynomial divisible by $h + v$. Then $\deg((uv^2)' - 1) \geq \deg(h + v)$; that is, $p - 2k - 1 \geq \deg(h) \geq k$. This establishes assertion a).

Assertion b) follows from the next lemma.

Lemma 9.3.4. *Suppose that $m = (p - 1)/2$ and*

$$q(x) = \sum_{k=1}^{p-1} \left(\frac{k}{p}\right) x^k = \sum_{i=1}^{m} (x^{s_i} - x^{t_i}),$$

where $(\frac{k}{p})$ is the Legendre symbol, $s_1, \dots, s_m$ are the quadratic residues in $\mathbb{F}_p^*$, and let $t_1, \dots, t_m$ be the quadratic nonresidues. Then multiplication by $q(x)$ gives an isomorphism of $\mathbb{Z}$-modules:

$$\Gamma_0^k \cong \Gamma_0^{k+m}, \ k > 0.$$

Moreover, $||Fq(x)||^2 = p||F||^2$ for every $F = F(x, y) \in \Gamma_0^k$.

Proof. Suppose that $F = F(x, y) = f(x)\Phi(y)$ is a non-zero element of Γ_0^k such that $(x - 1)^k | f(x)$, and write $f(x) = \sum_i a_i x^i$. We observe that $x^m - 1 = \prod_i (x - s_i)$ and $x^m + 1 = \prod_i (x - t_i)$, so, by (9.18), $(x - 1)^m | q(x)$ and $||q(x)\Phi(y)||^2 = p(p - 1)$.

Therefore, $qF \in \Gamma_0^{k+m}$. Let us establish the equality $||Fq(x)||^2 = p||F||^2$. To do this, we consider the function

$$f_l(x) = f(x)q(x^l) = \sum_{i=0}^{p-1}(a_{i-ls_1} + \cdots + a_{i-ls_m} - a_{i-lt_1} - \ldots - a_{i-lt_m})x^i,$$

where $1 \leq l \leq m$. Then

$$\sum_{l=1}^{m} ||f_l(x)\Phi(y)||^2 = p\sum_l \sum_i (a_{i-ls_1} + \cdots - a_{i-lt_m})^2.$$

In this sum, the term a_i^2 occurs $2pm^2$ times, and the term $a_i a_j$ with $i < j$ occurs with coefficient $-2pm$; so, taking account of the fact that $\sum_i a_i = 0$, we get that

$$\sum_{l=1}^{m} ||f_l(x)\Phi(y)||^2 = 2pm^2 \sum_i a_i^2 - 2pm \sum_{i<j} a_i a_j = p(2m^2 + m)\sum_i a_i^2 = pm||F||^2.$$

We also observe that $q(x^l) = (\frac{l}{p})q(x) = \pm q(x)$, and so $||f_l(x)\Phi(y)|| = ||Fq(x)||$.

From the equality just proved it follows that the linear map corresponding to multiplication by $q(x)$ is injective. It remains to verify that it is surjective. Suppose that $G = G(x, y) \in \Gamma_0^{k+m}$. Then $q(x)G \in \Gamma_0^{k+p-1} = p\Gamma_0^k$, that is, $q(x)G = pH$ for some $H \in \Gamma_0^k$. We observe that $p^2(p-1) = ||q(x)^2\Phi(y)||^2$ and $q^2\Phi(y) \in \Gamma_0^{p-1} = p\Gamma_0^0$. Hence, $q^2(x) = pt(x)$, where $t(x)\Phi(y) \in \Gamma_0^0$ and $||t(x)\Phi(y)||^2 = p - 1$. Consequently, $t(x) = \pm x^r$ and $q(x)G = (\pm px^r)(\pm x^{-r}H) = q^2(x)F$, where $F = \pm x^{-r}H$. Since the map $\Gamma_0^{k+m} \to \Gamma_0^{k+2m}$ corresponding to multiplication by $q(x)$ is injective, we have $G = q(x)F$. □

Remark 9.3.5. Calculations on a computer show that, for $p = 17$, there are no vectors $F(x, y)$ in the lattice Γ_0^5 with $||F||^2 = 17 \times 10$, even though $5 < (17-1)/3$. Hence estimate (9.14) can fail to be exact even for $k < (p-1)/3$.

Proposition 9.3.6. a) *Suppose that $F \in \Gamma^{k,l}$. Then either $F \in \Gamma^{k+1}$ or $||F||^2 \geq 2pk(p-k+1)$.*

b) *Suppose that $0 \neq F \in \Gamma^k \cap \tilde{\Gamma}$ and $k \geq 1$. Then $||F||^2 \geq 2p^2k$.*

Proof. a) We write $F \in \Gamma^{k,l}$ in the form

$$F = \sum_{i=0}^{k} a_i p L^{k-i}(x)L^i(y) + G,\ G \in \Gamma^{k+1}.$$

Suppose that $||F||^2 < 2pk(p-k+1)$. Since every nonzero element of Γ_σ^k has norm not less than $2pk$ (see (9.14)), we have $\mathrm{pr}_\sigma F = 0$ for at least $k+1$ points

$\sigma_1, \ldots, \sigma_{k+1}$ of $\mathbb{P}V$. Without loss of generality, we can assume that $(1, b_j) \in \sigma_j$, $1 \le j \le k+1$ (if $(0, b) \in \sigma_1$, apply a suitable automorphism $\hat{\varphi}$; this is possible, since $\hat{\varphi}$ is an isometry of Γ). We have

$$\mathrm{pr}_{\sigma_j} F = \Phi(x^{-b_j} y) \sum_{i=0}^{k} a_i L^{k-i}(x) L^i(y) \equiv L^k(x) \Phi(x^{-b_j} y) \sum_{i=0}^{k} a_i b_j^i \equiv$$

$$\equiv (x-1)^k \Phi(x^{-b_j} y) \sum_{i=0}^{k} a_i b_j^i \pmod{\Gamma^{k+1}}.$$

Here we have used Proposition 9.1.9. In accordance with our choice of σ_j we have the system $\sum_{i=0}^{k} a_i b_j^i = 0$, $1 \le j \le k+1$, from which we obtain $a_i = 0$ for $0 \le i \le k$. This means that $F \in \Gamma^{k+1}$.

b) By hypothesis, $F = \sum_{i,j=0}^{p-1} p a_{ij} x^i y^j$, $a_{ij} \in \mathbb{Z}$. Consider the expansion

$$pL^m(x)L^n(y) = \sum_{i,j} pL_{ij}(m, n) x^i y^j, \; m + n > 0.$$

Then clearly $\sum_{i,j} L_{ij}(m, n) = 0$. Therefore, we have $\sum_{i,j} a_{ij} = 0$ for $k \ge 1$. In view of (9.13), this gives $||F||^2 = p^2 \sum_{i,j} a_{ij}^2$. Since $F \ne 0$, F has at least one nonzero projection, say $\mathrm{pr}_\sigma F \ne 0$, $(1, b) \in \sigma$. This projection lies in Γ_σ^k and has the form

$$\mathrm{pr}_\sigma F = \Phi(x^{-b} y) \sum_{l=0}^{p-1} x^l \sum_{i+bj=l} a_{ij}.$$

Hence, in view of (9.17), we obtain the inequality

$$\sum_{l=0}^{p-1} \Big| \sum_{i+bj=l} a_{ij} \Big| \ge 2k,$$

which leads to the required estimate:

$$||F||^2 = p^2 \sum_{i,j} a_{ij}^2 \ge p^2 \sum_{i,j} |a_{ij}| \ge p^2 \sum_{l} \left| \sum_{i+bj=l} a_{ij} \right| \ge 2p^2 k. \qquad \square$$

9.4 The duality picture

We recall that by definition the *dual* of the lattice $(\Delta, (*|*))$, $\Delta \subset \mathbb{R}^n$, is the lattice

$$\Delta^* = \{x \in \mathbb{R}^n | (x|y) \in \mathbb{Z}, \forall y \in \Delta\}.$$

Clearly, G-invariance of Δ implies that of Δ^*.

Theorem 9.4.1. *For the indivisible invariant sublattices listed in Theorems* 9.2.4 *and* 9.2.7, *the dual lattices are the following:*

a) $(\Gamma^{k,l})^* = \frac{1}{p^2}\Gamma^{p-l,p-k}$, *where* $0 \le k \le l \le p$ *and* $(k,l) \ne (0,1)$; $(\Gamma^{0,1})^* = \frac{1}{p}\Gamma^w$, $(\Gamma^w)^* = \frac{1}{p}\Gamma^{0,1}$;

b) $(\Delta^{k,r})^* = \frac{1}{p^2}\Delta^{k,(-1)^{k-1}r}$, *where* $1 \le k \le (p-3)/2$ *and* $r \in \mathbb{F}_p^*$.

The proof splits into a number of steps.

a1) We claim that $(\Gamma^{k,k})^* = \frac{1}{p^2}\Gamma^{p-k,p-k}$. By definition, $\Gamma^{k,k} = \Gamma^k = \oplus_{\sigma\in\mathbb{P}V}\Gamma^k_\sigma$. Moreover, $(\Gamma^k_\sigma;\Gamma^l_{\sigma'}) = 0$ for $\sigma \ne \sigma'$. Therefore, it is sufficient to verify that $(\Gamma^k_0)^* = \frac{1}{p^2}\Gamma^{p-k}_0$. For this we first analyze the case $k = 0$, writing an arbitrary element $F \in (\Gamma^0_0)^*$ in the form $F = (\sum_{i=0}^{p-1} a_i x^i)\Phi(y)$. The fact that $F \in (\Gamma^0_0)^*$ is equivalent to the following chain of implications:

$$p(a_j - \frac{1}{p}\sum_i a_i) = (F; x^j\Phi(y)) \in \mathbb{Z},\ 0 \le j \le p-1,$$

$$\Leftrightarrow a_j = \frac{1}{p}(m + m_j),\ \text{where}\ m_j \in \mathbb{Z},\ m = \sum_i a_i,\ \sum_j m_j = 0,$$

$$\Leftrightarrow F = \frac{1}{p}(\sum_j m_j x^j)\Phi(y),\ \text{that is,}\ F \in \frac{1}{p}\Gamma^1_0 = \frac{1}{p^2}\Gamma^p_0.$$

We now turn to the case $0 < k < p$. We have $\Gamma^0_0 \supset \Gamma^k_0 \supset \Gamma^p_0$. Therefore,

$$\Gamma^p_0 = p^2(\Gamma^0_0)^* \subset p^2(\Gamma^k_0)^* \subset p^2(\Gamma^p_0)^* = \Gamma^0_0.$$

Moreover, $p^2(\Gamma^k_0)^*$ is a lattice invariant under G_0. By Theorem 9.1.3 we have $p^2(\Gamma^k_0)^* = \Gamma^l_0$ for some l with $0 < l < p$. But

$$\Gamma^l_0 = \Gamma^p_0 + < (x-1)^s\Phi(y) | l \le s \le p-1 >_{\mathbb{Z}},$$

$$\Gamma^k_0 = \Gamma^p_0 + < (x-1)^t\Phi(y) | k \le t \le p-1 >_{\mathbb{Z}}.$$

Therefore, by (9.12), $(\Gamma^k_0)^* = \frac{1}{p^2}\Gamma^l_0$ if and only if

$$(-1)^{s+t}\binom{s+t}{s} = \frac{1}{p}((x-1)^s\Phi(y); (x-1)^t\Phi(y)) \in p\mathbb{Z}$$

for the values of s and t given above. Hence, $s + t \ge p$, and so $l = p - k$.

a2) We claim that $(\Gamma^{0,1})^* = \frac{1}{p}\Gamma^w$. For this we consider the chain of inclusions

$$\Gamma = \Gamma^0 \supset \Gamma^{0,1} = \Gamma^1 + < p >_{\mathbb{Z}} \supset \Gamma^1.$$

Taking account of a1), we get that

$$\Gamma^1 = \frac{1}{p}\Gamma^p = p\Gamma^* \subset p(\Gamma^{0,1})^* \subset p(\Gamma^1)^* = \frac{1}{p}\Gamma^{p-1} = \Gamma.$$

It is therefore sufficient to determine which elements of Γ/Γ^1 belong to $p(\Gamma^{0,1})^*$, or equivalently, which elements $F \in \Gamma/\Gamma^1$ are such that $(p; F) \in p\mathbb{Z}$ (here and below, to simplify the notation we identify an element of the quotient lattice with a suitable representative of the coset). We introduce the notation

$$\Phi_\sigma = \Phi(x^a y^b);\ (a, b) \in \sigma, \tag{9.19}$$

and write the functions p and F in the form $p = \sum_{\sigma\in\mathbb{P}V} \Phi_\sigma$, $F = \sum_{\sigma\in\mathbb{P}V} a_\sigma\Phi_\sigma$. Taking account of Proposition 9.3.1, $(p; F) \in p\mathbb{Z}$ if and only if $(p-1)\sum_\sigma a_\sigma \in p\mathbb{Z}$, that is, $p|\sum_\sigma a_\sigma$. In other words, $F \in \pi_1^{-1}(W) = \Gamma^w$.

a3) We claim that $p^2(\Gamma^{k,l})^* = \Gamma^{p-l,p-k}$ for $1 \le k < l \le p-1$. Again by a1) we have $\Gamma^{p-k} \subset p^2(\Gamma^{k,l})^* \subset \Gamma^{p-l}$. From (9.13) it is obvious that $(px^s y^t; pL^a(x)L^b(y)) \in p^2\mathbb{Z}$ for $a+b > 0$. Moreover,

$$\Gamma^{k,l} = \Gamma^l + < pL^a(x)L^b(y) | k \le a+b < l >_{\mathbb{Z}} .$$

Therefore, $p^2(\Gamma^{k,l})^* \supset \Gamma^{p-l,p-k}$. It remains to show that $p^2(\Gamma^{k,l})^*/\Gamma^{p-k} = \Gamma^{p-l,p-k}/\Gamma^{p-k}$. We write an arbitrary element F of $\Gamma^{p-l}/\Gamma^{p-k}$ in the form

$$F = \sum_{i=p-l}^{p-1-k} \left\{ \left(\sum_{j=0}^{p-1} a_{ij} L^i(x)\Phi(x^{-j}y) \right) + a_{ip}L^i(y)\Phi(x) \right\}.$$

The condition $F \in p^2(\Gamma^{k,l})^*$ is equivalent to the inclusion

$$(F; pL^{t-s}(x)L^s(y)) \in p^2\mathbb{Z} \text{ for } 0 \le s \le t,\ k \le t \le l-1.$$

By Proposition 9.1.9 and (9.12), for $t = l-1$ we have the linear system

$$\sum_{j=0}^{p-1} a_{p-l,j} j^s \equiv 0 \,(\mathrm{mod}\, p);\ 0 \le s \le l-2;$$

$$\sum_{j=1}^{p-1} a_{p-l,j} j^{l-1} + a_{p-l,p} \equiv 0 \,(\mathrm{mod}\, p)$$

of rank l with respect to indeterminates $a_{p-l,j}$, $0 \le j \le p$. This means that the space

$$(p^2(\Gamma^{k,l})^* + \Gamma^{p-l+1})/\Gamma^{p-l+1}$$

has dimension $p+1-l$ over $\mathbb{F}_p$. On the other hand, it follows from the inclusion $p^2(\Gamma^{k,l})^* \supseteq \Gamma^{p-l,p-k}$ that

$$(p^2(\Gamma^{k,l})^* + \Gamma^{p-l+1})/\Gamma^{p-l+1} \supseteq (\Gamma^{p-l,p-k} + \Gamma^{p-l+1})/\Gamma^{p-l+1} \cong S^{p-l}(V^*),$$

and $\dim S^{p-l}(V^*) = p-l+1$. Therefore, $p^2(\Gamma^{k,l})^* + \Gamma^{p-l+1} = \Gamma^{p-l,p-k} + \Gamma^{p-l+1}$. Now the representatives of the coset in $\Gamma^{p-l}/\Gamma^{p-l+1}$ containing F are of the form

$$D = \sum_{s=0}^{p-l} b_s p L^{p-l-s}(x) L^s(y),$$

and therefore certainly $D \in p^2(\Gamma^{k,l})^*$. For this reason we ignore them.

Suppose now that $t = l-2$. Arguing in the same way, we see that the dimension of

$$((p^2(\Gamma^{k,l})^* \cap \Gamma^{p-l+1}) + \Gamma^{p-l+2})/\Gamma^{p-l+2}$$

over $\mathbb{F}_p$ is $p-l+2 = \dim_{\mathbb{F}_p} S^{p-l+1}(V^*)$, so that

$$(p^2(\Gamma^{k,l})^* \cap \Gamma^{p-l+1}) + \Gamma^{p-l+2} = (\Gamma^{p-l,p-k} \cap \Gamma^{p-l+1}) + \Gamma^{p-l+2}.$$

Continuing in the same spirit, we finally obtain that $p^2(\Gamma^{k,l})^* = \Gamma^{p-l,p-k}$.

The case $\Gamma^{0,l}$ with $2 \le l \le p-1$ is analyzed in exactly the same way.

Before proceeding to assertion b) of Theorem 9.4.1, we prove an auxiliary proposition about the explicit form of the lattices $\Delta^{k,r}$.

Proposition 9.4.2. *For* $0 < k < (p-1)/2$ *and* $r \in \mathbb{F}_p^*$,

$$\Delta^{k,r} = \Gamma^{k+1,p-k} + < B_r, C_{j,r} | 0 \le j \le k-1 >,$$

where

$$B_r = B(k,r) = L^{p-1-k}(y)\Phi(x) + r^{-1}pL^k(x);$$

$$C_{j,r} = C_j(k,r) = L^{p-1-k}(x)\Phi(x^{-j}y) + r^{-1}p(jL(x) - L(y))^k.$$

Proof. We consider the basis $\{u^{p-1-k-s}v^s | 0 \le s \le p-1-k\}$ of $S^{p-1-k}(V^*)$, which is at the same time a basis of the space of functions on the set $\{(1,k), \dots, (1,p-1)\}$. Consequently, for every coset in $\mathcal{F}^{p-1-k}(V)/S^{p-1-k}(V^*)$, there is a representative that vanishes at the points mentioned. This means that the quotient space

$$\bar{\mathcal{F}}^k(V) = \mathcal{F}^{p-1-k}(V)/S^{p-1-k}(V^*)$$

has a basis consisting of elements $u^{p-1-k} f_i$, $0 \le i \le k-1$, $v^{p-1-k} f_p$, where by $f_i = f_{<(1,i)>}$, $0 \le i \le k-1$, $f_p = f_{<(0,1)>}$, we understand the characteristic functions f_σ introduced in the course of the proof of Theorem 9.1.7.

We recall that there is a pairing $(g, h) \mapsto ((g, h)) = \pi(gh)$ between $\mathcal{F}^k(V)$ and $\mathcal{F}^{p-1-k}(V)$; π is the projection of $\mathcal{F}^0(V) = S^0(V^*) \oplus W$ along W. In our basis of $\bar{\mathcal{F}}^k(V)$ we have

$$\left(\left(\sum_{j=0}^{k} a_j u^{k-j} v^j, u^{p-1-k} f_i\right)\right) = \pi\left(\sum_{j=0}^{k} a_j u^{p-1} i^j f_i\right) = \sum_{j=0}^{k} a_j i^j;$$

$$\left(\left(\sum_{j=0}^{k} a_j u^{k-j} v^j, v^{p-1-k} f_p\right)\right) = \pi\left(a_k v^{p-1} f_p\right) = a_k.$$

Hence we arrive at the isomorphism $\bar{\mathcal{F}}^k(V) \cong (S^k(V^*))^*$.

We now construct the isomorphism $(S^k(V^*))^* \cong S^k(V^*)$ explicitly by using the skew-symmetric bilinear form $< *; * >$ on V^* (recall that

$$< au + bv; cu + dv >= ad - bc),$$

which carries over to $S^k(V^*)$ in the usual way:

$$< u^{k-i} v^i; u^{k-j} v^j >= \frac{1}{k!} \sum \underbrace{< u; * > \cdots < u; * >}_{k-i} \underbrace{< v; * > \cdots < v; * >}_{i} =$$

$$= \begin{cases} 0 & \text{if } \ j \neq k - i; \\ (-1)^i / \binom{k}{i} & \text{if } \ j = k - i. \end{cases}$$

By this means we set up a correspondence between the element $u^{k-i} v^i \in S^k(V^*)$, $0 \le i \le k$, and the function

$$\sum_j a_j u^{k-j} v^j \mapsto \frac{(-1)^i}{\binom{k}{i}} a_{k-i}.$$

The composition τ^{-1} (the notation τ was introduced just before Definition 9.2.3) of the two isomorphisms

$$\bar{\mathcal{F}}^k(V) \cong (S^k(V^*))^* \cong S^k(V^*)$$

has the form:

$$v^{p-1-k} f_p \mapsto u^k;$$

$$u^{p-1-k} f_i \mapsto \left\{ \sum_{j=0}^{k} a_j u^{k-j} v^j \mapsto \left(\left(\sum_j a_j u^{k-j} v^j; u^{p-1-k} f_i\right)\right) = \sum_j a_j i^j \right\} \mapsto$$

$$\mapsto \sum_{j=0}^{k} i^j(-1)^{k-j}\binom{k}{j} u^j v^{k-j} = (iu-v)^k.$$

Finally, we go over to the ring $\mathcal{A} = \mathbb{C}[x, y]/I$ using the isomorphism Θ, and obtain the explicit form of the isomorphism $\Theta^{-1} \circ \tau^{-1} \circ \Theta$:

$$L^{p-1-k}(y)\Phi(x) \mapsto pL^k(x);$$

$$L^{p-1-k}(x)\Phi(x^{-j}y) \mapsto p(jL(x) - L(y))^k.$$

By definition,

$$\begin{aligned} \Delta^{k,r} &= \Gamma^{k+1,p-k} + < f + r\tau(f) | f \in S^k(V^*), r \in \mathbb{F}_p^* >_{\mathbb{Z}} \\ &= \Gamma^{k+1,p-k} + < g + r^{-1}\tau^{-1}(g) | g \in \bar{\mathcal{F}}^k(V) >_{\mathbb{Z}} . \end{aligned} \qquad \square$$

Completion of the Proof of Theorem 9.4.1. b) Since $\Gamma^{k,p-1-k} \supset \Delta^{k,r} \supset \Gamma^{k+1,p-k}$, it follows that $\Gamma^{k+1,p-k} \subset p^2(\Delta^{k,r})^* \subset \Gamma^{k,p-1-k}$. We determine which elements of $\Gamma^{k,p-1-k}/\Gamma^{k+1,p-k}$ belong to $p^2(\Delta^{k,r})^*$. As before, we consider an arbitrary element $F \in \Gamma^{k,p-1-k}/\Gamma^{k+1,p-k}$. For any $s \in \mathbb{F}_p^*$, F can be written in the form

$$F = bB_s + \sum_{j=0}^{k-1} c_j C_{j,s} + dL^{p-1-k}(y)\Phi(x) + \sum_{j=0}^{k-1} a_j L^{p-1-k}(x)\Phi(x^{-j}y).$$

The condition $F \in p^2(\Delta^{k,r})^*$ is equivalent to the following system of inclusions:

$$(F; B_r) \in p^2\mathbb{Z},\ (F; C_{j,r}) \in p^2\mathbb{Z}.$$

Set

$$\mu = \left((x-1)^k\Phi(y); (x-1)^{p-1-k}\Phi(y)\right) = (-1)^k p\binom{p-1}{k} \equiv p \pmod{p^2}.$$

Then

$$(F; B_r) = \mu \sum_{j=0}^{k-1} \left\{c_j\left(r^{-1} + s^{-1}(-1)^k\right) + r^{-1}a_j\right\};$$

$$(F; C_{j,r}) = \mu\left\{b\left(s^{-1} + r^{-1}(-1)^k\right) + r^{-1}d(-1)^k + \right.$$

$$\left. + \sum_{j=0}^{k-1}\left(r^{-1}a_j + c_j\left(r^{-1} + s^{-1}(-1)^k\right)\right)(i-j)^k\right\}.$$

Now choose $s = (-1)^{k-1}r$. Then the condition $F \in p^2(\Delta^{k,r})^*$ takes the form of the following system of congruences:

$$\sum_{j=0}^{k-1} a_j \equiv 0 \,(\mathrm{mod}\, p);$$

$$(-1)^k d + \sum_{j=0}^{k-1} a_j (i-j)^k \equiv 0 \,(\mathrm{mod}\, p);\ 0 \le i \le k-1,$$

or, equivalently,

$$a_j \equiv 0 \,(\mathrm{mod}\, p),\ 0 \le j \le k-1;\ d \equiv 0 \,(\mathrm{mod}\, p).$$

Thus,

$$F \in p^2(\Delta^{k,r})^* \Leftrightarrow F = bB_s + \sum_{j=0}^{k-1} c_j C_{j,s},\ s = (-1)^{k-1}r.$$

In other words, $p^2(\Delta^{k,r})^* = \Delta^{k,(-1)^{k-1}r}$. □

In what follows it is convenient to use the notation $\frac{1}{p}\Delta = \{\frac{1}{p}\mathbf{x} | \mathbf{x} \in \Delta\}$, where Δ is an arbitrary lattice.

Corollary 9.4.3. *The lattices*

$$\frac{1}{p}\Gamma^{k,p-k},\ 1 \le k \le \frac{p-1}{2};$$

$$\frac{1}{p}\Delta^{k,r},\ 1 \le k \le \frac{p-3}{2},\ k \equiv 1 \,(\mathrm{mod}\, 2),$$

are even and unimodular.

Proof. The fact that the lattices are unimodular follows from Theorem 9.4.1. The fact that they are even is a general property of all sublattices of Γ. □

9.5 Study of unimodular invariant lattices

Let us agree to define the *norm* of a vector x of the lattice $(\Lambda, (*|*))$ as the quantity $(x|x) = N(x) = ||x||^2$. For various reasons there is special interest in vectors with least norm. We shall sometimes call them *minimal vectors*. As usual, a vector x in Λ with $(x|x) \le 2$ is called a *root*. The set of all roots of an (integral) lattice Λ forms a root system in the classical sense (see [Bour 2]).

Theorem 9.5.1. a) *For each* $r \in \mathbb{F}_p^*$, *the even unimodular lattice* $\frac{1}{p}\Delta^{1,r}$ *has a minimal vector with norm* 4.

b) *Minimal vectors of the even unimodular lattice* $\frac{1}{p}\Gamma^{1,p-1}$ *have norm* 2 *and form a root system of type* E_8 *if* $p = 3$, *and of type* A_{p^2-1} *if* $p > 3$.

c) *Minimal vectors of the even unimodular lattice* $\frac{1}{p}\Gamma^{(p-1)/2,(p+1)/2}$ *have norm* 2 *and form a root system of type* $(p+1)A_{p-1}$, *for any* $p > 3$.

Proof. a0) The existence in $\Delta^{1,r}$ of a vector (function) F with $(F;F) = 4p^2$ (or equivalently, the existence in $\frac{1}{p}\Delta^{1,r}$ of a vector with norm 4) is quite obvious: it is sufficient to take $F = p(x-1)(x^2-1)$. Suppose that there is an F in $\Delta^{1,r}$ with $(F;F) = 2p^2$. By Proposition 9.4.2, we have $\Delta^{1,r} = \Gamma^{2,p-1} + < B, C >_{\mathbb{Z}}$, where

$$B = rB_r = rL^{p-2}(y)\Phi(x) + pL(x),$$

$$C = -rC_{0,r} = -rL^{p-2}(x)\Phi(y) + pL(y).$$

Using the fact that $\Delta^{1,r}$ is $SL_2(p)$-invariant, and applying suitable automorphisms to F, we may assume without loss of generality that $F = bB + G$, $b \in \mathbb{Z}$, $G \in \Gamma^{2,p-1}$.

a1) Suppose that $b \equiv 0 \pmod p$. Then $bB \in \Gamma^{2,p-1}$, and we may suppose that $b = 0$ and $F \in \Gamma^{2,p-1}$. Since $2pk(p+1-k) > 2p^2$ for $1 < k < p-1$, successive application of Proposition 9.3.6a) gives $F \in \Gamma^{p-1}$. From Proposition 9.3.3b) we obtain $||F||^2 \geq p^2(p-1) > 2p^2$, which is a contradiction, since the lattices $\Delta^{1,s}$ are considered only for $p > 3$.

a2) Now we suppose that $b \not\equiv 0 \pmod p$. We may assume that

$$B = p(x-1) + r(y-1)^{p-2}\Phi(x)$$

modulo $\Gamma^{2,p-1}$. We rewrite F in the form

$$F = b\left\{p(x-1) + r(y-1)^{p-2}\Phi(x)\right\} + \sum_{j=2}^{p-2}\sum_{i=0}^{j} a_i^j pL^{j-i}(x)L^i(y) \pmod{p\Gamma}.$$

Set $\mathrm{pr}_{\sigma_i} F = f_i(x)\Phi(x^{-i}y)$, $0 \leq i \leq p-1$, $(1,i) \in \sigma_i$; $\mathrm{pr}_{\sigma_p} F = f_p(y)\Phi(x)$, $(0,1) \in \sigma_p$. It is immediately obvious that $(x-1)|f_i(x)$ and $(x-1)^2 \nmid f_i(x)$ for $0 \leq i \leq p-1$. From Proposition 9.3.3a) it follows that $||F||^2 \geq 2p(p+1) > 2p^2$ if $(y-1)^{p-1} \nmid f_p(y)$. It remains to suppose that $(y-1)^{p-1}|f_p(y)$, and this means that $a_j^j \equiv 0 \pmod p$ for $2 \leq j \leq p-3$ and

$$a_{p-2}^{p-2} \equiv -br \pmod p. \tag{9.20}$$

On the other hand, by (9.16) we have $f_i(x) = x^{s_i} - x^{t_i}$, $0 \leq s_i \neq t_i \leq p-1$. From the explicit expression for F we see that $x^{s_i} - x^{t_i} \equiv b(x-1) \pmod{(x-1)^2}$, so $s_i \equiv t_i + b \pmod p$. Thus,

$$F = (x^b - 1)\sum_{i=0}^{p-1} x^{t_i}\Phi(x^{-i}y).$$

We compare the resulting expression for F and the notation assumed for it modulo $(x-1)^3$:

$$b(x-1)+(a_0^2+ia_1^2)(x-1)^2 \equiv (x^b-1)x^{t_i} \pmod{(x-1)^3}.$$

Hence we have

$$a_0^2+ia_1^2 \equiv bt_i + b(b-1)/2 \pmod{p};$$

that is, all the t_i are determined mod p by specifying a_0^2, a_1^2, and b. We show by induction on j that all the a_i^j with $2 \le j \le p-3$ are also determined by specifying these three parameters. In fact,

$$\left(a_0^{j+1}+ia_1^{j+1}+\cdots+i^j a_j^{j+1}\right)(x-1)^j \equiv$$

$$\equiv x^{t_i}(x^b-1)-b(x-1)-\cdots \left(\bmod\,(x-1)^{j+1}\right).$$

The right-hand side of this congruence has the form

$$g_i(b,a_0^2,a_1^2)(x-1)^j \left(\bmod\,(x-1)^{j+1}\right),$$

by the inductive hypothesis. Putting $i=0,1,\ldots,j$, we obtain a linear system with given right-hand sides and nonzero (Vandermonde) determinant. Hence, $a_0^{j+1},\ldots,a_j^{j+1}$ are certain functions of a_0^2, a_1^2, and b. We recall that $a_{j+1}^{j+1} \equiv 0 \pmod{p}$ for $1 \le j \le p-4$.

Finally, comparison modulo $(x-1)^{p-1}$ of the two expressions for F gives

$$a_0^{p-2}+ia_1^{p-2}+\cdots+i^{p-2}a_{p-2}^{p-2} \equiv h_i(b,a_0^2,a_1^2) \pmod{p}, \tag{9.21}$$

for $0 \le i \le p-1$. We observe that h_i does not depend on r as a parameter of the lattice $\Delta^{1,r}$. Obviously, the system (9.21) of p linear equations with $p-1$ indeterminates can have at most one solution. We prove the existence of a solution by considering

$$F=(x^b-1)\sum_{i=0}^{p-1} x^{t_i}\Phi(x^{-i}y)$$

with

$$t_i=b^{-1}\left\{a_0^2+ia_1^2-\frac{b(b-1)}{2}\right\}=ia'+b'.$$

We have

$$\begin{aligned} F &= (x^b-1)\textstyle\sum_{i=0}^{p-1} x^{ia'+b'}\Phi(x^{-i}y) = (x^b-1)x^{b'}y^{a'}\sum_{i=0}^{p-1}\Phi(x^{-i}y) = \\ &= p(x^b-1)x^{b'}y^{a'} \in \Gamma^{1,p-1}=\Delta^{1,0}, \end{aligned}$$

and $||F||^2=2p^2$. Hence the system of coefficients a_i^{p-2} for F is the solution of (9.21). We now observe that the deduction of (9.20) did not depend on the value of r, and we get that $F \in \Delta^{1,0}$, that is, $a_{p-2}^{p-2} \equiv -b.0 \equiv 0 \pmod{p}$. We conclude that there cannot be roots in the lattice $\frac{1}{p}\Delta^{1,r}$ for $r \neq 0$.

b1) We assume first that $p > 3$. In the course of the arguments in a2), we showed that if $F \in \Delta^{1,0} = \Gamma^{1,p-1}$ and $||F||^2 = 2p^2$, then the action of some automorphism from $SL_2(p)$ reduces F to the form $F = p(x^b - 1)x^{b'}y^{a'}$. On the other hand, the G-orbit of the given element is the set

$$\mathcal{M} = \left\{p(x^a y^b - x^c y^d) | (c, d) \neq (a, b)\right\}.$$

Consequently, $\mathcal{M}$ is the set of minimal vectors in $\Gamma^{1,p-1}$, and it is easy to check that it is a root system of type A_{p^2-1}. The set of vectors $p(x^{a_i}y^{b_i} - x^{a_{i+1}}y^{b_{i+1}})$, $1 \leq i \leq p^2 - 1$, where $\{(a_1, b_1), \ldots, (a_{p^2}, b_{p^2})\} = \mathbb{F}_p^2$, serves as a system of simple roots.

b2) Next we treat the case $p = 3$. In accordance with the general theory [Ser 2], the lattice $\frac{1}{3}\Gamma^{1,2}$ must be isometric to the root lattice of type E_8.

The distinction between the cases $p > 3$ and $p = 3$ comes about because the relation $bt_i + b(b-1)/2 = a_0^2 + ia_1^2$ in a2), which also plays an essential role in b1), was obtained under the assumption that $2 < p - 2$; that is, $p \geq 5$.

c) In view of (9.15) and Proposition 9.3.6a), the conditions $F \in \Gamma^{(p-1)/2,(p+1)/2}$ and $||F||^2 = 2p^2$ imply that $F \in \Gamma_\sigma^{(p+1)/2} \backslash \Gamma_\sigma^{(p+3)/2}$ provided that $p > 3$. Hence, the set $\mathcal{N}$ of minimal vectors in $\Gamma^{(p-1)/2,(p+1)/2}$ is the union of the sets of minimal vectors in $\Gamma_\sigma^{(p+1)/2}$ with σ running over $\mathbb{P}V$. Applying Lemma 9.3.4, we find that the set of minimal vectors in $\Gamma_\sigma^{(p+1)/2}$ is a root system of type A_{p-1}. For $\sigma = \sigma_0$ this is simply the set

$$\{(x^a - x^b)q(x)\Phi(y) | 0 \leq a \neq b \leq p - 1\}.$$

Thus $\mathcal{N}$ is a root system of type $(p + 1)A_{p-1}$. □

Corollary 9.5.2. *The following assertions hold.*

a) *Among the invariant lattices of type* A_2 $(p = 3)$ *there is exactly one even unimodular lattice* E_8*; its automorphism group is the Weyl group* $W(E_8) = \mathbb{Z}_2.O_8^+(2)$.

b) *Among the invariant lattices of type* A_4 $(p = 5)$ *there are six even unimodular lattices:* $\frac{1}{5}\Gamma^{1,4}$ *of type* A_{24}*, with automorphism group* $\mathbb{Z}_2 \times \mathbb{S}_{25}$*;* $\frac{1}{5}\Gamma^{2,3}$ *of type* $6A_4$*, with automorphism group* $(\mathbb{S}_5)^6.(GL_2(5)/\mathbb{Z}_2)$*; and four lattices isometric to the Leech lattice* $(\frac{1}{5}\Delta^{1,r};\ r \in \mathbb{F}_5^*)$ *whose automorphism groups are central extensions* $\mathbb{Z}_2 \circ Co_1$ *of the Conway sporadic simple group* Co_1.

Proof. a) This assertion follows immediately from Theorem 9.5.1. For clarity we give the matrix realization of the system of simple roots of type E_8 (see Table 9.4).

b) For the 24-dimensional even unimodular lattices, Niemeier [Nie] and later using simpler arguments Venkov [Ven 1, 8] have proved that there are exactly 24 of them, and these lattices are uniquely determined by their root configurations (for the Leech lattice the configuration is empty). The automorphism group of the Leech lattice has been calculated by Conway [Con 2]. The isomorphism $\mathrm{Aut}(A_{24}) \cong \mathbb{Z}_2 \times \mathbb{S}_{25}$ is well known. Erokhin [Ero] has found that $\mathrm{Aut}(6A_4) \cong (\mathbb{S}_5)^6.H$, where H is a subgroup of order 240 of $(\mathbb{Z}_2)^6.\mathbb{S}_6$. We do not dwell on the proof of the isomorphism $H \cong GL_2(5)/\mathbb{Z}_2$.

The fact that all the $\Delta^{1,r}$ with $r \in \mathbb{F}_5^*$ are isomorphic to the Leech lattice follows from the Conway characterization of the Leech lattice [Con 1], but in our case we can arrive at this conclusion on the basis of the simple remark made at the end of §9.2.

Table 9.4
Lattice $\Gamma = \Gamma^{1,p-1}$ for $p = 3$

Dynkin scheme of simple roots		Normalization of the inner product	
e_1 — e_3 — e_4 — e_5 — e_6 — e_7 — e_8 (chain), with e_2 attached to e_4		$(x\|y) = \frac{1}{3}\mathrm{Tr}(xy)$ $\varepsilon = \exp(\frac{2\pi i}{3})$	
Simple root	$\det(e_i)$	Simple root	$\det(e_i)$
$e_1 = \begin{pmatrix} 0 & -\varepsilon^2 & -1 \\ -\varepsilon & 0 & -\varepsilon \\ -1 & -\varepsilon^2 & 0 \end{pmatrix}$	-2	$e_5 = \begin{pmatrix} 0 & -\varepsilon^2 & -\varepsilon^2 \\ -\varepsilon & 0 & -1 \\ -\varepsilon & -1 & 0 \end{pmatrix}$	-2
$e_2 = \begin{pmatrix} -2 & 0 & 0 \\ 0 & 1 & 0 \\ 0 & 0 & 1 \end{pmatrix}$	-2	$e_6 = \begin{pmatrix} 0 & -\varepsilon & -1 \\ -\varepsilon^2 & 0 & -\varepsilon^2 \\ -1 & -\varepsilon & 0 \end{pmatrix}$	-2
$e_3 = \begin{pmatrix} 0 & -\varepsilon & -\varepsilon \\ -\varepsilon^2 & 0 & -1 \\ -\varepsilon^2 & -1 & 0 \end{pmatrix}$	-2	$e_7 = \begin{pmatrix} 0 & -1 & 0 \\ -1 & -1 & \varepsilon^2 \\ 0 & \varepsilon & 1 \end{pmatrix}$	-1
$e_4 = \begin{pmatrix} 1 & -1 & 0 \\ -1 & 0 & 1 \\ 0 & 1 & -1 \end{pmatrix}$	0	$e_8 = \begin{pmatrix} 0 & -\varepsilon^2 & -\varepsilon \\ -\varepsilon & 0 & -\varepsilon^2 \\ -\varepsilon^2 & -\varepsilon & 0 \end{pmatrix}$	-2
Maximal root $e_{\max} = 2e_1 + 3e_2 + 4e_3 + 6e_4 + 5e_5 + 4e_6 + 3e_7 + 2e_8$ $= \begin{pmatrix} 0 & \varepsilon & \varepsilon^2 \\ \varepsilon^2 & 0 & \varepsilon \\ \varepsilon & \varepsilon^2 & 0 \end{pmatrix} = \varepsilon^2 J_{(0,1)} + \varepsilon J_{(0,2)}.$			

Theorem 9.5.3. *All unimodular G-invariant lattices in the Lie algebra of type A_{p-1} are indecomposable. All of them except $\frac{1}{p}\Gamma^{1,p-1}$ and $\frac{1}{p}\Gamma^{(p-1)/2,(p+1)/2}$ have no roots.*

Proof. 1) We show that if Δ is a G-invariant lattice and $\Delta = \oplus_{i \in I} \Delta_i$ is a (nontrivial) decomposition into a direct sum of indecomposable sublattices, then Δ is similar to one of the lattices Γ^k, $0 \leq k < p-1$ (and then I can be identified with the projective line $\mathbb{P}V$).

We should mention that since Δ is Euclidean, the decomposition into a direct sum of indecomposable sublattices is unique [Cas]. Hence, G fixes the decomposition $\Delta = \oplus_{i \in I} \Delta_i$. Because of the absolute irreducibility of G on $\mathcal{L}$, G acts transitively on the set $\{\Delta_i | i \in I\}$. Set $G_1 = St_G(\Delta_1)$. Since p and the index $(G : G_1) = |I|$ are coprime, by the Borel-Tits Theorem the subgroup G_1 contains the normal subgroup $\hat{V} = O_p(G)$ of G. Furthermore, the $\hat{V}$-character afforded by $\Delta_1 \otimes_{\mathbb{Z}} \mathbb{C}$ is rational. Finally, the transitive action of G on $\mathbb{P}V$ is primitive. These three remarks immediately give the required assertion.

2) Every unimodular G-invariant lattice Δ fails to be similar to Γ^k for any k. Therefore, in accordance with 1), Δ is indecomposable. Furthermore, the absence of roots has already been proved for $\frac{1}{p}\Delta^{1,r}$ (Theorem 9.5.1a). Turning to Proposition 9.3.3, we see that the minimum norm in $\Gamma^{k,l}$ is always at least

$$\min\{p^2(2l - (p-1)), 2pk'(p+1-k') | k \leq k' < l\}.$$

If $l = p - k$ and $1 < k < (p-1)/2$, then the lower bound indicated is obviously greater than $2p^2$, and so there are no roots in $\frac{1}{p}\Gamma^{k,p-k}$.

The same conclusion also applies to the lattice $\frac{1}{p}\Delta^{k,r}$ with k odd and $2 < k \leq (p-5)/2$, since $\Delta^{k,r} \subset \Gamma^{k,p-1-k}$. In the remaining case, that of $\frac{1}{p}\Delta^{(p-3)/2,r}$, it is convenient to make use of Proposition 9.3.6a). We omit the details. □

Remark 9.5.4. From the arguments used in the proof of Theorem 9.5.3 it follows immediately that *provided $k \approx \frac{p}{3}$, the minimum norm of vectors in the even unimodular lattices $\frac{1}{p}\Gamma^{k,p-k}$ and $\frac{1}{p}\Delta^{k,r}$ is at least $p/3$.* This means that as the dimension n of these even unimodular lattices tends to infinity, their minimum length grows like $n^{1/4}$. If $\mathbf{D}$ denotes the density of sphere packings corresponding to them, then

$$\frac{1}{n}\log_2 \mathbf{D} \approx -\frac{1}{4}\log_2 n \text{ as } n \to \infty.$$

9.6 Automorphism groups of projections of invariant lattices

Before studying the automorphism groups $\mathrm{Aut}(\Lambda)$ of invariant lattices Λ in the Lie algebra $\mathcal{L}$ of type A_{p-1}, we investigate the automorphism groups $\mathrm{Aut}(\mathrm{pr}_0\Lambda)$ of their projections onto the component $\mathcal{H}_0$ of the given decomposition $\mathcal{D}$.

By Theorem 9.1.3, the Cartan subalgebra $\mathcal{H}_0$ contains exactly $p-1$ indivisible invariant sublattices $\Gamma_k = \Gamma_0^k$, $0 \leq k \leq p-2$, and the projection $\mathrm{pr}_0\Lambda$ of an invariant lattice Λ of type A_{p-1} is similar to one of them. Furthermore, in view of Lemma 9.3.4 and Theorem 9.4.1 we have:

$$\mathrm{Aut}(\Gamma_k) \cong \mathrm{Aut}(\Gamma_{k+\frac{p-1}{2}}) \cong \mathrm{Aut}(\Gamma_{p-k}) \cong \mathrm{Aut}(\Gamma_{\frac{p+1}{2}-k}). \tag{9.22}$$

It is easy to see that

$$\mathrm{Aut}(\Gamma_0) = \mathrm{Aut}(\Gamma_1) = \mathbb{Z}_2 \times \mathbb{S}_p. \tag{9.23}$$

Our goal is to prove the following theorem, which completes the description of $\mathrm{Aut}(\Gamma_k)$ because of relations (9.22) and (9.23). Set $S = \{2, 3\}$.

Theorem 9.6.1. *Set $K = \mathrm{Aut}(\Gamma_k)$ for $1 < k < (p-1)/2$. Then*
(i) $K = \mathbb{Z}_2 \times (\mathbb{Z}_p.\mathbb{Z}_{p-1})$,
or
(ii) $p \equiv -1 \pmod{2s}$, *$k = (p+1)r/2s$ and $K = \mathbb{Z}_2 \times PGL_2(p)$ for some $s \in S$ and $r \in \{1, s-1\}$.*

Proof. In view of (9.22), one can suppose without loss of generality that

$$\frac{p+1}{4} \leq k < \frac{p-1}{2}. \tag{9.24}$$

In particular, $p \geq 7$. Let θ be a generator of the multiplicative group $\mathbb{F}_p^*$, $P = < g_{1,1} >\cong \mathbb{Z}_p$, $Q =< g_{0,\theta} >\cong \mathbb{Z}_{p-1}$, $Z =< \mathbf{t} : f(x) \mapsto -f(x) >\cong \mathbb{Z}_2$, $K_0 = Z \times (P.Q)$. Recall that the maps $g_{a,b}$ are defined in (9.1) and in the proof of Theorem 9.1.3; furthermore, we omit the symbol $\Phi(y)$ in the notation of elements from Γ_k. Clearly, $K_0 \subseteq K$. Our proof splits into several steps.

Lemma 9.6.2. *$N_K(P) = K_0$. In particular, P is a Sylow p-subgroup of K.*

Proof. Since K_0 is contained in $N_K(P)$ and acts on P as the whole group $\mathrm{Aut}(P)$ (with kernel $Z \times P$), it is sufficient to prove that $C_K(P) = Z \times P$. To this end we consider an element $\varphi \in C_K(P)$. Set $\varphi(p) = f(x) \in \Gamma_k$. Then

$$\varphi(px^l) = \varphi((g_{1,1})^l(p)) = (g_{1,1})^l(\varphi(p)) = x^l f(x).$$

In particular, $\varphi(p(x-1)^k) = (x-1)^k f(x)$. As $\varphi \in K$, φ fixes the lattice $\Gamma_{p-k} = p^2(\Gamma_k)^*$. But $p \in \Gamma_{p-k}$, and so $f(x) \in \Gamma_{p-k}$, that is,

$$f(x) \equiv \sum_{i=p-k}^{p-2} a_i(x-1)^i \pmod{\Gamma_{p-1}},$$

where $a_i \in \mathbb{F}_p$. Moreover, $p(x-1)^k \in p\Gamma_k$; therefore,

$$p\Gamma_k = \Gamma_{p-1+k} \ni (x-1)^k f(x) \equiv \sum_{i=p-k}^{p-2} a_i(x-1)^{i+k} \pmod{\Gamma_{p-1+k}}.$$

This means that $a_i = 0$ for $p-k \le i \le p-2$, that is, $f(x) \in \Gamma_{p-1}$, or equivalently, $f(x) = pg(x)$ for some $g(x) \in \Gamma_0$. Furthermore,

$$||g(x)||^2 = p^{-2}||f(x)||^2 = p^{-2}||p||^2 = p-1.$$

This implies that $g(x) = \pm x^n$, $0 \le n \le p-1$. Hence, we have $\varphi(px^l) = \pm x^{n+l}$, so that $\varphi \in Z \times P$. □

Lemma 9.6.3. *The* $\mathbb{F}_p$*-space*

$$W = \Gamma_k/\Gamma_{p-k}$$

is a faithful $(p-2k)$*-dimensional* K*-module. Moreover, the* P*-module* W *is indecomposable, and the linear transformation* $g_{1,1}$ *acting on* W *has minimal polynomial* $(t-1)^{p-2k}$*.*

Proof. 1) *Exactness.* Consider an element φ of K such that $\varphi(f(x)) \equiv f(x) \pmod{\Gamma_{p-k}}$ for each $f \in \Gamma_k$. We need to show that $\varphi = 1$. For $f(x)$ we take a minimal vector u of Γ_k. Then $v = \varphi(u) - u \in \Gamma_{p-k}$. By Proposition 9.3.3, we have $||\varphi(u)||^2 = ||u||^2 \le p(p-1)$ and $||v||^2 \ge 2p^2(p-k-(p-1)/2)$ if $v \ne 0$. But

$$||v||^2 \le 2(||u||^2 + ||\varphi(u)||^2) < 2p^2(p-k-(p-1)/2),$$

therefore we must have that $v = 0$ and $\varphi(u) = u$.

As $||ux^l|| = ||u||$, we have also $\varphi(ux^l) = ux^l$ for every l with $0 \le l \le p-1$. Since $< ux^l | 0 \le l \le p-1 >_{\mathbb{R}} = \mathbb{R}^{p-1}$, we conclude that $\varphi = 1$.

2) *Indecomposability.* In the basis $\{(x-1)^i | k \le i \le p-1-k\}$ of W, the element $g_{1,1}$ has matrix

$$\begin{pmatrix} 1 & 0 & 0 & \dots & 0 \\ 1 & 1 & 0 & \dots & 0 \\ 0 & 1 & 1 & \dots & 0 \\ \vdots & \vdots & \vdots & \ddots & \vdots \\ 0 & 0 & 0 & \dots & 1 \end{pmatrix}.$$

□

Using the inequality $p-2k < \frac{2}{3}(p-1)$, Lemmas 9.6.2, 9.6.3 and Feit's Theorem [Feit 1], we can conclude that any subgroup H with $K \supseteq H \supseteq P$ is a group of type $L_2(p)$, that is, each composition factor of H is $L_2(p)$, a p-group or a p'-group. Moreover, the following assertion holds.

Lemma 9.6.4. *Either* H *has (just one) composition factor* $L_2(p)$*, or else* $H \subseteq K_0$*.*

Proof. Suppose that H has no composition factors isomorphic to $L_2(p)$. Then H is p-soluble. If $O_p(H) = 1$, then by the Hall-Higman Theorem [HaH], $g_{1,1}$ must have minimal polynomial $(t-1)^l$ on W, where $l = p-1$ or $l = p$. But this is impossible, by Lemma 9.6.3. Therefore $O_p(H) \neq 1$, which implies that $P \lhd H$ and $H \subseteq N_K(P) = K_0$. □

By Lemma 9.6.4, K is either K_0 or a group with just one composition factor isomorphic to $L_2(p)$ (all other composition factors are p'-groups). From now on we shall suppose that $K \neq K_0$. It is easy to see that $Z(K) = Z$. Consider a composition series

$$K = K_1 \rhd \cdots \rhd K_l \rhd K_{l+1} \rhd \cdots \rhd K_n = Z \rhd \{1\},$$

where $K_l/K_{l+1} \cong L_2(p)$. We shall suppose that $P \subset K_l$. Clearly, K_{l+1} is a p'-group. Set $\bar{K} = K_l/K_{l+1}$, $\bar{P} = PK_{l+1}/K_{l+1}$. Then the isomorphism $\bar{K} \cong L_2(p)$ implies that $|N_{\bar{K}}(\bar{P})| = p(p-1)/2$. Let $\tilde{K}$ be the full preimage of $N_{\bar{K}}(\bar{P})$ in K_l. Then $|\tilde{K}| = \frac{p(p-1)}{2}|K_{l+1}|$ and $\tilde{K} \supseteq N_{K_l}(P)$. The normal series $\tilde{K} \rhd PK_{l+1} \rhd K_{l+1}$ shows that $\tilde{K}$ is a p-soluble group. In accordance with Lemma 9.6.4, we have $P \lhd \tilde{K}$, that is, $\tilde{K} = N_{K_l}(P)$.

It is easy to see that all Sylow p-subgroups of K are contained in K_l. Therefore $(K : N_K(P)) = (K_l : N_{K_l}(P))$, that is,

$$(K : K_l) = (N_K(P) : N_{K_l}(P)) = |K_0|/|\tilde{K}| = 4/|K_{l+1}|.$$

Consequently, $(K : K_l).(K_{l+1} : Z) = 2$, which implies that $l = 1$ and $(K_2 : Z) = 2$, or $l = 2$ and $K_3 = Z$. In the former case, the subgroup K_2 of order 4 would have been normalized by the subgroup $K_1 \supseteq P$ and $|P| = p \geq 7$. But this would have meant that P centralizes K_2. Meanwhile, the order $|C_K(P)| = |Z \times P| = 2p$ is not divisible by 4, a contradiction. Thus the latter case must occur, and K possesses a composition series

$$K = K_1 \rhd K_2 \rhd K_3 = Z(K) = Z,$$

where $K_1/K_2 \cong \mathbb{Z}_2$, $K_2/K_3 \cong L_2(p)$.

Lemma 9.6.5. *One of the following assertions holds.*

(i) $K_2 = \mathbb{Z}_2 \times L_2(p)$ *and* $K = \mathbb{Z}_2 \times PGL_2(p)$*;*

(ii) $K_2 = SL_2(p)$ *and*

$$K = SL_2^{\pm}(p) = \{X \in GL_2(p) | \det X = \pm 1\}.$$

Proof. 1) Consider the commutator subgroup K_2'. Then K_2/K_2' is abelian; but K_2 has only one abelian composition factor ($\mathbb{Z}_2$), therefore $K_2/K_2' \cong \mathbb{Z}_2$, or $K_2 = K_2'$. In the former case $K_2' \cong L_2(p)$ and $K_2 = Z \times L_2(p)$. In the latter case K_2 is the covering group of order $p(p^2-1)$ for $L_2(p)$, that is, $K_2 = SL_2(p)$.

2) Consider for instance the case $K_2 = SL_2(p)$. We may assume that the element $g_{1,1}$ is represented in K_2 by the matrix $\begin{pmatrix} 1 & 1 \\ 0 & 1 \end{pmatrix}$. Then

$$N = N_{K_2}(P) = \left\{ \begin{pmatrix} a & b \\ 0 & a^{-1} \end{pmatrix} | a \in \mathbb{F}_p^*, b \in \mathbb{F}_p \right\} = < g_{1,1} > . < h >,$$

where $h = \begin{pmatrix} \theta & 0 \\ 0 & \theta^{-1} \end{pmatrix}$ and $< \theta >= \mathbb{F}_p^*$. Since h is an element of order $p-1$ in $N_K(P) = Z \times (P.Q)$, the action of h on Γ_0 is given by

$$h : x^n \mapsto \delta x^{\alpha n+\beta},$$

where $\alpha \in \mathbb{F}_p^*$, $\beta \in \mathbb{F}_p$ and $\delta = \pm 1$. Here either $|< \alpha >| = p-1$; or $p \equiv -1 \pmod 4$, $| < \alpha > | = (p-1)/2$ and $\delta = -1$.

The first possibility leads to the equality $N = K_0$ (since $K_2 \supset Z$), a contradiction. Therefore, the second possibility must hold, and $N_K(P) =< N, \Xi >$, $K =< K_2, \Xi >$, where $\Xi : x^n \mapsto x^{\mu n}$ and μ is an arbitrary quadratic nonresidue in $\mathbb{F}_p$. One can take $\mu = -1$, for example. The additional element Ξ induces the automorphism $\Theta : g \mapsto \Xi^{-1} g \Xi$ of K_2. Here Ξ acts as follows on the parabolic subgroup N:

$$g_{1,1} \mapsto (g_{1,1})^{-1},\ h \mapsto (g_{1,1})^{-2\beta} h,$$

that is,

$$\Theta : \begin{pmatrix} 1 & 1 \\ 0 & 1 \end{pmatrix} \mapsto \begin{pmatrix} 1 & -1 \\ 0 & 1 \end{pmatrix};\ \begin{pmatrix} \theta & 0 \\ 0 & \theta^{-1} \end{pmatrix} \mapsto \begin{pmatrix} \theta & -2\beta\theta^{-1} \\ 0 & \theta^{-1} \end{pmatrix}.$$

Now we consider the element $\tilde{\Xi} = \begin{pmatrix} -1 & \xi \\ 0 & 1 \end{pmatrix}$ of $SL_2^{\pm}(p)$, where $\xi = 2\beta/(\theta^2 - 1)$. Then $\tilde{\Xi}$ induces some automorphism $\tilde{\Theta}$ of K_2; moreover, $\tilde{\Theta} = \Theta$ on N. In accordance with Theorem 5.6.6 from [O'Me], each automorphism of K_2 has the form: $g \mapsto A^{-1} g A$ for some $A \in GL_2(p)$. Using this description and the equality $\tilde{\Theta}|_N = \Theta|_N$, it is not difficult to show that $\tilde{\Theta} = \Theta$. Moreover, $| < \tilde{\Xi} > | = | < \Xi > | = 2$. Thus we have shown that

$$K =< K_2, \Xi > \cong < K_2, \tilde{\Xi} >= SL_2^{\pm}(p).$$

The case $K_2 = Z \times L_2(p)$ is treated similarly. □

Lemma 9.6.6. *Under the given hypotheses, we have $K = \mathbb{Z}_2 \times PGL_2(p)$.*

Proof. Using Lemma 9.6.5, it is sufficient to exclude the case $K = SL_2^{\pm}(p)$. Suppose the contrary. Consider the elements $A = \begin{pmatrix} 1 & 0 \\ 0 & -1 \end{pmatrix}$ and $J = \begin{pmatrix} 0 & 1 \\ -1 & 0 \end{pmatrix}$ of K. Observe that $J^{-1} A J = \mathbf{t} A$, where $\mathbf{t}$ is the (unique) central involution of K. Considering the action of the operator $J^{-1} A J$ on the module W, we arrive at the equality

$$\det(A|_W) = \det((J^{-1} A J)|_W) = \det((\mathbf{t} A)|_W) = (-1)^{p-2k} . \det(A|_W),$$

that is, $\det(A|_W) = 0$, a contradiction. □

Completion of the Proof of Theorem 9.6.1. From Lemmas 9.6.4 and 9.6.6, it follows that $K = K_0 = Z \times (P.Q)$ or $K = Z \times PGL_2(p)$.

Suppose that the second case occurs. Then $B = P.Q$ is a Borel subgroup of $L = PGL_2(p)$. Furthermore, L acts rationally and absolutely irreducibly on Γ_k. The character table of L is computed, for instance, in [Enn]. In particular, the irreducible characters $\chi^{(j)}$ of L of degree $p-1$ are parametrized by elements $j \in \mathbb{Z}/(p+1)\mathbb{Z}$ with $j \not\equiv 0 \,(\mathrm{mod}\, \frac{p+1}{2})$; furthermore, $\chi^{(j)} = \chi^{(-j)}$. Finally, the value field $\mathbb{Q}(\chi^{(j)})$ is equal to $\mathbb{Q}(\cos\frac{2\pi j}{p+1})$. Hence, $\chi^{(j)}$ can be rational-valued only in the case when $\cos(\frac{2\pi j}{p+1}) = 0, \pm\frac{1}{2}$, that is, when $p \equiv -1 \,(\mathrm{mod}\, 2s)$ for some $s \in S = \{2, 3\}$, $j = \pm\frac{r(p+1)}{2s}$ and $r = 1$ or $r = s-1$. Clearly, it is sufficient to restrict attention to the case $j = \frac{r(p+1)}{2s}$.

It is well known that in characteristic p, $SL_2(p)$ acts irreducibly on the space of homogeneous polynomials in 2 variables of degree i, for $i = 0, 1, \ldots, p-1$. Denote the corresponding p-Brauer character by β_{i+1}. Then

$$\hat{\chi}^{(j)} = \beta_{2j-1} + \beta_{p-2j},$$

where $\hat{}$ denotes the restriction to the p'-classes of $L_2(p)$. Now recall that $L_2(p)$ fixes the $\mathbb{F}_p$-space W of dimension $p-2k$. Comparing dimensions of $L_2(p)$-modules, we arrive at the conclusion that $p - 2k = 2j - 1$ or $p - 2k = p - 2j$. In other words, $k = \frac{p+1}{2} - j$ or $k = j$. In view of (9.24), we must have $k = \frac{(s-1)(p+1)}{2s}$.

Conversely, suppose that $p \equiv -1 \,(\mathrm{mod}\, 2s)$ for some $s \in S$. Consider the character $\chi = \chi^{(j)}$ of $L = PGL_2(p)$, where $j = \frac{r(p+1)}{2s}$ for $r = 1$ or $r = s-1$. Then the restriction of χ to a Borel subgroup B of L gives the unique faithful character of degree $p-1$ of B. Furthermore, $B \cong P.Q$ and the representation of $P.Q$ on $\mathcal{H}_0$ is faithful and can be written over $\mathbb{Q}$. Identifying B with $P.Q$, one can suppose that the representation of $P.Q$ on $\mathcal{H}_0$ extends to the whole group L. Since $\chi|_{P.Q}$ is irreducible and χ is rational-valued, χ is afforded by some rational representation, by Lemma 8.3.1. Applying the Deuring-Noether Theorem we conclude that L acts on some $P.Q$-invariant lattice Γ_k, and in this case $K = \mathrm{Aut}(\Gamma_k) = Z \times L$, by Lemmas 9.6.4 and 9.6.6. □

Corollary 9.6.7. *The total number of indivisible invariant sublattices Γ_k with* $\mathrm{Aut}(\Gamma_k) \cong \mathbb{Z}_2 \times PGL_2(p)$ *is* 0, 4, 2 *or* 6 *according as* $p \equiv 1, 5, 7$ *or* $11 \,(\mathrm{mod}\, 12)$, *respectively.* □

Remark 9.6.8. The above elegant argument giving explicit values of k with $\mathrm{Aut}(\Gamma_k) \cong \mathbb{Z}_2 \times PGL_2(p)$ is taken from [PlN]. However, this argument is not constructive. Below we explain another approach to $PGL_2(p)$, which is more constructive.

Proposition 9.6.9. *Under the assumptions of Theorem 9.6.1, the group $K = \mathrm{Aut}(\Gamma_k)$ is $\mathbb{Z}_2 \times PGL_2(p)$ if and only if there exists a vector v in Γ_k with the following properties:*
(i) $||v||^2 = p(p-1)$;
(ii) $v \equiv \sum_{i=k}^{(p-1)/2} a_i L^i(x)\Phi(y) \pmod{\Gamma_{p-k}}$, *where* $a_i \in \mathbb{F}_p^*$.

Proof. In view of Lemmas 9.6.4 – 9.6.6, we conclude that K is K_0 or $Z \times L$, where $L \cong PGL_2(p)$.

1) Firstly we suppose that $K = Z \times L$. One can suppose also that $g_{1,1}$ is represented in L by the matrix $\begin{pmatrix} 1 & 1 \\ 0 & 1 \end{pmatrix}$. But in this case, as an element of order $p-1$ in $N_K(P)$, $g_{0,\theta}$ must be represented by a matrix $\pm \begin{pmatrix} \alpha & \beta \\ 0 & 1 \end{pmatrix}$, where $\alpha, \beta \in \mathbb{F}_p$, $\alpha \neq 0, 1$. From this there follows the existence of an involution $J = \begin{pmatrix} -\zeta & 1-\zeta^2 \\ 1 & \zeta \end{pmatrix}$ that inverts the element $g_{0,\theta}$, where $\zeta = \beta/(\alpha - 1)$. In W we consider the eigenbasis $\{L^i(x) | k \leq i \leq p-1-k\}$ for $g_{0,\theta}$:

$$g_{0,\theta} : L^i(x) \mapsto \theta^{-i} L^i(x)$$

(we again omit the symbol $\Phi(y)$). Since J inverts $g_{0,\theta}$, modulo Γ_{p-k} we have

$$J : < L^i(x) >_{\mathbb{F}_p} \leftrightarrow < L^{p-1-i}(x) >_{\mathbb{F}_p},\ k \leq i \leq p-1-k.$$

Applying J to the vector

$$u = xq(x) = \sum_{i=1}^{p-1} i^{(p-1)/2} x^{i+1} \equiv \sum_{j=(p-1)/2}^{p-1-k} b_j L^j(x) \pmod{\Gamma_{p-k}}, \tag{9.25}$$

where $b_j \in \mathbb{F}_p^*$, we obtain the vector $v = Ju$ with the required properties (i), (ii).

2) Conversely, suppose that there exists a vector v in Γ_k with properties (i), (ii). Consider the vector u described in (9.25), and set

$$u_a = (-1)^a (g_{0,\theta})^a(u),\ v_a = (-1)^a (g_{0,\theta})^{-a}(v),\ 1 \leq a \leq p-1.$$

These vectors possess the following properties:

$$(u_a; u_a) = (v_a; v_a) = p(p-1),\ 1 \leq a \leq p-1, \tag{9.26}$$

$$(u_a; u_b) = (v_a; v_b) = -p,\ 1 \leq a \neq b \leq p-1, \tag{9.27}$$

$$(u_a; v_b) \equiv (u_b; v_a) \pmod{p^2},\ 1 \leq a, b \leq p-1. \tag{9.28}$$

In fact, relation (9.26) is obvious. Furthermore,

$$(u_a; u_b) = \left((-1)^a x^{\theta^{-a}} q(x^{\theta^{-a}}); (-1)^b x^{\theta^{-b}} q(x^{\theta^{-b}}) \right) =$$

$$= \left(x^{\theta^{-a}} q(x); x^{\theta^{-b}} q(x)\right) = -p,$$

because the following identities hold for $1 \le n \le p-1$:

$$q(x^n) = n^{(p-1)/2} q(x),$$

$$||(1-x^n)q(x)||^2 = p||1-x^n||^2 = 2p^2.$$

Applying $g_{0,\theta}$ to the quantity $(L^i(x); L^j(x))$, we come to the congruence

$$\left(L^i(x); L^j(x)\right) \equiv 0 \,(\operatorname{mod} p^2)$$

when $p-1$ does not divide $i+j$. Keeping in mind that

$$v \equiv \sum_{i=k}^{(p-1)/2} a_i L^i(x) \,(\operatorname{mod} \Gamma_{p-k}),$$

we have:

$$(v_a; v_b) \equiv (-1)^{a+b} \left(\sum_{i=k}^{(p-1)/2} a_i \theta^{ia} L^i(x); \sum_{j=k}^{(p-1)/2} a_j \theta^{jb} L^j(x) \right) \equiv$$

$$\equiv (a_{(p-1)/2})^2 \left(L^{(p-1)/2}(x); L^{(p-1)/2}(x)\right) \,(\operatorname{mod} p^2).$$

The last quantity does not depend on the subscripts a, b. In particular,

$$(v_a; v_b) \equiv (v_a; v_a) \equiv -p \,(\operatorname{mod} p^2).$$

But $|(v_a; v_b)| < ||v_a|| \cdot ||v_b|| = p(p-1)$, therefore we arrive at the equality $(v_a; v_b) = -p$. Finally, setting $\nu^{-1} = -\{\frac{p-1}{2}\}!$ we have

$$(u_a; v_b) \equiv (-1)^{a+b} \left(x^{\theta^{-a}} q(x^{\theta^{-a}}); \sum_{j=k}^{(p-1)/2} a_j L^j(x^{\theta^b}) \right) \equiv$$

$$\equiv (-1)^{a+b} \left(x^{\theta^{-a}} \theta^{-a(p-1)/2} q(x); \sum_{j=k}^{(p-1)/2} a_j \theta^{jb} L^j(x) \right) \equiv$$

$$\equiv (-1)^b \left(\nu L^{(p-1)/2}(x) \sum_{i=0}^{p-1} \frac{\theta^{-ia}}{i!} L^i(x); \sum_{j=k}^{(p-1)/2} a_j \theta^{jb} L^j(x) \right) \equiv$$

$$\equiv (-1)^{a+b} \nu \sum_{j=k}^{(p-1)/2} a_j \theta^{j(a+b)} \left(L^j(x); L^{p-1-j}(x)\right) \,(\operatorname{mod} p^2).$$

The last expression is symmetric in the parameters a, b. Hence, we arrive at relation (9.28).

Relations (9.26) and (9.27) show that the map J defined by

$$J : v_a \mapsto u_a,\ 1 \le a \le p-1, \tag{9.29}$$

is an isometry of $\mathbb{R}^{p-1}$. We claim that J preserves the lattice Γ_k. For this we consider $w_a = J(u_a) - v_a$, where $1 \le a \le p-1$. In view of (9.28), we have

$$(w_a; u_b) = (J(u_a); u_b) - (v_a; u_b) =$$

$$= (J(u_a); J(v_b)) - (v_a; u_b) = (u_a; v_b) - (v_a; u_b) \equiv 0\,(\mathrm{mod}\,p^2).$$

From this it follows that

$$w_a \in p^2\left(< u_1, \dots, u_{p-1} >_{\mathbb{Z}}\right)^* = p^2\left(\Gamma_{(p-1)/2}\right)^* = \Gamma_{(p+1)/2} \subset \Gamma_k.$$

This means simply that $J(u_a) \in \Gamma_k$ for all a, $1 \le a \le p-1$. But Γ_k is generated by vectors $u_1, \dots, u_{p-1}, v_1, \dots, v_{p-1}$, and so $J(\Gamma_k) = \Gamma_k$, as stated.

Thus, we have constructed the element $J \in K$ with $J(v) = u$. In particular, $J \notin K_0$. Hence, $K = Z \times L$. □

By Proposition 9.6.9, one can construct the group $K = \mathrm{Aut}(\Gamma_k) = \mathbb{Z}_2 \times PGL_2(p)$ as follows.

1) Find (for instance, using a computer) a vector v satisfying conditions (i) and (ii) of Proposition 9.6.9.

2) Define the map J by the rule (9.29), and then $K = < K_0, J >$.

Example 9.6.10. We indicate vectors v satisfying condition (i) and (ii) of Proposition 9.6.9 when $7 \le p \le 13$:

$$\begin{array}{lll}
p = 7, & k = 2, & v = \left(1 - 2x + x^2\right)\Phi(y); \\
p = 11, & k = 3, & v = \left(1 + 2x^4 - x^9 - 2x^5\right)\Phi(y); \\
 & k = 4, & v = \left(1 + 2x^4 + x^8 - x^2 - x^3 - x^5 - x^6\right)\Phi(y); \\
p = 13, & 2 \le k \le 5, & \text{there are no such vectors } v.
\end{array}$$

Furthermore, when $p = 11$ computer calculations show that $\mathrm{Aut}(\Gamma_3)$ is an insoluble group of order $2p(p^2 - 1) = 2640$ which acts transitively on the set consisting of 110 minimal vectors of Γ_3, with point stabilizer D_{24}.

As we have convinced ourselves, the automorphism groups of the lattices Γ_2, Γ_3 and Γ_4 are isomorphic to $\mathbb{Z}_2 \times PGL_2(11)$ if $p = 11$. In fact, there exist just three isomorphism classes of $\mathbb{Z}L$-lattices of dimension 10, where $L = PGL_2(11)$, and they are represented by the lattices Γ_k, $k = 2, 3, 4$. It is easy to show that $\mathrm{Aut}(\Gamma_2)$ and $\mathrm{Aut}(\Gamma_4)$ are $GL_{10}(\mathbb{Q})$-conjugate. Furthermore, the fact that all the groups $\mathrm{Aut}(\Gamma_1) = \mathbb{Z}_2 \times \mathbb{S}_{11}$, $\mathrm{Aut}(\Gamma_2)$ and $\mathrm{Aut}(\Gamma_3)$ contain the subgroup $B = \mathbb{Z}_{11}.\mathbb{Z}_{10}$ with

$$\mathrm{End}_B(\Gamma_i \otimes \mathbb{Q}) = \{\varphi \in \mathrm{End}_{\mathbb{Q}}(\Gamma_i \otimes \mathbb{Q}) | \forall g \in B, \varphi g = g\varphi\} \cong \mathbb{Q},$$

simply means that these groups are vertices of the unique 2-complex of the simplicial complex $M_{10}^{irr}(\mathbb{Q})$ (for the definitions of these notions, see [Ple 3]).

We conclude this section by stating (without proof) the following results, which are obtained using the classification of finite simple groups:

Theorem 9.6.11 [Blau]. *Let H be a finite group having a faithful rational representation with character χ such that $\chi(1) = p - 1$, where p is a prime dividing $|H|$. Then either H is a group of type $L_2(p)$, or H has a subgroup A of index p and $\chi = (\mathrm{Ind}_A^H(1_A) - 1_H)\lambda$, where λ is a linear character of H with $\lambda^2 = 1_H$.* □

Theorem 9.6.12 [PlN]. *Let p be a prime greater than 11, and H an irreducible maximal finite subgroup of $GL_{p-1}(\mathbb{Q})$ with $p||H|$. Then H is isomorphic to $\mathbb{Z}_2 \times \mathbb{S}_p$, $\mathbb{Z}_2 \times L_2(p)$ or $\mathbb{Z}_2 \times PGL_2(p)$.* □

9.7 On the automorphism groups of invariant lattices of type A_{p-1}

Postponing to Chapter 10 the complete description of automorphism groups of invariant lattices of type A_{p-1}, we restrict ourselves here to listing the most accessible groups among them. We begin by stating the following result without proof.

Theorem 9.7.1. *The following assertions hold.*

(i) $\mathrm{Aut}(\Delta) \cong \mathbb{Z}_2 \times AGL_2(p)$ *for* $\Delta = \Gamma^{0,l}$ $(1 < l < p - 1)$ *and* $\Delta = \Gamma^{1,l}$ $(2 < l < p - 1)$*;*

(ii) $\mathrm{Aut}(\Delta) \cong \mathbb{Z}_2 \times \mathbb{S}_{p^2}$ *for* $\Delta = \Gamma^{0,p-1}$ $(p \geq 3)$ *and* $\Delta = \Gamma^{1,p-1}$ $(p \geq 5)$*;*

(iii) $\mathrm{Aut}(\Gamma^{(p-1)/2,(p+1)/2}) \cong (\mathbb{S}_p)^{p+1}.(GL_2(p)/\mathbb{Z}_{(p-1)/2})$ $(p \geq 5)$*;*

(iv) $\mathrm{Aut}(\Gamma^{0,0}) \cong (\mathbb{Z}_2 \times \mathbb{S}_p) \wr \mathbb{S}_{p+1}$*, and more generally* $\mathrm{Aut}(\Gamma^{k,k}) = \mathrm{Aut}(\Gamma_0^k) \wr \mathbb{S}_{p+1}$*;*

(v) $\mathrm{Aut}(\Gamma^{0,1}) \cong \mathbb{Z}_2 \times (\mathbb{S}_p \wr \mathbb{S}_{p+1})$*;*

(vi) $\mathrm{Aut}(\Gamma^{1,2}) \cong (D_{2p})^{p+1}.GL_2(p)$ $(p \geq 5)$. □

The sets of minimal vectors of invariant lattices indicated in the formulation of this theorem are described very simply. Based on this, we obtain immediate information about the automorphism groups. It would have been dull to go into technical details.

We mentioned earlier that the series $\frac{1}{p}\Delta^{1,r}$, $r \in \mathbb{F}_p^*$, $p \geq 5$, of even unimodular root-free lattices involves the Leech lattice Λ_{24} if $p = 5$. Thus, it is interesting to look at the automorphism groups of these lattices when $p > 5$. Unfortunately, the following statement holds.

Theorem 9.7.2. *If $p > 5$, then*

$$\mathrm{Aut}(\Delta^{1,r}) = \mathbb{Z}_2 \times ASL_2(p).$$

Proof. 1) It is sufficient to treat the case $r = 1$. In the lattice $\Delta^{1,1}$ we consider the sublattice $\Gamma^{2,p-1}$ generated by its minimal vectors (of norm $4p^2$); by Theorem 9.7.1(i) its automorphism group is equal to $\mathbb{Z}_2 \times AGL_2(p)$. Clearly, the intersection of $\mathrm{Aut}(\Gamma^{2,p-1})$ with $\mathrm{Aut}(\Delta^{1,1})$ is $\mathbb{Z}_2 \times G = \mathbb{Z}_2 \times ASL_2(p)$. Consequently, all that is left is to show that every minimal vector of $\Delta^{1,1}$ has norm $4p^2$ and belongs to $\Gamma^{2,p-1}$.

Suppose that $X \in \Delta^{1,1} \backslash \Gamma^{2,p-1}$.

2) Applying a suitable automorphism from $\mathbb{Z}_2 \times G$ to X, one can suppose that $X \equiv B_1 \pmod{\Gamma^{2,p-1}}$, where the element $B_1 = B(1,1)$ is explained in Proposition 9.4.2. Therefore, X can be written in the form

$$X = \sum_{i,j=0}^{p} a_{ij} p x^i y^j - l(y)\Phi(y),$$

where $a_{ij} \in \mathbb{Z}$, $\sum_{i,j} a_{ij} = pa$ for some $a \in \mathbb{Z}$, $0 \le a < p$, and

$$l(y) = 1 + 2y + 3y^2 + \ldots + \frac{p-1}{2} y^{(p-3)/2} - \frac{p-1}{2} y^{(p-1)/2} - \ldots - y^{p-2} = \sum_{k=0}^{p-1} l_k y^k.$$

Taking a fixed, we give a lower bound for $||X||^2$. By Corollary 9.3.2b), we have

$$\left(px^i y^j; l(y)\Phi(x)\right) = pl_j.$$

Hence,

$$||X||^2 = p^2 \left(\sum_{i,j} (a_{ij})^2 - a^2 \right) + p \sum_j (l_j)^2 - 2p \sum_{i,j} a_{ij} l_j =$$

$$= \sum_{i,j} (pa_{ij} - l_j)^2 - p^2 a^2.$$

Now we claim that, under the assumption that $||X||^2$ takes the minimal value (among the vectors X of the form indicated with given a), all the coefficients a_{ij} are equal to 0 or 1. Indeed, if there exist indices i, j, m, n such that $a_{ij} > 0$, $a_{mn} < 0$, then

$$\left\{ (pa_{ij} - l_j)^2 + (pa_{mn} - l_n)^2 \right\} - \left\{ (p(a_{ij} - 1) - l_j)^2 + (p(a_{mn} + 1) - l_n)^2 \right\} =$$

$$= 2p^2(a_{ij} - 1) - 2p^2(a_{mn} + 1) + p(p - 2l_j) + p(p + 2l_n) > 0,$$

since $|l_k| \leq (p-1)/2$ for each k. Hence, we can suppose that $a_{ij} \geq 0$ for all i, j.

If $a_{ij} > 1$ for some pair (i, j), then there exists an element $a_{mn} = 0$, because $\sum_{m',n'} a_{m'n'} = pa < p^2$. In this case, we again have that

$$\left(pa_{ij} - l_j\right)^2 + (pa_{mn} - l_n)^2 > \left(p(a_{ij} - 1) - l_j\right)^2 + \left(p(a_{mn} + 1) - l_n\right)^2.$$

3) Recall that

$$l_j = \begin{cases} j+1 & \text{if} \quad 0 \leq j \leq (p-3)/2 \\ j+1-p & \text{if} \quad (p-1)/2 \leq j \leq p-1. \end{cases}$$

Replacing the index j with $(p-1)/2 \leq j \leq p-1$ by $j - p$, we obtain that

$$||X||^2 = \sum_{\substack{0 \leq i \leq p-1 \\ -\frac{p+1}{2} \leq j \leq \frac{p-3}{2}}} \left(pa_{ij} - (j+1)\right)^2 - p^2 a^2.$$

Clearly, under the minimality assumption on $||X||^2$, all the pa elements $a_{ij} = 1$ must be "arranged higher", that is, they are to have the maximal greatest possible indices j. This means that $||X||^2$ is minimal if and only if

$$a_{ij} = \begin{cases} 1 & \frac{p-3}{2} - a < j \leq \frac{p-3}{2} \\ 0 & \frac{p-3}{2} - a \geq j \geq -\frac{p+1}{2} \end{cases}.$$

Thus we have obtained the following estimate:

$$||X||^2 \geq \sum_{\substack{0 \leq i \leq p-1 \\ -\frac{p+1}{2} \leq j \leq \frac{p-3}{2} - a}} (j+1)^2 +$$

$$+ \sum_{\substack{0 \leq i \leq p-1 \\ -\frac{p-3}{2} - a + 1 \leq j \leq \frac{p-3}{2}}} (p-1-j)^2 - p^2 a^2 =$$

$$= p \sum_{-\frac{p-1}{2} \leq k \leq \frac{p-1}{2} - a} k^2 + p \sum_{\frac{p+1}{2} \leq k \leq \frac{p-1}{2} + a} k^2 - p^2 a^2 =$$

$$= p \sum_{0 \leq k \leq \frac{p-1}{2} - a} k^2 + p \sum_{0 \leq k \leq \frac{p-1}{2} + a} k^2 - p^2 a^2 = p^2 . \frac{p^2 - 1}{12}.$$

4) As a result of 1)–3) we get that $||X||^2 \geq p^2 . \frac{p^2-1}{12}$ for every $X \in \Delta^{1,1} \backslash \Gamma^{2,p-1}$. Now suppose that $||X||^2 \leq 4p^2$ (and $X \in \Delta^{1,1} \backslash \Gamma^{2,p-1}$). Then $p = 7$ and all the coefficients a_{ij} depend only on the subscript j, as we established above. But in

this case, all the projections of X to a Cartan subalgebra $\mathcal{H}_\sigma$, except at most one, are equal to 0. This means that $X \in \Gamma^{2,p-1}$, a contradiction. □

Unlike the case $p > 5$, when $p = 5$ the lattice $\Delta^{1,r}$ is generated by its minimal vectors (of norm $4p^2$).

Theorem 9.7.2 motivates the following

Definition 9.7.3. An invariant lattice Δ of type A_{p-1} is said to be *small* if every composition factor of $\mathrm{Aut}(\Delta)$ is either $L_2(p)$ or abelian.

Extending the statements of the previous two theorems, we prove, in particular, that the invariant lattices $\Gamma^{2,l}$, $\Delta^{2,r}$, $\Delta^{\frac{p-3}{2},r}$ and $\Delta^{\frac{p-5}{2},r}$ are small. For this we set $\mathbb{G} = \mathrm{Aut}(\Delta)$, where $\Delta = \Gamma^{k,l}$ or $\Delta = \Delta^{k,r}$, and

$$\mathbb{G}_0 = \{\varphi \in \mathbb{G} | \varphi \text{ preserves the set } \{\mathcal{H}_\sigma | \sigma \in \mathbb{P}V\}\}.$$

There arises a natural homomorphism $\tau : \mathbb{G}_0 \to \mathbb{S}_{p+1}$, namely that given by $\varphi(\mathcal{H}_\sigma) = \mathcal{H}_{\tau(\varphi)(\sigma)}$ for $\varphi \in \mathbb{G}_0$. Set $K = \mathrm{Ker}\tau$ and $H = \mathrm{Im}\tau$.

Lemma 9.7.4. *Suppose that* $0 \le m \ne n \le p-1$, $m \ne 0,\ 1,\ p-1,\ \frac{p\pm 1}{2},\ p-n$. *Then the intersection* $\mathrm{Aut}(\Gamma_0^m) \cap \mathrm{Aut}(\Gamma_0^n)$ *is* $K_0 = \mathbb{Z}_2 \times (\mathbb{Z}_p.\mathbb{Z}_{p-1})$.

Proof. An easy consequence of Theorem 9.6.1. □

Lemma 9.7.5. *If* $\Delta = \Gamma^{k,l}$ $(0 < k < l)$, $\Delta^{k,r}$ *(k even and* $r \in \mathbb{F}_p^*$*) or* $\Delta^{k,r}$ *(k odd and* $p \ne 11, 23, 2^d - 1$*), then*

$$L_2(p) \subseteq H \subseteq PGL_2(p).$$

Proof. It is not difficult to show that every element $\pi \in H\backslash\{1\}$ acting on the projective line $\mathbb{P}V$ has at most $k+1$ fixed points.

1) Suppose that $\Delta = \Gamma^{k,l}$. Then $\mathbb{G}_0$ contains the group $GL_2(p)$ of ring automorphisms (we mean here the ring $\mathcal{A} = \mathbb{C}[x,y]/I$). Hence $H \supseteq PGL_2(p)$, and so H is a faithful doubly transitive permutation group of degree $p+1$ containing a cycle of length $p+1$. Furthermore, H does not contain $\mathbb{A}_{p+1}$. Therefore, by Theorem 1.49 [Gor 2] we have $H = PGL_2(p)$.

2) If $\Delta = \Delta^{k,r}$, k even, then $\mathbb{G}$ preserves

$$\Delta^{k,r} \cap p^2(\Delta^{k,r})^* = \Delta^{k,r} \cap \Delta^{k,-r} = \Gamma^{k+1,p-k},$$

and so

$$L_2(p) \subseteq H = \mathrm{Im}\left(\tau|_{\mathbb{G}_0}\right) \subseteq \mathrm{Im}\left(\tau|_{(\mathrm{Aut}(\Gamma^{k+1,p-k}))_0}\right) = PGL_2(p).$$

If $\Delta = \Delta^{k,r}$, k odd, the condition $p \ne 2^d - 1$ implies that the socle of the doubly transitive permutation group H of degree $p+1$ is a nonabelian simple group. On scrutinizing the list of doubly transitive groups with nonabelian socle (see Table A2), we arrive at the required inclusion $L_2(p) \subseteq H \subseteq PGL_2(p)$. □

Theorem 9.7.6. *Suppose that*

(i) $\Delta = \Gamma^{k,l}$, *where* $1 \le k < l \le p-1$, $(k,l) \ne (1, p-1), (\frac{p-1}{2}, \frac{p+1}{2})$, *or* $\Delta = \Delta^{k,r}$, *where* $1 \le k \le \frac{p-3}{2}$, $r \in \mathbb{F}_p^*$;

(ii) $\mathbb{G} = \mathrm{Aut}(\Delta)$ *preserves the set* $\{\mathcal{H}_\sigma | \sigma \in \mathbb{P}V\}$.

Then Δ *is a small lattice.*

Proof. An immediate consequence of Lemmas 9.7.4, 9.7.5. □

Corollary 9.7.7. *The lattices* $\Gamma^{k,l}$, *where*

$$2 \le k < l \le \frac{p+5}{2};\ (k,l) \ne (\frac{p-1}{2}, \frac{p+1}{2}), (2, \frac{p+3}{2}), (2, \frac{p+5}{2}), (3, \frac{p+5}{2}),$$

are small.

Proof. Set

$$N_1 = \min\{||X||^2 | X \in \Gamma^{k,l} \backslash \Gamma^l\}, \quad N_2 = \min\{||X||^2 | X \in \Gamma^l \backslash \{0\}\}.$$

Under the hypotheses laid down, we have $N_1 > N_2$ (see §9.3), so that $\mathbb{G}$ preserves $\{\mathcal{H}_\sigma\}$ and $\Gamma^{k,l}$ is small. □

The following statement is proved similarly.

Corollary 9.7.8. *Suppose that* $p \ne 23, 2^d - 1$. *Then*

(i) *the lattices* $\Delta^{\frac{p-3}{2},r}$ *are small if* $p \ge 11$;

(ii) *the lattices* $\Delta^{\frac{p-5}{2},r}$ *are small if* $p \ge 13$. □

The proof of the next statement, which gives us some information about minimal vectors, is omitted.

Lemma 9.7.9. *Suppose that* $p \ge 11$ *and* $\frac{p+3}{2} < n \le p-1$. *Then all the minimal vectors of the lattice* $\Gamma^{2,n}$ *belong to* $\Gamma^{2,p-1}$ *and have norm* $4p^2$. □

Theorem 9.7.10. *The following assertions hold.*

(i) *Suppose that* $p \ge 11$ *and* $\frac{p+3}{2} < n \le p-1$. *Then the lattices* $\Gamma^{2,n}$ *are small:*

$$\mathrm{Aut}(\Gamma^{2,n}) = \mathbb{Z}_2 \times AGL_2(p).$$

(ii) *Suppose that* $p \ge 7$. *Then the lattices* $\Delta^{2,r}$ *are small.*

Proof. (i) Set

$$\mathcal{M} = \{X | X \in \Gamma^{2,n}, ||X||^2 = 4p^2\}.$$

Then by Lemma 9.7.9, $\tilde{\Gamma} =< \mathcal{M} >_{\mathbb{Z}}$ is an indivisible sublattice of $\Gamma^{2,p-1}$, and so $\tilde{\Gamma} = \Gamma^{2,p-1}$ or $\Gamma^{2,p}$. Hence, $\mathbb{G} = \mathrm{Aut}(\Gamma^{2,n})$ leaves fixed the lattice $p^2(\tilde{\Gamma})^* = \Gamma^{1,p-2}$ or $\Gamma^{0,p-2}$. But by Theorem 9.7.1,

$$\mathrm{Aut}(\Gamma^{0,p-2}) = \mathrm{Aut}(\Gamma^{1,p-2}) = \mathbb{Z}_2 \times AGL_2(p) \subseteq \mathbb{G}.$$

Hence, $\mathbb{G} = \mathbb{Z}_2 \times AGL_2(p)$.

(ii) Clearly, $\mathbb{G} = \mathrm{Aut}(\Delta^{2,r})$ leaves $\Delta^{2,r} \cap p^2(\Delta^{2,r})^* = \Gamma^{3,p-2}$ fixed. Hence,

$$\mathbb{G} \subseteq \mathrm{Aut}(\Gamma^{3,p-2}) = \mathrm{Aut}(\Gamma^{2,p-3}) = \mathbb{Z}_2 \times AGL_2(p).$$

From this it follows that $\mathbb{G} = \mathbb{Z}_2 \times ASL_2^{\pm}(p)$, where $ASL_2^{\pm}(p)$ consists of the two-dimensional affine transformations with determinant ± 1 over $\mathbb{F}_p$. □

In fact, assertion (i) of Theorem 9.7.10 is valid also when $n = \frac{p+3}{2}$.

The results of computing automorphism groups of invariant lattices of type A_{p-1} for $p \leq 11$ are summarized in Table 9.5.

To conclude this chapter, we construct an additional automorphism φ, which together with $G = \mathrm{Inn}(\mathcal{D})$ generates the whole group $\mathbb{G} = \mathrm{Aut}(\Delta)$, where $\Delta = \frac{1}{5}\Delta^{1,r}$ $(p = 5)$ is the Leech lattice. Recall that $G \cong ASL_2(5)$.

Theorem 9.7.11. *There exists an additional automorphism φ of order 5 such that*

$$< G, \varphi >= \mathbb{G} = \mathrm{Aut}(\Delta^{1,1}) = \mathbb{Z}_2 \circ Co_1.$$

Proof. 1) It is shown in [Cur] that $\mathbb{G}$ has a unique conjugacy class of elementary abelian subgroups of order 5^3 which contain representatives of all the three conjugacy classes $5A$, $5B$, $5C$ of elements of order 5 in $\mathbb{G}$. Using this fact, we shall find φ among the elements of $C_{\mathbb{G}}(B, C)$, where B and C are elements of the classes $5B$ and $5C$, respectively:

$$B(F(x, y)) = xF(x, y), \; C(F(x, y)) = F(x, xy).$$

Set $\mathcal{H}_i = \mathcal{H}_{<(1,i)>}$, $\mathcal{H}_\infty = \mathcal{H}_{<(0,1)>}$. Consider the following basis $\{F_{n,m}\}$ of the space $\oplus_{i=0}^4 \mathcal{H}_i$ that diagonalizes the operators B and C:

$$F_{n,m}(x, y) = \Phi(\varepsilon^{-m}x). \sum_{i=0}^{4} \varepsilon^{ni}\Phi(x^{-i}y), \; 1 \leq m \leq 4, 0 \leq n \leq 4,$$

Table 9.5
Automorphism groups of invariant lattices of type A_{p-1} for $p \leq 11$

	Δ	$\mathrm{Aut}(\Delta)$
$p=3$	$\Gamma^{0,0}, \Gamma^{1,1}$	$(\mathbb{Z}_2 \times \mathbb{S}_3) \wr \mathbb{S}_4$
	$\Gamma^{0,1}, \Gamma^{w}$	$\mathbb{Z}_2 \times (\mathbb{S}_3 \wr \mathbb{S}_4)$
	$\Gamma^{0,2}, \Gamma^{1,3}$	$\mathbb{Z}_2 \times \mathbb{S}_9$
	$\Gamma^{1,2}$	$W(E_8)$
$p=5$	$\Gamma^{0,0}, \Gamma^{1,1}, \Gamma^{2,2}, \Gamma^{3,3}$	$(\mathbb{Z}_2 \times \mathbb{S}_5) \wr \mathbb{S}_6$
	$\Gamma^{0,1}, \Gamma^{w}$	$\mathbb{Z}_2 \times (\mathbb{S}_5 \wr \mathbb{S}_6)$
	$\left\{ \Gamma^{0,2}, \Gamma^{0,3}, \Gamma^{1,3}, \Gamma^{2,4}, \Gamma^{2,5}, \Gamma^{3,5} \right.$	$\mathbb{Z}_2 \times AGL_2(5)$
	$\Gamma^{0,4}, \Gamma^{1,4}, \Gamma^{1,5}$	$\mathbb{Z}_2 \times \mathbb{S}_{25}$
	$\Gamma^{1,2}, \Gamma^{3,4}$	$(D_{10})^6.GL_2(5)$
	$\Gamma^{2,3}$	$(\mathbb{S}_5)^6.(GL_2(5)/\mathbb{Z}_2)$
	$\Delta^{1,r}\ (r \in \mathbb{F}_5^*)$	$\mathbb{Z}_2 \circ Co_1$
$p=7$	$\Gamma^{0,0}, \Gamma^{1,1}, \Gamma^{3,3}, \Gamma^{4,4}$	$(\mathbb{Z}_2 \times \mathbb{S}_7) \wr \mathbb{S}_8$
	$\Gamma^{2,2}, \Gamma^{5,5}$	$(\mathbb{Z}_2 \times PGL_2(7)) \wr \mathbb{S}_8$
	$\Gamma^{0,1}, \Gamma^{w}$	$\mathbb{Z}_2 \times (\mathbb{S}_7 \wr \mathbb{S}_8)$
	$\left\{ \Gamma^{0,2}, \Gamma^{0,3}, \Gamma^{0,4}, \Gamma^{0,5}, \Gamma^{1,3}, \Gamma^{1,4}, \Gamma^{1,5}, \Gamma^{2,4}, \Gamma^{2,5}, \Gamma^{2,6}, \Gamma^{2,7}, \Gamma^{3,5}, \Gamma^{3,6}, \Gamma^{3,7}, \Gamma^{4,6}, \Gamma^{4,7}, \Gamma^{5,7} \right.$	$\mathbb{Z}_2 \times AGL_2(7)$
	$\Gamma^{0,6}, \Gamma^{1,6}, \Gamma^{1,7}$	$\mathbb{Z}_2 \times \mathbb{S}_{49}$
	$\Gamma^{1,2}, \Gamma^{5,6}$	$(D_{14})^8.GL_2(7)$
	$\Gamma^{3,4}$	$(\mathbb{S}_7)^8.(GL_2(7)/\mathbb{Z}_3)$
	$\Delta^{1,r}\ (r \in \mathbb{F}_7^*)$	$\mathbb{Z}_2 \times ASL_2(7)$
	$\Delta^{2,r}\ (r \in \mathbb{F}_7^*)$	$\mathbb{Z}_2 \times ASL_2^{\pm}(7)$
	$\Gamma^{2,3}, \Gamma^{4,5}$	$\mathbb{Z}_2 \times \left\{ \left(\mathbb{Z}_7 \times (D_{14})^7 \right).GL_2(7) \right\}$

where $\varepsilon = \exp(\frac{2\pi i}{5})$. Here

$$B(F_{n,m}) = \varepsilon^m F_{n,m},\ C(F_{n,m}) = \varepsilon^n F_{n,m}.$$

Note that

$$\mathrm{Ker}(B-1) = \mathcal{H}_\infty,\ \mathrm{Ker}(C-1) = \mathcal{H}_\infty \oplus \mathcal{H}',$$

where $\mathcal{H}' = < F_{0,m} | 1 \leq m \leq 4 >_{\mathbb{C}}$. Since $\varphi \in C_{\mathbb{G}}(B, C)$ and $| < \varphi > | = 5$, we must have that

a) φ leaves $\mathcal{H}_\infty$ and $\mathcal{H}'$ fixed;

b) $\varphi(F_{n,m}) = \varepsilon^{A(n,m)} F_{n,m}$ for some $A(n,m) \in \mathbb{F}_5$.

Table 9.5 (continued)

	Δ	$\mathrm{Aut}(\Delta)$
$p = 11$	$\Gamma^{0,0}, \Gamma^{1,1}, \Gamma^{5,5}, \Gamma^{6,6}$	$(\mathbb{Z}_2 \times \mathbb{S}_{11}) \wr \mathbb{S}_{12}$
	$\left\{ \begin{array}{l} \Gamma^{2,2}, \Gamma^{3,3}, \Gamma^{4,4}, \\ \Gamma^{7,7}, \Gamma^{8,8}, \Gamma^{9,9} \end{array} \right.$	$(\mathbb{Z}_2 \times PGL_2(11)) \wr \mathbb{S}_{12}$
	$\Gamma^{0,1}, \Gamma^{w}$	$\mathbb{Z}_2 \times (\mathbb{S}_{11} \wr \mathbb{S}_{12})$
	$\Gamma^{0,2}, \Gamma^{0,3}, \Gamma^{0,4}, \Gamma^{0,5}, \Gamma^{0,6}, \Gamma^{0,7}, \Gamma^{0,8}, \Gamma^{0,9}, \Gamma^{1,3}, \Gamma^{1,4}, \Gamma^{1,5}, \Gamma^{1,6}, \Gamma^{1,7}, \Gamma^{1,8}, \Gamma^{1,9}, \Gamma^{2,3}, \Gamma^{2,4}, \Gamma^{2,5}, \Gamma^{2,6}, \Gamma^{2,7}, \Gamma^{2,8}, \Gamma^{2,9}, \Gamma^{2,10}, \Gamma^{2,11}, \Gamma^{3,4}, \Gamma^{3,5}, \Gamma^{3,6}, \Gamma^{3,7}, \Gamma^{3,9}, \Gamma^{3,10}, \Gamma^{3,11}, \Gamma^{4,5}, \Gamma^{4,6}, \Gamma^{4,7}, \Gamma^{4,8}, \Gamma^{4,9}, \Gamma^{4,10}, \Gamma^{4,11}, \Gamma^{5,7}, \Gamma^{5,8}, \Gamma^{5,9}, \Gamma^{5,10}, \Gamma^{5,11}, \Gamma^{6,7}, \Gamma^{6,8}, \Gamma^{6,9}, \Gamma^{6,10}, \Gamma^{6,11}, \Gamma^{7,8}, \Gamma^{7,9}, \Gamma^{7,10}, \Gamma^{7,11}, \Gamma^{8,9}, \Gamma^{8,10}, \Gamma^{8,11}, \Gamma^{9,11}, \Gamma^{3,8}, \Delta^{1,r}, \Delta^{2,r}, \Delta^{3,r}, \Delta^{4,r}\ (r \in \mathbb{F}_{11}^*)$	Δ are small
	$\Gamma^{0,10}, \Gamma^{1,10}, \Gamma^{1,11}$	$\mathbb{Z}_2 \times \mathbb{S}_{121}$
	$\Gamma^{1,2}, \Gamma^{9,10}$	$(D_{22})^{12}.GL_2(11)$
	$\Gamma^{5,6}$	$(\mathbb{S}_{11})^{12}.(GL_2(11)/\mathbb{Z}_5)$

Without loss of generality one can suppose that

$$\varphi(f(y)\Phi(x)) = y^t f(y)\Phi(x) \text{ for some } t \in \mathbb{F}_5 \tag{9.30}$$

and $\varphi|_{\mathcal{H}'} = 1$, that is,

$$A(0, m) = 0, \ 1 \leq m \leq 4. \tag{9.31}$$

The fact that φ is an isometry means simply that

$$A(-n, -m) = -A(n, m). \tag{9.32}$$

Now we make a reasonable assumption which is compatible with (9.31) and (9.32) (and prompted by considering a similar construction in the case $\Delta = \Gamma^{1,2}$, $p = 3$):

$$A(n, m) = nA(1, \frac{m}{n}), \ \forall m, n \in \mathbb{F}_5^*.$$

Clearly, φ belongs to $\mathbb{G}$ if and only if φ maps the generating vectors of $\Delta^{1,1}$, that is,

$$5(x-1) + L^3(y)\Phi(x), 5(y-1) - L^3(x)\Phi(y),$$

$$5x^k\Phi(x^{-l}y),\, 5y^k\Phi(x),\; 0 \le k, l \le 4,$$
$$5L^{i-j}(x)L^j(y),\; 2 \le i \le 3, 0 \le j \le i,$$

into vectors in the same lattice. This condition is equivalent to the following equation system:

$$\left\{ \begin{array}{l} \sum_{s=1}^4 \frac{a_s}{s} = 0,\; \sum_{s=1}^4 \frac{a_s}{s^2} = 0,\; 3\sum_{s=1}^4 \frac{a_s}{s^3} = t, \\ a_1 - 2\sum_{s=1}^4 \frac{(a_s)^2}{s^2} + \sum_{s=1}^4 \frac{a_s}{s^4} = 3t^2 + 3t \end{array} \right. ,$$

where $a_s = A(1, s)$ for $s \in \mathbb{F}_5^*$. Solving this system, we obtain

$$a_2 = a_1 + t,\; a_3 = a_1 - 2t,\; a_4 = a_1 - t. \tag{9.33}$$

Setting $t = 1$ and $a_1 = 0$ in (9.33), we get the required automorphism φ. Note that $< B, C, \varphi >\cong \mathbb{Z}_5^3$ and $\tilde{\varphi} = \varphi B$ is an element of class $5A$, since $\mathrm{Ker}(\tilde{\varphi} - 1) = 0$.

2) Now set $Z = Z(\mathbb{G}) \cong \mathbb{Z}_2$, $\mathbb{G}_1 =< G, \varphi >$, $\bar{\mathbb{G}} = \mathbb{G}/Z$ and $\bar{\mathbb{G}}_1 = Z\mathbb{G}_1/Z$, and suppose that

$$\mathbb{G}_1 \neq \mathbb{G}.$$

Note first that $5^4 \big| |\mathbb{G}_1|$. Indeed, $\mathbb{G}_1$ possesses two non-isomorphic subgroups of order 5^3: one of these, $< B, C, \varphi >$, is abelian, and the other, $< B, B', C >$, where $B'(F(x, y)) = yF(x, y)$, is non-abelian. Examining the list of maximal subgroups of the sporadic simple group Co_1 (see [ATLAS]), we see that there are only two maximal subgroups of order divisible by 5^4, namely,

$$\bar{M}_1 = 5_+^{1+2}.GL_2(5),\; \bar{M}_2 = \mathbb{Z}_5^3.(\mathbb{Z}_4 \times \mathbb{A}_5).\mathbb{Z}_2.$$

a) Suppose that $\bar{\mathbb{G}}_1 \subseteq \bar{M}_1$. Then the centre $Z_5 = Z(5_+^{1+2}) =< \bar{X} >$ of the normal extraspecial subgroup 5_+^{1+2} of $\bar{M}_1$ is also normal in $\bar{M}_1$. In particular, every element of odd order in $\bar{\mathbb{G}}_1$ centralizes $\bar{X}$. Lifting to $\mathbb{G}$, we observe that there exists an element $X \in \mathbb{G}$ of order 5 which is centralized by all elements of odd order in $\mathbb{G}_1$. In particular, $X \in \hat{C} = C_{\mathbb{G}}(B, B')$.

Consider the eigenbasis $\{F'_{n,m}\}$ for B, B':

$$F'_{n,m} = \left\{ \begin{array}{ll} \Phi(\varepsilon^{-n}x)\Phi(x^{-m}y) & 0 \le m \le 4, \\ \Phi(\varepsilon^{-n}y)\Phi(x) & m = \infty, \end{array} \right.$$

where $1 \le n \le 4$. Here

$$B(F'_{n,m}) = \left\{ \begin{array}{ll} \varepsilon^n F'_{n,m} & 0 \le m \le 4, \\ F'_{n,m} & m = \infty, \end{array} \right.$$

and

$$B'(F'_{n,m}) = \left\{ \begin{array}{ll} \varepsilon^{nm} F'_{n,m} & 0 \le m \le 4, \\ \varepsilon^n F'_{n,m} & m = \infty. \end{array} \right.$$

This basis gives a decomposition of $\mathcal{L}$ into a sum of twenty-four $< B, B' >$-eigensubspaces $\mathcal{L}_\rho$, where $\rho \in \mathrm{Irr}(< B, B' >)\backslash\{1_{<B,B'>}\}$. Furthermore, $\mathcal{H}_m = < F'_{n,m} | 1 \leq n \leq 4 >_{\mathbb{C}}$. Hence, every element of $\hat{C}$ leaves every Cartan subalgebra $\mathcal{H}_m$ fixed. Using this remark, it is not difficult to show that

$$\hat{C} = C_{\mathbb{G}}(B, B') = Z\times < B, B' > . \tag{9.34}$$

In view of (9.34), we have $X = B^k B'^l$ for some $k, l \in \mathbb{F}_5$, $(k, l) \neq (0, 0)$. If $l \neq 0$, then since $[\varphi, X] = [\varphi, B] = 1$, it follows immediately that $[\varphi, B'] = 1$, and so $\varphi \in \hat{C} \subset G$. But in this case, G has an abelian Sylow 5-subgroup $< B, C, \varphi >$, a contradiction. If $l = 0$, then $< B >=< X >$ and B centralizes every element of odd order in $\mathbb{G}_1$. However, B does not centralize the element $C' \in G$ of order 5, where $C'(F(x, y)) = F(xy, y)$, again a contradiction.

b) Next suppose that $\bar{\mathbb{G}}_1 \subseteq \bar{M}_2$. Denote $\bar{P} = O_5(\bar{M}_2) \cong \mathbb{Z}_5^3$, and consider a Sylow 5-subgroup $\bar{S}$ of $\bar{M}_2$ containing $\bar{P}$. Note that $\bar{S}$ has a unique elementary abelian subgroup of order 5^3. Indeed, suppose that $\bar{S} \supset \bar{P}_1 \cong \bar{P}$ and $\bar{P}_1 \neq \bar{P}$. Then $\bar{P}_0 = \bar{P}_1 \cap \bar{P} \cong \mathbb{Z}_5^2$ and $\bar{P}_0 \subseteq Z(\bar{S})$. Without loss of generality, one can suppose that $\bar{S}$ contains the images $\bar{B}$, $\bar{B}'$, $\bar{C}$ of the elements B, B', C in $\bar{\mathbb{G}}$. In this case, the full preimage P_0 of $\bar{P}_0$ in $\mathbb{G}$ centralizes B, B' and C. In view of (9.34) we must have $P_0 = Z\times < B, B' >$. From this it follows that $[C, B'] = 1$, a contradiction.

Now we set $\bar{Q} =< \bar{B}, \bar{C}, \bar{\varphi} >\cong \mathbb{Z}_5^3$, where $\bar{\varphi}$ is the image of φ in $\bar{\mathbb{G}}$. As we have established above, the subgroups $\bar{P}$ and $\bar{Q}$ are conjugate in $\bar{M}_2$. But $\bar{P} \lhd \bar{M}_2$, therefore in fact we have $\bar{P} = \bar{Q}$ and $\bar{Q} \lhd \bar{M}_2$. In particular, the image $\bar{C}'$ of C' in $\bar{\mathbb{G}}$ normalizes $\bar{Q}$. But $\bar{C}'\bar{B}(\bar{C}')^{-1} = \bar{B}\bar{B}'$, and so $\bar{Q} \ni \bar{B}, \bar{B}', \bar{C}$. However, $[\bar{C}, \bar{B}'] \neq 1$, that is, $\bar{Q}$ is nonabelian, again a contradiction.

The contradictions obtained in a) and b) show that in fact we have $\bar{\mathbb{G}}_1 = \bar{\mathbb{G}}$, that is, $\mathbb{G} = Z\mathbb{G}_1$. Remembering that the group $\mathbb{G} = \mathbb{Z}_2 \circ Co_1$ is non-split over its centre Z, we arrive at the required equality $\mathbb{G}_1 = \mathbb{G}$. □

Remark 9.7.12. A similar construction in the case $p = 3$ leads to the following result:

Suppose that $p = 3$, $\mathbb{G} = \mathrm{Aut}(\Gamma^{1,2}) = W(E_8)$ *and*

$$\mathbb{G}_0 = \{\psi \in \mathbb{G} | \psi \text{ preserves } \mathcal{D}\}.$$

Then there exists an additional automorphism φ *of order* 3 *such that*

$$\mathbb{G} =< \mathbb{G}_0, \varphi > .$$

Commentary

The presentation of the first five sections of this chapter follows, with insignificant improvements, the paper of Bondal and the authors [BKT], which was preceded by the notes [Bon 1] and [Tiep 1]. Theorem 9.7.2 was proved by Abdukhalikov [Abd 1]. All other material is taken from [Tiep 2, 3, 6, 18].

Note that the cases G_2, A_{p-1} (p a prime) and F_4 are the only ones when our programme of studying invariant lattices is fulfilled completely: all invariant lattices are classified, and their automorphism groups are computed. The type G_2 is considered in Chapter 8, type A_{p-1} in the present chapter, and type F_4 will be treated in Chapter 12.

As far as we know, at present there exist only six infinite series of even unimodular root-free lattices, as follows.

1) The Barnes-Wall series [BaW] of lattices of dimension 2^{2d+1}.

2) Lattices of dimensions $p+1$ ($p \equiv -1 \pmod 8$ a prime) and $2(p-1)$ ($p \equiv 1 \pmod 4$ a prime) with isometries of order p (see, for example, [Que 1,2]).

3) Our series of lattices of dimension $p^{2m}-1$ (p an odd prime and $p^m > 3$).

4) The Gow-Gross series [Gow 2], [Gro] of lattices of the same dimension $p^{2m}-1$.

5) The Gross series [Gro] of lattices of dimension $2p^2(p^2-1)$ (p a prime).

6) The Gow series [Gow 3] of lattices of dimension 2^n for specific values of n.

Lattices from series 1) and 3) are constructed explicitly, and there is more or less complete information about them. The series 3) with m arbitrary will be considered in Chapter 10, and the connection between series 1) and 6) in Chapter 14. The last three series 4) – 6) are derived implicitly, and only their automorphism groups are known (see Chapter 14). The same concernes the new series of even unimodular lattices of dimension $2(p^n-1) (p \equiv 1 \pmod 4$ a prime), which quite recently is discovered in [Tiep 20].

Proposition 9.3.3 has once more again been established by Bachoc and Batut [BaB]. It has also been conjectured by them that estimate (9.14) is exact for every k with $1 \leq k \leq \frac{p+1}{4}$. The exactness of (9.14) for $k = \frac{p+1}{4}$ has been shown by Elkies (see [Gro]).

The lattices Γ_k considered in Sections §§9.1, 9.6 were investigated independently by Bondal [Bon 1], and Craig (see [CoS 7]) (the notation there is $A_{p-1}^{(k)}$). If $p \equiv -1 \pmod 4$, then the lattice $\Gamma_{\frac{p+1}{4}}$ is directly related to the Mordell-Weil $SL_2(p)$-invariant lattices; this question will be touched upon in Chapter 14. We have mentioned in §9.6 the results of Blau [Blau] and of Plesken and Nebe [PlN] concerning the same lattices. Very interesting $2(p-1)$-dimensional analogues $B_{p-1}^{(k)}$ of the lattices Γ_k have recently been discovered by Quebbemann (private communication).

As mentioned in Theorem 9.5.1, when $p \geq 5$ the even unimodular lattice $\frac{1}{p}\Gamma^{(p-1)/2,(p+1)/2}$ admits the root system of type $(p+1)A_{p-1}$. In view of this remark, we state here the following two results.

(i) *In* $\mathbb{R}^{48}$, *there exist just four even unimodular lattices with root system* $8A_6$; *of them, just one is decomposable.*

(ii) *Suppose that* Δ *is an even unimodular lattice of dimension* $d(p-1)$ *with root system* dA_{p-1}*, where* $p \geq 7$ *is a prime and* $d \leq \frac{4}{3}(p-1)$ *(or* $d \leq \frac{3}{2}(p-1)$ *if* $p \geq 13$*). Then*

$$\mathrm{Aut}(\Delta) = (\mathbb{S}_p)^d.H.$$

Here, either H *is a* p'*-group, or* $d > p$ *and one of the following assertions holds.*

a) $H \subseteq (\mathbb{Z}_p.\mathbb{Z}_{p-1}) \times (\mathbb{Z}_2 \times (\mathbb{Z}_2 \wr \mathbb{S}_{d-p}))$.

b) H *is a group with one composition factor* $L_2(p)$ *and of order* $ap(p^2-1)/2$*, where the integer* a *divides* $2^{d-p+2} \cdot (d-p)!$.

It is worth mentioning that our classification of invariant lattices of type A_{p-1} is obtained by using functional means, and in terms that are far from the original matrix realization of the Lie algebra $\mathcal{L}$. It would be interesting to make the way back to $M_p(\mathbb{C})$, as demonstrated on the example of the lattice $\Gamma^{1,2} \cong E_8$ (see Table 9.4).

Finally, the results of this chapter settle, in particular, the questions posed by Hahn [Hahn] concerning so-called "*coset lattices*" of dimension $p-1$ and p^2-1.

Chapter 10

Invariant lattices of type A_{p^m-1}

Following [Abd 3], we suggest in this chapter an algorithm which solves in principle the classification problem for invariant lattices of type A_{p^m-1}, where p^m is an odd prime-power. We prove also a reduction theorem which reduces the description of the automorphism groups of invariant lattices to the similar problem for the projections of invariant lattices onto a single Cartan subalgebra. In particular, we complete the description of the automorphism groups for type A_{p-1} begun in Chapter 9.

Because of the cumbersome nature of the technical details, we restrict ourselves to a conceptual presentation, referring the reader to the original papers [Abd 3] and [Tiep 17].

10.1 Preliminaries

For the reader's convenience we recall the construction of the standard orthogonal decomposition of the complex simple Lie algebra $\mathcal{L} = sl_q(\mathbb{C})$ of type A_{q-1}, where $q = p^m$ and p a prime (see Chapter 1).

Let $V = \mathbb{F}_q \oplus \mathbb{F}_q$ be a 2-dimensional space over a field $\mathbb{F}_q$, and set $V_1 = (\mathbb{F}_q, 0) \subset V$, $V_2 = (0, \mathbb{F}_q) \subset V$. The additive groups V_1 and V_2 can be interpreted as vector spaces over $\mathbb{F}_p$. Furthermore, we choose bases $\{\rho_1, \ldots, \rho_m\}$ and $\{\tau_1, \ldots, \tau_m\}$ in V_1 and V_2 respectively, and then any element $u = \sum_i (a_i\rho_i + b_i\tau_i)$ can also be written in the form $u = (a_1, \ldots, a_m; b_1, \ldots, b_m)$. V is endowed with the following bilinear forms:

$$B(u, u') = -\sum_{i=1}^{m} b_i a'_i,$$

$$< u|u' >= B(u, u') - B(u', u) = \sum_{i=1}^{m} (a_i b'_i - a'_i b_i),$$

where $u' = (a'_1, \dots, a'_m; b'_1, \dots, b'_m)$. These formulae, as well as the matrices D, P, $J_{(a,b)} = D^a P^b$ and $J_u = J_{(a_1,b_1)} \otimes \cdots \otimes J_{(a_m,b_m)}$, were introduced in Chapter 1. We shall adopt the notation of Chapter 1 throughout. The decomposition

$$\mathcal{D} : \mathcal{L} = \bigoplus_{\sigma \in \mathbb{P}V} \mathcal{H}_\sigma, \ \mathcal{H}_\sigma = < J_u | u \in \sigma >_{\mathbb{C}}, \tag{10.1}$$

is an IOD.

In Chapter 5 it was shown that, provided $q \neq 27$, the decomposition (10.1) is the only IOD of $\mathcal{L}$ up to conjugacy. The automorphism group

$$\mathrm{Aut}(\mathcal{D}) = \mathrm{Inn}(\mathcal{D}).\mathbb{Z}_2$$

was described in Chapter 1. In particular, it contains the normal subgroup

$$\hat{V} = \{\hat{v} | v \in V\},$$

where $\hat{v} : J_u \mapsto \varepsilon^{\mathrm{tr}(v.u)} J_u$; $\varepsilon = \exp(\frac{2\pi i}{p})$; $v.u = ac + bd$ for $u = (a, b)$, $v = (c, d)$; and $\mathrm{tr}(\xi) = \mathrm{tr}_{\mathbb{F}_q/\mathbb{F}_p}(\xi)$ for $\xi \in \mathbb{F}_q$.

Lemma 10.1.1. *Suppose that* $q > 11$. *Then*

$$G = \mathbb{Z}_p^{2m}.SL_2(q) \cong V.SL_2(q)$$

is the unique minimal irreducible subgroup of $\mathrm{Aut}(\mathcal{D})$.

Proof. Clearly, G is irreducible on $\mathcal{L}$. Conversely, let H be an irreducible subgroup of $A = \mathrm{Aut}(\mathcal{D})$. Then

$$|H| > (\dim \mathcal{L})^2 = (q^2 - 1)^2,$$

and so

$$2mq^3(q^2 - 1) = |A| \geq |\hat{V}H| = \frac{|\hat{V}|.|H|}{|\hat{V} \cap H|} > \frac{q^2(q^2 - 1)^2}{|\hat{V} \cap H|}.$$

This means that $\hat{V} \cap H \neq 1$. From the irreducibility of H on $\mathcal{L}$ it follows that $\hat{V} \subset H$. It is not difficult to see now that H necessarily contains G when $q > 11$. □

Remark. Here we avoid the notations $G = \mathrm{Aut}(\mathcal{D})$ or $G = \mathrm{Inn}(\mathcal{D})$ used in Chapter 8 and Chapter 9, respectively. From now on the groups $\mathrm{Aut}(\mathcal{D})$ and $\mathrm{Inn}(\mathcal{D})$ play a secondary role in our investigation.

Recall that when $p > 2$, G is split over $\hat{V}$:

$$G = \{\hat{v}\hat{\varphi} | v \in V, \varphi \in SL_2(q)\},$$

where

$$\hat{\varphi} : J_u \mapsto \varepsilon^{\{B(\varphi(u),\varphi(u))-B(u,u)\}/2} J_{\varphi(u)}.$$

If $p = 2$, then G consists of the automorphisms $\Psi_{f,\varphi}$, where $f : V \to \mathbb{Z}_4$, $\varphi \in SL_2(q)$, and

$$\Psi_{f,\varphi}(J_u) = i^{f(u)} J_{\varphi(u)},\ i^2 = -1,$$

with

$$i^{f(u+v)} = i^{f(u)+f(v)}(-1)^{B(u,v)+B(\varphi(u),\varphi(v))}.$$

In this case, in view of Theorem 1.3.7, G is split over $\hat{V}$ if and only if $m \leq 2$.

Let T denote the subgroup of diagonal matrices and U the subgroup of unipotent upper-triangular matrices in $SL_2(q)$.

Lemma 10.1.2. *Suppose that* $p = 2$ *and* $u \in V_1$.

(i) *If* $\varphi \in U$, *then* $\Psi_{f,\varphi}(J_u) = \hat{w}(J_u)$ *for some* $w \in V_1$;

(ii) G *contains a subgroup* $\hat{T} = \{\hat{\varphi} | \varphi \in T\}$ *such that* $\hat{\varphi}(J_u) = J_{\varphi(u)}$.

Proof. (i) Under the given hypotheses, we have $\varphi(u) = u$ and, more generally, $\varphi(v) = v$ for all $v \in V_1$. Hence, $B(u, v) + B(\varphi(u), \varphi(v)) = 0$. This means that the functional f is additive: $f(u+v) = f(u) + f(v)$, and in fact f takes values only in the subgroup $\mathbb{Z}_2$ of $\mathbb{Z}_4$.

(ii) The group G contains a subgroup $\tilde{T}$ isomorphic to T since the orders of V and T are coprime. For $\hat{\varphi} \in \tilde{T}$ we have $\hat{\varphi}(J_u) = i^{f_\varphi(u)} J_{\varphi(u)}$, where f_φ is additive on V_1 and

$$f_{\psi\varphi}(u) = f_\varphi(u) + f_\psi(\varphi(u)),\ \forall \varphi, \psi \in T.$$

Next we consider a further embedding of T in G by setting

$$f'_\varphi = f_\varphi + l + l \circ \varphi.$$

Here the additive functional $l : V \to \mathbb{Z}_4$ is chosen so that

$$l|_{V_1} = \sum_{\psi \in T} f_\psi|_{V_1}.$$

Then

$$f'_\varphi(u) = f_\varphi(u) + \sum_{\psi \in T} f_\psi(u) + \sum_{\psi \in T} f_\psi(\varphi(u)) =$$

$$= f_\varphi(u) + \sum_{\psi \in T} f_\psi(u) + \sum_{\psi \in T} \{f_{\psi\varphi}(u) - f_\varphi(u)\} = 0.$$ □

Once again, we fix again the following notation: $(1, 0) \in \sigma_0$, $\mathcal{H}_0 = \mathcal{H}_{\sigma_0}$, $G_0 = St_G(\mathcal{H}_0)$. More generally, the subscript σ_0 is replaced by 0 everywhere. Furthermore, a character χ of a finite group H is said to be *rational* (respectively *real*), if χ is afforded by a rational (respectively real) representation.

In every Cartan subalgebra $\mathcal{H}_\sigma$ we choose the $\mathbb{Z}$-module Λ_σ spanned by the basis dual to the system of simple roots. Then the lattice $(\Lambda, (X|Y))$, where

$$\Lambda = \bigoplus_{\sigma \in \mathbb{P}V} \Lambda_\sigma,$$

$$(X|Y) = \frac{1}{2q^2} K(X, Y) = \frac{1}{q} \mathrm{Tr} XY,$$

is said to be a *basic lattice.*

Lemma 10.1.3. *Suppose that q is odd. Then G has a unique real absolutely irreducible representation of degree $q^2 - 1$. More precisely, G has exactly q complex irreducible characters $\rho_0, \rho_1, \dots, \rho_{q-1}$ of degree $q^2 - 1$, where ρ_0 is rational, and*

$$\mathbb{Q}(\rho_1) = \dots = \mathbb{Q}(\rho_{q-1}) = \mathbb{Q}(\varepsilon).$$

If q is even, then $ASL_2(q)$ has exactly q complex irreducible characters of degree $q^2 - 1$, and all of them are rational.

Proof. 1) Let $c(X)$ denote the number of conjugacy classes of a finite group X. Then $c(G) = q + c(S)$, where throughout this proof G denotes the special affine group $ASL_2(q) = V.S$, $V = O_p(G)$ and $S = SL_2(q)$. Furthermore, G has precisely $c(S)$ complex irreducible characters that are trivial on V. Let I_0 denote the set consisting of such characters and $I' = \mathrm{Irr}(G) \backslash I_0$. Clearly, for all $\rho \in I_0$,

$$\deg \rho < \sqrt{|S|} < q^2 - 1$$

On the other hand, $|I'| = q$, and by Clifford's Theorem, $(q^2 - 1) | \deg \rho$ for all $\rho \in I'$. Finally,

$$\sum_{\rho \in I'} (\deg \rho)^2 = |G| - \sum_{\rho \in I_0} (\deg \rho)^2 = |G| - |S| = q(q^2 - 1)^2.$$

We arrive at the conclusion that

$$I' = \{\rho_0, \dots, \rho_{q-1}\},$$

where $\deg \rho_i = q^2 - 1$.

Now we rename the characters ρ_i in such a way that the character $1_G + \rho_0$ corresponds to the natural doubly transitive permutation representation of G. Clearly, $\rho_0 = \delta_0|_G$ for some irreducible character δ_0 of $F = AGL_2(q)$. If we denote

$$\mathrm{Irr}(F/G) = \{\varsigma_0 = 1_{F/G}, \varsigma_1, \dots, \varsigma_{q-2}\},$$

then $\delta_i = \delta_0 \varsigma_i$, $0 \leq i \leq q - 2$, are distinct irreducible characters of F. Moreover, F possesses a further complex irreducible character δ, which is of degree $(q-1)(q^2-1)$ and nontrivial on V (this is the *pivotal* character in the terminology of Faddeev [Fad]). It is not difficult to see that

$$\delta|_G = \rho_1 + \cdots + \rho_{q-1}.$$

By Clifford's Theorem, the characters $\rho_1, \ldots, \rho_{q-1}$ are conjugate under F. In particular, they have the same Schur-Frobenius indicator: $\mathrm{ind}(\rho_i) = r$ for some $r \in \{0, \pm 1\}$. Moreover, $\mathrm{ind}(\rho_0) = 1$.

2) Next, we denote by ρ_G and $\rho_{G/V}$ the regular characters of G and $G/V \cong S$, respectively. Then

$$\rho_G - \rho_{G/V} = \sum_{\zeta \in \mathrm{Irr}(G)} (\deg \zeta)\zeta - \sum_{\zeta \in I_0} (\deg \zeta)\zeta = (q^2-1)\sum_{i=0}^{q-1} \rho_i.$$

This implies that

$$(q-1)(q^2-1)r = \mathrm{ind}\left((q^2-1)\sum_{i=1}^{q-1}\rho_i\right) =$$

$$= \mathrm{ind}(\rho_G) - \mathrm{ind}(\rho_{G/V}) - (q^2-1) =$$

$$= \{1 + \text{the number of involutions in } G\} -$$

$$- \{1 + \text{the number of involutions in } S\} - \left(q^2-1\right) =$$

$$= \begin{cases} (q^2+1) - 2 - (q^2-1) = 0 & \text{if } \ 2 \nmid q; \\ (q^2 + (q^2-1)q) - q^2 - (q^2-1) = (q-1)(q^2-1) & \text{if } \ 2|q. \end{cases}$$

Hence,

$$r = \begin{cases} 0 & \text{if } \ 2 \nmid q; \\ 1 & \text{if } \ 2|q. \end{cases}$$

Finally, the equalities $\mathbb{Q}(\rho_1) = \ldots = \mathbb{Q}(\rho_{q-1}) = \mathbb{Q}(\varepsilon)$ valid when q is odd follow from the fact that all the characters ρ_i can be obtained by inducing characters of a subgroup $\mathbb{Z}_p^m.(\mathbb{Z}_p^m \times \mathbb{Z}_p^m)$. □

In view of Lemma 10.1.3 and the Deuring-Noether Theorem [CuR], one can suppose that every G-invariant lattice is isometrically embedded in the basic lattice Λ. Hence, in what follows, by an *invariant lattice of type* A_{q-1} we mean an arbitrary G-invariant sublattice $\Lambda' \subseteq \Lambda$. A sublattice $\Lambda' \subseteq \Lambda$ is said to be *divisible* if there exists a natural $n > 1$ such that $\Lambda' \subseteq n\Lambda$, and *indivisible* otherwise.

The following statement is obvious.

Lemma 10.1.4. (i) *$G_0 = \hat{V}.(T.U)$ and*

$$G_0|_{\mathcal{H}_0} = \left\{ g_{v,t} : J_{(a,0)} \mapsto \varepsilon^{\mathrm{tr}(v.(a,0))} J_{(ta,0)} | v \in V_1, t \in \mathbb{F}_q^* \right\}.$$

(ii) *If* $Z = \frac{1}{q}\sum_{a\in\mathbb{F}_q^*} J_{(a,0)}$, *then* $\Lambda_0 =< \hat{v}(Z) | v \in V_1 >_{\mathbb{Z}}$.

(iii) *All linear relations between the elements* $\hat{v}(Z)$, $v \in V_1$, *are proportional to the relation* $\sum_{v\in V_1} \hat{v}(Z) = 0$.

(iv) *The* $\mathbb{Z}G$*-module* Λ *is isomorphic to the module*

$$\mathrm{Ind}_{G_0}^{G}(\Lambda_0) = \mathbb{Z}G \otimes_{\mathbb{Z}G_0} \Lambda_0. \qquad \square$$

Next, we consider a free (q^2-1)-dimensional $\mathbb{Z}$-module $\mathcal{A}$ generated over $\mathbb{Z}$ by the formal symbols $x^\alpha y^\beta$, $\alpha, \beta \in \mathbb{F}_q$, with a single relation

$$\sum_{\alpha,\beta\in\mathbb{F}_q} x^\alpha y^\beta = 0.$$

Clearly, $\mathcal{A}$ is a commutative ring under the product

$$x^\alpha y^\beta . x^\gamma y^\delta = x^{\alpha+\gamma} y^{\beta+\delta}.$$

Note that the choice of basis $\{\alpha_1, \ldots, \alpha_m\}$ in $\mathbb{F}_q$, regarded as a space over $\mathbb{F}_p$, defines a surjective ring homomorphism

$$\Psi : \mathbb{Z}[X_1, \ldots, X_m, Y_1, \ldots, Y_m] \to \mathcal{A}, \quad \Psi(X_i Y_j) = x^{\alpha_i} y^{\alpha_j}.$$

Here KerΨ is the ideal generated by the polynomials

$$(X_i)^p - 1,\ (Y_i)^p - 1, \quad \sum_{0\le i_1,\ldots,j_m\le p-1} (X_1)^{i_1}\cdots(X_m)^{i_m}(Y_1)^{j_1}\cdots(Y_m)^{j_m}.$$

We endow the ring $\mathcal{A}$ with the structure of a G-module by setting

a) $\hat{v}(x^\alpha y^\beta) = x^\gamma y^\delta . x^\alpha y^\beta = x^{\gamma+\alpha} y^{\delta+\beta}$ for $v = (\gamma, \delta) \in V$;

b) $\hat{\varphi}(x^\alpha y^\beta) = x^{a\alpha+c\beta} y^{b\alpha+d\beta}$ for $\begin{pmatrix} a & b \\ c & d \end{pmatrix} = \varphi^{-1} \in SL_2(q)$.

Here we mean that $(x^a y^b)^\alpha = x^{a\alpha} y^{b\alpha}$. The definition is correct, since $\sum_{\alpha,\beta} x^\alpha y^\beta$ is invariant under the actions introduced. Clearly, the G-module $\mathcal{A}$ is an exact analogue of the ring $\mathbb{Z} < x, y >$ introduced in §9.1.

For convenience we set $x^0 = y^0 = x^0 y^0 = 1$ and denote

$$\Phi_\sigma = 1 + \sum_{(\beta,-\alpha)\in\sigma} x^\alpha y^\beta,\ \Phi(x^\alpha y^\beta) = \sum_{\gamma\in\mathbb{F}_q} (x^\alpha y^\beta)^\gamma.$$

In particular, $\Phi(x) = \sum_\gamma x^\gamma$, $\Phi(y) = \sum_\gamma y^\gamma$.

The following statement is trivial to verify.

Lemma 10.1.5. (i) $\Phi_\sigma . \Phi_{\sigma'}$ *is equal to* 0 *if* $\sigma \neq \sigma'$; *and* $q\Phi_\sigma$ *if* $\sigma = \sigma'$.

(ii) *The element* $\Phi(x^\alpha y^\beta)$ *does not depend on the choice of representative* $(\beta, -\alpha) \in \sigma$.

(iii) *The element* $x^\gamma y^\delta \Phi(x^\alpha y^\beta)$ *does not depend on the choice of representative* $(\delta, -\gamma) \in V/\sigma$ *on an affine line* V/σ *parallel to* σ *in* V. □

Next, we consider the $\mathbb{Z}G$-submodule Γ of $\mathcal{A}$ generated by $\Phi(y)$.

Theorem 10.1.6. *The* $\mathbb{Z}G$*-modules* Γ *and* Λ *are isomorphic for* $p > 2$.

Proof. Set $\Gamma_0 = \{(\sum_{\alpha\in\mathbb{F}_q} a_\alpha x^\alpha)\Phi(y) | a_\alpha \in \mathbb{Z}\} \subset \Gamma$. Clearly, $St_G(\Gamma_0) = G_0$. We claim that the G_0-modules Γ_0 and Λ_0 are isomorphic. Indeed, Γ_0 is generated over $\mathbb{Z}$ by the elements $\{\hat{v}(\Phi(y)) | v \in V_1\} = \{x^\alpha \Phi(y) | \alpha \in \mathbb{F}_q\}$, and the unique linear relation between them is $\sum_\alpha x^\alpha \Phi(y) = 0$. Clearly, $St_{G_0}(\Phi(y)) = \hat{V}_2.(T.U)$. Next, define a map

$$\chi : \Lambda_0 \to \Gamma_0$$

by setting $\chi(\hat{v}(Z)) = \hat{v}(\Phi(y))$. Then χ is defined correctly, by Lemma 10.1.4, and is a G_0-homomorphism:

$$\chi(\hat{\varphi}\hat{v}(Z)) = \chi(\hat{\varphi}\hat{v}\hat{\varphi}^{-1}(Z)) = \hat{\varphi}\hat{v}\hat{\varphi}^{-1}(\Phi(y)) = \hat{\varphi}\hat{v}(\Phi(y))$$

for $\varphi \in T.U$, since $\hat{\varphi}(Z) = Z$ and $\hat{\varphi}\hat{v}\hat{\varphi}^{-1} \in \hat{V}_1$. From Lemma 10.1.5 it follows that $\Gamma \cong \mathrm{Ind}_{G_0}^G(\Gamma_0)$. Hence the isomorphism χ can be extended to the whole of Λ. □

Corollary 10.1.7. *For any* $\sigma \in \mathbb{P}V$ *we have*

$$\chi(\Lambda_\sigma) = \sum_{\alpha,\beta} \mathbb{Z}x^\alpha y^\beta \Phi_\sigma.$$ □

Theorem 10.1.6 allows us to go over from the lattice Λ to the lattice Γ. Thus, when $p > 2$, the description of indivisible G-invariant sublattices of Λ is equivalent to the description of indivisible ideals of the ring $\mathcal{A}$ that are contained in Γ and invariant under $SL_2(q)$. On the other hand, Lemma 10.1.2 tells us that, when $p = 2$, the classification of G_0-invariant sublattices of Λ_0 is equivalent to the classification of G_0-invariant sublattices of Γ_0.

Furthermore, let $\mathbb{K}$ be an algebraic number field, and R a maximal order in $\mathbb{K}$. Consider the RG-modules $\mathcal{A}^R = \mathcal{A} \otimes_\mathbb{Z} R$, $\Gamma^R = \Gamma \otimes_\mathbb{Z} R$. The factor-lattice $\Gamma/p\Gamma$ is a vector space over $\mathbb{F}_p$, and we wish to choose R such that $\Gamma^R/p\Gamma^R$ is a vector space over $\mathbb{F}_q$.

We claim that there exists an algebraic number field $\mathbb{K}$ such that $R/pR \cong \mathbb{F}_q$. Indeed, let $f(t) \in \mathbb{Z}[t]$ be a monic polynomial of degree m whose image $\bar{f}(t)$ under the homomorphism $\mathbb{Z}[t] \to \mathbb{F}_p[t]$ is irreducible over $\mathbb{F}_p$. Then the field $\mathbb{K} = \mathbb{Q}(\theta)$, where θ is a root of $f(t)$, satisfies the required condition [BoS]. Fix a maximal order R in $\mathbb{K}$, and from now on identify R/pR with $\mathbb{F}_q$. The Galois group Gal of the extension $\mathbb{K}/\mathbb{Q}$ is cyclic of order m, and there exists a generator g of Gal that induces the Frobenius automorphism $\beta \mapsto \beta^p$ of $\mathbb{F}_q$. Identifying Γ with

$\Gamma\otimes\mathbb{Z}\subset\Gamma\otimes R=\Gamma^R$, one can suppose that $\Gamma\subset\Gamma^R$. Set $\Gamma^R_\sigma=\Gamma_\sigma\otimes R$. The group *Gal* acts on Γ^R by the rule: $g(F\otimes r)=F\otimes g(r)$ for $F\in\Gamma$ and $r\in R$. Then

$$\Gamma=\{a\in\Gamma^R|g(a)=a\}.$$

Note that if $I\lhd R$ and Γ' is a RG-submodule of Γ^R, then $I.\Gamma'=\{ra|r\in I,a\in\Gamma'\}$ is also an RG-submodule. In this connection we introduce the following definition: a G-module $\Gamma'\subseteq\Gamma^R$ is said to be *divisible* if there exists a proper ideal $I\lhd R$ such that $\Gamma'\subseteq I\Gamma^R$, and *indivisible* otherwise.

By analogy with the case of $\mathbb{Z}$-lattices, one can try to describe (up to similarity) all the G-invariant R-submodules (sublattices) of Γ^R. Clearly, this problem is equivalent to the classification of $SL_2(q)$-invariant indivisible ideals of the R-algebra $\mathcal{A}$ that are contained in Γ^R.

The following statement is obvious.

Lemma 10.1.8. *Let $\Gamma'\subseteq\Gamma^R$ be an indivisible invariant sublattice. Then*

(i) *the projection $\Gamma'_0=\mathrm{pr}_0\Gamma'$ of Γ' onto Γ^R_0 is an indivisible G_0-invariant sublattice;*

(ii) $\Gamma'_0\supset q\Gamma^R_0$;

(iii) $\Gamma'\supset q\Gamma'_0$ *and* $\Gamma'\supset q^2\Gamma^R$. □

The following theorem explains the connection between Γ and Γ^R.

Theorem 10.1.9. *The map*

$$\Gamma'\mapsto\Gamma'\cap\Gamma$$

defines a bijective correspondence between the set of indivisible R-submodules of Γ^R that are invariant under G and Gal, and the set of indivisible $\mathbb{Z}$-submodules in Γ that are invariant under G.

Proof. Let $\{\beta,\beta^p,\dots,\beta^{p^{m-1}}\}$ be a normal basis of $\mathbb{F}_q$ over $\mathbb{F}_p$. Note that the matrix

$$\begin{pmatrix}\beta & \beta^p & \beta^{p^2} & \dots & \beta^{p^{m-1}}\\ \beta^p & \beta^{p^2} & \beta^{p^3} & \dots & \beta\\ & & \dots & & \\ \beta^{p^{m-1}} & \beta & \beta^{p^2} & \dots & \beta^{p^{m-2}}\end{pmatrix}$$

is non-singular in this situation.

Now we suppose that Γ' is an indivisible sublattice invariant under G and *Gal*. Set $\Gamma''=(\Gamma'\cap\Gamma)R$. We shall prove by induction on $i=0,1,\dots,2m$ that

$$\Gamma'\cap p^{2m-i}\Gamma^R=\Gamma''\cap p^{2m-i}\Gamma^R. \tag{10.2}$$

Indeed, if $i=0$, then

$$\Gamma'\cap q^2\Gamma^R=q^2\Gamma^R=\Gamma''\cap q^2\Gamma^R.$$

Assume that $j > 0$ and that (10.2) holds for $i = j-1$. Set

$$C_1 = (\Gamma' \cap p^{2m-j}\Gamma^R)/(\Gamma' \cap p^{2m-j+1}\Gamma^R),$$

$$C_2 = (\Gamma'' \cap p^{2m-j}\Gamma^R)/(\Gamma'' \cap p^{2m-j+1}\Gamma^R).$$

Then C_1 and C_2 are spaces over R/pR invariant under Gal. We show that $C_1 = C_2$. Clearly, $C_1 \supseteq C_2$. We must represent an arbitrary element $f \in C_1$ as an $\mathbb{F}_q$-linear combination of elements of C_2. This is possible since the equation

$$f = \lambda_0 \sum_{k=0}^{m-1} g^k(\beta f) + \lambda_1 \sum_{k=0}^{m-1} g^k(\beta^p f) + \ldots + \lambda_{m-1} \sum_{k=0}^{m-1} g^k(\beta^{p^{m-1}} f)$$

has a solution, or equivalently, that the following equation system has a solution:

$$\begin{cases} \lambda_0\beta + \lambda_1\beta^p + \ldots + \lambda_{m-1}\beta^{p^{m-1}} & = & 1, \\ \lambda_0\beta^p + \lambda_1\beta^{p^2} + \ldots + \lambda_{m-1}\beta & = & 0, \\ \ldots & & \\ \lambda_0\beta^{p^{m-1}} + \lambda_1\beta + \ldots + \lambda_{m-1}\beta^{p^{m-2}} & = & 0. \end{cases}$$

The matrix of this last system is non-singular, as we mentioned above.

This proves (10.2). In particular, when $i = 2m$ we obtain

$$\Gamma' = (\Gamma' \cap \Gamma)R.$$

Now the map $\Gamma' \mapsto \Gamma' \cap \Gamma$ is injective, since whenever $\Gamma'_1 \cap \Gamma = \Gamma'_2 \cap \Gamma$, we have that

$$\Gamma'_1 = (\Gamma'_1 \cap \Gamma)R = (\Gamma'_2 \cap \Gamma)R = \Gamma'_2.$$

This map is also surjective, since the module $\Gamma_1.R$ corresponds to the sublattice $\Gamma_1 \subseteq \Gamma$. □

Consequently, our classification problem is equivalent to that of describing ideals of $\mathcal{A}$ that are contained in Γ^R and invariant under $SL_2(q)$ and Gal.

In view of Lemma 10.1.8, first of all we have to study G_0-invariant sublattices of Γ_0^R.

10.2 Classification of projections of invariant sublattices to a Cartan subalgebra

Recall that

$$\Gamma_0^R = \{\sum_\alpha a_\alpha x^\alpha \Phi(y) | a_\alpha \in R\}.$$

For convenience we introduce the module

$$\Delta = \{\sum_{\alpha} a_\alpha x^\alpha | a_\alpha \in R, \sum_{\alpha} x^\alpha = 0\}.$$

Clearly, Δ is a G_0-module isomorphic to Γ_0^R. Elements of Γ_0^R are obtained by attaching the symbol $\Phi(y)$ to elements of Δ. Furthermore, classification of G_0-invariant sublattices in Γ_0^R is equivalent to the description of all ideals of the ring Δ that are invariant under the subgroup T of $SL_2(q)$, that is, under the automorphisms $x^\alpha \mapsto x^{\alpha t}$ with $t \in \mathbb{F}_q^*$.

We define the ideal Δ^k of Δ to be that generated by the elements of the form

$$\prod_{i=1}^{k}(1 - x^{\gamma_i}),\ \gamma_i \in \mathbb{F}_q.$$

Lemma 10.2.1. *The following assertions hold.*

(i) *For all $\alpha, \beta \in \mathbb{F}_q$ and $n \in \mathbb{Z}$,*

$$(1 - x^\alpha) + (1 - x^\beta) \equiv (1 - x^{\alpha+\beta}),\ (1 - x^{n\alpha}) \equiv n(1 - x^\alpha) \pmod{\Delta^2}.$$

(ii) *For all $\alpha \in \mathbb{F}_q$,*

$$(1 - x^\alpha)^p \in p\Delta^1,\ (1 - x^\alpha)^p \equiv -p(1 - x^\alpha) \pmod{p\Delta^2}.$$

(iii) *If $\{\alpha_1, \dots, \alpha_m\}$ is a basis of $\mathbb{F}_q$ over $\mathbb{F}_p$, then the elements*

$$(1 - x^{\alpha_1})^{n_1} \cdots (1 - x^{\alpha_m})^{n_m},\ 0 \le n_i < p, \sum_{i=1}^{m} n_i < m(p-1),$$

form a basis of Δ over R.

(iv) $(1 - x^{\alpha_1})^{p-1} \cdots (1 - x^{\alpha_m})^{p-1} \in p\Delta.$ □

Next, we study the structure of the module

$$\Delta/p\Delta = \{\sum_{\alpha} a_\alpha x^\alpha | a_\alpha \in \mathbb{F}_q, \alpha \in \mathbb{F}_q\},$$

which is a $(q-1)$-dimensional space over $R/pR = \mathbb{F}_q$. For this we introduce the following generalizations of the function $L(x)$ considered in Chapter 9:

$$\bar{L}_0(x) = \sum_{\alpha \ne 0} \frac{1}{\alpha} x^\alpha,$$

$$\bar{L}_i(x) = g^i(\bar{L}_0(x)) = \sum_{\alpha \ne 0} \frac{1}{\alpha^{p^i}} x^\alpha;\ 0 \le i \le m-1.$$

Observe that

$$\bar{L}_0(x^t) = \sum_{\alpha \neq 0} \frac{1}{\alpha} x^{\alpha t} = t \sum_{\alpha \neq 0} \frac{1}{\alpha t} x^{\alpha t} = t.\bar{L}_0(x).$$

Similarly, $\bar{L}_i(x^t) = t^{p^i}.\bar{L}_i(x)$. This means that $\bar{L}_i(x)$ is an eigenvector for T. Note also that the elements $(\bar{L}_0(x))^p$ and $\bar{L}_1(x)$ are proportional. Set

$$\mathcal{K} = \{k = (k_0, \ldots, k_{m-1}) = \sum_{t=0}^{m-1} k_t p^t | 0 \leq k_t < p, \sum_t k_t < m(p-1)\}.$$

Lemma 10.2.2. *The space $\bar{\Delta} = \Delta / p\Delta$ has as basis the set*

$$\bar{L}^k(x) = (\bar{L}_0(x))^{k_0} \cdots (\bar{L}_{m-1}(x))^{k_{m-1}};\ k \in \mathcal{K}.$$

Proof. The ideals $\bar{\Delta}^i = (\Delta^i + p\Delta)/p\Delta$ are invariant under T and $Gal =< g >$. Since

$$\bar{\Delta}^1 = \{\sum_\alpha a_\alpha x^\alpha | \sum_\alpha a_\alpha = 0\},$$

and $\sum_{\alpha \neq 0} \frac{1}{\alpha} = 0$, we have $\bar{L}_0(x) \in \bar{\Delta}^1$. Similarly, $\bar{L}_i(x) \in \bar{\Delta}^1$, and we show that $\bar{L}_i(x) \notin \bar{\Delta}^2$. To this end we consider the linear map

$$\partial : \bar{\Delta} \to \mathbb{F}_q,$$

given by $\partial(\sum_\alpha a_\alpha x^\alpha) = \sum_\alpha \alpha a_\alpha$. Then $\bar{\Delta}^2 \subseteq \mathrm{Ker}\partial$ because

$$\partial((1 - x^\alpha)(1 - x^\beta)x^\gamma) = 0.$$

But $\partial(\bar{L}_0(x)) = \sum_{\alpha \neq 0} \frac{1}{\alpha}.\alpha = -1$, and so $\bar{L}_0(x) \notin \bar{\Delta}^2$ and $\bar{L}_i(x) \notin \bar{\Delta}^2$. Meanwhile, we have that $\partial(\bar{L}_i(x)) = 0$ for $1 \leq i \leq m-1$.

Furthermore, by Lemma 10.2.1(iii) the element

$$\bar{M}(x) = \bar{L}_0(x) + \ldots + \bar{L}_{m-1}(x)$$

can be written in the form

$$\bar{M}(x) = \beta_1(1 - x^{\alpha_1}) + \ldots + \beta_m(1 - x^{\alpha_m}) + F(x),$$

where $F(x) \in \bar{\Delta}^2$. Since $g(\bar{M}(x)) = \bar{M}(x)$, we have $\beta_i \in \mathbb{F}_p$, and so by Lemma 10.2.1(i) we can suppose that

$$\bar{M}(x) = (1 - x^{\beta_1\alpha_1 + \ldots + \beta_m\alpha_m}) + F(x).$$

This implies that $\partial(\bar{M}(x)) = -\sum_i \beta_i \alpha_i$, on the one hand. On the other hand,

$$\partial(\bar{M}(x)) = \partial(\bar{L}_0(x) + \ldots + \bar{L}_{m-1}(x)) = -1,$$

hence,

$$\bar{M}(x) = (1-x) + F(x),\ F(x) \in \bar{\Delta}^2.$$

It is now clear that $\bar{M}(x^\alpha) = (1-x^\alpha)+F(x^\alpha)$, $F(x^\alpha) \in \bar{\Delta}^2$. By Lemma 10.2.1(iii), every element of $\bar{\Delta}$ can be expressed in terms of the elements $\bar{L}^k(x)$, $k \in \mathcal{K}$. Now just recall that $|\mathcal{K}| = q-1 = \dim_{\mathbb{F}_q} \Delta$. □

Consequently, we have decomposed the T-module $\bar{\Delta}$ into a direct sum of pairwise non-isomorphic T-submodules:

$$\bar{\Delta} = \bigoplus_{k \in \mathcal{K}} < \bar{L}^k(x) >_{\mathbb{F}_q} .$$

According to Lemma 10.1.8, every indivisible invariant sublattice $\Gamma' \subseteq \Gamma^R$ contains $q^2\Gamma^R$. Therefore, we have to study the structure of the module $\Delta/q^2\Delta$. Denote by φ_t the element $\begin{pmatrix} t^{-1} & 0 \\ 0 & t \end{pmatrix}$ of T, where $t \in \mathbb{F}_q^*$.

Proposition 10.2.3. *There exist elements $L_i(x) \in \Delta/q^2\Delta$, $0 \le i \le m-1$ with the following properties:*

(i) $\varphi_t(L_i(x)) \in (R/q^2R).L_i(x)$;

(ii) $\varphi_t(L_i(x)) \equiv t^{p^i} L_i(x) \pmod{p(R/q^2R).L_i(x)}$;

(iii) $L_i(x) = g^i(L_0(x))$;

(iv) $(L_i(x))^p = -pL_{i+1}(x)$;

(v) $L_0(x)^{p-1} \cdots L_{m-1}(x)^{p-1} = q(-1)^m$;

(vi) $\Delta/q^2\Delta = \oplus_{k \in \mathcal{K}} < L^k(x) >_{R/q^2R}$ *is a direct decomposition of the T-module $\Delta/q^2\Delta$, where*

$$L^k(x) = L_0(x)^{k_0} \cdots L_{m-1}(x)^{k_{m-1}}$$

for $k = (k_0, \ldots, k_{m-1}) \in \mathcal{K}$.

Proof. It is not difficult to see that

$$\mathrm{Rad}(\Delta/q^2\Delta) = p\Delta/q^2\Delta, \quad (\Delta/q^2\Delta)/\mathrm{Rad}(\Delta/q^2\Delta) \cong \Delta/p\Delta.$$

Furthermore, the (R/q^2R)-module $\Delta/q^2\Delta$ is free:

$$\Delta/q^2\Delta = \bigoplus_{\alpha \neq 0} (R/q^2R)x^\alpha,$$

and artinian. Hence, $\Delta/q^2\Delta$ is semiperfect and projective. In this case it is known [CaE] that the direct decomposition

$$(\Delta/q^2\Delta)/\mathrm{Rad}(\Delta/q^2\Delta) = \bigoplus_{k \in \mathcal{K}} < \bar{L}^k(x) >_{R/pR}$$

of the semisimple module $(\Delta/q^2\Delta)/\mathrm{Rad}(\Delta/q^2\Delta)$ can be lifted to a direct decomposition

$$\Delta/q^2\Delta = \bigoplus_{k\in\mathcal{K}} \mathcal{P}(k) \tag{10.3}$$

of the module $\Delta/q^2\Delta$ such that

$$\mathcal{P}(k)/p\mathcal{P}(k) \cong < \bar{L}^k(x) >_{\mathbb{F}_q} . \tag{10.4}$$

This decomposition is unique, since $\Delta/p\Delta$ is a direct sum of non-isomorphic irreducible modules. Let $L_0(x)$ denote the generator of $\mathcal{P}(1, 0, \ldots, 0)$ whose image in $\Delta/p\Delta$ is equal to $\bar{L}_0(x)$. Set

$$L_i(x) = g^i(L_0(x)).$$

Then from the uniqueness of the decomposition (10.3), and condition (10.4), it follows that

$$\mathcal{P}(k) = < L^k(x) >_{R/q^2R} .$$

Thus we have constructed elements $L_i(x)$ satisfying conditions (i) – (iii) and (vi).

Now we "adjust" the elements $L_i(x)$ so that, in addition to conditions (i) – (iii), (vi), they satisfy condition (iv). From (10.3) and (10.4) it follows that $L_0(x)^p \in p\mathcal{P}(0, 1, 0, \ldots, 0)$. Furthermore, if we write

$$L_0(x) \equiv \sum_{i=0}^{m-1} \lambda_i(1 - x^{\alpha_i}) \,(\mathrm{mod}\, \Delta^2/q^2\Delta)$$

for $\lambda_i \in \mathbb{F}_q$, and $\{\alpha_i | 0 \le i \le m-1\}$ is a basis of $\mathbb{F}_q$ over $\mathbb{F}_p$, then

$$L_0(x)^p \equiv \sum_{i=0}^{m-1} (\lambda_i)^p (1 - x^{\alpha_i})^p \equiv$$

$$\equiv -p \sum_{i=0}^{m-1} (\lambda_i)^p (1 - x^{\alpha_i}) \equiv -pL_1(x) \,(\mathrm{mod}\, p\Delta^2/q^2\Delta).$$

Here we have used Lemma 10.2.1(ii) and the equality $L_1(x) = g(L_0(x))$. Furthermore,

$$(p\Delta^2/q^2\Delta) \cap p\mathcal{P}(0, 1, 0, \ldots, 0) = p^2\mathcal{P}(0, 1, 0, \ldots, 0).$$

Hence,

$$L_0(x)^p \equiv -pL_1(x) \,(\mathrm{mod}\, p^2\mathcal{P}(0, 1, 0, \ldots, 0)),$$

that is, $L_0(x)^p = -p(1 + pb)L_1(x)$ for some $b \in R/q^2R$. We shall prove by induction on $n \ge 2$ that $L_0(x)$ can be chosen so that

$$L_0(x)^p = -p(1 + p^{n-1}c)L_1(x)$$

for some $c \in R/q^2R$. To begin with, the basis of induction $n = 2$ has been just established. The induction step: set

$$L_0'(x) = (1 + p^{n-1}a)L_0(x),\ L_1'(x) = g(L_0'(x)) = (1 + p^{n-1}g(a))L_1(x),$$

where $a = g^{-1}(c)$. Then

$$(L_0'(x))^p = (1 + p^{n-1}a)^p L_0(x)^p = (1 + p^{n-1}a)^p(-p)(1 + p^{n-1}c)L_1(x) =$$
$$= -p(1 + p^{n-1}a)^p(1 + p^{n-1}c)(1 + p^{n-1}g(a))^{-1}L_1'(x) =$$
$$= -p(1 + p.p^{n-1}a + \ldots)L_1'(x) = -p(1 + p^n c')L_1'(x).$$

Thus, setting $n = 2m$ we can choose $L_0'(x)$ such that $(L_0'(x))^p = -pL_1'(x)$, that is, (iv) is satisfied.

Finally, we prove (v). From (10.3) and (10.4) it follows that

$$L_0(x)^{p-1} \cdots L_{m-1}(x)^{p-1} \in \mathcal{P}(0, \ldots, 0),$$

that is, $L_0(x)^{p-1} \cdots L_{m-1}(x)^{p-1} = a \in R/q^2R$. Multiplying both of parts of this equality by $L_0(x)$, we have

$$aL_0(x) = L_0(x)^p \cdots L_{m-1}(x)^{p-1} =$$
$$= -pL_1(x)^p \cdots L_{m-1}(x)^{p-1} = \ldots =$$
$$= (-p)^{m-1}L_{m-1}(x)^p = (-p)^m L_0(x),$$

that is, $a = q(-1)^m$. □

Recall that $\mathcal{P}(k) =< L^k(x) = L_0(x)^{k_0} \ldots L_{m-1}(x)^{k_{m-1}} >_{R/q^2R}$ for $k \in \mathcal{K}$.

Theorem 10.2.4. *A sublattice $\Delta' \subseteq \Delta$ is an indivisible sublattice invariant under G_0 and Gal if and only if*

$$\Delta' \supset q^2\Delta \text{ and } \Delta'/q^2\Delta = \bigoplus_{k\in\mathcal{K}} p^{t(k)}\mathcal{P}(k),$$

where $0 \le t(k) \le 2m$ and

(a) $t(k_0, k_1, \ldots, k_{m-1}) = t(k_{m-1}, k_0, k_1, \ldots, k_{m-2})$*;*

(b) *if $p^t\mathcal{P}(k) \subseteq \Delta'/q^2\Delta$, then*

$$p^t\mathcal{P}(k_0, \ldots, k_{i-1}, k_i + 1, k_{i+1}, \ldots, k_{m-1}) \subseteq \Delta'/q^2\Delta;$$

(here $\mathcal{P}(\ldots, p, k_i, \ldots) = p\mathcal{P}(\ldots, 0, k_i + 1, \ldots)$ and $\mathcal{P}(p-1, \ldots, p-1) = q\mathcal{P}(0, \ldots, 0)$);

(c) $t(p-2, p-1, \ldots, p-1) = 0$.

Proof. The existence of the required decomposition for $\Delta'/q^2\Delta$ follows by Proposition 10.2.3. Condition (a) is equivalent to invariance under *Gal*. Furthermore, condition (b) means that Δ' is fixed under multiplication by the $L_i(x)$, or equivalently, $\Delta' \lhd \Delta$. Finally, condition (c) is equivalent to the indivisibility of Δ'. □

Example 10.2.5. Let us consider the case $q = 8$. We introduce the following submodules of $\Delta/8\Delta$:

$$\bar{P}_0 =< 1 >_{\mathbb{Z}/8\mathbb{Z}},$$

$$\bar{P}_1 = \langle L_0(x^\alpha) + L_1(x^\alpha) + L_2(x^\alpha) | \alpha \in \mathbb{F}_8 \rangle_{\mathbb{Z}/8\mathbb{Z}},$$

$$\bar{P}_3 = \left\langle \frac{1}{2}(L_0(x^\alpha))^3 + L_1(x^\alpha)^3 + L_2(x^\alpha)^3 | \alpha \in \mathbb{F}_8 \right\rangle_{\mathbb{Z}/8\mathbb{Z}}.$$

Furthermore, let P_i denote the full preimages of these submodules in Δ. Then Δ *has just nine indivisible invariant sublattices, as follows:*

a) *the sublattices that contain* 2Δ*:*

$$\Delta_0 = \Delta,$$

$$\Delta_1 = 2P_0 + P_1 + P_3,$$

$$\Delta_2 = 2P_0 + 2P_1 + P_3;$$

b) *the sublattices that contain* 4Δ *but not* 2Δ*:*

$$\Delta_3 = 4P_0 + P_1 + P_3,$$

$$\Delta_4 = 4P_0 + 2P_1 + P_3,$$

$$\Delta_5 = 4P_0 + 4P_1 + P_3;$$

c) *the sublattices that do not contain* 4Δ*:*

$$\Delta_6 = P_1 + P_3,$$

$$\Delta_7 = 2P_1 + P_3,$$

$$\Delta_8 = 4P_1 + P_3.$$

In particular, Δ_8 is isometric to the root lattice of type E_7.

The mutual disposition of these sublattices is given in Figure 10.1.

Example 10.2.6. Next we consider the case $q = 9$, and introduce the following submodules of $\Delta/9\Delta$:

$$\bar{P}_0 =< 1 >_{\mathbb{Z}/9\mathbb{Z}},$$

$$\bar{P}_1 = \langle L_0(x^\alpha) + L_1(x^\alpha) | \alpha \in \mathbb{F}_9 \rangle_{\mathbb{Z}/9\mathbb{Z}},$$

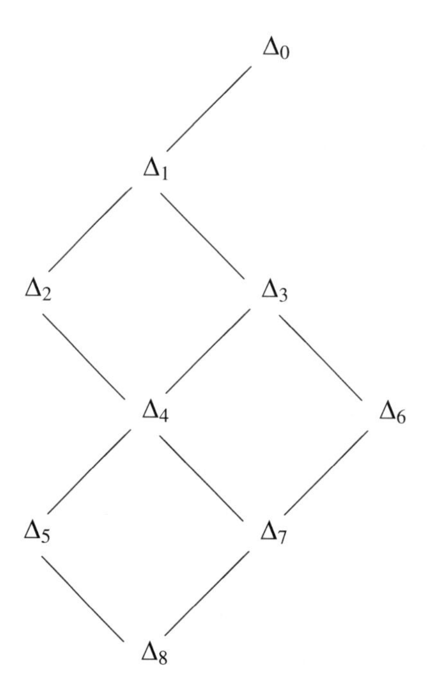

Figure 10.1

$$\bar{P}_2 = \left\langle L_0(x^\alpha)^2 + L_1(x^\alpha)^2 | \alpha \in \mathbb{F}_9 \right\rangle_{\mathbb{Z}/9\mathbb{Z}},$$

$$\bar{P}_4 = \left\langle \frac{1}{3} L_0(x^\alpha)^4 | \alpha \in \mathbb{F}_9 \right\rangle_{\mathbb{Z}/9\mathbb{Z}},$$

$$\bar{P}_5 = \left\langle \frac{1}{3} L_0(x^\alpha)^5 + L_1(x^\alpha)^5 | \alpha \in \mathbb{F}_9 \right\rangle_{\mathbb{Z}/9\mathbb{Z}}.$$

Furthermore, let P_i denote the full preimages of these submodules in Δ. Then Δ *has just fourteen indivisible invariant sublattices, as follows:*

a) *the sublattices that contain* 3Δ*:*

$$\Delta_0 = \Delta,$$

$$\Delta_1 = 3P_0 + P_1 + P_2 + P_4 + P_5,$$

$$\Delta_2 = 3P_0 + 3P_1 + P_2 + P_4 + P_5,$$

$$\Delta_3 = 3P_0 + 3P_1 + 3P_2 + P_4 + P_5,$$

$$\Delta_4 = 3P_0 + 3P_1 + P_2 + 3P_4 + P_5,$$

$$\Delta_5 = 3P_0 + 3P_1 + 3P_2 + 3P_4 + P_5;$$

b) *the sublattices that do not contain* 3Δ*:*

$$\Delta_6 = P_1 + P_2 + P_4 + P_5,$$

$$\Delta_7 = 3P_1 + P_2 + P_4 + P_5,$$

$$\Delta_8 = 3P_1 + 3P_2 + P_4 + P_5,$$

$$\Delta_9 = 3P_1 + P_2 + 3P_4 + P_5,$$

$$\Delta_{10} = 3P_1 + 3P_2 + 3P_4 + P_5,$$

$$\Delta_{11} = 3P_2 + P_4 + P_5,$$

$$\Delta_{12} = 3P_2 + 3P_4 + P_5,$$

$$\Delta_{13} = 3P_2 + P_5.$$

In particular, Δ_{12} is isometric to the root lattice of type E_8.

The mutual disposition of these sublattices is given in Figure 10.2.

10.3 The structure of the $SL_2(q)$-module $\Gamma^R/p\Gamma^R$

In this section we deal with the (q^2-1)-dimensional $\mathbb{F}_q SL_2(q)$-module $\Gamma^R/p\Gamma^R$.

First of all we consider the space

$$\mathcal{F} = \{f : V\backslash\{0\} \to \mathbb{F}_q\}$$

of functions endowed with the natural module structure with respect to $SL_2(q)$ and Gal:

$$(\varphi \circ f)(X, Y) = f(aX + bY, cX + dY),\ \varphi = \begin{pmatrix} d & -b \\ -c & a \end{pmatrix} \in SL_2(q),$$

$$g(f) = f^p,\ Gal = < g > .$$

The following statement is obvious.

Proposition 10.3.1. *If* $p > 2$*, then the representations of* $SL_2(q)$ *and* Gal *on* $\Gamma^R/p\Gamma^R$ *are equivalent to their representations on* $\mathcal{F}$. □

Clearly,

$$\mathcal{F} = \bigoplus_{k=0}^{q-2} \mathcal{F}^k,$$

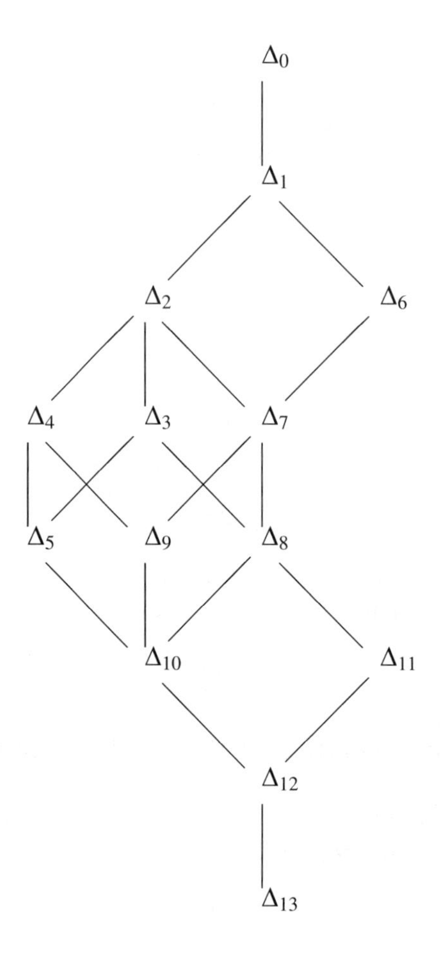

Figure 10.2

where

$$\mathcal{F}^k = \{f \in \mathcal{F} | f(\lambda X, \lambda Y) = \lambda^k f(X, Y), \forall \lambda \in \mathbb{F}_q^*\}.$$

We denote

$$\Omega = \{0, 1, \ldots, m-1\},$$

and set

$$\mathcal{M}_\omega(k) = \Big\langle X^{i_0+pi_1+\ldots+p^{m-1}i_{m-1}} Y^{j_0+pj_1+\ldots+p^{m-1}j_{m-1}} \in \mathcal{F}^k |$$
$$i = (i_0, \ldots, i_{m-1}) \in \mathcal{K}, j = (j_0, \ldots, j_{m-1}) \in \mathcal{K},\ i_t + j_t \le k_t, \forall t \in \Omega \backslash \omega \Big\rangle_{\mathbb{F}_q}$$

for any subset $\omega \subseteq \Omega$ and for any index $k = (k_0, k_1, \ldots, k_{m-1}) \in \mathcal{K}$.

As always, we agree that the notation $\omega \subset \nu$ for $\omega, \nu \subseteq \Omega$ denotes proper inclusion. Then $\mathcal{M}_\omega(k) \subseteq \mathcal{M}_\nu(k)$ if $\omega \subset \nu$; and $\mathcal{M}_\Omega(k) = \mathcal{F}^k$ if we identify the index $k = (k_0, k_1, \ldots, k_{m-1}) \in \mathcal{K}$ with the integer $k = k_0 + pk_1 + \ldots + p^{m-1}k_{m-1}$. We claim that $\mathcal{M}_\omega(k)$ is an $SL_2(q)$-submodule. Indeed, it is sufficient to verify the invariance of this space under the subgroup T and the elements $\vartheta = \begin{pmatrix} 0 & 1 \\ -1 & 0 \end{pmatrix}$, $\varsigma = \begin{pmatrix} 1 & -1 \\ 0 & 1 \end{pmatrix}$. Invariance under T and ϑ is obvious. Furthermore,

$$\varsigma(X^{i_0+pi_1+\ldots}Y^{j_0+pj_1+\ldots}) = (X+Y)^{i_0+pi_1+\ldots}Y^{j_0+pj_1+\ldots} =$$

$$= (X+Y)^{i_0}(X^p+Y^p)^{i_1}\cdots(X^{p^{m-1}}+Y^{p^{m-1}})^{i_{m-1}}Y^{j_0+pj_1+\ldots}.$$

The last element is a linear combination of elements like

$$X^{r_0+pr_1+\ldots}Y^{(i_0-r_0+j_0)+p(i_1-r_1+j_1)+\ldots},$$

where $0 \le r_t \le i_t$. Furthermore, for $t \in \Omega\backslash\omega$ we have $i_t + j_t = k_t$ or $i_t + j_t = k_t - 1$, since $i_t + j_t \le k_t$ and

$$(i_0 + pi_1 + \ldots) + (j_0 + pj_1 + \ldots) \equiv k = (k_0 + pk_1 + \ldots) \pmod{(q-1)}.$$

But in this case, the elements indicated belong to $\mathcal{M}_\omega(k)$ as well.

We shall see below that every $SL_2(q)$-submodule of $\mathcal{F}^k$ is equal, as a rule, to a sum of modules of type $\mathcal{M}_\omega(k)$.

Note that $\mathcal{M}_{\{0,1,\ldots,m-2\}}(k)$ coincides with the space $S^k(V^*)$ of homogeneous polynomials of degree k. Set

$$W_k = \mathcal{M}_\emptyset(k) = \langle X^{i_0+pi_1+\ldots}Y^{j_0+pj_1+\ldots} | i_t + j_t = k_t, \forall t \in \Omega \rangle.$$

According to [Ste 3], the module W_k is absolutely irreducible, and every irreducible $\mathbb{F}_q SL_2(q)$-module is isomorphic to some module of this kind.

Clearly, $\mathcal{F}^0 = \mathcal{F}^{q-1} = S^0(V^*) \oplus S^{q-1}(V^*)$ is a direct sum of irreducible submodules.

There is a non-degenerate $SL_2(q)$-invariant pairing between $\mathcal{F}^k$ and $\mathcal{F}^{q-1-k}$. It is sufficient to set

$$((f, h)) = \pi(f.h),$$

where π is the projection of $\mathcal{F}^0$ onto $S^0(V^*)$ along $S^{q-1}(V^*)$.

Lemma 10.3.2. *Suppose that $0 < k < q-1$. Then W_k is the unique minimal submodule of $\mathcal{F}^k$.*

Proof. Suppose that $0 \ne \mathcal{M} \subset \mathcal{F}^k$; we have to show that $W_k \subseteq \mathcal{M}$. The finite p-group U acts on the point set of $\mathcal{M}$, and every U-orbit has length a power of p.

Meanwhile, the length of the trivial orbit $\{0\}$ is 1. Hence, $\mathcal{M}$ contains a nonzero element f that is invariant under U. Write f in the form

$$f = \sum_{\gamma \in \mathbb{F}_q} c_\gamma X^k f_{<(1,\gamma)>} + cY^k f_{<(0,1)>}, \tag{10.5}$$

where, as in Chapter 9, f_σ denotes the characteristic function of the line σ in $\mathbb{P}V$. Now the U-invariance of f means simply that $c_\gamma = c\gamma^k$ for all $\gamma \neq 0$. Hence, $f = c_0 X^k f_0 + cY^k$. Furthermore, $\sum_\gamma \hat{\varphi}_\gamma(f) = c_0 X^k$, where $\varphi_\gamma = \begin{pmatrix} 1 & 0 \\ \gamma & 1 \end{pmatrix}$ for $\gamma \in \mathbb{F}_q$. We arrive at the conclusion that $X^k \in \mathcal{M}$ or $Y^k \in \mathcal{M}$. In particular, $W_k \subseteq \mathcal{M}$. □

Lemma 10.3.3. *W_k possesses an invariant bilinear form that is unique up to multiplication by an element of $\mathbb{F}_q$. This form is symmetric if k is even, and skew-symmetric if k is odd.*

Proof. First we prove the existence of an invariant form. Suppose that $k < p$, that is, $k = (k_0, 0, \dots, 0)$. Then $W_k = S^k(V^*)$, and the natural skew-symmetric bilinear form

$$< \alpha X + \beta Y; \gamma X + \delta Y >= \alpha\delta - \beta\gamma$$

on V^* can be extended in the usual way to $S^k(V^*)$:

$$< X^i Y^{k-i}; X^j Y^{k-j} >= \frac{1}{k!} \sum \underbrace{< X; * > \cdots < X; * >}_{i} \underbrace{< Y; * > \cdots < Y; * >}_{k-i} =$$

$$= \frac{1}{\begin{pmatrix} k \\ i \end{pmatrix}} (-1)^{k-i} \delta_{k-i,j}.$$

Furthermore, let Ψ_ℓ denote the representation of $SL_2(q)$ on $S^\ell(V^*)$, $\ell < p$, and Fr the (Frobenius) automorphism $A = (a_{ij}) \mapsto ((a_{ij})^p)$ of $SL_2(q)$. It is not difficult to see that the representation Ψ of $SL_2(q)$ on W_k is isomorphic to the representation $\otimes_{t=0}^{m-1} \Psi_{k_t} \circ \mathrm{Fr}^t$ on $\otimes_{t=0}^{m-1} S^{k_t}(V^*)$ (*the Steinberg Tensor Product Theorem*), if $k = \sum_{t=0}^{m-1} k_t p^t$. Set

$$< a_0 \otimes \cdots \otimes a_{m-1}; b_0 \otimes \cdots \otimes b_{m-1} >=< a_0; b_0 > \cdots < a_{m-1}; b_{m-1} >$$

for $a_t, b_t \in S^{k_t}(V^*)$, $0 \leq t \leq m-1$. Then $< *; * >$ is a nonzero invariant form, and for $i = \sum_t i_t p^t$, $j = \sum_t j_t p^t$, $0 \leq i_t, j_t \leq k_t$, we have

$$< X^i Y^{k-i}; X^j Y^{k-j} >= \prod_{t=0}^{m-1} \left\{ \begin{pmatrix} k_t \\ i_t \end{pmatrix}^{-1} (-1)^{k_t - i_t} \delta_{k_t - i_t, j_t} \right\} =$$

$$= \binom{k}{i}^{-1} (-1)^{k-i} \delta_{k-i,j}.$$

Here we should mention that, when $p > 2$,

$$k = \sum_t k_t p^t \equiv \sum_t k_t \pmod 2.$$

Furthermore, we call a form B' symmetric (respectively skew-symmetric), if

$$\exists \kappa,\ B'(b, a) = \kappa B'(a, b),\ \forall a, b,$$

and $\kappa = 1$ (respectively $\kappa = -1$). In particular, when $p = 2$ the notions of symmetric and skew-symmetric forms (in our sense) are identical.

Finally, the uniqueness of the invariant form follows from the absolute irreducibility of W_k. □

From now on we call the index ω of a module $\mathcal{M}_\omega(k)$ *uncancellable* if $\mathcal{M}_\nu(k) \neq \mathcal{M}_\omega(k)$ for every $\nu \subset \omega$. The uncancellability of the index ω of $\mathcal{M}_\omega(k)$ means that for any $t \in \omega$ there exists an element

$$X^{i_0+pi_1+\dots} Y^{j_0+pj_1+\dots} \in \mathcal{M}_\omega(k)$$

with $i_t + j_t > k_t$. Furthermore, the index ω of $\mathcal{M}_\omega(k)$ is uncancellable if and only if

a) the conditions $t \in \omega$, $t+1 \notin \omega$ imply that $k_{t+1} > 0$;

b) the conditions $t \in \omega$, $t-1 \notin \omega$ imply that $k_t < p-1$.

(Here addition of subscripts t is carried out modulo m).

Lemma 10.3.4. *Suppose that the index ω of a module $\mathcal{M}_\omega(k)$ is uncancellable. Then the factor module $\mathcal{M}_\omega(k)/\sum_{\nu\subset\omega}\mathcal{M}_\nu(k)$ is isomorphic to W_ℓ, where $\ell = \sum_{t=0}^{m-1} \ell_t p^t$ and*

$$\ell_t = \begin{cases} k_t & \text{if } t \notin \omega,\ t-1 \notin \omega; \\ k_t - 1 & \text{if } t \notin \omega,\ t-1 \in \omega; \\ p-2-k_t & \text{if } t \in \omega,\ t-1 \notin \omega; \\ p-1-k_t & \text{if } t \in \omega,\ t-1 \in \omega. \end{cases} \tag{10.6}$$

Proof. Consider elements like $X^{s_0+ps_1+\dots} Y^{r_0+pr_1+\dots} \in \mathcal{M}_\omega(k)$, where

$$\begin{array}{ll} 0 \le s_t \le k_t & \text{if } t \notin \omega,\ t-1 \notin \omega; \\ 0 \le s_t \le k_t - 1 & \text{if } t \notin \omega,\ t-1 \in \omega; \\ k_t + 1 \le s_t \le p-1 & \text{if } t \in \omega,\ t-1 \notin \omega; \\ k_t \le s_t \le p-1 & \text{if } t \in \omega,\ t-1 \in \omega. \end{array} \tag{10.7}$$

The images of these elements under factoring by $\sum_{\nu\subset\omega}\mathcal{M}_\nu(k)$ form a basis of the factor-module $\mathcal{M}=\mathcal{M}_\omega(k)/\sum_{\nu\subset\omega}\mathcal{M}_\nu(k)$. Let $\mathcal{M}'$ be a minimal submodule of $\mathcal{M}$. We find a highest vector, that is, a vector v that is invariant under U. It is not difficult to see that v is proportional to $X^{s_0+ps_1+\cdots}Y^{r_0+pr_1+\cdots}$, where s_t takes the least value satisfying condition (10.7). But the weight of a highest vector determines the irreducible representation uniquely (see e.g. [Ste 3]). Therefore, comparison of the weights of v and of a highest vector of W_ℓ, where ℓ is defined in (10.6), shows that $\mathcal{M}'\cong W_\ell$. Finally, by comparing dimensions we see that $\mathcal{M}\cong W_\ell$. □

Suppose that $\mathcal{M}_\omega(k)$ is a module with uncancellable index ω. We restrict the pairing

$$((*,*)):\mathcal{F}^k\times\mathcal{F}^{q-1-k}\to\mathbb{F}_q$$

to $\mathcal{M}_\omega(k)\times\mathcal{M}_{\omega^*}(k^*)$, where

$$\omega^*=\Omega\backslash\omega,\ k^*=q-1-k=\sum_{t=0}^{m-1}(p-1-k_t)p^t.$$

It is not difficult to see that the left and right kernels of this pairing are $\sum_{\nu\subset\omega}\mathcal{M}_\nu(k)$ and $\sum_{\nu\subset\omega^*}\mathcal{M}_\nu(k^*)$ respectively. Applying Lemmas 10.3.3 and 10.3.4, we obtain an isomorphism

$$\mathcal{M}_\omega(k)/\sum_{\nu\subset\omega}\mathcal{M}_\nu(k)\cong\mathcal{M}_{\omega^*}(k^*)/\sum_{\nu\subset\omega^*}\mathcal{M}_\nu(k^*).$$

In particular, when $k=\frac{q-1}{2}$ we get an isomorphism

$$\tau_\omega:\mathcal{M}_\omega(k)/\sum_{\nu\subset\omega}\mathcal{M}_\nu(k)\cong\mathcal{M}_{\omega^*}(k)/\sum_{\nu\subset\omega^*}\mathcal{M}_\nu(k).$$

For $\lambda\in\mathbb{F}_q^*$ and $\omega\neq\emptyset,\Omega$ we denote by $\mathcal{M}_\omega^\lambda(\frac{q-1}{2})$ the full preimage of the module

$$\left\{\bar{x}+\lambda\tau_\omega(\bar{x})|\bar{x}\in\mathcal{M}_\omega(k)/\sum_{\nu\subset\omega}\mathcal{M}_\nu(k)\right\}$$

under the factorization

$$(\mathcal{M}_\omega(k)+\mathcal{M}_{\omega^*}(k))/\left(\sum_{\nu\subset\omega}\mathcal{M}_\nu(k)+\sum_{\nu\subset\omega^*}\mathcal{M}_\nu(k)\right)\cong$$

$$\cong\left(\mathcal{M}_\omega(k)/\sum_{\nu\subset\omega}\mathcal{M}_\nu(k)\right)\bigoplus\left(\mathcal{M}_{\omega^*}(k)/\sum_{\nu\subset\omega^*}\mathcal{M}_\nu(k)\right).$$

Clearly,

$$\mathcal{M}^{\lambda}_{\omega}(\frac{q-1}{2}) = \mathcal{M}^{\lambda^{-1}}_{\omega^*}(\frac{q-1}{2}).$$

Proposition 10.3.5. *The following assertions hold.*

(i) *If $k = 0$, then the proper $SL_2(q)$-submodules of $\mathcal{F}^k$ are $S^0(V^*)$ (the space of constants) and $S^{q-1}(V^*)$ (the space of homogeneous polynomials of degree $q-1$).*

(ii) *If $0 < k < q-1$ and $k \neq \frac{q-1}{2}$, then the submodules of $\mathcal{F}^k$ are precisely the modules $\mathcal{M}_{\omega}(k)$ with ω uncancellable, together with their sums.*

(iii) *If $k = \frac{q-1}{2}$, then the submodules of $\mathcal{F}^k$ are modules of the form $\mathcal{M}^{\lambda}_{\omega}(\frac{q-1}{2})$, where $\lambda \in \mathbb{F}^*_q$, ω is uncancellable and $\omega \neq \emptyset, \Omega$; $\mathcal{M}_{\omega}(\frac{q-1}{2})$, where ω is uncancellable; and their sums.*

Proof. 1) Assertion (i) has been already established. Suppose that $0 < k < q-1$.

2) By Lemma 10.3.2, every nonzero submodule $\mathcal{M}$ of $\mathcal{F}^k$ contains W_k. Therefore, one can suppose that

$$W_k \subset \mathcal{M} \subset \mathcal{F}^k.$$

The space $\mathcal{F}^k$ is a direct sum of 1-dimensional T-modules that are generated by elements

$$e_i = X^i Y^{k-i},\ 0 \leq i \leq q-1;\ e_q = X^k Y^{q-1}.$$

3) Suppose that $p = 2$. Then the T-modules $< e_i >_{\mathbb{F}_q}$ are isomorphic to each other only in the following cases: $< e_0 > \cong < e_{q-1} >$, $< e_k > \cong < e_q >$. Note that $e_0, e_k \in W_k$. Hence, if $f = \sum_i a_i e_i \in \mathcal{M}$, we have $a_i e_i \in \mathcal{M}$.

4) Suppose that $p > 2$. Then the T-modules $< e_i >_{\mathbb{F}_q}$ are isomorphic to each other only in the following cases:

$$< e_0 > \cong < e_{(q-1)/2} > \cong < e_{q-1} >,$$

$$< e_k > \cong < e_{k+(q-1)/2} > \cong < e_q >,$$

$$< e_i > \cong < e_{i+(q-1)/2} >,\ 0 < i < (q-1)/2, i \neq k.$$

Hence, if $f = \sum_i a_i e_i \in \mathcal{M}$, then

$$a_i e_i + a_{i+(q-1)/2} e_{i+(q-1)/2} \in \mathcal{M},\ 0 < i \leq (q-1)/2, i \neq k;$$

$$a_{k+(q-1)/2} e_{k+(q-1)/2} + a_q e_q \in \mathcal{M}.$$

5) Suppose that the indices ω and μ of $\mathcal{M}_{\omega}(k)$ and $\mathcal{M}_{\mu}(k)$ are uncancellable, and

$$\mathcal{M}_{\omega}(k)/\sum_{\nu \subset \omega} \mathcal{M}_{\nu}(k) \cong W_{\ell},\ \mathcal{M}_{\mu}(k)/\sum_{\nu \subset \mu} \mathcal{M}_{\nu}(k) \cong W_s.$$

Suppose in addition that $W_{\ell} \cong W_s$, that is, $\ell = s$. We claim that in this case we have $\omega = \mu$, or $\omega = \mu^*$ and $k = \frac{q-1}{2}$. For, suppose that $\omega \neq \mu$. If $r \in \mu \backslash \omega$, then from (10.6) and the condition $\ell_r = s_r$, it follows that one of the following holds:

a) $r-1 \in \omega \Rightarrow \ell_r = k_r - 1 \Rightarrow s_r = p-2-k_r,\ r-1 \notin \mu,\ k_r = (p-1)/2$;
b) $r-1 \notin \omega \Rightarrow \ell_r = k_r \Rightarrow s_r = p-1-k_r,\ r-1 \in \mu,\ k_r = (p-1)/2$.
Hence, $r-1 \in (\omega+\mu) = (\omega\backslash\mu) \cup (\mu\backslash\omega)$. Similarly, $r-2 \in (\omega+\mu)$, and so on.

6) Suppose that $f = ae_i + be_j \in \mathcal{M}$, $< e_i > \cong < e_j >$. Choose a module $\mathcal{M}_\omega(k)$ containing e_i and a module $\mathcal{M}_\mu(k)$ containing e_j such that $|\omega|$ and $|\mu|$ take the least possible value.

If the modules $\mathcal{M}_\omega(k)/\sum_{\nu\subset\omega}\mathcal{M}_\nu(k)$ and $\mathcal{M}_\mu(k)/\sum_{\nu\subset\mu}\mathcal{M}_\nu(k)$ are not isomorphic, then the element ae_i belongs to $\mathcal{M}$ modulo

$$\left(\sum_{\nu\subset\omega}\mathcal{M}_\nu(k) + \sum_{\nu\subset\mu}\mathcal{M}_\nu(k)\right).$$

But in this case, we have $ae_i \in \mathcal{M}$ because of 4).

Furthermore, any element $h \in \mathcal{M}_\omega(k)\backslash\sum_{\nu\subset\omega}\mathcal{M}_\nu(k)$ generates the whole module $\mathcal{M}_\omega(k)$. To see this, let $\mathcal{M}_1$ be the submodule generated by h. By 4), 5) and Lemma 10.3.4, $\mathcal{M}_1$ contains all the monomials that belong to $\mathcal{M}_\omega(k)\backslash\sum_{\nu\subset\omega}\mathcal{M}_\nu(k)$. Similarly, if an element $h_1 = \sum_i b_i e_i$ is obtained by applying $\varphi \in SL_2(q)$ to these monomials, then $b_i e_i \in \mathcal{M}_1$. Hence, $\mathcal{M}_1 = \mathcal{M}_\omega(k)$. In particular, ae_i generates the whole of the module $\mathcal{M}_\omega(k)$, and $\mathcal{M} = \mathcal{M}_\omega(k)$.

If $\mathcal{M}_\omega(k)/\sum_{\nu\subset\omega}\mathcal{M}_\nu(k)$ and $\mathcal{M}_\mu(k)/\sum_{\nu\subset\mu}\mathcal{M}_\nu(k)$ are isomorphic, then $\mu = \omega^*$, $k = \frac{q-1}{2}$ and $f \in \mathcal{M}_\omega^\lambda(k)$ for some $\lambda \in \mathbb{F}_q^*$. In this case, f generates $\mathcal{M}_\omega^\lambda(k)$, since ae_i generates $\mathcal{M}_\omega(k)$, be_j generates $\mathcal{M}_{\omega^*}(k)$, $\mathcal{M}_\omega(k) \cap \mathcal{M}_{\omega^*}(k) = W_k$, and the modules

$$\left(\sum_{\nu\subset\omega}\mathcal{M}_\nu(k)\right)/W_k,\ \left(\sum_{\nu\subset\omega^*}\mathcal{M}_\nu(k)\right)/W_k$$

have no common composition factors. This means that $\mathcal{M} = \mathcal{M}_\omega^\lambda(k)$. □

Remark. It is helpful to compare Proposition 10.3.5 with Proposition 9.1.11, in which assertion (iii) is absent for natural reasons.

We shall need the following statement. Recall that

$$\mathcal{M}_\omega(k) = \mathcal{M}_\omega(k_0, \ldots, k_{m-1}),$$

if $k = \sum_{t=0}^{m-1} k_t p^t = (k_0, \ldots, k_{m-1}) \in \mathcal{K}$.

Lemma 10.3.6. *If the index ω of a module $\mathcal{M}_\omega(k)$ is uncancellable, then*
(i) $< X^{p^i}, Y^{p^i} > \cdot \mathcal{M}_\omega(k) = \mathcal{M}_\omega(k_0, \ldots, k_i + 1, \ldots, k_{m-1})$, *where*

$$\mathcal{M}_\omega(k_0, \ldots, \underbrace{p}_{i}, k_{i+1}, \ldots, k_{m-1}) = \mathcal{M}_{\omega\cup\{i\}}(k_0, \ldots, \underbrace{0}_{i}, k_{i+1}+1, \ldots, k_{m-1}),$$

$$\mathcal{M}_\omega(q-1) = \begin{cases} S^{q-1}(V^*) & \text{if } \ \omega \neq \Omega, \\ \mathcal{M}_\Omega(0) & \text{if } \ \omega = \Omega. \end{cases}$$

(ii) $< X^{p^i}, Y^{p^i} > \cdot \mathcal{M}^{\lambda}_{\omega}(\frac{q-1}{2}) = \mathcal{M}_{\omega}(\frac{q-1}{2} + p^i) + \mathcal{M}_{\omega^*}(\frac{q-1}{2} + p^i)$.
(iii) $< X^{p^i}, Y^{p^i} > \cdot S^{q-1}(V^*) = \mathcal{M}_{\Omega}(p^i)$.
(iv) $g(\mathcal{M}_{\omega}(k)) = \mathcal{M}_{\omega+1}(k_{m-1}, k_0, k_1, \dots, k_{m-2})$, *and*

$$g(\mathcal{M}^{\lambda}_{\omega}(k)) = \mathcal{M}^{\lambda p}_{\omega+1}(k_{m-1}, k_0, k_1, \dots, k_{m-2}),$$

where $\omega + 1 = \{t_1 + 1, t_2 + 1, \dots, t_s + 1\}$ *for* $\omega = \{t_1, t_2, \dots, t_s\}$.

Proof. This is an immediate consequence of the definitions. □

10.4 The structure of the $SL_2(q)$-module $\Gamma^R/q^2\Gamma^R$

For $k \in \mathcal{K}$ we set

$$\mathcal{N}(k) = SL_2(q)(\mathcal{P}(k)),$$

where the submodule $\mathcal{P}(k)$ of $\Gamma^R_0/q^2\Gamma^R_0$ is defined in Proposition 10.2.3. Clearly,

$$\mathcal{N}(k) = \sum_{i=1}^{q+1} \varphi_i(\mathcal{P}(k)),$$

where elements φ_i of $SL_2(q)$ map the point σ_0 of $\mathbb{P}V$ into distinct points σ in $\mathbb{P}V$. Furthermore, $\mathcal{N}(k)$ is $SL_2(q)$-invariant,

$$\mathcal{N}(k)/p\mathcal{N}(k) \cong \mathcal{F}^k,$$

and

$$\Gamma^R/q^2\Gamma^R = \bigoplus_{k \in \mathcal{K}} \mathcal{N}(k).$$

Recall that the $\hat{V}$-invariance of a $SL_2(q)$-module $\mathcal{N}$ in $\Gamma^R/q^2\Gamma^R$ is equivalent to the conditions $L_i(x^a y^b) \cdot \mathcal{N} \subseteq \mathcal{N}$, where $0 \le i \le m-1$ and $(a, b) = (0, 1), (1, 0)$. From Proposition 10.2.3 it follows that

$$< L_i(x), L_i(y) > \cdot \mathcal{N}(k) = \mathcal{N}(k_0, \dots, k_{i-1}, k_i + 1, k_{i+1}, \dots, k_{m-1}), \tag{10.8}$$

where

$$\mathcal{N}(k_0, \dots, p, k_{i+1}, \dots, k_{m-1}) = p\mathcal{N}(k_0, \dots, 0, k_{i+1} + 1, \dots, k_{m-1}), \tag{10.9}$$

$$\mathcal{N}(q-1) = q\mathcal{N}(0). \tag{10.10}$$

At this point we describe the $SL_2(q)$-submodules of $\mathcal{N}(k)$. First of all, we show that

$$\mathcal{N}(0) = \mathcal{N}_{\emptyset}(0) \oplus \mathcal{N}^w(0)$$

is a direct sum of $(R/q^2R)SL_2(q)$-modules, where

$$\mathcal{N}_{\emptyset}(0)/p\mathcal{N}_{\emptyset}(0) \cong \mathcal{M}_{\emptyset}(0) =< 1 >_{R/q^2R},$$

$$\mathcal{N}^w(0)/p\mathcal{N}^w(0) \cong S^{q-1}(V^*).$$

We have,

$$\mathcal{N}_{\emptyset}(0) =< \sum_{\sigma\in\mathbb{P}V} \Phi_\sigma >_{R/q^2R},$$

$$\mathcal{N}^w(0) = \{\sum_{\sigma\in\mathbb{P}V} a_\sigma\Phi_\sigma | a_\sigma \in R/q^2R, \sum_\sigma a_\sigma = 0\}.$$

Furthermore, every $SL_2(q)$-module $\mathcal{N}'$ in $\mathcal{N}(0)$ is of the form $p^r\mathcal{N}_{\emptyset}(0) + p^s\mathcal{N}^w(0)$ for some non-negative integers r, s.

Next, we denote by $\mathcal{E}_\omega(k)$ the full preimage of $\mathcal{M}_\omega(k)$ under the natural homomorphism

$$\mathcal{N}(k) \to \mathcal{N}(k)/p\mathcal{N}(k) \cong \mathcal{F}^k.$$

The submodules $\mathcal{E}^w(0)$ and $\mathcal{E}^\lambda_\omega(\frac{q-1}{2})$ corresponding to the modules $S^{q-1}(V^*)$ and $\mathcal{M}^\lambda_\omega(\frac{q-1}{2})$ respectively are defined similarly. Furthermore, we introduce the (R/q^2R)-modules

$$\mathcal{N}'_\omega(k) = \big\langle qL_0(x)^{i_0} \cdots L_{m-1}(x)^{i_{m-1}} L_0(y)^{j_0} \cdots L_{m-1}(y)^{j_{m-1}} |$$

$$|X^{i_0+pi_1+\ldots+p^{m-1}i_{m-1}} Y^{j_0+pj_1+\ldots+p^{m-1}j_{m-1}} \in \mathcal{M}_\omega(k),\ 0 \le i_s, j_s < p \big\rangle_{R/q^2R}.$$

Then

$$\mathcal{E}_\omega(k) = \left\{F + p\mathcal{N}(k) | F \in \mathcal{N}'_\omega(k)\right\}.$$

Lemma 10.4.1. *If $\mathcal{M}_\omega(k)$ is a module with ω uncancellable, then*

(i) $< L_i(x), L_i(y) > \cdot\mathcal{E}_\omega(k) = \mathcal{E}_\omega(k_0, \ldots, k_{i-1}, k_i + 1, k_{i+1}, \ldots, k_{m-1})$, *where*

$$\mathcal{E}_\omega(k_0, \ldots, k_{i-1}, p, k_{i+1}, \ldots, k_{m-1}) = p\mathcal{E}_{\omega\cup\{i\}}(k_0, \ldots, k_{i-1}, 0, k_{i+1} + 1, \ldots, k_{m-1}),$$

$$\mathcal{E}_\omega(q-1) = \begin{cases} q\mathcal{E}^w(0) & \textit{if} \quad \omega \neq \Omega, \\ q\mathcal{E}_\Omega(0) & \textit{if} \quad \omega = \Omega. \end{cases}$$

(ii) $< L_i(x), L_i(y) > \cdot\mathcal{E}^w(0) = \mathcal{E}_\Omega(p^i)$.

Proof. An immediate consequence of relations (10.8) – (10.10) and Lemma 10.3.6. □

We introduce the following indivisible invariant sublattice:

$$\tilde{\Gamma}^1 = \left\langle (1 - x^\alpha y^\beta)\Phi_\sigma | \sigma \in \mathbb{P}V;\ \alpha, \beta \in \mathbb{F}_q \right\rangle_R =$$

$$= \langle L_i(x)\Phi_\sigma, L_i(y)\Phi_\sigma | 0 \le i \le m-1; \sigma \in \mathbb{P}V\rangle + q^2\Gamma^R.$$

Clearly,

$$\tilde{\Gamma}^1/q^2\Gamma^R = q\mathcal{N}(0) + \bigoplus_{k\in\mathcal{K}, k>0} \mathcal{N}(k).$$

Furthermore, the module

$$q\mathcal{N}_\emptyset(0) \bigoplus \mathcal{N}^w(0) \bigoplus \left(\bigoplus_{k\in\mathcal{K}, k>0} \mathcal{N}(k) \right) \tag{10.11}$$

is a G-invariant submodule of $\Gamma^R/q^2\Gamma^R$. The full preimage of the module (10.11) under the natural homomorphism $\Gamma^R \to \Gamma^R/q^2\Gamma^R$ is denoted by Γ^w.

Proposition 10.4.2. *Every indivisible invariant sublattice Γ' contains $q\Gamma^w$ and $q\tilde{\Gamma}^1$.*

Proof. Because $\Gamma^w \supset \tilde{\Gamma}^1$, it is sufficient to show that $\Gamma' \supseteq q\Gamma^w$. Since Γ' is indivisible, $(\Gamma' + p\Gamma^R)/q^2\Gamma^R$ contains $\mathcal{E}_\emptyset(q-1-p^i)$ for some i. Then $\Gamma'/q^2\Gamma^R$ contains an element $F = F_1 + F_2$ such that

$$F_1 = qL_0(x)^{p-1} \cdots L_i(x)^{p-2} \cdots L_{m-1}(x)^{p-1}, \; F_2 \in p\Gamma^R/q^2\Gamma^R.$$

Hence,

$$L_i(x)F_1 \in q\mathcal{N}^w(0) \setminus (p\tilde{\Gamma}^1/q^2\Gamma^R), \; L_i(x)F_2 \in p\tilde{\Gamma}^1/q^2\Gamma^R.$$

Since $p\tilde{\Gamma}^1/p^{m+1}\Gamma^R \cong \Gamma^1/q\Gamma^R$ does not have composition factors isomorphic to $S^{q-1}(V^*)$, we obtain the inclusion

$$(\Gamma' + p^{m+1}\Gamma^R)/q^2\Gamma^R \supseteq q\mathcal{N}^w(0).$$

By Lemma 10.4.1, for $k > 0$ we have

$$(\Gamma' + p^{m+1}\Gamma^R)/q^2\Gamma^R \supseteq q\mathcal{N}(k).$$

Hence,

$$\Gamma'/q^2\Gamma^R \supseteq q\mathcal{N}^w(0) \bigoplus \left(\bigoplus_{k\in\mathcal{K}, k>0} q\mathcal{N}(k) \right).$$ □

The next step in our investigation is to consider the lattice

$$\hat{\Gamma} = \left\langle q(1 - x^\alpha y^\beta) | \alpha, \beta \in \mathbb{F}_q \right\rangle_R.$$

Note that $\hat{\Gamma} \subset \tilde{\Gamma}^1 \subset \Gamma^R$, because $q = \sum_{\sigma\in\mathbb{P}V} \Phi_\sigma$.

Proposition 10.4.3. (i) *The lattice* $\hat{\Gamma}$ *is indivisible and invariant, and it contains* $q^2\Gamma^R$ *but not* $p^{2m-1}\Gamma^R$;

(ii) *the factor-lattice* $\hat{\Gamma}/q^2\Gamma^R$ *is homogeneous with respect to* $\mathcal{N}(k)$, *that is,*

$$\sum_{k\in\mathcal{K}}\left\{(\hat{\Gamma}/q^2\Gamma^R)\cap\mathcal{N}(k)\right\}=\hat{\Gamma}/q^2\Gamma^R;$$

(iii) $(\hat{\Gamma}/q^2\Gamma^R)\cap\mathcal{N}(0)=q\mathcal{N}^w(0)$;

(iv) *if* $k\in\mathcal{K}$ *and* $k>0$, *then*

$$\left((\hat{\Gamma}/q^2\Gamma^R)\cap p^j\mathcal{N}(k)\right)+p^{j+1}\mathcal{N}(k)=\sum_{|\omega|\leq j}p^j\mathcal{E}_\omega(k).$$

Proof. (i) Suppose that $\hat{\Gamma}\supseteq p^{2m-1}\Gamma^R$. Then $p^{2m-1}.q=p^{2m-1}.\sum_\sigma\Phi_\sigma\in\hat{\Gamma}$, that is,

$$p^{2m-1}.q=\sum_{\alpha,\beta}a_{\alpha\beta}q(1-x^\alpha y^\beta)$$

for some $a_{\alpha\beta}\in R$. Going over to the ring

$$\mathcal{C}=\left\langle\xi^\alpha\eta^\beta|\xi^\alpha\eta^\beta.\xi^\gamma\eta^\delta=\xi^{\alpha+\gamma}\eta^{\beta+\delta}\right\rangle_R,$$

where $\mathcal{C}/R(\sum_{\alpha,\beta}\xi^\alpha\eta^\beta)\cong\mathcal{A}$, and $\mathcal{A}$ is as defined in §10.1, we have

$$p^{2m-1}.q=\sum_{\alpha,\beta\in\mathbb{F}_q}a_{\alpha\beta}q(\xi^0\eta^0-\xi^\alpha\eta^\beta)+d\sum_{\alpha,\beta\in\mathbb{F}_q}\xi^\alpha\eta^\beta.$$

Comparing the sums of coefficients of $\xi^\alpha\eta^\beta$, we obtain that $p^{2m-1}q=dq^2$. This means that $d=p^{m-1}$, and

$$q.q=\sum_{\alpha,\beta\in\mathbb{F}_q}a_{\alpha\beta}p(\xi^0\eta^0-\xi^\alpha\eta^\beta)+\sum_{\alpha,\beta\in\mathbb{F}_q}\xi^\alpha\eta^\beta,$$

which is a contradiction.

Assertion (ii) follows from the fact that

$$(1-x^\alpha y^\beta)=(1-x^\alpha)+(1-y^\beta)-(1-x^\alpha)(1-y^\beta)$$

and

$$\hat{\Gamma}=\left\langle q(1-x^\alpha y^\beta)\right\rangle_R=$$

$$=\langle q(1-(x_1)^{\alpha_1})(1-(x_2)^{\alpha_2})\cdots(1-(x_n)^{\alpha_n})|\alpha_t\in\mathbb{F}_q,\ x_t\in\{x,y\},\ n\geq 1\rangle_R=$$

$$=\langle qL_{i_1}(x_1)L_{i_2}(x_2)\cdots L_{i_n}(x_n)|0\leq i_t\leq m-1,\ x_t\in\{x,y\},\ n\geq 1\rangle_R+q^2\Gamma^R.$$

(iii) In view of Proposition 10.4.2, we have

$$(\hat{\Gamma}/q^2\Gamma^R) \cap \mathcal{N}(0) \supseteq q\mathcal{N}^w(0).$$

The inclusion $\tilde{\Gamma}^1 \supset \hat{\Gamma}$ implies that

$$q\mathcal{N}(0) \supseteq (\hat{\Gamma}/q^2\Gamma^R) \cap \mathcal{N}(0).$$

We claim that

$$(\hat{\Gamma}/q^2\Gamma^R) \cap \mathcal{N}(0) = q\mathcal{N}^w(0).$$

Indeed, in the contrary case we would have had

$$p^t\mathcal{N}_\emptyset(0) \subseteq (\hat{\Gamma}/q^2\Gamma^R) \cap \mathcal{N}(0)$$

for some t, $m \leq t < 2m$. But

$$p^t\mathcal{N}_\emptyset(0) + p^t\mathcal{N}^w(0) = p^t\mathcal{N}(0),$$

and so $p^t\Gamma^R \subseteq \hat{\Gamma}$, contrary to (i).

(iv) If $j \geq 2m$, then both sides of the required equality are equal to zero. If $2m > j \geq m$, then the required equality follows from Proposition 10.4.2. Suppose that $j < m$. We shall prove by induction on $s = |\omega|(p-1) + \sum_{i=0}^{m-1} k_i > 0$ that

$$\left((\hat{\Gamma}/q^2\Gamma^R) \cap p^j\mathcal{N}(k)\right) + p^{j+1}\mathcal{N}(k) \supseteq \sum_{|\omega|\leq j} p^j\mathcal{E}_\omega(k). \tag{10.12}$$

The basis of induction ($s = 1$) is obvious:

$$\left((\hat{\Gamma}/q^2\Gamma^R) \cap \mathcal{N}(p^i)\right) + p\mathcal{N}(p^i) =$$

$$= \langle qL_i(x) + p\mathcal{N}(p^i), qL_i(y) + p\mathcal{N}(p^i)\rangle_{R/q^2R} = \mathcal{E}_\emptyset(p^i).$$

The induction step: suppose that $s > 1$ and that (10.12) has been established for all cases with $|\omega|(p-1) + \sum_{i=0}^{m-1} k_i < s$. It is sufficient to prove that

$$\left((\hat{\Gamma}/q^2\Gamma^R) \cap p^j\mathcal{N}(k)\right) + p^{j+1}\mathcal{N}(k) \supseteq p^j\mathcal{E}_\omega(k)$$

for all uncancellable indices ω with $|\omega| = j$. Without loss of generality, one can suppose that $m-1 \notin \omega$ and $0 \in \omega$. The following two cases arise:

1) there exists an integer t such that $t-1 \notin \omega$ and $k_t > 0$, or $t-1 \in \omega$ and $k_t > 1$, or $t-1, t \in \omega$ and $k_t > 0$;

2) $\{0, 1, \ldots, t-1\} \subseteq \omega$, $t \notin \omega$, $k_0 = k_1 = \ldots = k_{t-1} = 0$ and $k_t = 1$.

In the former case, by the induction hypothesis we have

$$\left((\hat{\Gamma}/q^2\Gamma^R) \cap p^j\mathcal{N}(\ldots, k_{t-1}, k_t - 1, k_{t+1}, \ldots)\right) +$$

$$+p^{j+1}\mathcal{N}(\ldots, k_{t-1}, k_t - 1, k_{t+1}, \ldots) \supseteq$$
$$\supseteq p^j\mathcal{E}_\omega(\ldots, k_{t-1}, k_t - 1, k_{t+1}, \ldots).$$

Multiplying by $L_t(x)$ and $L_t(y)$ we obtain

$$\left((\hat{\Gamma}/q^2\Gamma^R) \cap p^j\mathcal{N}(k)\right) + p^{j+1}\mathcal{N}(k) \supseteq p^j\mathcal{E}_\omega(k).$$

In the latter case, by the induction hypothesis we have

$$\left((\hat{\Gamma}/q^2\Gamma^R) \cap p^{j-t}\mathcal{N}(p-1, \ldots, p-1, 0, k_{t+1}, \ldots)\right) +$$

$$+p^{j-t+1}\mathcal{N}(p-1, \ldots, p-1, 0, k_{t+1}, \ldots) \supseteq$$
$$\supseteq p^{j-t}\mathcal{E}_{\omega\setminus\{0,1,\ldots,t-1\}}(p-1, \ldots, p-1, 0, k_{t+1}, \ldots).$$

Again multiplying by $L_0(x)$ and $L_0(y)$ we obtain

$$\left((\hat{\Gamma}/q^2\Gamma^R) \cap p^j\mathcal{N}(0, \ldots, 0, 1, k_{t+1}, \ldots)\right) + p^{j+1}\mathcal{N}(0, \ldots, 0, 1, k_{t+1}, \ldots) \supseteq$$

$$\supseteq p^j\mathcal{E}_\omega(0, \ldots, 0, 1, k_{t+1}, \ldots).$$

This proves relation (10.12).

Now suppose that for some uncancellable index ω, $|\omega| = j + 1$, we have:

$$\left((\hat{\Gamma}/q^2\Gamma^R) \cap p^j\mathcal{N}(k)\right) + p^{j+1}\mathcal{N}(k) \supseteq p^j\mathcal{E}_\omega(k).$$

Using the same arguments as above, we obtain for some integer t:

$$\left((\hat{\Gamma}/q^2\Gamma^R) \cap p^{m-1}\mathcal{N}(p-1, \ldots, \underbrace{p-2}_{t}, \ldots, p-1)\right) +$$

$$+p^m\mathcal{N}(p-1, \ldots, \underbrace{p-2}_{t}, \ldots, p-1) \supseteq$$

$$\supseteq p^{m-1}\mathcal{E}_\Omega(p-1, \ldots, \underbrace{p-2}_{t}, \ldots, p-1),$$

$$\left((\hat{\Gamma}/q^2\Gamma^R) \cap p^{2m-1}\mathcal{N}(0)\right) + p^{2m}\mathcal{N}(0) \supseteq p^{2m-1}\mathcal{N}(0).$$

But $q^2\mathcal{N}(0) = 0$, and so

$$\hat{\Gamma}/q^2\Gamma^R \supseteq p^{2m-1}\mathcal{N}(0),$$

contrary to (iii). □

Now define a map $\Theta : \hat{\Gamma}/p\hat{\Gamma} \to \mathcal{F}$,

$$\Theta\left(q(1-x^\alpha y^\beta)+p\hat{\Gamma}\right) = f_{\beta,-\alpha}(X,Y),$$

where

$$f_{\gamma,\delta}(X,Y) = \begin{cases} 1 & \text{if} \quad (X,Y) = (\gamma,\delta), \\ 0 & \text{if} \quad (X,Y) \neq (\gamma,\delta). \end{cases}$$

Lemma 10.4.4. *The following assertions hold.*

(i) *The map Θ is a module homomorphism over $SL_2(q)$ and Gal.*

(ii) $\Theta(qL_t(x)+p\hat{\Gamma}) = Y^{q-1-p^t} f_{<(0,1)>}$.

(iii) *Suppose that $i = i_0+pi_1+\ldots+p^{m-1}i_{m-1}$ and $j = j_0+pj_1+\ldots+p^{m-1}j_{m-1}$ with $0 \leq i_t, j_t < p$, $(i,j) \neq (0,0), (q-1,q-1)$. Suppose in addition that $\Theta(F+p\hat{\Gamma}) = X^iY^j$ for some $F \in \hat{\Gamma}$. Then*

$$\Theta(L_t(x).F+p\hat{\Gamma}) = -j_tX^iY^{j-p^t}, \quad \Theta(L_t(y).F+p\hat{\Gamma}) = i_tX^{i-p^t}Y^j.$$

(iv) *For $k \in \mathcal{K}$, $k > 0$, we have*

$$\left((\hat{\Gamma}/q^2\Gamma^R)\cap\mathcal{N}(k)\right)/p\left((\hat{\Gamma}/q^2\Gamma^R)\cap\mathcal{N}(k)\right) \cong \mathcal{M}_\Omega(q-1-k).$$

Proof. (i) is immediately verified.

(ii) Recall that

$$\bar{L}_0(x) = \sum_{\alpha\neq 0}\frac{1}{\alpha}x^\alpha = -\sum_{\alpha\neq 0}\frac{1}{\alpha}(1-x^\alpha).$$

Hence,

$$\Theta(qL_0(x)+p\hat{\Gamma}) = \Theta(q\bar{L}_0(x)+p\hat{\Gamma}) =$$
$$= \sum_{\alpha\neq 0}\frac{1}{\alpha}f_{0,-\alpha}(X,Y) = Y^{q-2}f_{<(0,1)>}(X,Y),$$

and after this one can use the action of *Gal*.

(iii) If $F = \sum_{\alpha,\beta} a_{\alpha\beta}q(1-x^\alpha y^\beta)$, then

$$\Theta(F+p\hat{\Gamma}) = \sum_{\alpha,\beta} a_{\alpha\beta}f_{\beta,-\alpha}(X,Y) = X^iY^j,$$

$$\Theta(x^\gamma.F+p\hat{\Gamma}) = \sum_{\alpha,\beta} a_{\alpha\beta}f_{\beta,-\alpha-\gamma}(X,Y) = X^i(Y+\gamma)^j.$$

This implies that

$$\Theta\left(L_0(x).F+p\hat{\Gamma}\right) = \Theta\left(\sum_{\gamma\neq 0}\frac{1}{\gamma}(x^\gamma-1)F+p\hat{\Gamma}\right) =$$

$$= \sum_{\gamma \neq 0} \frac{1}{\gamma} \Theta(-F + x^\gamma F + p\hat{\Gamma}) = \sum_{\gamma \neq 0} \frac{1}{\gamma} (X^i(Y+\gamma)^j - X^i Y^j) =$$

$$= \sum_{\gamma \neq 0} \frac{1}{\gamma} \left\{ X^i (Y+\gamma)^{j_0} (Y^p + \gamma^p)^{j_1} \cdots (Y^{p^{m-1}} + \gamma^{p^{m-1}})^{j_{m-1}} - X^i Y^j \right\} =$$

$$= -j_0 X^i Y^{j-1}.$$

The proofs for $L_t(x)$, $0 < t < m$, and $L_t(y)$, $0 \leq t < m$, are similar.

(iv) We have the decomposition

$$\hat{\Gamma}/p\hat{\Gamma} \cong \sum_{k \in \mathcal{K}} \left((\hat{\Gamma}/q^2\Gamma^R) \cap \mathcal{N}(k) \right) / p \left((\hat{\Gamma}/q^2\Gamma^R) \cap \mathcal{N}(k) \right).$$

By the Krull-Schmidt Theorem, decomposition of the $SL_2(q)$-module

$$\hat{\Gamma}/p\hat{\Gamma} \cong \mathcal{F} \cong \Gamma^R \big/ p\Gamma^R$$

into a sum of indecomposable submodules is unique up to isomorphism. On the other hand,

$$\left\{ \left((\hat{\Gamma}/q^2\Gamma^R) \cap \mathcal{N}(k) \right) + p\mathcal{N}(k) \right\} / p\mathcal{N}(k) \cong \mathcal{M}_{\emptyset}(k) \cong$$

$$\cong \mathcal{M}_{\Omega}(q-1-k) \Big/ \sum_{|\nu|=m-1} \mathcal{M}_{\nu}(q-1-k),$$

and so the required isomorphism follows immediately. □

We introduce here an $SL_2(q)$-submodule

$$\mathcal{N}_{\omega}(k) = SL_2(q) \cdot \mathcal{N}'_{\omega}(k)$$

of $\mathcal{N}(k)$, where the $\mathcal{N}'_{\omega}(k)$ are as defined at the beginning of this section.

Lemma 10.4.5. *Suppose that the index ω of $\mathcal{M}_{\omega}(k)$ is uncancellable. Then the following assertions hold.*

(i) $\mathcal{N}_{\omega}(k) = \sum_{\nu \supseteq \omega} p^{|\nu|-|\omega|} \mathcal{N}'_{\omega}(k)$. *Furthermore,*

$$\mathcal{N}_\omega(k) = \Big\langle qL_0(x)^{i_0}L_1(x)^{i_1}\cdots L_{m-1}(x)^{i_{m-1}}L_0(y)^{j_0}L_1(y)^{j_1}\cdots L_{m-1}(y)^{j_{m-1}}|$$

$$|i_0 + \ldots + p^{m-1}i_{m-1} + j_0 + \ldots + p^{m-1}j_{m-1} \equiv k \,(\mathrm{mod}\, q-1)\Big\rangle_R.$$

(ii) $\hat{\Gamma}/q^2\Gamma^R = \sum_{k\in\mathcal{K},k>0}\mathcal{N}_\emptyset(k) + q\mathcal{N}^w(0)$. *Furthermore, for* $k > 0$ *we have*

$$\mathcal{N}_\emptyset(k)/p\mathcal{N}_\emptyset(k) \cong \mathcal{M}_\Omega(q-1-k).$$

(iii) *If* $\omega \neq \emptyset$, *then*

$$\mathcal{N}_\omega(k)/\sum_{\nu\subset\omega}\mathcal{N}_\nu(k) \cong \mathcal{M}_{\Omega\backslash\omega}(q-1-k).$$

Proof. (i) First of all we study the modules

$$\mathcal{N}_\emptyset(1) = SL_2(q)\cdot < qL_0(x) >_R = \Big\langle qL_0(x^\alpha y^\beta)|\alpha,\beta\in\mathbb{F}_q\Big\rangle_R,$$

where $L_0(x^\alpha y^\beta) = \hat{\varphi}(L_0(x))$ if $\varphi \in SL_2(q)$ and $\varphi(x) = x^\alpha y^\beta$. As

$$qL_0(x) \in (\hat{\Gamma}/q^2\Gamma^R)\cap\mathcal{N}(1),$$

we have

$$\mathcal{N}_\emptyset(1) \subseteq (\hat{\Gamma}/q^2\Gamma^R)\cap\mathcal{N}(1).$$

We claim that in fact equality occurs. Indeed, by Lemma 10.4.4(ii), the image of $qL_0(x)$ under Θ is

$$Y^{q-2}f_{<(0,1)>} = \sum_{s=0}^{q-3} X^sY^{q-2-s} + X^{q-2}Y^{q-1},$$

and the last element generates the module $\mathcal{M}_\Omega(q-2)$ (over $SL_2(q)$) (see Proposition 10.3.5). Hence, in view of Lemma 10.4.4(iv),

$$\mathcal{N}_\emptyset(1) = (\hat{\Gamma}/q^2\Gamma^R)\cap\mathcal{N}(1).$$

Applying Proposition 10.4.3(iv), we obtain assertion (i) for $\omega = \emptyset$. In particular,

$$\mathcal{N}_\emptyset(1) = \Big\langle qL_0(x)^{i_0}L_1(x)^{i_1}\cdots L_{m-1}(x)^{i_{m-1}}L_0(y)^{j_0}L_1(y)^{j_1}\cdots L_{m-1}(y)^{j_{m-1}}|$$

$$|i_0 + \ldots + p^{m-1}i_{m-1} + j_0 + \ldots + p^{m-1}j_{m-1} \equiv 1 \,(\mathrm{mod}\, q-1)\Big\rangle_R,$$

since $\mathcal{N}_\emptyset(1) = (\hat{\Gamma}/q^2\Gamma^R)\cap\mathcal{N}(1)$. Using the action of the group Gal, we have

$$\begin{aligned}\mathcal{N}_\emptyset(p^i) &= \langle qL_i(x^\alpha y^\beta)|\alpha,\beta\in\mathbb{F}_q\rangle_R = \\ &= \langle qL_0(x)^{i_0}L_1(x)^{i_1}\cdots L_{m-1}(x)^{i_{m-1}}L_0(y)^{j_0}L_1(y)^{j_1}\cdots L_{m-1}(y)^{j_{m-1}}| \\ &|i_0 + \ldots + p^{m-1}i_{m-1} + j_0 + \ldots + p^{m-1}j_{m-1} \equiv p^i \,(\mathrm{mod}\, q-1)\rangle_R,\end{aligned} \quad (10.13)$$

for $0 \le i < m$.

Now assertion (i) for $\omega \neq \emptyset$ is proved using (10.13) and induction processes similar to those in the proof of Proposition 10.4.3(iv).

(ii) According to (i) we have

$$\mathcal{N}_{\emptyset}(k) = (\hat{\Gamma}/q^2\Gamma^R) \cap \mathcal{N}(k),$$

and so we can apply Lemma 10.4.4(iv).

(iii) From (i) it follows that $p^{|\omega|}\mathcal{N}_{\omega}(k) \supseteq q\mathcal{N}(k)$. Hence,

$$\mathcal{N}_{\omega}(k)/\sum_{\nu\subset\omega}\mathcal{N}_{\nu}(k) \cong p^{|\omega|}\mathcal{N}_{\omega}(k)/\sum_{\nu\subset\omega}p^{|\omega|}\mathcal{N}_{\nu}(k) =$$

$$= p^{|\omega|}\mathcal{N}_{\omega}(k)/(p^{|\omega|}\mathcal{N}_{\omega}(k) \cap p\hat{\Gamma}) \cong (p^{|\omega|}\mathcal{N}_{\omega}(k) + p\hat{\Gamma})/p\hat{\Gamma}.$$

The last module is embedded in

$$\mathcal{N}_{\emptyset}(k)/p\mathcal{N}_{\emptyset}(k) \cong \mathcal{M}_{\Omega}(q-1-k).$$

Furthermore, elements of $p^{|\omega|}\mathcal{N}_{\omega}(k)$ are obtained from some module $\mathcal{N}_{\emptyset}(p^i)$ on multiplying by elements like $L_j(x^{\alpha}y^{\beta})$. Therefore, our assertion follows from relation (10.13) and Lemma 10.4.4 (ii), (iii). □

Next we consider the case $k = \frac{q-1}{2}$. Then for $t = 0$ and $\omega \neq \emptyset, \Omega$, and for $0 < t < m - |\omega|$, we have

$$(\mathcal{N}_{\omega}(k) + p^t\mathcal{N}_{\omega^*}(k))/\mathcal{N}_t \cong \left(\mathcal{M}_{\omega}(k)/\sum_{\nu\subset\omega}\mathcal{M}_{\nu}(k)\right) \bigoplus \left(\mathcal{M}_{\omega^*}(k)/\sum_{\nu\subset\omega^*}\mathcal{M}_{\nu}(k)\right),$$

where we recall that $\omega^* = \Omega\backslash\omega$; furthermore,

$$\mathcal{N}_t = \sum_{\nu\subset\omega}\mathcal{N}_{\nu}(k) + (\mathcal{N}_{\omega}(k) \cap p\mathcal{N}(k)) +$$

$$+ \sum_{\nu\subset\omega^*}p^t\mathcal{N}_{\nu}(k) + (p^t\mathcal{N}_{\omega^*}(k) \cap p^{t+1}\mathcal{N}(k)).$$

Denote by $\mathcal{N}_{\omega}^{\lambda,t}(k)$ the full preimage of the module

$$\left\{\bar{a} + \lambda\tau_{\omega}(\bar{a}) | \bar{a} \in \mathcal{M}_{\omega}(k)/\sum_{\nu\subset\omega}\mathcal{M}_{\nu}(k)\right\}$$

on factoring $\mathcal{N}_{\omega}(k) + p^t\mathcal{N}_{\omega^*}(k)$ by $\mathcal{N}_t$, where τ_{ω} is defined in §10.3.

Theorem 10.4.6. *The following assertions hold.*

(i) *Every* $SL_2(q)$*-submodule of* $\mathcal{N}(0)$ *has the form*

$$p^i\mathcal{N}_\emptyset(0) \oplus p^j\mathcal{N}_\emptyset(0).$$

(ii) *If* $k \neq 0, \frac{q-1}{2}, q-1$, *then all* $SL_2(q)$*-submodules of* $\mathcal{N}(k)$ *are precisely the* $p^i\mathcal{N}_\omega(k)$ *with* ω *uncancellable, together with their sums.*

(iii) *If* $k = \frac{q-1}{2}$, *then the* $SL_2(q)$*-submodules of* $\mathcal{N}(k)$ *are the modules* $p^i\mathcal{N}(k)$, $p^j\mathcal{N}^{\lambda,t}_\omega(k)$ *(where* $\lambda \in \mathbb{F}^*_q$, *and either* $t = 0$ *and* $\omega \neq \emptyset, \Omega$, *or* $0 < t < m - |\omega|$*), together with their sums.*

(iv) $g(\mathcal{N}_\omega(k)) = \mathcal{N}_{\omega+1}(k_{m-1}, k_0, \ldots, k_{m-2})$,

$$g\left(\mathcal{N}^{\lambda,t}_\omega(\frac{q-1}{2})\right) = \mathcal{N}^{\lambda^p,t}_{\omega+1}(\frac{q-1}{2}).$$

(v) $\sum_{\alpha,\beta} < L_i(x^\alpha y^\beta) \cdot \mathcal{N}^w(0) >= \mathcal{N}(p^i)$;

$$\sum_{\alpha,\beta}\left\langle L_i(x^\alpha y^\beta) \cdot \mathcal{N}^{\lambda,t}_\omega(\frac{q-1}{2})\right\rangle = \mathcal{N}_\omega(\frac{q-1}{2} + p^i) + p^t\mathcal{N}_{\Omega\setminus\omega}(\frac{q-1}{2} + p^i),$$

$$\sum_{\alpha,\beta}\left\langle L_i(x^\alpha y^\beta) \cdot \mathcal{N}_\omega(k)\right\rangle = \mathcal{N}_\omega(k_0, \ldots, k_i + 1, \ldots, k_{m-1}),$$

where for $k_i + 1 = p$ *we set by definition*

$$\mathcal{N}_\omega(k_0, \ldots, p, k_{i+1}, \ldots) = p\mathcal{N}_{\omega\cup\{i\}}(k_0, \ldots, 0, k_{i+1} + 1, \ldots),$$

$$\mathcal{N}_\omega(q-1) = \begin{cases} q\mathcal{N}^w(0), & \text{if } \omega \neq \Omega, \\ q\mathcal{N}(0), & \text{if } \omega = \Omega. \end{cases}$$

Proof. Assertion (i) has been mentioned above.

(ii) According to Lemma 10.4.5, every minimal $SL_2(q)$-submodule $\mathcal{N}$ of $\mathcal{N}(k)$ such that

$$(\mathcal{N} + p\mathcal{N}(k))/p\mathcal{N}(k) \cong \mathcal{M}_\omega(k)$$

is isomorphic to $\mathcal{N}_\omega(k)$. All composition factors of the $SL_2(q)$-module $\mathcal{N}(k)/p\mathcal{N}(k) \cong \mathcal{F}^k$ are pairwise non-isomorphic. Hence, every submodule of $\mathcal{N}(k)$ is isomorphic to a sum of modules $p^i\mathcal{N}_\omega(k)$.

Assertion (iii) is proved similarly, but the presence of isomorphic composition factors

$$\mathcal{M}_\omega(k)/\sum_{\nu\subset\omega}\mathcal{M}_\nu(k) \cong \mathcal{M}_{\omega^*}(k)/\sum_{\nu\subset\omega^*}\mathcal{M}_\nu(k)$$

leads to the additional modules $p^i\mathcal{N}^{\lambda,t}_\omega(k)$.

Assertions (iv), (v) follow from our construction. □

Theorem 10.4.6 settles our classification problem *in principle*. To construct invariant sublattices, we first choose an $SL_2(q)$-submodule of $\Gamma^R/q^2\Gamma^R$ that is preserved under multiplications by $L_i(x)$ and $L_i(y)$, and invariant under the action of Gal. Next, we consider its full preimage $\Gamma' \subseteq \Gamma^R$ and take the intersection $\Gamma' \cap \Gamma$, and this consists of Gal-invariant elements.

An invariant submodule of $\Gamma^R/q^2\Gamma^R$ can be chosen as a sum of modules $p^i\mathcal{N}(k)$, $p^i\mathcal{N}^{\lambda,t}_{\omega}(k)$ and modules obtained by "intertwining" isomorphic factors of these modules, rather like in the construction of $\mathcal{M}^{\lambda}_{\omega}(k)$ and $\mathcal{N}^{\lambda,t}_{\omega}(k)$. For every given odd prime-power q, this process can be carried out by simple enumeration using Theorem 10.4.6 and Lemma 10.4.5. An example of the construction of sublattices is given in the next section. On the other hand, in the simplest case $q = 9$, Abdukhalikov has described 164 indivisible invariant sublattices Γ_i, $1 \leq i \leq 164$, such that every indivisible invariant sublattice is a sum of several Γ_i !

10.5 A series of unimodular invariant lattices

We carry over the inner product defined on Λ to Γ by means of the isomorphism χ (see Theorem 10.1.6):

$$\Big(x^{\alpha}\Phi(y); x^{\beta}\Phi(y)\Big) = \Big(\chi^{-1}(x^{\alpha}\Phi(y))|\chi^{-1}(x^{\beta}\Phi(y))\Big) =$$

$$= (\hat{v}_{\alpha}(Z)|\hat{v}_{\beta}(Z)) = \frac{1}{q}\mathrm{Tr}(\hat{v}_{\alpha}(Z).\hat{v}_{\beta}(Z)).$$

Here, for $\alpha \neq \beta$, $\hat{v}_{\alpha}(Z)$ and $\hat{v}_{\beta}(Z)$ are distinct diagonal matrices with one entry $\frac{q-1}{q}$ and all other equal to $-\frac{1}{q}$. Hence,

$$\Big(x^{\alpha}\Phi(y); x^{\beta}\Phi(y)\Big) = \frac{q.\delta_{\alpha,\beta} - 1}{q^2}.$$

Proposition 10.5.1. *The following assertions hold.*

(i) $\Big(\sum_{\alpha} a_{\alpha}x^{\alpha}\Phi(y); \sum_{\beta} b_{\beta}x^{\beta}\Phi(y)\Big) = \frac{1}{q^2}(q\sum_{\alpha} a_{\alpha}b_{\alpha} - \sum_{\alpha} a_{\alpha}.\sum_{\alpha} b_{\alpha})$. *In particular, if* $\sum_{\alpha} a_{\alpha} = 0$, *then*

$$\left(\sum_{\alpha} a_{\alpha}x^{\alpha}\Phi(y); \sum_{\beta} b_{\beta}x^{\beta}\Phi(y)\right) = \frac{1}{q}\sum_{\alpha} a_{\alpha}b_{\alpha}.$$

(ii) $(x^{\alpha}y^{\beta}\Phi_{\sigma}; x^{\gamma}y^{\delta}\Phi_{\sigma'}) = 0$, *if* $\sigma \neq \sigma'$.

(iii) *If* $\{\alpha_1, \ldots, \alpha_m\}$ *is a basis of* $\mathbb{F}_q$ *over* $\mathbb{F}_p$, $0 \leq k_t, \ell_t \leq p-1$, *and* $\sum_t k_t > 0$, *then*

$$\Big((1 - x^{\alpha_1})^{k_1} \cdots (1 - x^{\alpha_m})^{k_m}\Phi(y); (1 - x^{\alpha_1})^{\ell_1} \cdots (1 - x^{\alpha_m})^{\ell_m}\Phi(y)\Big) =$$

$$= \frac{1}{q}\binom{k_1+\ell_1}{k_1}\cdots\binom{k_m+\ell_m}{k_m}.$$

(iv)

$$(\sum_{\alpha,\beta} a_{\alpha\beta} q x^\alpha y^\beta; \sum_{\alpha,\beta} b_{\alpha\beta} q x^\alpha y^\beta) = \sum_{\alpha,\beta} a_{\alpha\beta} b_{\alpha\beta} - \frac{1}{q^2}\sum_{\alpha,\beta} a_{\alpha\beta}. \sum_{\alpha,\beta} b_{\alpha\beta}.$$

Proof. This is a direct translation of Proposition 9.3.1 and Corollary 9.3.2. □

Recall that the lattice

$$\Delta^* = \{a \in \mathbb{Q}^n | (a;\Delta) \subseteq \mathbb{Z}\}$$

is said to be dual to the lattice $(\Delta, (*;*))$ in $\mathbb{Q}^n$. Similarly, the R-lattice

$$\tilde{\Delta}^* = \{a \in \mathbb{K}^n | (a;\tilde{\Delta}) \subseteq R\}$$

is said to be *dual* to the R-lattice $(\tilde{\Delta}, (*;*))$ in $\mathbb{K}^n$.

Computation of dual lattices is based on the following two simple statements.

Lemma 10.5.2. *Let H be a finite group acting on R-lattices $\tilde{\Delta}_1$ and $\tilde{\Delta}_2$, where $\tilde{\Delta}_1 \supseteq \tilde{\Delta}_2 \supseteq p\tilde{\Delta}_2$. Then the $\mathbb{F}_pH$-modules $\tilde{\Delta}_1/\tilde{\Delta}_2$ and $\tilde{\Delta}_2^*/\tilde{\Delta}_1^*$ are dual to each other.* □

Lemma 10.5.3. *Let $\tilde{\Delta}$ be an invariant R-sublattice in Γ^R and $\Delta = \tilde{\Delta} \cap \Gamma$. Then*

$$\Delta^* = \tilde{\Delta}^* \cap \Gamma.$$ □

Now set

$$\Gamma_\sigma^n = \left\langle (1 - x^{\alpha_1} y^{\beta_1}) \cdots (1 - x^{\alpha_i} y^{\beta_i})\Phi_\sigma | i \geq n \right\rangle_{\mathbb{Z}}, \ \tilde{\Gamma}_\sigma^n = R\Gamma_\sigma^n,$$

$$\Gamma^n = \bigoplus_{\sigma \in \mathbb{P}V} \Gamma_\sigma^n, \ \tilde{\Gamma}^n = \bigoplus_{\sigma \in \mathbb{P}V} \tilde{\Gamma}_\sigma^n.$$

Clearly, Γ^n (respectively $\tilde{\Gamma}^n$) is an invariant lattice, indivisible if $n < m(p-1)$, and divisible if $n \geq m(p-1)$. Furthermore,

$$\tilde{\Gamma}^n = \langle L_0(x)^{i_0} \cdots L_{m-1}(x)^{i_{m-1}} L_0(y)^{j_0} \cdots L_{m-1}(y)^{j_{m-1}} \Phi_\sigma |$$

$$| i_0 + \ldots + i_{m-1} + j_0 + \ldots + j_{m-1} \geq n, \sigma \in \mathbb{P}V\rangle_R + q^2\Gamma^R.$$

Lemma 10.5.4. *If $1 \leq n < m(p-1)$, then*

$$(\tilde{\Gamma}_0^n + p\tilde{\Gamma}_0^1)^* = q\tilde{\Gamma}_0^0 + p^{m-1}\tilde{\Gamma}_0^{m(p-1)-n+1}.$$

Proof. We prove the lemma first for $n = 1$. For any element $F = \sum_\alpha a_\alpha x^\alpha \Phi(y)$ the following chain of implications holds:

$$F \in (\tilde{\Gamma}_0^1)^* \Leftrightarrow R \ni \left(\sum_\alpha a_\alpha x^\alpha \Phi(y); (x^\beta - x^\gamma)\Phi(y)\right) = \frac{1}{q}(a_\beta - a\gamma) \Leftrightarrow$$

$$\Leftrightarrow a_\alpha = c_\alpha + a,\ c_\alpha \in qR \Leftrightarrow F = \sum_\alpha c_\alpha x^\alpha \Phi(y) \in q\Gamma_0^R = q\tilde{\Gamma}_0^0.$$

Now suppose that $n = 2$. The inclusion $(\tilde{\Gamma}_0^2 + p\tilde{\Gamma}_0^1)^* \supseteq q\tilde{\Gamma}_0^0$ is obvious. Furthermore, by Lemma 10.5.2 an isomorphism

$$\tilde{\Gamma}_0^1/(\tilde{\Gamma}_0^2 + p\tilde{\Gamma}_0^1) \cong \bigoplus_{s=0}^{m-1} < X^{p^s} >_{\mathbb{F}_q},$$

of T-modules implies the following isomorphism of T-modules:

$$(\tilde{\Gamma}_0^2 + p\tilde{\Gamma}_0^1)^*/q\tilde{\Gamma}_0^0 = (\tilde{\Gamma}_0^2 + p\tilde{\Gamma}_0^1)^*/(\tilde{\Gamma}_0^1)^* \cong \bigoplus_{s=0}^{m-1} < X^{q-1-p^s} >_{\mathbb{F}_q}.$$

Hence,

$$(\tilde{\Gamma}_0^2 + p\tilde{\Gamma}_0^1)^* = q\tilde{\Gamma}_0^0 + \sum_{s=0}^{m-1} \left\langle p^{m-1} L_0(x)^{p-1} \cdots L_s(x)^{p-2} \cdots L_{m-1}(x)^{p-1}\Phi(y)\right\rangle_R =$$

$$= q\tilde{\Gamma}_0^0 + p^{m-1}\tilde{\Gamma}_0^{m(p-1)-n+1}.$$

The case $n > 2$ is treated similarly. □

Corollary 10.5.5. *If* $1 \leq n < m(p-1)$*, then*

$$(\Gamma_0^n + p\Gamma_0^1)^* = q\Gamma_0 + p^{m-1}\Gamma_0^{m(p-1)-n+1}. \qquad \square$$

Corollary 10.5.6. *If* $1 \leq n < m(p-1)$*, then*

$$(\tilde{\Gamma}^n + p\tilde{\Gamma}^1)^* = q\Gamma^R + p^{m-1}\tilde{\Gamma}^{m(p-1)-n+1},$$

$$(\Gamma^n + p\Gamma^1)^* = q\Gamma + p^{m-1}\Gamma^{m(p-1)-n+1}. \qquad \square$$

Now we construct some unimodular lattices. To this end we consider a lattice

$$\hat{\Delta} = \left\langle q(1 - x^\alpha y^\beta) | \alpha, \beta \in \mathbb{F}_q \right\rangle_{\mathbb{Z}}.$$

Then $\hat{\Delta} + q\Gamma$ is unimodular. For, $\hat{\Delta}$ is isometric to the root lattice of type A_{q^2-1}. Furthermore, the inner product on $\hat{\Delta} + q\Gamma = \hat{\Delta} + \mathbb{Z}q^2$ takes only integer values, and the additional element q^2 has order q in the factor-group $(\hat{\Delta}+q\Gamma)/\hat{\Delta}$. In particular, when $q=3$, the lattice $\hat{\Delta}+q\Gamma$ is a root lattice of type E_8 (see also Theorem 9.5.1).

Applying Lemma 10.5.3, we see that the R-lattice $\hat{\Gamma}+q\Gamma^R$ is also unimodular. Set

$$\tilde{\Gamma}^{n,\ell} = (\hat{\Gamma}+q\Gamma^R) \cap (\tilde{\Gamma}^n + p\tilde{\Gamma}^1) + p^{m-1}\tilde{\Gamma}^\ell.$$

By Lemma 10.4.5 we have

$$\tilde{\Gamma}^{n,\ell}/q^2\Gamma^R = \\ = \sum_{k\in\mathcal{K},\sum_i k_i \geq n} \mathcal{N}_\emptyset(k) + \sum_{k\in\mathcal{K},\omega\neq\emptyset} p^{|\omega|}\mathcal{N}_\omega(k) + \sum_{k\in\mathcal{K},\sum_i k_i\geq \ell} p^{m-1}\mathcal{N}_\Omega(k).$$

Therefore, in view of Lemma 10.5.2 and Corollary 10.5.6, we get that the lattices

$$\tilde{\Gamma}^{n,p^*-n},\ 1\leq n< m(p-1) \text{ if } m>1, \text{ and } 1\leq n\leq \frac{p-1}{2} \text{ if } m=1,$$

where $p^*=m(p-1)+1$, are unimodular.

Now we "adjust" these lattices to get new unimodular lattices. For $q>3$ we have

$$\tilde{\Gamma}^{n,p^*-n-1}/\tilde{\Gamma}^{n+1,p^*-n} = \tilde{\Gamma}^{n+1,p^*-n-1}/\tilde{\Gamma}^{n+1,p^*-n} \oplus \tilde{\Gamma}^{n,p^*-n}/\tilde{\Gamma}^{n+1,p^*-n}, \tag{10.14}$$

$$\tilde{\Gamma}^{n,p^*-n}/\tilde{\Gamma}^{n+1,p^*-n} \cong \bigoplus_{k\in\mathcal{K},\sum_i k_i=n} \mathcal{M}_\emptyset(k), \tag{10.15}$$

$$\tilde{\Gamma}^{n+1,p^*-n-1}/\tilde{\Gamma}^{n+1,p^*-n} \cong \bigoplus_{k\in\mathcal{K},\sum_i k_i=n} \left(\mathcal{M}_\Omega(q-1-k)/\sum_{|\nu|=m-1}\mathcal{M}_\nu(q-1-k)\right).$$

Let τ_k denote the isomorphism

$$\mathcal{M}_\emptyset(k) \cong \mathcal{M}_\Omega(q-1-k)/\sum_{|\nu|=m-1}\mathcal{M}_\nu(q-1-k).$$

One can suppose that τ_k is compatible with Gal: $g\circ\tau_k=\tau_k\circ g$. For convenience we set

$$\mathcal{K}_n = \{k\in\mathcal{K} | \sum_{i=0}^{m-1} k_i = n\}.$$

Furthermore, suppose that we are given a functional

$$\mathbf{r}:\mathcal{K}_n\to\mathbb{F}_q$$

which is invariant under $Gal=<g>$, that is,

$$\mathbf{r}(k)^p = \mathbf{r}(kp), \tag{10.16}$$

where kp is naturally identified with $(k_{m-1}, k_0, \dots, k_{m-2})$, if

$$k = (k_0, \dots, k_{m-1}) = \sum_t k_t p^t \in \mathcal{K}.$$

We define also the functional *dual* to $\mathbf{r}$:

$$\mathbf{r}^* : \mathcal{K}_n \to \mathbb{F}_q, \ \mathbf{r}^*(k) = (-1)^{p(n-1)}\mathbf{r}(k).$$

Finally, we set

$$\mathcal{S}^{n,\mathbf{r}} = \{a + \mathbf{r}(k)\tau_k(a) | a \in \mathcal{M}_\emptyset(k), \ k \in \mathcal{K}_n\}.$$

Then the module $\mathcal{S}^{n,\mathbf{r}}$ is isomorphic to the $SL_2(q)$-module figuring in (10.15), and, by (10.16), it is invariant under Gal. Let $\tilde{\Delta}^{n,\mathbf{r}}$ denote the full preimage of $\mathcal{S}^{n,\mathbf{r}}$ under the factorization (10.14), where $1 \le n < m(p-1)$ if $m > 1$, and $1 \le n < \frac{p-1}{2}$ if $m = 1$. Then $\tilde{\Delta}^{n,\mathbf{r}}$ is an R-lattice invariant under G and Gal. Note also that

$$\tilde{\Delta}^{n,\mathbf{0}} = \tilde{\Gamma}^{n,p^*-n},$$

where $\mathbf{0}(n) : \mathcal{K}_n \ni k \mapsto 0$.

Before finding dual lattices for $\tilde{\Delta}^{n,\mathbf{r}}$, we prove some auxiliary statements concerning the module

$$\mathcal{W} = \mathcal{M}_\emptyset(k_0, k_1, \dots, k_{m-1}) \oplus \mathcal{M}_\emptyset(k_{m-1}, k_0, \dots, k_{m-2}) \oplus$$

$$\oplus \cdots \oplus \mathcal{M}_\emptyset(k_1, \dots, k_{m-1}, k_0),$$

which is a sum of irreducible $SL_2(q)$-modules.

Lemma 10.5.7. *Suppose that all the components of $\mathcal{W}$ are pairwise non-isomorphic. Then every $SL_2(q)$-invariant bilinear form on $\mathcal{W}$ with values in $\mathbb{F}_q$ is symmetric if $2|k$, and skew-symmetric if $2 \not| k$.*

Proof. This follows from Lemma 10.3.3 and the fact that there does not exist a non-degenerate pairing between distinct modules W_k and W_ℓ. □

Proposition 10.5.8. *The following formula holds:*

$$(\tilde{\Delta}^{n,\mathbf{r}})^* = \tilde{\Delta}^{n,\mathbf{r}^*}.$$

Proof. We give the necessary arguments for the case $n = 1$. By Lemma 10.5.2 and Corollary 10.5.6, we have

$$(\tilde{\Delta}^{1,\mathbf{r}})^* = \tilde{\Delta}^{1,\mathbf{s}}$$

for some Gal-invariant functional $\mathbf{s} : \mathcal{K}_1 \to \mathbb{F}_q$. Note that

$$\mathcal{K}_1 = \{p^i | 0 \le i \le m-1\},$$

so that it follows from condition (10.16) that $\mathbf{r}$ is uniquely determined by its value $r = \mathbf{r}(1)$:

$$\mathbf{r}(p^i) = r^{p^i}.$$

Similarly, $\mathbf{s}(p^i) = s^{p^i}$ when $s = \mathbf{s}(1)$. As $n = 1$, $\mathbf{r}^* = \mathbf{r}$, and so we have to show that $s = r$.

Thus, we must prove that $(F; H) \in R$ for all $F, H \in \tilde{\Delta}^{1,\mathbf{r}}$. It is sufficient to prove that

$$A = (u + r^{p^i}\tau(u); v + r^{p^j}\tau(v)) \in R$$

for all $u \in \mathcal{M}_\emptyset(p^i)$ and $v \in \mathcal{M}_\emptyset(p^j)$, $0 \le i, j \le m-1$. Here, for brevity, we identify elements of the factor-lattice with suitable representatives of the cosets; furthermore, we omit the subscript k in the notation for τ_k. Clearly, $pA \in R$. We claim that $pA \in pR$. We have

$$pA = p(u + r^{p^i}\tau(u); v + r^{p^j}\tau(v)) =$$

$$= p(u; v) + p(r^{p^i}\tau(u); v) + p(u; r^{p^j}\tau(v)) + p(r^{p^i}\tau(u); r^{p^j}\tau(v)) \equiv$$

$$\equiv p(u; r^{p^j}\tau(v)) + p(r^{p^i}\tau(u); v) \pmod{pR}.$$

Since the form $B'(u, v) = p(r^{p^i}\tau(u); v)$ is invariant bilinear, it is skew-symmetric by Lemma 10.5.7. Hence,

$$pA \equiv p(u; r^{p^j}\tau(v)) + p(r^{p^i}\tau(u); v) \equiv$$

$$\equiv p(u; r^{p^j}\tau(v)) - p(u; r^{p^j}\tau(v)) \equiv 0 \pmod{pR}.$$

The case $n > 1$ is treated similarly. □

Now we go over to invariant $\mathbb{Z}$-lattices:

$$\Gamma^{n,\ell} = (\hat{\Delta} + q\Gamma) \cap (\Gamma^n + p\Gamma^1) + p^{m-1}\Gamma^\ell,$$

$$\Delta^{n,\mathbf{r}} = \tilde{\Delta}^{n,\mathbf{r}} \cap \Gamma.$$

The lattices $\Delta^{n,\mathbf{r}}$ (where $p(n-1)$ is even) and Γ^{n,p^*-n} are direct analogues of the unimodular lattices $\Delta^{k,r}$ and $\Gamma^{n,p-n}$ in the case $m = 1$ (see §9.4). They are also indecomposable, as is shown by the following result.

Lemma 10.5.9. *An invariant lattice $\Gamma' \subseteq \Gamma$ is decomposable if and only if*

$$\Gamma' = \bigoplus_{\sigma \in \mathbb{P}V} (\Gamma' \cap \Gamma_\sigma).$$

Proof. This is a direct translation of the proof of Theorem 9.5.3. □

A verbatim repeat of the proof of Theorems 9.5.1 and 9.7.2 yields the following result.

Theorem 10.5.10. *Suppose that* $q \geq 5$ *and* $r = \mathbf{r}(1) \neq 0$. *Then*
(i) *the lattice* $\Delta^{1,\mathbf{r}}$ *is a unimodular root-free lattice with minimum norm* 4;
(ii)

$$\mathrm{Aut}(\Delta^{1,\mathbf{r}}) = \begin{cases} \mathbb{Z}_2 \times \mathbb{Z}_p^{2m}.Sp_{2m}(p) & for \quad q > 5, \\ \mathbb{Z}_2 \circ Co_1 & for \quad q = 5. \end{cases}$$ □

Remark. When $q = 4$, the lattice $\Delta^{1,\mathbf{r}}$, where $r = \mathbf{r}(1) \neq 0$, is also unimodular. It is isometric to the unique indecomposable (odd) unimodular lattice of dimension 15 (see [Kne]). In particular, its minimal vectors form a root system of type A_{15}, and its automorphism group is $\mathbb{Z}_2 \times \mathbb{S}_{16}$. However, recall that when $q = 2^m \geq 8$ the lattices $\Delta^{n,\mathbf{r}}$ and $\Gamma^{n,\ell}$ considered above *are not* invariant lattices of type A_{q-1}: they are invariant under the *split* extension $\mathbb{Z}_p^{2m}.SL_2(p^m)$, but not under $G = \hat{V} \circ SL_2(q)$ (see Theorem 1.3.7).

10.6 Reduction theorem: statement of results

The rest of the chapter is devoted to a study of automorphism groups of invariant lattices of type A_{q-1} for an arbitrary prime power $q = p^m$. Our analysis does not require the classification of lattices given in §§10.1 – 10.5, but it does use the classification of finite simple groups. Except in a few cases, the question about the structure of the automorphism group $\mathbb{G} = \mathrm{Aut}(\Lambda)$ for an invariant lattice Λ of type A_{q-1}, $\dim \Lambda = p^{2m} - 1$, can be reduced to the same question for lattices Λ_1 of dimension $p^s - 1$, where $s|m$. This is the case that we shall refer to as *good reduction*. Moreover, Λ_1 is a projection of some invariant lattice of type A_{p^s-1} (onto a Cartan subalgebra). The outline of our arguments is as follows. Supposing that $\mathbb{G}$ is an imprimitive complex linear group (on $\mathcal{L}$), we prove in §10.7 that Λ has a good reduction. If $\mathbb{G}$ is a primitive linear group, then the methods of representation theory allow us to determine $\mathbb{G}$ in §10.8. From now on we denote by χ the $\mathbb{G}$-character afforded by $\mathcal{L}$, and $\mathcal{D}$ the standard OD of $\mathcal{L}$. By Lemma 10.1.1, we can assume that $\mathbb{G} \supseteq G = \mathbb{Z}_p^{2m}.SL_2(q)$.

Definition 10.6.1. We say that an invariant lattice Λ of type A_{q-1} has a *good reduction* if $\mathcal{L}$ admits a $\mathbb{G}$-invariant decomposition $\mathcal{P} : \mathcal{L} = \oplus_{i=1}^N V_i$ with the following properties.

(i) There exist natural numbers s, t such that

$$m = st,\ N = \frac{q^2-1}{p^s-1},\ \dim V_i = p^s - 1,$$

and V_i is a Cartan subalgebra that occurs in some IOD of a Lie algebra of type A_{p^s-1}.

(ii) $\mathbb{G}$ permutes the N components V_i of $\mathcal{P}$ primitively. Furthermore, the kernel $\mathbb{K}$ of this action is a subgroup of the direct power $(\mathrm{Aut}(\Lambda_1))^N$, where $\Lambda_1 = \Lambda \cap V_1$ is a (p^s-1)-dimensional lattice. Finally, if $\bar{\mathbb{G}} = \mathbb{G}/\mathbb{K}$, and $T = \mathrm{soc}(\bar{\mathbb{G}})$, then one of the following assertions holds:

a) $T = \mathbb{A}_N$;

b) $T = L_{2t}(p^s)$ or $T = PSp_{2t}(p^s)$;

c) the triple $(q, N, \bar{\mathbb{G}})$ is one of the following:

$$(7, 8, ASL_3(2)),\ (11, 12, M_{12}),\ (23, 24, M_{24}),\ (4, 15, \mathbb{A}_7).$$

Excluding the case of the sporadic IOD of the Lie algebra of type A_{26} as in §§10.1 – 10.5 (see Chapter 1), we prove the following reduction theorem that describes the automorphism groups of all invariant lattices of type A_n.

Theorem 10.6.2 (Reduction Theorem). *Let Λ be an invariant lattice of type A_{q-1}, $q = p^m$, $\mathbb{G} = \mathrm{Aut}(\Lambda)$. Then one of the following assertions holds.*

(i) *Λ has a good reduction.*

(ii) *$\mathbb{G}$ has a normal subgroup F with the following properties:*

a) *$F \subseteq (\mathrm{Aut}(\Gamma))^N$ for some d-dimensional lattice Γ, where $d|(q-1)$, $(p-1)|d$ and $N = \frac{q^2-1}{d}$;*

b) *$\mathbb{A}_N \lhd \mathbb{G}/F \subseteq \mathbb{S}_N$.*

(iii) *$q = 3$, Λ is isometric to the root lattice of type E_8 and $\mathbb{G} = \mathbb{Z}_2 \circ O_8^+(2)$.*

(iv) *$q = 5$, Λ is isometric to the Leech lattice and $\mathbb{G} = \mathbb{Z}_2 \circ Co_1$.*

(v) *$\mathbb{Z}_2 \times \mathbb{A}_{q^2} \lhd \mathbb{G} \subseteq \mathbb{Z}_2 \times \mathbb{S}_{q^2}$ and $q \notin \{2^a | a \geq 3\}$.*

As a consequence of this theorem, we obtain a complete description of the automorphism groups of invariant lattices of type A_{p-1}.

Corollary 10.6.3. *Let Λ be an invariant lattice of type A_{p-1}, p a prime, $\mathbb{G} = \mathrm{Aut}(\Lambda)$. Then one of the following assertions holds.*

1) *$p = 3$, Λ is isometric to the root lattice of type E_8 and $\mathbb{G} = \mathbb{Z}_2 \circ O_8^+(2)$.*

2) *$p = 5$, Λ is isometric to the Leech lattice and $\mathbb{G} = \mathbb{Z}_2 \circ Co_1$.*

3) *$\mathbb{G} = \mathbb{Z}_2 \times \mathbb{S}_{p^2}$.*

4) *$\mathbb{G}$ has a normal subgroup F with the following properties:*

a) *$F \subseteq H^{p+1}$, $H \in \{\mathbb{Z}_2 \times \mathbb{S}_p, \mathbb{Z}_2 \times PGL_2(p), \mathbb{Z}_2 \times (\mathbb{Z}_p.\mathbb{Z}_{p-1})\}$;*

b) *either $\mathbb{G}/F \in \{L_2(p), PGL_2(p), \mathbb{A}_{p+1}, \mathbb{S}_{p+1}\}$, or $(\mathbb{G}/F, p)$ is one of $(ASL_3(2), 7)$, $(M_{12}, 11)$, $(M_{24}, 23)$.*

Proof. Apply Theorem 10.6.2. Here possibilities (i) and (ii) mean simply that $\mathbb{G}$ preserves the given IOD $\mathcal{D}$, that is, we are in the situation considered in §9.7. Furthermore, the lattices Λ_1 mentioned in (i), and the lattices Γ mentioned in (ii), are nothing but projections of Λ onto a Cartan subalgebra $\mathcal{H}_\sigma$. Hence, they are isometric to the lattices Γ_k considered in §9.6. In particular, their automorphism groups are as computed in Theorem 9.6.1. Thus we arrive at assertion 4). In case (v) of Theorem 10.6.2, it is not difficult to convince oneself that $\mathbb{G} = \mathbb{Z}_2 \times \mathbb{S}_{p^2}$. □

The remaining part of this chapter is devoted to proving Theorem 10.6.2. We shall suppose that $q \geq 4$, since cases A_1 and A_2 have been already studied in Chapter 9. First of all, in this section we establish some auxiliary statements concerning the subgroup G of $\mathbb{G}$.

From now on to the end of the chapter we fix the following notation: $E = O_p(G) \cong \mathbb{Z}_p^{2m}$, $S = G/E \cong SL_2(q)$. If $p > 2$, or $q = 4$ we can assume that $S \hookrightarrow G$ (see Theorem 1.3.7).

The following statement is obvious.

Lemma 10.6.4. *E is the unique minimal normal subgroup of G. Furthermore, G acts (in the usual way) transitively on the sets $E\backslash\{1\}$ and $\mathrm{Irr}(E)\backslash\{1_E\}$. Finally, $C_G(E) = E$ and $[G,G] = G$.* □

Lemma 10.6.5. *Let $\mathbb{F}$ be a field of characteristic other than p and U a finite-dimensional faithful projective $\mathbb{F}G$-module. Then $\dim U \geq q$, or $q = 9$ and $\dim U \geq 6$. Moreover, if in addition a certain central extension $\hat{G} = \mathbb{Z}_\ell.G$, where $p \not| \ell$, is faithfully represented on U, then $\dim U \geq q^2 - 1$.*

Proof. By [KlL 2, Lemma 5.5.3] we have

$$\dim U \geq \min\{q, P(S)\},$$

where $P(X)$ denotes the smallest index of proper subgroups in a finite group X. Hence, we can use Table A6.

Next we suppose that $\hat{G} = \mathbb{Z}_\ell.G \hookrightarrow GL(U)$, where $p \not| \ell$. Setting $\hat{E} = \mathbb{Z}_\ell.E$ and using Lemma 10.6.4, one can show that $E \cong O_p(\hat{E}) \lhd \hat{G}$, and $\hat{G}$ acts transitively on $E^*\backslash\{0\}$, where $E^* = \mathrm{Hom}(E, \mathbb{F}_p)$. By Maschke's Theorem, the E-module

$$\bar{U} = U \otimes_{\mathbb{F}} \bar{\mathbb{F}},$$

where $\bar{\mathbb{F}}$ is the algebraic closure of $\mathbb{F}$, can be decomposed into a direct sum of submodules $\bar{U}_f$, $f \in E^*$. Here $e(v) = \zeta^{f(e)}v$ for $e \in E$, $v \in \bar{U}_f$, $f \in E^*$, and ζ is a p-th primitive root of unity in $\bar{\mathbb{F}}$. Clearly, the set

$$\Omega = \{f \in E^*\backslash\{0\} | \bar{U}_f \neq 0\}$$

is $\bar{G}$-invariant. Hence,

$$\dim U \geq |\Omega| = q^2 - 1.$$ □

As an immediate consequence we obtain:

Corollary 10.6.6. *Every linear representation π of G over a field of characteristic other than p is trivial on E if $\deg \pi < q^2 - 1$. Every permutation representation π of G is trivial on E if $\deg \pi < q^2$.* □

We omit the proof of the following statement.

Lemma 10.6.7. *The centralizer $C_{\mathbb{G}}(E)$ is an abelian group of exponent $2p$.* □

10.7 Invariant lattices of type A_n: the imprimitive case

Throughout this section we shall suppose that the complex linear group $\mathbb{G} = \mathrm{Aut}(\Lambda)$ is imprimitive on $\mathcal{L}$. This means that $\mathcal{L}$ can be decomposed into a direct sum of subspaces V_i, $1 \leq i \leq N$, $N > 1$, which are permuted transitively by $\mathbb{G}$. In fact, in the presence of a good reduction for Λ (see Definition 10.6.1), we can impose much more rigid restrictions on the decomposition $\mathcal{P} : \mathcal{L} = \oplus_{i=1}^{N} V_i$. As can be seen from this definition, $\mathcal{P}$ is obtained by stratifying the space $W = \mathbb{F}_q^2 = (\mathbb{F}_{p^s})^{2t}$ of indices of the matrices J_u (see Chapter 1) into $\mathbb{F}_{p^s}$-lines. Meanwhile, the given IOD $\mathcal{D}$ is obtained by stratifying the space $W = \mathbb{F}_q^2$ into $\mathbb{F}_q$-lines. Moreover, the question about the structure of the automorphism group $\mathbb{G}$ of the (q^2-1)-dimensional lattice Λ can be reduced to the similar question for the automorphism group of a certain (p^s-1)-dimensional $St_G(V_1)$-invariant lattice Λ_1, which is a projection of some invariant lattice of type A_{p^s-1}.

The main result of this section is the following theorem.

Theorem 10.7.1. *Suppose that $\mathbb{G} = \mathrm{Aut}(\Lambda)$ is an imprimitive complex linear group on $\mathcal{L}$. Then one of the following assertions holds.*

(i) *Λ has a good reduction.*

(ii) *$\mathbb{G}$ has a normal subgroup F with the following properties:*

a) *$F \subseteq (\mathrm{Aut}(\Gamma))^N$ for some d-dimensional lattice Γ, where $d|(q-1)$, $(p-1)|d$ and $N = \frac{q^2-1}{d}$;*

b) $\mathbb{A}_N \lhd \mathbb{G}/F \subseteq \mathbb{S}_N$.

To prove this theorem we choose a (nontrivial) $\mathbb{G}$-invariant decomposition $\mathcal{P}$ with the least possible number N of components V_i.

Lemma 10.7.2. *The following assertions hold.*

(i) $\mathbb{G}$ *acts primitively on the set* $\{V_i\}_{i=1}^N$.

(ii) $St_G(V_1) = E.S_1$, *where* $S_1 = \mathbb{Z}_p^m.\mathbb{Z}_{c(p-1)}$ *is a subgroup of* S, $c|\frac{q-1}{p-1}$, *and* $N = \frac{q^2-1}{c(p-1)}$.

(iii) $\Lambda_1 = \Lambda \cap V_1$ *is a* $c(p-1)$*-dimensional* $St_{\mathbb{G}}(V_1)$*-invariant lattice.*

Proof. First we suppose that $\mathbb{G}_1 = St_{\mathbb{G}}(V_1)$ is not maximal in $\mathbb{G}$, so that $\mathbb{G}_1 \subset M \subset \mathbb{G}$ for some subgroup M. Let ς be the $\mathbb{G}_1$-character afforded by V_1. Then

$$\chi = \mathrm{Ind}_{\mathbb{G}_1}^{\mathbb{G}}(\varsigma) = \mathrm{Ind}_M^{\mathbb{G}}(\varsigma'),$$

where $\varsigma' = \mathrm{Ind}_{\mathbb{G}_1}^M(\varsigma)$ is the M-character afforded by $V_1' = M.V_1$. Hence, the decomposition

$$\mathcal{L} = \bigoplus_{i=1}^{N'} g_i V_1',$$

where $\{g_1, \ldots, g_{N'}\}$ is a complete system of representatives of the cosets of $\mathbb{G}$ modulo M, is $\mathbb{G}$-invariant. Since $N' < N$, we get a contradiction, and this establishes (i).

Now we set $G_1 = St_G(V_1)$. Then $N = (G : G_1)$ divides $q^2 - 1$, and so is coprime to p. Hence, $G_1 \supseteq O_p(G) = E$, that is, $G_1 = E.S_1$, where S_1 is a subgroup of index N in $S = SL_2(q)$. By the Borel-Tits Theorem, S_1 is contained in some parabolic subgroup $P_1 = \mathbb{Z}_p^m.\mathbb{Z}_{q-1}$ of S. Therefore we have $S = \mathbb{Z}_p^m.\mathbb{Z}_d$ for some d dividing $q-1$. Namely,

$$S_1 = \left\{ \begin{pmatrix} \alpha & \beta \\ 0 & \alpha^{-1} \end{pmatrix} \middle| \alpha \in \mathbb{Z}_d \subseteq \mathbb{Z}_{q-1} = \mathbb{F}_q^*, \beta \in \mathbb{F}_q \right\}.$$

Setting $\Omega = \{V_1, \ldots, V_N\}$, we show that there exists a unique minimal subset $\Omega_1 \subseteq \Omega$ such that

a) $\Omega \ni V_1$;

b) $[\Omega_1] \cap \Lambda \neq 0$,

where $[\Omega'] = \sum_{U \in \Omega'} U$ for $\Omega' \subseteq \Omega$. Recall that

$$\chi|_E = \sum_{u \in W \setminus \{0\}} \bar{u},$$

where $\bar{u}(e) = \varepsilon^{u(e)}$ for $e \in E$, $u \in W \cong E^* = \mathrm{Hom}(E, \mathbb{F}_p)$ and $\varepsilon = \exp(\frac{2\pi i}{p})$. Here $\bar{u}$ is the E-character afforded by the line $< J_u >_{\mathbb{C}}$. From the explicit formula for G_1 it follows that

$$V_1 = \langle J_{\alpha a} | \alpha \in \mathbb{Z}_d \rangle_{\mathbb{C}}$$

for some $a \in W \setminus \{0\}$. Now suppose that $\Omega \supseteq \Omega' \ni V_1$, and $[\Omega'] \cap \Lambda \neq 0$. Because $[\Omega'] \cap \Lambda$ is a nonzero $\mathbb{Z}E$-module, it must afford all the characters $\overline{\lambda \alpha a}$, where $\lambda \in \mathbb{F}_p^*$, $\alpha \in \mathbb{Z}_d$. This means that

$$[\Omega'] \supseteq \tilde{V}_1 = \left\langle J_{\lambda \alpha a} | \lambda \in \mathbb{F}_p^*, \alpha \in \mathbb{Z}_d \right\rangle_{\mathbb{C}}.$$

It is not difficult to see that $\tilde{V}_1 \cap \Lambda \neq 0$ and $\tilde{V}_1 = [\Omega_1]$ for some $\Omega_1 \subseteq \Omega$. Thus the required subset Ω_1 exists and is unique.

From the uniqueness of Ω_1, it now follows that for any $g \in \mathbb{G}$, the intersection $g[\Omega_1] \cap [\Omega_1]$ is either 0 or $[\Omega_1]$. In other words, the decomposition

$$\mathcal{L} = \bigoplus_{i=1}^{N''} h_i[\Omega_1],$$

where $\{h_1, \ldots, h_{N''}\}$ is a complete system of representatives of the cosets of $\mathbb{G}$ modulo $St_{\mathbb{G}}([\Omega_1])$, is $\mathbb{G}$-invariant. By the minimality of N, we must have that $N'' = N$, $[\Omega_1] = V_1$. This proves assertions (ii) and (iii). □

Denote by $\mathbb{K}$ and $\bar{\mathbb{G}}$ the kernel and image of $\mathbb{G}$ in its permutation action on the set $\Omega = \{V_1, \ldots, V_N\}$. Clearly, $\mathbb{K}$ leaves every component $\Lambda \cap V_i$ of the lattice

$$(\Lambda \cap V_1) \oplus \cdots \oplus (\Lambda \cap V_N) \cong (\Lambda_1)^N$$

fixed. Hence, $\mathbb{K} \subseteq (\mathrm{Aut}(\Lambda_1))^N$. Set $T = \mathrm{soc}(\bar{\mathbb{G}})$.

Lemma 10.7.3. *Either T is a nonabelian simple group and $L_2(q) \hookrightarrow T$, or $q = p = 7$ and $\bar{\mathbb{G}} = ASL_3(2)$.*

Proof. By the O'Nan-Scott Theorem, T is either an elementary abelian group and then $|T| = N$, or a direct power $L^k = L \times \cdots \times L$ for some nonabelian simple group L, in which case $k < r(\bar{\mathbb{G}})$. Here $r(\bar{\mathbb{G}})$ denotes the rank of the primitive permutation group $\bar{\mathbb{G}}$. By Lemma 10.7.2, we have $G \cap \mathbb{K} = E.Z(S)$, hence

$$\bar{\mathbb{G}} \supseteq \bar{S} = S/Z(S) = L_2(q). \tag{10.17}$$

Assume first that $T = \mathbb{Z}_h^{\ell}$, where h is prime. Then $q + 1 = p^m + 1$ divides $N = h^{\ell}$, hence $p^m + 1 = h^{\ell'}$ for some integer ℓ'. It is not difficult to see that one of the following assertions holds:

a) $m = 1$, $h = 2$, $p = 2^{\ell'} - 1$, $\bar{\mathbb{G}} \subseteq ASL_{\ell'}(2)$;

b) $p = 2$, $\ell' = 1$, $h = 2^m + 1$, $\bar{\mathbb{G}} \subseteq AGL_1(h)$;

c) $p = 2$, $m = 3$, $h = 3$, $\ell' = 2$, $\bar{\mathbb{G}} \subseteq AGL_2(3)$.

But $\bar{\mathbb{G}}$ is insoluble, therefore case a) occurs, and $\bar{S} \hookrightarrow SL_{\ell'}(2)$. According to Table A7, we have

$$\ell' \geq R_{p'}(S) = \frac{q-1}{2} = 2^{\ell'-1} - 1,$$

so that, $\ell' = 3$, $q = p = 7$, and $\bar{\mathbb{G}} = ASL_3(2)$.

Now suppose that $T = L^k = L_1 \times \cdots \times L_k$, where $L_i \cong L$ for some nonabelian simple group L, and $k > 1$. Using the the action of $\bar{S}$ on Ω described in Lemma 10.7.2, one can show that

$$r = r(\bar{\mathbb{G}}) \leq r(\bar{S}) = \frac{2(q-1)}{p-1}.$$

Clearly, $\bar{S}$ permutes the k components $L_1, \ldots, L_k$ of T.

If $q = 9$, then by [DiM] we have $N = 10$, 20 or 40, and $k = 1$, a contradiction. If $p > 2$ and $q \neq 9$, then $k \leq r - 1 \leq q - 2 < P(\bar{S})$ (see Table A6). If $p = 2$, then the number N divides $q^2 - 1$ and this is odd. As $k > 1$, $\bar{\mathbb{G}}$ must be a primitive permutation group of type $3B$ (so-called *product action*, see [LPS]). Moreover, $\bar{\mathbb{G}} \subseteq H \wr \mathbb{S}_k$, where H is a primitive permutation group of degree n, $N = n^k$, and $L \lhd H \subseteq \mathrm{Aut}(L)$. Hence, $k = \log_n N \leq \log_5(q^2 - 1) < q < P(\bar{S})$.

Thus we have $k < P(\bar{S})$ in all cases. Therefore, $\bar{S}$ leaves every component L_i fixed. If, in addition, $\bar{S}$ centralizes every L_i, then $\bar{S} \subseteq C_{\bar{\mathbb{G}}}(T) = 1$, a contradiction. So, $\bar{S}$ can be embedded in $\mathrm{Aut}(L_i) \cong L.\mathrm{Out}(L)$ for some i, $1 \leq i \leq k$. But in accordance with the "Schreier Conjecture", $\mathrm{Out}(L)$ is soluble. Hence, $\bar{S} \hookrightarrow L$. In particular, $P(L) \geq P(\bar{S}) \geq q$. If $\bar{\mathbb{G}}$ is a primitive permutation group of type $3A$ or $3C$ as described in [LPS], then

$$\frac{q^2 - 1}{p - 1} \geq N \geq |L| \geq |\bar{S}|,$$

a contradiction. Therefore $\bar{\mathbb{G}}$ is a group of type $3B$: $\bar{\mathbb{G}} \subseteq H \wr \mathbb{S}_k$ for some primitive permutation group H of degree n, where $N = n^k$, and $L \lhd H \subseteq \mathrm{Aut}(L)$. In particular, L has a subgroup of index $n \leq \sqrt{N} < q$, contrary to the inequality $P(L) \geq q$. □

Lemma 10.7.4. *Theorem* 10.7.1 *holds if* $q \leq 11$ *or* $q = p$.

Proof. Recall that we are assuming that $q \geq 4$. By Lemma 10.7.3 and [DiM] we obtain the following possibilities when $q \leq 11$:

1) $q = 4$, and either $N = 5$, $T = \mathbb{A}_5$, or $N = 15$, $T \in \{\mathbb{A}_6 \cong Sp_4(2)', \mathbb{A}_7, \mathbb{A}_8 \cong SL_4(2), \mathbb{A}_{15}\}$;

2) $q = 5$, and $N = 6$, $T \in \{\mathbb{A}_5 \cong L_2(5), \mathbb{A}_6\}$;

3) $q = 7$, and $N = 8$, $T \in \{\mathbb{Z}_2^3, L_2(7), \mathbb{A}_8\}$;

4) $q = 8$, and either $N = 9$, $T \in \{L_2(8), \mathbb{A}_9\}$, or $N = 63$, $T \in \{Sp_6(2), SL_6(2), \mathbb{A}_{63}\}$;

5) $q = 9$, and either $N = 10$, $T \in \{L_2(9), \mathbb{A}_{10}\}$, or $N = 20$, $T = \mathbb{A}_{20}$, or $N = 40$, $T \in \{PSp_4(3), L_4(3), \mathbb{A}_{40}\}$;

6) $q = 11$, and $N = 12$, $T \in \{L_2(11), M_{11}, M_{12}, \mathbb{A}_{12}\}$.

If $q = p \geq 13$, then $N = p + 1$. As p divides $|\bar{\mathbb{G}}|$, $\bar{\mathbb{G}}$ is doubly transitive. In accordance with Lemma 10.7.3 and Table A2, we have either $T \in \{L_2(p), \mathbb{A}_{p+1}\}$, or $p = 23$ and $T = M_{24}$. We have to exclude the case $(p, N, T) = (11, 12, M_{11})$. To that end, observe that no transitive permutation representation of $\bar{S} = L_2(11)$ of degree 12 can be extended to a transitive permutation representation of M_{11} of the same degree. □

For the rest of the section we shall suppose that $q \geq 13$ and $q > p$. In this case $T = \mathrm{soc}(\bar{\mathbb{G}})$ is a nonabelian simple group. Using the classification of finite simple groups, we find all possibilities for T.

Lemma 10.7.5. *If $T = \mathbb{A}_\ell$ is the alternating group on ℓ symbols, then $\ell = N$.*

Proof. Clearly, $\ell \geq P(S) = q + 1$. Furthermore, $q + 1 \leq N \leq \frac{q^2-1}{p-1}$. One can suppose that $\bar{\mathbb{G}} \hookrightarrow \mathrm{Sym}(\Xi)$, where $\Xi = \{1, \ldots, \ell\}$. Setting $T_1 = St_T(V_1)$, we consider the action of T_1 on Ξ.

Assume firstly that T_1 acts primitively on Ξ. Then by Bochert's Theorem (see [Wie, Theorem 14.2]) we have

$$2N = 2(T : T_1) = (\mathbb{S}_\ell : T_1) \geq [\frac{\ell+1}{2}]! \geq (1 + [\frac{q}{2}])! > 2(q^2 - 1)$$

since $q > 13$, a contradiction.

Next, assume that T_1 acts transitively but imprimitively on Ξ. Then there exist integers $a, b \geq 2$ such that $\ell = ab$ and $T_1 \subseteq (\mathbb{S}_a \wr \mathbb{S}_b) \cap \mathbb{A}_\ell$. Hence

$$N = (T : T_1) \geq (\mathbb{S}_\ell : (\mathbb{S}_a \wr \mathbb{S}_b)) = \frac{(ab)!}{(a!)^b b!} > ab(ab-1) = \ell(\ell-1) > q^2 - 1,$$

a contradiction.

Thus T_1 acts intransitively on Ξ. This means that there exists an integer a, $1 \leq a \leq \frac{\ell}{2}$, such that $T_1 \subseteq (\mathbb{S}_a \times \mathbb{S}_{\ell-a}) \cap \mathbb{A}_\ell$. If $a \geq 3$, then

$$N = (T : T_1) \geq (\mathbb{S}_\ell : (\mathbb{S}_a \times \mathbb{S}_{\ell-a})) = \binom{\ell}{a} \geq \binom{\ell}{3} > \ell(\ell-1) > q^2,$$

a contradiction. If $a = 1$, then $T_1 \subseteq \mathbb{A}_{\ell-1}$. If, in addition, $T_1 \subset \mathbb{A}_{\ell-1}$, then

$$N = (T : T_1) = \ell(\mathbb{A}_{\ell-1} : T_1) \geq \ell(\ell-1) > q^2,$$

again a contradiction. Hence, $T_1 = \mathbb{A}_{\ell-1}$ and $\ell = N$ as stated.

Finally, suppose that $a = 2$. If $T_1 \subset (\mathbb{S}_2 \times \mathbb{S}_{\ell-2}) \cap \mathbb{A}_\ell$, then

$$N = (T : T_1) \geq 2(\mathbb{S}_\ell : (\mathbb{S}_2 \times \mathbb{S}_{\ell-2})) = \ell(\ell-1) > q^2,$$

a contradiction. Hence,

$$T_1 = (\mathbb{S}_2 \times \mathbb{S}_{\ell-2}) \cap \mathbb{A}_\ell = St_T(\{1, 2\}).$$

Consider the induced action of T on the set $\Xi(2) = \big\{\{i, j\} | 1 \leq i < j \leq \ell\big\}$. Since

$$(\bar{S} : \bar{S} \cap T_1) = N = \frac{\ell(\ell-1)}{2} = |\Xi(2)|,$$

$\bar{S}$ acts transitively on $\Xi(2)$. This means that $\bar{S}$ is a doubly homogeneous permutation group of even order on Ξ. In other words, $\bar{S}$ is doubly transitive. But when $q \geq 13$ the group $\bar{S} = L_2(q)$ has a unique doubly transitive permutation representation, and this representation is of degree $q+1$. Therefore we have $\ell = q+1$, and so $N = \frac{q(q+1)}{2}$ divides q^2-1, again a contradiction. □

Lemma 10.7.6. *T cannot be a finite group of Lie type of characteristic other than p.*

Proof. Suppose the contrary: T is a finite group of Lie type defined over a field $\mathbb{F}_{\hat{q}}$ of characteristic $r \neq p$, $\hat{q} = r^{t'}$. As $\bar{S} \hookrightarrow T$, we have on the one hand

$$\ell = R_r(T) \geq R_{p'}(S) = \frac{q-1}{\gcd(2, q-1)}. \tag{10.18}$$

On the other hand, it is clear that

$$P(T) \leq N \leq \frac{q^2-1}{p-1}. \tag{10.19}$$

Our aim is to deduce a contradiction from (10.18) and (10.19). Since $q > 13$ and $q > p$, it follows from (10.18) that $\ell \geq 12$. In particular, T does not belong to the types ${}^2B_2$, G_2, ${}^2G_2$ or ${}^3D_4$.

First, suppose that T is a classical finite simple group, for instance, $T = L_{\ell'}(\hat{q})$. Then, by Table A3 and (10.18), we have $\ell' = \ell \geq 12$. In this case, using Table A6 we obtain

$$P(T) = \frac{\hat{q}^\ell - 1}{\hat{q}-1}.$$

Hence, $P(T) > q^2$, contrary to (10.19). The remaining classical finite simple groups are considered in a similar way.

Now let T be an exceptional finite group of Lie type, for example, $T = F_4(\hat{q})$ or ${}^2F_4(\hat{q})$. Then, by Table A3 and (10.18), we obtain

$$\frac{q-1}{\gcd(2, q-1)} \leq \ell \leq 26.$$

This means that $q \in \{16, 25, 27, 49\}$, and so

$$N \leq \frac{q^2-1}{p-1} \leq 400.$$

Clearly, $P(T) \geq R_{r'}(T) + 1$. If $T = F_4(\hat{q})$ and $r > 2$, then

$$P(T) > R_{r'}(T) \geq \hat{q}^6(\hat{q}^2-1) \geq 8.3^6.$$

If $T = F_4(2^{t'})$ and $t' > 1$, then

$$P(T) > R_{2'}(T) \geq \frac{\hat{q}^7(\hat{q}^3 - 1)(\hat{q} - 1)}{2} > 2^{14}.$$

Furthermore, $P(F_4(2)) \geq P({}^2F_4(2)') = 1600$. Finally, if $\hat{q} = 2^{t'} > 2$, then

$$P({}^2F_4(\hat{q})) > R_{2'}({}^2F_4(\hat{q})) \geq \hat{q}^4(\hat{q} - 1)\sqrt{\frac{\hat{q}}{2}} \geq 7.2^{13}.$$

In any case we have $P(T) > N$, a contradiction. The cases of the other exceptional finite groups of Lie type are treated similarly. □

Lemma 10.7.7. *T cannot be a sporadic finite simple group.*

Proof. Suppose the contrary: T is one of the 26 sporadic finite simple groups. Let $T_1 = St_T(V_1)$ be a point stabilizer. Since the index $(T : T_1) = N$ is coprime to p, T_1 is an overgroup of some Sylow p-subgroup P in T. Moreover, $|P| \geq q > p$. Maximal overgroups of Sylow subgroups in sporadic simple groups have been classified by Aschbacher [Asch 2]. Let M be a maximal subgroup of T that contains T_1. Then the index $\ell = (T : M)$ must divide $q^2 - 1$. For example, we consider the cases when T is the Janko group J_4 or the Fischer-Griess Monster F_1.

If $T = J_4$, then

$$|T| = 2^{21}.3^3.5.7.11^3.23.29.31.37.43,$$

and $p = 2$, 3, or 11. Moreover,

$$\frac{q^2 - 1}{p - 1} \geq N \geq P(T) > d(T) = 1333.$$

(We adopt the notation of Tables A4 – A6). If $p = 3$, then $q^2 - 1 > 2666$, and $q > 3^3$, a contradiction. If $p = 11$, then $q^2 - 1 > 13330$, and $q \geq 11^2$. But by [ATLAS], $P = 11_+^{1+2}$. Moreover, a Sylow p-subgroup P_1 of $\bar{S}$ is an elementary abelian group of order q, and $P_1 \subseteq P$. Hence, $q = 11^2$, and so J_4 has a subgroup of index N dividing $\frac{q^2-1}{p-1} = 1464$, a contradiction. If $p = 2$, then by [Asch 2], there are three possibilities for M:

$$(2_+^{1+12}.3M_{22}).\mathbb{Z}_2,\ 2^{3+12}.(\mathbb{S}_5 \times SL_3(2)),\ \mathbb{Z}_2^{11}.M_{24}.$$

In particular, the index $\ell = (T : M)$ is divisible by 31.43. Hence, $2^{2m} - 1 = q^2 - 1$ is divisible by 31.43. This implies that m is divisible by 35, and so $q \geq 2^{35}$, a contradiction.

If $T = F_1$, then

$$|T| = 2^{46}.3^{20}.5^9.7^6.11^2.13^3.17.19.23.29.31.41.47.59.71,$$

and $p = 2, 3, 5, 7, 11$ or 13. Moreover,

$$\frac{q^2-1}{p-1} \geq N \geq P(T) > d(T) = 196883.$$

If $p = 13$, then $P = 13_+^{1+2}$ by [ATLAS], and a Sylow p-subgroup P_1 of $\bar{S}$ is elementary abelian: $P_1 = \mathbb{Z}_{13}^m$. Hence, $q \leq 13^2$, and

$$\frac{q^2-1}{p-1} < 3000 < P(T),$$

a contradiction. If $p = 11$, then $q^2 - 1 > 1968830$, and $q > 11^2$, again a contradiction. If $p = 7$, then by [Asch 2], there are two possibilities for M:

$$7^{1+4}.(\mathbb{Z}_3 \times 2\mathbb{S}_7),\ [7^5].GL_2(7),$$

where $[n]$ denotes a soluble group of order n. In particular, the index $\ell = (T : M)$ is divisible by 59, so that $7^{2m} - 1 = q^2 - 1$ is divisible by 59. This implies that m is divisible by 29, and so $q \geq 7^{29}$, a contradiction. If $p = 5$, then by [Asch 2], there are three possibilities for M:

$$(5^{1+6}.4J_2).\mathbb{Z}_2,\ [5^8].(\mathbb{S}_3 \times GL_2(5)),\ 5^{3+3}.(\mathbb{Z}_2 \times SL_3(5)).$$

Again 59 divides ℓ and $q^2 - 1 = 5^{2m} - 1$, that is, 29 divides m and so $q \geq 5^{29}$, a contradiction. If $p = 3$, then by [Asch 2], there are four possibilities for M:

$$(3_+^{1+12}.2Suz).\mathbb{Z}_2,\ [3^{17}].(M_{11} \times 2\mathbb{S}_4),\ [3^{17}].(SL_3(3) \times [16]),\ (\mathbb{Z}_3^8.P\Omega_8^-(3)).\mathbb{Z}_2.$$

Hence, 59 divides ℓ and $q^2 - 1 = 3^{2m} - 1$, that is, 29 divides m and so $q \geq 3^{29}$, a contradiction. If $p = 2$, then by [Asch 2], there are five possibilities for M:

$$2_+^{1+24}.Co_1,\ [2^{35}].(\mathbb{S}_3 \times M_{24}),\ [2^{26}].\Omega_{10}^+(2),$$

$$[2^{39}].(3\mathbb{S}_6 \times SL_3(2)),\ [2^{35}].(\mathbb{S}_3 \times SL_5(2)).$$

Hence, 47.59 divides ℓ and $q^2 - 1 = 2^{2m} - 1$, that is, 23.29 divides m and so $q \geq 2^{23.29}$, a contradiction.

The remaining sporadic simple groups are excluded using similar arguments.
□

To complete the proof of Theorem 10.7.1, we have to consider the case $T \in Chev(p)$. Let T be a finite group of Lie type defined over a field $\mathbb{F}_{\hat{q}}$ of characteristic

p, where $\hat{q} = p^s$, and let $T_1 = St_T(V_1)$ be a point stabilizer. Since the index $N = (T : T_1)$ is coprime to p, by the Borel-Tits Theorem T_1 is contained in some maximal parabolic subgroup P of T. We choose some primitive prime divisor ℓ of $p^{2m} - 1$, that is, ℓ divides $p^{2m} - 1$, but not $\prod_{i<2m}(p^i - 1)$. The existence of such a primitive prime divisor ℓ, which is guaranteed by [Zsi], enables us to find a lower bound for $R_p(T)$ and $P(T)$.

First suppose that T is an exceptional finite group of Lie type. For example, we consider the case when T is of type E_8 or ${}^2E_6$. If $T = E_8(\hat{q})$, then the factor-group $P/O_p(P)$ is a Chevalley group of type E_7, $A_1 + E_6$, $A_2 + D_5$, $A_3 + A_4$, $A_1 + A_2 + A_4$, A_7, $A_1 + A_6$, or D_7. It is not difficult to show that $(T : P) > \hat{q}^{30} - 1$ in all cases. Hence,

$$q^2 - 1 \geq N = (T : T_1) \geq (T : P) > \hat{q}^{30} - 1,$$

and so $q > \hat{q}^{15}$. But in this case, the primitive prime divisor ℓ of $q^2 - 1$ cannot divide $|T|$, a contradiction. If $T = {}^2E_6(\hat{q})$, then the factor-group $P/O_p(P)$ is a Chevalley group of type ${}^2D_4(\hat{q})$, $A_1(\hat{q}^2) + A_2(\hat{q})$, $A_1(\hat{q}) + A_2(\hat{q}^2)$, or ${}^2A_5(\hat{q})$. In any case we have

$$q^2 - 1 \geq N = (T : T_1) \geq (T : P) > \hat{q}^{18} - 1,$$

and so $q > \hat{q}^9$. But in this case, the primitive prime divisor ℓ of $q^2 - 1$ cannot divide $|T|$, again a contradiction.

The remaining exceptional finite groups of Lie type are considered in a similar way.

As regards the cases when T is a classical finite simple group, we shall consider them individually.

Lemma 10.7.8. *Suppose that* $T = L_k(\hat{q})$, $\hat{q} = p^s$. *Then the conclusion* (i) *of Theorem* 10.7.1 *holds.*

Proof. A primitive prime divisor ℓ of $q^2 - 1$ divides $|T|$, therefore $ks \geq 2m$. Furthermore, P is the stabilizer (in T) of some i-dimensional subspace with $1 \leq i \leq k - 1$, if T is projectively represented on the space $\mathbb{F}_{\hat{q}}^k$. If $2 \leq i \leq k - 2$, then

$$(T : P) > \hat{q}^k - 1 \geq q^2 - 1 > N,$$

a contradiction. Therefore $i = 1$ or $i = q - 1$, and so $p^{2m} - 1 = q^2 - 1$ is divisible by $(T : P) = \frac{\hat{q}^k - 1}{\hat{q} - 1}$. Using the presence of a primitive prime divisor of $\hat{q}^k - 1$, we arrive at the conclusion that ks divides $2m$. Hence, $ks = 2m$, and $\hat{q}^k = q^2$. Now we set

$$\bar{\mathbb{G}}_1 = St_{\bar{\mathbb{G}}}(V_1),\ T_1 = \bar{\mathbb{G}}_1 \cap T.$$

Then $(P : T_1)$ divides $\hat{q} - 1$, and from this it follows that $T_1 \lhd P$. If $\bar{\mathbb{G}}_1 \supseteq P$, then $T_1 = P$. If $\bar{\mathbb{G}}_1 \not\supseteq P$, then by the maximality of $\bar{\mathbb{G}}_1$ we have $< \bar{\mathbb{G}}_1, P >= \bar{\mathbb{G}}$.

But $T_1 \lhd < \bar{\mathbb{G}}_1, P >$, and so $T_1 \lhd T$. Since T is simple, we have $T_1 = 1$, a contradiction. This means that $T_1 = P$. Recall that $p^m + 1 = q + 1$ divides $N = (T : T_1) = (T : P)$, so that $s|m$ and $m = st$ for $t \in \mathbb{N}$. Conclusion (i) of Theorem 10.7.1 is now immediate. □

Similar arguments give

Lemma 10.7.9. *Suppose that $T = PSp_{2t}(\hat{q})$, $\hat{q} = p^s$ and $t \geq 2$. Then conclusion* (i) *of Theorem* 10.7.1 *holds.* □

Lemma 10.7.10. *T cannot belong to the types $PSU_k(\hat{q})$ $(k \geq 3)$; $\Omega_{2k+1}(\hat{q})$ $(k \geq 3, p > 2)$; and $P\Omega^{\pm}_{2k}(\hat{q})$ $(k \geq 4)$, where $\hat{q} = p^s$.*

Proof. Suppose that $T = PSU_k(\hat{q})$ with $k \geq 3$. Then P is one of the stabilizers P_i (in T) of i-dimensional totally singular subspaces with $1 \leq i \leq t = [\frac{k}{2}]$, if T is projectively represented on the space $\mathbb{F}^k_{\hat{q}^2}$. Setting $q_j = \hat{q}^j - (-1)^j$ for $1 \leq j \leq k$, we have

$$n_i = (T : P_i) = \frac{q_k q_{k-1} \cdots q_{k-2i+1}}{q_2 q_4 \cdots q_{2i}}.$$

In particular, $n_i \geq \min\{n_2, n_t\}$, if $2 \leq i \leq t$. Using the fact that $|T|$ is divisible by the primitive prime divisor ℓ of $q^2 - 1$, one can show that $m \leq ks$ if k is odd; and $m \leq (k-1)s$ if k is even. If $k \geq 7$, then

$$n_2, n_t \geq \hat{q}^{2k} \geq p^{2ks} \geq q^2,$$

on the one hand. On the other hand, $n_1 = q_k q_{k-1}/q_2$ does not divide $q^2 - 1$. Hence, $i \neq 1$ and $3 \leq k \leq 6$. These remaining possibilities are excluded individually.

Suppose now that $T = \Omega_{2k+1}(\hat{q})$, where $k \geq 3$ and $p > 2$. Then P is one of the stabilizers P_i (in T) of i-dimensional totally isotropic subspaces in T with $1 \leq i \leq k$, if T is projectively represented on the space $U = \mathbb{F}^{2k+1}_{\hat{q}}$. Using primitive prime divisors, one can show that $q = \hat{q}^k$. But $(T : P_i) > \hat{q}^{2k}$ for $2 \leq i \leq k$, hence $i = 1$ and $P = P_1$. Following the proof of Lemma 10.7.8, we get that $T_1 = P$ and $N = \frac{q^2-1}{\hat{q}-1}$. Thus we can identify the set $\Omega = \{V_1, \ldots, V_N\}$ with the set of isotropic points of the projective space $\mathbb{P}U$. In particular, T has the following subdegrees in its action on Ω:

$$1, \frac{\hat{q}(\hat{q}^{2k-2} - 1)}{\hat{q} - 1}, \hat{q}^{2k-1}.$$

On the other hand, it is known (from Lemma 10.7.2 with $c(p-1) = \hat{q} - 1$) that $\bar{S}$ has the following subdegrees in its action on Ω:

$$\underbrace{\hat{q}^k = q, q, \ldots, q}_{\kappa \text{ times}}, \underbrace{1, 1, \ldots, 1}_{\kappa \text{ times}},$$

where $\kappa = \frac{q-1}{\hat{q}-1}$. Hence, there exist integers a and b with $0 \leq a, b \leq \kappa$ such that

$$\hat{q}^{2k-1} = a\hat{q}^k + b,$$

a contradiction.

The case $T = P\Omega_{2k}^{\pm}(\hat{q})$, $k \geq 4$, is treated similarly. □

This last lemma completes the proof of Theorem 10.7.1. □

10.8 Invariant lattices of type A_n: the primitive case

Throughout this section we shall suppose that the complex linear group $\mathbb{G} = \mathrm{Aut}(\Lambda)$ is primitive on $\mathcal{L}$, and deduce conclusions (iv) and (v) of Theorem 10.6.2 from this assumption. Recall that $q = p^m \geq 4$.

The following statement is obvious.

Lemma 10.8.1. *$\mathbb{G}$ has a unique maximal abelian normal subgroup* abel($\mathbb{G}$). *Moreover,*

$$\mathrm{abel}(\mathbb{G}) = Z(\mathbb{G}) = Z = <\mathbf{t} : x \mapsto -x> = \mathbb{Z}_2. \qquad \square$$

Lemma 10.8.2. *The maximal soluble normal subgroup* sol($\mathbb{G}$) *is precisely Z.*

Proof. Suppose that $R = \mathrm{sol}(\mathbb{G}) \supset Z$. Taking the derived series

$$R = R^{(1)} \supset R^{(2)} \supset \cdots \supset R^{(\ell)} \supset R^{(\ell+1)} = 1,$$

where $R^{(i+1)} = [R^{(i)}, R^{(i)}]$, we set $Q = O_2(R^{(\ell-1)})$. Then $Q \lhd \mathbb{G}$, and $[Q, Q] = Z(Q) = Z$, that is, Q is a normal nilpotent subgroup of class 2 of $\mathbb{G}$. Choose a subgroup P of Q such that

$$P \supseteq Z,\ P/Z = \Omega_2(Q/Z).$$

Then $P \lhd \mathbb{G}$, and $[P, P] = \Phi(P) = Z(P) = Z$, so that P is a normal extraspecial 2-subgroup of $\mathbb{G}$: $P = 2_{\pm}^{1+2k}$ for some natural k.

By Clifford's Theorem and the primitivity of $\mathbb{G}$, we have $\chi|_P = e\varsigma$, where $e \in \mathbb{N}$ and ς is the unique faithful irreducible complex character of P. In particular, $2^k = \deg\varsigma$ divides $\deg\chi = q^2 - 1$. This implies that $p > 2$ and $2^k \leq 8[\frac{q+1}{4}]$. Hence, $2k < q^2 - 1$. Applying Lemma 10.6.5, we arrive at the conclusion that the $\mathbb{F}_2G$-module $P/Z = \mathbb{Z}_2^{2k}$ is not faithful, and this means that $E = \mathrm{soc}(G)$ centralizes P/Z. But as $p > 2$, $\mathrm{Hom}(E, Z) = 1$, and so $[E, P] = 1$, $P \subseteq C_{\mathbb{G}}(E)$. This last inclusion contradicts Lemma 10.6.7. □

Proposition 10.8.3. *The socle* $T = \mathrm{soc}(\mathbb{G}/Z)$ *is a nonabelian simple group containing* G.

Proof. Set $\bar{\mathbb{G}} = \mathbb{G}/Z$, and more generally, set $\bar{K} = KZ/Z$ for any subgroup $K \subseteq \mathbb{G}$. Observe that

$$C_{\bar{\mathbb{G}}}(\bar{K}) = C_{\mathbb{G}}(K)/Z \tag{10.20}$$

for every perfect subgroup $\bar{K}$ of $\bar{\mathbb{G}}$. Identity (10.20) will be used frequently in the proof of this proposition.

Let $\bar{M}$ be a minimal normal subgroup of $\bar{\mathbb{G}}$. As $\mathrm{sol}(\mathbb{G}) = Z$, $\bar{M}$ is insoluble, that is, $\bar{M} = \bar{L}_1 \times \ldots \times \bar{L}_k$, where $\bar{L}_1 \cong \ldots \cong \bar{L}_k \cong L$ for some nonabelian simple group L. Let $M, L_1, \ldots, L_k$ denote the full preimages of $\bar{M}, \bar{L}_1, \ldots, \bar{L}_k$ in $\mathbb{G}$. In view of (10.20), M is a central product of subgroups $L_1, \ldots, L_k$:

$$M = L_1 * \ldots * L_k.$$

Let $d = d(L)$ denote the smallest degree of faithful projective complex representations of L. Then

$$k \leq \log_d(q^2 - 1) \leq \log_2(q^2 - 1). \tag{10.21}$$

For, by the primitivity of $\mathbb{G}$, we have $\chi|_M = e\varsigma$ for some $e \in \mathbb{N}$ and $\varsigma \in \mathrm{Irr}(M)$. Moreover, there exist characters $\rho_i \in \mathrm{Irr}(L_i)$, $1 \leq i \leq k$, such that

$$\varsigma(x_1 x_2 \ldots x_k) = \rho_1(x_1)\rho_2(x_2)\ldots\rho_k(x_k)$$

for all $x_i \in L_i$. In particular,

$$\chi|_{L_i} = \frac{\deg \chi}{\deg \rho_i}\rho_i.$$

This means that ρ_i is faithful, and in particular $\deg \rho_i \geq d = d(L) \geq 2$. Hence, $q^2 - 1 = \deg \chi \geq d^k$, as stated.

Clearly, G can be embedded in $\bar{\mathbb{G}}$. Now G, as a subgroup of $\bar{\mathbb{G}}$, permutes the k components $\bar{L}_1, \ldots, \bar{L}_k$ of the normal subgroup $\bar{M}$. From (10.21) it follows that $k < q$. Hence, by Corollary 10.6.6, $E = \mathrm{soc}(G)$ normalizes every subgroup $\bar{L}_i$, $1 \leq i \leq k$. Furthermore, when $q \neq 9$ we have $k < q \leq P(S)$. If $q = 9$, then the integer $\deg \chi = q^2 - 1 = 80$ can be decomposed into a product of at most five divisors, each of which is greater than 1; therefore, $k \leq 5 < P(S) = 6$. The inequality $k < P(S)$ thus obtained means simply that G normalizes each of the subgroups $\bar{L}_i$, $1 \leq i \leq k$. Next we consider the homomorphism that arises here:

$$\pi : G \to \mathrm{Aut}(\bar{L}_1) = \mathrm{Aut}(L) = L.\mathrm{Out}(L).$$

We have

$$\mathrm{Ker}\pi = G \cap C_{\bar{\mathbb{G}}}(\bar{L}_1) = G \cap (C_{\mathbb{G}}(L_1)/Z).$$

If $\mathrm{Ker}\pi \neq 1$, then $\mathrm{Ker}\pi \supseteq E$ by Lemma 10.6.4, which implies that $E \subseteq C_{\mathbb{G}}(L_1)$ and so $L_1 \subseteq C_{\mathbb{G}}(E)$, contrary to Lemma 10.6.7. Hence, $\mathrm{Ker}\pi = 1$. Moreover, $\mathrm{Out}(L)$ is soluble and G is perfect, therefore $G \hookrightarrow L$. In particular, $d(L) \geq d(G)$. Furthermore, $d(G) \geq q$ by Lemma 10.6.5. Hence from (10.21) it follows that $k = 1$. This means that every minimal normal subgroup $\bar{M}$ of $\bar{\mathbb{G}}$ is a nonabelian simple group containing G. It is not difficult to see that such a subgroup $\bar{M}$ is unique. □

Keeping in mind Proposition 10.8.3, we use the classification of finite simple groups to find all possibilities for $T = \mathrm{soc}(\mathbb{G}/Z)$.

Lemma 10.8.4. *Suppose that T is the alternating group on N symbols: $T = \mathbb{A}_N$. Then the following assertions hold.*

(i) $N = q^2$;

(ii) $q \notin \{2^a | a \geq 3\}$;

(iii) $\mathbb{Z}_2 \times \mathbb{A}_N \lhd \mathbb{G} \subseteq \mathbb{Z}_2 \times \mathbb{S}_N$.

Proof. Since $G \hookrightarrow T$, we have $N \geq q^2$ by Corollary 10.6.6, and in particular, $N \geq 16$. In this case we have also that

$$\dim \mathcal{L} = q^2 - 1 \geq d(\mathbb{A}_N) = N - 1,$$

that is, $N = q^2$. Now as a subgroup of $T = \mathbb{A}_{q^2}$, G has a faithful permutation representation Ψ of degree q^2 with character ς. Decompose ς into a sum of irreducible constituents:

$$\varsigma = e.1_G + \rho_1 + \ldots + \rho_s,$$

where $e, s \in \mathbb{N}$, $\rho_i \in \mathrm{Irr}(G)$, $\deg \rho_i > 1$. Here we have used the fact that $[G, G] = G$. Clearly,

$$\deg \rho_i = (\deg \varsigma - e) - \sum_{j \neq i} \deg \rho_j \leq q^2 - (e + 2(s - 1)).$$

If $e + 2(s-1) > 1$, then for any i with $1 \leq i \leq s$ we have $\deg \rho_i < q^2 - 1$; thus, by Corollary 10.6.6, all the characters ρ_i restricted to E are trivial, that is, $\mathrm{Ker}\varsigma \supseteq E$, a contradiction. Hence, $e + 2(s - 1) = 1$, so that $e = s = 1$, $\varsigma = 1_G + \rho_1$, and Ψ is doubly transitive. But $E = \mathrm{soc}(G)$ is abelian, therefore the permutation representation $\Psi|_E$ is regular. For a point stabilizer G_1 we have $G = E.G_1$, that is, G is split over E. By Theorem 1.3.7, this fact gives assertion (ii).

Next we consider the full preimage $\tilde{T} = Z.T$ of T in $\mathbb{G}$. By Clifford's Theorem, we have $\chi_{\tilde{T}} = r\vartheta$ for some natural number r and some faithful irreducible complex character ϑ of $\tilde{T}$. If $\tilde{T}$ is not split over Z, then $\tilde{T}$ is a double covering group for $\mathbb{A}_N$, and so, by [Wag 3],

$$\deg \vartheta \geq 2^{[\frac{N}{2}]-1} > N = q^2 > \dim \mathcal{L},$$

a contradiction. Hence, $\tilde{T} = Z \times T$.

Suppose that $\mathbb{G} \neq \tilde{T}$. Then

$$\mathbb{S}_N = \mathrm{Aut}(\mathbb{A}_N) \supseteq \mathbb{G}/Z \supset \tilde{T}/Z = \mathbb{A}_N.$$

In other words, $\mathbb{G} = \mathbb{Z}_2.\mathbb{S}_N$. Take an element $g \in \mathbb{G}$ whose image in $\mathbb{G}/Z$ is a transposition. Clearly, $g \notin \tilde{T}$ and $g^2 \in Z$. Recall that $Z =< \mathbf{t} >\cong \mathbb{Z}_2$. Suppose that $g^2 = \mathbf{t}$. As $\chi(g) \in \mathbb{Z}$, we have

$$\mathcal{L} = \mathcal{L}_+ \oplus \mathcal{L}_-,$$

where $\mathcal{L}_+ = \mathrm{Ker}(g - i.1_{\mathcal{L}})$, $\mathcal{L}_- = \mathrm{Ker}(g + i.1_{\mathcal{L}})$, $i = \sqrt{-1}$, $\dim \mathcal{L}_+ = \dim \mathcal{L}_- = \frac{q^2-1}{2}$. Observe that $\bar{g} = gZ$ centralizes a certain perfect subgroup $T_1 \cong \mathbb{A}_{N-2}$ with full preimage $\tilde{T}_1 = Z.T_1$. Clearly,

$$\bar{g} \in C_{\bar{\mathbb{G}}}(T_1) = C_{\mathbb{G}}(\tilde{T}_1)/Z,$$

that is, g centralizes $\tilde{T}_1$. This means that $\tilde{T}_1$ leaves both of the subspaces $\mathcal{L}_+$, $\mathcal{L}_-$ fixed. But

$$\dim \mathcal{L}_\pm = \frac{q^2-1}{2} < q^2 - 3 = d(\tilde{T}_1),$$

therefore $\tilde{T}_1 = Z \times T_1$ and T_1 acts trivially on $\mathcal{L}$, a contradiction. Consequently, $g^2 = 1$ and

$$\mathbb{G} =< Z, T, g >= Z\times < T, g >\cong \mathbb{Z}_2 \times \mathbb{S}_N. \qquad \square$$

Lemma 10.8.5. *T cannot be a finite group of Lie type of characteristic other than p: $T \notin Chev(p')$.*

Proof. Suppose the contrary, so that T is a finite group of Lie type defined over a field $\mathbb{F}_{\hat{q}}$ of characteristic $r \neq p$, $\hat{q} = r^t$. The outline of our arguments is as follows. As $G \hookrightarrow T$, we have the obvious estimate

$$\ell = R_r(T) \geq R_{p'}(G). \tag{10.22}$$

Using (10.22) and Table A7, we can give a lower bound for $d(T)$, on the one hand. On the other hand,

$$d(T) \leq q^2 - 1.$$

More precisely,

the central extension $\tilde{T} = Z.T$ has a faithful
$\mathbb{Q}$-valued irreducible complex character (10.23)
ϑ of degree dividing $q^2 - 1$.

Conditions (10.22) and (10.23) are incompatible, and we get the required contradiction.

To begin with, suppose that T is a classical finite simple group. Then T is the factor-group of some linear group $\hat{T}$ defined over $\mathbb{F}_{\hat{q}}$ by its centre $Z(\hat{T})$. As an example, we consider the case when T is a group of type A or C. The remaining classical finite simple groups are treated similarly.

1) Suppose that $T = L_k(\hat{q})$. Setting $e = \gcd(k, \hat{q} - 1)$, we have

$$\hat{T} = SL_k(\hat{q}) = \mathbb{Z}_e.T \supseteq \mathbb{Z}_e.G.$$

If p does not divide e, then it follows from (10.22) and Lemma 10.6.5 that $k \geq q^2 - 1 \geq 15$. In this case, by Table A7 we have

$$d(T) \geq \hat{q}^{k-1} - 1 \geq 2^{q^2-2} - 1 > q^2,$$

contrary to (10.23). Hence, p divides e. Suppose that $q = 9$. Then in view of (10.22) and Lemma 10.6.5, we have $k = \ell \geq R_{p'}(G) \geq 6$. But $p = 3$ and p divides e, therefore $3|(\hat{q} - 1)$, $\hat{q} \geq 4$, and, by Table A7,

$$d(T) \geq \hat{q}^{k-1} - 1 \geq 4^5 - 1 > 80,$$

a contradiction. Thus one can suppose that $q \neq 9$. By Lemma 10.6.5, we have $k = \ell \geq R_{p'}(G) \geq q \geq 4$. Moreover, p divides $\hat{q} - 1$, which implies that $\hat{q} \geq 3$. Hence,

$$d(T) \geq \hat{q}^{k-1} - 1 \geq 3^{q-1} - 1 > q^2,$$

again a contradiction.

2) Next, suppose that $T = PSp_{2k}(\hat{q})$, $k \geq 2$. Setting $e = \gcd(2, \hat{q} - 1)$, we have

$$\hat{T} = Sp_{2k}(\hat{q}) = \mathbb{Z}_e.T \supseteq \mathbb{Z}_e.G.$$

If $r = 2$, then $e = 1$ and $2k \geq q^2 - 1 \geq 15$, by Lemma 10.6.5. Using Table A7, we get the estimate

$$d(T) \geq \frac{\hat{q}^{k-1}(\hat{q}^{k-1} - 1)(\hat{q} - 1)}{2} \geq 2^{k-2}(2^{k-1} - 1) > q^2,$$

which is contrary to (10.23). Thus we can suppose that $r > 2$, $e = 2$. If p does not divide e, then again we have $2k \geq q^2 - 1$. In this case, Table A7 gives that

$$d(T) = \frac{\hat{q}^k - 1}{2} \geq \frac{3^{\frac{q^2-1}{2}} - 1}{2} > q^2,$$

a contradiction. Hence, one can put $p = 2$, so in particular, $q \neq 9$. Applying Lemma 10.6.5, we have $2k \geq R_{p'}(G) \geq q$; furthermore, $d(T) = \frac{\hat{q}^k-1}{2}$ by Table A7. If $\hat{q} \geq 7$, then

$$d(T) \geq \frac{7^{q/2} - 1}{2} > q^2,$$

a contradiction. If $\hat{q} = 5$ and $q > 4$, then

$$d(T) \geq \frac{5^{q/2} - 1}{2} > q^2,$$

again a contradiction. If $\hat{q} = 3$ and $q > 8$, then

$$d(T) \geq \frac{3^{q/2} - 1}{2} > q^2,$$

yet again a contradiction.

We still have to exclude the cases $(\hat{q}, q) = (5, 4), (3, 4), (3, 8)$. If $(\hat{q}, q) = (5, 4)$, then condition (10.23) implies that $15 = \dim \mathcal{L} \geq d(T) = \frac{5^k-1}{2}$, that is, $k = 2$ and $T = PSp_4(5)$. But, in accordance with [ATLAS], the group $T = PSp_4(5)$ cannot satisfy condition (10.23). If $(\hat{q}, q) = (3, 4)$ or $(3, 8)$, then (10.23) implies that $63 \geq \dim \mathcal{L} \geq d(T) = \frac{3^k-1}{2}$, that is, $2 \leq k \leq 4$. Note that $Sp_6(3)$ has no irreducible characters of degree dividing 15 or 63 [ATLAS], and therefore $k \neq 3$. Suppose that $k = 2$, that is, $T = PSp_4(3) \cong PSU_4(2)$. As $G = \mathbb{Z}_p^{2m}.SL_2(q)$ is embedded in T, we must have $q = 4$. Moreover, as a subgroup of index 27 in T, G is conjugate to the subgroup

$$\left\{ \begin{pmatrix} A & X \\ 0 & (A^*)^{-1} \end{pmatrix} | A \in SL_2(4), X \in M_2(\mathbb{F}_4), AX^* + XA^* = 0 \right\},$$

where $A^* = {}^t\bar{A}$. In particular, G has orbits of lengths 1, 5, and 10 in its action by conjugations on $E = O_2(G)$, contrary to Lemma 10.6.4.

Finally, suppose that $k = 4$, $T = PSp_8(3)$, $q = 8$. Set $\hat{T} = Sp_8(3)$, and let $U = \mathbb{F}_3^8$ be the natural symplectic module for $\hat{T}$, and $\hat{E}$ the full preimage of E in $\hat{T}$. Then one can show that $\hat{E}$ is $2_\pm^{1+6}$ or $\mathbb{Z}_2^7$. If $\hat{E} = 2_\pm^{1+6}$, then $\hat{E}$ acts absolutely irreducibly on U, $C_{\hat{T}}(\hat{E}) = Z(\hat{E}) = \mathbb{Z}_2$, and $S = SL_2(8)$ can be embedded in $\mathrm{Out}(\hat{E}) = O_6^\pm(2)$, a contradiction. If $\hat{E} = \mathbb{Z}_2^7$, then by Maschke's Theorem, $\hat{E}$ is completely reducible on U, and so $\hat{E}$ consists of diagonal matrices in some basis $\{e_1, \dots, e_8\}$ of U:

$$\hat{E} \subseteq \{\mathrm{diag}(a_1, \dots, a_8) | a_i = \pm 1\}.$$

Using the fact that $|\hat{E}| = 2^7$, one can now show that U has a 6-dimensional totally isotropic subspace, contrary to the non-degeneracy of U.

Now we suppose that T is an exceptional finite group of Lie type, for instance, $T = E_6(\hat{q})$. Then T has Schur multiplier $\mathbb{Z}_e$, where $e = \gcd(3, \hat{q}-1)$. Furthermore, from Table A3 we have $\mathbb{Z}_e.T \hookrightarrow GL_{27}(\hat{q})$. Hence, by Lemma 10.6.5, $q \leq 27$. But in this case we get

$$d(T) \geq \hat{q}^9(\hat{q}^2 - 1) \geq 3.2^9 > q^2,$$

contrary to (10.23). The remaining cases are excluded in a similar way. □

Lemma 10.8.6. *Suppose that T is a sporadic finite simple group. Then $q = 5$, $T = Co_1$, $\mathbb{G} = \mathbb{Z}_2 \circ Co_1$, and Λ is isometric to the Leech lattice.*

Proof. Our arguments will be based on (10.23) and the following obvious observation:

$$q^3 = p^{3m} \text{ divides } |T|. \tag{10.24}$$

As examples we consider the cases $T = Co_1$ and $T = F_1$. The other sporadic simple groups are treated similarly.

1) If $T = Co_1$, then it follows from (10.24) that $q = 5$, 9, 27, or 2^m for $2 \leq m \leq 7$. Furthermore, condition (10.23) implies that $q \neq 9, 27, 4, 8, 16, 64$. Moreover, the condition $(q-1)||T|$ implies that $q \neq 32, 128$, so that $q = 5$. Since $T = Co_1$ has no irreducible complex characters of degree ≤ 24, $\tilde{T}$ is not split over Z. Finally, $\mathrm{Out}(T) = 1$ [ATLAS]. We arrive at the conclusion that $\mathbb{G} = \tilde{T} \cong \mathbb{Z}_2 \circ Co_1$. It is well known (see Chapter 9) that $\mathbb{Z}_2 \circ Co_1$ can be realized as the automorphism group of invariant lattices $\Delta^{1,r}$, $r \in \mathbb{F}_5^*$. Moreover, every $(\mathbb{Z}_2 \circ Co_1)$-invariant lattice of dimension 24 is isometric to the Leech lattice. The last statement is an immediate consequence of the observation that $R_r(Co_1) = 24$ for all primes r.

2) Suppose finally that $T = F_1$. Then $d(T) = 196883$, and it follows from (10.23) and (10.24) that $q = 3^6$, or $q = 2^m$ for $9 \leq m \leq 15$. If $q = 3^6$, then 73 divides $q+1$ and $|G|$, but not $|T|$, a contradiction. If $q = 2^9$, 2^{11}, 2^{13}, or 2^{14}, then $q-1$ and $|G|$ are divisible by $r = 73$, 89, 8191, or 127, respectively. Meanwhile $|T|$ is not divisible by this prime r. If $q = 2^{12}$ or $q = 2^{15}$, then $q+1$ and $|G|$, but not $|T|$, are divisible by 241 or 331. Finally, the case $q = 2^{10}$ is impossible by (10.23). □

Proposition 10.8.7. *Suppose that T is a finite group of Lie type of characteristic p: $T \in Chev(p)$. Then one of the following assertions holds.*

(i) *$T = L_\ell(\hat{q})$, $\hat{q} = p^t$, $t(\ell - 1) = 2m$, and $\ell \geq 3$.*

(ii) *$T = PSp_{2k}(\hat{q})$, $\hat{q} = p^t$, $p > 2$, and either*

a) *$2m = kt$,*

or

b) *there exists a natural number j such that $m = jt$ and $\frac{k}{2} \leq j < k$.*

(iii) *$T = Sp_{2(m+1)}(2)$ and $q = 4$ or $q = 2^m \geq 16$.*

(iv) *$T = PSU_\ell(\hat{q})$, $\hat{q} = p^t$, $m = t[\frac{\ell}{2}]$, and $\ell \geq 3$.*

(v) *$T = PSU_\ell(\hat{q})$, $\hat{q} = p^t$, and $m = tj$ for some odd integer j, $1 \leq \frac{\ell-1}{2} \leq j \leq \ell - 2$.*

(vi) *$T = \Omega_{2k+1}(\hat{q})$, $\hat{q} = p^t$, $k \geq 3$, $p > 2$, and $m = t(k-1)$.*

(vii) *$T = P\Omega^-_{2k}(\hat{q})$, $\hat{q} = p^t$, $k \geq 4$, $2 \nmid k$, and $m = t(k-1)$.*

Proof. Suppose that T is a finite group of Lie type defined over the field $\mathbb{F}_{\hat{q}}$, $\hat{q} = p^t$. As $G = E.S$ is a p-local subgroup of T, by the Borel-Tits Theorem G is contained in some maximal parabolic subgroup P of T. Using this inclusion and primitive prime divisors of $q^2-1=p^{2m}-1$ (if they exist), we can give an upper bound for the Lie rank of the group T. Whereupon, using the Landázuri-Seitz-Zalesskii lower bound for the degrees of projective representations of the Chevalley groups (Table A7), we are able to deduce the required contradiction in most cases.

1) $T=L_\ell(\hat{q})$, $\ell \geq 2$.

a) If $\ell=2$, then $\hat{q}=p^t \geq q^3=p^{3m} \geq 8$, and by Table A7,

$$d(T) \geq \frac{\hat{q}-1}{\gcd(2,\hat{q}-1)} \geq \frac{q^3-1}{\gcd(2,q-1)} > q^2,$$

contrary to (10.23). If $\ell=3$, then $\hat{q}^3=p^{3t} \geq q^3=p^{3m}$, and so $\hat{q} \geq q$. If, in addition, $\hat{q}>q$, then $\hat{q} \geq 8$ and $d(T) \geq \hat{q}^2-1>q^2$, a contradiction. So $\hat{q}=q$, and we arrive at conclusion (i) with $\ell=3$.

b) We can now suppose that $\ell \geq 4$. By (10.23) and Table A7, we have $p^{2m}-1 \geq d(T) \geq \hat{q}^{\ell-1}-1=p^{t(\ell-1)}-1$, that is,

$$t(\ell-1) \leq 2m. \tag{10.25}$$

In particular, $m>1$. Suppose that $q=8$. Then inequality (10.25) implies that $6 \geq t(\ell-1) \geq 3t$, that is, $t=1$ or $t=2$. If $t=2$, then $\ell=4$ and $T=SL_4(4)$. Hence, as a maximal parabolic subgroup of T, P can have only the following composition factors: $SL_2(4) \cong \mathbb{A}_5$, and $L_3(4)$. Recall that $G \leq P$, and so $S=SL_2(8)$ is a section of $SL_2(4)$ or $L_3(4)$, a contradiction. Therefore $t=1$, $\hat{q}=2$, $3<\ell \leq 7$. Since $S=SL_2(8)$ is not a section of $SL_5(2)$, we get that $\ell=7$. This means that possiblity (i) is realized with $(t,\ell,m)=(1,7,3)$.

Finally, suppose that $q \neq 8$, and consider a primitive prime divisor κ of $p^{2m}-1$. If $U=\mathbb{F}_{\hat{q}}^{\ell}$ denotes the natural (projective) T-module, then P is the stabilizer of some i-dimensional subspace of U, where $1 \leq i \leq \ell-1$. Since $G \subseteq P$, $S=SL_2(q)$ is a section of one of the nonabelian composition factors $L_i(\hat{q})$ and $L_{\ell-i}(\hat{q})$ of P. Hence, κ divides $\hat{q}^j-1$ for some j, $1 \leq j \leq \ell-1$. In particular, $jt \geq 2m$. From this and (10.25) it follows that $t(\ell-1)=2m$, as stated in (i). Note also that $i=1$ or $i=\ell-1$.

2) $T=PSp_{2k}(\hat{q})$, $k \geq 2$.

Let $U=\mathbb{F}_{\hat{q}}^{2k}$ denotes the natural (projective) symplectic $\mathbb{F}_{\hat{q}}T$-module. Then P is the stabilizer of some i-dimensional totally isotropic subspace of U, $1 \leq i \leq k$. Furthermore, every nonabelian composition factor of P is $A_i=L_i(\hat{q})$ or $B_i=PSp_{2k-2i}(\hat{q})$. Hence, $S=SL_2(q)$ is involved in A_i or B_i.

a) Suppose first that $p>2$. Then, from (10.23) and Table A7, it follows that

$$p^{2m}-1 \geq d(T)=\frac{\hat{q}^k-1}{2},\ p^{2m} \geq \frac{\hat{q}^k+1}{2} > p^{kt-1},$$

and so

$$2m \geq kt. \tag{10.26}$$

If $q = p$, then $m = 1$, $k = 2$, $t = 1$, and we obtain possibility (ii) a). If $q > p$, then we take a primitive prime divisor κ of $p^{2m} - 1$. Furthermore, if S is involved in A_i, then κ divides $\hat{q}^j - 1$ for some j, $1 \le j \le i \le k$. Hence, $2m$ divides it, and from (10.26) it follows that $2m = kt$ as stated in (ii). If S is involved in B_i, then κ divides $\hat{q}^{2j} - 1$ for some j, $1 \le j \le k - i \le k - 1$. Hence, m divides jt, and by (10.26) we have $m = jt$, $\frac{k}{2} \le j \le k - 1$. In other words, possibility (ii) b) is realized.

b) Now suppose that $p = 2$. If $q = 4$, then $d(T) \le 15$ by (10.23). If, in addition, $\hat{q} \ge 4$, or $\hat{q} = 2$ and $k \ge 4$, then

$$d(T) \ge \frac{\hat{q}^{k-1}(\hat{q}^{k-1} - 1)(\hat{q} - 1)}{2} > 15,$$

a contradiction. Furthermore,

$$G = \mathbb{Z}_2^4.SL_2(4) \not\to Sp_4(2).$$

Therefore $k = 3$ and $T = Sp_6(2)$, as stated in (iii).

If $q = 8$, then $d(T) \le 63$ by (10.23). If, in addition, $\hat{q} \ge 8$, or $\hat{q} = 4$ and $k \ge 3$, or $\hat{q} = 2$ and $k \ge 5$, then

$$d(T) \ge \frac{\hat{q}^{k-1}(\hat{q}^{k-1} - 1)(\hat{q} - 1)}{2} > 63,$$

a contradiction. Furthermore,

$$G = \mathbb{Z}_2^6 \circ SL_2(8) \not\to Sp_4(2), Sp_4(4), Sp_6(2).$$

Finally, the last case $T = Sp_8(2)$ is impossible in view of (10.23). Therefore $q \ne 8$.

We can now suppose that $q > 8$, and take a primitive prime divisor κ of $q^2 - 1$. Moreover, from (10.23) and Table A7 it follows that

$$2^{2m} - 1 \ge d(T) \ge \frac{\hat{q}^{k-1}(\hat{q}^{k-1} - 1)(\hat{q} - 1)}{2} \ge 2^{t(2k-1)-3},$$

that is,

$$2m \ge t(2k - 1) - 2. \tag{10.27}$$

Recall that $S = SL_2(q)$ is involved in A_i or B_i. In the former case, κ divides $\hat{q}^j - 1$ for some j, $1 \le j \le i \le k$. Hence, $2m$ divides jt, contrary to (10.27). In the latter case, κ divides $\hat{q}^{2j} - 1$ for some j, $1 \le j \le k - i \le k - 1$. Hence, m divides jt, and from (10.23) and (10.27) it follows that $t = 1$, $j = k - 1$, $k = m + 1$. Thus we arrive at conclusion (iii). Note that in this case we also have $i = 1$.

3) $T = PSU_\ell(\hat{q})$, $\ell \ge 3$; $\Omega_{2k+1}(\hat{q})$, $k \ge 3$; $P\Omega_{2k}^{\pm}(\hat{q})$, $k \ge 4$.

These cases are treated like 2).

4) $T = {}^2B_2(\hat{q})$, ${}^2G_2(\hat{q})$.

In these cases, all the parabolic subgroups are soluble, and so the inclusion $G \subseteq P$ is impossible.

5) $T = {}^3D_4(\hat{q})$.

Here, T has two conjugacy classes of maximal parabolic subgroups P, namely, $P/O_p(P)$ is isomorphic to one of the groups

$$(SL_2(\hat{q}^3) * \mathbb{Z}_{\hat{q}-1}).\mathbb{Z}_e,\ (SL_2(\hat{q}) * \mathbb{Z}_{\hat{q}^3-1}).\mathbb{Z}_e,$$

where $e = \gcd(2, \hat{q}-1)$. Furthermore, by [DeM] we have

$$p^{2m} - 1 \geq d(T) = \hat{q}(\hat{q}^4 - \hat{q}^2 + 1) \geq \frac{\hat{q}^5}{2} \geq p^{5t-1},$$

that is, $2m \geq 5t$. Hence, $L_2(q)$ must be involved in $SL_2(\hat{q}^3)$, and so $q = \hat{q}^3$. However, according to [DeM], T has no nontrivial irreducible characters of degree dividing $q^2 - 1 = \hat{q}^6 - 1$, and this contradicts (10.23).

6) $T = G_2(\hat{q})$: similar to 5).

7) $T = {}^2E_6(\hat{q})$.

From (10.23) and Table A7 it follows that

$$p^{2m} - 1 \geq d(T) \geq \hat{q}^9(\hat{q}^2 - 1) \geq \frac{\hat{q}^{11}}{2} \geq p^{11t-1},$$

that is,

$$2m \geq 11t. \tag{10.28}$$

Furthermore, T has four conjugacy classes of maximal parabolic subgroups P, namely, $P/O_p(P)$ is isomorphic to one of the groups

$${}^2D_4(\hat{q}),\ A_2(\hat{q}) + A_1(\hat{q}^2),\ A_2(\hat{q}^2) + A_1(\hat{q}),\ {}^2A_5(\hat{q}).$$

Now let κ be a primitive prime divisor of $q^2 - 1$. Then κ divides $\hat{q}^j - 1$ for some j, $j \leq 10$, since $G \subseteq P$. Hence, $2m$ divides jt, contradicting (10.28).

8) $T = F_4(\hat{q})$, ${}^2F_4(\hat{q})$, $E_6(\hat{q})$, $E_7(\hat{q})$, $E_8(\hat{q})$.

These cases are considered like 7), and this completes the proof of Proposition 10.8.7. □

Below we analyze in detail the possibilities mentioned in the last proposition.

Lemma 10.8.8. *Case* (i) *formulated in Proposition* 10.8.7 *cannot occur.*

Proof. Suppose the contrary: $T = L_\ell(\hat{q})$, $\hat{q} = p^t$, $t(\ell - 1) = 2m$, $\ell \geq 3$. If $\ell = 3$, then $\hat{q} = q$, and, by [SiF], we have

$$d(T) = q(q+1) > q^2,$$

contrary to (10.23). If $\ell = 4$, then $2m = 3t$, and so m is divisible by 3. In particular, $T \neq L_4(2), L_4(3)$. Hence, when $\ell \geq 4$ we can use the improved lower bound for $d(T)$ found by Seitz and Zalesskii [SeZ] and reproduced in Table A7:

$$d(T) \geq \frac{\hat{q}^{\ell} - 1}{\hat{q} - 1} - \ell \geq \hat{q}^{\ell-1} + \hat{q}^{\ell-2} - \ell > \hat{q}^{\ell-1} = q^2,$$

again contrary to (10.23). □

Lemma 10.8.9. *Possibility* (ii) *formulated in Proposition* 10.8.7 *cannot occur.*

Proof. In the notation of Proposition 10.8.7, suppose that $T = PSp_{2k}(\hat{q})$, $\hat{q} = p^t$ and $p > 2$. It is well known (see e.g. [War]) that $d(T) = \frac{\hat{q}^k - 1}{2}$. Namely, $\hat{T} = Sp_{2k}(\hat{q})$ has a certain irreducible complex representation Ψ of degree $\frac{\hat{q}^k - 1}{2}$ whose kernel is contained in $Z(\hat{T}) = \mathbb{Z}_2$.

Let $\hat{G}$ be the full preimage of G in $\hat{T}$: $\hat{G} = \mathbb{Z}_2.G$. Recall that $G = E.S$ and that the Schur multiplier of S is

$$Mult(S) = \begin{cases} 1 & \text{if} \quad q \neq 4, 9, \\ \mathbb{Z}_p & \text{if} \quad q = 4, 9, \end{cases}$$

(see [Gor 2]). Denote by $\hat{E}$ and $\hat{S}$ the full preimages of E and S in $\hat{T}$. Since $p > 2$, $\hat{S} = \mathbb{Z}_2 \times S$. Furthermore, by Sylow's Theorem, $\hat{E} = \mathbb{Z}_2 \times E$, so that $\hat{G} = \mathbb{Z}_2 \times G$. In particular, $\Psi|_G$ is a faithful complex representation of G. Applying Lemma 10.6.5 we obtain on the one hand that

$$\frac{\hat{q}^k - 1}{2} = \deg \Psi \geq q^2 - 1.$$

On the other hand, (10.23) means that

$$\frac{\hat{q}^k - 1}{2} = d(T) \leq q^2 - 1.$$

Hence,

$$\frac{p^{kt} - 1}{2} = \frac{\hat{q}^k - 1}{2} = q^2 - 1 = p^{2m} - 1, \; p^{kt} - 2p^{2m} = -1,$$

a contradiction. □

Lemma 10.8.10. *Possibility* (iii) *formulated in Proposition* 10.8.7 *cannot occur.*

Proof. Suppose that $T = Sp_{2(m+1)}(2)$. Recall that, in the notation of the proof of Proposition 10.8.7,

$$G = E.S \subseteq P = Q.T_1,$$

where $Q = \mathbb{Z}_2^{1+2m}$, $T_1 = Sp_{2m}(2)$, and P is split over Q.

Suppose first that $q = 4$. Then $T = Sp_6(2)$, and so T has a faithful irreducible complex representation Ψ of degree 7. Clearly, Ψ_G is also faithful, so by Lemma 10.6.5 we get

$$\deg \Psi \geq q^2 - 1 = 15,$$

a contradiction.

Thus, we can suppose that $q = 2^m \geq 16$. In particular, one can take a primitive prime divisor κ of $q^2 - 1$, and $Q \cap G = Q \cap E$ is a normal 2-subgroup of G. So, by the minimality of E, $Q \cap G$ is 1 or E. In the former case,

$$G \cong QG/Q \subseteq P/Q = T_1,$$

that is, G can be regarded as a 2-local subgroup of $T_1 = Sp_{2m}(2)$. Using the Borel-Tits Theorem, one can show that κ divides $\prod_{1 \leq j \leq m-1}(2^{2j} - 1)$, a contradiction. Hence, $Q \cap G = E$. It is easy to see that P contains some central transvection $\mathbf{j}$ such that $\mathbb{Z}_2 \cong J =< \mathbf{j} > \subseteq Q$. Thus we can consider the subgroup $JG = J \times G$. As $Q \subseteq JG \subseteq P = Q.T_1$, JG is split over Q, say, $JG = Q.S_1$. Here

$$S_1 = JG/Q \cong (JG/J)/(Q/J) \cong G/(Q \cap G) = G/E = S.$$

Furthermore, $S_1 = [S_1, S_1] \subseteq [JG, JG] = [G, G] = G$. Thus, G contains subgroup S_1 isomorphic to $S = G/E$, that is, G is split over E, contrary to Theorem 1.3.7. □

Lemma 10.8.11. *Possibilities* (iv) *and* (v) *formulated in Proposition* 10.8.7 *cannot occur.*

Proof. Suppose that $T = PSU_\ell(\hat{q})$, $\hat{q} = p^t$, $\ell \geq 3$. First we assume in addition that $(\ell, \hat{q}) \neq (4, 2), (4, 3)$. Let ς denote the unipotent irreducible complex character of T that is parametrized by the partition $(1, \ell - 1)$ of the number ℓ (see [Car]). Then

$$\deg \varsigma = \frac{\hat{q}(\hat{q}^{\ell-1} - (-1)^{\ell-1})}{\hat{q} + 1},$$

and $\varsigma|_G$ is faithful. By Lemma 10.6.5 we have

$$p^{2m} - 1 \leq \frac{\hat{q}(\hat{q}^{\ell-1} - (-1)^{\ell-1})}{\hat{q} + 1} < \hat{q}^{\ell-1} - 1,$$

that is, $p^{2m} < \hat{q}^{\ell-1}$, $p^{2m+1} \leq \hat{q}^{\ell-1}$. But in this case, by Table A7 we have

$$d(T) \geq \frac{\hat{q}(\hat{q}^{\ell-1} - 1)}{\hat{q} + 1} \geq \frac{\hat{q}(p^{2m+1} - 1)}{\hat{q} + 1} > \frac{p\hat{q}(p^{2m} - 1)}{\hat{q} + 1} > p^{2m} - 1,$$

contrary to (10.23).

The case $(\ell, \hat{q}) = (4, 2)$ is excluded by Proposition 10.8.5, since $PSU_4(2) \cong PSp_4(3)$. Finally, if $(\ell, \hat{q}) = (4, 3)$, then

$$G \subseteq T = PSU_4(3) \hookrightarrow GL_{21}(\mathbb{C}).$$

By Lemma 10.6.5, we must have that $q^2 - 1 = 3^{2m} - 1 \leq 21$, $m = 1$, $q = 3$, which contradicts our assumption that $q \geq 4$. □

Before analyzing possibilities (vi) and (vii) of Proposition 10.8.7, we establish several auxiliary statements.

Lemma 10.8.12. *Let L be a finite simple group of type D_n or ${}^2D_n$ defined over a field $\mathbb{F}_s$, and $\nu = 1$ or -1, respectively. Furthermore, let $M = SO^{\pm}_{2n}(s)$ denote the corresponding special orthogonal group. Then the following assertions hold.*

(i) *If $2|s$ and $M \neq SO^+_4(2)$, then $M = L.\mathbb{Z}_2$;*

(ii) *if $2 \not| \, s$ and $n(s-1) \equiv \nu + 1 \pmod 4$, then $M = L \times \mathbb{Z}_2$. If $2 \not| \, s$ and $n(s-1) \equiv \nu - 1 \pmod 4$, then $M = (\mathbb{Z}_2.L).\mathbb{Z}_2$.*

Proof. Use the following equality [KlL 2]:

$$|M| : |L| = 2\frac{\gcd(4, s^n - \nu)}{\gcd(2, s-1)}.$$

□

Lemma 10.8.13. *Suppose that $q = \hat{q}^k = p^{kt}$, where p is an odd prime and $q \geq 5$. Then the group $S = SL_2(q)$ cannot be embedded in the orthogonal groups $O^{\pm} = O^{\pm}_{2k}(\hat{q})$.*

Proof. If $k = 1$, then p divides $|S|$, but not $|O^{\pm}|$, therefore $S \not\hookrightarrow O^{\pm}$. If $(k, t) = (2, 1)$, then

$$|S| = p^2(p^4 - 1), \; |O^+| = 2p^2(p^2 - 1),$$

and so $S \not\hookrightarrow O^+$. Furthermore, as $p > 2$ we have

$$O^- = O^-_4(p) = (\mathbb{Z}_2 \times L_2(p^2)).\mathbb{Z}_2, \; S = SL_2(p^2),$$

and so $S \not\hookrightarrow O^-$.

We can therefore suppose that $k \geq 2$ and $(k, t) \neq (2, 1)$. In particular, there exist primitive prime divisors κ (of $\hat{q}^{2k} - 1$) and κ' (of $\hat{q}^k - 1$). Since κ does not divide $|O^+|$, $S \not\hookrightarrow O^+$. Suppose that $S \hookrightarrow O^-$. Then $\hat{q}^k - 1$ divides $\prod_{1 \leq j \leq k-1}(\hat{q}^{2j} - 1)$, and so κ' divides $\hat{q}^{2j} - 1$ for some j, $1 \leq j \leq k-1$. This means that $k = 2j$, and, in particular, k is even. In view of Lemma 10.8.12, we have $S \subseteq SO^-_{2k}(\hat{q}) = \mathbb{Z}_2 \times L$, where $L = P\Omega^-_{2k}(\hat{q})$ is a finite simple group. Let $U = \mathbb{F}^{2k}_{\hat{q}}$ denote the natural $\mathbb{F}_{\hat{q}}O^-$-module and $\mathbf{t}'$ the central involution in S. Then $U = U_+ \oplus U_-$, where the subspaces $U_+ = \mathrm{Ker}(\mathbf{t}' - 1_U)$ and $U_- = \mathrm{Ker}(\mathbf{t}' + 1_U)$ are S-invariant. Clearly,

$U_+ \neq U$. In fact, using κ one can show that $U_+ = 0$, $U_- = U$. This means that $\mathbf{t}'$ is a central involution in $SO^-_{2k}(\hat{q})$, and $SO^-_{2k}(\hat{q}) =< \mathbf{t}' > \times L$. But in this case, $S = SL_2(q)$ is split over its centre $< \mathbf{t}' >= \mathbb{Z}_2$, a contradiction. □

Lemma 10.8.14. *Case* (vi) *formulated in Proposition* 10.8.7 *cannot occur.*

Proof. Suppose that $T = \Omega_{2k+1}(\hat{q})$, $k \geq 3$, $p > 2$, in the notation of Proposition 10.8.7. Recall that we showed in the proof of this proposition that $i = 1$ and $q = \hat{q}^{k-1}$. In fact, one can see that $G \subseteq Q.T_1$, where Q is an elementary abelian group of order $\hat{q}^{2k-1}$, $T_1 = SO_{2k-1}(\hat{q})$, and the extension $Q.T_1$ is split. Let κ denote a primitive prime divisor of $q^2 - 1$. Then by definition κ does not divide $\prod_{1 \leq j < k-1}(\hat{q}^{2j} - 1)$, and so $G \not\hookrightarrow T_1$, $Q \cap G = E$, and

$$S = G/E = G/(Q \cap G) \cong QG/Q \subseteq T_1.$$

Consider again the central involution $\mathbf{t}'$ of S. If $U = \mathbb{F}_{\hat{q}}^{2k-1}$ denotes the natural $\mathbb{F}_{\hat{q}}T_1$-module, then $U = U_+ \oplus U_-$, where both of the subspaces $U_+ = \mathrm{Ker}(\mathbf{t}' - 1_U)$, $U_- = \mathrm{Ker}(\mathbf{t}' + 1_U)$ are S-invariant. Clearly, $U_+ \neq U$. Furthermore,

$$(-1)^{\dim U_-} = \det(\mathbf{t}'|_U) = 1,$$

that is, $\dim U_- = 2s$, $1 \leq s \leq k-1$. Now $\dim U_+ = 2k - 1 - 2s \leq 2k - 3$, and κ does not divide $|O_{2k-3}(\hat{q})|$, so that S acts trivially on U_+. But S acts faithfully on U, hence S acts faithfully on U_-. This means that $S \hookrightarrow O(U_-) = O_{2s}(\hat{q})$. The last inclusion is impossible, by Lemma 10.8.13. □

Lemma 10.8.15. *Case* (vii) *formulated in Proposition* 10.8.7 *cannot occur.*

Proof. Suppose that $T = P\Omega^-_{2k}(\hat{q})$, $q = \hat{q}^{k-1}$, $k \geq 4$ in the notation of Proposition 10.8.7. Recall that in the proof of this proposition we showed that $G \subseteq P$ and $i = 1$. Let $\hat{P}$ and $\hat{G}$ denote the full preimages of P and G in $\hat{T} = SO^-_{2k}(\hat{q})$. Then $\hat{P} = Q.T_1$, where Q is an elementary abelian group of order $\hat{q}^{2(k-1)}$, $T_1 = \mathbb{Z}_{\hat{q}-1} \times SO^-_{2k-2}(\hat{q})$, and $\hat{P}$ is split over Q.

Assume first that $p > 2$. Then $P = \hat{P}/ < \mathbf{t}' >$, where $\mathbf{t}'$ is a central involution of $\hat{T}$. But the Schur multiplier of S is contained in $\mathbb{Z}_p$, therefore $\hat{G} =< \mathbf{t}' > \times G$. Furthermore, $Q \cap \hat{G}$ is a normal p-subgroup of $\hat{G}$; hence, $Q \cap \hat{G}$ is 1 or E. In the former case,

$$G \subseteq \hat{G} = \hat{G}/(Q \cap \hat{G}) \cong Q\hat{G}/Q \subseteq \hat{P}/Q = T_1.$$

This implies that

$$S \subset G = [G, G] \subseteq [T_1, T_1] \subseteq SO^-_{2k-2}(\hat{q}).$$

In the latter case,

$$S \subseteq \hat{G}/E = \hat{G}/(Q \cap \hat{G}) \cong Q\hat{G}/Q \subseteq \hat{P}/Q = T_1.$$

This implies that

$$S = [S, S] \subseteq [T_1, T_1] \subseteq SO^-_{2k-2}(\hat{q}).$$

In all cases, we obtain an embedding $SL_2(\hat{q}^{k-1}) \hookrightarrow SO^-_{2k-2}(\hat{q})$, contrary to Lemma 10.8.13.

Now suppose that $p = 2$. Then G and P are contained in $Q.T_2$, where $T_2 = \mathbb{Z}_{\hat{q}-1} \times O^-_{2k-2}(\hat{q})$, and $Q.T_2$ is split over Q. Furthermore, $q = \hat{q}^{k-1} \geq 8$, therefore $G = E.S$ is non-split over E. As $Q \cap G$ is a normal 2-subgroup of G, $Q \cap G$ is 1 or E. If $Q \cap G = E$, then in fact $Q = E$, $G = E.S \subseteq E.T_2$, and so G is split over E, a contradiction. Therefore, $Q \cap G = 1$, and

$$G = G/(Q \cap G) \cong QG/Q \subseteq P/Q = T_2.$$

This implies that

$$G = [G, G] \subseteq [T_2, T_2] \subseteq O^-_{2k-2}(\hat{q}).$$

If in addition $q = 8$, then $\hat{q} = 2$, $k = 4$. In this case, 7 divides $|G|$, but not $|O^-_{2k-2}(\hat{q})|$, a contradiction. Assume that $q > 8$. Then there exists a primitive prime divisor κ of $q^2 - 1$. Since G is a 2-local subgroup of $O^-_{2k-2}(\hat{q})$, by the Borel-Tits Theorem G is contained in some maximal parabolic subgroup of $O^-_{2k-2}(\hat{q})$. As a consequence of this fact we obtain that κ divides $\prod_{1 \leq j \leq k-2}(\hat{q}^{2j} - 1)$, again a contradiction. □

Lemma 10.8.15 completes the proof of Theorem 10.6.2. □

Commentary

1) If difficulties arise in the first reading of the first five sections, the reader is recommended to return to Chapter 9 and project the general construction on the special case $m = 1$. Generally speaking, it should be pointed out that the material of §§10.1 – 10.5 can be regarded as an attempt to carry over the results of Chapter 9, which themselves require some "polishing", to the case A_{p^m-1}. Of course, in the case $m > 1$ we dealt with many additional obstacles, which were overcome to a large extent by Abdukhalikov, whose results [Abd 1, 3] are reproduced in this Chapter.

2) As mentioned in the text (see §10.2), the classification of projections of invariant lattices (Theorem 10.2.4) is valid for all primes p. Examples 10.2.5 and 10.2.6 are due to the efforts of Abdukhalikov. Meanwhile, Theorem 10.4.6 enables us to classify invariant lattices (in principle) *only in the case* $p > 2$, since when $p = 2$ the group $\mathrm{Inn}(\mathcal{D}) = \mathbb{Z}_p^{2m}.\Sigma L_2(p^m)$ is non-split (Theorem 1.3.7). However, the results concerning $SL_2(q)$-modules explained in §§10.3, 10.4 might be useful

for the case $p = 2$. We do not know nontrivial examples of groups $\mathbb{G}$ such that $\mathbb{Z}_2^{2m} \circ SL_2(2^m) \subseteq \mathbb{G} \subset GL_{2^{2m}-1}(\mathbb{Z})$.

We remark also that the description of projections of invariant lattices given in §10.2 is very inconvenient in practice, because of the implicit construction of the "truncated logarithms" $L_i(x)$. It would be alluring to fill all the gaps mentioned above.

The reason for not unifying the approaches to the cases $m = 1$ and $m > 1$ was that we wanted to keep the style of the original papers [BKT], [Abd 3].

3) In §10.5 several unimodular lattices are constructed that lie in the upper "floors" of the "tower" of $SL_2(q)$-modules (recall Figures 9.1 and 9.2!). Of course, they do not exhaust the set of all unimodular invariant lattices. On the other hand, they are *new* examples of root-free indecomposable unimodular lattices of dimension $p^{2m} - 1$ ($m > 1$): by Theorem 10.6.2, they are isometric neither to the Gow-Gross lattices [Gow 2], [Gro], nor to any other known lattices.

4) As mentioned at the beginning of §10.6, the Reduction Theorem (Theorem 10.6.2) highlights the importance of studying the automorphism groups $\mathrm{Aut}(\Lambda_1)$, where Λ_1 is a projection of an invariant lattice onto a Cartan subalgebra. In the case of A_{p-1}, this problem is completely settled in §9.6. In the general case of Lie algebras of type A_{p^m-1}, this problem is still open, in spite of the classification given in §10.2. It seems almost impossible that this problem will be solved in the near future.

As a nontrivial example, we consider the case $q = p^2$. It is well known that the Frobenius group $H = \mathbb{Z}_p^2.\mathbb{Z}_{p^2-1}$ has a unique absolutely irreducible rational representation of degree $p^2 - 1$: $H \hookrightarrow GL_{p^2-1}(\mathbb{Q})$. The problem mentioned above is reduced to that of finding all intermediate groups $\mathbb{H}$: $H \subseteq \mathbb{H} \subset GL_{p^2-1}(\mathbb{Z})$. All the automorphism groups $\mathrm{Aut}(\Lambda_1)$ occur as groups of this type.

The nontriviality of the given example is clear from the observation that it includes all the invariant lattices $\Gamma^{k,l}$ considered in Chapter 9.

Chapter 11

The types B_{2^m-1} and D_{2^m}

In this chapter we study the automorphism groups of invariant lattices arising in connection with multiplicative orthogonal decompositions (MODs) of Lie algebras $\mathcal{L}$ of type B_{2^m-1} and D_{2^m} $(m \geq 2)$. The similar problem involving classification of lattices is completely settled for the Lie algebra of type G_2 (see Chapter 8). Regarding the two other Lie algebras that admit MODs, the case of A_1 is trivial, and the case of E_8 will be considered in Chapter 13. Unlike the case G_2, where all the invariant lattices are classified, it is difficult to give a complete classification of invariant lattices for B_{2^m-1} and D_{2^m}. However, we give a complete description of their aumorphism groups that requires only a rough partition of the set of invariant lattices into the so-called lattices of *root-type* and *nonroot-type* (precise definitions will be given later).

The main results of the chapter are the following theorems.

Theorem 11.0.1. *Let $\mathbb{G}$ be the automorphism group of an invariant lattice of root-type in the Lie algebra $\mathcal{L}$ of type D_{2^m} $(m \geq 2)$, and set $N = \dim \mathcal{L}$. Then one of the following assertions holds.*

(i) $m = 3$ *and* $\mathbb{G} \subseteq W(E_8) \wr \mathbb{S}_{15}$.

(ii) $\mathbb{G} = E_0.(E_1.\mathbb{G}_1)$, *where* E_0 *is an elementary abelian normal* 2*-subgroup if* $m > 2$, *and a soluble normal* $\{2,3\}$*-group if* $m = 2$; $E_1 \subseteq (\mathbb{S}_{2^m})^{2^{m+1}-1}$; *and* $\mathbb{G}_1$ *is one of* $SL_{m+1}(2)$, $\mathbb{A}_{2^{m+1}-1}$, *and* $\mathbb{S}_{2^{m+1}-1}$.

(iii) $\mathbb{G} = E_0.\mathbb{G}_1$, *where* E_0 *is an elementary abelian normal* 2*-subgroup and* $\mathbb{G}_1$ *satisfies one of the following conditions:*

a) $\mathbb{A}_N \lhd \mathbb{G}_1 \subseteq \mathbb{S}_N$,

b) $\mathbb{A}_{2^{m+1}} \lhd \mathbb{G}_1 \subseteq \mathbb{S}_{2^{m+1}}$,

c) $m = 2$ *and* $\mathbb{G}_1 = Sp_6(2)$.

All these possibilities occur.

Theorem 11.0.2. *Let $\mathbb{G}$ be the automorphism group of an invariant lattice of root-type in the Lie algebra $\mathcal{L}$ of type B_{2^m-1} $(m \geq 2)$, and set $N = \dim \mathcal{L}$. Then one of the following assertions holds.*

(i) $m = 3$ *and* $\mathbb{G} \subseteq W(E_7) \wr \mathbb{S}_{15}$.

(ii) $\mathbb{G} = E_0.(E_1.\mathbb{G}_1)$, *where* E_0 *is an elementary abelian normal 2-subgroup;* $E_1 \subseteq (\mathbb{S}_{2^m-1})^{2^{m+1}-1}$; *and* $\mathbb{G}_1$ *is one of* $SL_{m+1}(2)$, $\mathbb{A}_{2^{m+1}-1}$, *and* $\mathbb{S}_{2^{m+1}-1}$.

(iii) $\mathbb{G} = E_0.\mathbb{G}_1$, *where* E_0 *is a soluble normal* $\{2,3\}$*-subgroup and* $\mathbb{G}_1$ *satisfies one of the following conditions:*

a) $\mathbb{A}_{\frac{N}{3}} \lhd \mathbb{G}_1 \subseteq \mathbb{S}_{\frac{N}{3}}$,

b) $m = 2$ *and* $L_2(7) \cong SL_3(2) \lhd \mathbb{G}_1 \subseteq PGL_2(7)$,

c) $m = 3$ *and* $\mathbb{A}_8 \cong SL_4(2) \lhd \mathbb{G}_1 \subseteq \mathbb{S}_8$,

d) $m > 3$ *and* $\mathbb{G}_1 = SL_{m+1}(2)$.

(iv) $\mathbb{G} = E_0.\mathbb{G}_1$, *where* E_0 *is an elementary abelian normal 2-subgroup and* $\mathbb{G}_1$ *is one of* $\mathbb{A}_N$, $\mathbb{S}_N$, $\mathbb{A}_{2^{m+1}-1}$, *and* $\mathbb{S}_{2^{m+1}-1}$.

All these possibilities occur.

Theorem 11.0.3. *The Lie algebra* $\mathcal{L}$ *of type* B_{2^m-1} *admits an invariant lattice of nonroot-type if and only if* $m = 2$. *Furthermore, the automorphism group of every invariant lattice of nonroot-type is isomorphic to* $\mathbb{Z}_2 \times Sp_6(2)$.

Theorem 11.0.4. *The Lie algebra* $\mathcal{L}$ *of type* D_{2^m} *admits an invariant lattice of nonroot-type if and only if* $m = 2$. *Furthermore, the automorphism group of every invariant lattice of nonroot-type is isomorphic to* $\mathbb{Z}_2 \times (D_4(2).\mathbb{Z}_2)$ *or* $\mathbb{Z}_2.\mathrm{Aut}(D_4(2))$.

11.1 Preliminaries

Let us recall the construction of MODs of the Lie algebra $\mathcal{L}$ of types D_{2^m} and B_{2^m-1} ($m \geq 2$) (see Chapter 2). To this end, we realize the Lie algebra $\mathcal{L}$ of type D_{2^m} by skew-symmetric matrices of degree ℓ, where $\ell = 2^{m+1}$:

$$\mathcal{L} = \langle E(i,j) | i, j = 1, 2, \ldots, \ell; i \neq j \rangle_{\mathbb{C}}.$$

Here $E(i,j) = E_{ij} - E_{ji}$, and $E_{ij} = (\delta_{r,i}\delta_{s,j})_{1 \leq r,s \leq \ell}$ are the standard matrix units. Set $V = \mathbb{F}_2^{m+1}$, $V^0 = V$ and identify the set $\{1, 2, \ldots, \ell\}$ with V^0. Then the MOD $\mathcal{D}$ of $\mathcal{L}$ is of the form

$$\mathcal{D} : \mathcal{L} = \bigoplus_{z \in V \setminus \{0\}} \mathcal{H}_z,\ \mathcal{H}_z = \langle E(u,v) | u + v = z \rangle_{\mathbb{C}}.$$

The automorphism group of $\mathcal{D}$ is $G = E.G_0$, where

$$E = \left\{ \bar{f} : E(u,v) \mapsto (-1)^{f(u)+f(v)} E(u,v) | f : V^0 \to \{0,1\} \right\},$$

$$G_0 = \{ \bar{\varphi} : E(u,v) \mapsto E(\varphi(u), \varphi(v)) | \varphi \in ASL(V) \},$$

and $ASL(V) \cong ASL_{m+1}(2)$ denotes the group of affine transformations of the (affine) space V. (If $m = 2$, then $\mathrm{Aut}(\mathcal{D}) = \langle G, T \rangle$, where T is an outer automorphism of order 3 of $\mathcal{L}$). By an *invariant lattice (of type D_{2^m})* we mean any $\mathbb{Z}G$-module Γ in $\mathcal{L}$ such that the Killing form K restricted to Γ is integer-valued and (positive or negative) definite.

In the case where $\mathcal{L}$ is of type B_{2^m-1}, we have to set $\ell = 2^{m+1} - 1$, $V^0 = V\backslash\{0\}$, $G_0 = \{\bar{\varphi}|\varphi \in SL(V)\}$. In this way we obtain an embedding of the Lie algebra of type B_{2^m-1} in the Lie algebra of type D_{2^m}.

Set $\Omega = \{\{a, b\}|a, b \in V^0, a \neq b\}$. Consider the automorphisms $\bar{f}_a \in E$, $a \in V^0$ given by $f_a(v) = \delta_{a,v}$, and let Γ be an arbitrary invariant lattice. Then

$$\pi_a = 1_{\mathcal{L}} - \bar{f}_a \in \mathrm{End}_{\mathbb{Z}}(\Gamma).$$

Furthermore,

$$\pi_a(E(u, v)) = \begin{cases} 2E(u, v) & \text{if} \quad a \in \{u, v\} \\ 0 & \text{if} \quad a \notin \{u, v\} \end{cases},$$

$$\pi_a\pi_b(E(u, v)) = \begin{cases} 4E(u, v) & \text{if} \quad \{a, b\} = \{u, v\} \\ 0 & \text{if} \quad \{a, b\} \neq \{u, v\} \end{cases},$$

where $a, b \in V^0$ and $a \neq b$. We introduce the set

$$I_{ab} = \left\{\lambda \in \mathbb{C}\big|\exists X = \sum_{c,d} r_{cd}E(c, d) \in \Gamma,\ r_{ab} = \lambda\right\}.$$

Since $E(a, b) = -E(b, a)$ and G_0 acts transitively on Ω, we have $I_{ab} = I$ whenever $a \neq b$. Set

$$\Lambda(I) = \left\{X = \sum_{a,b} r_{ab}E(a, b)\big|r_{ab} \in I\right\}.$$

Clearly, $\pi_a\pi_b(X) = 4r_{ab}E(a, b) \in \Gamma$ for all $X = \sum_{a,b} r_{ab}E(a, b) \in \Gamma$. Hence, $4\Lambda(I) \subseteq \Gamma \subseteq \Lambda(I)$.

The Killing form K of $\mathcal{L}$ is proportional to the form $(*|*)$, where

$$(E(a, b)|E(c, d)) = \begin{cases} 0 & \text{if} \quad \{a, b\} \neq \{c, d\} \\ 1 & \text{if} \quad a = c, b = d \end{cases}.$$

As Γ is a lattice, the $\mathbb{Z}$-module I satisfies the following condition:

$$(4rE(a, b)|4rE(a, b)) = 16r^2 \in \mathbb{Z}_+,\ \forall r \in I,$$

where $\mathbb{Z}_+ = \{n \in \mathbb{Z}|n \geq 0\}$. The following simple assertion holds for such $\mathbb{Z}$-modules I.

Lemma 11.1.1. *Let J be a $\mathbb{Z}$-module in $\mathbb{C}$ such that $r^2 \in \mathbb{Z}_+$ for all $r \in J$. Then $I = \sqrt{n}\mathbb{Z}$ for some $n \in \mathbb{Z}_+$.* □

Set

$$\Lambda = \langle E(u, v) | \{u, v\} \in \Omega \rangle_{\mathbb{Z}} .$$

We have shown that up to similarity one can suppose that any invariant lattice Γ satisfies the inclusion $4\Lambda \subseteq \Gamma \subseteq \Lambda$. For this reason, we call Λ a *basic lattice*. From now on, by an *invariant lattice* we mean any G-invariant lattice Γ with $4\Lambda \subseteq \Gamma \subseteq \Lambda$. Furthermore, an invariant lattice Γ is said to be *divisible* if $2\Lambda \supseteq \Gamma$, and *indivisible* otherwise.

We introduce at this point the following definition playing a major role in our further investigation.

Definition 11.1.2. An invariant lattice Γ is said to be a lattice of *root-type* (briefly: $\Gamma \in \mathcal{R}$), if Γ contains an Aut(Γ)-invariant sublattice Γ', which is a root lattice in the classical sense. Otherwise Γ is called a lattice of *nonroot-type* (briefly: $\Gamma \in \overline{\mathcal{R}}$).

Note that if $\Gamma \in \mathcal{R}$, then the sublattice Γ' mentioned in the definition is a root lattice of maximal rank. In fact, Aut(Γ') $\supseteq G$, so that Γ' is G-invariant, which implies that $\Gamma' \supseteq r\Lambda$ for some $r \in \mathbb{Q}^*$.

Lemma 11.1.3. *Let* Γ_1 *be an* Aut(Γ)*-invariant sublattice of* Γ, *and suppose that* $\Gamma_1 \in \mathcal{R}$. *Then* $\Gamma \in \mathcal{R}$.

Proof. Take an Aut(Γ_1)-invariant root sublattice Γ_2 in Γ_1. Then $\Gamma_2 \subseteq \Gamma$, and Γ_2 is clearly Aut(Γ)-invariant. Hence, $\Gamma \in \mathcal{R}$. □

Now in an arbitrary indivisible sublattice Γ we set

$$\mathcal{U} = \{X \in \Gamma \mid (X|X) = 16,\ (X|\Gamma) \subseteq 4\mathbb{Z}\},\ \Gamma_r = < \mathcal{U} >_{\mathbb{Z}} .$$

In what follows, the sublattice Γ_r will be called the *reduced sublattice* of Γ. Clearly, Γ_r contains 4Λ, is Aut(Γ)-invariant, and satisfies the following conditions:

P1) $(X|Y) \in 4\mathbb{Z}$ for all $X, Y \in \Gamma_r$;

P2) $(X|X) \in 8\mathbb{Z}$ for all $X \in \Gamma_r$.

Next, one can distinguish the following cases

P3) $\exists X \in \Gamma_r,\ (X|X) = 8$;

P4) $\nexists X \in \Gamma_r,\ (X|X) = 8$.

Case 4) splits into the two following subcases:

P4a) $\mathcal{U} \subset 2\Lambda$;

P4b) $\mathcal{U} \not\subseteq 2\Lambda$.

Denote by $\mathcal{R}_1$ (respectively, $\mathcal{R}_2$, $\mathcal{R}_3$) the set of indivisible sublattices whose reduced sublattice satisfies condition P3) (respectively, conditions P4), P4a), and P4), P4b)).

Lemma 11.1.4. $\overline{\mathcal{R}} \subseteq \mathcal{R}_3$.

Proof. Suppose that $\Gamma \notin \mathcal{R}_3$, so that $\Gamma \in \mathcal{R}_1$ or $\Gamma \in \mathcal{R}_2$. If $\Gamma \in \mathcal{R}_1$, then $\Gamma_2 = \langle X \in \Gamma_r | (X|X) = 8\rangle_{\mathbb{Z}}$ is an $\mathrm{Aut}(\Gamma_r)$-invariant root lattice. This means that $\Gamma_r \in \mathcal{R}$, and so $\Gamma \in \mathcal{R}$ by Lemma 11.1.3. If $\Gamma \in \mathcal{R}_2$, then $\Gamma_r \subseteq 2\Lambda$, and so the set

$$\mathcal{V} = \{X \in \Gamma_r | (X|X) = 16, (X|\Gamma_r) \subseteq 8\mathbb{Z}\}$$

is not empty. Hence, $\Gamma_3 = \langle \mathcal{V} \rangle_{\mathbb{Z}}$ is an $\mathrm{Aut}(\Gamma_r)$-invariant root lattice, so that Γ_r and Γ belong to $\mathcal{R}$. □

We claim that $\mathcal{R}_3$, and a fortiori $\overline{\mathcal{R}}$, is empty when $m > 2$. Indeed, suppose that $\Gamma \in \mathcal{R}_3$ and $X = \sum_{\omega \in \Omega} x_\omega E_\omega \in \mathcal{U} \backslash 2\Lambda$. Here and in what follows we identify each element $\omega \in \Omega$ with some ordered pair (a, b) and set $E_\omega = E(a, b)$. Clearly, the set

$$S(X) = \{\omega \in \Omega | x_\omega \equiv 1 \pmod 2\}$$

is not empty. Furthermore, $16 = (X|X) \equiv |S(X)| \pmod 4$, therefore

$$4 \big| |S(X)|, \ |S(X)| \leq 16. \tag{11.1}$$

For $v \in V^0$ we set

$$S_v(X) = \{\omega \in S(X) | v \in \omega\}.$$

Then

$$\Gamma_r \ni \pi_v(X) \equiv 2 \sum_{\omega \in S_v(X)} E_\omega \pmod{4\Lambda},$$

and so $Y = 2\sum_{\omega \in S_v(X)} E_\omega \in \Gamma_r$. Because of condition P1), we have $4\mathbb{Z} \ni (X|Y) \equiv 2|S_v(X)| \pmod 4$. Moreover, if $S_v = S_v(X) \neq \emptyset$, then

$$8 < (Y|Y) = 4|S_v|.$$

Hence,

$$|S_v| \in 2\mathbb{Z}, \ |S_v| \neq 2. \tag{11.2}$$

Lemma 11.1.5. *Suppose that $X \in \mathcal{U} \backslash 2\Lambda$.*

(i) *If $\bar{\varphi} \in G_0$, $\varphi(v) = v$, then*

$$a_v(\varphi) = |S_v \cap \varphi(S_v)|$$

is even;

(ii) *there exists $u \in V^0$ such that $|S_u| = 4$; moreover, if $\mathcal{L}$ is of type D_{2^m}, then*

$$S_u = \big\{\{a, u\}, \{b, u\}, \{c, u\}, \{d, u\}\big\}$$

with $a + b + c + d = 0$.

Proof. (i) Write

$$X \equiv \sum_{\omega \in S_v} E_\omega + \sum_{\omega \in \Omega'} E_\omega \,(\mathrm{mod}\, 2\Lambda),$$

where $\Omega' = S(X)\backslash S_v$. For brevity, we identify $\bar{\varphi} \in G_0$ with φ from now on. Then $\varphi(Y) = 2\sum_{\omega \in S_v} \pm E_{\varphi(\omega)}$, and $a_v(\varphi)$ is even, since $4|(X|Y)$.

(ii) Suppose the contrary: $|S_v| \neq 4$ for every v. Then, by (11.2) we have $|S_v| \geq 6$ for some v, that is, $S_v \ni \{v, u_i\}$, $1 \leq i \leq 6$. As $S_{u_i} \ni \{v, u_i\}$, $|S_{u_i}|$ is also not less than 6. Hence,

$$16 \geq |S(X)| \geq \frac{1}{2}(|S_v| + \sum_i |S_{u_i}|) \geq 21,$$

a contradiction.

Finally, suppose that $|S_u| = 4$ and $\mathcal{L}$ is of type D_{2^m}. Applying the translation $t_u : x \mapsto x + u$ from G_0, one can suppose that $u = 0$, and so $S_0 = \big\{\{a, 0\}, \{b, 0\}, \{c, 0\}, \{d, 0\}\big\}$. One can suppose also that a, b, c are linearly independent. If $d \notin < a, b, c >_{\mathbb{F}_2}$, then $a_0(\varphi) = 3$ for the element φ of $SL(V)$ such that

$$\varphi : a \mapsto a, b \mapsto b, c \mapsto c, d \mapsto a + d,$$

contrary to (i). If $d \in < a, b, c >_{\mathbb{F}_2}$, but $d \neq a+b+c$, then we can set $d = a+b$, and so $a_0(\varphi) = 3$ for φ in $SL(V)$ such that

$$\varphi : a \mapsto a, b \mapsto b, c \mapsto a + c,$$

again contrary to (i). Hence, $a + b + c + d = 0$. □

Theorem 11.1.6. *$\mathcal{R}_3 = \emptyset$ if $m > 2$.*

Proof. We consider the case D_{2^m} as an example. Suppose that $\Gamma \in \mathcal{R}_3$, and take $X \in \mathcal{U}\backslash 2\Lambda$. By Lemma 11.1.5, we can assume that $S_0 = S_0(X) = \big\{\{a, 0\}, \{b, 0\}, \{c, 0\}, \{d, 0\}\big\}$ with $a + b + c + d = 0$. Now we use the condition $m > 2$, and choose $e \in V\backslash < a, b, c >_{\mathbb{F}_2}$. Then $a_0(\varphi) = 1$ for the element φ of $SL(V)$ such that

$$\varphi : a \mapsto a, b \mapsto e, c \mapsto b + c,$$

contrary to Lemma 11.1.5 (i).

The case B_{2^m-1} is settled in a similar way. □

Invariant lattices in $\mathcal{R}_1$ and $\mathcal{R}_2$ are studied in Sections §§11.2, 11.3. The class $\mathcal{R}_3$ is considered in § 11.4 (for type B_3) and §11.5 (for type D_4). It will be shown, in particular, that both the sets $\overline{\mathcal{R}}$ and $\mathcal{R}_3\backslash\overline{\mathcal{R}}$ are not empty for these types. The last section, §11.6, is devoted to the relation between invariant lattices and $\mathbb{Z}$-forms of Lie algebras.

11.2 Possible configurations of root systems

Suppose that $\Gamma \in \mathcal{R}_1 \cup \mathcal{R}_2$. Then a certain Aut($\Gamma$)-invariant sublattice Γ' of Γ is spanned by some root system of maximal rank. We shall find all possible configurations of this root system and thereby extract useful information about the group $\mathbb{G} = \mathrm{Aut}(\Gamma)$.

A. The Case D_{2^m}
First of all we prove a statement that gives better information on the transitivity of G_0 on Ω:

Lemma 11.2.1. *Let $\Omega = \bigcup_{\alpha\in\mathcal{A}} \Omega_\alpha$ be a nontrivial G_0-invariant partition of Ω. Then $|\mathcal{A}| = 2^{m+1} - 1$, and every Ω_α is of the form*

$$\ell(a) = \big\{\{x, y\} \in \Omega | x + y = a\big\}$$

for some $a \in V\backslash\{0\}$.

Proof. 1) First we assume that some block Ω_α contains two elements like $\{a, b\}$, $\{a, c\}$. Applying the translation t_a, one can suppose that $a = 0$. Clearly, $c \in V\backslash < b >_{\mathbb{F}_2}$. For each $d \in V\backslash < b >_{\mathbb{F}_2}$ there exists a transformation φ in $SL(V)$,

$$\varphi : b \mapsto b, c \mapsto d,$$

that maps $\{0, b\}$, $\{0, c\}$ into $\{0, b\}$, $\{0, d\}$. Hence, $\{0, d\} \in \Omega_\alpha$, so that $\{0, d\} \in \Omega_\alpha$ for all $d \in V\backslash\{0\}$. Take an arbitrary pair $\{d, e\} \in \Omega$ with $d, e \neq 0$. Then $\{0, d\}$, $\{0, d+e\} \in \Omega_\alpha$. But the translation t_d leaves $\{0, d\}$ fixed and maps $\{0, d+e\}$ into $\{d, e\}$, and hence $\{d, e\} \in \Omega_\alpha$. Thus we have $\Omega_\alpha = \Omega$, a contradiction.

2) We have shown that every two distinct elements of Ω_α are mutually disjoint, that is, look like $\{a, b\}$ and $\{c, d\}$. As above, one can suppose that $a = 0$. We claim that $d = b + c$. Indeed, if $d \neq b + c$, then $\dim_{\mathbb{F}_2} < b, c, d >= 3$. In this case a transformation φ in $SL(V)$ such that

$$\varphi : b \mapsto b, c \mapsto c, d \mapsto b + d$$

leaves $\{0, b\}$ fixed and maps $\{c, d\}$ into $\{c, b + d\}$. This means that $\{c, d\}$, $\{c, b + d\} \in \Omega_\alpha$, a contradiction. Hence, $a + b = c + d$. We have just shown that for every $\alpha \in \mathcal{A}$ there exists $a \in V\backslash\{0\}$ such that $\Omega_\alpha \subseteq \ell(a)$. In fact equality holds, because $St_{G_0}(a)$ acts doubly transitively on the (affine) line $\ell(a)$. □

For $\Omega' \subseteq \Omega$ we set

$$\mathcal{M}(\Omega') = \{\pm 2E_{\omega_1} \pm 2E_{\omega_2} | \omega_i \in \Omega', \omega_1 \neq \omega_2\}.$$

In particular, $\mathcal{M}_1 = \mathcal{M}(\Omega)$ is a root system of type D_N, where $N = \dim \mathcal{L}$. Furthermore,

$$\mathcal{M}_2 = \bigoplus_{a \in V \setminus \{0\}} \mathcal{M}(\ell(a))$$

is a root system of type $(2^{m+1} - 1)D_{2^m}$.

Proposition 11.2.2. *Suppose that* $\Gamma \in \mathcal{R}_1$, $\Gamma_r =< \mathcal{U} >_{\mathbb{Z}}$, $\mathcal{M} = \{X \in \Gamma_r | (X|X) = 8\}$. *Then* $\mathcal{M} = \mathcal{M}_1$ *or* $\mathcal{M} = \mathcal{M}_2$.

Proof. 1) To begin with, suppose that $\mathcal{M} \cap 2\Lambda \neq \emptyset$. Clearly, every element $X \in \mathcal{M} \cap 2\Lambda$ is of the form $X = 2\varepsilon_1 E_{\omega_1} + 2\varepsilon_2 E_{\omega_2}$ for some $\varepsilon_i = \pm 1$. In this case, all the vectors $\pm 2E_{\omega_1} \pm 2E_{\omega_2}$ belong also to $\mathcal{M} \cap 2\Lambda$, because $4\Lambda \subseteq \Gamma_r$. We introduce the following relation $\sim$ on Ω:

$$\omega, \omega' \in \Omega, \ \omega \sim \omega' \Leftrightarrow 2E_\omega + 2E_{\omega'} \in \Gamma_r.$$

Clearly, $\sim$ is an equivalence relation, and so it splits Ω into blocks Ω_α, $\alpha \in \mathcal{A}$. Furthermore, if $\varphi \in G_0$ and $\omega \sim \omega'$, then $\varphi(\omega) \sim \varphi(\omega')$, which means that the partition $\Omega = \bigcup_{\alpha \in \mathcal{A}} \Omega_\alpha$ is G_0-invariant. Applying Lemma 11.2.1 we obtain that $\mathcal{M} \cap 2\Lambda$ is $\mathcal{M}_1$ or $\mathcal{M}_2$.

2) Recall that $\mathcal{M} \neq \emptyset$ when $\Gamma \in \mathcal{R}_1$. Suppose that $\mathcal{M} \not\subseteq 2\Lambda$, and take $X \in \mathcal{M} \setminus 2\Lambda$. Write

$$X \equiv \sum_{\omega \in A} E_\omega \ (\text{mod}\, 2)$$

for some $A = S(X) \subseteq \Omega$. Since $8 = (X|X) \equiv |A| \ (\text{mod}\, 4)$, we have $|A| = 4$ or 8. For $a \in V^0$ we set $m_a = |S_a(X)|$. Consider also $\bar{f}_a \in G$ defined by the equalities $f_a(v) = \delta_{a,v}$. Then

$$Y = X - \bar{f}_a(X) \equiv 2E(a, b_1) + \ldots + 2E(a, b_{m_a}) \ (\text{mod}\, 4\Lambda).$$

In particular, $Z = 2E(a, b_1) + \ldots + 2E(a, b_{m_a}) \in \Gamma_r$. By property P1) of Γ_r, 4 divides $(X|Z) \equiv 2m_a \ (\text{mod}\, 4)$, that is, m_a is even. Suppose that $m_a \neq 2$ for all $a \in V^0$. Then $|A| > 4$, and so $|A| = 8$. But with $n = |\{a \in V^0 | m_a \neq 0\}|$, we have

$$4n \leq 2|A| = 16 \leq n(n-1),$$

a contradiction. Hence, there exists $a \in V^0$ with $m_a = 2$. In this case $Z = 2E(a, b_1) + 2E(a, b_2) \in \mathcal{M} \cap 2\Lambda$. In view of 1) we have $\mathcal{M} \cap 2\Lambda = \mathcal{M}_1$. In particular, there is a vector $T = 2E_{\omega_1} + 2E_{\omega_2}$ in Γ_r with $\omega_1 \in A$ but $\omega_2 \notin A$. Then $(X|T) \equiv 2 \ (\text{mod}\, 4)$, contrary to P1). Consequently, $\mathcal{M} \subseteq 2\Lambda$, and so $\mathcal{M} = \mathcal{M} \cap 2\Lambda = \mathcal{M}_1$ or $\mathcal{M}_2$. $\square$

Next, suppose that $\Gamma \in \mathcal{R}_2$, and set

$$\mathcal{W} = \{X \mid 2X \in \Gamma_r, \ (X|X) = 4, \ (X|\Gamma_r) \subseteq 4\mathbb{Z}\}; \ \Gamma_2 =< \mathcal{W} >_{\mathbb{Z}}.$$

Then $2E_\omega \in \mathcal{W}$ for all $\omega \in \Omega$, and so

a) $\Lambda \supseteq \Gamma_2 \supseteq 2\Lambda$;

b) $(X|Y) \in 2\mathbb{Z}$ and $(X|X) \in 4\mathbb{Z}$ for all $X, Y \in \Gamma_2$.

Set

$$\mathcal{M} = \{X \in \Gamma_2 | (X|X) = 4\}.$$

In view of b), $\mathcal{M}$ is a root system; moreover, $\mathcal{M} \supseteq \mathcal{M}_0$, where

$$\mathcal{M}_0 = \{\pm 2E_\omega \mid \omega \in \Omega\}.$$

Proposition 11.2.3. $\mathcal{M} = \mathcal{M}_0$, *except in the following two cases:*

(i) $m = 2$ *and* $\mathcal{M}$ *is a root system of type* $7D_4$*;*

(ii) $m = 3$ *and* $\mathcal{M}$ *is a root system of type* $15E_8$.

In these exceptional cases, $\mathcal{M}$ *is also uniquely determined.*

Proof. 1) Consider a decomposition of $\mathcal{M}$ into a sum of indecomposable components: $\mathcal{M} = \oplus_{i=1}^n \mathcal{M}_i$. As G is a subgroup of $\mathrm{Aut}(\Gamma_2)$, it leaves $\mathcal{M}$ fixed, and by [Cas], G fixes the decomposition. But G is absolutely irreducible on $\mathcal{L}$, therefore all the components $\mathcal{M}_i$ are isomorphic, and one can write $\mathcal{M} = n\mathcal{M}_1$. Here $\mathcal{M}_1$ is a root system of type A_k, D_k, or E_k. Furthermore, set

$$\Omega_i = \{\omega \in \Omega \mid \pm 2E_\omega \in \mathcal{M}_i\}.$$

Clearly, the partition $\Omega = \bigcup_{i=1}^n \Omega_i$ is G_0-invariant. Applying Lemma 11.2.1 we obtain $k = 1$, $k = 2^m$, or $k = |\Omega| = N = 2^m(2^{m+1} - 1)$.

If $k = 1$, then $\mathcal{M} = \mathcal{M}_0$. In what follows we shall assume that $k > 1$. Note that in this case $\mathcal{M}_1 \cong D_k$ or E_8. For, we have either $k = |\Omega| \geq 28$, or $k = 2^m$. In particular, $k \neq 6, 7$, and so $\mathcal{M}_1 \neq E_6, E_7$. Furthermore, it is easy to see that there are at most $[\frac{k+1}{2}]$ mutually orthogonal vectors in a root system of type A_k. This means that when $\mathcal{M}_1 = A_k$, we would have $nk = |\Omega| \leq n[\frac{k+1}{2}]$, a contradiction.

2) Consider the case $k = 2^m$. Then by Lemma 11.2.1 we have $\Omega_i = \ell(a)$ for some $a \in V\backslash\{0\}$. It is easy to see that $\mathcal{M}_1$ is a root system in

$$\mathbb{R}^k = \langle e_x = E(x, a + x) | x \in < a_1, \ldots, a_m >_{\mathbb{F}_2} \rangle_{\mathbb{R}},$$

invariant under the subgroup $E.(\mathbb{Z}_2^{m+1}.St_{SL(V)}(a))$, where $V = < a = a_0, a_1, \ldots, a_m >_{\mathbb{F}_2}$. Consider an arbitrary vector $X = e_x + e_y + e_z + e_t \in \mathcal{M}_1$. Signs are unimportant, because $\Gamma_2 \supseteq 2\Lambda$. We claim that $x + y + z + t = 0$. Suppose the contrary. Then applying the translation $v \mapsto v + t$ one can assume that $t = 0$, $x, y, z \neq 0$ and $x + y + z \neq 0$. Now apply the map

$$\varphi : a \mapsto a, x \mapsto x, y \mapsto y, z \mapsto x + z$$

in $SL(V)$ to X and obtain $Y = e_0 + e_x + e_y + e_{x+z} \in \mathcal{M}_1$. Hence, $e_z + e_{x+z} \in \Gamma_2$, contrary to the definition of the class $\mathcal{R}_2$. Conversely, the affine group $St_{SL(V)}(a)$

acts transitively on the set of nonordered quadruples (x, y, z, t), where $x, y, z, t \in\ < a_1, \dots, a_m >_{\mathbb{F}_2}$ and $x+y+z+t=0$. This means that $\pm e_x \pm e_y \pm e_z \pm e_t \in \mathcal{M}_1$ for any such quadruple (x, y, z, t).

Note that $m \le 3$. For, if $m \ge 4$, then we can choose four linearly independent vectors a, b, c, d in $< a_1, \dots, a_m >_{\mathbb{F}_2}$. As we have shown above, Γ_2 contains vectors

$$X' = e_0 + e_a + e_b + e_{a+b},\ Y' = e_0 + e_c + e_d + e_{c+d}.$$

But then $(X'|Y') = 1$, contrary to the definition of $\mathcal{R}_2$.

If $m = 2$, then $\mathcal{M}_1$ is a root system of type D_4:

$$\mathcal{M}_1 = \left\{\pm 2e_{x_1},\ \pm 2e_{x_2},\ \pm 2e_{x_3},\ \pm 2e_{x_4},\ \pm e_{x_1} \pm e_{x_2} \pm e_{x_3} \pm e_{x_4}\right\},$$

where $< a_1, a_2 >_{\mathbb{F}_2} = \{x_1, x_2, x_3, x_4\}$.

If $m = 3$, we obtain an affine realization of the root system of type E_8:

$$\mathcal{M}_1 = \left\{\pm 2e_x,\ \pm e_x \pm e_y \pm e_z \pm e_t \right|$$

$$\left| x, y, z, t \in < a_1, a_2, a_3 >_{\mathbb{F}_2}, x+y+z+t=0\right\}.$$

3) Finally, consider the possibility $\mathcal{M} = D_N$, where $N = |\Omega|$. Suppose that $X = \sum_{i=1}^{4} \pm E(a_i, b_i) \in \mathcal{M}$. Using the condition $(X|\varphi(X)) \in 2\mathbb{Z}$ valid for all $\varphi \in G_0$, one can show that

$$a_1 + b_1 = a_2 + b_2 = a_3 + b_3 = a_4 + b_4.$$

This means that $\mathcal{M}$ can be split into orthogonal components

$$\mathcal{M} = \bigoplus_{a \in V \setminus \{0\}} \mathcal{M}(a),\ \mathcal{M}(a) \subset \langle E(x, a+x) | x \in V \rangle_{\mathbb{C}}.$$

But this contradicts the indecomposability of $\mathcal{M}$. □

Corollary 11.2.4. *If* $\Gamma \in \mathcal{R}_1 \cup \mathcal{R}_2$, *then*

$$\mathrm{Aut}(\Gamma) \subseteq \mathrm{Aut}(\Lambda) = \mathbb{Z}_2 \wr \mathbb{S}_N$$

except in the following two cases:

(i) $m = 2$, $\mathrm{Aut}(\Gamma) \subseteq A(D_4) \wr \mathbb{S}_7$, *where* $A(D_4)$ *denotes the full automorphism group of a root system of type* D_4;

(ii) $m = 3$, $\mathrm{Aut}(\Gamma) \subseteq W(E_8) \wr \mathbb{S}_{15}$.

Proof. The statement is obvious for $\Gamma \in \mathcal{R}_2$ by Proposition 11.2.3. Now we assume that $\Gamma \in \mathcal{R}_1$. Then $\mathrm{Aut}(\Gamma) \subseteq \mathrm{Aut}(\Gamma_r) \subseteq \mathrm{Aut}(\mathcal{M})$ (see Proposition 11.2.2). If $\mathcal{M} = \mathcal{M}_1$, then $\mathrm{Aut}(\mathcal{M}) = \mathbb{Z}_2 \wr \mathbb{S}_N = \mathrm{Aut}(\Lambda)$. If $\mathcal{M} = \mathcal{M}_2$ and $m > 2$, then $\mathrm{Aut}(\mathcal{M}) \subset \mathrm{Aut}(\Lambda)$. Finally, if $\mathcal{M} = \mathcal{M}_2$ and $m = 2$, then $\mathrm{Aut}(\mathcal{M}) = A(D_4) \wr \mathbb{S}_7$. □

B. The Case B_{2^m-1}
Here the scheme of the arguments remains the same as in the case of D_{2^m}. Therefore, we shall restrict ourselves to formulating the principal statements. Set

$$\Omega = \{\{u, v\} | u, v \in V \backslash \{0\}, u \neq v\}.$$

Lemma 11.2.5. *$G_0 = SL(V)$ acts transitively but imprimitively on Ω. Moreover, the possible sets of imprimitivity are the following:*

(i) $t(W) = \{\{u, v\} \in \Omega | u, v \in W\}$, where $W \in \mathcal{V}_2$ and $\mathcal{V}_2$ is the variety of 2-dimensional subspaces in V;

(ii) $\ell(a) = \{\{u, v\} \in \Omega | u + v = a\}$, where $a \in V \backslash \{0\}$. □

Set

$$\mathcal{M}(\Omega') = \{\pm 2E_{\omega_1} \pm 2E_{\omega_2} | \omega_i \in \Omega', \omega_1 \neq \omega_2\},$$

for $\Omega' \subseteq \Omega$. In particular, $\mathcal{M}_1 = \mathcal{M}(\Omega)$ is a root system of type D_N, where $N = \dim \mathcal{L}$. Furthermore,

$$\mathcal{M}_2 = \bigoplus_{a \in V \backslash \{0\}} \mathcal{M}(\ell(a))$$

is a root system of type $(2^{m+1} - 1)D_{2^m-1}$. Finally,

$$\mathcal{M}_3 = \bigoplus_{W \in \mathcal{V}_2} \mathcal{M}(t(W))$$

is a root system of type $\frac{N}{3}D_3$.

Proposition 11.2.6. *Suppose that $\Gamma \in \mathcal{R}_1$, $\Gamma_r =< \mathcal{U} >_{\mathbb{Z}}$, $\mathcal{M} = \{X \in \Gamma_r \mid (X|X) = 8\}$. Then $\mathcal{M} = \mathcal{M}_1$, $\mathcal{M}_2$, or $\mathcal{M}_3$.* □

Next, we suppose that $\Gamma \in \mathcal{R}_2$. Set

$$\mathcal{W} = \{X \mid 2X \in \Gamma_r, \ (X|X) = 4, \ (X|\Gamma_r) \subseteq 4\mathbb{Z}\}; \Gamma_2 =< \mathcal{W} >_{\mathbb{Z}}.$$

Then $2E_\omega \in \mathcal{W}$ for all $\omega \in \Omega$, and so

a) $\Lambda \supseteq \Gamma_2 \supseteq 2\Lambda$;

b) $(X|Y) \in 2\mathbb{Z}$ and $(X|X) \in 4\mathbb{Z}$ for all $X, Y \in \Gamma_2$.

Set

$$\mathcal{M} = \{X \in \Gamma_2 | (X|X) = 4\}.$$

In view of b), $\mathcal{M}$ is a root system; moreover, $\mathcal{M} \supseteq \mathcal{M}_0$, where

$$\mathcal{M}_0 = \{\pm 2E_\omega | \omega \in \Omega\}.$$

Proposition 11.2.7. $\mathcal{M} = \mathcal{M}_0$, *except when* $m = 3$ *and* $\mathcal{M}$ *is a root system of type* $15E_7$. *In this exceptional case* $\mathcal{M}$ *is also uniquely determined.* □

Corollary 11.2.8. *If* $\Gamma \in \mathcal{R}_1 \cup \mathcal{R}_2$, *then*

$$\mathrm{Aut}(\Gamma) \subseteq \mathrm{Aut}(\Lambda) = \mathbb{Z}_2 \wr \mathbb{S}_N$$

except in the case $m = 3$, $\mathrm{Aut}(\Gamma) \subseteq W(E_7) \wr \mathbb{S}_{15}$. □

11.3 Automorphism groups of lattices: the classes $\mathcal{R}_1$ and $\mathcal{R}_2$

The results of §11.2 actually reduce the determination of the automorphism groups of lattices $\Gamma \in \mathcal{R}_1 \cup \mathcal{R}_2$ to the investigation of primitive permutation groups of specific types. More precisely, Corollaries 11.2.4 and 11.2.8 show that, except for a few cases, the group $\mathbb{G} = \mathrm{Aut}(\Gamma)$ is embedded in $\mathrm{Aut}(\Lambda) = \mathbb{Z}_2 \wr \mathbb{S}_N$, where $N = |\Omega|$. In particular, $\mathbb{G}$ acts on the set $\{\pm E_\omega | \omega \in \Omega\}$, and also on Ω. Hence, we get a homomorphism

$$\tau_0 : \mathbb{G} \to \mathbb{S}_N$$

whose kernel $\mathrm{Ker}\tau_0$ is obviously an elementary abelian 2-group $E_0 = \mathbb{Z}_2^e$, where $\ell \le e \le N$. Recall that $\ell = 2^{m+1}$ for $\mathcal{L}$ of type D_{2^m}, and $\ell = 2^{m+1} - 1$ for $\mathcal{L}$ of type B_{2^m-1}. Furthermore,

$$\mathrm{Ker}\left(\tau_0|_{\mathbb{Z}_2 \times G}\right) = \mathbb{Z}_2 \times \mathbb{Z}_2^{\ell-1} = \mathbb{Z}_2^{\ell}.$$

It remains to determine the structure of $\mathbb{G}_0 = \mathrm{Im}\tau_0$. Note that $\mathbb{G}_0$ acts transitively on Ω. Here the following two cases arise:

1) $\mathbb{G}_0$ is primitive on Ω, that is, $\mathbb{G}_0 = \mathbb{G}_1$ is a primitive permutation group of degree

$$N = \begin{cases} N_1 = 2^m(2^{m+1} - 1) & \text{if } \mathcal{L} \text{ is of type } D_{2^m}, \\ N_2 = (2^m - 1)(2^{m+1} - 1) & \text{if } \mathcal{L} \text{ is of type } B_{2^m-1}. \end{cases}$$

2) $\mathbb{G}_0$ is imprimitive on Ω. Let $\{\Omega_\alpha | \alpha \in \mathcal{A}\}$ be a nontrivial imprimitivity system for $\mathbb{G}_0$ on Ω.

a) Suppose that $\mathcal{L}$ is of type D_{2^m}. Then, since $G_0 \subseteq \mathbb{G}_0$, it follows from Lemma 11.2.1 that

$$|\mathcal{A}| = 2^{m+1} - 1 = N_3,\ \Omega_\alpha = \ell(a),\ a \in V \backslash \{0\}.$$

The action of $\mathbb{G}_0$ on the blocks Ω_α induces a homomorphism

$$\tau_1 : \mathbb{G}_0 \to \mathbb{S}_{N_3}$$

with

$$\mathbb{Z}_2^{m+1} \subseteq E_1 = \mathrm{Ker}\tau_1 \subseteq (\mathbb{S}_{2^m})^{N_3}.$$

Furthermore,

$$\mathbb{G}_1 = \mathrm{Im}\tau_1 \supseteq SL_{m+1}(2)$$

is a primitive permutation group of degree N_3.

b) Suppose that $\mathcal{L}$ is of type B_{2^m-1}. Then, since $G_0 \subseteq \mathbb{G}_0$, it follows from Lemma 11.2.5 that

$$|\mathcal{A}| = 2^{m+1} - 1 = N_3,\ \Omega_\alpha = \ell(a),\ a \in V\backslash\{0\},$$

or

$$|\mathcal{A}| = \frac{1}{3}(2^m - 1)(2^{m+1} - 1) = N_4,\ \Omega_\alpha = t(W),\ W \in \mathcal{V}_2.$$

In the former case, the action of $\mathbb{G}_0$ on the blocks Ω_α induces a homomorphism

$$\tau_1 : \mathbb{G}_0 \to \mathbb{S}_{N_3}$$

with

$$E_1 = \mathrm{Ker}\tau_1 \subseteq (\mathbb{S}_{2^m-1})^{N_3}.$$

Furthermore,

$$\mathbb{G}_1 = \mathrm{Im}\tau_1 \supseteq SL_{m+1}(2)$$

is a primitive permutation group of degree N_3.

In the latter case, the action of $\mathbb{G}_0$ on the blocks Ω_α leads to a homomorphism

$$\tau_2 : \mathbb{G}_0 \to \mathbb{S}_{N_4}$$

with

$$E_1 = \mathrm{Ker}\tau_2 \subseteq (\mathbb{S}_3)^{N_4};$$

in particular, E_1 is a soluble $\{2, 3\}$-group. Furthermore,

$$\mathbb{G}_1 = \mathrm{Im}\tau_2 \supseteq SL_{m+1}(2)$$

is a primitive permutation group of degree N_4.

Thus, we have to find primitive overgroups $\mathbb{G}_1$ of $ASL(V)$ or $SL(V)$, of degrees $N' = N_1, N_2, N_3$, or N_4.

Lemma 11.3.1. *If* $\mathbb{G}_1$ *is doubly transitive, then* $T = \mathrm{soc}(\mathbb{G}_1)$ *is a nonabelian finite simple group.*

Proof. By the O'Nan-Scott Theorem, it is sufficient to show that every degree $N' \in \{N_1, N_2, N_3, N_4\}$ is not a proper power x^n, $x, n \in \mathbb{N}$, $n > 1$. Clearly, the last assertion reduces to establishing the unsolvability of the equation

$$x^n = 2^d - 1, \text{ where } x, d, n \in \mathbb{N}, \text{ and } n, d > 1.$$

Suppose the contrary, so that $2^d = x^n + 1$. Then both x and n are odd; moreover, $x \equiv -1 \pmod 4$. Rewrite x in the form $x = 2^r y - 1$, where $r, y \in \mathbb{N}$, $r \geq 2$, 2 nvy. Then the number

$$2^d = x^n + 1 = (2^r y - 1)^n + 1 = 2^{2r} z - 2^r ny$$

is not divisible by 2^{r+1}. This implies that $d \leq r$ and $n = 1$, a contradiction. □

Proposition 11.3.2. *The permutation group* $\mathbb{G}_1$ *of degree* $N_3 = 2^{m+1} - 1$ *is one of* $SL_{m+1}(2)$, $\mathbb{A}_{2^{m+1}-1}$ *and* $\mathbb{S}_{2^{m+1}-1}$.

Proof. Recall that $\mathbb{G}_1$ contains $S = SL(V)$ and acts faithfully and primitively on the set of blocks $\ell(a)$, $a \in V \backslash \{0\}$. Here the action of S is induced by its natural action on $V \backslash \{0\}$. By [Ust], we have either $\mathbb{G}_1 = S$, or $\mathbb{A}_{N_3} \lhd \mathbb{G}_1 \subseteq \mathbb{S}_{N_3}$. □

Proposition 11.3.3. *Either the permutation group* $\mathbb{G}_1$ *of degree* $N_1 = 2^m(2^{m+1} - 1)$ *is doubly transitive, or it has rank* 3 *and then* $\mathbb{A}_{2^{m+1}} \lhd \mathbb{G}_1 \subseteq \mathbb{S}_{2^{m+1}}$.

Proof. It is known that $\mathbb{G}_1 \supseteq G_0 = ASL(V)$ and that it acts primitively on Ω. Here, G_0 acts on Ω as a permutation group of rank 4, that is, the stabilizer $G_0(\omega)$ of the point $\omega = \{0, a\}$ has three orbits on $\Omega \backslash \omega$:

$$\mathcal{A} = \big\{\{0, b\}, \{a, b\} | b \in V \backslash < a >\big\},$$

$$\mathcal{B} = \big\{\{x, y\} | x, y \in V \backslash < a >, x + y = a\big\},$$

$$\mathcal{C} = \big\{\{x, y\} | x, y, x + y \in V \backslash < a >\big\}.$$

Furthermore, $|\mathcal{A}| = 2^{m+2} - 4$, $|\mathcal{B}| = 2^m - 1$, $|\mathcal{C}| = 2^{2m+1} - 3.2^{m+1} + 4$. Suppose that $\text{rank}(\mathbb{G}_1) \geq 3$. Then at least one of the sets $\mathcal{A}$, $\mathcal{B}$, $\mathcal{C}$ is a $\mathbb{G}_1(\omega)$-orbit, where $\mathbb{G}_1(\omega) = St_{\mathbb{G}_1}(\omega)$.

1) Suppose that $\mathcal{A}$ is a $\mathbb{G}_1(\omega)$-orbit. We claim that

$$|\varphi(\omega_1) \cap \varphi(\omega_2)| = |\omega_1 \cap \omega_2|, \ \forall \varphi \in \mathbb{G}_1, \ \forall \omega_1, \omega_2 \in \Omega.$$

To see this, note that

$$\mathcal{A} = \{\omega' \in \Omega | |\omega \cap \omega'| = 1\}.$$

Consider a decomposition $\varphi = \theta_1 \psi \theta_2$, where $\theta_1, \theta_2 \in G_0$, $\theta_1(\omega) = \varphi(\omega_1)$, $\theta_2(\omega_1) = \omega$, $\psi \in \mathbb{G}_1(\omega)$. Since elements of G_0 possess the property indicated, we have

$$|\omega_1 \cap \omega_2| = |\theta_2(\omega_1) \cap \theta_2(\omega_2)| = |\omega \cap \omega'|,$$

(where $\omega' = \theta_2(\omega_2)$), and

$$|\psi(\omega) \cap \psi(\omega')| = |\psi\theta_2(\omega_1) \cap \psi\theta_2(\omega_2)| = |\varphi(\omega_1) \cap \varphi(\omega_2)|.$$

It remains to show that $|\omega \cap \omega'| = |\psi(\omega) \cap \psi(\omega')|$ for $\psi \in \mathbb{G}_1(\omega)$. But this is in fact the case, since

$$|\omega \cap \omega'| = 2 \Leftrightarrow \omega = \omega' \Leftrightarrow \psi(\omega) = \psi(\omega') \Leftrightarrow |\psi(\omega) \cap \psi(\omega')| = 2;$$

$$|\omega \cap \omega'| = 1 \Leftrightarrow \omega' \in \mathcal{A} \Leftrightarrow \psi(\omega') \in \mathcal{A} \Leftrightarrow |\psi(\omega) \cap \psi(\omega')| = 1.$$

For $a \in V$ we define a sheaf

$$\Pi(a) = \{\omega \in \Omega | \omega \ni a\},\ |\Pi(a)| = 2^{m+1} - 1.$$

We claim that $\Pi = \Pi(b)$ for some $b \in V$, if $\Pi = \varphi(\Pi(a))$ and $\varphi \in \mathbb{G}_1$. Note that

$$\omega_1, \omega_2 \in \Pi(a) \Rightarrow |\omega_1 \cap \omega_2| = 1;$$

$$\omega \notin \Pi(a) \Rightarrow \exists \omega' \in \Pi(a), \omega \cap \omega' = \emptyset.$$

As we have shown above, the same condition is satisfied by Π. Choose two elements $\omega_1 = \{b, c\}, \omega_2 = \{b, d\} \in \Pi$. If $\omega \ni b$ for all $\omega \in \Pi$, then $\Pi = \Pi(b)$. Suppose that there exists $\omega_3 \in \Pi$ such that $\omega_3 \not\ni b$. Then, since $|\omega_1 \cap \omega_3| = |\omega_2 \cap \omega_3| = 1$, it follows that $\omega_3 = \{c, d\}$. Because $|\Pi| = |\Pi(a)| \geq 4$, Π contains an element ω_4 different from ω_1, ω_2, ω_3. It is easy to see that the condition

$$|\omega_1 \cap \omega_4| = |\omega_2 \cap \omega_4| = |\omega_3 \cap \omega_4| = 1$$

cannot be satisfied, and this means that $\varphi(\Pi(a)) = \Pi(b)$.

We have thereby established the existence of a homomorphism

$$\pi : \mathbb{G}_1 \to \mathrm{Sym}(V) = \mathbb{S}_{2^{m+1}}$$

such that $\varphi(\Pi(a)) = \Pi(\pi(\varphi)(a))$ for $\varphi \in \mathbb{G}_1$. In particular,

$$\varphi : \{a, b\} = \Pi(a) \cap \Pi(b) \mapsto \{\pi(\varphi)(a), \pi(\varphi)(b)\},$$

that is, π is injective. Hence, $ASL(V) = G_0 \subseteq \mathbb{G}_1 \hookrightarrow \mathrm{Sym}(V)$. Moreover, $\mathbb{G}_1 \neq G_0$, since G_0 acts imprimitively on Ω. If we denote by $\mathbb{G}_1(0)$ the stabilizer of the point $\Pi(0)$, $0 \in V$, then

$$SL(V) = G_0(0) \subset \mathbb{G}_1(0) \subseteq \mathrm{Sym}(V \backslash \{0\}).$$

By Proposition 11.3.2 we have

$$\mathbb{A}_{2^{m+1}-1} \lhd \mathbb{G}_1(0) \subseteq \mathbb{S}_{2^{m+1}-1}.$$

Hence,

$$\mathbb{A}_{2m+1} \lhd \mathbb{G}_1 \subseteq \mathbb{S}_{2m+1}.$$

Note that both $\mathbb{A}_{2m+1}$ and $\mathbb{S}_{2m+1}$ act primitively on Ω as groups of rank 3.

2) Suppose that $\mathcal{B}$ is a $\mathbb{G}_1(\omega)$-orbit. We claim that

$$\forall \varphi \in \mathbb{G}_1,\ \forall \omega_1, \omega_2 \in \Omega,\ [\omega_1] = [\omega_2] \Rightarrow [\varphi(\omega_1)] = [\varphi(\omega_2)],$$

where $[\omega] = a + b$ for $\omega = \{a, b\}$. Using a decomposition $\varphi = \theta_1 \psi \theta_2$ with $\theta_1, \theta_2 \in G_0$, $\psi \in \mathbb{G}_1(\omega)$, it is sufficient to establish the implication

$$[\omega] = [\omega'] \Rightarrow [\omega] = [\psi(\omega')]$$

for every $\psi \in \mathbb{G}_1(\omega)$. But

$$[\omega'] = [\omega] \Rightarrow \omega' \in \mathcal{B} \cup \{\omega\} \Rightarrow \psi(\omega') \in \mathcal{B} \cup \{\omega\} \Rightarrow [\psi(\omega')] = [\omega].$$

The property established above means simply that $\mathbb{G}_1$ has $\{\ell(a) | a \in V \backslash \{0\}\}$ as an imprimitivity set, since

$$\ell(a) = \{\omega \in \Omega | [\omega] = a\}.$$

This contradicts our assumption.

3) Finally, suppose that $\mathcal{C}$ is a $\mathbb{G}_1(\omega)$-orbit. For $\omega' = \{a, b\}$ we set

$$\mathcal{S}(\omega') = \big\{\{x, y\} \in \Omega | x, y \neq a, b;\ x + y \neq a + b\big\}.$$

Then $\mathcal{C} = \mathcal{S}(\omega)$. Using (as above) a decomposition $\varphi = \theta_1 \psi \theta_2$, where $\theta_1, \theta_2 \in G_0$, $\psi \in \mathbb{G}_1(\omega)$, one can show that $\mathbb{G}_1$ acts on the set $\{\mathcal{S}(\omega') | \omega' \in \Omega\}$. Namely,

$$\varphi(\mathcal{S}(\omega')) = \mathcal{S}(\varphi(\omega'))$$

for $\varphi \in \mathbb{G}_1$. In particular,

$$|\mathcal{S}(\omega_1) \cap \mathcal{S}(\omega_2)| = |\mathcal{S}(\varphi(\omega_1)) \cap \mathcal{S}(\varphi(\omega_2))|.$$

Moreover,

$$|\mathcal{S}(\omega_1) \cap \mathcal{S}(\omega_2)| = \begin{cases} r = (2^m - 2)(2^{m+1} - 5) & \text{if} \quad |\omega_1 \cap \omega_2| = 1, \\ s = r - (2^m - 2) & \text{if} \quad |\omega_1 \cap \omega_2| = 0,\ [\omega_1] = [\omega_2], \\ t = r - (2^{m+1} - 6) & \text{if} \quad |\omega_1 \cap \omega_2| = 0,\ [\omega_1] \neq [\omega_2]. \end{cases}$$

Here $r > s \geq t$. Hence, φ preserves $|\omega_1 \cap \omega_2|$. But

$$\mathcal{A} = \{\omega' \in \Omega | |\omega' \cap \omega| = 1\},$$

therefore $\mathbb{G}_1(\omega)$ has $\mathcal{A}$ as an orbit on Ω. Furthermore, we assumed that $\mathcal{C}$ is a $\mathbb{G}_1(\omega)$-orbit, so that, $\mathcal{B}$ is also a $\mathbb{G}_1(\omega)$-orbit, and thus $\mathbb{G}_1$ is imprimitive on Ω, a contradiction. □

Proposition 11.3.4. *The permutation group $\mathbb{G}_1$ of degree $N_4 = \frac{1}{3}(2^m-1)(2^{m+1}-1)$ is one of $SL(V)$, $\mathbb{A}_{N_4}$ and $\mathbb{S}_{N_4}$, or else $m = 3$ and $\mathbb{G}_1 = \mathbb{S}_8$.*

Proof. Recall that $\mathbb{G}_1$ contains $S = SL(V)$ and acts faithfully and primitively on the set of blocks $t(W)$, $W \in \mathcal{V}_2$. The action of S here is that induced by its natural action on $\mathcal{V}_2$. Applying [Ust], we see that $\mathbb{G}_1 \in \{SL(V), \mathbb{A}_{N_4}, \mathbb{S}_{N_4}\}$. The unique exception is the case $\frac{m+1}{2} = 2$, that is, $m = 3$, when there arises the overgroup $\mathbb{S}_8$ of S in its primitive permutation representation of degree 35. □

Using the same arguments as in the proof of Proposition 11.3.3, one can prove the following statement.

Proposition 11.3.5. *If $m > 2$, then either the permutation group $\mathbb{G}_1$ of degree $N_2 = (2^m-1)(2^{m+1}-1)$ is doubly transitive, or it has rank 3 and $\mathbb{A}_{2^{m+1}-1} \lhd \mathbb{G}_1 \subseteq \mathbb{S}_{2^{m+1}-1}$.* □

The case $m = 2$ and $N' = N_2$ is treated in the following statement.

Proposition 11.3.6. *When $m = 2$ the permutation group $\mathbb{G}_1$ of degree 21 satisfies one of the following conditions:*

(i) $\mathbb{A}_{21} \lhd \mathbb{G}_1 \subseteq \mathbb{S}_{21}$,

(ii) $\mathbb{A}_7 \lhd \mathbb{G}_1 \subseteq \mathbb{S}_7$,

(iii) $\mathbb{G}_1 = PGL_2(7)$.

Proof. It is shown in [Pog] that the only primitive permutation groups $\mathbb{G}_1$ of degree 21 are (i) – (iii) and

(iv) $T = L_3(4) \lhd \mathbb{G}_1 \subseteq \mathrm{Aut}(T) = P\Gamma L_3(4)$.

We claim that case (iv) cannot occur. Indeed, suppose the contrary. Clearly, $S \subseteq T$, and T is a transitive permutation group of degree 21. Let σ denote the corresponding character. By [ATLAS], we have $\sigma = 1_T + \sigma_1$, where $\sigma_1 \in \mathrm{Irr}(T)$. Furthermore, T has a unique conjugacy class of elements of order 3, and $\sigma_1(x) = 2$ if $x \in T$, $| < x > | = 3$. However, if we take $x \in S$ with $| < x > | = 3$, then x acts fixed-point-freely on Ω, that is, $\sigma(x) = 0$, a contradiction.

The possibilities (i) and (ii) are realized by a general construction described below. Regarding possibility (iii), we remark that the groups $S = SL_3(2) \cong L_2(7)$ and $PGL_2(7)$ have unique transitive permutation representations of degree 21. Hence, the transitive action of S on Ω can be extended to a primitive permutation representation of $PGL_2(7)$. □

Next, we consider the case when the permutation group $\mathbb{G}_1$ is doubly transitive.

Proposition 11.3.7. *If the permutation group $\mathbb{G}_1$ of degree $N' \neq N_3$ is doubly transitive, then $T = \mathrm{soc}(\mathbb{G}_1)$ satisfies one of the following conditions:*
(i) $T = \mathbb{A}_{N'}$,
(ii) $m = 2$, $N' = N_4 = 7$, *and* $T = SL_3(2)$,
(iii) $m = 2$, $N' = N_1 = 28$, *and* $T = Sp_6(2)$.

Proof. By Lemma 11.3.1, T is a nonabelian finite simple group, so we can use Table A2.

1) It should be pointed out that $(T, N') \neq (L_2(11), 11)$, $(\mathbb{A}_7, 15)$, $(HS, 176)$, $(M_{11}, 11)$, $(M_{11}, 12)$, $(M_{12}, 12)$, $(M_{22}, 22)$, $(M_{23}, 23)$, $(M_{24}, 24)$, $(PSU_3(q), q^3 + 1)$, $({}^2B_2(q), q^2 + 1)$, $({}^2G_2(q), q^3 + 1)$, since in these cases $N' \notin \{N_1, N_2, N_3, N_4\}$. A possible exception is the case $m = 2$, $N' = N_1 = 28$, $T = PSU_3(3)$.

2) We claim that $T \neq L_n(q)$, except, perhaps, in the cases
a) $m = 2$, $N' = N_4 = 7$, $T = SL_3(2)$;
b) $m = 2$, $N' = N_1 = 28$, $T = L_2(27)$ or $T = L_2(8)$.
For, if $N' = N_1$, then since

$$2^m(2^{m+1} - 1) = \frac{q^n - 1}{q - 1},$$

it follows that $m = 2$, $n = 2$, $q = 27$, that is, $T = L_2(27)$. Another possible exception is $N' = N_1 = 28$, $T = L_2(8)$.

Suppose that $N' = N_2$ or $N' = N_4$. First assume that q is even: $q = 2^r$. If $N' = N_2$, then because $(N_2 - 1)/2^m$ is odd, we have $r = m$. This means that

$$2q^2 - 3q = N_2 - 1 = q + q^2 + \cdots + q^{n-1}.$$

In particular, $n \leq 3$. If $n = 3$, then we arrive at the possibility

$$q = 4,\ N' = N_2 = 21,\ T = L_3(4),$$

which is excluded by Proposition 11.3.6. If $n = 2$, then $q = 2$ and $T = SL_2(2)$ is soluble, a contradiction. If $N' = N_4$, then $(N' - 1)/2$ is odd, therefore $r = 1$ and $q = 2$. Furthermore, T is simple, which implies that $n \geq 3$. Hence, $N' + 1 = 2^n$ is divisible by 8. But $N_4 + 1$ is not divisible by 8 when $m \geq 3$. We arrive at the conclusion that $m = 2$, $n = 3$, $T = L_3(2)$.

Lastly, we consider the case where q is odd. Since both N_2 and N_4 are odd, $n \equiv 1 \pmod 2$. Clearly, $S \subseteq T$. We wish to determine the number of fixed points (on Ω) for any involution in T. It is known that T has two inequivalent doubly transitive permutation representations: on lines, and on hyperplanes of the space $U = \mathbb{F}_q^n$. The latter representation is contragradient to the former; therefore, the number of fixed points is the same for these representations. We shall restrict ourselves to a consideration of the action of T on points of $\mathbb{P}U$. Suppose that

$\varphi \in SL_n(q)$ and $| < \varphi > | = 2$ in $L_n(q)$. Then $\varphi^2 = \lambda.1_U$, where $\lambda = \theta^r \in < \theta > = \mathbb{F}_q^*$. In particular,

$$1 = (\det \varphi)^2 = \lambda^n = \theta^{rn}.$$

This means that $(q-1)|rn$; however, $2 \not| \, n$, therefore r is even, and $\lambda = \mu^2$ for some $\mu \in \mathbb{F}_q^*$. Taking $\psi = \mu^{-1}\varphi$, we have $\psi \in GL_n(q)$ and $\psi^2 = 1_U$. As q is odd, we have the decomposition

$$U = U_+ \oplus U_-,$$

where $U_+ = \mathrm{Ker}(\psi - 1_U)$, $U_- = \mathrm{Ker}(\psi + 1_U)$. Furthermore, $U_\pm \neq 0$. Setting $\{\dim U_+, \dim U_-\} = \{k, n-k\}$ for $1 \leq k \leq [\frac{n}{2}]$, we see that the number of fixed points of any involution in T is

$$N(k) = |\mathbb{P}U_+| + |\mathbb{P}U_-| = \frac{q^k + q^{n-k} - 2}{q-1}.$$

Clearly,
α) $N(1) > N(2) > \cdots > N([\frac{n}{2}])$;
β) $N(2) < \frac{N'}{7} = \frac{N(0)}{7}$ if $n \geq 4$;
γ) if $q > 5$, then $N(1) < \frac{N'}{5}$; if $q = 5$, then $\frac{N'}{5} < N(1) < \frac{3N'}{13}$; if $q = 3$, then $\frac{N'}{3} < N(1) < \frac{N'}{2}$.
Now we consider the following involution in S:

$$\mathbf{t} = \begin{pmatrix} 1 & 0 & 0 & \dots & 0 & 1 \\ 0 & 1 & 0 & \dots & 0 & 0 \\ 0 & 0 & 1 & \dots & 0 & 0 \\ & & & \dots & & \\ 0 & 0 & 0 & \dots & 0 & 1 \end{pmatrix}.$$

In the representation of S of degree N_2, $\mathbf{t}$ has

$$N_2' = (2^m - 1)(2^{m-1} - 1) + 2^{m-1}$$

fixed points. In the representation of S of degree N_4, $\mathbf{t}$ has

$$N_4' = 2^{m-1} + \frac{(2^m - 1)(2^{m-1} - 1)}{3}$$

fixed points. Furthermore,
δ) if $m \geq 2$, then $N_2/4 > N_2' > 3N_2/13$; if $m \geq 3$, then $N_4/3 > N_4' > N_4/4$.
From α) – δ), it now follows that

$$N' = N_4, \; N_4' = N(1), \; m = 2, \; q = 3,$$

a contradiction. Thus we have excluded the case $T = L_n(q)$.

3) Next, suppose that $N' = N_1 = 28$, and exclude the cases

$$T = PSU_3(3),\ L_2(8),\ L_2(27).$$

As always, $S \subseteq T$. But $|S| = 2^3.3.7$, therefore S cannot be embedded in $L_2(27)$. Furthermore, $|L_2(8)| = 3|S|$, and so $S \not\hookrightarrow L_2(8)$. Finally, suppose that $T = PSU_3(3)$. Let σ denote the character of the unique doubly transitive permutation representation of T. In accordance with [ATLAS], $\sigma(x) = 4$ for every involution $x \in T$. Meanwhile, every involution of S has exactly 8 fixed points on Ω, a contradiction.

4) We still have to consider the last (and most interesting) possibility that $T = Sp_{2(m+1)}(2)$. As $\mathrm{Out}(T) = 1$, we have also $\mathbb{G}_1 = T$, hence $T \supseteq G_0 = ASL(V)$. Denote $O_2(G_0)$ by the same symbol V. Since G_0 is a 2-local subgroup of T, by the Borel-Tits Theorem G_0 is contained in some maximal parabolic subgroup

$$P_j = (\mathbb{Z}_2^{j(j+1)/2}.\mathbb{Z}_2^{2j(m+1-j)}).(SL_j(2) \times Sp_{2(m+1-j)}(2)),$$

where $1 \le j \le m+1$.

Of course, we can apply the same arguments as those carried out at the beginning of this section with respect to $\mathbb{G}_0$, to the group P_j acting on Ω. Hence, one of the following assertions holds.

α) P_j is imprimitive on Ω, and then, by Proposition 11.3.2, P_j has $SL_{m+1}(2)$ or $\mathbb{A}_{2^{m+1}-1}$ as composition factor.

β) P_j is primitive of rank 3 on Ω, and then, by Proposition 11.3.3, P_j has $\mathbb{A}_{2^{m+1}}$ as composition factor.

Note that P_j cannot be doubly transitive on Ω, since $O_2(P_j) \ne 1$ (see Lemma 11.3.1). Recall that P_j has the following composition factors: $SL_j(2)$, $Sp_{2(m+1-j)}(2)$, and $\mathbb{Z}_2$. Comparing with α), β), we arrive at the conclusion that $j = m+1$. This means that $G_0 = V.S \subseteq P = P_{m+1}$, where P is the stabilizer of some maximal totally isotropic subspace in T, that is,

$$P = \left\{ \begin{pmatrix} A & B \\ 0 & {}^tA^{-1} \end{pmatrix} \middle| A \in SL_{m+1}(2), B \in M_{m+1}(\mathbb{F}_2), B.{}^tA + A.{}^tB = 0 \right\}.$$

We can write $P = Q.S'$, where

$$Q = \left\{ \begin{pmatrix} E_{m+1} & X \\ 0 & E_{m+1} \end{pmatrix} \middle| X + {}^tX = 0 \right\} \cong \mathbb{Z}_2^{(m+1)(m+2)/2},$$

$$S' = \left\{ \begin{pmatrix} A & 0 \\ 0 & {}^tA^{-1} \end{pmatrix} \middle| A \in SL_{m+1}(2) \right\} \cong SL_{m+1}(2).$$

It is easy to see that $Q \cap G_0 = V$. Moreover, $V \lhd P$, $P = Q.S$, and P and G_0 act (by conjugation) transitively on $V \backslash \{0\}$. But Q centralizes V, therefore in fact S acts on V, and we rewrite this condition in terms of matrices. For

$$f = \begin{pmatrix} E_{m+1} & X \\ 0 & E_{m+1} \end{pmatrix} \in Q,\ g = \begin{pmatrix} A & 0 \\ 0 & {}^tA^{-1} \end{pmatrix} \in S$$

we have

$$gfg^{-1} = \begin{pmatrix} E_{m+1} & AX.{}^tA \\ 0 & E_{m+1} \end{pmatrix}.$$

Thus the space Q of symmetric matrices of degree $m+1$ over $\mathbb{F}_2$ contains some $(m+1)$-dimensional subspace V such that the group $S = SL_{m+1}(2)$ acts transitively on $V\backslash\{0\}$ according to the rule $X \mapsto AX.{}^tA$, where $X \in V$, $A \in S$. We claim that this can occur only when $m = 2$.

First, we convince ourselves that every matrix X in V has zero principal diagonal. To see this, suppose that some element X has nonzero principal diagonal $\bar{X} = [x_{11}, \ldots, x_{m+1,m+1}]$. If $Y = AX.{}^tA$, then by the symmetry of X the principal diagonal $\bar{Y}$ is $A\bar{X}$ (here we view $\bar{X}, \bar{Y}$ as columns over $\mathbb{F}_2$). But S acts transitively on $\mathbb{F}_2^{m+1}\backslash\{0\}$, therefore one can suppose that $\bar{X} = [1, 0, \ldots, 0]$. Set

$$S(X) = \{A \in S | AX.{}^tA = X\}.$$

Since S is transitive on $V\backslash\{0\}$, $(S : S(X)) = 2^{m+1} - 1$. Furthermore, the inclusion $A \in S(X)$ implies that $\bar{X} = \overline{AX.{}^tA} = A\bar{X}$, that is, A belongs to the stabilizer of $\bar{X}$ in S (here we mean the action of S on the natural module $\mathbb{F}_2^{m+1}$). Since this subgroup is of the same index $2^{m+1} - 1$ in S, we arrive at the equality

$$S(X) = \{A \in S | A\bar{X} = \bar{X}\}.$$

Recall that $\bar{X} = [1, 0, \ldots, 0]$. Hence,

$$S(X) = \left\{ \begin{pmatrix} 1 & a \\ 0 & B \end{pmatrix} | a \in \mathbb{F}_2^m, B \in SL_m(2) \right\}.$$

From this it is not difficult to deduce that $X = \mathrm{diag}(1, 0, \ldots, 0)$. In this case, the S-orbit of X contains the matrix

$$\mathrm{diag}(0, \ldots, 0, \underbrace{1}_{i}, 0, \ldots, 0),$$

for each $i = 1, \ldots, m+1$. But V is an $(m+1)$-dimensional linear space, therefore V coincides with the space of diagonal matrices. We arrive at the contradiction: if Y is a diagonal matrix, then $AY.{}^tA$ is also a diagonal matrix for all $A \in S$.

Thus, we can suppose that every matrix X in V has zero principal diagonal. The group S acts on the set $\mathcal{O}$ of symmetric matrices with zero principal diagonal with orbits $\mathcal{O}_i$, $0 \le i \le [\frac{m+1}{2}]$. Namely,

$$\mathcal{O}_i = \{X \in \mathcal{O} | \mathrm{rank}(X) = 2i\}.$$

In other words, $\mathcal{O}_i$ is the set of alternating bilinear forms of rank $2i$. Take X in $\mathcal{O}_r$, and let $< *|* >$ denote the form

$$(u, u') \mapsto {}^tuXu'.$$

Clearly,

$$A \in S(X) \Leftrightarrow AX.{}^tA = X \Leftrightarrow$$

$$\Leftrightarrow < {}^tAu|{}^tAv >= {}^tuAX.{}^tAv = {}^tuXv =< u|v > \Leftrightarrow {}^tA \in O(X),$$

where $O(X)$ denotes the orthogonal group corresponding to X. In particular,

$$|S(X)| = |O(X)| = 2^{2rs}.|SL_s(2)|.|Sp_{2r}(2)| = 2^{\frac{m(m+1)}{2} - r(r-1)}.t,$$

where $s = m + 1 - 2r$ and $2 \not| \, t$. Hence, $(S : S(X))$ is even if $r > 1$, and $(S : S(X)) = \frac{(2^m-1)(2^{m+1}-1)}{3}$ if $r = 1$. So, the equality

$$(S : S(X)) = 2^{m+1} - 1$$

implies that $r = 1$ and $m = 2$. Conversely, when $m = 2$,

$$V = \left\{ \begin{pmatrix} 0 & a & b \\ a & 0 & c \\ b & c & 0 \end{pmatrix} \Bigg| a, b, c \in \mathbb{F}_2 \right\}$$

is the desired S-invariant subspace. □

We have finished the "negative" part of the section: all the unsuitable permutation groups are excluded. The rest of this section is devoted to the "positive" activity: find invariant lattices Γ where the "components" $\mathbb{G}_1$ of $\mathbb{G} = \mathrm{Aut}(\Gamma)$ are as listed above.

Before starting on this activity, we recall some definitions from coding theory (see [McS]). By a *linear binary code* (in other words, linear code over $\mathbb{F}_2$) of length n we mean any subspace $\mathcal{C}$ of the space $V = \mathbb{F}_2^n =< e_1, \ldots, e_n >_{\mathbb{F}_2}$ endowed with the standard bilinear form

$$< \sum_i x_i e_i, \sum_i y_i e_i >= \sum_i x_i y_i.$$

The code

$$\mathcal{C}^\perp = \{v \in V | < v, \mathcal{C} >= 0\}$$

is said to be *dual* to $\mathcal{C}$. $\mathcal{C}$ is called *self-orthogonal* if $\mathcal{C} \subseteq \mathcal{C}^\perp$, and *self-dual* if $\mathcal{C} = \mathcal{C}^\perp$. A *code word*, that is, an element $c = \sum_i x_i e_i$ of $\mathcal{C}$, has *weight* $w(c) = |\{i | x_i \neq 0\}|$. $\mathcal{C}$ is said to be *doubly even* if $4 | w(c)$ for all $c \in \mathcal{C}$. Finally, the automorphism group of $\mathcal{C}$ is

$$\mathrm{Aut}(\mathcal{C}) = \{\varphi \in \mathbb{S}_n | \varphi(\mathcal{C}) = \mathcal{C}\},$$

where $\mathbb{S}_n$ is the symmetric group on the n symbols $e_1, \ldots, e_n$.

The outline of our arguments is as follows. Let $\mathbb{G}_1$ be one of the primitive permutation groups of degree $N' = N_1$ or N_2 listed above. We shall construct a G_0-invariant linear code $\mathcal{C}$ of length N' over $\mathbb{F}_2$ with automorphism group $\mathrm{Aut}(\mathcal{C}) = \mathbb{G}_1$ and with minimum weight greater than 4. Next, we take a lattice Γ, $\Lambda \supset \Gamma \supset 2\Lambda$, such that

$$\Gamma/2\Lambda \cong \mathcal{C}.$$

Here, we identify $\Lambda/2\Lambda$ with the underlying space $\mathbb{F}_2^{N'}$ for $\mathcal{C}$. Then it is easy to see that

$$\mathrm{Aut}(\Gamma) = \mathbb{Z}_2^{N'}.\mathbb{G}_1.$$

In the case $N' = N_3$ or N_4, the imprimitivity blocks $\ell(a)$ (or $t(W)$) are taken as basis vectors e_i of the underlying space $\mathbb{F}_2^{N'}$ for $\mathcal{C}$. Moreover, one can avoid the condition $w(c) > 4$ for $c \in \mathcal{C}$. The case $T = \mathrm{soc}(\mathbb{G}_1) = \mathbb{A}_{N'}$ is trivial: it is sufficient to set $\mathcal{C} = \mathbb{F}_2^{N'}$.

We start with the following simple statement.

Proposition 11.3.8. (i) *There exists a G_0-invariant linear code $\mathcal{C}$ of length $N_3 = 2^{m+1} - 1$ over $\mathbb{F}_2$ with* $\mathrm{Aut}(\mathcal{C}) = SL_{m+1}(2)$.

(ii) *For each n there exists a binary linear code $\mathcal{C}_n$ of length $\frac{n(n-1)}{2}$ with minimum weight $n-1$ and automorphism group* $\mathrm{Aut}(\mathcal{C}_n) = \mathbb{S}_n$. *Moreover, if we identify basis vectors of the underlying space $\mathbb{F}_2^{\frac{n(n-1)}{2}}$ with elements of Ω, where $n = 2^{m+1}$ for type D_{2^m}, and $n = 2^{m+1} - 1$ for type B_{2^m-1}, then the code $\mathcal{C}_n$ is G_0-invariant.*

Proof. (i) We take

$$\mathbb{F}_2^{N_3} = \langle e_v | v \in V \backslash \{0\} \rangle_{\mathbb{F}_2},$$

where $V = \mathbb{F}_2^{m+1}$. The desired code $\mathcal{C}$ is spanned by the words $e_W = \sum_{v \in V \backslash W} e_v$, where W runs over the variety $\mathcal{V}_m$ of all hyperplanes in V. It is easy to see that

$$e_W + e_{W'} = e_{W''},$$

where

$$W =< a_1, \dots, a_{m-1}, b >_{\mathbb{F}_2},\ W' =< a_1, \dots, a_{m-1}, b' >_{\mathbb{F}_2},$$

$$W'' =< a_1, \dots, a_{m-1}, b + b' >_{\mathbb{F}_2}.$$

Hence,

$$\mathcal{C} = \{0, e_W | W \in \mathcal{V}_m\},\ |\mathcal{C}| = 2^{m+1}.$$

Moreover, the group $G_0 = SL(V)$ acts faithfully on $\mathcal{C}$. In fact, we have $\mathrm{Aut}(\mathcal{C}) = SL(V)$.

(ii) Set

$$\mathbb{F}_2^{\frac{n(n-1)}{2}} = \langle e_\omega | \omega \in \Omega_n \rangle_{\mathbb{F}_2},$$

where $\Omega_n = \{\{i,j\} | 1 \leq i \neq j \leq n\}$. Then $\mathcal{C}_n$ is spanned by the words

$$A_i = \sum_{\omega \in \Omega_n, \omega \ni i} e_\omega,\ 1 \leq i \leq n.$$

Note that if one identifies $\{1, 2, \ldots, n\}$ with (the point set of) $V = \mathbb{F}_2^{m+1}$ if $n = 2^{m+1}$, or $V \backslash \{0\}$ if $n = 2^{m+1} - 1$; and makes G_0 act in the natural way on $\mathbb{F}_2^{n(n-1)/2}$:

$$\varphi : e_\omega \mapsto e_{\varphi(\omega)},$$

then $\mathcal{C}_n$ is G_0-invariant. Furthermore, if $1 \leq i_1 < \ldots < i_k \leq n$, then

$$w(A_{i_1} + \cdots + A_{i_k}) = k(n-k).$$

In particular, $w(c) \geq n - 1$ for all c in $\mathcal{C}_n \backslash \{0\}$; moreover,

$$w(c) = n - 1 \Leftrightarrow c = A_i \text{ for some } i,\ 1 \leq i \leq n.$$

Hence, for any ψ in $\mathrm{Aut}(\mathcal{C}_n)$ there exists π in $\mathbb{S}_n$ such that $\psi(A_i) = A_{\pi(i)}$. Moreover,

$$\psi(e_{\{i,j\}}) = e_{\{\pi(i),\pi(j)\}}.$$

This means that $\mathrm{Aut}(\mathcal{C}_n) \hookrightarrow \mathbb{S}_n$. On the other hand, $\mathbb{S}_n$ is embedded in $\mathrm{Aut}(\mathcal{C}_n)$, so that $\mathrm{Aut}(\mathcal{C}_n) = \mathbb{S}_n$. □

Proposition 11.3.9. *Suppose that $d \geq 3$. There exist doubly even self-orthogonal binary codes $\mathcal{C}^\pm$ of dimension $2d + 1$ and length $2^{d-1}(2^d \mp 1)$, and with automorphism groups $\mathrm{Aut}(\mathcal{C}^\pm) = Sp_{2d}(2)$. Furthermore, their minimum weight is $4^{d-1} - 2^{d-1}$ for $\mathcal{C}^+$, and 4^{d-1} for $\mathcal{C}^-$. Moreover, if $d = 3$, then the transitive permutation representation of degree 28 of $G_0 = ASL_3(2)$ on Ω can be extended to a transitive representation of degree 28 of $Sp_6(2)$ such that $\mathcal{C}^+$ is G_0-invariant.*

Proof. 1) Let us recall some facts about classical groups (see [Dieu]). Suppose that $V = \mathbb{F}_q^{2d}$, where $q = 2^h$, is endowed with a non-degenerate bilinear alternating form $< *|* >$, and that $\{u_1, \ldots, u_d, v_1, \ldots, v_d\}$ is a symplectic basis. Consider the set Ξ of all quadratic forms Q on V that are compatible with $< *|* >$, that is, are such that

$$Q(x + y) = Q(x) + Q(y) + < x|y > .$$

The *index* ν of the element Q of Ξ is defined to be the dimension of the maximal totally Q-singular subspaces of V. It is known that there exists a basis $\{g_i\}$ of V such that

$$Q(\sum_{i=1}^{2d} z_i g_i) = z_1 z_{d+1} + \cdots + z_{d-1} z_{2d-1} + (\zeta z_d^2 + z_d z_{2d} + \zeta z_{2d}^2),$$

where either $\nu = d$ and $\zeta = 0$, or $\nu = d-1$, $\zeta \neq 0$ and the polynomial $\zeta X^2 + X + \zeta$ is irreducible over $\mathbb{F}_q$. Let $O_{2d}(Q)$ denote the orthogonal group corresponding to Q and $\Omega_{2d} = \Omega_{2d}(Q) = [O_{2d}(Q), O_{2d}(Q)]$. The group $Sp_{2d}(q)$ acts on Ξ; moreover, the stabilizer of a form Q is $O_{2d}(Q)$. The so-called Dickson invariant $D \in \mathrm{Hom}(O_{2d}(Q), \mathbb{F}_2)$ is such that the rotation subgroup $O^+_{2d} = O^+_{2d}(Q) = \{f \in O_{2d}(Q) | D(f) = 0\}$ is a subgroup of index 2 in $O_{2d}(Q)$. If $d \geq 2$, then $\Omega_{2d} = [O^+_{2d}, O^+_{2d}]$. If $d \geq 1$ and $\nu \geq 1$, then except in the case $(d, q, \nu) = (2, 2, 2)$, we have $O^+_{2d}/\Omega_{2d} \cong \mathbb{F}_q/\mathbb{F}^{*2}_q$, where $\mathbb{F}^{*2}_q = \{x^2 | x \in \mathbb{F}_q\}$. If $d \geq 3$ and $\nu \geq 1$, Ω_{2d} is simple. As $q = 2^h$, we have $\Omega_{2d} = O^+_{2d}$ and

$$|\Omega_{2d}| = q^{d(d-1)/2}(q^d - (-1)^{\mu(Q)}) \prod_{i=1}^{d-1} (q^{2i} - 1),$$

where $\mu(Q) = 0$ if $\nu = d$, and $\mu(Q) = 1$ if $\nu = d - 1$. In the meantime,

$$Sp_{2d}(q) = q^{d^2} \prod_{i=1}^{d} (q^{2i} - 1).$$

This means that $Sp_{2d}(q)$ acts on Ξ with the two orbits

$$\Xi^+ = \{Q \in \Xi | \nu(Q) = d - 1\}, |\Xi^+| = \frac{1}{2} q^d (q^d - 1),$$

and

$$\Xi^- = \{Q \in \Xi | \nu(Q) = d\}, |\Xi^-| = \frac{1}{2} q^d (q^d + 1).$$

2) Next, consider the case $q = 2$. For $Q \in \Xi$ we have

$$\mu(Q) = \sum_{i=1}^{d} Q(u_i)Q(v_i).$$

For $a \in V \backslash \{0\}$ we set

$$\Theta^+(a) = \{Q \in \Xi^+ | Q(a) = 0\}, \ \Theta^-(a) = \{Q \in \Xi^- | Q(a) = 0\}.$$

Direct calculations show that

a) $|\Theta^+(a)| = 2N_{d-1}$, $|\Theta^-(a)| = 2M_{d-1}$;

b)

$$|\Theta^+(a) \cap \Theta^+(b)| = \begin{cases} 4N_{d-2} & \text{if } a \neq b, < a|b >= 0; \\ N_{d-1} & \text{if } a \neq b, < a|b >= 1; \end{cases}$$

$$|\Theta^-(a) \cap \Theta^-(b)| = \begin{cases} 4M_{d-2} & \text{if } a \neq b, < a|b >= 0; \\ M_{d-1} & \text{if } a \neq b, < a|b >= 1; \end{cases}$$

c)

$$\Theta^{\pm}(a)+\Theta^{\pm}(b)=\begin{cases}\overline{\Theta^{\pm}(a+b)} & \text{if } a\neq b,\ <a|b>=0;\\ \Theta^{\pm}(a+b) & \text{if } a\neq b,\ <a|b>=1.\end{cases}$$

Here $M_d = 2^{d-1}(2^d+1)$, $N_d = 2^{d-1}(2^d-1)$; $\bar{X}$ is equal to $\Xi^+\backslash X$ if $X\subseteq\Xi^+$, and $\Xi^-\backslash X$ if $X\subseteq\Xi^-$; finally, $X+Y=(X\backslash Y)\cup(Y\backslash X)$ for all $X, Y\subseteq\Xi$.

3) We shall view the set $2^{\Xi^{\pm}}$ of all subsets of $\Xi^{\pm}$ as an $\mathbb{F}_2$-space, where the addition operation is defined above. In this space we distinguish the basis consisting of the 1-element subsets. The subspaces

$$\mathcal{C}^+ = \left\langle\Theta^+(a)|a\in V\backslash\{0\}\right\rangle_{\mathbb{F}_2},\ \mathcal{C}^- = \langle\Theta^-(a)|a\in V\backslash\{0\}\rangle_{\mathbb{F}_2},$$

can then be considered as binary codes. From a) – c) it follows that, when $d\geq 3$, the code $\mathcal{C}^{\pm}$ is a doubly even self-orthogonal code of dimension $2d+1$. Clearly, $Sp_{2d}(2)\subseteq \mathrm{Aut}(\mathcal{C}^{\pm})$, and we have to prove the reverse inclusion. We introduce the following form on $\mathcal{C}^{\pm}$:

$$(A,B)=\frac{1}{2^{d-2}}|A\cap B|\ (\mathrm{mod}\, 2).$$

It is obvious that $(*,*)$ is a bilinear alternating form. Furthermore, in the basis

$$\left\{\Xi^{\pm},\Theta^{\pm}(u_i),\Theta^{\pm}(v_i)|1\leq i\leq d\right\}$$

of $\mathcal{C}^{\pm}$, $(*,*)$ has matrix

$$\begin{pmatrix}0&0&0\\0&0&E_d\\0&E_d&0\end{pmatrix}.$$

Since $C=\mathrm{Aut}(\mathcal{C}^{\pm})$ preserves $(*,*)$, we obtain thereby a homomorphism

$$\vartheta: C\to Sp_{2d}(2).$$

Suppose that $\varphi\in\mathrm{Ker}\vartheta$, that is, $\varphi=1$ on $\mathcal{C}^{\pm}/\mathrm{Ker}(*,*)$. As $\mathrm{Ker}(*,*)=<\Xi^{\pm}>$, for all $A\in\mathcal{C}^{\pm}$ we have

$$\varphi(A)=A \text{ or } \bar{A}.$$

But $|A|\neq|\bar{A}|$, so in fact we have $\varphi(A)=A$ for all $A\in\mathcal{C}^{\pm}$, that is, $\varphi=1$. Hence ϑ embeds C in $Sp_{2d}(2)$, and we conclude that $\mathrm{Aut}(\mathcal{C}^{\pm})=Sp_{2d}(2)$.

4) Finally, we consider the case $d=2$ in detail.

It is known [ATLAS] that the irreducible complex characters of $Sp_6(2)$ have degrees 1, 7, 15, 21, 27, or greater than 27. Moreover, $Sp_6(2)$ has a unique irreducible character of degree 27. Therefore, every transitive permutation representation of degree 28 of $Sp_6(2)$ is equivalent (as complex representation) to the doubly transitive permutation representation of $Sp_6(2)$ on Ξ^+. But $Sp_6(2)$ has a unique conjugacy class of maximal subgroups of index 28 [ATLAS], hence, in fact we have an equivalence of permutation representations.

Now we realize the root system of type E_8 in the form

$$R = \left\{\pm 2e_a, \pm e_a \pm e_b \pm e_c \pm e_d | a, b, c, d \in \mathbb{F}_2^3, a+b+c+d=0\right\},$$

where $\{e_a | a \in V = \mathbb{F}_2^3\}$ is an orthonormal basis for the space $U = \mathbb{R}^8$. Then the set $R' = \{r \in R \mid r \perp e_0\}$ is a root system of type E_7. Furthermore, the stabilizer

$$St_{W(E_8)}(e_0) = W(E_7) = \mathbb{Z}_2 \times Sp_6(2)$$

acts transitively on the set

$$\{r \in R \mid (r, e_0) = 1\} = \left\{e_0 \pm e_a \pm e_b \pm e_{a+b} \mid a, b \in V, < a, b >_{\mathbb{F}_2} \in \mathcal{V}_2\right\}.$$

This means that $Sp_6(2)$ acts transitively on the set Σ consisting of the 28 pairs $\pm(e_a \pm e_b \pm e_{a+b})$, where $< a, b >_{\mathbb{F}_2} \in \mathcal{V}_2$. As we have pointed out above, Ξ^+ can be identified with Σ.

Now we identify Σ with Ω, and embed $G_0 = ASL_3(2)$ in $Sp_6(2)$. To this end, consider the subgroup

$$G_1 = E.S \subset \mathbb{Z}_2^6.SL_3(2) \subset Sp_6(2),$$

where

$$S = \left\{\bar{\varphi} : e_a \mapsto e_{\varphi(a)} \mid \varphi \in SL(V)\right\},$$

$$E = \left\{\bar{f} : e_a \mapsto (-1)^{f(a)} e_a \mid f \in \mathcal{F}_2\right\},$$

$$\mathcal{F}_2 = \left\{f : V \to \{0, 1\} \mid f^{-1}(0) \in \{V\} \cup \mathcal{V}_2\right\}.$$

Furthermore, we identify V with $\{0, 1, \ldots, 7\}$ by the rule:

$$\updownarrow \begin{matrix} 0 & a & b & a+b & c & a+c & b+c & a+b+c \\ 0 & 1 & 2 & 3 & 4 & 5 & 6 & 7 \end{matrix}.$$

Correspondingly, the elements of $\mathcal{V}_2$ are indexed as follows:

$$W_1 =< a, b >_{\mathbb{F}_2}= \{0, 1, 2, 3\},\ W_2 =< a, c >_{\mathbb{F}_2}= \{0, 1, 4, 5\},$$

$$W_3 =< a, b+c >_{\mathbb{F}_2}= \{0, 1, 6, 7\},\ W_4 =< b, c >_{\mathbb{F}_2}= \{0, 2, 4, 6\},$$

$$W_5 =< b, a+c >_{\mathbb{F}_2}= \{0, 2, 5, 7\},\ W_6 =< a+b, c >_{\mathbb{F}_2}= \{0, 3, 4, 7\},$$

$$W_7 =< a+b, a+c >_{\mathbb{F}_2}= \{0, 3, 5, 6\}.$$

Note that

$$W_i \cap W_j \cap W_k \neq \{0\} \Leftrightarrow i+j+k = 0 \text{ in } V.$$

Next, identify Σ with $\Omega = \{\{i, j\} | 0 \leq i \neq j \leq 7\}$ as follows:

$$\pm(e_a + e_b + e_{a+b}) \rightsquigarrow \{0, i\} \in \Omega, \text{ if } W_i =< a, b >_{\mathbb{F}_2};$$

$$\pm(-e_a + e_b + e_{a+b}) \rightsquigarrow \{j, k\} \in \Omega, \text{ if } W_i =< a, b >_{\mathbb{F}_2} \text{ and } a \in W_i \cap W_j \cap W_k.$$

Finally, we describe the action of G_1 on Σ. To each element $\varphi \in SL(V)$ there corresponds a pair of permutations ψ, $\pi \in \mathbb{S}_7$; namely, those given by

$$\varphi(a) = \psi(a), \ \forall a \in V \backslash \{0\},$$

$$\varphi(W_i) = W_{\pi(i)} \text{ for } 1 \leq i \leq 7.$$

We claim that $\psi, \pi \in SL(V)$. Indeed, this is obvious for $\psi = \varphi$. Furthermore, the inclusion

$$a \in W_i \cap W_j \cap W_k$$

implies that

$$\varphi(a) \in W_{\pi(i)} \cap W_{\pi(j)} \cap W_{\pi(k)}.$$

In other words,

$$\pi(i + j) = \pi(i) + \pi(j),$$

as required.

Next, we have

$$\varphi(\{0, i\}) = \varphi(e_a + e_b + e_{a+b}) = e_{\varphi(a)} + e_{\varphi(b)} + e_{\varphi(a+b)} = \{0, \pi(i)\},$$

$$\varphi(\{j, k\}) = \varphi(-e_a + e_b + e_{a+b}) = -e_{\varphi(a)} + e_{\varphi(b)} + e_{\varphi(a+b)} = \{\pi(j), \pi(k)\},$$

because

$$W_i =< a, b >_{\mathbb{F}_2}, \ W_j =< a, c >_{\mathbb{F}_2}, \ W_k =< a, b + c >_{\mathbb{F}_2},$$

$$W_{\pi(i)} =< \varphi(a), \varphi(b) >_{\mathbb{F}_2}, \ W_{\pi(j)} =< \varphi(a), \varphi(c) >_{\mathbb{F}_2}, \ldots .$$

Write E in the form $E = \{1_U, \delta_i | 1 \leq i \leq 7\}$, where

$$\delta_i(e_a) = \begin{cases} e_a & \text{if } a \in W_i, \\ -e_a & \text{if } a \notin W_i. \end{cases}$$

Clearly, $\varphi \delta_i \varphi^{-1} = \delta_{\pi(i)}$. Furthermore, if we fix the notation

$$W_i =< a, b >_{\mathbb{F}_2}, \ W_j =< a, c >_{\mathbb{F}_2},$$

$$W_{i+j} =< a, b + c >_{\mathbb{F}_2}, \ W_k =< b, c >_{\mathbb{F}_2},$$

then δ_i acts on Σ as follows:

$$\{0, i\} = e_a + e_b + e_{a+b} \mapsto e_a + e_b + e_{a+b} = \{0, i\},$$

$$\{j, i+j\} = -e_a + e_b + e_{a+b} \mapsto -e_a + e_b + e_{a+b} = \{j, i+j\},$$

$$\{0, j\} = e_a + e_c + e_{a+c} \leftrightarrow -(-e_a + e_c + e_{a+c}) = \{i, i+j\},$$

$$\{k, j+k\} = -e_c + e_a + e_{a+c} \leftrightarrow -e_{a+c} + e_a + e_c = \{i+k, i+j+k\}.$$

Here, we have used the facts that

$$W_{i+k} =< b, a+c >_{\mathbb{F}_2}, W_{j+k} =< c, a+b >_{\mathbb{F}_2}, W_{i+j+k} =< a+b, a+c >_{\mathbb{F}_2} .$$

Consequently, we have shown that

$$\varphi(\{k,\ell\}) = \{\pi(k), \pi(\ell)\},\ \delta_i(\{k,\ell\}) = \{i+k, i+\ell\}.$$

This means that the representations of G_1 and G_0 on Σ are the same, and this completes the proof of Proposition 11.3.9. □

Proof of Theorems 11.0.1 *and* 11.0.2. Suppose that $\Gamma \in \mathcal{R}_1 \cup \mathcal{R}_2$. Then Corollaries 11.2.4 and 11.2.8 enable us to suppose that $\mathbb{G} = \mathrm{Aut}(\Gamma) \subseteq \mathrm{Aut}(\Lambda) = \mathbb{Z}_2 \wr \mathbb{S}_N$. The arguments carried out at the beginning of this section reduce the description of $\mathbb{G}$ to the problem of finding some primitive permutation groups $\mathbb{G}_1$. All the possibilities for these permutation groups $\mathbb{G}_1$ are determined in Propositions 11.3.2 – 11.3.7. Finally, by Propositions 11.3.8 and 11.3.9, one shows that the possibilities indicated are actually realized. Thus we have proved Theorems 11.0.1 and 11.0.2 for Γ in $\mathcal{R}_1 \cup \mathcal{R}_2$. However, as will be shown in §§11.4, 11.5, the lattices that belong to $\mathcal{R} \cap \mathcal{R}_3$ also satisfy the conclusion of these theorems. □

Remark 11.3.10. When $m > 2$, $\mathcal{R}_3$ is empty, therefore in Theorems 11.0.1 and 11.0.2 one can remove the words "of root-type", that is, these theorems give a description of the automorphism groups of *all* invariant lattices.

11.4 Lattices of nonroot-type: the case B_3

As mentioned in §11.1, invariant lattices of nonroot-type (see Definition 11.1.2) can exist only in the case of Lie algebras of types B_3 and D_4. It is convenient to consider these lattices in the general context of the class $\mathcal{R}_3$ introduced in §11.1. This section is devoted to the case of B_3.

Suppose that $\mathcal{L}$ is a Lie algebra of type B_3, and take Γ in $\mathcal{R}_3$. One can show that in our case there exist only two possibilities, denoted by Λ^0, Λ^1, for the reduced sublattice Γ_r of Γ. To describe them we identify $\Lambda/2\Lambda$ with the $\mathbb{F}_2$-space 2^Ω as usual:

$$E_\omega \rightsquigarrow \omega.$$

Set

$$\mathcal{A} = (\Gamma_r + 2\Lambda)/2\Lambda, \ \mathcal{B} = (\Gamma_r \cap 2\Lambda)/4\Lambda.$$

Furthermore, for $v \in V\backslash\{0\}$ and $A \subseteq \Omega$ we set

$$S_v(A) = \{\omega \in A | \omega \ni v\},$$

$$\mathcal{S}(v) = \big\{\{x, y\} \in \Omega | x, y, x + y \neq v\big\}.$$

The following simple statement holds.

Lemma 11.4.1. (i) $\dim_{\mathbb{F}_2} \mathcal{A} = 3, \ \mathcal{A} = \{\emptyset, \mathcal{S}(v) | v \in V\backslash\{0\}\}$.
(ii) $\dim_{\mathbb{F}_2} \mathcal{B} = 14, \ \mathcal{B} = \langle A, S_v(A) | A \in \mathcal{A}, v \in V\backslash\{0\}\rangle_{\mathbb{F}_2}$. □

As in §11.3, we realize the root system of type E_7 in the form

$$R = \{\pm 2e_a, \pm e_a \pm e_b \pm e_c \pm e_d | a, b, c, d \in V\backslash\{0\}, a + b + c + d = 0\},$$

where $\{e_a | a \in V\backslash\{0\}\}$ is an orthonormal basis of the space $U = \mathbb{C}^7$. The exterior square $U \wedge U$ admits the following inner product induced from U:

$$(x \wedge y, z \wedge t) = 2\{(x, z)(y, t) - (x, t)(y, z)\}, \ x, y, z, t \in U.$$

Moreover, there exists an isomorphism

$$\Phi : U \wedge U \cong \mathcal{L}, \ \Phi(e_i \wedge e_j) = E(i, j),$$

which is "isometric" in the sense that

$$(\Phi(w) | \Phi(w)) = \frac{1}{2}(w, w).$$

Now we set

$$\Lambda_0 = \langle \alpha \wedge \beta | \alpha, \beta \in R, \alpha \perp \beta\rangle_{\mathbb{Z}}.$$

Then $\Lambda^0 = \Phi(\Lambda_0)$. Furthermore, one can choose a vector $X^0 \in \Lambda^0$ with $(X^0 | X^0) = 16$, of the form

$$X^0 = 2E(2,3) + E(2,4) + E(2,5) + E(2,6) + E(2,7) + E(3,4) + E(3,5) +$$
$$+ E(4,6) + E(4,7) + E(5,6) + E(5,7) - E(3,6) - E(3,7),$$

and set

$$X^1 = X^0 - (-2E(3,6) - 2E(3,7) + 2E(4,7) + 2E(5,6)).$$

Then $\Lambda^1 = \big\langle g(X^1) | g \in G\big\rangle_{\mathbb{Z}}$. It is shown later that

$$\Lambda^0 \in \overline{\mathcal{R}}, \ \Lambda^1 \in \mathcal{R}_3 \backslash \overline{\mathcal{R}}.$$

In what follows, we identify the point set of V with $\{0, 1, \dots, 7\}$ according to the rule (see §11.3)

$$\updownarrow \begin{array}{cccccccc} 0 & a & b & a+b & c & a+c & b+c & a+b+c \\ 0 & 1 & 2 & 3 & 4 & 5 & 6 & 7 \end{array}.$$

Recall that Γ_r is Λ^0 or Λ^1. Now we set

$$\Gamma^* = \{X \in \mathcal{L} | (X|\Gamma) \subseteq 4\mathbb{Z}\},$$

$$\mathcal{X}^* = \left\{Y \in 2^{\Omega} | \forall X \in \mathcal{X}, |X \cap Y| \equiv 0 \,(\mathrm{mod}\, 2)\right\},$$

where $\Lambda \supseteq \Gamma \supseteq 4\Lambda$, $\mathcal{X} \subseteq 2^{\Omega}$. The following statement is obvious.

Lemma 11.4.2. (i) $(\Gamma_r^* + 2\Lambda)/2\Lambda = \mathcal{B}^*$, $(\Gamma_r^* \cap 2\Lambda)/4\Lambda = \mathcal{A}^*$.

(ii) *If* $A \in \mathcal{A}^* \backslash \{\emptyset\}$, *then* $|A| \geq 2$; *moreover,* $|A| = 2$ *if and only if* $A = \{\{a, b\}, \{a, a+b\}\}$ *for some* $a, b \in V \backslash \{0\}$ *with* $a \neq b$.

(iii) *If* $A \in \mathcal{B}^* \backslash \{\emptyset\}$, *then* $|A| \geq 6$. □

Set $\mathbb{G}_r = \mathrm{Aut}(\Gamma_r)$, $Z =< \mathbf{t} = -1_{\mathcal{L}} >= \mathbb{Z}_2$. First we determine $\mathbb{G}_r \cap \mathrm{Aut}(\Lambda)$. Clearly, if $\varphi \in \mathrm{Aut}(\Lambda)$, then φ preserves the set $\{\pm E_\omega | \omega \in \Omega\}$.

Lemma 11.4.3. *Suppose that* $\varphi \in \mathbb{G}_r \cap \mathrm{Aut}(\Lambda)$, *and, in addition, that* φ *preserves every triple* $\tilde{t}(W) = \{E(a, b), E(a, a+b), E(b, a+b)\}$ *modulo* 2Λ, *where* $W = < a, b > \in \mathcal{V}_2$. *Then* $\varphi \in Z \times G$.

Proof. 1) We claim that there exist $\lambda_{ij} = \pm 1$ such that $\varphi(E(i, j)) = \lambda_{ij} E(i, j)$. Indeed, suppose that φ acts nontrivially on Ω, for example, φ acts nontrivially on the set $t(W_1) = \{\{1, 2\}, \{1, 3\}, \{2, 3\}\}$. We consider all the possible cases.

a) $\varphi : \{1, 2\} \leftrightarrow \{1, 3\}$, or $\varphi : \{1, 2\} \mapsto \{2, 3\} \mapsto \{1, 3\} \mapsto \{1, 2\}$.

Taking a vector $X = 2(E(1, 2) + E(2, 3) + E(2, 5) + E(2, 7))$ in Γ_r, we get

$$\Gamma_r \ni Y = \varphi(X) = 2(E(1, 3) + E(2, 3) + E(a, b) + E(a, c)),$$

where $\{a, b\}, \{a, c\} \in \{\{2, 5\}, \{2, 7\}, \{5, 7\}\}$, which is impossible by Lemma 11.4.1.

b) $\varphi : \{1, 2\} \leftrightarrow \{2, 3\}$ or $\varphi : \{1, 3\} \leftrightarrow \{2, 3\}$.

This case is excluded similarly to a).

2) Next we determine the constants λ_{ij}. "Adjusting" φ by means of the subgroup $E = O_2(G) = \mathbb{Z}_2^6$, one can suppose that

$$\lambda_{24} = \lambda_{25} = \lambda_{26} = \lambda_{27} = \lambda_{14} = \lambda_{34} = 1.$$

Recall that $\Gamma_r = \langle g(X^i) | g \in G \rangle_{\mathbb{Z}}$, $i = 0, 1$. Setting $Z = X^i$, we have

$$\varphi(Z) - Z \equiv 2 \sum_{\{i,j\} \in A} E(i, j) \,(\mathrm{mod}\, 4\Lambda),$$

where $A \subseteq \{\{3,5\},\{3,6\},\{3,7\},\{4,6\},\{4,7\},\{5,6\},\{5,7\}\}$. Since $A \in \mathcal{B}$, we get from Lemma 11.4.1 that $A = \emptyset$ or

$$A = \{\{4,6\},\{4,7\},\{5,6\},\{5,7\}\}.$$

This means that

$$\lambda_{35} = \lambda_{36} = \lambda_{37} = 1,\ \lambda_{46} = \lambda_{47} = \lambda_{56} = \lambda_{57} = \lambda.$$

Taking $Z = \psi(X^i)$, where $\psi = (12)(56) \in S$ (recall that $S = SL_3(2)$ acts faithfully on $V\backslash\{0\} = \{1,2,\ldots,7\}$), and using the same arguments, one can see that

$$\lambda_{14} = \lambda_{15} = \lambda_{16} = \lambda_{17} = 1,\ \lambda_{45} = \lambda_{67} = \lambda.$$

Finally, on considering $Z = \theta(X^i)$, where $\theta = (14)(36) \in S$, we get that

$$\lambda_{12} = \lambda_{23} = \lambda_{13} = \lambda.$$

Now we simply note that $\varphi = 1_{\mathcal{L}}$ if $\lambda = 1$, and $\varphi = -\bar{f}$ if $\lambda = -1$, where

$$f : V\backslash\{0\} \to \{0,1\},\ f(1) = f(2) = f(3) = 0,\ f(4) = f(5) = f(6) = f(7) = 1.$$

In other words, $\varphi \in Z \times G$. □

Proposition 11.4.4. $\mathbb{G}_r \cap \mathrm{Aut}(\Lambda) = Z \times G$.

Proof. Obviously, $\mathbb{G}_r \cap \mathrm{Aut}(\Lambda) \supseteq Z \times G$; we prove the reverse inclusion. Suppose that $\varphi \in \mathbb{G}_r \cap \mathrm{Aut}(\Lambda)$.

1) As usual, let $\mathcal{V}_2 = \{W_1,\ldots,W_7\}$ denote the variety of 2-dimensional subspaces in V, so that $\Omega = \cup_{i=1}^7 t(W_i)$. We claim that φ leaves this decomposition fixed. For, suppose that

$$\varphi : \{i,j\} \mapsto \{k,\ell\},\ \{i,i+j\} \mapsto \{m,n\}.$$

Since

$$X = 2E(i,j) + 2E(i,i+j) \in \Gamma_r^*$$

by Lemma 11.4.2, we have

$$\varphi(X) = 2E(k,\ell) + 2E(m,n) \in \Gamma_r^*.$$

Again applying Lemma 11.4.2, we obtain $\{m,n\} = \{k,k+\ell\}$ or $\{\ell,k+\ell\}$. This means that φ maps the triple $t(W)$ into the triple $t(W')$, where $W =< i,j >$, $W' =< k,\ell >$.

2) We claim that there exists an element $\psi \in S$ such that φ and ψ act identically on the set of triples $t(W_i)$, $1 \le i \le 7$. To this end, we fix the notation

$$W_1 =< a,b >_{\mathbb{F}_2},\ W_2 =< a,c >_{\mathbb{F}_2},\ W_3 =< a,b+c >_{\mathbb{F}_2},$$

$$W_4 =< b, c >_{\mathbb{F}_2}, \, W_5 =< b, a+c >_{\mathbb{F}_2}, \, W_6 =< c, a+b >_{\mathbb{F}_2},$$

$$W_7 =< a+b, b+c >_{\mathbb{F}_2} .$$

Note that S acts doubly transitively on $\mathcal{V}_2$. Set $S_2 = \{\theta \in S | \theta(W_1) = W_1, \theta(W_2) = W_2\}$. Then S_2 acts on $\mathcal{V}_2$ with orbits

$$\{W_1\}, \, \{W_2\}, \, \{W_3\}, \, \{W_4, W_5, W_6, W_7\}.$$

Furthermore, $W' = W_3$ is the unique element of $\mathcal{V}_2 \backslash \{W_1, W_2\}$ with the property that no element $\mathcal{S}(v)$, $v \in V \backslash \{0\}$, contains all the triples $t(W_1)$, $t(W_2)$, $t(W')$. The triple $t(W_3)$ is said to be *complementary* to the couple $(t(W_1), t(W_2))$.

As φ preserves Γ_r and Λ, it leaves the set $\mathcal{A} \backslash \{\emptyset\} = \{\mathcal{S}(v) | v \in V \backslash \{0\}\}$ fixed. Hence, if the triple $t(W_3)$ is complementary to the couple $(t(W_1), t(W_2))$, then the triple $\varphi(t(W_3))$ is complementary to the couple $(\varphi(t(W_1)), \varphi(t(W_2)))$. In view of this remark, we can "adjust" φ by means of S and suppose that

$$\varphi : t(W_1) \mapsto t(W_1), \; t(W_2) \mapsto t(W_2), \; t(W_3) \mapsto t(W_3).$$

"Adjusting" φ by means of S_2, one can suppose that $\varphi(t(W_4)) = t(W_4)$. Now $t(W_5)$ is complementary to the couple $(t(W_1), t(W_4))$, therefore $\varphi(t(W_5)) = t(W_5)$. Similarly, φ leaves $t(W_i)$, $i = 6, 7$, fixed.

3) By 2), we can assume that φ leaves every triple $t(W_i)$, $1 \le i \le 7$, fixed. By Lemma 11.4.3, $\varphi \in Z \times G$. □

Next, we consider a further $\mathbb{G}_r$-invariant lattice:

$$\Gamma_e = \left\{ X \in \Gamma_r^* | (X|X) \in 2\mathbb{Z} \right\} .$$

Clearly, $[\Gamma_r^* : \Gamma_e] = 2$, and $\Gamma_e \supseteq \Gamma_r^* \cap 2\Lambda$. Furthermore, it is not difficult to see that

$$\mathcal{B}_1 = \mathcal{B} \cap \mathcal{B}^*$$

admits a basis consisting of elements A_i, $1 \le i \le 6$, where

$$A_1 = \{\{1,2\}, \{1,5\}, \{1,7\}, \{2,6\}, \{2,7\}, \{5,6\}, \{5,7\}, \{6,7\}\} ,$$

$$A_2 = \{\{1,3\}, \{1,5\}, \{1,6\}, \{3,6\}, \{3,7\}, \{5,6\}, \{5,7\}, \{6,7\}\} ,$$

$$A_3 = \{\{1,4\}, \{1,5\}, \{1,6\}, \{1,7\}, \{4,6\}, \{4,7\}, \{5,6\}, \{5,7\}\} ,$$

$$A_4 = \{\{2,3\}, \{2,5\}, \{2,6\}, \{3,5\}, \{3,7\}, \{5,6\}, \{5,7\}, \{6,7\}\} ,$$

$$A_5 = \{\{2,4\}, \{2,5\}, \{2,6\}, \{2,7\}, \{4,5\}, \{4,7\}, \{5,6\}, \{6,7\}\} ,$$

$$A_6 = \{\{3,4\}, \{3,5\}, \{3,6\}, \{3,7\}, \{4,5\}, \{4,6\}, \{5,7\}, \{6,7\}\} ,$$

and

$$\left(|A_i \cap A_j|\right)_{1\le i,j\le 6} = \begin{pmatrix} 8 & 4 & 4 & 4 & 4 & 2 \\ 4 & 8 & 4 & 4 & 2 & 4 \\ 4 & 4 & 8 & 2 & 2 & 2 \\ 4 & 4 & 2 & 8 & 4 & 4 \\ 4 & 2 & 2 & 4 & 8 & 2 \\ 2 & 4 & 2 & 4 & 2 & 8 \end{pmatrix}. \tag{11.3}$$

Moreover,

$$A_1 + A_2 + A_4 = \mathcal{S}(4),\ A_3 + A_5 = \mathcal{S}(3),\ A_5 + A_6 = \mathcal{S}(1).$$

It is obvious that $(\Gamma_e + 2\Lambda)/2\Lambda = \mathcal{B}_1$. We note an important property of Γ_e, which is a consequence of (11.3):

$$\forall X, Y \in \Gamma_e,\ (X|X) \in 4\mathbb{Z},\ (X|Y) \in 2\mathbb{Z}.$$

Finally, $\Gamma_e \supseteq \Gamma_r \supseteq 2\Gamma_e$. So, we can consider the $\mathbb{F}_2\mathbb{G}_r$-module $\Delta = \Gamma_e/\Gamma_r$.

a) $\dim_{\mathbb{F}_2} \Delta = 7$. Indeed,

$$[\Gamma_e : \Gamma_r] = \frac{[\Lambda : \Gamma_r]}{[\Lambda : \Gamma_e]} = \frac{2^{(21-3)+(21-14)}}{2^{1+(21-18)+(21-7)}} = 2^7.$$

b) We define the bilinear form

$$B(\bar{X}, \bar{Y}) = \frac{1}{2}(X|Y) \pmod 2,$$

on Δ, where $\bar{X} = X + \Gamma_r$, $\bar{Y} = Y + \Gamma_r$. Note that B is correctly defined, since for $Z \in \Gamma_r$ we have

$$B(\bar{X}, \overline{Y+Z}) - B(\bar{X}, \bar{Y}) = \frac{1}{2}(X|Z) \pmod 2 = 0.$$

As $(X|X) \in 4\mathbb{Z}$ for all $X \in \Gamma_e$, B is alternating. It is not difficult to see that

$$\mathrm{Ker} B = \Gamma_e^*/\Gamma_r$$

has dimension 1. Thus we obtain a homomorphism

$$\vartheta : \mathbb{G}_r \to Sp(\bar{B}) = Sp_6(2),$$

where $\bar{B}$ is a non-degenerate bilinear alternating form defined on Γ_e/Γ_e^*.

The proof of the following statement is omitted.

Lemma 11.4.5. $(X|X) \geq 12$ *for all* $X \in \Gamma_e^* \backslash \{0\}$. *If in addition* $\Gamma_r = \Lambda^0$, *then*

$$\Gamma_e^* = \Phi\left(\langle \alpha \wedge \beta | \alpha, \beta \in R \rangle_{\mathbb{Z}}\right). \qquad \square$$

Using Proposition 11.4.4 it is not difficult to prove:

Lemma 11.4.6. $\mathrm{Ker}\vartheta = Z$. □

By Lemma 11.4.6,

$$G = \mathbb{Z}_2^6.SL_3(2) \subseteq \mathrm{Im}\vartheta \subseteq Sp_6(2).$$

Furthermore, G is a maximal subgroup of $Sp_6(2)$. Hence, either $\mathrm{Im}\vartheta = G$ and then $\mathbb{G}_r = Z \times G$, or $\mathrm{Im}\vartheta = Sp_6(2)$ and then $\mathbb{G}_r = \mathbb{Z}_2.Sp_6(2)$.

One can see that the lattices Λ^0 and Λ^1 are very much alike. At first sight, it seems that there exists an element $f \in \mathbb{Z}_2^{21} = O_2(\mathrm{Aut}(\Lambda))$ such that $f(\Lambda^0) = \Lambda^1$. Unfortunately, direct calculations show that there is no such element. In fact, the internal geometries of these lattices are quite different.

Theorem 11.4.7. *The following equalities hold.*

(i) $\mathrm{Aut}(\Lambda^0) = \mathbb{Z}_2 \times Sp_6(2) \cong W(E_7)$.

(ii) $\mathrm{Aut}(\Lambda^1) = \mathbb{Z}_2 \times (\mathbb{Z}_2^6.SL_3(2)) = Z \times G$.

Proof. (i) Clearly, $\mathrm{Aut}(R) = W(E_7)$ acts on the lattice $\Lambda_0 = \langle \alpha \wedge \beta | \alpha, \beta \in R,\ \alpha \perp \beta \rangle_{\mathbb{Z}}$ with kernel $\mathbb{Z}_2$. Correspondingly,

$$Sp_6(2) = W(E_7)/\mathbb{Z}_2 \hookrightarrow \mathrm{Aut}(\Lambda^0).$$

The restriction $\vartheta|_{Sp_6(2)}$ is obviously injective, so that $\mathrm{Im}\vartheta = Sp_6(2)$. Thus we obtain that

$$\mathrm{Aut}(\Lambda^0) = \mathbb{Z}_2 \times Sp_6(2).$$

(ii) First, we formulate some auxiliary statements.

Lemma 11.4.8. *If $1 \subset K \subset E = O_2(G)$ and $K \lhd G$, then $K \cong \mathbb{Z}_2^3$. In particular, G has no proper subgroups of index less than 7.* □

Lemma 11.4.9. *Let C_E and N_E denote the centralizer and normalizer of $E = O_2(G)$ in $GL(\mathcal{L}) = GL_{21}(\mathbb{C})$. Then*

1) *C_E consists of diagonal elements $E(i,j) \mapsto a_{ij}E(i,j)$;*

2) *N_E consists of monomial elements*

$$E(i,j) \mapsto a_{ij}E(\pi(i),\pi(j));\ \pi \in \mathbb{S}_7,\ a_{ij} \in \mathbb{C}^*.$$ □

Lemma 11.4.10. *Let C and N denote the centralizer and normalizer of G in $GL(\mathcal{L}) = GL_{21}(\mathbb{C})$. Then $N = C \times G$, and C consists of diagonal elements $E(i,j) \mapsto aE(i,j)$, $a \in \mathbb{C}^*$.* □

Completion of the proof of Theorem 11.4.7. By (i), we have

$$\mathbb{G}^0 = \mathrm{Aut}(\Lambda^0) = \mathbb{Z}_2 \times Sp_6(2).$$

Suppose that

$$\mathbb{G}^1 = \mathrm{Aut}(\Lambda^1) \supset Z \times G.$$

As we have shown above, under this assumption, $\mathbb{G}^1 = \mathbb{Z}_2.Sp_6(2)$. If $\mathbb{G}^1$ is perfect, then $\mathbb{G}^1$ is a double cover for $Sp_6(2)$. But in this case $\mathbb{G}^1$ has no faithful irreducible complex representations of degree 21 (see [ATLAS]), a contradiction. Hence, $\mathbb{G}^1$ is not perfect, and so $\mathbb{G}^1 = \mathbb{Z}_2 \times Sp_6(2)$. It is convenient to regard $\mathbb{G}^0$ and $\mathbb{G}^1$ as two faithful irreducible complex representations of degree 21 of the group $\mathbb{Z}_2 \times T$, where $T = Sp_6(2) \supset \mathbb{Z}_2^6.SL_3(2) = G_0$:

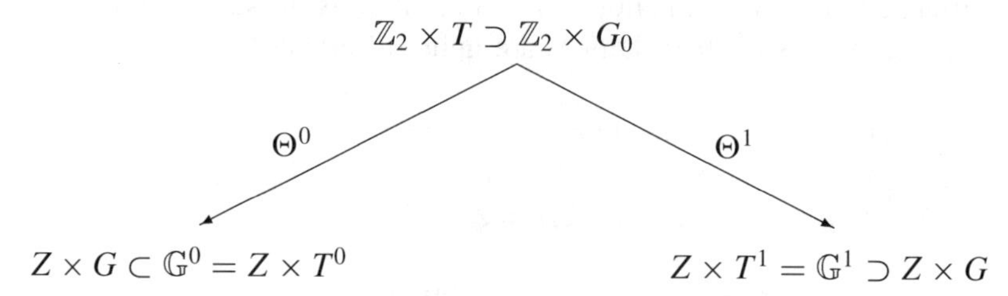

Clearly, Θ^i maps $\mathbb{Z}_2$ onto Z, and T onto T^i, where $i = 0, 1$. Next we consider the subgroups G and $\Theta^0(G_0)$, which are maximal subgroups of type $\mathbb{Z}_2^6.SL_3(2)$ of T^0. As $Sp_6(2)$ has a unique conjugacy class of maximal subgroups of this type, there exists an element g of $\mathbb{G}^0$ such that $\Theta^0(G_0) = gGg^{-1}$. Replacing Θ^0 by the representation $x \mapsto g^{-1}\Theta^0(x)g$, one can suppose that $\Theta^0(G_0) = G$. Similarly, $\Theta^1(G_0) = G$.

In accordance with [ATLAS], the group $T = Sp_6(2)$ has just two irreducible complex characters σ_1, σ_2 of degree 21. Moreover, if x is an involution in T, then $\sigma_1(x) \in \{9, 1, -3\}$, and $\sigma_2(x) \in \{-11, -3, 5\}$. In the meantime, the involution $\bar{f}_a \in E$ has trace 9 on $\mathcal{L}$. Hence, the characters $\mathrm{Tr}\Theta^0$, $\mathrm{Tr}\Theta^1$ are precisely σ_1, and the representations Θ^0, Θ^1 are equivalent:

$$\forall x \in \mathbb{Z}_2 \times T,\ \Theta^1(x) = A\Theta^0(x)A^{-1}$$

for some $A \in GL_{21}(\mathbb{C})$. In particular,

$$AGA^{-1} = A\Theta^0(G_0)A^{-1} = \Theta^1(G_0) = G,$$

that is, A normalizes G. By Lemma 11.4.10, $A = \lambda h$ for some $\lambda \in \mathbb{C}^*$, $h \in G$. But in that case,

$$\forall x \in \mathbb{Z}_2 \times T,\ \Theta^1(x) = A\Theta^0(x)A^{-1} = h\Theta^0(x)h^{-1} \in Z \times T^0,$$

that is, $\mathbb{G}^0 = \mathbb{G}^1$. From this it follows that $\mathbb{G}^0$ leaves the sublattice $\Lambda^0 \cap \Lambda^1 = \Lambda^0 \cap 2\Lambda$ fixed, on the one hand. On the other hand, some $\psi \in \mathbb{G}^0$ maps the generating vector X^0 of Λ^0 (see the beginning of the section) into a vector lying in 2Λ. In other words, ψ^{-1} is an element of $\mathbb{G}^0$ that does not preserve $\Lambda^0 \cap 2\Lambda$. This contradiction completes the proof of Theorem 11.4.7. □

Proof of Theorem 11.0.3. Suppose that $m = 2$ and $\Gamma \in \mathcal{R}_3$. Then

$$Z \times G \subseteq \mathbb{G} = \mathrm{Aut}(\Gamma) \subseteq \mathbb{G}_r = \mathrm{Aut}(\Gamma_r).$$

Furthermore, by Theorem 11.4.7, either $\mathbb{G}_r = \mathbb{Z}_2 \times Sp_6(2)$ and then $Z \times G$ is a maximal subgroup in $\mathbb{G}_r$, or $\mathbb{G}_r = Z \times G$. Hence,

$$\mathbb{G} \in \{\mathbb{Z}_2 \times Sp_6(2), Z \times G\}.$$

Next, we note that if $\mathbb{G} = Z \times G$, then $\mathbb{G} \subset \mathrm{Aut}(\Lambda)$ and so 4Λ is a $\mathbb{G}$-invariant root sublattice (of type $21A_1$) of Γ. This means that $\Gamma \in \mathcal{R}$. For example, $\Lambda^1 \in \mathcal{R}_3 \cap \mathcal{R}$. This implies also that

$$\forall \Gamma' \in \overline{\mathcal{R}},\ \mathrm{Aut}(\Gamma') = \mathbb{Z}_2 \times Sp_6(2).$$

Conversely, suppose that $\mathbb{G} = \mathbb{Z}_2 \times Sp_6(2)$, but $\Gamma \in \mathcal{R}$. Let Γ_1 be a $\mathbb{G}$-invariant root sublattice of Γ: $\Gamma_1 =< \mathcal{M} >_{\mathbb{Z}}$, where $\mathcal{M}$ is a root system. Clearly,

$$\mathcal{M} = \mathcal{M}_1 \oplus \cdots \oplus \mathcal{M}_n \cong n\mathcal{M}_1,$$

where $\mathcal{M}_1$ is an indecomposable root system, and $n|21$. Furthermore, $Sp_6(2)$ leaves this decomposition fixed. But $Sp_6(2)$ has no proper subgroups of index less than 28, so in fact $n = 1$ and $\mathcal{M}_1$ is of type A_{21}, B_{21}, C_{21}, or D_{21}. This implies that

$$Sp_6(2) \subseteq \mathrm{Aut}(\mathcal{M}_1) \subseteq \mathbb{Z}_2 \wr \mathbb{S}_{22}.$$

In particular, $Sp_6(2)$ has a proper subgroup of index less than 22, a contradiction. Hence, $\Gamma \in \overline{\mathcal{R}}$. For instance, $\Lambda^0 \in \overline{\mathcal{R}}$. □

11.5 Lattices of nonroot-type: the case D_4

Suppose that $\mathcal{L}$ is a Lie algebra of type D_4, and take Γ in $\mathcal{R}_3$. First we determine the structure of the reduced sublattice Γ_r. Set

$$\mathcal{A} = (\Gamma_r + 2\Lambda)/2\Lambda,\ \mathcal{B} = (\Gamma_r \cap 2\Lambda)/4\Lambda.$$

Other notation is taken from §§11.1, 11.4. By Lemma 11.1.5, for any $u \in \mathcal{U}\backslash 2\Lambda$, $S(u)$ has cardinality not less than $\frac{1}{2}(4 + 4.4) = 10$. This means that instead of (11.1) and (11.2), the following conditions hold here:

$$|S(u)| = 12 \text{ or } 16;$$

$$|S_v(u)| = 0,\ 4 \text{ or } 6.$$

In fact, it is not difficult to see that $S(u)$ belongs to one of the following G_0-types:

$$\mathcal{S}(a, b) = \big\{\{x, y\} \in \Omega | x, y \neq a, b;\ x + y \neq a + b\big\}, \tag{11.4}$$

where $\{a, b\} \in \Omega$;

$$\mathcal{T}(W) = \big\{\{x, y\} \in \Omega | x \in W, y \notin W\big\}, \tag{11.5}$$

where $W \in \mathcal{V}_2$; and

$$\begin{array}{l} \big\{\{0, 4\}, \{0, 5\}, \{0, 6\}, \{0, 7\}, \{1, 4\}, \{2, 4\}, \{3, 4\}, \{1, 2\}, \\ \{1, 3\}, \{1, 5\}, \{2, 3\}, \{2, 6\}, \{3, 7\}, \{5, 6\}, \{5, 7\}, \{6, 7\}\big\}. \end{array} \tag{11.6}$$

Moreover, $G_0 = ASL_3(2)$ acts transitively on the sets $\mathcal{M}_i$, $i = 1, 2, 3$, of all subsets $S(u)$ of types (11.4), (11.5), (11.6), respectively. Furthermore,

$$|\mathcal{M}_1| = |\mathcal{M}_3| = 28,\ |\mathcal{M}_2| = 7.$$

Thus, the following statement holds.

Lemma 11.5.1. *If* $u \in \mathcal{U}\backslash 2\Lambda$, *then* $S(u) \in \mathcal{M}_1 \cup \mathcal{M}_2 \cup \mathcal{M}_3$. □

As $\Gamma_r =< \mathcal{U} >_{\mathbb{Z}}$, we have $\mathcal{A} = \langle S(u) | u \in \mathcal{U}\backslash 2\Lambda\rangle_{\mathbb{F}_2}$. Therefore, as an immediate consequence of Lemma 11.5.1, we get

Lemma 11.5.2. $\mathcal{A}$ *is either* $\mathcal{A}_1 = \{\emptyset\} \cup \mathcal{M}_1 \cup \mathcal{M}_2 \cup \mathcal{M}_3$, *or* $\mathcal{A}_2 = \{\emptyset\} \cup \mathcal{M}_2$. □

We mention the identity

$$\mathcal{S}(a, b) + \mathcal{S}(a, c) = \mathcal{S}(a, a + b + c)$$

valid for distinct $a, b, c \in V$, and the fact that

$$\mathcal{A}_1 = \langle \mathcal{S}(0, 1), \mathcal{S}(0, 2), \mathcal{S}(0, 4), \mathcal{S}(1, 3), \mathcal{S}(1, 5), \mathcal{S}(2, 6)\rangle_{\mathbb{F}_2}.$$

In the first instance we are interested in lattices $\Gamma \in \mathcal{R}_3$ that satisfy the additional condition that

$$\dim_{\mathbb{F}_2} \mathcal{A} = 6,$$

that is, $\mathcal{A} = \mathcal{A}_1$. The remaining lattices of $\mathcal{R}_3$, as will be shown later, are not of great interest.

We omit the proof of the following statement.

Lemma 11.5.3. *If* $\dim_{\mathbb{F}_2} \mathcal{A} = 6$, *then* $\dim_{\mathbb{F}_2} \mathcal{B} = 21$, *and*

$$\mathcal{B} = \langle A, S_v(A) | A \in \mathcal{A}, v \in V\rangle_{\mathbb{F}_2}.$$

□

Recall that

$$\Gamma'^{*} = \{X \in \mathcal{L} | (X|\Gamma') \subseteq 4\mathbb{Z}\}.$$

Corollary 11.5.4. *If* $\dim_{\mathbb{F}_2} \mathcal{A} = 6$, *then* $[\Gamma_r^* : \Gamma_r] = 4$, *that is,* Γ_r *is almost unimodular.* □

Proposition 11.5.5. *There exist just two possibilities,* Γ^0 *and* Γ^1, *for the reduced sublattice* Γ_r *of a lattice* $\Gamma \in \mathcal{R}_3$ *that satisfies the additional condition* $\dim_{\mathbb{F}_2} \mathcal{A} = 6$.

Proof. Suppose that $\Gamma \in \mathcal{R}_3$ and $\mathcal{A} = \mathcal{A}_1$. Note that every vector $u \in \Gamma_r \backslash 2\Lambda$ such that

$$u + 2\Lambda \notin \mathcal{A}_2$$

determines Γ_r uniquely in the sense that

$$\Gamma_r = \langle g(u) | g \in G \rangle_{\mathbb{Z}}.$$

Without loss of generality, one can suppose that

$$(u|u) = 16, \; u + 2\Lambda = \mathcal{S}(0, 1).$$

This means that

$$u = 2a_{k\ell} E(k, \ell) + \sum_{\omega \in \mathcal{S}(0,1)} a_\omega E_\omega,$$

where $\{k, \ell\} \in \Omega \backslash \mathcal{S}(0, 1)$; $a_{k\ell}$, $a_\omega = \pm 1$. "Adjusting" u by means of vectors $v \in \Gamma_r \cap 2\Lambda$, we can reduce u to the form

$$u = 2E(2, 3) + \sum_{\omega \in \mathcal{S}(0,1)} a_\omega E_\omega. \tag{11.7}$$

Applying a suitable automorphism $\bar{f}$ from $E = \mathbb{Z}_2^7$ to u, and one more "adjusting" $\bar{f}(u)$ by means of vectors $v \in \Gamma_r \cap 2\Lambda$, we can suppose that

$$a_{24} = a_{25} = a_{26} = a_{27} = a_{34} = a_{46} = 1. \tag{11.8}$$

But $(u|g(u)) \in 4\mathbb{Z}$ for all $g \in G$, therefore there exist at most two additional vectors u of the form (11.7) satisfying condition (11.8). Namely, if u and u_1 are different vectors of that kind, then

$$v_1 = u_1 - u \equiv 2 \sum_{\omega \in A} E_\omega \,(\mathrm{mod}\, 4\Lambda),$$

where $A = \{\{3, 6\}, \{3, 7\}, \{4, 7\}, \{5, 6\}\}$. Moreover, if u is an additional vector, and $u_1 \equiv u + v_1 \,(\mathrm{mod}\, 4\Lambda)$, then u_1 also satisfies (11.7) and (11.8). Note that $A \notin \mathcal{B}$, so that

$$< g(u) | g \in G >_{\mathbb{Z}} \neq < g(u_1) | g \in G >_{\mathbb{Z}}.$$

Finally, we construct a vector of the required kind. Realize the root system R of type E_8 in the form

$$R = \{\pm 2e_a,\ \pm e_a \pm e_b \pm e_c \pm e_d | a, b, c, d \in V, a + b + c + d = 0\},$$

where $\{e_a | a \in V\}$ is an orthonormal basis for $U = \mathbb{C}^8$. The construction described at the beginning of §11.4, with the root system of type E_7 replaced by one of type E_8, leads to

$$\Gamma^0 = \left\langle g(X^0) | g \in G \right\rangle_{\mathbb{Z}}$$

with $u = X^0$. Any other possibility for Γ_r is denoted by Γ^1. □

Using the description of $\mathcal{A}$ and $\mathcal{B}$, one can prove the following two lemmas.

Lemma 11.5.6. *For all* $X \in (\Gamma^i)^* \backslash \{0\}$ *we have*

$$4 | (X|X) \text{ and } (X|X) \geq 12.$$

If $(X|X) = 12$, *then*

$$X = \pm 2E(a, b) \pm 2E(a, c) \pm 2E(a, d)$$

for some $a, b, c, d \in V$ *such that* $a + b + c + d = 0$. □

Lemma 11.5.7. (i) *If* $A \in \mathcal{B}^*$, $|A| = 16$, *then* $A \in \mathcal{M}_2 \cup \mathcal{M}_3$ *or* $\Omega \backslash A \in \mathcal{M}_1$.

(ii) *If* $A \in \mathcal{A}^*$ *and* $|A| = 4$, *then* $A \in \mathcal{B}$ *and* A *belongs to one of the following five* G_0*-types:*

$$\Omega_1 = \big\{\{0, 1\}, \{2, 3\}, \{4, 5\}, \{6, 7\}\big\},$$

$$B_0 = \big\{\{0, 7\}, \{2, 7\}, \{4, 7\}, \{6, 7\}\big\},\ B_1 = \big\{\{0, 7\}, \{1, 2\}, \{1, 4\}, \{6, 7\}\big\},$$

$$C_0 = \big\{\{0, 1\}, \{2, 3\}, \{0, 3\}, \{1, 2\}\big\},\ C_1 = \big\{\{0, 1\}, \{2, 3\}, \{4, 7\}, \{5, 6\}\big\}.$$ □

Corollary 11.5.8. *There are invariant lattices* $\tilde{\Gamma}^i$, $i = 0, 1$, *such that*

(i) $(\Gamma^i)^* \supset \tilde{\Gamma}^i \supset \Gamma^i$, $[(\Gamma^i)^* : \tilde{\Gamma}^i] = [\tilde{\Gamma}^i : \Gamma^i] = 2$;

(ii) $\frac{1}{2}\tilde{\Gamma}^i$ *is an odd unimodular lattice with minimum norm* 3.

Proof. Set

$$\tilde{\Gamma}^i = \langle \Gamma^i, 2(E(0, 1) + E(0, 2) + E(0, 3)) \rangle_{\mathbb{Z}}.$$

Then

$$(\tilde{\Gamma}^i + 2\Lambda)/2\Lambda = \mathcal{A},\ (\tilde{\Gamma}^i \cap 2\Lambda)/4\Lambda = \mathcal{A}^*,$$

and so $\frac{1}{2}\tilde{\Gamma}^i$ is an unimodular lattice. Furthermore, by Lemma 11.5.6, the minimum norm of $\frac{1}{2}\tilde{\Gamma}^i$ is 3. Note also that

$$\Gamma^i = \big\{X \in \tilde{\Gamma}^i | (X|X) \in 8\mathbb{Z}\big\}.$$ □

Proposition 11.5.9. *Suppose that* $\Gamma \in \mathcal{R}_3$ *and* $\dim_{\mathbb{F}_2} \mathcal{A} = 6$, *and set* $\mathbb{G}_r = \mathrm{Aut}(\Gamma_r)$. *Then* $\mathbb{G}_r \cap \mathrm{Aut}(\Lambda) = Z \times G$.

Proof. If $\varphi \in \mathbb{G}_r \cap \mathrm{Aut}(\Lambda)$, then φ leaves the set

$$\{X \in \Gamma_r^* \cap 2\Lambda | (X|X) = 12\} =$$

$$= \{X = \pm 2E(a,b) \pm 2E(a,c) \pm 2E(a,d) | a,b,c,d \in V, a+b+c+d=0\}$$

fixed. Now by changing slightly the arguments used in the proof of Proposition 11.4.4, we can establish the desired inclusion: $\varphi \in Z \times G$. □

Our next goal is to describe the action of $\mathbb{G}_r$ on the set of minimal vectors of Γ_r^*. For this we set

$$\mathcal{N} = \{X \in \Gamma_r^* | (X|X) = 12\}.$$

By Lemma 11.5.6, $\mathcal{N} = \bigcup_{j=0}^3 \mathcal{N}_j$, where

$$\mathcal{N}_0 = \mathcal{N} \cap 2\Lambda =$$

$$= \{\pm 2E(a,b) \pm 2E(a,c) \pm 2E(a,d) | a,b,c,d \in V, a+b+c+d=0\},$$

$$\mathcal{N}_1 = \left\{X = \sum_{\omega \in A} \pm E_\omega \in \mathcal{N} | A \in \mathcal{M}_1\right\},$$

$$\mathcal{N}_j = \left\{X = \sum_{\omega \in A} \pm E_\omega \in \mathcal{N} | \Omega \backslash A \in \mathcal{M}_j\right\},\ j = 2, 3.$$

Note that

$$|\mathcal{N}_0| = 2^6.7,\ |\mathcal{N}_1| = 2^8.7,\ |\mathcal{N}_2| = 2^7.7,\ |\mathcal{N}_3| = 2^9.7. \qquad (11.9)$$

The following statement is obvious.

Lemma 11.5.10. $Z \times G$ *acts transitively on each of the sets* $\mathcal{N}_j$, $0 \le j \le 3$. □

In view of Proposition 11.5.9, we first assume that

$$\mathbb{G}_r \supset Z \times G. \qquad (11.10)$$

Lemma 11.5.11. *Under assumption* (11.10), $\mathcal{N}_0$ *and* $\mathcal{N}_2$ *cannot be* $\mathbb{G}_r$*-orbits.*

Proof. 1) Suppose that $\mathcal{N}_0$ is a $\mathbb{G}_r$-orbit. Then $\mathbb{G}_r$ fixes both of the sublattices

$$\Gamma_1 = \langle \mathcal{N}_0 \rangle_{\mathbb{Z}},\ \langle X \in \Gamma_1^* | (X|X) = 4 \rangle_{\mathbb{Z}} = 2\Lambda.$$

Hence, $\mathbb{G}_r = \mathbb{G}_r \cap \mathrm{Aut}(\Lambda) = Z \times G$, contrary to (11.10).

2) Suppose that $\mathcal{N}_2$ is a $\mathbb{G}_r$-orbit. Then $\mathbb{G}_r$ fixes $\Gamma_2 = \langle \mathcal{N}_2 \rangle_{\mathbb{Z}}$ and

$$\{X \in \Gamma_2^* | (X|X) = 8\} =$$

$$= \{\pm 2E(a, b) \pm 2E(c, d) | a, b, c, d \in V,\ a + b = c + d\}.$$

In particular, $\mathbb{G}_r$ preserves the given MOD $\mathcal{D}$. From this it is not difficult to see that $\mathbb{G}_r = \mathbb{G}_r \cap \mathrm{Aut}(\Lambda) = Z \times G$. □

From now on we set

$$\mathbb{G}^i = \mathrm{Aut}(\Gamma^i),\ \tilde{\mathbb{G}}^i = \mathrm{Aut}(\tilde{\Gamma}^i),\ i = 0, 1.$$

Recall that

$$\Gamma^0 = \Phi\left(\langle \alpha \wedge \beta | \alpha, \beta \in R, \alpha \perp \beta \rangle_{\mathbb{Z}}\right).$$

Therefore, the group $\mathrm{Aut}(R) = W(E_8)$ acts on Γ^0 with kernel $\mathbb{Z}_2 = Z(W(E_8))$. In particular, $\mathbb{G}^0$ and $\tilde{\mathbb{G}}^0$ both contain the subgroup

$$\mathbb{G}_0^0 = Z \times (\mathrm{Aut}(R)/Z(\mathrm{Aut}(R))) \cong \mathbb{Z}_2 \times O_8^+(2).$$

Moreover, the following simple statement holds.

Lemma 11.5.12. (i) *For $i = 0, 1$ we have $\mathbb{G}^i \supseteq \tilde{\mathbb{G}}^i$.*
(ii) *$\mathbb{G}_0^0$ acts on $\mathcal{N}$ with two orbits, namely $\mathcal{N}_0 \cup \mathcal{N}_1$ and $\mathcal{N}_2 \cup \mathcal{N}_3$.* □

Before determining the groups $\mathbb{G}^i$, we study some of their subgroups. We shall use the following embeddings:

$$U' = \langle e_a | a \in V \backslash \{0\} \rangle_{\mathbb{C}} \subset U,$$

$$\mathcal{L}' = \langle E(i, j) | 1 \le i \ne j \le 7 \rangle_{\mathbb{C}} \subset \mathcal{L},$$

$$G' = \left(St_G(\mathcal{L}')\right)|_{\mathcal{L}'},$$

$$R' = \{r \in R | r \perp e_0\},$$

where $\mathcal{L}'$ is a Lie algebra of type B_3, and R' is a root system of type E_7. Set also

$$\mathcal{E} = \langle E(0, j) | 1 \le j \le 7 \rangle_{\mathbb{C}},$$

$$\mathbb{G}_7^i = \{\varphi \in \mathbb{G}^i | \varphi(\mathcal{E}) = \mathcal{E}\}.$$

Proposition 11.5.13. *The following equalities hold:*

$$\mathbb{G}_7^0 = Z \times \left((\mathrm{Aut}(R')/Z(\mathrm{Aut}(R'))) \times < -\bar{f}_0 >\right) \cong \mathbb{Z}_2 \times (Sp_6(2) \times \mathbb{Z}_2) \subset \mathbb{G}_0^0,$$

$$\mathbb{G}_7^1 = Z \times (G' \times < -\bar{f}_0 >) \cong \mathbb{Z}_2 \times (\mathbb{Z}_2^6.SL_3(2) \times \mathbb{Z}_2) \subset Z \times G.$$

Proof. Recall that the automorphism $\bar{f}_0$ acts as follows:

$$E(i,j) \mapsto (-1)^{\delta_{i,0}} E(i,j),\ 0 \le i < j \le 7.$$

Clearly, every element φ of $\mathbb{G}_7^i$ fixes the sublattices

$$\Delta^i = \Gamma^i \cap \mathcal{E},\ \Lambda^i = \Gamma^i \cap \mathcal{L}'$$

Here, the lattices Λ^i, $i = 0, 1$, are nothing but the two possilibities for the reduced sublattice considered in §11.4. Hence, by Theorem 11.4.7 we have

$$\varphi|_{\mathcal{L}'} \in \begin{cases} \mathrm{Aut}(\Lambda^0) = (\mathrm{Aut}(R')/Z(\mathrm{Aut}(R'))) \times < -\bar{f}_0 > & \text{if} \quad i = 0, \\ \mathrm{Aut}(\Lambda^1) = G' \times < -\bar{f}_0 > & \text{if} \quad i = 1. \end{cases}$$

"Adjusting" φ using $\mathbb{G}_0^0$ if $i = 0$, and G if $i = 1$, we can suppose that $\varphi|_{\mathcal{L}'} = 1_{\mathcal{L}'}$. Furthermore,

$$\Delta^0 = \Delta^1 = \{2 \sum_{j=1}^{7} a_j E(0,j) | a_j \in \mathbb{Z},$$

$$a_1 \equiv a_6 + a_7,\ a_2 \equiv a_5 + a_7,\ a_3 \equiv a_5 + a_6,\ a_4 \equiv a_5 + a_6 + a_7 \pmod 2\},$$

is a root lattice of type E_7. Hence, $\varphi|_{\mathcal{E}} \in W(E_7)$. From this and the condition $\varphi|_{\mathcal{L}'} = 1_{\mathcal{L}'}$ it is not difficult to show that in fact $\varphi = 1_{\mathcal{L}}$ or $\varphi = \bar{f}_0$. □

Proposition 11.5.14. *Set*

$$\mathbb{G}_1^i = \{\varphi \in \mathbb{G}^i | \varphi(E(0,1)) = \pm E(0,1)\}.$$

Then $\mathbb{G}_1^0 \subseteq \mathbb{G}_0^0$, $\mathbb{G}_1^1 \subseteq Z \times G$.

Proof. First, by Lemma 11.5.6, every element φ of $\mathbb{G}_1^i$ preserves the set

$$\mathcal{K} = \{\pm 2E(0,1) \pm 2E(k,j) \pm 2E(k,j+1) | k = 0, 1,\ j = 2, 4, 6\}.$$

Furthermore,

$$\mathcal{L}_1 = \langle \mathcal{K} \rangle_{\mathbb{C}} = \langle E(0,1), E(0,j), E(1,j) | 2 \le j \le 7 \rangle_{\mathbb{C}}.$$

Because $\varphi(\mathcal{L}_1) = \mathcal{L}_1$, φ preserves the sublattice $\Delta_1 = \mathcal{L}_1 \cap (\Gamma^i)^*$. Setting $\Lambda_1 = \mathcal{L}_1 \cap \Lambda$, we have

$$4\Lambda_1 \subset \Delta_1 \subset 2\Lambda_1,\ \dim_{\mathbb{F}_2} \Delta_1/4\Lambda_1 = 8.$$

Hence, for

$$\Delta_2 = \{X \in \mathcal{L}_1 | (X|\Delta_1) \subseteq 4\mathbb{Z},\ (X|X) \in 2\mathbb{Z}\}$$

we get

$$\varphi(\Delta_2) = \Delta_2,\ 2\Lambda_1 \subset \Delta_2 \subset \Lambda_1,\ \dim_{\mathbb{F}_2} \Delta_2/2\Lambda_1 = 4.$$

Namely, Δ_2 is generated modulo $2\Lambda_1$ by the four vectors

$$E(0,2) + E(0,3) + E(0,4) + E(0,5),\ E(0,2) + E(0,3) + E(0,6) + E(0,7),$$

$$E(1,2) + E(1,3) + E(1,4) + E(1,5),\ E(1,2) + E(1,3) + E(1,6) + E(1,7).$$

In particular, the set

$$\mathcal{K}' = \{X \in \Delta_2 | (X|X) = 4\}$$

is nothing but a root system of type $A_1 \oplus D_6 \oplus D_6$, where the first component A_1 lies in $\mathcal{L}_{10} =< E(0,1) >_{\mathbb{C}}$, and the second and the third components are contained in

$$\mathcal{L}_{11} =< E(0,2), \ldots, E(0,7) >_{\mathbb{C}}$$

and

$$\mathcal{L}_{12} =< E(1,2), \ldots, E(1,7) >_{\mathbb{C}},$$

respectively. Clearly, φ preserves A_1, and permutes the components of type D_6 among themselves. The translation

$$t_1 : E(a,b) \mapsto E(a+1, b+1)$$

lies in $\mathbb{G}_1^i$ and transposes the components of type D_6. Hence, either φ, or φt_1 preserves the subspace $\mathcal{L}_{10} \oplus \mathcal{L}_{11} = \mathcal{E}$. Now simply apply Proposition 11.5.13. □

As to the relation between the groups $\mathbb{G}^i$ and $\tilde{\mathbb{G}}^i$, one can prove the following statement.

Proposition 11.5.15. *One of the following assertions holds.*
(i) $\mathbb{G}^i = \tilde{\mathbb{G}}^i$. *If in addition* $\mathbb{G}^i \neq Z \times G$, *then* $\mathbb{G}^i$ *acts on the set*

$$\mathcal{N} = \{X \in (\Gamma^i)^* | (X|X) = 12\}$$

of minimal vectors of $(\Gamma^i)^*$ *with just two orbits,* $\mathcal{N}_0 \cup \mathcal{N}_1$ *and* $\mathcal{N}_2 \cup \mathcal{N}_3$.
(ii) $(\mathbb{G}^i : \tilde{\mathbb{G}}^i) = 3$. □

The internal geometry of the lattices $\tilde{\Gamma}^i$, $i = 0, 1$, is elucidated in the following statement.

Proposition 11.5.16. *For any* X *in* $\tilde{\Gamma}^i$ *with* $(X|X) = 16$, *we set*

$$p(X) = |\{Y \in \tilde{\Gamma}^i | (Y|Y) = 12,\ (X|Y) = 8\}|.$$

Then

(i) *$p(X) \leq 24$;*

(ii) *if $i = 0$, then $p(X) = 24$ if and only if X belongs to the $\mathbb{G}_0^0$-orbit of the vector $X_0 = 4E(0, 1)$;*

(iii) *if $i = 1$, then $p(X) = 24$ if and only if X belongs to the $(Z \times G)$-orbit of one of the following vectors:*

$$X_0 = 4E(0, 1),$$

$$X_1 = 2\left(E(0, 2) + E(0, 3) + E(0, 4) + E(0, 5)\right),$$

$$X_2 = 2\left(E(0, 2) + E(0, 3) + E(1, 2) + E(1, 3)\right),$$

$$X_3 = (E(2, 4) + E(2, 5) + E(2, 6) + E(2, 7)) +$$
$$+ (E(0, 4) - E(0, 5) + E(0, 6) - E(0, 7)) +$$
$$+ (E(1, 4) - E(1, 5) - E(1, 6) + E(1, 7)) +$$
$$+ (E(3, 4) + E(3, 5) - E(3, 6) - E(3, 7)).$$

□

We are now able to determine the groups $\tilde{\mathbb{G}}^i$, $i = 0, 1$.

Proposition 11.5.17. *The automorphism group* $\mathrm{Aut}(\tilde{\Gamma}^0)$ *of the odd unimodular lattice $\tilde{\Gamma}^0$ is $\mathbb{G}_0^0 = \mathbb{Z}_2 \times O_8^+(2)$.*

Proof. We know that $\tilde{\mathbb{G}}^0 \supseteq \mathbb{G}_0^0$. Conversely, take φ in $\tilde{\mathbb{G}}^0$. Then by Proposition 11.5.16, the vector $\varphi(X_0)$, where $X_0 = 4E(0, 1)$, belongs to the $\mathbb{G}_0^0$-orbit of X_0, that is, $\varphi(X_0) = \psi(X_0)$ for some $\psi \in \mathbb{G}_0^0$. In this case,

$$\psi^{-1}\varphi \in \mathbb{G}_1^0 \subseteq \mathbb{G}_0^0,$$

by Proposition 11.5.14. Hence, $\varphi \in \mathbb{G}_0^0$. □

Proposition 11.5.18. *The automorphism group* $\mathrm{Aut}(\tilde{\Gamma}^1)$ *of the odd unimodular lattice $\tilde{\Gamma}^1$ is $Z \times G$.*

Proof. Set $\Gamma_2 =< \mathcal{X} >_{\mathbb{Z}}$, where

$$\mathcal{X} = \left\{X \in \tilde{\Gamma}^1 | (X|X) = 16,\ p(X) = 24\right\}.$$

Then Γ_2 is a $\tilde{\mathbb{G}}^1$-invariant lattice, $\Lambda \supset \Gamma_2 \supset 4\Lambda$, and

$$(\Gamma_2 + 2\Lambda)/2\Lambda = \mathcal{A}_2,\ (\Gamma_2 \cap 2\Lambda)/4\Lambda = \mathcal{B}_2,$$

where

$$\mathcal{B}_2 = \langle S_v(A) \mid A \in \mathcal{A}_2,\ v \in V\rangle_{\mathbb{F}_2}.$$

Furthermore, we set

$$\mathcal{K} = \left\{X \in \Gamma_2^* | (X|X) = 8\right\}.$$

Then $\mathcal{K} \subset 2\Lambda$. More precisely,

$$\mathcal{K} = \{\pm 2E(a, b) \pm 2E(c, d) | a, b, c, d \in V, a + b = c + d\}$$

is a root system of type $7D_4$. But $\tilde{\mathbb{G}}^1$ preserves $\mathcal{K}$, so that $\tilde{\mathbb{G}}^1$ is contained in $\mathrm{Aut}(\mathcal{K})$ and acts on the set $\{\mathcal{H}_v | v \in V\backslash\{0\}\}$ of Cartan subalgebras that occur in our MOD. Now, among those vectors $X \in \mathcal{X}$, only those of the form $\pm 4E_\omega$, $\omega \in \Omega$, have just one nonzero projection onto Cartan subalgebras $\mathcal{H}_v$, $v \in V\backslash\{0\}$. This means that $\tilde{\mathbb{G}}^1$ preserves the lattice

$$\Lambda = \langle E_\omega | \omega \in \Omega\rangle_{\mathbb{Z}}.$$

To conclude the proof, apply Proposition 11.5.9. □

The proof of the following result is omitted.

Lemma 11.5.19. *If $\varphi \in \mathbb{G}^i\backslash\tilde{\mathbb{G}}^i$, then φ normalizes neither G nor $E = \mathbb{Z}_2^7$. In fact,*

$$\varphi E \varphi^{-1} \not\subset \tilde{\mathbb{G}}^i.$$

□

Corollary 11.5.20. *Suppose that $\mathbb{G}^i \neq \tilde{\mathbb{G}}^i$. Then*

$$\mathbb{G}^i = \begin{cases} \mathbb{Z}_2.\mathrm{Aut}(\Omega_8^+(2)) & \text{if } i = 0, \\ E_1.SL_3(2) & \text{if } i = 1, \end{cases}$$

where E_1 is a certain soluble group of order $2^{11}.3$.

Proof. 1) First we assume that $i = 0$. By Proposition 11.5.17, we have $\tilde{\mathbb{G}}^0 = Z \times (L.\mathbb{Z}_2)$, where $L = \Omega_8^+(2)$ is a simple group and $L.\mathbb{Z}_2 = O_8^+(2)$. In view of Proposition 11.5.15 one can suppose that $(\mathbb{G}^0 : \tilde{\mathbb{G}}^0) = 3$. Consider the permutation representation of $\mathbb{G}^0$ on the cosets of $\tilde{\mathbb{G}}^0$, with kernel K, say. If $(\mathbb{G}^0 : K) = 3$, then $\tilde{\mathbb{G}}^0 = K \lhd \mathbb{G}^0$, contrary to Lemma 11.5.19. Therefore, $\mathbb{G}^0/K \cong \mathbb{S}_3$, $(\tilde{\mathbb{G}}^0 : K) = 2$. If $K \not\supseteq Z$, then $\tilde{\mathbb{G}}^0 =< K, Z >$ is a normal subgroup of $\mathbb{G}^0$, again contrary to Lemma 11.5.19. Hence, $K = Z \times L$. Next we consider the natural homomorphism

$$\mathbb{G}^0 \to \mathrm{Aut}(K) = \mathrm{Aut}(L)$$

with kernel $C = C_{\mathbb{G}^0}(K)$. Clearly, $C \cap K = Z$. But $C, K \lhd \mathbb{G}^0$, and so

$$\mathbb{S}_3 = \mathbb{G}^0/K \rhd KC/K \cong C/(K \cap C).$$

This means that $|C| = 2$, 6, or 12. It is easy to see that C has no elements of order 3, so that $|C| = 2$ and $C = Z$. Thus we arrive at the conclusion that $\mathbb{G}^0/Z = \mathrm{Aut}(L)$.

2) Now suppose that $i = 1$. By Proposition 11.5.18, $\tilde{\mathbb{G}}^1 = Z \times G$. In particular, $\tilde{\mathbb{G}}^1$ has four orbits $\mathcal{N}_j$, $0 \leq j \leq 3$, on $\mathcal{N}$. Consider any X in $\mathcal{N}_0$. As $\mathbb{G}^1 \supset \tilde{\mathbb{G}}^1 = Z \times G$, we have $X^{\mathbb{G}^1} \neq \mathcal{N}_0$ by Lemma 11.5.11. Hence,

$$|\mathcal{N}_0| < |X^{\mathbb{G}^1}| = \frac{|\mathbb{G}^1|}{|St_{\mathbb{G}^1}(X)|} \leq \frac{3|\tilde{\mathbb{G}}^1|}{|St_{\tilde{\mathbb{G}}^1}(X)|} = 3|\mathcal{N}_0|.$$

In view of (11.9), $X^{\mathbb{G}^1} = \mathcal{N}_0 \cup \mathcal{N}_2$. Following the proof of Lemma 11.5.11, we arrive at the conclusion that $\mathbb{G}^1$ preserves the given MOD $\mathcal{D}$, that is, $\mathbb{G}^1$ acts on the set $\{\mathcal{H}_v | v \in V\backslash\{0\}\}$. Consider the homomorphism ϑ arising here:

$$\vartheta : \mathbb{G}^1 \to \mathbb{S}_7 = \mathrm{Sym}(V\backslash\{0\}).$$

Clearly, $\mathrm{Im}\vartheta \supseteq SL_3(2)$. In fact, equality occurs. To see this, note that if Y is a minimal vector of $(\Gamma^1)^*$ with just three nonzero projections to Cartan subalgebras $\mathcal{H}_a$, $\mathcal{H}_b$, $\mathcal{H}_c$, then $Y \in \mathcal{N}_0 \cup \mathcal{N}_2$, and $a + b + c = 0$. This means that

$$\forall \varphi \in \mathbb{G}^1,\ a + b + c = 0 \Rightarrow \vartheta(\varphi)(a) + \vartheta(\varphi)(b) + \vartheta(\varphi)(c) = 0.$$

In other words, $\mathrm{Im}\vartheta = SL(V)$. Finally,

$$|\mathrm{Ker}\vartheta| = \frac{3|Z \times G|}{|\mathrm{Im}\vartheta|} = 2^{11}.3. \qquad \square$$

Remark 11.5.21. Under the assumption of Corollary 11.5.20, it is possible that some element φ in $\mathbb{G}^i \backslash \tilde{\mathbb{G}}^i$ is induced by a so-called *triality automorphism*, that is, an outer automorphism of order 3 of the Lie algebra $\mathcal{L}$ of type D_4. This does not contradict Lemma 11.5.19, because *G is not a normal subgroup* of $Aut_{MOD}(D_4) = < G, T >$, where T is a triality automorphism of $\mathcal{L}$.

The automorphism groups of the remaining lattices in $\mathcal{R}_3$ are described by the following statement.

Proposition 11.5.22. *Suppose that* $\Gamma \in \mathcal{R}_3$, *and* $\dim_{\mathbb{F}_2}(\Gamma_r + 2\Lambda)/2\Lambda = 3$. *Then* $\mathbb{G} = \mathrm{Aut}(\Gamma) = E_1.\mathbb{G}_1$, *where* E_1 *is a soluble* $\{2, 3\}$*-group and* $\mathbb{G}_1 \in \{SL_3(2), \mathbb{A}_7, \mathbb{S}_7\}$. *Moreover,* $\Gamma \in \mathcal{R}$.

Proof. By Lemma 11.5.2, we have $(\Gamma + 2\Lambda)/2\Lambda = \mathcal{A}_2$. In particular, $\mathcal{B} = (\Gamma \cap 2\Lambda)/4\Lambda \supseteq \mathcal{B}_2$, where

$$\mathcal{B}_2 = \langle S_v(A) | A \in \mathcal{A}_2, v \in V \rangle_{\mathbb{F}_2}$$

is a 12-dimensional subspace of 2^{Ω}. By Lemma 11.4.2 we have

$$(\Gamma^* + 2\Lambda)/2\Lambda = \mathcal{B}^* \subseteq \mathcal{B}_2^*,\ (\Gamma^* \cap 2\Lambda)/4\Lambda = \mathcal{A}_2^*.$$

It is not difficult to prove the following assertions:

1) $|A| \geq 2$ for all $A \in \mathcal{A}_2^* \backslash \{\emptyset\}$. Moreover, $|A| = 2$ if and only if $A = \{\{a,b\},\{c,d\}\}$ for some $a,b,c,d \in V$ with $a+b = c+d$.

2) For every $B \in \mathcal{B}_2^* \backslash \{\emptyset\}$ we have $|A| = 6$ or $|A| \geq 9$.

Set $\mathcal{K} = \{X \in \Gamma^* | (X|X) = 8\}$. In view of 2), $\mathcal{K} \subset 2\Lambda$. Furthermore, from 1) it follows that

$$\mathcal{K} = \bigoplus_{a \in V \backslash \{0\}} \mathcal{K}(a)$$

is a root system of type $7D_4$, where

$$\mathcal{K}(a) = \left\{ \pm 2E_{\omega_1} \pm 2E_{\omega_2} | \omega_i \in \ell(a) \right\}.$$

Now $\mathbb{G}$ preserves the root sublattice $2 < \mathcal{K} >_{\mathbb{Z}}$, so that $\Gamma \in \mathcal{R}$. The remaining assertions in Proposition 11.5.22 follow since

$$G \subseteq \mathbb{G} \subseteq \mathrm{Aut}(\mathcal{K}) = A(D_4) \wr \mathbb{S}_7. \qquad \square$$

We omit the proof of the following statement, which is based on [ATLAS].

Lemma 11.5.23. *The group* $G = \mathbb{Z}_2^7.ASL_3(2)$ *is maximal in* $O_8^+(2)$. *Furthermore, if*

$$Z \times G \subset H \subset \mathbb{G}^0 = \mathrm{Aut}(\Gamma^0),$$

then either $H = \mathbb{G}_0^0 = \mathbb{Z}_2 \times O_8^+(2)$, *or* $H \subseteq A(D_4) \wr \mathbb{S}_7$ *(and then* H *fixes a certain root system of type* $7D_4$ *in* 4Λ*).* $\square$

Theorem 11.5.24. *In the Lie algebra* $\mathcal{L}$ *of type* D_4, *the sets* $\overline{\mathcal{R}}$ *and* $\mathcal{R}_3 \backslash \overline{\mathcal{R}}$ *are not empty. In fact,* $\Gamma^0 \in \overline{\mathcal{R}}$ *and* $\Gamma^1 \in \mathcal{R}_3 \backslash \overline{\mathcal{R}}$. *For every invariant lattice* Γ *of nonroot-type, the group* $\mathbb{G} = \mathrm{Aut}(\Gamma)$ *is* $\mathbb{Z}_2.O_8^+(2)$ *or* $\mathbb{Z}_2.\mathrm{Aut}(\Omega_8^+(2))$. *If* $\Gamma \in \mathcal{R}_3 \backslash \overline{\mathcal{R}}$, *then* $\mathrm{Aut}(\Gamma) = E_1.\mathbb{G}_1$, *where* E_1 *is a soluble* $\{2,3\}$*-group and* $\mathbb{G}_1 \in \{SL_3(2), \mathbb{A}_7, \mathbb{S}_7\}$.

Proof. Suppose that $\Gamma \in \mathcal{R}_3$ and $\mathbb{G} = \mathrm{Aut}(\Gamma)$. By Proposition 11.5.22, one can suppose that $\dim_{\mathbb{F}_2}(\Gamma_r + 2\Lambda)/2\Lambda = 6$. Then $\Gamma_r = \Gamma^i$, and $\mathbb{G} \subseteq \mathbb{G}^i$, where i is 0 or 1.

1) If $i = 1$, then by Corollary 11.5.20 we have $\mathbb{G} = E_1.\mathbb{G}_1$, where E_1 is a $\{2,3\}$-group of order 2^{11} or $2^{11}.3$, and $\mathbb{G}_1 = SL_3(2)$. Moreover, $\mathbb{G}$ preserves the root system

$$\mathcal{K} = \bigoplus_{a \in V \backslash \{0\}} \mathcal{K}(a)$$

of type $7D_4$, where

$$\mathcal{K}(a) = \left\{ \pm 2E_{\omega_1} \pm 2E_{\omega_2} | \omega_i \in \ell(a) \right\}.$$

In other words, $\Gamma \in \mathcal{R}$.

2) If $i = 0$, then we can use Corollary 11.5.20 and Lemma 11.5.23. If in addition $\mathbb{G} \subseteq A(D_4) \wr \mathbb{S}_7$, then $\Gamma \in \mathcal{R}$. If $\mathbb{G} \supseteq \mathbb{G}_0^0$, then $\mathbb{G}$ is $\mathbb{Z}_2.O_8^+(2)$ or $\mathbb{Z}_2.\mathrm{Aut}(\Omega_8^+(2))$. Using the same arguments as in the proof of Theorem 11.0.3 (see §11.4), we conclude that $\Gamma \in \overline{\mathcal{R}}$. □

Theorem 11.0.4 is now an immediate consequence of Theorems 11.1.6 and 11.5.24. □

11.6 $\mathbb{Z}$-forms of Lie algebras of types G_2, B_3 and D_4

We start with the following notion.

Definition 11.6.1. By a *$\mathbb{Z}$-form* of a Lie algebra $\mathcal{L}$ we understand any $\mathbb{Z}$-module $\Lambda \subset \mathcal{L}$ with the following properties:
(i) $\Lambda \otimes_{\mathbb{Z}} \mathbb{C} = \mathcal{L}$;
(ii) Λ is closed under the Lie product in $\mathcal{L}$;
(iii) (Λ, K) is a definite integral lattice, where K is the Killing form.

In this section we show that the most interesting invariant lattices, namely, lattices of nonroot-type in Lie algebras of types B_3, D_4, and deep lattices in the Lie algebra of type G_2 (see Chapter 8), arise from $\mathbb{Z}$-forms.

Take a $\mathbb{C}$-space U with a basis $\{e_i | 1 \le i \le \ell\}$ and standard form $(e_i|e_j) = \delta_{i,j}$. Here $\ell = 7$ if the Lie algebra $\mathcal{L}$ belongs to type B_3, and $\ell = 8$ for type D_4. Using the form $(*|*)$ we can identify $U^* = \mathrm{Hom}(U, \mathbb{C})$ with U, $U^* \otimes_{\mathbb{C}} U = \mathrm{End}_{\mathbb{C}}(U)$ with $U \otimes_{\mathbb{C}} U$, and, finally,

$$\mathcal{L} = \{A \in \mathrm{End}_{\mathbb{C}}(U) | A + {}^tA = 0\}$$

with $U \wedge U$. Let R denote a root lattice of type E_ℓ lying in U, and $R^* = \{r \in U | (r|R) \subseteq \mathbb{Z}\}$ the dual lattice. Recall that in §§11.4, 11.5 we dealt with the following lattices of nonroot-type, which will be denoted below as Δ_j:
a) $\mathcal{L}$ of type B_3:

$$\Delta_0 = \Lambda^0,\ \Delta_1 = R \wedge R,\ \Delta_2 = R \wedge R^* = R^* \wedge R^*,\ \Delta_3 = (\Lambda^0)^*;$$

moreover, $\Delta_0 = \{X \in \Delta_1 | (X|X) \in 8\mathbb{Z}\}$, and $\Delta_0 \subset \Delta_1 \subset \Delta_2 \subset \Delta_3$.
b) $\mathcal{L}$ of type D_4:

$$\Delta_0 = \Gamma^0,\ \Delta_1 = R \wedge R,\ \Delta_2 = (\Lambda^0)^*;$$

moreover, $\Delta_0 = \{X \in \Delta_1 | (X|X) \in 8\mathbb{Z}\}$, and $\Delta_0 \subset \Delta_1 \subset \Delta_2$.
It is clear that all the lattices Δ_j can be uniquely reconstructed from the lattice $\Delta_1 = R \wedge R$.

Proposition 11.6.2. Δ_1 *is precisely*

$$\{X \in \mathcal{L} \subset \mathrm{End}_{\mathbb{C}}(U) | X(R^*) \subseteq R\},$$

and is a $\mathbb{Z}$-form. In the case of type B_3, $2\Delta_2$ is also a $\mathbb{Z}$-form.

Proof. 1) For any lattices $M, N \subset U$, the equalities

$$\mathrm{Hom}_{\mathbb{Z}}(M, N) = M^* \otimes_{\mathbb{Z}} N, \ (M \otimes_{\mathbb{Z}} M) \cap (U \wedge U) = M \wedge M,$$

are obvious. Hence,

$$\{X \in \mathcal{L} | X(R^*) \subseteq R\} = \mathrm{Hom}_{\mathbb{Z}}(R^*, R) \cap (U \wedge U) =$$

$$= (R \otimes_{\mathbb{Z}} R) \cap (U \wedge U) = R \wedge R = \Delta_1.$$

Furthermore, suppose that $A, B \in \Delta_1$. Then $A(R^*)$ and $B(R^*)$ are contained in R. But $R^* \supseteq R$, therefore,

$$[A, B](R^*) \subseteq A(R) + B(R) \subseteq R.$$

This means that $[A, B] \in \Delta_1$, that is, Δ_1 is a $\mathbb{Z}$-form.

2) We have to show that $[2A, 2B] \in 2\Delta_2$, that is, $2[A, B] \in \Delta_2$ for any $A, B \in \Delta_2$. As in 1) we have

$$\Delta_2 = \{X \in \mathcal{L} | X(R) \subseteq R^*\}.$$

Hence,

$$2AB(R) \subseteq 2A(R^*) = A(2R^*) \subseteq A(R) \subseteq R^*.$$

Similarly, $2BA(R) \subseteq R^*$, and so $2[A, B] \in \Delta_2$. □

Recall that one can embed the Lie algebra $\mathcal{L}'$ of type G_2 in the Lie algebra $\mathcal{L}$ of type B_3 in such a way that, if

$$\mathcal{D} : \mathcal{L} = \oplus_{i=1}^{7} \mathcal{H}_i$$

is an MOD for $\mathcal{L}$, then

$$\mathcal{D}' : \mathcal{L}' = \oplus_{i=1}^{7} (\mathcal{H}_i \cap \mathcal{L}')$$

is an MOD for $\mathcal{L}'$. Moreover, the following statement holds.

Proposition 11.6.3. *The lattice $\Delta = 2\Delta_2 \cap \mathcal{L}'$ is the (up to similarity and duality) unique invariant lattice of type G_2 with automorphism group $\mathbb{Z}_2 \times G_2(3)$. Moreover, Δ is a $\mathbb{Z}$-form.*

Proof. If Λ is a basic lattice in $\mathcal{L}$, then $\Gamma = \Lambda \cap \mathcal{L}'$ is basic in $\mathcal{L}'$. Furthermore, it is not difficult to verify that

$$\Gamma \supset \Delta \supset 4\Gamma, \ \Delta \not\supseteq 2\Gamma, \ 2\Gamma \not\supseteq \Delta.$$

In other words, Δ is a deep lattice. Our assertions now follow from the results of Chapter 8. □

Propositions 11.6.2 and 11.6.3 expose the natural nature of the most interesting invariant lattices of types D_4, B_3, and G_2: they arise from $\mathbb{Z}$-forms of the corresponding Lie algebras.

We now treat this problem in the general case.

Question 11.6.4. *Suppose that a Lie algebra $\mathcal{L}$ admits an OD. Are invariant lattices of type $\mathcal{L}$ similar to $\mathbb{Z}$-forms of $\mathcal{L}$?*

Definition 11.6.5. We say that an OD $\mathcal{D}$ of a Lie algebra $\mathcal{L}$ *satisfies condition* ($\mathcal{B}$) if there exists a G-invariant lattice Λ_B, where $G = \mathrm{Aut}(\mathcal{D})$, with the following properties:

(i) for any G-invariant (positive or negative) definite lattice $\Gamma \subset \mathcal{L}$ there is $\lambda \in \mathbb{R} \cup i\mathbb{R}$ such that $\Gamma \subseteq \lambda\Lambda_B$ ($i = \sqrt{-1}$);

(ii) there is a natural number n such that $\Lambda_B \supseteq \Gamma \supset n\Lambda_B$ for any indivisible G-invariant sublattice $\Gamma \subseteq \Lambda_B$.

A lattice like the Λ_B mentioned above is called a *base lattice*.

Proposition 11.6.6. *The standard OD of the Lie algebras of type A_{p^m-1} (p a prime) and also the MODs of Lie algebras of types B_{2^m-1}, D_{2^m}, and G_2, satisfy condition* ($\mathcal{B}$).

Proof. In each of these cases, the basic lattice Λ introduced in §10.1 (for type A_{p^m-1}), §11.1 (for types B_{2^m-1} and D_{2^m}), and [KKU 5] (for type G_2), is a base lattice. □

Proposition 11.6.7. *Suppose that an OD $\mathcal{D}$ of a Lie algebra $\mathcal{L}$ satisfies condition* ($\mathcal{B}$). *Then the following assertions are equivalent:*

(i) *each G-invariant lattice $\Gamma \subset \mathcal{L}$ is similar to some $\mathbb{Z}$-form of $\mathcal{L}$;*

(ii) *among the G-invariant lattices there exists at least one lattice Γ that is similar to some $\mathbb{Z}$-form of $\mathcal{L}$;*

(iii) *every base lattice Λ_B satisfies the inclusion*

$$[\Lambda_B, \Lambda_B] \subseteq r\Lambda_B$$

for some

$$r \in \tilde{\mathbb{Q}} = \{\pm\sqrt{\pm s} | s \in \mathbb{Q}, s \geq 0\}.$$

Proof. 1) (i) $\Rightarrow$ (ii): this is obvious.

2) (ii) $\Rightarrow$ (iii). One can suppose that $\Lambda_B \supseteq \Gamma \supset n\Lambda_B$ for some $n \in \mathbb{N}$. We know that $\lambda\Gamma$ is a $\mathbb{Z}$-form for some $\lambda \in \mathbb{C}$. Since (Γ, K) and $(\lambda\Gamma, K)$ are definite

lattices, $\lambda \in \tilde{\mathbb{Q}}$. Furthermore, $[\lambda\Gamma, \lambda\Gamma] \subseteq \lambda\Gamma$. Hence,

$$[\Lambda_B, \Lambda_B] \subseteq [\frac{1}{n}\Gamma, \frac{1}{n}\Gamma] = \frac{1}{n^2\lambda^2}[\lambda\Gamma, \lambda\Gamma] \subseteq \frac{1}{n^2\lambda}\Gamma \subseteq r\Lambda_B,$$

where $r = \frac{1}{n^2\lambda} \in \tilde{\mathbb{Q}}$.

3) (iii) $\Rightarrow$ (i). Suppose again that $\Lambda_B \supseteq \Gamma \supset n\Lambda_B$ for some $n \in \mathbb{N}$. Then

$$[\Gamma, \Gamma] \subseteq [\Lambda_B, \Lambda_B] \subseteq r\Lambda_B \subseteq \frac{r}{n}\Gamma.$$

Setting $\Gamma_0 = \frac{n}{r}\Gamma$, we have $[\Gamma_0, \Gamma_0] \subseteq \Gamma_0$. This means that Γ_0 is a full $\mathbb{Z}$-module in $\mathcal{L}$ closed under the Lie product. In particular, the Killing form K is integral on Γ_0. As (Γ, K) is a definite lattice, and $\frac{n}{r} \in \tilde{\mathbb{Q}}$, (Γ_0, K) is also a (positive or negative) definite lattice. In other words, Γ_0 is a $\mathbb{Z}$-form of $\mathcal{L}$. □

Corollary 11.6.8. *Let $\mathcal{D}$ be an MOD of a Lie algebra $\mathcal{L}$ of type B_{2^m-1}, D_{2^m} or G_2. Then, every invariant lattice of type $\mathcal{L}$ is similar to some $\mathbb{Z}$-form of $\mathcal{L}$.*

Proof. In view of Propositions 11.6.6 and 11.6.7, it is sufficient to show that some invariant lattice of type $\mathcal{L}$ is similar to a $\mathbb{Z}$-form of $\mathcal{L}$. But the last assertion follows from Propositions 11.6.2 and 11.6.3. □

Corollary 11.6.9. *Let $\mathcal{D}$ denote the standard IOD of the Lie algebra $\mathcal{L}$ of type A_{p-1}, p a prime, $p \geq 5$. No invariant lattice of type $\mathcal{L}$ is similar to a $\mathbb{Z}$-form of $\mathcal{L}$.*

Proof. By Propositions 11.6.6 and 11.6.7, it is sufficient to show that there does not exist $r \in \tilde{\mathbb{Q}}$ such that $[\Lambda, \Lambda] \subseteq r\Lambda$, where Λ is a basic lattice as introduced in Chapter 9. Recall that

$$\Lambda \cong \Lambda_0 \oplus \cdots \oplus \Lambda_{p-1} \oplus \Lambda_p = SL_2(p)(\Lambda_0),$$

where Λ_0 is spanned by the vectors

$$X_0^n = \frac{1}{p}\sum_{a=1}^{p-1} \varepsilon^{na} J_{(a,0)},\ 0 \leq n \leq p-1,\ \varepsilon = \exp(\frac{2\pi i}{p}).$$

The automorphism $\hat{\varphi}$ with $\varphi = \begin{pmatrix} 0 & 1 \\ -1 & 0 \end{pmatrix} \in SL_2(p)$ maps Λ_0 onto Λ_p, and X_0^n onto

$$X_p^n = \frac{1}{p}\sum_{a=1}^{p-1} \varepsilon^{na} J_{(0,a)},\ 0 \leq n \leq p-1.$$

Furthermore, when $1 \leq k \leq p-1$ the automorphism $\hat{\varphi}_k$ with $\varphi_k = \begin{pmatrix} 1 & 0 \\ k & 1 \end{pmatrix} \in SL_2(p)$ maps Λ_0 onto Λ_k, and X_0^n onto

$$X_k^n = \frac{1}{p}\sum_{a=1}^{p-1} \varepsilon^{na - \frac{ka^2}{2}} J_{(a,ka)},\ 0 \leq n \leq p-1.$$

Suppose that $[\Lambda, \Lambda] \subseteq r\Lambda$ for some $r \in \tilde{\mathbb{Q}}$. Then

$$r\Lambda \ni p[X_0^0, X_p^0] = \frac{1}{p} \sum_{a,b\neq 0} [J_{(a,0)}, J_{(0,b)}] = \frac{1}{p} \sum_{a,b\neq 0} (1 - \varepsilon^{-ab}) J_{(a,b)}.$$

The projection of the last vector into a Cartan subalgebra $\mathcal{H}_k$ lies in $r\Lambda_k$. In particular, for $k = 2$ we can find integers $k_0, \ldots, k_{p-1}$ such that

$$\frac{1}{p} \sum_{a\neq 0} (1 - \varepsilon^{-2a^2}) J_{(a,2a)} = r \sum_{n=0}^{p-1} k_n X_2^n =$$

$$= \frac{r}{p} \sum_{n=0}^{p-1} \left(k_n \sum_{a\neq 0} \varepsilon^{na-a^2} J_{(a,2a)} \right) = \frac{r}{p} \sum_{a\neq 0} \left(\sum_{n=0}^{p-1} k_n \varepsilon^{na-a^2} \right) J_{(a,2a)},$$

that is,

$$r \sum_{n=0}^{p-1} k_n \varepsilon^{na} = \varepsilon^{a^2} - \varepsilon^{-a^2}, \; 1 \le a \le p-1. \tag{11.11}$$

Applying complex conjugation to (11.11), we obtain

$$\bar{r} \sum_{n=0}^{p-1} k_n \varepsilon^{-na} = \varepsilon^{-a^2} - \varepsilon^{a^2} = -r \sum_{n=0}^{p-1} k_n \varepsilon^{-na}.$$

Hence,

$$0 = (r + \bar{r}) \sum_{n=0}^{p-1} k_n \varepsilon^n = (r + \bar{r})(\varepsilon - \varepsilon^{-1}) r^{-1},$$

that is, $\bar{r} = -r$. This means that $r = \sqrt{-s}$ for some $s \in \mathbb{Q}$, $s > 0$. Squaring (11.11), we get

$$s \left(\sum_{n=0}^{p-1} k_n \varepsilon^{na} \right)^2 = 2 - \varepsilon^{2a^2} - \varepsilon^{-2a^2}, \; 1 \le a \le p-1. \tag{11.12}$$

Set $\eta = \exp(\frac{2\pi i}{4p})$. It is not difficult to see that there exists an automorphism $\psi \in Gal(\mathbb{Q}(\eta)/\mathbb{Q})$ such that

$$\psi : \varepsilon \mapsto \varepsilon^2, \; i = \sqrt{-1} \mapsto i.$$

Taking $a = 1$ in (11.12) and applying ψ, we obtain that

$$s \left(\sum_{n=0}^{p-1} k_n \varepsilon^{2n} \right)^2 = 2 - \varepsilon^4 - \varepsilon^{-4}.$$

On the other hand, when $a = 2$ the relation (11.12) gives us that

$$s\left(\sum_{n=0}^{p-1} k_n \varepsilon^{2n}\right)^2 = 2 - \varepsilon^8 - \varepsilon^{-8}.$$

Hence, $\varepsilon^4 + \varepsilon^{-4} = \varepsilon^8 + \varepsilon^{-8}$, that is, $(\varepsilon^{12} - 1)(\varepsilon^4 - 1) = 0$, which is impossible if $p \geq 5$. □

Commentary

The idea of the rough partition of the set of invariant lattices into lattices of root-type and nonroot-type was suggested by Burichenko. The construction of the lattices Λ^0, Λ^1, Γ^0, Γ^1 (see §§11.4, 11.5) is also due to him. As the material of the chapter shows, this idea is very fruitful for the description of automorphism groups of invariant lattices. In principle, the same approach can be applied to the study of invariant lattices associated with other IODs of Lie algebras of types B_n and D_n. Other results of the chapter are taken from [Tiep 4, 5].

In the course of the investigation of the case D_4 we obtained two odd unimodular root-free lattices of rank 28, namely $\tilde{\Gamma}^0$ and $\tilde{\Gamma}^1$. Unfortunately, the authors were forced to make large omissions, especially in §11.5. However, we hope that a persistent reader will be able to reconstruct the omitted details by using the outline of the arguments expounded in the text.

To conclude this chapter, we point out the methodological significance of Corollary 11.6.9. In the Workshop on Finite Groups (May 1988, Jaroslavl', Russia), Borovik posed a question as to whether invariant lattices and $\mathbb{Z}$-forms of the corresponding Lie algebras coincide (see Question 11.6.4). As is clear from Corollary 11.6.9, the answer to this question can sometimes be in the negative. For instance, the Leech lattice $\Lambda_{24} \cong \Delta^{1,1}$ is not similar to any $\mathbb{Z}$-form of a Lie algebra of type A_4.

Chapter 12

Invariant lattices of types F_4 and E_6, and the finite simple groups $L_4(3), \Omega_7(3), Fi_{22}$

This chapter is devoted to a description of the automorphism groups of all invariant lattices of types F_4 and E_6. The main results are summarized in the following theorems.

Theorem 12.0.1. *Up to $\mathbb{Z}$-similarity, there are precisely* 82 *invariant lattices of type F_4. If Λ is such a lattice, then its automorphism group $\mathbb{G} = \mathrm{Aut}(\Lambda)$ is one of the following groups:*
(i) $(\mathbb{Z}_3^k.SL_3(3)) \times \mathbb{Z}_2$, $k = 3, 10, 13$;
(ii) $W(F_4) \wr \mathbb{S}_{13}$;
(iii) $\mathbb{Z}_2 \times L_4(3), \mathbb{Z}_2 \times (L_4(3).\mathbb{Z}_2)$.

Theorem 12.0.2. *Let Λ be an invariant lattice of type E_6, and set $\mathbb{G} = \mathrm{Aut}(\Lambda)$. Then one of the following assertions hold:*
(i) $\mathbb{Z}_2 \times \Omega_7(3) \lhd \mathbb{G} \subseteq (\mathbb{Z}_2 \times \Omega_7(3)).\mathbb{Z}_2$;
(ii) $\mathbb{Z}_2 \times Fi_{22} \lhd \mathbb{G} \subseteq (\mathbb{Z}_2 \times Fi_{22}).\mathbb{Z}_2$;
(iii) $\mathbb{G} = \mathbb{E}.H$, *where* $\mathbb{E} \subseteq (A(E_6))^{13}$ *and* $H = \mathbb{G}/\mathbb{E}$ *is one of the groups* $SL_3(3)$, $\mathbb{A}_{13}$ *and* $\mathbb{S}_{13}$.
All possibilities occur.

In Section §12.1 some consequences of Theorem 12.0.1 and certain details related to invariant lattices of type F_4 are stated. However, we omit the proof of this theorem, referring the reader to [Bur 2]. We remark simply that the methods of the proof of Theorem 12.0.2 explained here can be applied to obtain complete information about the automorphism groups of invariant lattices of type F_4, but not about the classification of these lattices up to $\mathbb{Z}$-similarity. The main result of Chapter 12 can be viewed as the "Lie" realization, in the context of orthogonal decompositions and integral invariant lattices, of three finite simple groups: $L_4(3) = PSL_4(3), \Omega_7(3)$ (the commutator subgroup of the special orthogonal group $SO_7(3)$) and the Fischer sporadic simple group Fi_{22}.

12.1 On invariant lattices of type F_4

In this section we give a sketch of the proof of Theorem 12.0.1. Details can be found in [Bur 2].

Recall (see Chapter 4) that the Lie algebra $\mathcal{L}$ of type F_4 has a unique IOD:

$$\mathcal{D} : \mathcal{L} = \bigoplus_{i=1}^{h+1} \mathcal{H}_i. \tag{12.1}$$

Here $h = 12$ and $G = \mathrm{Aut}(\mathcal{D}) \cong E.S$, where $E \cong \mathbb{Z}_3^3$ is a Jordan subgroup of the complex Lie group $\mathcal{G} = F_4(\mathbb{C})$, and $S \cong SL_3(3)$. The extension G of E by S is split since $H^2(SL_3(3), \mathbb{F}_3^3) = 0$ (see [Sah]). Set $G_1 = St_G(\mathcal{H}_1)$. Then G_1 acts on $\mathcal{H}_1$ like $\mathbb{Z}_3.GL_2(3)$. It can be shown that G is the only subgroup of G that is irreducible on $\mathcal{L}$. So, every invariant lattice of type F_4 (see Preamble to Part II) is a G-invariant lattice of dimension 52.

Let Λ be such a lattice and Λ_i its projection to $\mathcal{H}_i$. Then $\Lambda_i \cong \Lambda_1$ is a $\mathbb{Z}_3.GL_2(3)$-invariant lattice of dimension 4. Moreover,

$$3^2(\bigoplus_{i=1}^{13} \Lambda_i) \subseteq \Lambda \subseteq \bigoplus_{i=1}^{13} \Lambda_i.$$

So, the question reduces to the classification of submodules in the induced module

$$\tilde{\Gamma} = \mathrm{Ind}_{G_1}^{G}(\Gamma/3^2\Gamma),$$

where Γ is some 4-dimensional G_1-invariant lattice.

One can show that, up to similarity, there exist only four G_1-invariant lattices in $\mathcal{H}_1$ (moreover, all of them are isometric to the root lattice of type F_4). If Γ is one of these lattices, then

$$\bar{\Gamma} = \Gamma/(1-E)\Gamma \cong \mathbb{Z}_3^2,$$

where $(1-E)\Gamma =< (1-f)x | f \in E, x \in \Gamma >_{\mathbb{Z}}$; and G_1 acts irreducibly on $\bar{\Gamma}$.

Now let M be a G-submodule in $\tilde{\Gamma}$. Clearly,

$$(1-E)^d \tilde{\Gamma} \subseteq M$$

for some non-negative integer d, since

$$(1-E)^4 \tilde{\Gamma} \subseteq \mathrm{Ind}_{G_1}^{G}((1-E)^4(\Gamma/3^2\Gamma)) = \mathrm{Ind}_{G_1}^{G}(3^2\Gamma/3^2\Gamma) = 0.$$

The least integer $d = d(M)$ with this property is said to be the *depth* of the module M. Clearly, $d(M) = 0$ if and only if $M = \tilde{\Gamma}$. If $d(M) = 1$, then it is enough to study G-submodules of $\mathrm{Ind}_{G_1}^{G}(\bar{\Gamma})$. Here we shall deal with the following simple $SL_3(3)$-modules. Let $V =< e_1, e_2, e_3 >_{\mathbb{F}_3}$ denote the natural module of column-vectors over $\mathbb{F}_3$ for $SL_3(3)$, and $V^* =< x_1, x_2, x_3 >_{\mathbb{F}_3}$ its dual module: $x_i(e_j) = \delta_{ij}$. Furthermore, let U denote the factor-module of the submodule of $V^* \otimes_{\mathbb{F}_3} V$ consisting of tensors with zero trace by the 1-dimensional submodule $< x_1 \otimes e_1 + x_2 \otimes e_2 + x_3 \otimes e_3 >_{\mathbb{F}_3}$. Finally, let W be the submodule of $V^* \otimes_{\mathbb{F}_3} S^2(V)$ consisting of tensors with zero convolution on the first two indices. As composition factors of the S-module $\tilde{\Gamma}$, we encounter the following simple modules: the trivial module, 3-dimensional modules V and V^*, a 7-dimensional module U and 15-dimensional modules W and W^*. The Loewy structure of projective indecomposable modules for $SL_3(3)$, in particular, the functor Ext^1, is studied in [Kosh].

Here we formulate some consequences of Theorem 12.0.1.

1) *Up to* $\mathbb{Z}$*-equivalence, there exist precisely* 24 *absolutely irreducible integral representations of degree* 52 *of* $L_4(3)$.

This assertion follows from the complete description of invariant lattices.

2) *In* $\mathbb{R}^{52}$ *there exists a configuration consisting of* 1560 *lines making angles*

$$0, \arccos\frac{1}{6}, \arccos\frac{1}{3}, \frac{\pi}{3}, \text{ or } \frac{\pi}{2}$$

with each other, such that the group of line permutations preserving angles is $L_4(3)$.

12.2 Invariant lattices of type E_6: the imprimitive case

Recall (see Chapter 4) that the Lie algebra $\mathcal{L}$ of type E_6 has a unique IOD $\mathcal{D}$ of form (12.1). Here $h = 12$ and $G = \mathrm{Aut}(\mathcal{D}) \cong C.S$, where

$$C = C_{\mathcal{G}}(E) \cong 3^{3+3}$$

is a special 3-group of order 3^6, $Z(C) = [C, C] = \Phi(C) = E \cong \mathbb{Z}_3^3$ is a Jordan subgroup of the complex Lie group $\mathcal{G} = E_6(\mathbb{C})$, and $S = SL_3(3)$. Remark (see Proposition 14.1.7) that G is the unique extension of C by S with the property that $[C, C]$ and $C/[C, C]$ are irreducible S-modules. Furthermore, it is not difficult to show that G is the only subgroup of G that is irreducible on $\mathcal{L}$. So, every invariant lattice of type E_6 is a G-invariant lattice of dimension 78.

Keeping alternative $(\mathcal{A})$ in mind (see Preamble to Part II), throughout this section we shall suppose that Λ is an invariant lattice of type E_6 such that $\mathbb{G} = \mathrm{Aut}(\Lambda)$ is an imprimitive complex linear group on $\mathcal{L}$. From now on we denote the character of $\mathbb{G}$ on $\mathcal{L}$ by χ.

Proposition 12.2.1. *Under the given hypotheses,* $\mathbb{G}$ *preserves the decomposition* $\mathcal{D}$.

Proof. It is known that

$$\chi|_E = 3 \sum_{1_E \neq \sigma \in \mathrm{Irr}\,(E)} \sigma.$$

If

$$\mathcal{L}_\sigma = \{x \in \mathcal{L} | \forall f \in E, f(x) = \sigma(f)x\}$$

is the σ-eigensubspace for E on $\mathcal{L}$, then

$$\mathcal{H}_i = \mathcal{L}_{\sigma_i} \oplus \mathcal{L}_{\bar{\sigma}_i}$$

for some nontrivial character $\sigma_i \in \mathrm{Irr}\,(E)$. As usual, $\bar{\sigma}$ denotes the complex conjugate of σ.

1) We claim that

$$\chi|_C = \sum_{i=1}^{13} (\alpha_i + \bar{\alpha}_i), \tag{12.2}$$

where $\{\alpha_1, \bar{\alpha}_1, \dots, \alpha_{13}, \bar{\alpha}_{13}\}$ is the set of 26 distinct irreducible characters of degree 3 of C. Moreover, $\mathcal{H}_i$ affords the C-character $\alpha_i + \bar{\alpha}_i$ and

$$\alpha_i|_E = 3\sigma_i, \; \bar{\alpha}_i|_E = 3\bar{\sigma}_i.$$

Indeed, by Clifford's Theorem,

$$\chi|_C = e \sum_{i=1}^{t} \beta_i, \; \beta_i \in \mathrm{Irr}\,(C).$$

As $\deg \beta_i$ divides $3 = \gcd(78, 3^6) = \gcd(|C|, \dim \mathcal{L})$ and C is nonabelian, we have $\deg \beta_i = 3$. In particular, $et = 26$. If $e > 1$, then on restricting χ to E we obtain that

$$3 \sum_{1_E \neq \sigma \in \mathrm{Irr}\,(E)} \sigma = \chi|_E = e \sum_i \beta_i|_E,$$

where $e = 2$, 13 or 26, a contradiction.

Hence, $e = 1$ and

$$\chi|_C = \sum_{i=1}^{26} \beta_i,$$

where $\beta_1, \dots, \beta_{26}$ are distinct characters from $\mathrm{Irr}\,(C)$. As $\mathcal{H}_i$ is C-invariant, it affords some C-character, say, $\beta_1 + \beta_2$. But $E = Z(C)$, so by Schur's Lemma E acts scalarly on the β_i-eigensubspaces for C, $i = 1, 2$. Because E has character $3\sigma_1 + 3\bar{\sigma}_1$ on $\mathcal{H}_1$, one can suppose that $\beta_1|_E = 3\sigma_1$, $\beta_2|_E = 3\bar{\sigma}_1$. In particular, $\beta_1 \neq \bar{\beta}_1$, $\beta_2 \neq \bar{\beta}_2$. But $\beta_1 + \beta_2$ is real (the group $G_1 = St_G(\mathcal{H}_1)$ acts on $\mathcal{H}_1^*$ as a subgroup of the Weyl group), so $\beta_2 = \bar{\beta}_1$.

2) Next we show that if C_1 is the kernel of the representation of C on $\mathcal{H}_1$, then $C/C_1 = 3_+^{1+2}$. In fact, since C has character $\alpha_1 + \bar{\alpha}_1$ on $\mathcal{H}_1$, we have

$C_1 = \mathrm{Ker}\,\alpha_1$. Furthermore, α_1 is a faithful irreducible character of C/C_1 and $\deg\alpha_1 > 1$, so that C/C_1 is nonabelian. In particular, $|C/C_1| \geq 3^3$, $|C_1| \leq 3^3$. Moreover, $E \cap C_1 = \mathbb{Z}_3^2$. Thus

$$[C/C_1, C/C_1] = [C, C]C_1/C_1 = EC_1/C_1 \cong E/(E \cap C_1) = \mathbb{Z}_3.$$

By Schur's Lemma, the centre $Z(C/C_1)$ acts scalarly on the α_1-eigensubspace for C. But α_1 is faithful on C/C_1, so $Z(C/C_1)$ is cyclic. Recall that $\exp(C) = 3$, and hence

$$\exp(C/C_1) = 3, Z(C/C_1) = \mathbb{Z}_3, Z(C/C_1) = [C/C_1, C/C_1] = \mathbb{Z}_3.$$

The factor-group $(C/C_1)/\mathbb{Z}_3$ is an elementary abelian 3-group, which implies that $\Phi(C/C_1) = \mathbb{Z}_3$. This means that $C/C_1 = 3_+^{1+2}$.

3) Now choose a $\mathbb{G}$-imprimitivity system $\mathcal{L} = \oplus_{i=1}^N V_i$ with maximal dimension of components (and with $N > 1$). Set $H_1 = St_G(V_1)$. Clearly, $(G : H_1) = N$ divides 78. Here we show that $|C \cap H_1| \geq 3^5$ and that N is divisible by 13.

Indeed, the equality $|CH_1| = |C|.|H_1|/|C \cap H_1|$ implies that $(C : C \cap H_1) = (CH_1 : H_1)$ divides $\gcd(3^6, (G : H_1))$. Of course $\gcd(3^6, 78) = 3$, therefore $|C \cap H_1| \geq 3^5$. In particular, $C \cap H_1 \lhd C$. But clearly $C \cap H_1 \lhd H_1$, so $C \cap H_1 \lhd CH_1$. Using this fact we show that $CH_1 \neq G$. Suppose the contrary: $CH_1 = G$. Because $H_1 \neq G$, we have $C \cap H_1 \neq C$ and $|C \cap H_1| = 3^5$. As we have shown above, $C \cap H_1 \lhd CH_1 = G$. Furthermore, since $C/(C \cap H_1) = \mathbb{Z}_3$, it follows that $C \cap H_1 \supseteq [C, C] = E$ and so

$$\mathbb{Z}_3^2 = (C \cap H_1)/E \lhd G/E = \mathbb{Z}_3^3.SL_3(3),$$

a contradiction. Hence, $CH_1 \neq G$. This implies that

$$CH_1/C \subset G/C = S = SL_3(3).$$

The group $SL_3(3)$ has maximal subgroups $\mathbb{Z}_3^2.2\mathbb{S}_4$, $\mathbb{Z}_{13}.\mathbb{Z}_3$ and $\mathbb{S}_4$ of index 13, 144 and 234, respectively. But the index $(G/C : CH_1/C) = (G : CH_1)$ divides 78, so that

$$CH_1/C \subseteq \mathbb{Z}_3^2 \cdot 2\mathbb{S}_4 = P,$$

and in this case N is divisible by $(G : CH_1)$ and by $(S : P) = 13$.

4) At this point we show that in fact $H_1 \supset C$. Suppose the contrary: $|C \cap H_1| = 3^5$. As mentioned above, $C \cap H_1 \supseteq E$. Hence $(C \cap H_1)/E$ is a H_1-invariant subspace of codimension 1 in the $SL_3(3)$-module $C/E \cong E^*$. It was established in 3) that CH_1/C, in its action on E, is contained in $P \cong \mathbb{Z}_3^2.2\mathbb{S}_4$, where P is either a line stabilizer or a plane stabilizer and E is considered as a 3-dimensional $\mathbb{F}_3$-space. Here we have $(CH_1 : H_1) = 3$, which implies that the index

$$(P : CH_1/C) = \frac{1}{13}(S : CH_1/C) = \frac{1}{13}(G : CH_1) = \frac{N}{13.(CH_1 : H_1)} = \frac{N}{39}$$

divides $78/39 = 2$. But $P = \mathbb{Z}_3^2.GL_2(3)$, so that

$$P \supseteq CH_1/C \supseteq \mathbb{Z}_3^2.SL_2(3).$$

Recall that H_1 fixes some plane in E^*, therefore H_1 fixes some line in E.

We have shown that H_1 has either three orbits of length $8, 9, 9$ or two orbits of length $8, 18$ on $E^* \backslash \{1\}$, on the one hand. On the other hand, $E \lhd H_1$ and H_1 acts on the subspace V_1 of dimension $78/N \leq 6$. Hence, E acts trivially on V_1, a contradiction.

5) As $H_1 \supseteq C$, it follows from decomposition (12.2) and the fact that V_1 is C-invariant that 3 divides $\dim V_1$. But 13 divides $N = \dim \mathcal{L}/\dim V_1 = 78/\dim V_1$, so $\dim V_1 = 3$ or 6.

5a) First assume that $\dim V_1 = 6$. Then $N = 13$, which implies that $CH_1/C = P$. If P is a line stabilizer (in the action on E), then P has two orbits of length 8 and 18 on $E^* \backslash \{1\}$, and consideration of the action of the normal subgroup E of H_1 on V_1 gives a contradiction. Thus P is a plane stabilizer. But in that case, P has two orbits of length 2 and 24 on $E^* \backslash \{1\}$. Considering the action of H_1 on V_1 again, we come to the conclusion that V_1 affords the E-character $3\sigma_1 + 3\bar{\sigma}_1$, where $\{\sigma_1, \bar{\sigma}_1\}$ is a P-orbit. In other words, $V_1 = \mathcal{H}_1$.

5b) Finally, assume that $\dim V_1 = 3$. Then $N = 26$ and we can suppose that V_1 affords the C-character α_1. In that case, V_1 is $\mathcal{L}_\sigma$ for some $\sigma \in \mathrm{Irr}\,(E) \backslash \{1_E\}$. Consider a subspace $[V_1]$ of minimal dimension with the following properties:

(i) $[V_1]$ is a sum of certain of the components V_i;

(ii) $[V_1] \supseteq V_1$;

(iii) $[V_1] \cap \Lambda \neq 0$, where Λ is the given invariant lattice.

It is easy to see that the subspace $\mathcal{H}_1 = \mathcal{L}_\sigma \oplus \mathcal{L}_{\bar{\sigma}}$ possesses properties (i) – (iii), on the one hand. On the other hand, let δ denote the character of the $\mathbb{Z}E$-module $[V_1] \cap \Lambda$. If σ does not enter into δ, then we can move V_1 from $[V_1]$, which contradicts the minimality of $[V_1]$. So σ enters into δ. In that case, since δ is rational, it follows that $\bar{\sigma}$ also enters into δ. In other words, $[V_1] \cap \mathcal{L}_{\bar{\sigma}} \neq 0$ and $[V_1] \supseteq \mathcal{H}_1$. By the minimality of $[V_1]$, we must have $[V_1] = \mathcal{H}_1$.

Using the uniqueness of $[V_1]$, it is easy to show that $\{g[V_1] | g \in \mathbb{G}\}$ constitutes a $\mathbb{G}$-imprimitivity system on $\mathcal{L}$. The maximality of the decomposition $\mathcal{L} = \oplus_i V_i$ implies that $V_1 = [V_1] = \mathcal{H}_1$, and this contradicts the fact that $\dim V_1 = 3$. □

Lemma 12.2.2. *The lattice $\Lambda_1 = \Lambda \cap \mathcal{H}_1$ is isometric to the root lattice E_6, the dual lattice E_6^* or the root lattice $3A_2$.*

Proof. 1) Recall that the group $Q = C/C_1 = 3_+^{1+2}$ acts faithfully on Λ_1 with character $\alpha + \bar{\alpha}$, where $\alpha \in \mathrm{Irr}\,(Q)$ and $\deg \alpha = 3$. In particular, a certain central element $\varphi \in Z(Q)$ of order 3 acts fixed-point-freely. Set $\varepsilon = \exp(2\pi i/3)$. We define the structure of a $\mathbb{Z}[\varepsilon]$-lattice Γ_1 on Λ_1 by setting

$$\varepsilon.u = \varphi(u),$$

$$u \circ v = (u|v) + \bar{\varepsilon}(u|\varphi(v)) + \varepsilon(u|\varphi^2(v)).$$

Then

$$(\varepsilon u) \circ v = \bar{\varepsilon}(u \circ v),\ v \circ u = \overline{u \circ v},\ (u|v) = \frac{1}{3}\mathrm{Tr}_{\mathbb{Q}(\varepsilon)/\mathbb{Q}}(u \circ v).$$

In such a case we shall write

$$\Lambda_1 = (\Gamma_1)_{\mathbb{Z}},\ \Gamma_1 = (\Lambda_1)^{\mathbb{Z}[\varepsilon]}.$$

It is enough to classify 3-dimensional Q-invariant $\mathbb{Z}[\varepsilon]$-lattices Γ.

2) Suppose that the group $Q =< D, P >$ acts on $\mathbb{C}^3 =< e_1, e_2, e_3 >_{\mathbb{C}}$ as follows:

$$D \mapsto \begin{pmatrix} 1 & & \\ & \varepsilon & \\ & & \varepsilon^2 \end{pmatrix},\ P \mapsto \begin{pmatrix} 0 & 0 & 1 \\ 1 & 0 & 0 \\ 0 & 1 & 0 \end{pmatrix}.$$

Since this action of Q on $\mathbb{C}^3$ is irreducible and the Hermitian form defined by the relation $e_i * e_j = \delta_{ij}$ is Q-invariant, we can suppose that $e_i \circ e_j = e_i * e_j = \delta_{ij}$, that is, the given basis is orthonormal. Set

$$I_i = \{x | \exists y = (y_1, y_2, y_3) \in \Gamma,\ y_i = x\},\ i = 1, 2, 3.$$

By considering the action of the element P, one can see that $I_1 = I_2 = I_3 = I$ is a $\mathbb{Z}[\varepsilon]$-module in $\mathbb{C}$. Set

$$\Gamma(I) = \{x = (x_1, x_2, x_3) | x_i \in I\}.$$

Then $\Gamma \ni y = x + Dx + D^2x = (3x_1, 0, 0)$, that is,

$$\Gamma(I) \supseteq \Gamma \supseteq 3\Gamma(I).$$

Furthermore, $9|x_1|^2 = y \circ y \in \mathbb{Z}$, that is, $9|z|^2 \in \mathbb{Z}$ for all $z \in I$. It is easy to see that

$$I = \lambda\mathbb{Z}[\varepsilon]$$

for some $\lambda \in \mathbb{C}$ in this case. Up to similarity one can suppose that

$$\Gamma_0 \supseteq \Gamma \supseteq 3\Gamma_0,$$

where $\Gamma_0 = \{x = (x_1, x_2, x_3) | x_i \in \mathbb{Z}[\varepsilon]\}$.

3) Now we introduce the following sublattices of Γ_0, where $\theta = \varepsilon - \bar{\varepsilon}$:

$$\Gamma_1 = \{(x, y, z) \in \Gamma_0 | x + y + z \equiv 0 \pmod{\theta}\},$$

$$\Gamma_2 = \{(x, y, z) \in \Gamma_0 | x \equiv y \equiv z \pmod{\theta}\},$$

$$\Gamma_3 = \theta\Gamma_0,\ \Gamma_4 = \theta\Gamma_1,\ \Gamma_5 = \theta\Gamma_2,\ \Gamma_6 = 3\Gamma_0,$$

$$\Gamma_7 = \{(x, y, z) \in \Gamma_0 | x \equiv y \equiv z \pmod{\theta}, x + y + z \equiv 0 \pmod{3}\},$$

$$\Gamma_8 = \{(x, y, z) \in \Gamma_0 | x \equiv y \equiv z \pmod{\theta}, x + y + \varepsilon z \equiv 0 \pmod{3}\},$$

$$\Gamma_9 = \{(x, y, z) \in \Gamma_0 | x \equiv y \equiv z \pmod{\theta}, x + y + \varepsilon^2 z \equiv 0 \pmod{3}\}.$$

We show that Γ is one of $\Gamma_0, \ldots, \Gamma_9$. Observe that Γ_0/Γ_3 (and in general Γ_i/Γ_{i+3}, $0 \le i \le 3$) is isomorphic to $\mathbb{F}_3^3$. On this factor-lattice, D acts trivially and P acts as the cyclic permutation $(x, y, z) \mapsto (y, z, x)$. So Γ_0/Γ_3 has only two nontrivial submodules, namely Γ_i/Γ_3 with $i = 1, 2$. Suppose now that $\Gamma_0 \supseteq \Gamma \supseteq \Gamma_6$ but $\Gamma \ne \Gamma_i$ for any i, $0 \le i \le 6$. In particular, one can find an element $u \in \Gamma$ of the form $u = (x, y, z)$ such that y does not belong to $\theta\mathbb{Z}[\varepsilon]$. Then

$$\Gamma \ni u - Du = (0, y(1 - \varepsilon), z(1 - \varepsilon^2)) = \theta(0, \varepsilon^2 y, -\varepsilon z).$$

We have used only the condition that Γ is not contained in Γ_3, and shown that $\theta v \in (\Gamma \cap \Gamma_3)/\Gamma_6$, where $v = (0, \varepsilon^2 y, -\varepsilon z)$. Since Γ does not contain Γ_3, we must have $(\Gamma \cap \Gamma_3)/\Gamma_6 \subseteq \Gamma_4/\Gamma_6$. In particular, $y \equiv z \pmod{\theta}$. But in that case, θ does not divide $\varepsilon^2 y$, and so $(\Gamma \cap \Gamma_3)/\Gamma_6 \ne \Gamma_5/\Gamma_6$. In other words, $(\Gamma \cap \Gamma_3)/\Gamma_6 = \Gamma_4/\Gamma_6$ and $\Gamma \cap \Gamma_3 = \Gamma_4$. Applying the same arguments to the element $Du = (y, z, x)$ (which is possible because $z \equiv y \pmod{\theta}$), we obtain that $z \equiv x \pmod{\theta}$. Hence, if θ does not divide y, then $x \equiv y \equiv z \pmod{\theta}$ and $u \in \Gamma_2$. If $x, y, z \equiv 0 \pmod{\theta}$, then $u \in \Gamma \cap \Gamma_3 = \Gamma_4 \subset \Gamma_2$. In all cases we have $\Gamma_2 \supset \Gamma \supset \Gamma_4$.

Now we recall that $\dim_{\mathbb{F}_3} \Gamma_2/\Gamma_4 = 2$, and therefore $\dim_{\mathbb{F}_3} \Gamma/\Gamma_4 = 1$ and $\Gamma = \, < \Gamma_4, u >_{\mathbb{F}_3}$, with

$$u = (x_1 + x_2\varepsilon, y_1 + y_2\varepsilon, z_1 + z_2\varepsilon), \ x_i, y_i, z_i \in \mathbb{F}_3, x_1 + x_2 \equiv y_1 + y_2 \equiv z_1 + z_2 \pmod{3}.$$

Of course, we can suppose that

$$(x_1, x_2) = (x, 1 - x), (y_1, y_2) = (y, 1 - y), (z_1, z_2) = (z, 1 - z).$$

Setting $t = x + y + z \in \mathbb{F}_3$, we obtain that

$$\varepsilon^2 u \equiv (1, 1, 1), (1, 1, \varepsilon^2) \text{ or } (1, 1, \varepsilon) \pmod{\Gamma_4},$$

according as $t = 0, 1$ or -1, respectively. Hence, Γ is Γ_7, Γ_8 or Γ_9.

4) Finally, we study the integral lattices $\Delta_i = (\Gamma_i)_{\mathbb{Z}}$. Clearly, Γ' and $\theta\Gamma'$ are isometric: $\theta u \circ \theta v = 3(u \circ v)$, so we have to consider the lattices Γ_i with $i = 0, 1, 2, 7, 8, 9$.

4a) Observe that Γ_0 is a root $\mathbb{Z}[\varepsilon]$-lattice of type $3A_1$. Namely, the set $\{\pm\varepsilon^i e_k | 0 \le i \le 2\}$, where $k = 0, 1, 2$, constitutes a root system of type A_1 (over $\mathbb{Z}[\varepsilon]$). Hence, $\Delta_0 = (\Gamma_0)_{\mathbb{Z}}$ is a root lattice of type $3A_2$ (over $\mathbb{Z}$). Indeed,

$$(\varepsilon^i e_1 | \varepsilon^j e_1) = \frac{1}{3} \mathrm{Tr}_{\mathbb{Q}(\varepsilon)/\mathbb{Q}}(\varepsilon^{i-j}) = \begin{cases} 2/3 & \text{if } i = j, \\ -1/3 & \text{if } i \ne j. \end{cases}$$

After the identification

$$e_1 \rightsquigarrow f_1 - f_2,\ \varepsilon e_1 \rightsquigarrow f_2 - f_3,\ \varepsilon^2 e_1 \rightsquigarrow f_3 - f_1,$$

where $(f_i|f_j) = \delta_{ij}$, the root system of type A_1 over $\mathbb{Z}[\varepsilon]$ becomes a root system of type A_2 over $\mathbb{Z}$.

4b) The lattices Γ_7, Γ_8, Γ_9 are isometric: Γ_8 and Γ_9 are obtained from Γ_7 on replacing e_3 by $\varepsilon^2 e_3$ and εe_3, respectively. Furthermore, Γ_7 is generated by 18 minimal vectors of norm 3 which constitute a root system of type $3A_1$ over $\mathbb{Z}[\varepsilon]$. They are

$$\pm\varepsilon^k(1, 1, 1),\ \pm\varepsilon^k(1, \varepsilon, \varepsilon^2),\ \pm\varepsilon^k(1, \varepsilon^2, \varepsilon);\ k = 0, 1, 2.$$

This means that $\Delta_i \cong 3A_2$ for $i = 7, 8, 9$.

4c) The lattice Γ_2 is generated by 72 minimal vectors of norm 3:

$$\pm(\varepsilon^i, \varepsilon^j, \varepsilon^k)\ (54\ \text{vectors}),$$

$$\pm\theta\varepsilon^i(1, 0, 0),\ \pm\theta\varepsilon^i(0, 1, 0),\ \pm\theta\varepsilon^i(0, 0, 1)\ (18\ \text{vectors}).$$

Let $\mathcal{M}$ denote the set of these vectors. It is easy to see that

$$\Gamma_2^* = \Gamma_4,$$

where $\Gamma'^* = \{u | u \circ \Gamma' \subseteq 3\theta\mathbb{Z}[\varepsilon]\}$. In particular, $(u|u) = (v|v) = 6/3 = 2$ and $(u|v) \in \mathbb{Z}$ for all $u, v \in \mathcal{M}$. Thus, $\mathcal{M}$ is a classical root system in $\mathbb{R}^6$. But $|\mathcal{M}| = 72$, so $\mathcal{M}$ is a root system of type E_6 and Δ_2 is a root lattice of type E_6. Furthermore, observe that

$$\Delta_2^* = \frac{1}{3}\Delta_4,$$

where, as usual, $\Lambda'^* = \{v | (v|\Lambda') \subseteq \mathbb{Z}\}$ for an integral lattice Λ'. Provided $a, b \in 3\mathbb{Z}$, we have

$$a + b\varepsilon \in 3\theta\mathbb{Z}[\varepsilon] \Rightarrow 9|(a + b) \Rightarrow (a + b\varepsilon) + (a + b\bar{\varepsilon}) = 2a - b \in 9\mathbb{Z}.$$

Hence, $3\Delta_2^* \supseteq \Delta_4$. Conversely, suppose that $v \in 3\Delta_2^*$ and consider an arbitrary vector u in Δ_2. Set $v \circ u = a + b\varepsilon$ for suitable $a, b \in \mathbb{R}$. Then $9\mathbb{Z} \ni (v \circ u + \overline{v \circ u}) = 2a - b$. But $v \circ u\varepsilon = a\varepsilon + b\bar{\varepsilon}$, therefore $9\mathbb{Z} \ni ((a\varepsilon + b\bar{\varepsilon}) + (a\bar{\varepsilon} + b\varepsilon)) = -(a + b)$. This means that $a, b \in 3\mathbb{Z}$ and $a + b \in 9\mathbb{Z}$. Hence, $a + b\varepsilon \in 3\theta\mathbb{Z}[\varepsilon]$ and $v \in \Delta_4$. Thus Δ_1 and Δ_4 are isometric to the dual lattice E_6^*. □

Now we show that the case $3A_2$ listed in the preceding lemma cannot occur.

Lemma 12.2.3. *The automorphism group A of the root lattice Δ of type $3A_2$ cannot contain a subgroup B satisfying the following conditions:*

(i) $B \cong 3_+^{1+2}.SL_2(3)$;

(ii) *B centralizes some element $\varphi \in A$ of order* 3 *with* $\mathrm{Tr}_{\Delta}\varphi = -3$.

Proof. Suppose the contrary, so that A possesses such a subgroup B.

1) We claim that $C = C_A(\varphi)$ can be identified with the automorphism group of some Hermitian $\mathbb{Z}[\varepsilon]$-lattice Γ of type $3A_1$ and, in particular, $C = \mathbb{Z}_6 \wr \mathbb{S}_3$. Here $\varepsilon = \exp(2\pi i/3)$ as above.

To this end, we write Δ in the form of an orthogonal sum of nonzero sublattices: $\Delta = \Delta_1 \oplus \Delta_2 \oplus \Delta_3$. Then, like every element of A, φ permutes the components Δ_i. If it permutes them cyclically, then $\mathrm{Tr}_{\Delta}\varphi = 0$, a contradiction. So, $\varphi(\Delta_i) = \Delta_i$ for every i. Consider Δ_1, for example. As usual, we can write Δ_1 in the form

$$\Delta_1 =< e_i - e_j | 1 \le i \ne j \le 3 >_{\mathbb{Z}},$$

and then φ permutes the pairs $\pm e_j$, $1 \le j \le 3$. If φ leaves every pair fixed, then, as $|\varphi| = 3$, we must have that $\varphi(e_j) = e_j$ for all j and $\mathrm{Tr}_{\Delta_1}\varphi = 2$. If it permutes them cyclically, then without loss of generality we can suppose that $\varphi : e_1 \mapsto e_2 \mapsto e_3 \mapsto e_1$ and so $\mathrm{Tr}_{\Delta_1}\varphi = -1$. Thus we have shown that $\mathrm{Tr}_{\Delta_i}\varphi \in \{-1, 2\}$. But $\sum_{i=1}^{3} \mathrm{Tr}_{\Delta_i}\varphi = \mathrm{Tr}_{\Delta}\varphi = -3$, therefore $\mathrm{Tr}_{\Delta_i}\varphi = -1$ for all i.

Set $a = e_1 - e_2$ and consider the cyclic module $\mathbb{Z}[\varepsilon]a$ with the Hermitian form $a \circ a = 1$. Viewing $\mathbb{Z}[\varepsilon]a$ as a $\mathbb{Z}$-lattice, we define the Euclidean form $(x, y) = \mathrm{Tr}_{\mathbb{Q}(\varepsilon)/\mathbb{Q}}(x \circ y)$ on it. Then

$$(a, a) = 2 = (e_1 - e_2 | e_1 - e_2),$$

$$(a, \varphi(a)) = (a, \varepsilon a) = \mathrm{Tr}\,(\varepsilon + \bar{\varepsilon}) = -1 = (e_1 - e_2 | e_2 - e_3) = (e_1 - e_2 | \varphi(e_1 - e_2)),$$

$$(a, \varphi^2(a)) = (a, \varepsilon^2 a) = \mathrm{Tr}\,(\bar{\varepsilon} + \varepsilon) = -1 = (e_1 - e_2 | e_3 - e_1) = (e_1 - e_2 | \varphi^2(e_1 - e_2)).$$

Thus Δ can be considered as the $\mathbb{Z}$-form of the Hermitian root lattice $\Gamma = < a, b, c >_{\mathbb{Z}[\varepsilon]}$ of type $3A_1$, on which φ acts as multiplication by ε.

Now let h be any element of $C_A(\varphi)$. Note that the subgroup $C_1 = \mathbb{Z}_6 \wr \mathbb{S}_3$, where $\mathbb{Z}_6$ is generated by the map $a \mapsto -\varepsilon a$ and $\mathbb{S}_3$ permutes the a, b, c, is contained in $C_A(\varphi)$. So we can suppose without loss of generality that $h(a) = a$. In that case we have

$$h(\pm\varepsilon^i a) = \pm h\varphi^i(a) = \pm\varphi^i h(a) = \pm\varepsilon^i a.$$

This means that $h(b) \in \{\pm\varepsilon^j b, \pm\varepsilon^k c\}$. As $St_{C_1}(a)$ acts transitively on the last set, we can suppose that $h(b) = b$. From this it follows that $h(\pm\varepsilon^j b) = \pm\varepsilon^j b$ and $h(c) = \pm\varepsilon^k c$, that is, $h \in C_1$. Hence, $C_A(\varphi) = C_1 = \mathbb{Z}_6 \wr \mathbb{S}_3$.

2) We consider the subgroup $C' = (\mathbb{Z}_6)^3.\mathbb{A}_3 = \mathbb{Z}_6 \wr \mathbb{A}_3$ of index 2 in C. Then $|C'| = 3.6^3 = |B|$. Furthermore, the Sylow 2-subgroup of B is isomorphic to the quaternion group Q_8, and the Sylow 2-subgroup of C' is elementary abelian. So $C' \ne B$ and

$$BC' = C,\ B/(B \cap C') \cong BC'/C' = C/C' \cong \mathbb{Z}_2.$$

Clearly, $B \cap C'$ contains the subgroup $Q = 3_+^{1+2}$ of B. Therefore the factor-group $B/Q \cong SL_2(3)$ has the subgroup $(B \cap C')/Q$ of index 2, a contradiction. □

We are now in a position to prove the expected strengthening of Lemma 12.2.2.

Lemma 12.2.2'. *The lattice $\Lambda_1 = \Lambda \cap \mathcal{H}_1$ is isometric to the root lattice E_6 or to its dual E_6^*.*

Proof. By Lemma 12.2.2, it is enough to consider the case when Λ_1 is isometric to the root lattice $3A_2$. Clearly, Λ_1 is fixed by some group $B \cong 3_+^{1+2}.SL_2(3)$, which is simply the image of the group $K_1 = [N_G(E_1), N_G(E_1)]$ in its action on $\mathcal{H}_1 = C_{\mathcal{L}}(E_1)$, where $(E : E_1) = 3$. Consider a central element φ of order 3 of the subgroup $Q = 3_+^{1+2}$. Note that $\mathcal{H}_1$ is a sum of two faithful irreducible B-submodules, U_+ and U_- say. By Schur's Lemma, φ acts on U_+ and on U_- as multiplications by ε^k and ε^l, where $k, l \in \{1, 2\}$. But B acts rationally on $\mathcal{H}_1$, so $\mathbb{Q} \ni \mathrm{Tr}_{\mathcal{H}_1}\varphi = 3(\varepsilon^k + \varepsilon^l)$, that is, $\mathrm{Tr}_{\mathcal{H}_1}\varphi = -3$. However, by Lemma 12.2.3 the group $A = \mathrm{Aut}(\Lambda_1)$ cannot contain such a subgroup B, a contradiction. □

Corollary 12.2.4. *Suppose that Λ is an invariant lattice of type E_6 such that the group $\mathbb{G} =$ Aut (Λ) is imprimitive on $\mathcal{L}$. Then $\mathbb{G}$ satisfies conclusion* (iii) *of Theorem* 12.0.2.

Proof. By Proposition 12.2.1, $\mathbb{G}$ preserves the decomposition $\mathcal{D}$ and so permutes the set $\{\mathcal{H}_i\}_{i=1}^{13}$ of its components transitively. Let $\mathbb{E}$ and H denote the kernel and image of this permutation representation. Then $\mathbb{E} \subseteq (\mathrm{Aut}(\Lambda_1))^{13}$, where $\Lambda_1 = \Lambda \cap \mathcal{H}_1$ and $\mathrm{Aut}(\Lambda_1) = A(E_6)$. Furthermore, $SL_3(3) \subseteq H \subseteq \mathbb{S}_{13}$, therefore H is one of the groups $SL_3(3)$, $\mathbb{A}_{13}$, $\mathbb{S}_{13}$. □

12.3 Character computation

The aim of the section is to prove that if we embed the group

$$G = 3^{3+3}.SL_3(3) = \mathrm{Aut}(\mathcal{D})$$

in $\Omega = \Omega_7(3)$ as a maximal subgroup, then

$$\nu|_G = \chi,$$

where ν is the unique irreducible character of degree 78 of Ω (see [ATLAS]), and χ is the G-character afforded by $\mathcal{L}$.

The group G has the factorization

$$G = CD \tag{12.3}$$

in which $C = O_3(G) = 3^{3+3}$ and $D = E.S = \mathbb{Z}_3^3.SL_3(3)$. In fact, one can choose a Lie subalgebra $\mathcal{L}_0$ of type F_4 in $\mathcal{L}$ such that the decomposition

$$\mathcal{D}_0 : \mathcal{L}_0 = \bigoplus_{i=1}^{13} (\mathcal{H}_i \cap \mathcal{L}_0)$$

is an IOD of $\mathcal{L}_0$, and then $D = \mathrm{Aut}\,(\mathcal{D}_0)$ [Ale 1]. Furthermore, one can embed D in $L = L_4(3)$ as follows:

$$D = \left\{ \begin{pmatrix} A & a \\ 0 & 1 \end{pmatrix} \mid A \in SL_3(3),\, a \in \mathbb{F}_3^3 \right\} \hookrightarrow L. \tag{12.4}$$

As $(L : D) = 40$, D is a maximal subgroup of L. Moreover, L has two conjugacy classes of subgroups of index 40, and they are interchanged by an outer automorphism 2_2 (in the notation of [ATLAS]) of L. Furthermore,

$$L. < 2_2 > \cong P\Omega_6^+(3) \hookrightarrow \Omega.$$

For, consider the natural representation of Ω on the 7-dimensional $\mathbb{F}_3$-space $< u_1, u_2, u_3, v_1, v_2, v_3, w >$ endowed with the following quadratic form q and symmetric bilinear form b:

$$q(u_i) = q(v_i) = 0,\; q(w) = 1,\; q(x+y) = q(x) + q(y) + b(x, y);$$

$$b(u_i, u_j) = b(v_i, v_j) = b(u_i, w) = b(v_i, w) = 0;$$

$$b(u_i, v_j) = \delta_{i,j},\; b(w, w) = 1.$$

It is known that $G = St_\Omega(< u_1, u_2, u_3 >_{\mathbb{F}_3}) \cong 3^{3+3}.SL_3(3)$. Set $P = St_\Omega(< w >_{\mathbb{F}_3}) \cong L_4(3). < 2_2 >$. Note that $D = G \cap P \cong \mathbb{Z}_3^3.SL_3(3)$. Indeed, consider the matrix

$$[X, Y, Z, a, \lambda] := \begin{pmatrix} X & Y & 0 \\ 0 & Z & 0 \\ 0 & {}^t a & \lambda \end{pmatrix};\; \lambda \in \mathbb{F}_3,\; a \in \mathbb{F}_3^3,\; X, Y, Z \in M_3(\mathbb{F}_3)$$

(in the given basis) of any element of $St_{SO_7(3)}(< u_1, u_2, u_3 >, < w >)$. Then $Z = {}^t X^{-1}$, $\lambda = 1$, $a = 0$, $Y.{}^t X + X.{}^t Y = 0$. If we add the condition that $\det X = 1$, then we obtain a perfect subgroup $D' = \mathbb{Z}_3^3.SL_3(3)$, where

$$\mathbb{Z}_3^3 = \{[E_3, Y, E_3, 0, 1]\},\; SL_3(3) = \{[X, 0, {}^t X^{-1}, 0, 1]\}.$$

As usual, E_m denotes the unit matrix of degree m. Clearly, $1 \le (D : D') \le 2$. But $(G : D') = 27$, so $D = D'$. Note also that $C = O_3(G)$ can be identified with the subgroup

$$\left\{ (a, X) := \begin{pmatrix} E_3 & X & -a \\ 0 & E_3 & 0 \\ 0 & {}^t a & 1 \end{pmatrix} | a \in \mathbb{F}_3^3,\; X \in M_3(\mathbb{F}_3),\; X + {}^t X + a.{}^t a = 0 \right\}$$

with multiplication

$$(a, X).(b, Y) = (a + b, X + Y - a.^t b).$$

Hence, the subgroup $\mathbb{Z}_3^3 = O_3(D)$ coincides with $E = [C, C]$, that is, $D = E.S$.

Now suppose that $\rho \in \mathrm{Irr}\,(D)\backslash\mathrm{Irr}\,(D/E)$ and $\deg \rho \leq 52$. Clearly, 26 divides $\deg \rho$ and $\rho = \mathrm{Ind}_B^D(\alpha)$, where $B = \mathbb{Z}_3^3.(\mathbb{Z}_3^2.SL_2(3))$ is the commutator subgroup of the stabilizer $\tilde{B} = St_D(< e_1, e_2 >)$ with $E =< e_1, e_2, e_3 >$, and $\alpha \in \mathrm{Irr}\,(B)$. Moreover, $\mathrm{Ker}\,\alpha \supseteq \mathbb{Z}_3^2.\mathbb{Z}_3^2$. This means that α is in fact an irreducible character of $\mathbb{Z}_3 \times SL_2(3)$ (here one can identify the direct factor $\mathbb{Z}_3$ with $< e_3 >$). One can suppose also that

$$\alpha(e_3) = \varepsilon. \deg \alpha \tag{12.5}$$

where, as above, $\varepsilon = \exp(2\pi i/3)$. If $\deg \rho = 26$, then $\deg \alpha = 1$ and there exist precisely three possibilities α_k, $k = 0, 1, 2$, for a 1-dimensional character α of B with property (12.5), where

$$\alpha_k(x_0) = \varepsilon^k \text{ for } x_0 = \begin{pmatrix} 1 & 1 \\ 0 & 1 \end{pmatrix} \in SL_2(3). \tag{12.6}$$

Correspondingly, we obtain three 26-dimensional irreducible characters $\rho_k = \mathrm{Ind}_B^D(\alpha_k)$ of D, $k = 0, 1, 2$.

Now suppose that $\deg \rho = 52$ and so $\deg \alpha = 2$. The group $SL_2(3)$ has seven conjugacy classes; $\{1\}$, $\{-1\}$, $\{y||y| = 4\}$, and four classes with representatives x_0, x_0^{-1}, $-x_0$ and $-x_0^{-1}$. Furthermore, $SL_2(3)$ has precisely three irreducible characters β_i of degree 2, $i = 0, 1, 2$, with

$$\beta_i(x_0) = -\varepsilon^i \tag{12.7}$$

Correspondingly, D has three irreducible characters $\mu_i = \mathrm{Ind}_B^D(\beta_i)$ of degree 52, $i = 0, 1, 2$.

Let X_0 denote the element

$$\begin{pmatrix} 1 & 1 & 0 & 0 \\ 0 & 1 & 0 & 0 \\ 0 & 0 & 1 & 0 \\ 0 & 0 & 0 & 1 \end{pmatrix}$$

of D (see embedding (12.4)).

Lemma 12.3.1. *Of the six characters ρ_i, μ_j, only ρ_1, ρ_2 and μ_0 can be extended to the group L; moreover, $\rho_1(X_0) = \rho_2(X_0) = -1$, $\mu_0(X_0) = -2$.*

Proof. 1) We view each element

$$\begin{pmatrix} Y & y \\ 0 & 1 \end{pmatrix}$$

of D as an affine transformation:

$$v \mapsto Yv + y,\ v \in E = < e_1, e_2, e_3 >_{\mathbb{F}_3}.$$

Then $X_0(v) = Av$, where

$$A = \begin{pmatrix} 1 & 1 & 0 \\ 0 & 1 & 0 \\ 0 & 0 & 1 \end{pmatrix}.$$

For every element $g : v \mapsto Yv + y$ we have

$$gX_0g^{-1} : v \mapsto YAY^{-1}v + (E_3 - YAY^{-1})y.$$

In the induction from B to D we are interested only in those elements g with $gX_0g^{-1} \in B$, that is, such that YAY^{-1} fixes the plane $W = < e_1, e_2 >_{\mathbb{F}_3}$ (and has determinant 1 on W). Furthermore,

$$YAY^{-1}(W) = W \Leftrightarrow AY^{-1}(W) = Y^{-1}(W) \Leftrightarrow (A - E_3) \text{ fixes } Y^{-1}(W).$$

So we obtain the following two possibilities: either

$$Y^{-1}(W) = < e_1, e_3 > \tag{12.8}$$

or

$$Y^{-1}(W) = < e_1, e_2 + \lambda e_3 >,\ \lambda \in \mathbb{F}_3. \tag{12.9}$$

In case (12.8), we have

$$Y^{-1} = \begin{pmatrix} a_1 & b_1 & c_1 \\ 0 & 0 & c_2 \\ a_3 & b_3 & c_3 \end{pmatrix}, \quad \text{where } c_2(a_1b_3 - a_3b_1) = -1.$$

Correspondingly,

$$YAY^{-1} = \begin{pmatrix} 1 & 0 & -b_3 \\ 0 & 1 & a_3 \\ 0 & 0 & 1 \end{pmatrix}.$$

Hence, $gX_0g^{-1}(v) = YAY^{-1}v + y_3(b_3e_1 - a_3e_2)$, where $y = [y_1, y_2, y_3]$. Provided $\alpha \in \{\alpha_i, \beta_j | 0 \leq i, j \leq 2\}$ we have (see (12.6), (12.7)):

$$\alpha(gX_0g^{-1}) = \deg \alpha.$$

Thus, in the sum

$$\rho(X_0) = \mathrm{Ind}_B^D(\alpha)(X_0) = \frac{1}{|B|} \sum_{g \in D, gX_0g^{-1} \in B} \alpha(gX_0g^{-1})$$

terms of type (12.8) enter

$$\frac{1}{|B|}.3^3.3^2.48$$

times.

In case (12.9), we have

$$Y^{-1} = \begin{pmatrix} a_1 & b_1 & c_1 \\ a_2 & b_2 & c_2 \\ \lambda a_2 & \lambda b_2 & c_3 \end{pmatrix}, \text{ where } (c_3 - \lambda c_2)(a_1 b_2 - a_2 b_1) = 1.$$

Correspondingly,

$$YAY^{-1} = \begin{pmatrix} 1 + a_2 b_2 (c_3 - \lambda c_2) & b_2^2 (c_3 - \lambda c_2) & b_2 c_2 (c_3 - \lambda c_2) \\ -a_2^2 (c_3 - \lambda c_2) & 1 - a_2 b_2 (c_3 - \lambda c_2) & -a_2 c_2 (c_3 - \lambda c_2) \\ 0 & 0 & 1 \end{pmatrix}.$$

From this it follows that

$$\alpha(gX_0 g^{-1}) = \begin{cases} \alpha(x_0) & \text{if} \quad c_3 - \lambda c_2 = 1, \\ \overline{\alpha(x_0)} & \text{if} \quad c_3 - \lambda c_2 = -1, \end{cases}$$

where x_0 is as described in (12.6). Here we have used the fact that the following pairs of elements are conjugate in $SL_3(3)$:

$$\begin{pmatrix} 1 & 0 \\ 1 & 1 \end{pmatrix} \sim \begin{pmatrix} 1 & -1 \\ 0 & 1 \end{pmatrix},$$

$$\begin{pmatrix} 1+t & t \\ -t & 1-t \end{pmatrix} \sim \begin{pmatrix} 1 & t \\ 0 & 1 \end{pmatrix}, \begin{pmatrix} 1-t & t \\ -t & 1+t \end{pmatrix} \sim \begin{pmatrix} 1 & t \\ 0 & 1 \end{pmatrix}.$$

Uniting cases (12.8) and (12.9), we have:

$$\rho(X_0) = \frac{1}{3^3.3^2.24}(\deg \alpha.3^3.3^2.48 + \alpha(x_0).3^3.3^3.24 + \overline{\alpha(x_0)}.3^3.3^3.24),$$

that is,

$$\rho(X_0) = 2.\deg \alpha + 6.\Re \alpha(x_0) \tag{12.10}$$

2) Now let $\rho = \rho_i$ or μ_j. Then from (12.10) it follows that

$$\rho_i(X_0) = 2 + 6.\Re \varepsilon^i = \begin{cases} 8 & \text{if} \quad i = 0 \\ -1 & \text{if} \quad i = 1, 2 \end{cases} \tag{12.11}$$

$$\mu_j(X_0) = 4 - 6.\Re \varepsilon^j = \begin{cases} -2 & \text{if} \quad j = 0 \\ 7 & \text{if} \quad j = 1, 2 \end{cases} \tag{12.12}$$

The group $L = L_4(3)$ has two irreducible characters χ_2, χ_3 of degree 26, and one irreducible character χ_5 of degree 52 (this is the notation of [ATLAS]). Moreover,

$$\chi_2(X_0) = \chi_3(X_0) = -1,\ \chi_5(X_0) = -2,$$

since X_0 is an element of class $3A$ in L. Furthermore, all of these characters can be extended to the group $L. < 2_2 >$, and thus their restrictions to a subgroup D' of L of index 40 do not depend on the choice of D'. Hence, ρ_0, μ_1 and μ_2 cannot be extended to L. But $\chi_2|_D$ and $\chi_3|_D$ are faithful characters of degree 26, so they belong to $\mathrm{Irr}\,(D)\backslash\mathrm{Irr}\,(D/E)$, which implies that $\chi_2|_D, \chi_3|_D \in \{\rho_1, \rho_2\}$. It is not difficult to see that in fact $\{\chi_2|_D, \chi_3|_D\} = \{\rho_1, \rho_2\}$.

Suppose that $\chi_5|_D$ is reducible. From the exactness of χ_5 and the formula $\chi_5(X_0) = -2$ it follows that $\chi_5|_D \in \{\rho_1 + \rho_2, 2\rho_1, 2\rho_2\}$. Using the same arguments as above, we obtain that

$$\rho(Y_0) = 2.\deg\alpha + 6.\Re(\varepsilon.\alpha(x_0)),$$

where

$$Y_0 = \begin{pmatrix} 1 & 1 & 0 & 0 \\ 0 & 1 & 0 & 0 \\ 0 & 0 & 1 & 1 \\ 0 & 0 & 0 & 1 \end{pmatrix}$$

is an element of D of class $3B$ in L (in the notation of [ATLAS]). In particular, $\rho_1(Y_0) = -1$, $\rho_2(Y_0) = 8$, $\chi_5(Y_0) = 7$, so that $\chi_5|_D = \rho_1 + \rho_2$. Now for the element

$$Z_0 = \mathrm{diag}\,(-1, -1, 1, 1)$$

in D of class $2B$ in L, we have

$$\chi_5(Z_0) = -4,\ \rho_1(Z_0) + \rho_2(Z_0) = \chi_2(Z_0) + \chi_3(Z_0) = 4,$$

a contradiction. Hence, $\chi_5|_D$ is irreducible, and so it is clear that $\chi_5|_D = \mu_0$. $\square$

It is known that the group $SL_3(3)$ has the following irreducible characters:

$$\overline{1},\ \overline{12},\ \overline{13},\ \underbrace{\overline{16_1},\ \overline{16_2},\ \overline{16_3},\ \overline{16_4}}_{\text{algebr. conj.}},\ \overline{26_1},\quad \underbrace{\overline{26_2},\ \overline{26_3}}_{\text{complex conj.}},\quad \overline{27},\ \overline{39},$$

(here, for brevity, we are denoting characters by their degrees).

Lemma 12.3.2. *Suppose that a subgroup $S = SL_3(3) \subset \mathcal{G}_0 = F_4(\mathbb{C}) = \mathrm{Aut}\,(\mathcal{L}_0)$ acts on the Lie algebra $\mathcal{L}_0$ of type F_4 with integer-valued character τ. Then*

$$\tau = \overline{26_2} + \overline{26_3}.$$

In particular, $\tau(x) = -2$ *for every element* x *of* $SL_3(3)$ *of order* 3. *Furthermore, if in addition* S *acts on* $\mathcal{L}$ *with integer-valued character* τ' *in the embeddings* $\mathcal{L}_0 \hookrightarrow \mathcal{L}$, $\mathcal{G}_0 \hookrightarrow \mathcal{G}$, *where* $\mathcal{L}$ *is a Lie algebra of type* E_6 *and* $\mathcal{G} = \mathrm{Aut}(\mathcal{L})$, *then*

$$\tau' = \overline{26_1} + \overline{26_2} + \overline{26_3}.$$

Proof. It is known that every involution of $\mathcal{G}_0$ has trace -4 or 20 on $\mathcal{L}_0$, and every element of order 3 has trace 7 or -2 (see [CoG]).

a) If some $\overline{26_i}$ with $i \geq 2$ enters τ, then $\tau = \overline{26_2} + \overline{26_3}$ because τ is integer-valued.

b) Each character $\overline{16_j}$, $1 \leq j \leq 4$, does not enter τ. Otherwise $\overline{16_k}$ would enter τ for all k, and so $\deg \tau \geq 64$, a contradiction.

c) Now suppose that $\tau = a.\overline{1} + b.\overline{12} + c.\overline{13} + d.\overline{26_1} + e.\overline{27} + f.\overline{39}$, where $a, \ldots, f \in \mathbb{Z}_0^+$. Then

$$\tau(2A) = a + 4b - 3c + 2d + 3e - f \in \{-4, 20\},$$

$$\left.\begin{array}{rcl} \tau(3A) & = & a + 3b + 4c - d + 3f \\ \tau(3B) & = & a + c - d \end{array}\right\} \in \{7, -2\},$$

(the symbols $2A$, $3A$ and $3B$ denote representatives of the corresponding conjugacy classes). If $\tau(3B) = -2$, then $a + c - d = -2$, $2 = 52/26 \leq d = a + c + 2$, and so $a = c = 0$, $d = 2$, that is, $\tau = 2.\overline{26_1}$. But in that case $\tau(2A) = 4$, a contradiction. Hence, $\tau(3B) = 7$. But $\tau(3A) \geq \tau(3B)$, so that $\tau(3A) = \tau(3B) = 7$, $b = c = f = 0$, $a = d + 7$. In particular,

$$\tau(2A) = a + 2d + 3e = 7 + 3(d + e) \equiv 1 \pmod 3,$$

that is, $\tau(2A)$ cannot belong to $\{-4, 20\}$, again a contradiction.

d) Finally, suppose that $\tau - \tau' \neq \overline{26_1}$. As τ and τ' are integer-valued, we must have

$$\tau' = \overline{26_2} + \overline{26_3} + a'.\overline{1} + b'.\overline{12} + c'.\overline{13}$$

for some non-negative integers a', b', c' with $a' + 12b' + 13c' = 26$. In particular,

$$\tau'(2A) = a' + 4b' - 3c' - 4, \ \tau'(3B) = a' + c' - 2.$$

Furthermore, every automorphism of order 3 of $\mathcal{L}$ has trace ± 3, 6, 15 or 30 on $\mathcal{L}$ (use the description of automorphisms of simple Lie algebras explained in [Kac]). But direct calculations show that $\tau'(3B) \in \{0, 12, 24\}$ in our case, a contradiction. □

Definition 12.3.3. An irreducible character ζ of degree 78 of $G = 3^{3+3}.SL_3(3)$ is said to be *D-regular* if

$$\zeta|_D = \mu_0 + \rho_1 \text{ or } \mu_0 + \rho_2,$$

where D is a subgroup of type $\mathbb{Z}_3^3.SL_3(3)$ in G and the D-characters μ_0, ρ_1 and ρ_2 are as defined in Lemma 12.3.1.

Recall that all the subgroups of type $\mathbb{Z}_3^3.SL_3(3)$ are conjugate in G.

Lemma 12.3.4. *Suppose that a subgroup $D = \mathbb{Z}_3^3.SL_3(3)$ of $\mathcal{G}_0$ acts on $\mathcal{L}_0$ with integer-valued character η. Then $\eta \in \mathrm{Irr}(D)$ and $\eta = \mu_0$. Furthermore, if in addition D acts on $\mathcal{L}$ with integer-valued character η' in the embedding $\mathcal{G}_0 \hookrightarrow \mathcal{G}$, then*

$$(\eta' - \mu_0) \in \{\rho_1, \rho_2\}.$$

Proof. Embed $S = SL_3(3)$ in D. By Lemma 12.3.2, $\eta|_S = \overline{26_2} + \overline{26_3}$. Suppose that η is reducible. Then $\eta = \alpha + \beta$, where $\alpha, \beta \in \mathrm{Irr}(D)$ and $\deg\alpha = \deg\beta = 26$. Furthermore, $\alpha(X) = \beta(X) = -1$ for every element X of S of order 3. In the proof of Lemma 12.3.1, we showed that $\alpha \in \{\rho_i | 0 \leq i \leq 2\}$ and $\rho_i(X_0) = 9\delta_{i,0} - 1$ (see (12.11)). Hence, $\alpha = \rho_1$ or $\alpha = \rho_2$. By Lemma 12.3.1, α can be extended to a unique irreducible character α' of degree 26 of $L = L_4(3)$. In particular,

$$\mathbb{Q}(i) = \mathbb{Q}(\overline{26_2}) \subseteq \mathbb{Q}(\alpha') = \mathbb{Q},$$

a contradiction, and so η is irreducible. In particular, $\eta \in \{\mu_0, \mu_1, \mu_2\}$. Applying (12.12) and the equality $\eta(X_0) = -2$ (see Lemma 12.3.2), we conclude that $\eta = \mu_0$.

Suppose now that $\lambda = \eta' - \mu_0$ is neither ρ_1 nor ρ_2. Set $E = O_3(D)$, as usual. If $\lambda|_E$ is trivial, then we have $\mu_0(e) = -2$ and $\lambda(e) = 26$ for every element $e \in E \backslash \{1\}$, that is, $\eta'(e) = 24$, a contradiction (see the proof of Lemma 12.3.2). Thus $\lambda|_E$ is non-trivial, λ is faithful and so $\lambda = \rho_0$. In particular, $\lambda(X_0) = 8$ (see (12.11)), on the one hand. On the other hand, by Lemma 12.3.2 we have $\lambda(X_0) = \overline{26_1}(X_0) = -1$, a contradiction. □

Lemma 12.3.5. *The character χ of $G = \mathrm{Aut}(\mathcal{D})$ afforded by $\mathcal{L}$ is D-regular.*

Proof. Clearly, $\chi \in \mathrm{Irr}(G)$ and $\deg\chi = 78$. For D we again take the subgroup $D = \mathrm{Aut}(\mathcal{D}_0)$ for the given embedding $\mathcal{L}_0 \hookrightarrow \mathcal{L}$. Then D acts faithfully and irreducibly on $\mathcal{L}_0$, with integral character. Now simply apply Lemma 12.3.4. □

Recall that the finite simple group $\Omega = \Omega_7(3) = [SO_7(3), SO_7(3)]$ has a unique irreducible character ν of degree 78 [ATLAS]. Moreover, $\mathbb{Q}(\nu) = \mathbb{Q}$, and G can be embedded in Ω as a maximal subgroup.

Lemma 12.3.6. *The character $\nu|_G$ is D-regular.*

Proof. 1) Firstly we determine the restriction $\nu|_L$ to the subgroup $[P, P] = L = L_4(3)$. It is known that

$$\mathrm{Irr}(L) = \{\overline{1}, \overline{26_1}, \overline{26_2}, \overline{39}, \overline{52}, \overline{65_1}, \overline{65_2}, \overline{90}, ...\}$$

(again we denote the characters by their degrees), and $\nu|_L$ is faithful. Write

$$\nu|_L = a.\overline{1} + b_1.\overline{26_1} + b_2.\overline{26_2} + c.\overline{39} + d.\overline{52} + e_1.\overline{65_1} + e_2.\overline{65_2}$$

for some non-negative integers $a, b_1, \ldots, e_2$. Then

$$\begin{array}{rclcl} 78 & = & \nu(1) & = & a + 26(b_1 + b_2) + 39c + 52d + 65(e_1 + e_2), \\ 0 & = & \nu(13A) & = & a, \\ 3 & = & \nu(5A) & = & a + b_1 + b_2 - c + 2d, \\ -1 & = & \nu(20A) & = & a - b_1 - b_2 - c. \end{array}$$

The names of conjugacy classes of L and the values of ν restricted from Ω to L are taken from [ATLAS]. Solving this equation system, we find that $a = c = e_1 = e_2 = 0$ and $d = b_1 + b_2 = 1$, that is, $\nu|_L = \overline{52} + \overline{26_i}$ for some $i = 1, 2$.

2) Clearly, D can be chosen to be a subgroup of L. Furthermore, $\nu|_D$ is faithful. By Lemma 12.3.1, the character μ_0 of D can be extended to the character $\overline{52}$ of L. Hence, μ_0 occurs in $\nu|_D$. One can suppose that

$$\overline{26_j}|_D = \rho_j \text{ for } j = 1, 2.$$

Hence, $\nu|_D = \mu_0 + \rho_i$, and we show that $\nu|_G$ is irreducible. Suppose the contrary. Because of the decomposition $\nu|_D = \mu_0 + \rho_i$, we must have that $\nu|_G = \alpha + \beta$, where $\alpha, \beta \in \mathrm{Irr}\,(G)$, $\deg\alpha = 52$, $\deg\beta = 26$. Restrict α to the subgroup $C = O_3(G)$:

$$\alpha|_C = e\sum_{k=1}^{t} \lambda_k,\ e, t \in \mathbb{N},\ \lambda_1, \ldots, \lambda_t \in \mathrm{Irr}\,(C).$$

If $\deg\lambda_1 > 1$, then 3 divides $\deg\lambda_k$ for all k, and so 3 divides 52, a contradiction. Therefore $\deg\lambda_k = 1$ and $\lambda_k = 1$ on $E = [C, C]$ for all k, that is, $\mathrm{Ker}\,\alpha \supseteq E$. Similarly, $\mathrm{Ker}\,\beta \supseteq E$. Thus we obtain the contradiction that $\mathrm{Ker}\,\nu \supseteq E$. □

Proposition 12.3.7. *The group $G = 3^{3+3}.SL_3(3)$ has exactly one D-regular character.*

Proof. 1) The existence of D-regular characters follows from Lemma 12.3.5. Let $\rho \in \mathrm{Irr}\,(G)$ be any D-regular character. Then $(\rho|_D, \mu_0)_D > 0$, and therefore E cannot be contained in $\mathrm{Ker}\,\rho$. Considering the restriction of ρ to E and applying Clifford's Theorem, we get that $\rho = \mathrm{Ind}_H^G(\alpha)$, where α is an irreducible character of degree 3 of the subgroup $H = C.H_0$ with $C = O_3(G)$ and $H_0 = \mathbb{Z}_3^2.SL_2(3)$. The subgroup H is characterized by the fact that it fixes the plane $< e_1, e_2 >$ and the point e_3 modulo $< e_1, e_2 >$ in its action on $E =< e_1, e_2, e_3 >_{\mathbb{F}_3}$. Since $\deg\alpha = 3 < 8$, $\mathrm{Ker}\,\alpha$ contains the normal subgroup $\mathbb{Z}_3^2 = O_3(H_0)$. Furthermore, $\mathrm{Ker}\,\alpha$ contains another elementary abelian subgroup, namely $< e_1, e_2 >$. One can suppose also that

$$\alpha(e_3) = 3\varepsilon, \tag{12.13}$$

where $\varepsilon = \exp(2\pi i/3)$.

2) We set $K = \operatorname{Ker}\alpha \cap C$ and determine C/K. Note that C/K is nonabelian. In fact, if C/K is abelian, then $K \supseteq [C, C] = E$ and so $e_3 \in K$, contrary to (12.13). In particular, $|C/K| \geq 3^3$, $|K| \leq 3^3$. Recall that $K \cap E =< e_1, e_2 >\cong \mathbb{Z}_3^2$, which implies that

$$[C/K, C/K] = [C, C]K/K = EK/K \cong E/(E \cap K) = \mathbb{Z}_3 =< e_3 > .$$

Furthermore, $\alpha|_{C/K}$ is irreducible. Indeed, by Clifford's Theorem, we have

$$\alpha|_{C/K} = e \sum_{i=1}^{t} \alpha_i,\ \alpha_i \in \operatorname{Irr}(C/K).$$

Therefore, in the case when $\deg \alpha_1 = 1$ we have $\deg \alpha_i = \deg \alpha_1 = 1$, which implies that $\operatorname{Ker}\alpha_i \supseteq [C/K, C/K] =< e_3 >$ and so $\operatorname{Ker}\alpha \ni e_3$, a contradiction. Hence, $\deg \alpha_1 = 3$ and $\alpha|_{C/K} = \alpha_1$.

Thus, α is a faithful irreducible character of C/K. By Schur's Lemma, the centre $Z(C/K)$ acts scalarly (and faithfully). This means that $Z(C/K) \hookrightarrow \mathbb{C}^*$. In particular, $Z(C/K)$ is cyclic. Moreover, $\exp(C/K) = \exp(C) = 3$ and $e_3 \in Z(C/K)$. Consequently,

$$Z(C/K) = \mathbb{Z}_3 =< e_3 >= [C/K, C/K].$$

Finally, $(C/K)/(EK/K) \cong C/EK$ is elementary abelian, therefore $\Phi(C/K) = \mathbb{Z}_3$. We have shown that C/K is an extraspecial group of exponent 3 and of order not greater than 3^4. In other words, $C/K = 3_+^{1+2}$.

3) More concretely, we write $E = F \wedge F$, where $F = E^* = C/E = < f_1, f_2, f_3 >_{\mathbb{F}_3}$. The identification of the $SL_3(3)$-modules E and $F \wedge F$ is given by

$$e_1 \rightsquigarrow f_2 \wedge f_3,\ e_2 \rightsquigarrow f_3 \wedge f_1,\ e_3 \rightsquigarrow f_1 \wedge f_2,$$

and the convolution is as follows:

$$\delta_{i,j} = f_j(e_i) = f_j \wedge e_i$$

(we are identifying $\wedge^3(F)$ with $\mathbb{F}_3$). It is now clear that if g is an element of the subgroup $St_{SL_3(3)}(f_3)$ of index 26 in $SL_3(3)$, then the action of g on F has matrix of the form

$$X = \begin{pmatrix} a & c & 0 \\ b & d & 0 \\ x & y & 1 \end{pmatrix}$$

in the basis $\{f_1, f_2, f_3\}$, and the action of g on E has matrix

$${}^tX^{-1} = \begin{pmatrix} d & -b & by - dx \\ -c & a & cx - ay \\ 0 & 0 & \underbrace{ad - bc}_{1} \end{pmatrix}$$

in the basis $\{e_1, e_2, e_3\}$. This means that $St(f_3)$ is contained in $St(< e_1, e_2 >, e_3 \pmod{< e_1, e_2 >})$. This last subgroup has also index 26, so in fact we have

$$St(f_3) = St(< e_1, e_2 >, e_3 \pmod{< e_1, e_2 >}).$$

Furthermore, the group $\mathbb{Z}_3 = KE/E \cong K/(K \cap E)$ is normal in H, so that $St(f_3)$ fixes some subgroup $\mathbb{Z}_3$ in F, which implies that $KE/K =< f_3 >$. Hence, we have found that

$$K =< e_1, e_2, f_3 >, \; C/K =< e_3, f_1, f_2 > .$$

4) The product in the group $C = \{(u, x) | u \in F, x \in E\}$ is given by the rule

$$(u, x).(v, y) = (u + v, x + y + u \wedge v), \; u, v \in F, \; x, y \in E.$$

In particular,

$$[f_1, f_2] = f_1 f_2 f_1^{-1} f_2^{-1} = (f_1, 0).(f_2, 0).(-f_1, 0).(-f_2, 0) = (0, -e_3) = e_3^{-1}.$$

Consider again the matrices

$$\delta = \begin{pmatrix} 1 & & \\ & \varepsilon & \\ & & \varepsilon^2 \end{pmatrix}, \; \pi = \begin{pmatrix} 0 & 1 & 0 \\ 0 & 0 & 1 \\ 1 & 0 & 0 \end{pmatrix}$$

(which were denoted by D and P^{-1} in §12.2 !). Then $[\delta, \pi] = \varepsilon^2 E_3$. Hence we can suppose without loss of generality that f_1 acts as δ, and f_2 as π in the representation with character α.

5) Consider again the element $X_0 \in D = \mathbb{Z}_3^3.SL_3(3)$ introduced just before Lemma 12.3.1. Since the matrices

$$\begin{pmatrix} 1 & 1 & \\ & 1 & \\ & & 1 \end{pmatrix}, \begin{pmatrix} 1 & & \\ 1 & 1 & \\ & & 1 \end{pmatrix}$$

are conjugate in $S = SL_3(3)$, in computing character values at the point X_0 we can replace X_0 by

$$X_1 = \begin{pmatrix} 1 & & & \\ 1 & 1 & & \\ & & 1 & \\ & & & 1 \end{pmatrix}.$$

Clearly, $X_1 \in S$ and X_1 acts on $E = O_3(D)$ with matrix

$$Y_1 = \begin{pmatrix} 1 & & \\ 1 & 1 & \\ & & 1 \end{pmatrix}$$

in the basis $\{e_1, e_2, e_3\}$. In this case X_1 acts on C/E with matrix ${}^tY_1^{-1}$ in the basis $\{f_1, f_2, f_3\}$. Hence, we have the following relations for the matrix φ of X_1 in the representation with character α:

(i) $|\varphi| = 3$;

(ii) $\varphi\delta\varphi^{-1} = \zeta\delta$, $\varphi\pi\varphi^{-1} = \eta\delta^{-1}\pi$.

Here $\zeta, \eta \in \{1, \varepsilon, \varepsilon^2\}$. Using conditions (i) and (ii), we shall next determine $\operatorname{Spec}\varphi$.

5a) Firstly, suppose that $\zeta = 1$. Since $\delta = \operatorname{diag}(1, \varepsilon, \varepsilon^2)$, we must have $\varphi = \operatorname{diag}(a, b, c)$ for some $a, b, c \in \mathbb{C}^*$. From this it follows that $[\varphi, \pi] = \operatorname{diag}(ab^{-1}, bc^{-1}, ca^{-1})$. When $\eta = 1$ we have $a = b = \varepsilon^2 c$. If $\eta = \varepsilon$, then $b = c = \varepsilon^2 a$. Finally, if $\eta = \varepsilon^2$, then $c = a = \varepsilon^2 b$. In all cases we have

$$\operatorname{Spec}\varphi = \{d, \varepsilon^2 d, \varepsilon^2 d\}$$

(counting multiplicities) for some $d \in \mathbb{C}$, $d^3 = 1$. In particular, $\alpha(X_1) = d(1 + 2\varepsilon^2) = \varepsilon^i(1 + 2\varepsilon^2)$ for some $i = 0, 1, 2$.

5b) Secondly, suppose that $\zeta = \varepsilon$, and set $\psi = \varphi\pi^{-1}$. Then

$$\psi\delta\psi^{-1} = \varphi\pi^{-1}\delta\pi\varphi^{-1} = \varepsilon^2\varphi\delta\varphi^{-1} = \delta,$$

and

$$\psi\pi\psi^{-1} = \varphi\pi^{-1}\pi\pi\varphi^{-1} = \varphi\pi\varphi^{-1} = \eta\delta^{-1}\pi.$$

This means that ψ plays the same role as φ in 5a). Hence, $\psi = \operatorname{diag}(a, b, c)$, where $\{a, b, c\} = \{d, \varepsilon^2 d, \varepsilon^2 d\}$, and

$$\varphi = \psi\pi = \begin{pmatrix} 0 & a & 0 \\ 0 & 0 & b \\ c & 0 & 0 \end{pmatrix}.$$

But $\varphi^3 = 1$, so that $abc = 1$, $d^3 = \varepsilon^{-1}$, $\alpha(X_1) = 0$ and

$$\operatorname{Spec}\varphi = \{1, \varepsilon, \varepsilon^2\}.$$

5c) Finally, suppose that $\zeta = \varepsilon^2$. Using the same arguments as in 5b) we obtain that $\alpha(X_1) = 0$ and $\operatorname{Spec}\varphi = \{1, \varepsilon, \varepsilon^2\}$.

6) Fix a representation Φ affording the character α of H, which gives a D-regular character in inducing it from H to $G = 3^{3+3}.SL_3(3)$ and satisfies condition (12.13). Let Ψ be a representation of H with some character β, which also leads to a D-regular character of G and is such that $\beta(e_3) = 3\varepsilon$. In this case $\Phi, \Psi \in \operatorname{Hom}(3_+^{1+2}.SL_2(3), GL_3(\mathbb{C}))$. Set $Q = 3_+^{1+2}$, $R = SL_2(3)$. Since $\alpha|_Q = \beta|_Q$, we can suppose that $\Phi|_Q = \Psi|_Q$. When $x \in R$ and $q \in Q$ we have

$$\Phi(x)\Phi(q)\Phi(x)^{-1} = \Phi(xqx^{-1}) = \Psi(xqx^{-1}) = \Psi(x)\Psi(q)\Psi(x)^{-1} =$$

$$= \Psi(x)\Phi(q)\Psi(x)^{-1},$$

which implies that $[\Psi(x)^{-1}\Phi(x), \Phi(q)] = 1$. By Schur's Lemma, $\Psi(x)^{-1}\Phi(x) = \lambda = \lambda(x) \in \mathbb{C}^*$. Furthermore, for $x, y \in R$ we have

$$\lambda(xy) = \Psi(xy)^{-1}\Phi(xy) = \Psi(y)^{-1}.(\Psi(x)^{-1}\Phi(x)).\Phi(y) =$$

$$= \lambda(x).\Psi(y)^{-1}\Phi(y) = \lambda(x).\lambda(y).$$

This means that

$$\lambda \in \mathrm{Hom}\,(R, \mathbb{C}^*) = \mathrm{Hom}\,(R/[R, R], \mathbb{C}^*) = \mathrm{Hom}\,(\mathbb{Z}_3, \mathbb{C}^*) = \mathbb{Z}_3.$$

Thus $\Psi(x) = \xi^i(x)\Phi(x)$, where $\xi \in \mathrm{Hom}\,(\mathbb{Z}_3, \mathbb{C}^*)\backslash\{1\}$ and $i = 0, 1, 2$. Set

$$\alpha^{(i)} = \alpha.\xi^i.$$

We have shown that G has at most three irreducible characters of degree 78 nontrivial on E; namely, $\rho^{(i)} = \mathrm{Ind}\,_H^G(\alpha^{(i)})$ for $i = 0, 1, 2$.

7) Recall that $G = CD$, where $D = E.S$. Set $B = H \cap D$ and $\alpha' = \alpha|_B$. We claim that

$$\rho|_D = \mathrm{Ind}\,_B^D(\alpha')$$

if $\rho = \mathrm{Ind}\,_H^G(\alpha)$. For, we have $G = CD = HD$, $|G| = |H|.|D|/|B|$, and so $(G : H) = (D : B)$. Consider a decomposition $D = \bigcup_{k=1}^n d_kB$ of D into cosets modulo B. Suppose that $d_kH \cap d_lH$ is not empty, so that $d_k = d_lh$ for some $h \in H$. Then $h = d_l^{-1}d_k \in D \cap H = B$ and $d_kB = d_lB$, $k = l$. Therefore, we obtain a decomposition $G = \bigcup_{k=1}^n d_kH$ of G into cosets modulo H with the same representatives. For x in D we have

$$\rho(x) = \sum_{1 \leq k \leq n,\ d_kxd_k^{-1} \in H} \alpha(d_kxd_k^{-1}).$$

Here

$$d_kxd_k^{-1} \in H \Leftrightarrow d_kxd_k^{-1} \in D \cap H = B.$$

Hence,

$$\rho(x) = \sum_{1 \leq k \leq n,\ d_kxd_k^{-1} \in B} \alpha'(d_kxd_k^{-1}) = \mathrm{Ind}\,_B^D(\alpha')(x)$$

as stated.

8) Finally, take for ρ the D-regular character which is the restriction of ν to G (see Lemma 12.3.6). Then $\rho|_D = \mu_0 + \rho_i$ for $i = 1$ or 2. By Lemma 12.3.1, we have $\rho(X_0) = -2 - 1 = -3$, on the one hand. On the other hand, $\rho|_D = \mathrm{Ind}\,_B^D(\alpha')$, where α' is actually a character of $\mathbb{Z}_3 \times SL_2(3)$. Hence, according to (12.10) we must have $-3 = \rho(X_0) = 2.\deg\alpha + 6.\Re\alpha(X_0)$, that is, $\Re\alpha(X_0) = -3/2$. As we have shown in 5), $\alpha(X_0) \in \{0, 1 + 2\varepsilon^2, \varepsilon(1 + 2\varepsilon^2), \varepsilon^2(1 + 2\varepsilon^2)\}$. Comparing these facts, we conclude that $\alpha(X_0) = 2\varepsilon + \varepsilon^2$.

Conversely, if ρ' is an arbitrary D-regular character of G, then, as was shown in 6), $\rho' = \rho^{(i)} = \mathrm{Ind}_H^G(\alpha^{(i)})$ for some $i \in \{0, 1, 2\}$. One can suppose that $\alpha^{(i)} = \xi^i\alpha$, where $\alpha(X_0) = 2\varepsilon + \varepsilon^2$. Again we have

$$-3 = \rho'(X_0) = 6 + 6.\Re\{\varepsilon^i(2\varepsilon + \varepsilon^2)\}$$

and so $i = 0$. Consequently, $\rho' = \rho^{(0)}(= \rho)$. □

Corollary 12.3.8. *Let ν be the unique irreducible complex character of degree* 78 *of the finite simple group $\Omega_7(3)$. Then the restriction $\nu|_G$ to the maximal subgroup $G = \mathrm{Aut}\,(\mathcal{D}) = 3^{3+3}.SL_3(3)$ is also irreducible, and is the G-character afforded by $\mathcal{L}$.* □

Corollary 12.3.9. *In the notation of Corollary* 12.3.8, *the character ν is rational. More precisely, ν is character of some integral representation of the group $\Omega_7(3)$.*

Proof. Note that ν is integer-valued (see [ATLAS]). According to Corollary 12.3.8, $\nu|_G$ is irreducible and rational. It is well-known (see Lemma 8.3.1) that ν is also rational in this case. □

Corollary 12.3.10. *The unique irreducible complex character σ of degree* 78 *of the Fischer sporadic simple group Fi_{22} is rational.*

Proof. By [ATLAS], σ is integer-valued. So it is enough to note that $\Omega = \Omega_7(3)$ can be embedded in Fi_{22} (as a maximal subgroup), and then $\sigma|_\Omega = \nu$. □

12.4 Invariant lattices of type E_6: the primitive case

Throughout this section we shall suppose that Λ is an invariant lattice of type E_6 such that the group $\mathbb{G} = \mathrm{Aut}\,(\Lambda)$ is a primitive complex linear group over $\mathcal{L} = \Lambda \otimes_{\mathbb{Z}} \mathbb{C}$. Let χ denote the character of this representation of $\mathbb{G}$ on $\mathcal{L}$. Furthermore, if A is a finite group, then
– $\mathrm{soc}\,(A)$ and $\mathrm{sol}\,(A)$ denote the socle and the maximal soluble normal subgroup of A;
– $P(A)$ denotes the minimum degree of nontrivial permutation representations of A;
– $P_f(A)$ denotes the minimum degree of faithful permutation representations of A;
– $R_3(A)$ denotes the minimum degree of faithful 3-modular projective representations of A;
– $e_{3'}(A)$ denotes the minimum degree of nontrivial projective representations of A over a field of characteristic $\neq 3$.

Lemma 12.4.1. *The group* $G = \mathrm{Aut}\,(\mathcal{D})$ *has the following properties:*
(i) $\mathrm{soc}\,(G) = E = \mathbb{Z}_3^3$, $\mathrm{sol}\,(G) = C = 3^{3+3}$*;*
(ii) $P(G) = 13$, $P_f(G) \geq 81$*;*
(iii) $e_{3'}(G) \geq 10$*;*
(iv) $R_3(G) \geq 5$*; more precisely, G cannot be embedded in any of the finite simple groups* ${}^3D_4(q)$, $A_3(q)$, $B_2(q)$, ${}^2A_3(q)$, $G_2(q)$, ${}^2G_2(q)$, $C_3(3)$ *and* $C_4(3)$.

Proof. (i) Clearly, $C \trianglelefteq \mathrm{sol}\,(G)$. Hence, $\mathrm{sol}\,(G)/C = \mathrm{sol}\,(G/C) = \mathrm{sol}\,(S) = 1$, that is, $\mathrm{sol}\,(G) = C$. Furthermore, it is clear that E is a minimal normal subgroup of G. Let H be another minimal normal subgroup of G. Then $H \cap E = 1$, $[H, E] = 1$, $H \subseteq C_G(E) = C$. This implies that $1 \neq H \cong HE/E \subseteq C/E$ and HE/E is a $SL_3(3)$-submodule of C/E. In that case, $HE/E = C/E$, $C = HE = H \times E$, $[C, C] \cap E = [H, H] \cap E \subseteq H \cap E = 1$, a contradiction. Thus $\mathrm{soc}\,(G) = E$.

(ii) Let $\pi : G \to \mathbb{S}_n$ be a nontrivial homomorphism with $n \leq 12$. Then 13 divides $|\mathrm{Ker}\,\pi|$ but not $|\mathrm{Im}\,\pi|$, and so $K = \mathrm{Ker}\,\pi \neq 1$. As $\mathrm{soc}\,(G) = E$, we have $K \supseteq E$ and $13||K/E|$. This means that $1 \neq K/E \trianglelefteq G/E = (C/E).S$. It is easy to see that $\mathrm{soc}\,(G/E) = C/E$, and therefore $K/E \supseteq C/E$, $K \supseteq C$. Now $1 \neq K/C \trianglelefteq S = SL_3(3)$, that is, $K/C = S$, $K = G$ and π is trivial, a contradiction. Hence, $P(G) \geq 13$. But G has a subgroup of index 13, so that $P(G) = 13$.

Suppose now that $\pi : G \to \mathbb{S}_n$ is an embedding and that $n \leq 80$. Consider the corresponding action of G on the set $X = \{1, 2, \ldots, n\}$ with permutation character

$$\tau = e.1_G + \tau_1 + \cdots + \tau_t,$$

where $\tau_i \in \mathrm{Irr}\,(G) \backslash \{1_G\}$; $\tau_1, \ldots, \tau_s$ are not faithful and $\tau_{s+1}, \ldots, \tau_t$ are faithful, $0 \leq s \leq t$ and $t \geq 1$, $e \geq 1$. Note that $\deg \tau_i < 78$ for all i. For, this condition holds if $n - e \leq 77$, in particular if $n \leq 78$. Suppose that $n = 79$ and $e = 1$. This means that G acts transitively on the set X of cardinality 79; in particular, 79 divides $|G|$, a contradiction. Finally, suppose that $n = 80$ and $n - e \geq 78$. As 80 does not divide $|G|$, G cannot be transitive on X, that is, $e = 2$. If $t > 1$, then clearly $\deg \tau_i < 78$. Finally, we consider the case $t = 1$, that is, $\tau = 2.1_A + \tau_1$. Here G has two orbits on X, say, X_1 and X_2; moreover, one can suppose that $|X_1| = 79$. We again come to the contradiction that 79 divides $|G|$.

Hence, $\deg \tau_i < 78$ for all i. Furthermore, for $i \leq s$ we have $1 \neq \mathrm{Ker}\,\tau_i \trianglelefteq G$, that is, $\mathrm{Ker}\,\tau_i \supseteq E$. As τ is faithful, we must have $s < t$. Therefore τ_t is a faithful irreducible character of G of degree less than 78. Following the proof of Proposition 12.3.7, we have $\tau_t = \mathrm{Ind}_H^G(\alpha)$, where α is an irreducible character of $H = C.(\mathbb{Z}_3^2.SL_2(3))$ with $\deg \alpha < 3$. In this case, it is not difficult to see that $\mathrm{Ker}\,\alpha \supseteq [C, C] = E$ and so $\mathrm{Ker}\,\tau_t \supseteq E$, a contradiction.

(iii) Let $\pi : G \to PGL_n(\mathbb{F})$ be a nontrivial homomorphism with $n < 10$ and $\mathrm{char}\mathbb{F} \neq 3$. As $S = SL_3(3) \hookrightarrow G$ and

$$d_{3'}(S) \geq \frac{3^3 - 1}{3 - 1} - 3 = 10,$$

(see [SeZ]), we must have that $\mathrm{Ker}\,\pi \supseteq S$. But then $\mathrm{Ker}\,\pi = G$, contrary to the nontriviality of π. Hence, $d_{3'}(G) \geq 10$.

(iv) It is enough to show that G cannot be embedded in $A_3(q)$, ${}^3D_4(q)$, $G_2(q)$ (where $3|g$) or $C_4(3)$. Firstly, by the Borel-Tits Theorem, if G is a subgroup of a Chevalley group $X(q)$ defined over the field $\mathbb{F}_q$, then G is contained in some maximal parabolic subgroup P of $X(q)$, because G is 3-local. Therefore, if G is embedded in ${}^3D_4(q)$ or $G_2(q)$, we have $G \hookrightarrow P = O_3(P).SL_2(q)$. In particular, $S \hookrightarrow SL_2(q)$, that is, $R_3(SL_3(3)) \leq 2$, a contradiction.

Suppose next that G can be embedded in $X = PGL_4(\mathbb{F})$, where $\mathrm{char}\mathbb{F} = 3$. By considering the natural projective representation of X on

$$V =< e_1, e_2, e_3, e_4 >_{\mathbb{F}},$$

one can suppose that $G \hookrightarrow P = St_X(V_i)$, where $V_i =< e_1, \ldots, e_i >_{\mathbb{F}}$ and $1 \leq i \leq 2$. In particular, we have a homomorphism $\pi : G \to PGL(V_i) = PGL_i(\mathbb{F})$. As $S \hookrightarrow G$ and $R_3(S) = 3 > i$, S is contained in $\mathrm{Ker}\,\pi$ and so $\mathrm{Ker}\,\pi = G$, that is, G acts trivially on V_i. Similarly, if $i = 2$, then G acts trivially on V/V_i, which implies that G is soluble, a contradiction. Therefore $i = 1$ and one can suppose that $G \subseteq F.T$, where F is an abelian 3-group and $T = GL_3(\mathbb{F})$. Clearly, $F \cap G$ is an abelian normal subgroup of G, so that $F \cap G = 1$ or $F \cap G = E$. Correspondingly, either G or $G/E = (C/E).S$ can be embedded in T. From this it is not difficult to see that $S = SL_3(3)$ is embedded in $GL_2(\mathbb{F})$, which is false since $R_3(S) = 3$.

Finally, suppose that $G \hookrightarrow X = PSp_8(3)$. The group X has a faithful complex representation of degree $(3^4+1)/2 = 41$ (see e.g. [War]), which is a constituent of the Weil representation. Meanwhile, in (ii) we have shown that the degree of every faithful irreducible representation of G is not less than 78, again a contradiction. □

Lemma 12.4.2. *The following equalities hold:* $\mathrm{sol}\,(\mathbb{G}) = Z(\mathbb{G}) = Z \cong \mathbb{Z}_2$.

Proof. 1) Firstly suppose that H is a non-trivial abelian normal subgroup of $\mathbb{G}$. Then by Clifford's Theorem, we have $\chi|_H = e\sum_{i=1}^{t} \tau_i$, $\tau_i \in \mathrm{Irr}\,(H)$. The primitivity of χ implies that $t = 1$, so that τ_1 is a faithful $\mathbb{Q}$-valued irreducible character of the abelian group H. Hence, $H = Z(\mathbb{G}) = Z =< \mathbf{t} >\cong \mathbb{Z}_2$, where $\mathbf{t}$ acts on $\mathcal{L}$ according to the rule $\mathbf{t}(x) = -x$.

2) Suppose next that $R = \mathrm{sol}\,(\mathbb{G}) \supset Z$, and consider the derived series

$$R = R^{(1)} \supset R^{(2)} \supset \ldots \supset R^{(l+1)} = 1,$$

where $R^{(i+1)} = [R^{(i)}, R^{(i)}]$ as usual. Set $T = R^{(l-1)}$. Then $T \supset T' = [T,T] \supset [T',T'] = 1$. This means that T' is a nontrivial abelian normal subgroup of G. By 1), we have $T' = Z$. In particular, $[T,T'] = 1$, that is, T is nilpotent of class 2. Set $Q = O_2(T)$, so that $Q' = O_2(T') = Z$. Finally, with $P = Z.\Omega_2(Q/Z)$ it is clear $Z \subset P \lhd \mathbb{G}$. Furthermore, $Z(P) = P' = Z$, and P/Z is elementary abelian. In other words, P is an extraspecial normal subgroup of G: $P = 2_{\pm}^{1+2m}$ for some m.

By Clifford's Theorem and the primitivity of χ, we have $\chi|_P = e\tau$, where τ is a faithful irreducible character of P. In particular, $78 = e.\deg\tau = e.2^m$, which implies that $m = 1$ and $P = 2^{1+2}_{\pm}$. Furthermore, G normalizes P, so we obtain a homomorphism π from G into the soluble group $\mathrm{Aut}\,(P) = \mathbb{S}_4$. In particular, $G/\mathrm{Ker}\,\pi$ is soluble. It is easy to see that $\mathrm{Ker}\,\pi = G$ in this case, so that $[G, P] = 1$, $P \subseteq C_{\mathbb{G}}(G)$. By Schur's Lemma, P acts scalarly (and faithfully) on $\mathcal{L}$. In other words, P is abelian, a contradiction. □

Lemma 12.4.3. *The socle L of $\mathbb{G}/Z$ is a nonabelian simple group. Moreover, G can be embedded in L.*

Proof. 1) Consider a minimal normal subgroup

$$\bar{M} = \bar{L}_1 \times \ldots \times \bar{L}_m \cong L^m$$

of $\bar{\mathbb{G}} = \mathbb{G}/Z$, where L is a nonabelian simple group. Let $M, L_1, \ldots, L_m$ denote the full preimages of $\bar{M}, \bar{L}_1, \ldots, \bar{L}_m$ in $\mathbb{G}$. Then, as is easy to see, $M = L_1 * \ldots * L_m$. By Clifford's Theorem and the primitivity of χ, we have $\chi|_M = e\rho = e\rho_1 \cdots \rho_m$, where $\rho_i \in \mathrm{Irr}\,(L_i)$. In particular,

$$\chi|_{L_i} = \frac{\deg\chi}{\deg\rho_i}\rho_i,$$

that is, ρ_i is a faithful irreducible character of L_i. Hence,

$$m \le \log_d 78,$$

where $d = d(L)$ denotes the minimum degree of faithful projective complex representations of L. In particular, $m \le \log_2 78 < 7$.

2) Like $\mathbb{G}$, the group G normalizes $\bar{M}$, and so it permutes the m minimal normal subgroups $\bar{L}_i$ of $\bar{M}$. But $m \le 6$ and $P(G) = 13$, so that G leaves every subgroup $\bar{L}_i$ fixed. Consider the homomorphism $\pi : G \to \mathrm{Aut}\,(\bar{L}_i) = \mathrm{Aut}\,(L)$ arising here. If $\pi = 1$, then $G \subseteq C_{\bar{\mathbb{G}}}(\bar{L}_i) = C_{\mathbb{G}}(L_i)/Z$ and $L_i \subseteq C_{\mathbb{G}}(G) = Z$, a contradiction. Hence $\pi \ne 1$, $\mathrm{Ker}\,\pi \ne G$. In particular, S cannot be contained in $\mathrm{Ker}\,\pi$. But in this case $S \cap \mathrm{Ker}\,\pi = 1$, and $\pi|_S : S \to \mathrm{Aut}\,(L)$ is an embedding. Furthermore, $\mathrm{Out}\,(L)$ is soluble and S is simple, hence in fact π embeds S in L. In particular, $d = d(L) \ge d(S) \ge 10$. From this it follows that $m \le \log_{10} 78 < 2$. In other words, $\bar{M} = L$ and $M = Z.L$.

3) Now let $\bar{N}$ denote another minimal normal subgroup of $\bar{\mathbb{G}}$, and set $N = Z.\bar{N}$. Then $\bar{M} \cap \bar{N} = 1$, and the full preimage K of $\bar{M} \times \bar{N}$ in $\mathbb{G}$ is a central product $M * N$. Moreover, $K \trianglelefteq \mathbb{G}$. Hence, $\chi|_K = e\alpha\beta$, where $\alpha \in \mathrm{Irr}\,(M)$ and $\beta \in \mathrm{Irr}\,(N)$. As we have mentioned above, $\deg\alpha \ge d(L) \ge d(S) \ge 10$. Similarly, $\deg\beta \ge 10$, and this gives a contradiction: $78 = \deg\chi \ge 100e$.

Consequently, $\operatorname{soc}(\bar{\mathbb{G}}) = L$ is simple. As $G \cap Z = 1$, G is embedded in $\bar{\mathbb{G}}$. But G is perfect and $\bar{\mathbb{G}}/L$ is soluble since it is a subgroup of $\operatorname{Out}(L)$, so $G \hookrightarrow L$. □

Thus we have established that $G \subseteq \tilde{L} = \mathbb{Z}_2.L \subseteq \mathbb{G}$, and that $\tilde{L}$ has a faithful irreducible integral character χ of degree 78. Using the classification of finite simple groups, we determine all the possibilities for L.

Lemma 12.4.4. *L is not an alternating group.*

Proof. Assume the contrary: $L = \mathbb{A}_n$ for some n. Since $G \hookrightarrow L$ and $P_f(G) \geq 81$ (Lemma 12.4.1), we must have $n \geq 81$. But in this case $d(L) = n - 1 \geq 80 > 78$ (see [Wag 3]), a contradiction. □

Lemma 12.4.5. *L does not belong to the set $Chev(3')$ of finite simple Chevalley groups defined over fields of characteristic $\neq 3$.*

Proof. Suppose that L is a Chevalley group of type X defined over the field $\mathbb{F}_q$ of characteristic $p \neq 3$, $q = p^s$. Let l be the minimum dimension of faithful projective L-modules over $\mathbb{F}_q$. Then $G \hookrightarrow L \hookrightarrow PGL_l(\mathbb{F})$, and therefore

$$l \geq e_{3'}(G) \geq 10$$

(see Lemma 12.4.1). In particular, $X \neq {}^2B_2, G_2, {}^2G_2, {}^3D_4$. We shall consider the remaining cases separately.

1) $L = PSL_l(q)$. Then $l \geq 10$, $d(L) \geq q^{l-1} - 1 \geq 2^9 - 1 = 511 > 78$, a contradiction. Here and below we frequently use the lower bound for the minimum degrees of projective representations for Chevalley groups determined in [SeZ] and reproduced in Table A7.

2) $L = PSp_{2m}(q)$. Here $10 \leq l = 2m$, so that $m \geq 5$. If $p = 2$, then

$$d(L) \geq \frac{1}{2}q^{m-1}(q^{m-1} - 1)(q - 1) \geq 2^3(2^4 - 1) = 240 > 78.$$

If $p \geq 5$, then

$$d(L) = \frac{1}{2}(q^m - 1) \geq \frac{5^5 - 1}{2} = 1562 > 78.$$

3) $L = PSU_l(q)$. Here $l \geq 10$ and

$$d(L) \geq \frac{q(q^{l-1} - 1)}{q + 1} \geq \frac{2^{10} - 1}{3} - 1 = 340 > 78.$$

4) $L = P\Omega^{\pm}_{2m}(q)$. Here $10 \leq l = 2m$, so that $m \geq 5$. Hence,

$$d(L) \geq (q^{m-1} - 1)(q^{m-2} - 1) \geq (2^4 - 1)(2^3 - 1) = 105 > 78.$$

5) $L = \Omega_{2m+1}(q)$. Here $p > 3$, $10 \leq l = 2m+1$, so that $m \geq 5$. Hence,

$$d(L) \geq q^{m-1}(q^{m-1}-1) \geq 5^4(5^4-1) > 78.$$

6) $L = {}^2E_6(q)$, $E_6(q)$, $E_7(q)$ or $E_8(q)$. Here

$$d(L) \geq q^9(q^2-1) \geq 3.2^9 > 78.$$

7) $L = {}^2F_4(q)$ or $F_4(q)$. If $q = 2$, then $|{}^2F_4(2)'|$ and $|F_4(2)|$ are not divisible by $|G|$, that is, G cannot be embedded in any of these groups. If $q > 2$ and $p = 2$, then

$$d({}^2F_4(q)) \geq q^4\sqrt{q/2}(q-1) \geq 7.2^{13} > 78;$$

$$d(F_4(q)) \geq \frac{1}{2}q^7(q^3-1)(q-1) > 2^{13} > 78.$$

If $p \geq 5$, then $d(F_4(q)) \geq q^6(q^2-1) \geq 5^6(5^2-1) > 78$. □

Lemma 12.4.6. *Suppose that L is a Chevalley group of characteristic 3. Then $L = \Omega_7(3)$.*

Proof. Suppose that L is a Chevalley group of type X defined over the field $\mathbb{F}_q$, $q = 3^s$. According to Lemma 12.4.1, $X \neq {}^2B_2, G_2, {}^2G_2, {}^3D_4$.

1) $X = F_4, {}^2F_4, E_6, {}^2E_6, E_7, E_8$. For these types we have $d(L) \geq q^6(q^2-1) \geq 8.3^6 > 78$.

2) $L = PSL_n(q)$. By Lemma 12.4.1 we have $n \geq 5$. Hence, $d(L) \geq q^{n-1}-1 \geq 3^4-1 = 80 > 78$.

3) $L = PSp_{2m}(q)$. By Lemma 12.4.1 we have $m \geq 3$ (and $m \geq 5$, if $q = 3$). Therefore

$$d(L) = \frac{q^m-1}{2}\left\{\begin{array}{lll} \geq & (3^5-1)/2 = 121 & \text{if } \ q=3 \\ \geq & (9^3-1)/2 = 364 & \text{if } \ q \geq 9 \end{array}\right\} > 78.$$

4) $L = PSU_n(q)$. Again by Lemma 12.4.1, we have $n \geq 5$. If $n \geq 6$, then

$$d(L) \geq \frac{q^6-1}{q+1} \geq \frac{3^6-1}{4} = 182 > 78.$$

If $n = 5$ and $q \geq 9$, then

$$d(L) \geq \frac{q(q^4-1)}{q+1} > 656 > 78.$$

Finally, if $n = 5$ and $q = 3$, then $|L|$ is not divisible by $|G|$.

5) $L = P\Omega^{\pm}_{2m}(q)$, $m \geq 4$. Then

$$d(L) \geq (q^{m-2}-1)(q^{m-1}-1) \geq (3^2-1)(3^3-1) = 208 > 78.$$

6) $L = \Omega_{2m+1}(q)$, $m \geq 3$. Suppose that $L \neq \Omega_7(3)$. Then either $q \geq 9$ and $d(L) \geq q^{2(m-1)} - 1 \geq 6560 > 78$; or $q = 3$, $m \geq 4$ and $d(L) \geq q^{m-1}(q^{m-1} - 1) \geq 27.26 > 78$. Consequently, $L = \Omega_7(3)$. □

Lemma 12.4.7. *Suppose that L is one of the 26 sporadic simple groups. Then* $L = Fi_{22}$.

Proof. Because $|L|$ is divisible by $|G| = 3^9.5.13$, we have $L \neq M_{11}, M_{12}, M_{22}, M_{23}, M_{24}, J_1, J_2, J_3, J_4, HS, Suz, McL, Ru, He, Ly, O'N, Co_2, Co_3, HN$. This means that $L \in \{Co_1, Fi_{22}, Fi_{23}, Fi'_{24}, F_3, F_2, F_1\}$. Furthermore, $d(Fi_{23}) = 782$, $d(Fi'_{24}) = 783$, $d(F_3) = 248$, $d(F_2) = 4371$, $d(F_1) = 196883$ (see [ATLAS]). Finally, suppose that $L = Co_1$. Recall that our group L is such that $\tilde{L} = \mathbb{Z}_2.L$ has a faithful irreducible character of degree 78. Meanwhile the group Co_1 does not satisfy this condition (see [ATLAS]), a contradiction. Consequently, $L = Fi_{22}$. □

It now follows from Lemmas 12.4.4 – 12.4.7 that, under our hypotheses, the following embeddings hold:

$$\mathbb{Z}_2.L \trianglelefteq \mathbb{G} \subseteq \mathbb{Z}_2.\mathrm{Aut}\,(L),$$

where $L = \Omega_7(3)$ or $L = Fi_{22}$. Both of these simple groups have Schur multiplier $\mathbb{Z}_6$ and outer automorphism group $\mathbb{Z}_2$. Furthermore, if we denote irreducible characters by their degrees, then

$$\mathrm{Irr}\,(\mathbb{Z}_2 \circ L) \backslash \mathrm{Irr}\,(L) = \begin{cases} \{520, 560, \ldots\} & \text{if} \quad L = \Omega_7(3); \\ \{352, 2080, \ldots\} & \text{if} \quad L = Fi_{22}. \end{cases}$$

Hence, $\mathbb{Z}_2 \times L \trianglelefteq \mathbb{G} \subseteq (\mathbb{Z}_2 \times L).\mathbb{Z}_2$.

Proof of Theorem 12.0.2. As we have shown above, only possibilities (i) – (iii) indicated in the statement of Theorem 12.0.2 can occur. It remains to prove that every possibility is in fact realized. Clearly, (iii) is realized for the lattice $\Lambda = \oplus\Lambda_i$, where $\Lambda_i \subset \mathcal{H}_i$ and Λ_i is isometric to the root lattice E_6:

$$\mathrm{Aut}\,(\Lambda) = A(E_6) \wr \mathbb{S}_{13}.$$

Furthermore, the following embeddings hold:

$$G \subset \Omega_7(3) \subset Fi_{22}.$$

The groups Fi_{22} and $\Omega_7(3)$ each has a unique irreducible complex character of degree 78, say, σ and ν, respectively. If χ denotes the character of G afforded by $\mathcal{L}$, then $\sigma|_G = \nu|_G = \chi$ (see Section 12.3). By Corollaries 12.3.9 – 12.3.10, σ and

ν are rational. Moreover, σ can be extended to integer-valued characters σ^1, σ^2 of $\mathrm{Aut}\,(Fi_{22})$. Hence, there exists an $\mathrm{Aut}\,(Fi_{22})$-invariant lattice, Λ_F say. Clearly,

$$\mathrm{Aut}\,(\Lambda_F) = \mathbb{Z}_2 \times \mathrm{Aut}\,(Fi_{22}).$$

Similarly, ν can be extended to integer-valued characters ν^1, ν^2 of the group $SO_7(3) = \mathrm{Aut}\,(\Omega_7(3))$. This means that one can find an $\mathrm{Aut}\,(\Omega_7(3))$-invariant lattice, Λ_Ω say. We claim that

$$\mathrm{Aut}\,(\Lambda_\Omega) = \mathbb{Z}_2 \times \mathrm{Aut}\,(\Omega_7(3)). \tag{12.14}$$

Indeed, suppose that $\mathbb{G} = \mathrm{Aut}\,(\Lambda_\Omega) \supset \mathbb{Z}_2 \times \mathrm{Aut}\,(\Omega)$, where $\Omega = \Omega_7(3)$. Then $\mathbb{G}$ satisfies conclusion (ii) of Theorem 12.0.2, so that $\mathbb{Z}_2 \times F \trianglelefteq \mathbb{G} \subseteq (\mathbb{Z}_2 \times F).\mathbb{Z}_2$, where $F = Fi_{22}$. In particular, $\Omega \subseteq [[\mathbb{G}, \mathbb{G}], [\mathbb{G}, \mathbb{G}]] = F \triangleleft \mathbb{G}$. But Ω is a maximal subgroup of F, so we must have that $\mathbb{G} = (\mathbb{Z}_2 \times F).\mathbb{Z}_2$. One can suppose that $\mathrm{Aut}\,(\Omega) =< \Omega, x >$, where x in an involution of type $2D$ in the notation of [ATLAS], and the character ν^1 of $\mathrm{Aut}\,(\Omega)$ takes the value 26 at x. Since x normalizes F but does not centralize F, we have $< F, x >\cong \mathrm{Aut}\,(F)$. Thus, we have shown that $\mathrm{Aut}\,(F)$ is contained in $\mathbb{G}$, on the one hand. On the other hand, the characters σ^i, $i = 1, 2$ of $\mathrm{Aut}\,(F)$ can take only the values -34, 14, -2, ± 6, ± 22 at involutions; this is a contradiction.

Another explanation for equality (12.14) is given by the fact that $\mathrm{Aut}\,(\Omega_7(3))$ cannot be embedded in $\mathrm{Aut}\,(Fi_{22})$. This completes the proof of Theorem 12.0.2. □

Commentary

Probably, several details of the proofs explained here will seem unnecessary to the reader. However, our intention was to make the material related to the case E_6 maximally transparent. All the groups $L_4(3)$, $\Omega_7(3)$ and Fi_{22} are remarkable in themselves, and it would perhaps be interesting to investigate geometric characteristics of the corresponding invariant lattices. Furthermore, by analogy with the $G_2(3)$-invariant lattice in the Lie algebra of type G_2 (see Chapter 8), it seems important to find an explanation in terms of our lattices of the well known embeddings $L_4(3) \hookrightarrow F_4(2)$ and $\Omega_7(3) \hookrightarrow Fi_{22} \hookrightarrow^2 E_6(2)$ (see [Coo 2] and [ATLAS]). Another approach to these embeddings (but omitting the intermediate term Fi_{22}) was suggested by Burichenko [Bur 3] (see also §14.1), who used loops and Hermitian lattices for this purpose. By the way, Theorem 12.0.1 forms the main result of Burichenko's nice M. Sc. Thesis [Bur 2]. Other results of this chapter are taken from [BuT].

Chapter 13

Invariant lattices of type E_8 and the finite simple groups F_3, $L_4(5)$

Among the sporadic finite simple groups, the Monster F_1 is the most mysterious occurrence. Meanwhile, the class of complex simple Lie algebras contains the Lie algebra of type E_8 which is, among other things, directly related to the "Lie" realization of F_1 (see [FLM 1, 2]). As mentioned above, the starting points of the study of orthogonal decompositions are the Lie algebra $\mathcal{L}$ of type E_8 and the Thompson sporadic simple group F_3, which is a direct factor of the centralizer of an element of order 3 in F_1 (see [Tho 2, 3]). Here, in Section §13.1, we reproduce several exciting fragments from Thompson's report in Sapporo [Tho 3] and Smith's Ph. D. Thesis [Smi 1].

Complete classification of integral invariant lattices of type E_8 is still out of sight, on the one hand. On the other hand, a complete description of their automorphism groups will be given in the subsequent sections. As is known (see Chapter 4), the Lie algebra $\mathcal{L}$ of type E_8 admits two irreducible orthogonal decompositions — the *Dempwolff decomposition* (in the terminology of Thompson [Tho 2], see also [Ale 1]) and the *Borovik decomposition* (see [Bor 3]). The former led to an even unimodular lattice of dimension 248 with automorphism group $\mathbb{Z}_2 \times F_3$; and the latter to an integral realization of the finite simple Chevalley group $A_3(5) \cong L_4(5) = PSL_4(5)$. This fact is apparently connected with the embedding $L_4(5) \hookrightarrow E_8(4)$, which was first discovered in [CLSS]. The statement of the main results of this chapter is given in Section §13.2. Some details of the proofs are omitted, and the reader can find them in [Tiep 16].

13.1 The Thompson-Smith lattice

One of the remarkable steps in the process of classifying the finite simple groups was the construction by Thompson and Smith [Tho 2, 3], [Smi 1, 2] of the sporadic simple group F_3, whose existence would have followed from the (at that time hypothetical) existence of the Monster F_1. To reproduce more precisely the circumstances surrounding the construction of F_3 (which Thompson denoted by E and many subsequent authors denoted by Th or F_3), we shall quote, with Thompson's permission, several fragments from his report [Tho 3].

"**Theorem.** *There is precisely one group E with the following properties:*

(a) *All involutions of E are conjugate.*

(b) *If z is an involution of E,* $C = C_E(z)$ *and* $H = O_2(C)$*, then H is extra-special of order* 2^9 *and* $C/H \cong \mathbb{A}_9$.

Additional properties of E:

(c) $|E| = 2^{15}.3^{10}.5^3.7^2.13.19.31$.

(d) E has just one irreducible character χ of degree 248, and χ is the character of a rational representation.

Let V be a $\mathbb{C}E$-module which affords χ.

(e) V can be given the structure of a Lie algebra of type E_8 in such a way that E preserves the Killing from. If Λ is a lattice in V stable under E, then $\Lambda/p\Lambda$ is irreducible for all primes p. Λ can be chosen so that the Killing form is negative definite on Λ.[1]

(f) If D is the largest subgroup of E which preserves the Lie multiplication, then D is a non splitting extension of an elementary abelian group of order 2^5 by $L_5(2)$.

(g) If

$$R = \{\frac{a}{2^n} | a \in \mathbb{Z}, n = 0, 1, ...\}$$

and $M = R\Lambda$, then $M = [M, M]$, and E preserves the Lie multiplication of $M/3M$.
It is this last property which explains the title[2].

It took several months' work to unravel the preceding results, although it is fairly easy to guess that these things are true. I became convinced that $E \subseteq E_8(p)$ for some prime p, and one of the main pleasures for me was to see $p = 3$ emerge.

... So I explicitly constructed the relevant automorphisms of E_8, and left it to Peter Smith and the IBM 370 to check that I had indeed constructed D. In particular, at this stage we had established the existence of Dempwolff's group.

Now $D \cap C = C_0 \supseteq H$ with $C_0/H \cong \mathbb{A}_8$. Also, $\chi|_{C_0} = \alpha_0 + \beta_0$, where $\alpha_0, \beta_0 \in \mathrm{Irr}(C_0)$, $\alpha_0(1) = 120$, $\beta_0(1) = 128$. Since $\chi|_C = \alpha + \beta$ with $\alpha|_{C_0} = \alpha_0$, $\beta|_{C_0} = \beta_0$, the construction of C is unique... Once again the machine was used to show that the explicit automorphisms constructed (above) stabilize a lattice Λ in V, and it can be shown that the automorphism group of the lattice is $\mathbb{Z}_2 \times E$. It should be understood that the embedding $D \subseteq E_8(\mathbb{C})$ gives us the "correct" quadratic form, which is fixed by E, namely the Killing form.

The final twist which gives $E \subseteq E_8(3)$ comes from the fact that there are two irreducible characters β, β' of C which restrict to β_0. If one uses β' in place of β, the resulting group E^* is infinite, but acts on $R\Lambda = M$, agreeing with E on $M/3M$. And in addition, this infinite group is in $E_8(\mathbb{C})$, as was shown by the IBM 370.

One side result which emerged from this work is the classification of what I call Dempwolff decompositions of E_8.

Definition. Let $\mathcal{L}$ be a Lie algebra $/\mathbb{C}$ of type E_8. A *Dempwolff decomposition* of $\mathcal{L}$ is a family $\mathcal{F}$ of Cartan subalgebras of $\mathcal{L}$ such that

---

[1]Clearly, multiplication by $i = \sqrt{-1}$ turns Λ into a Euclidean lattice.

[2]*A simple subgroup of* $E_8(3)$

(a)

$$\mathcal{L} = \sum_{F \in \mathcal{F}} F$$

(b) If F_1, F_2 are distinct members of $\mathcal{F}$, then $[F_1, F_2] \in \mathcal{F}$.

Theorem. *$E_8(\mathbb{C})$ is transitive on Dempwolff decompositions. If K is the largest subgroup of $E_8(\mathbb{C})$ which fixes the Dempwolff decomposition $\mathcal{F}$, and L is the largest subgroup of K which fixes every member of $\mathcal{F}$, then $|L| = 2^{15}$ and $K/L \cong L_5(2)$.*

Since $|L| = 2^{15}$ (and not 2^5), I wonder if there might be a new simple finite group in $E_8(\mathbb{C})$. I don't recommend hunting for such a thing, but if one happens to come your way, you might want to examine it."

The rationality of (the unique irreducible character of degree 248) χ is, of course, implied by Lemma 8.3.1 and the rationality of the character $\chi|_D$. The same assertion also follows from the following elegant argument due to Conway. We have to show that the Schur index $m = m_{\mathbb{Q}}(\chi)$ is equal to 1. Clearly, m divides the multiplicity $(\chi, \xi)_E$ for any permutation character ξ. If for ξ one takes the regular character, then this multiplicity is equal to 248. Now if ξ corresponds to the action by conjugations of E on itself:

$$\xi(x) = |C_E(x)|, \ \forall x \in E,$$

then the multiplicity $(\chi, \xi)_E$ is

$$\sum_{i=1}^{48} \chi(x_i) = 275,$$

where $x_1 = 1, x_2, \ldots, x_{48}$ are representatives of the 48 conjugacy classes of E. Since 248 and 275 are coprime, we conclude that $m = 1$.

Now for some remarks about the computer computation of Thompson and Smith. Regarding the group E, it was well known by Thompson that the centralizer $C = C_E(\underline{z})$ of any involution $\underline{z} \in E$ is isomorphic to $2_+^{1+8}.\mathbb{A}_9$. Furthermore, if $\underline{z}'$ is a non-central involution in $H = O_2(C) = 2_+^{1+8}$, then $O_2(C_E(\underline{z})) \cap O_2(C_E(\underline{z}')) = F$ is an elementary abelian group of order 2^5. Moreover, the group $D = N_E(F)$ must be a non-split extension of $F = \mathbb{Z}_2^5$ by $D/F \cong SL_5(2)$, which is unique if it exists (see [Demp 1]). Using the table of structure constants for the Lie algebra $\mathcal{L}$ of type E_8 in the Chevalley basis $\{h_i, e_k, e_{-k} | 1 \leq i \leq 8, 1 \leq k \leq 120\}$ calculated by Smith on a computer, Thompson determined certain elements $\underline{a}, \underline{b}, \underline{c}, \underline{d}, \underline{r}, \underline{s}, \underline{x}, \underline{z} \in \mathrm{Aut}(\mathcal{L})$. It was verified that these elements together do indeed generate the group $D \cong \mathbb{Z}_2^5 \circ SL_5(2)$, where $F = O_2(D) = < \underline{a}^2, \underline{b}^2, \underline{c}^2, \underline{d}^2, \underline{z} >$. The set of indices $\{1, 2, \ldots, 120\}$ splits into 15 disjoint octads O_j, $1 \leq j \leq 15$, such that the Cartan subalgebras

$$\mathcal{H}_0 = < h_i | 1 \leq i \leq 8 >_{\mathbb{C}},$$

$$\mathcal{H}_{-j} = < e_k + e_{-k} | k \in O_j >_{\mathbb{C}}, \ \mathcal{H}_j = < e_k - e_{-k} | k \in O_j >_{\mathbb{C}},$$

which are eigensubspaces for F, constitute a multiplicative orthogonal decomposition called by Thompson a Dempwolff decomposition. As we see easily, the embedding $\{MOD\} \subset \{ROD\}$ is used here.

Thus, the subgroup D and, together with it, the subgroup

$$C_0 = C \cap D = < \underline{a}^2, \underline{b}^2, \underline{c}^2, \underline{d}^2, \underline{s}, \underline{x} > \cong 2_+^{1+8}.\mathbb{A}_8$$

are constructed. The next step was the (computer) construction of an involutory linear map $\underline{t}$ of the space $\mathcal{L}$, which together with C_0 generates the whole group C. Firstly, Thompson and Smith built C as an automorphism group of the Lie algebra

$$\mathcal{L}_+ = < e_k - e_{-k} | 1 \le k \le 120 >_{\mathbb{C}} = \sum_{j=1}^{15} \mathcal{H}_j$$

of type D_8. It remained to define the action of $\underline{t}$ on

$$\mathcal{L}_- = < h_i, e_k + e_{-k} >_{\mathbb{C}} = \sum_{0}^{-15} \mathcal{H}_j.$$

Just here we treat one of the most delicate aspects of the process: the character β_0 of $C_0 = 2_+^{1+8}.\mathbb{A}_8$ afforded by $\mathcal{L}_-$ can be extended to two distinct irreducible characters β and β' of $C = 2_+^{1+8}.\mathbb{A}_9$ of degree 128; moreover, $\beta \equiv \beta' (\mathrm{mod}\, 3)$. Corresponding to them, we obtain two endomorphisms $\underline{t}$ and $\underline{t}'$; moreover, $\underline{t}'$ is an automorphism of $\mathcal{L}$. This fact led to the embedding of the group $E^* = < \underline{a}, \underline{s}, \underline{t}' >$ in the Lie group $E_8(\mathbb{C})$ as an infinite subgroup. The right choice to achieve our goal is in fact the endomorphism $\underline{t}$, which is *not* contained in $\mathrm{Aut}(\mathcal{L})$.

The last step was to produce an Euclidean lattice $\Lambda_{TS} = \Lambda = (\Lambda, K)$ (where K is the Killing form of $\mathcal{L}$), stable under the group $E = < C, D > = < \underline{a}, \underline{s}, \underline{t} >$. The required lattice is constructed (using a computer) as a sublattice of index 4^{248} in the lattice $\bigoplus_{j=-15}^{15} \Lambda_j$, where each sublattice $\Lambda_j \subset \mathcal{H}_j$ is dual to the system of simple roots corresponding to the Cartan subalgebra $\mathcal{H}_j$. Moreover, $\mathrm{Aut}(\Lambda) = \mathbb{Z}_2 \times F_3$; and the lattice $\frac{1}{4}\Lambda$ is even and unimodular.

Unfortunately, the explicit presentation of generating vectors of the lattice $\Lambda = \Lambda_{TS}$, which occupies 12 pages of [Smi 1], makes this lattice visual, but not at all easy to understand.

13.2 Statement of results

As mentioned above, the program of studying invariant lattices associated with IODs of Lie algebras $\mathcal{L}$ is fully realized for the types G_2 (see Chapter 8) and F_4

(see Chapter 12). The results of this investigation can be stated as follows. If Λ is an invariant lattice of type G_2 (respectively, F_4), then the following alternative $(\mathcal{A})$ holds.

Either $\mathbb{G} = \mathrm{Aut}(\Lambda)$ *is an imprimitive complex linear group on* $\mathcal{L}$ *and then* $\mathbb{G}$ *has well-described composition structure, or* $\mathbb{G}/Z(\mathbb{G})$ *is an almost simple group:* $\mathbb{G} = \mathbb{Z}_2 \times G_2(3)$ *(respectively,* $\mathbb{G} = \mathbb{Z}_2 \times L_4(3)$ *or* $\mathbb{G} = \mathbb{Z}_2 \times (L_4(3).\mathbb{Z}_2)$*).*

By Theorem 4.5.1, the Lie algebra $\mathcal{L}$ of type E_8 admits just two IODs up to $\mathrm{Aut}(\mathcal{L})$-conjugacy. One of these is the MOD constructed by Thompson (see Chapter 3 and also [Tho 2]). The other was first constructed in [Bor 3]. The corresponding automorphism groups $G = \mathrm{Aut}(\mathcal{D})$ are $2^{5+10} \circ SL_5(2)$ and $\mathbb{Z}_5^3.SL_3(5)$.

In the remaining sections of the chapter, we shall show that, as with the cases G_2 and F_4, invariant lattices of type E_8 satisfy alternative $(\mathcal{A})$, where the role of the exceptional automorphism group $\mathrm{Aut}(\Lambda)$ is played by F_3 or $L_4(5)$. More precisely, it was established in Chapter 4 that $2^{5+10} \circ SL_5(2)$ has exactly two minimal irreducible subgroups, $D = \mathbb{Z}_2^5 \circ SL_5(2)$ and $I = 2^{5+10}.\mathbb{Z}_{31}$. Meanwhile, the only irreducible subgroup of $B = \mathbb{Z}_5^3.SL_3(5)$ is B itself. Correspondingly, the integral lattices of type E_8 invariant under these subgroups will be called D-, I- and B-lattices. The main result of this chapter is the following

Theorem 13.2.1. *Let* Λ *be an invariant lattice of type* E_8*, and set* $\mathbb{G} = \mathrm{Aut}(\Lambda)$*. Then one of the following assertions holds.*

(i) Λ *is the Thompson-Smith lattice and* $\mathbb{G} = \mathbb{Z}_2 \times F_3$*;*

(ii) $\mathbb{Z}_2 \circ L_4(5) \lhd \mathbb{G} \subseteq \mathbb{Z}_2.\mathrm{Aut}(L_4(5))$*;*

(iii) $\mathbb{G}$ *acts transitively on the set of Cartan subalgebras constituting the given IOD. Moreover, if* K *is the kernel of this permutation action and* $\bar{\mathbb{G}} = \mathbb{G}/K$*, then either*

a) Λ *is a D-lattice,* $K \subseteq (W(E_8))^{31}$ *and* $\bar{\mathbb{G}} \in \{SL_5(2), \mathbb{A}_{31}, \mathbb{S}_{31}\}$*;*

or

b) Λ *is a B-lattice,* $K \subseteq (G_1)^{31}$*,* $G_1 \in \{W(E_8), (SL_2(5) * SL_2(5)).\mathbb{Z}_2\}$ *and* $\bar{\mathbb{G}} \in \{SL_3(5), \mathbb{A}_{31}, \mathbb{S}_{31}\}$.

(iv) Λ *is an I-lattice and* $\mathbb{G}$ *admits a normal subgroup* K *such that* $K \subseteq (G_1)^N$*,* $N = 248/d$*,* $d|8$*,* $G_1 \hookrightarrow GL_d(\mathbb{C})$*, and* $\bar{\mathbb{G}} = \mathbb{G}/K$ *satisfies one of the following conditions:*

a) $\mathbb{A}_N \lhd \bar{\mathbb{G}} \subseteq \mathbb{S}_N$*;*

b) $d = 8$ *and either* $\mathbb{Z}_{31} \lhd \bar{\mathbb{G}} \subseteq \mathbb{Z}_{31}.\mathbb{Z}_{30}$ *or* $\bar{\mathbb{G}} \in \{SL_3(5), SL_5(2)\}$*;*

c) $d = 1$ *and* $\bar{\mathbb{G}} = L_2(31)$*;*

d) $d = 4$ *and* $\bar{\mathbb{G}} \in \{L_2(61), PGL_2(61)\}$.

The result related to $SL_4(5)$-invariant lattice described in assertion (ii) of Theorem 13.2.1 can in fact be generalized to any group $SL_4(q)$ with odd q as follows.

Theorem 13.2.2. *Set* $q = p^f$*, where* p *is an odd prime. Then* $SL_4(q)$ *has exactly two irreducible complex representations of degree* $N_q = (q-1)(q^3-1)/2$*, and both* *of them are rational. If $q = p$ and Λ is an $SL_4(q)$-invariant lattice of dimension N_q, then $\mathbb{Z}_2.L_4(q) \lhd \mathrm{Aut}(\Lambda) \subseteq \mathbb{Z}_2.\mathrm{Aut}(L_4(q))$.*

13.3 The imprimitive case

We start this section by recalling certain properties of the automorphism group $G = \mathrm{Aut}_{MOD}(\mathcal{L})$ of the unique MOD of the Lie algebra $\mathcal{L}$ of type E_8. It is known [Gri 2] that $G = N_{E_8(\mathbb{C})}(E) = C.S$, where $E = \mathbb{Z}_2^5$ and $S = SL_5(2)$. (We have avoided the notation of §13.1). Furthermore, $C = C_{E_8(\mathbb{C})}(E)$ is a special 2-group with the following properties:

(1) $[C,C] = Z(C) = \Phi(C) = E$;

(2) $G = C.S$ is a non-split extension;

(3) E and C/E are irreducible $SL_5(2)$-modules;

(4) G admits a factorization $G = C.D$, where $C \cap D = E$ and $D = E.S$ is a non-split extension.

Finally, E acts on $\mathcal{L}$ with character

$$8 \sum_{\rho \in \mathrm{Irr}(E) \setminus \{1_E\}} \rho,$$

and the MOD of $\mathcal{L}$ is of the form

$$\mathcal{L} = \bigoplus_{\rho} \mathcal{H}_\rho,$$

where $\mathcal{H}_\rho$ affords the E-character 8ρ.

We remind the reader of some notation. For a finite group X let $e_0(X)$ (respectively, $e_{p'}(X)$) denote the least degree of nontrivial projective representations of X over the field $\mathbb{C}$ (respectively, over fields of characteristic other than p). Furthermore, $P(X)$ (respectively, $P_f(X)$, $P_f^t(X)$) denotes the least degree of transitive (respectively, faithful, faithful transitive) permutation representations of X.

Lemma 13.3.1. *The following estimates hold.*

(i) $e_0(D) = 30$, $e_{2'}(D) \geq 26$;

(ii) $e_0(B) = 30$, $e_{5'}(B) \geq 28$.

Proof. Denote one of D and B by X, and set $E = \mathrm{sol}(X) = \mathrm{soc}(X)$, $\bar{X} = X/E$. Each projective representation of X over a field $\mathbb{F}$ is a homomorphism $\pi : X \to PGL_n(\mathbb{F})$, and nontriviality means that $\mathrm{Ker}\pi \neq X$.

1) Firstly, suppose that $\mathrm{Ker}\pi \neq 1$. Then $\mathrm{Ker}\pi = E$ and $\pi : \bar{X} \hookrightarrow PGL_n(\mathbb{F})$ is an embedding. But the Schur multiplier of $\bar{X}$ is trivial (see [ATLAS]), so π can

be lifted to a faithful n-dimensional representation of $\bar{X}$. Hence, if $\mathbb{F} = \mathbb{C}$, then $n \geq 30$ (see [ATLAS]). If char$\mathbb{F} \neq 2$ and $\bar{X} = SL_5(2)$, then by [SeZ], we have $n \geq (2^5 - 1) - 5 = 26$. If char$\mathbb{F} \neq 5$ and $\bar{X} = SL_3(5)$, then again according to [SeZ] we have $n \geq \frac{5^3-1}{5-1} - 3 = 28$.

2) Suppose now that Ker$\pi = 1$, so that $\pi : X \hookrightarrow PGL_n(\mathbb{F}) \hookrightarrow PGL_n(\bar{\mathbb{F}})$. Following [KlL 1], we set

$$R_{\bar{\mathbb{F}}}(Z) = \min\{m | Z \hookrightarrow PGL_m(\bar{\mathbb{F}})\},$$

$$M_{\bar{\mathbb{F}}}(Z) = \min\{m \mid \text{ some finite extension } Y.Z \text{ can be embedded in } PGL_m(\bar{\mathbb{F}})\}.$$

By [KlL 1], we have $M_{\bar{\mathbb{F}}}(\bar{X}) = R_{\bar{\mathbb{F}}}(\bar{X})$, if $\bar{X} = SL_3(5)$ or $SL_5(2)$. Hence, $n \geq R_{\bar{\mathbb{F}}}(\bar{X})$ and one can apply the result of 1). □

Lemma 13.3.2. *The following estimates hold.*

(i) $P(D) = 31$, $P_f^t(D) = P_f(D) \geq 62$;

(ii) $P(B) = 31$, $P_f^t(B) = P_f(B) = 125$.

Proof. As above, we denote one of D and B by X, and set $E = \mathrm{soc}(X)$, $\bar{X} = X/E$.

1) Let Y be a maximal subgroup of X. If $Y \supseteq E$, then $(X : Y) \geq P(\bar{X}) = 31$, by [ATLAS]. If Y does not contain E, then $E \cap Y \lhd EY = X$. By the minimality of E, this means that $E \cap Y = 1$ and so $X = E.Y$. Hence $X = B$, because D is non-split over E. Thus $(B : Y) = |E| = 125 > 31$ and $P(X) = 31$.

2) It is not difficult to show that $P_f^t(X) = P_f(X)$.

Now assume that $X = D$ and that the permutation representation of D on the cosets modulo a subgroup Z is faithful. Choose a maximal subgroup Y of X containing Z. In 1) we have shown that $Y \supseteq E$; furthermore, from the exactness of the representation it follows that E is not contained in Z. Hence, $(Y : Z) \geq 2$, $(D : Z) \geq 62$, $P_f(D) \geq 62$. Similarly, $P_f(B) = 125$. □

Lemma 13.3.3. *The following assertions hold for the group $I = 2^{5+10}.\mathbb{Z}_{31}$.*

(i) *If H is a normal subgroup of I and $31 || H|$, then $H = I$.*

(ii) $P(I) = 31$, $P_f(I) \geq 62$.

(iii) *Suppose that H is a non-trivial elementary abelian normal 2-subgroup of I. Then $H = E = Z(O_2(I))$.*

Proof. Recall that $C = 2^{5+10} = O_2(I)$ and $E = Z(C) = \mathbb{Z}_2^5$. Set $\mathbb{Z}_{31} =< \varphi >$.

1) Suppose that $H \lhd I$, $31 || H|$ but $H \neq I$. If $H \cap C = 1$, then $I = H \times C$, a contradiction. Therefore $1 \neq H \cap C \lhd C$, which implies that $1 \neq H \cap E = (H \cap C) \cap Z(C)$. But φ permutes the elements of $E \backslash \{1\}$ transitively, so $H \supseteq E$. As $HC = I$, we must have that $|H \cap C| = |H|/31 < |I|/31 = |C|$, that is, $E \subseteq H \cap C \subset C$. Setting $\bar{C} = C/E = \mathbb{Z}_2^{10}$, $\bar{I} = I/E = \bar{C}. < \varphi >$, $\bar{H} = H/E$, we see that φ acts fixed-point-freely on $\bar{C} \backslash \{1\}$. In particular, $\bar{H} \cap \bar{C}$ is 1 or $\mathbb{Z}_2^5$. In the

former case we would have $\bar{I} = \bar{H} \times \bar{C}$, a contradiction. Hence, $\bar{H} \cap \bar{C} = \mathbb{Z}_2^5$, and thus $\bar{I}/(\bar{H} \cap \bar{C}) \cong \mathbb{Z}_2^5.\mathbb{Z}_{31}$. But $\bar{H}/(\bar{H} \cap \bar{C}) \cong H/(H \cap C) \cong HC/C = I/C = \mathbb{Z}_{31}$ and $\bar{C}/(\bar{H} \cap \bar{C}) \cong \mathbb{Z}_2^5$. So $\bar{I}/(\bar{H} \cap \bar{C}) \cong \mathbb{Z}_2^5 \times \mathbb{Z}_{31}$, again a contradiction. Consequently, $H = I$, that is, (i) holds.

2) Let X be a proper subgroup in I with $(I : X) < 31$. Setting $H = \cap_{x \in I} xXx^{-1}$, we have $I/H \hookrightarrow \mathbb{S}_{30}$. In particular, $H \lhd I$ and $31 || H|$. By 1), $H = I$ and $X = I$, a contradiction. Hence, $P(I) \geq 31$, on the one hand. On the other hand, $(I : C) = 31$, so that $P(I) = 31$. Observe that C is the only subgroup of I of index 31. Furthermore, suppose that $H \subset I$ and $31 < (I : H) < 62$. Then $(I : H) = 32$. If $E \cap H = 1$, then $I = E.H$, $C = E.(H \cap C) = E \times (H \cap C)$, a contradiction. Therefore $E \cap H \neq 1$. But $E \cap H \lhd H$ and so $H \supseteq E$, showing that every subgroup of index less than 62 in I contains E. This means that $P_f(I) \geq 62$.

3) Suppose that $1 \neq H \lhd I$, $\exp(H) = 2$ but $H \neq E$. It is not difficult to see that H is $\mathbb{Z}_2^{15}$ or $\mathbb{Z}_2^{10}$ in this case, on the one hand. On the other hand, as is shown in [Ada], the rank of every elementary abelian subgroup of $E_8(\mathbb{C})$ does not exceed 9, a contradiction. Hence $H = E$. □

Lemma 13.3.4. *The group $I = 2^{5+10}.\mathbb{Z}_{31}$ cannot be embedded in $SL_8(2)$ nor in $Sp_{10}(2)$.*

Proof. 1) Suppose that $I \hookrightarrow T = T_n = SL_n(2)$ for some $n \leq 8$. Note at once that $n > 6$. In fact, if $n = 6$ we would have $C \in \mathrm{Syl}_2(T)$, which implies that $x^4 = 1$ for every 2-element $x \in SL_6(2)$, a contradiction. Consider the natural $\mathbb{F}_2T$-module $W =< e_1, \ldots, e_n >_{\mathbb{F}_2}$. By the Borel-Tits Theorem, the 2-local subgroup I is contained in some maximal parabolic subgroup P. We shall suppose that $I \subseteq P = St_T(< e_1, \ldots, e_i >)$, where $1 \leq i \leq n/2$ and i is the largest integer with this property. Setting $W_i =< e_1, \ldots, e_i >_{\mathbb{F}_2}$, we consider the natural homomorphism $\pi : I \to GL(W_i) = T_i$. Since $i \leq 4$, 31 does not divide $|T_i|$, and so $31||\mathrm{Ker}\pi|$. By Lemma 13.3.3, $\mathrm{Ker}\pi = I$ and $\pi = 1$. This means that $I \subseteq J.T_{n-i}$, where $J = \mathbb{Z}_2^{i(n-i)}$. Clearly, $i \leq n - 5$. Furthermore, $I \cap J$ is a normal subgroup of exponent not greater than 2 in I. Therefore by Lemma 13.3.3, we have $I \cap J = 1$ or $I \cap J = E$. Correspondingly, either

$$I \hookrightarrow T_{n-i} \tag{13.1}$$

or

$$\mathbb{Z}_2^{10}.\mathbb{Z}_{31} = I/E \hookrightarrow T_{n-i}. \tag{13.2}$$

Suppose now that $n = 7$, so that $1 \leq i \leq 2$. In case (13.1), we would have $I \hookrightarrow T_6$, which is impossible. So case (13.2) occurs. Since I/E is 2-local, we obtain the embedding $\mathbb{Z}_2^{10}.\mathbb{Z}_{31} = I/E \hookrightarrow \mathbb{Z}_2^5.T_5$. It is not difficult to see that this embedding cannot occur, and this means that I cannot be embedded in T_7.

Finally, suppose that $n = 8$, so that $1 \leq i \leq 3$. In case (13.1), we would obtain that I can be embedded in T_7, a contradiction. We mentioned above that

I/E cannot be embedded in T_6. Therefore, case (13.2) must occur with $i = 1$: $I \subseteq J.T_7 \cong St_{T_8}(W/U)$, where $U =< e_2, \dots, e_8 >_{\mathbb{F}_2}$ and $I \cap J = E$. Every element j of J can be written in the form $j(u) = u + f(u)e_1$ for some $f \in U^* = \mathrm{Hom}(U, \mathbb{F}_2)$. Hence, the subgroup E corresponds to some subspace of dimension 5 of U^*, which means that $\dim(W' \cap U) = 7 - 5 = 2$ and $\dim W' = 3$, where $W' = \{v \in W | \forall e \in E, e(v) = v\}$. Moreover, $E \lhd I$. Therefore I fixes the 3-dimensional subspace W', and so $I \subseteq P_3 = St_{T_8}(W')$, contrary to the maximality of i.

2) Finally, suppose that $I \hookrightarrow Sp_{10}(2)$. The 2-localization of I implies the existence of an index i, $1 \leq i \leq 5$, and a homomorphism $\pi : I \to SL_i(2) \times Sp_{10-2i}(2)$, such that $\mathrm{Ker}\pi$ has exponent not greater than 2. By Lemma 13.3.3, $\mathrm{Ker}\pi \subseteq E$. In particular, $31 || I|$, $i = 5$ and $I/\mathrm{Ker}\pi \hookrightarrow SL_5(2)$, a contradiction. □

We shall require the two following statements.

Lemma 13.3.5. *The group $D = \mathbb{Z}_2^5 \circ SL_5(2)$ cannot be embedded in $SL_8(2)$ nor in $Sp_{10}(2)$.*

Proof. 1) First note that D cannot be embedded in a finite group X with the following properties:

(a) $X = Y.S$ is a split extension, where $S = SL_5(2)$;

(b) 31 does not divide $|Y|$ and $|Y| < 2^{15}.63$.

For, suppose the contrary. Without loss of generality one can assume that $S \cap D \supseteq P \in \mathrm{Syl}_{31}(S)$. Clearly, either $S \subseteq D$ or $S \cap D \subseteq N_S(P) = P.\mathbb{Z}_5$ (see [ATLAS]). In the former case, D would be split over $O_2(D)$, a contradiction. Therefore $|S \cap D| \leq 31.5$, which implies that

$$|Y|.|S| = |X| \geq |SD| = \frac{|S|.|D|}{|S \cap D|} \geq \frac{2^{15}.|S|.3^2.5.7.31}{5.31},$$

that is, $|Y| \geq 2^{15}.63$, contrary to assumption (b).

2) Suppose that $D \hookrightarrow T = Sp_{10}(2)$. Then by the Borel-Tits Theorem, D can be embedded in a maximal parabolic subgroup $P = \mathbb{Z}_2^{15}.SL_5(2)$, contrary to the results of 1).

As in the proof of the previous lemma we set $T_n = SL_n(2)$.

3) Suppose that $D \hookrightarrow T_7$. By the Borel-Tits Theorem, there exists an index i, $1 \leq i \leq 2$, such that D is embedded in a maximal parabolic subgroup $P_i = \mathbb{Z}_2^{i(7-i)}.(T_i \times T_{7-i})$. Here $P_2 = (\mathbb{Z}_2^{10}.\mathbb{S}_3).T_5$, so that D cannot be embedded in P_2, because of 1). This means that $D \hookrightarrow P_1 = F.T_6$, where $F = \mathbb{Z}_2^6$. If $D \cap F = 1$, then $D \cong DF/F \subseteq T_6$, a contradiction. Hence $D \cap F = E = O_2(D)$ and so $D \subseteq N_{P_1}(E) = (\mathbb{Z}_2^6.\mathbb{Z}_2^5).T_5$, contrary to 1).

4) Finally, suppose that $D \hookrightarrow T_8$. As above, there exists an index i, $1 \leq i \leq 3$, such that $D \hookrightarrow P_i = \mathbb{Z}_2^{i(8-i)}.(T_i \times T_{8-i})$. From the results of 2) it follows that D cannot be embedded in $T_i \times T_{8-i}$. Therefore, if one sets $E_i = \mathbb{Z}_2^{i(8-i)}$,

then $E_i \cap D \neq 1$ and $E_i \cap D = E = O_2(D)$. Hence, for $i = 1$ we have $D \subseteq N_{P_1}(E) = E_1.(\mathbb{Z}_2^{10}.(\mathbb{S}_3 \times T_5)) = (2^{7+10}.\mathbb{S}_3).T_5$, contrary to 1). Now let $i = 3$. Then $E_3D/E_3 \cong D/(E_3 \cap D) = T_5$ is embedded in $T_3 \times T_5$. But $\mathrm{Hom}(T_5, T_3) = 1$, thus in fact E_3D/E_3 coincides with T_5. Hence, $D \subseteq E_3D = E_3.T_5 = \mathbb{Z}_2^{15}.T_5$, which again contradicts the results of 1). Lastly, suppose that $i = 2$. Then $E_2D/E_2 \cong T_5$ is embedded in $T_2 \times T_6$, and so this factor-group is embedded in a maximal parabolic subgroup $\mathbb{Z}_2^5.T_5$ of T_6. This means that D can be embedded in $E_2.(\mathbb{Z}_2^5.T_5) = 2^{12+5}.T_5$, again a contradiction. □

Lemma 13.3.6. *Suppose that $n \geq 9$. Then $d(\mathbb{A}_n) = n - 1$. Furthermore, $\mathbb{A}_n$ has just one irreducible complex character of degree $n - 1$, and $\mathbb{S}_n$ has just two.*

Proof. The fact that $d(\mathbb{A}_n) = n - 1$ is well known (see e.g. [Wag 3]). Now suppose that $\sigma \in \mathrm{Irr}(\mathbb{A}_n)$ and $\deg \sigma = n - 1$. Let δ_n denote the character afforded by the natural doubly transitive permutation representation of $\mathbb{A}_n$. From [ATLAS] it follows that σ is $\delta_n - 1_{\mathbb{A}_n}$ if $n = 7$, 8 or 9. We shall prove by induction on n that $\sigma = \delta_n - 1_{\mathbb{A}_n}$ for $n \geq 10$. Consider the standard subgroup $Y = \mathbb{A}_{n-1}$ in $X = \mathbb{A}_n$. As $d(Y) = n - 2 > (n-1)/2$ and Y has no irreducible characters of degree $n - 1$, the induction hypothesis implies that $\sigma|_Y = \delta_{n-1} = \delta_n|_Y - 1_Y$. Thus $\delta_n(x) = \sigma(x) + 1$ whenever $x \in X$ and $\delta_n(x) > 0$. From this it follows that

$$\sum_{x \in X} |\sigma(x) + 1|^2 = |X|(\sigma + 1_X, \sigma + 1_X)_X = 2|X| = |X|(\delta_n, \delta_n)_X =$$

$$= \sum_{x \in X} |\delta_n(x)|^2 = \sum_{x \in X, \delta_n(x) > 0} |\delta_n(x)|^2 = \sum_{x \in X, \delta_n(x) > 0} |\sigma(x) + 1|^2.$$

Hence, for each x in X with $\delta_n(x) = 0$ we have $\sigma(x) + 1 = 0$. In other words, $\sigma(x) + 1 = \delta_n(x)$ for all x in X, and $\sigma = \delta_n - 1_X$ as required.

Finally, suppose that $\sigma \in \mathrm{Irr}(\mathbb{S}_n)$, $\deg \sigma = n - 1$ and $n \geq 7$. Denote the non-linear irreducible constituent of the natural doubly transitive permutation character of $\mathbb{S}_n$ by δ. Furthermore, set $S = \mathbb{S}_n$ and $A = \mathbb{A}_n$. Then, as shown above, $\sigma|_A = \delta|_A$ and $\mathrm{Ind}_A^S(\delta|_A) = \delta + \delta.\varepsilon$, where $\varepsilon(x) = \mathrm{sgn}(x)$ for all x in S. Furthermore,

$$(\sigma, \mathrm{Ind}_A^S(\delta|_A))_S = (\sigma|_A, \delta|_A)_A = 1.$$

This means that $\sigma = \delta$ or $\sigma = \delta.\varepsilon$. □

In the rest of this section we shall suppose that the following condition holds: $(*)$ *the automorphism group* $\mathbb{G} = \mathrm{Aut}(\Lambda)$ *of an invariant lattice* Λ *is an imprimitive complex linear group on* $\mathcal{L}$.

Proposition 13.3.7. *Suppose that Λ is a D-lattice satisfying condition* $(*)$. *Then conclusion* (iii) *of Theorem* 13.2.1 *holds.*

Proof. From Lemma 4.5.3 it follows that $\mathbb{G}$ preserves the given MOD: $\mathcal{L} = \oplus_i \mathcal{H}_i$; furthermore, the subgroup $D_1 = St_D(\mathcal{H}_1)$ acts on $\mathcal{H}_1$ like the covering group $2\mathbb{A}_8$ on its basic spin module. Hence (see the more general statement in Chapter 14), the $2\mathbb{A}_8$-invariant lattice $\Lambda_1 = \Lambda \cap \mathcal{H}_1$ is isometric to the root lattice of type E_8. Therefore, if K denotes the kernel of the permutation representation of $\mathbb{G}$ on the set $\{\mathcal{H}_i\}_{i=1}^{31}$ and $\bar{\mathbb{G}} = \mathbb{G}/K$, then $K \subseteq (\mathrm{Aut}(\Lambda_1))^{31} = (W(E_8))^{31}$ and $SL_5(2) \subseteq \bar{\mathbb{G}} \subseteq \mathbb{S}_{31}$, that is, $\bar{\mathbb{G}} \in \{SL_5(2), \mathbb{A}_{31}, \mathbb{S}_{31}\}$ (see [DiM]). □

Lemma 13.3.8. *Suppose that* Λ *is a B-lattice satisfying condition* $(*)$. *Then the group* $\mathbb{G} = \mathrm{Aut}(\Lambda)$ *preserves the given IOD* $\mathcal{D}$.

Proof. Choose a $\mathbb{G}$-invariant decomposition $\mathcal{L} = \oplus_{i=1}^N V_i$ with the smallest number of components $N > 1$, and set $B_1 = St_B(V_1)$.

1) As $N|248$, we have $\gcd((B : B_1), 5) = \gcd(248, 5) = 1$. This means that $B_1 \supseteq E = O_5(B)$, $B_1 = E.S_1$, where S_1 is a subgroup of index N in $S = SL_3(5)$. By the Borel-Tits Theorem, S_1 is contained in some parabolic subgroup $P = F.T$, where $F = \mathbb{Z}_5^2$ and $T = GL_2(5)$. Here $S_1 \supseteq F$, $S_1 = F.T_1$, where $T_1 \subseteq T$ and the index $(T : T_1) = (P : S_1) = N/31$ divides 8. It is not difficult to show that $U \subseteq T_1 \subseteq T$, where $U = [T, T] = SL_2(5)$.

2) We claim that, in its action on E, P fixes a subgroup E_1 of index 5. Indeed, suppose the contrary. Then P fixes a subgroup of order 5 of E:

$$P = \left\{ \begin{pmatrix} A & 0 \\ a & \det A^{-1} \end{pmatrix} \mid A \in GL_2(5), a \in \mathbb{F}_5^2 \right\},$$

and

$$S_1 \supseteq \left\{ \begin{pmatrix} A & 0 \\ a & 1 \end{pmatrix} \mid A \in SL_2(5), a \in \mathbb{F}_5^2 \right\}.$$

In particular, $[S_1, S_1]$ acts on $\mathrm{Irr}(E)\backslash\{1_E\}$ with orbits of lengths 24, 25, 25, 25 and 25. Hence, by Clifford's Theorem, E must be trivial on the ES_1-module V_1, since $\dim V_1 \leq 8$, on the one hand. On the other hand, it is known that E acts on $\mathcal{L}$ with character

$$2 \sum_{\rho \in \mathrm{Irr}(E)\backslash\{1_E\}} \rho,$$

which is a contradiction.

Thus, we can suppose that $P = St_B(E_1)$, where $E_1 = \langle e_2, e_3 \rangle_{\mathbb{F}_5}$ and $E = \langle e_1, e_2, e_3 \rangle_{\mathbb{F}_5}$. In this case, $[S_1, S_1]$ acts on $\mathrm{Irr}(E)\backslash\{1_E\}$ with orbits of lengths 120, 1, 1, 1 and 1. Hence, E_1 acts trivially on V_1 and $V_1 \subseteq C_{\mathcal{L}}(E_1) = \mathcal{H}_1$. Moreover, the subgroup $F = O_5(S_1)$ is also trivial on V_1, because S_1 acts transitively on $F\backslash\{1\}$. Thus in fact V_1 is acted on by the group $\bar{B_1} = B_1/K_1 = \mathbb{Z}_5.(SL_2(5).\mathbb{Z}_d)$, where $K_1 = \mathbb{Z}_5^4$, $\mathbb{Z}_5 = \langle e_1 \rangle$, $d = 124/N$ and d divides 4. It is not difficult to verify that $\mathcal{H}_1$ affords the character $\alpha_1 + \alpha_2 + \alpha_3 + \alpha_4$ of the commutator subgroup

$[\bar{B_1}, \bar{B_1}] = \mathbb{Z}_5 \times SL_2(5)$ of $\bar{B_1}$, where $\alpha_1, \ldots, \alpha_4$ are distinct irreducible characters of degree 2. Consider now the character σ of E afforded by V_1:

$$\sigma = \sum_{i=1}^{t} e_i \sigma_i,\ \sigma_i \in \mathrm{Irr}(E),\ \sum_{i=1}^{t} e_i = 2d,\ 1 \leq e_i \leq 2.$$

Clearly, every σ_i-eigensubspace of E on V_1 is $[\bar{B_1}, \bar{B_1}]$-invariant, and so its dimension e_i is not less than $2 = \dim \alpha_j$. This means that $t = d$ and $e_i = 2$. Hence, if one decomposes $\mathcal{L}$ into a sum of eigensubspaces for E:

$$\mathcal{L} = \sum_{\rho \in \mathrm{Irr}(E) \setminus \{1_E\}} \mathcal{L}_\rho,$$

where $\mathcal{L}_\rho$ affords the E-character 2ρ, then each V_i is a sum of d such components $\mathcal{L}_\rho$.

3) Based on the results of 2), it is not difficult to show that there exists a unique minimal subspace $[V_1]$ with the following properties:

(a) $[V_1] \supseteq V_1$;

(b) $[V_1]$ is a sum of some components V_i;

(c) $[V_1] \cap \Lambda \neq 0$.

Namely, $[V_1] = \mathcal{H}_1$. Furthermore, the set $\{g[V_1] | g \in \mathbb{G}\}$ constitutes a $\mathbb{G}$-imprimitivity set. By the minimality of the number $N = \dim \mathcal{L} / \dim V_1$, we must have $V_1 = [V_1] = \mathcal{H}_1$. This means that $\mathbb{G}$ preserves the given IOD $\mathcal{L} = \oplus \mathcal{H}_i$. □

Corollary 13.3.9. *Let Λ be a B-lattice satisfying condition* $(*)$. *Then conclusion* (iii) *of Theorem* 13.2.1 *holds.*

Proof. From the previous lemma it follows that the group $\mathbb{G} = \mathrm{Aut}(\Lambda)$ fixes the lattice $\Lambda_1 \oplus \ldots \oplus \Lambda_{31} \cong (\Lambda_1)^{31}$, where $\Lambda_i = \Lambda \cap V_i \cong \Lambda_1$ is an 8-dimensional $\mathbb{Z}_5.GL_2(5)$-invariant lattice. The maximal finite subgroups in $GL_8(\mathbb{Z})$ are classified in [PlP 1]. From this it follows in particular that $\mathrm{Aut}(\Lambda_1)$ is contained either in $W(E_8)$ or in $(SL_2(5) * SL_2(5)).\mathbb{Z}_2$. Furthermore, if K is the kernel of the permutation action of $\mathbb{G}$ on the set $\{\mathcal{H}_i\}_{i=1}^{31}$ and $\bar{\mathbb{G}} = \mathbb{G}/K$, then $K \subseteq (\mathrm{Aut}(\Lambda_1))^{31}$ and $SL_3(5) \subseteq \bar{\mathbb{G}} \subseteq \mathbb{S}_{31}$, that is, $\bar{\mathbb{G}} \in \{SL_3(5), \mathbb{A}_{31}, \mathbb{S}_{31}\}$ (see [DiM]). □

Finally, we consider imprimitive I-lattices.

Proposition 13.3.10. *Let Λ be an I-lattice satisfying condition* $(*)$. *Then conclusion* (iv) *of Theorem* 13.2.1 *holds.*

Proof. Condition $(*)$ means that the group $\mathbb{G} = \mathrm{Aut}(\Lambda)$ leaves some decomposition $\mathcal{L} = \oplus_{i=1}^{N} V_i$ fixed, where $1 < N$ and $N|248$. From Lemma 13.3.3 it follows that $N \geq 31$, and this implies that 31 divides N. Set $I_1 = St_I(V_1)$, and let χ, ξ be the characters of I on $\mathcal{L}$ and I_1 on V_1. As $|I_1| = |I|/N$ divides 2^{15}, we have that $I_1 \subseteq O_2(I) = C = 2^{5+10}$. Hence, $\chi = \mathrm{Ind}_{I_1}^{I}(\xi) = \mathrm{Ind}_{C}^{I}(\xi')$, where $\xi' = \mathrm{Ind}_{I_1}^{C}(\xi)$ is an irreducible C-character. In the proof of Lemma 4.5.2 we established that $\chi|_C = \sum_{i=1}^{31} \delta_i$, where δ_i is the C-character afforded by $\mathcal{H}_i$, $\delta_i \in \mathrm{Irr}(C)$ and $\delta_1, \ldots, \delta_{31}$ are distinct. Hence, without loss of generality, one can suppose that $\xi' = \delta_1$, $\mathcal{H}_1 = CV_1$, that is, every Cartan subalgebra $\mathcal{H}_i$ is a sum of some of the components V_j.

Now we choose a $\mathbb{G}$-invariant decomposition $\mathcal{L} = \oplus_{i=1}^{N} V_i$ in which the components have highest dimensions. Then the action of $\mathbb{G}$ on the set $\{V_i\}$ is primitive. As we have shown above, N is 31, 62, 124 or 248. Since C acts on $\mathcal{H}_1$ like the extraspecial 2-group 2_+^{1+6} on its unique faithful 8-dimensional module, all these possibilities for N can be realized. Furthermore, if K denotes the kernel of this permutation action and $\bar{\mathbb{G}} = \mathbb{G}/K$, then $K \subseteq (G_1)^N$, where $G_1 \hookrightarrow GL_d(\mathbb{C})$ and $d = 248/N$, on the one hand. On the other hand, $\bar{\mathbb{G}}$ is a primitive permutation group of degree N. Now simply apply the results of [DiM]. □

13.4 The primitive case

Throughout this section, we shall suppose that the following condition holds.
$(**)$: *The automorphism group* $\mathbb{G} = \mathrm{Aut}(\Lambda)$ *of an invariant lattice* Λ *is a primitive complex linear group over* $\mathcal{L}$.

Let χ be the $\mathbb{G}$-character afforded by $\mathcal{L}$, and $\mathbf{t}$ the central involution: $\mathbf{t}(x) = -x$ for all x in $\mathcal{L}$. Furthermore, $H = B$, D or I according as Λ is a B-, D- or I-lattice respectively.

Lemma 13.4.1. *Under hypothesis* $(**)$, *we have* $\mathrm{sol}(\mathbb{G}) = Z(\mathbb{G}) = Z =< \mathbf{t} >\cong \mathbb{Z}_2$.

Proof. This uses standard arguments (see e.g. §12.4). □

Following the proof of Lemma 12.4.3 and using the estimates stated in Lemmas 13.3.1 – 13.3.3, one can readily check the folowing result.

Lemma 13.4.2. *Under hypothesis* $(**)$, *the socle* $L = \mathrm{soc}(\mathbb{G}/Z)$ *is a nonabelian simple group. Furthermore,* $H \subseteq \tilde{L} = Z.L = \mathbb{Z}_2.L$. □

Using the classification of finite simple groups, we determine the group $L = \mathrm{soc}(\mathbb{G}/Z)$, which clearly satisfies the following conditions:

$$d(L) \leq 248, \tag{13.3}$$

more precisely, $\tilde{L} = \mathbb{Z}_2.L$ *has a faithful rational absolutely irreducible character* χ *of degree* 248, and

$$H \hookrightarrow L. \tag{13.4}$$

Lemma 13.4.3. *L cannot be an alternating group.*

Proof. Suppose the contrary: $L = \mathbb{A}_n$ for some $n \geq 5$. From (13.4) and Lemmas 13.3.2 – 13.3.3, it follows that $n \geq P_f(H) \geq 62$. Furthermore, if the full preimage $\tilde{L}$ of L in $\mathbb{G}$ is perfect, then $\deg \chi \geq 2^{[n/2]-1} \geq 2^{30} > 248$, contrary to (13.3). Therefore $\tilde{L} = Z \times L$ and $\chi|_L \in \mathrm{Irr}(L)$. Since $H \subseteq \tilde{L}$ and H has no subgroups of index 2, we have $H \subseteq L$. We claim that $\deg \chi = n-1$. For, set $M = \mathbb{S}_n = L.\mathbb{Z}_2$ and $\xi = \mathrm{Ind}_L^M(\chi|_L)$. Then either $\xi = \xi_1 \in \mathrm{Irr}(M)$ or $\xi = \xi_1 + \xi_2$, where $\xi_i \in \mathrm{Irr}(M)$ and $\deg \xi_1 = \deg \xi_2$. Hence, $248 = \deg \chi \leq \deg \xi_1 \leq \deg \xi = 2\deg \chi = 496 < n(n-3)/2$. By [Ras], we must have that $\deg \xi_1 = n-1$. From Lemma 13.3.6 it follows that $\xi_1|_L \in \mathrm{Irr}(L)$. But $\chi|_L$ enters $\xi_1|_L$, so $\chi|_L = \xi_1|_L$ and $248 = \deg \chi = \deg \xi_1 = n-1$, that is, $n = 249$. Thus H is a subgroup of $L = \mathbb{A}_{249}$ such that the restriction of the non-linear irreducible constituent of the natural doubly transitive permutation character of L to H is also irreducible. In other words, H has a doubly transitive permutation representation of degree 249, a contradiction. □

Lemma 13.4.4. *L cannot be an exceptional group of Lie type.*

Proof. Suppose the contrary: L is an exceptional group of Lie type defined over the field $\mathbb{F}_q$, $q = r^f$, r prime.

If L is of types E_6, ${}^2E_6$, E_7 or E_8, then by the Landázuri-Seitz-Zalesskii lower bound for the minimum degrees of projective representations of finite Chevalley groups [LaS], [SeZ] (reproduced in Table A7 and used frequently here), $d(L) \geq q^9(q^2-1) \geq 3.2^9 > 248$, contrary to (13.3).

Suppose that L is of type F_4 or ${}^2F_4$. If $r > 2$, then $d(L) \geq q^6(q^2-1) > 3^6 > 248$, contrary to (13.3). If $r = 2$, then $5|f$ because $31||L|$; hence $q \geq 32$ and $d(L) \geq q^4(q-1)\sqrt{q/2} > 2^{20} > 248$, again a contradiction.

Now suppose that L is of type ${}^2B_2$. Then (13.3) implies that $248 \geq (q-1)\sqrt{q/2}$, that is, $q \leq 32$. Furthermore, $31||L|$, therefore in fact $q = 32$. But none of the groups B, D and I can be embedded in ${}^2B_2(32)$, a contradiction.

If L is of type ${}^3D_4$, then (13.3) implies that $248 \geq q^3(q^2-1)$, that is, $q \leq 3$. But in this case 31 cannot divide $|L|$, a contradiction.

Next, suppose that L is of type G_2. If $q \geq 7$, then $d(L) \geq q(q^2-1) \geq 336 > 248$, contrary to (13.3). If $q < 7$, then none of the groups B, D and I can be embedded in L.

Finally, L is of type ${}^2G_2(q)$. Then $q \geq 27$ and $d(L) \geq q(q-1) > 248$, again a contradiction. □

Following the proof of Lemma 13.4.4 and using Table A7 and Lemmas 13.3.4, 13.3.5, we obtain:

Lemma 13.4.5. *Let Λ be an invariant lattice of type E_8 satisfying condition $(**)$, and suppose that $L = \mathrm{soc}(\mathbb{G}/Z)$ is a classical group of Lie type. Then Λ is a B-lattice and $L = L_4(5)$. Moreover,*

$$\mathbb{Z}_2 \circ L \lhd \mathbb{G} \subseteq \mathbb{Z}_2.\mathrm{Aut}(L). \qquad \square$$

Lemma 13.4.6. *Let Λ be an invariant lattice of type E_8 satisfying condition $(**)$, and suppose that $L = \mathrm{soc}(\mathbb{G}/Z)$ is a sporadic simple group. Then*

(i) *Λ is a D-lattice;*

(ii) *$L = F_3$, $\mathbb{G} = \mathbb{Z}_2 \times F_3$;*

(iii) *Λ is isometric to the Thompson-Smith lattice.*

Proof. 1) Firstly, suppose that Λ is a B-lattice. Then the order of the sporadic simple group L is divisible by $5^6.31$, which implies that $L \in \{Ly, F_2, F_1\}$. But in this case condition (13.3) is not satisfied, because $d(Ly) = 2480$, $d(F_2) = 4371$ and $d(F_1) = 196883$.

2) Now let Λ be a D- or I-lattice. Then the order of L is divisible by $2^{15}.31$, and so $L \in \{J_4, F_3, F_2, F_1\}$. As in 1) we must have that $L \neq F_2, F_1$. Furthermore, $d(J_4) = 1333$, therefore $L \neq J_4$. Finally, F_3 has trivial Schur multiplier and trivial outer automorphism group. Hence $\mathbb{G} = \mathbb{Z}_2 \times F_3$ when $L = F_3$. Note that I cannot be embedded in F_3, since Sylow 2-subgroups of I have exponent 4 while Sylow 2-subgroups of F_3 have the same order 2^{15} but exponent 8. Thus, when $L = F_3$, Λ is a D-lattice. For example, if $\Lambda = \Lambda_{TS}$ is the Thompson-Smith lattice, then $\mathrm{Aut}(\Lambda) = \mathbb{Z}_2 \times F_3$.

Conversely, let Γ be a D-lattice with $\mathrm{Aut}(\Gamma) = \mathbb{Z}_2 \times F_3$. Since F_3 has a unique irreducible complex representation of degree 248, by the Deuring-Noether Theorem [CuR] one can suppose that $\Gamma \otimes_{\mathbb{Z}} \mathbb{Q} = \Lambda \otimes_{\mathbb{Z}} \mathbb{Q}$ and $\Gamma \subseteq \Lambda$. Clearly, there exist a greatest m and a least k such that $m\Lambda \supseteq \Gamma \supseteq k\Lambda$. Of course, m divides k, that is, $k = mn$ for some integer n. Suppose that $n > 1$, so that n is divisible by some prime number p. By the maximality of m, Γ is not contained in $mp\Lambda$ and so $\Gamma + mp\Lambda \neq mp\Lambda$. Moreover, $(\Gamma + mp\Lambda)/mp\Lambda$ is an $\mathbb{F}_p F_3$-submodule of $m\Lambda/mp\Lambda \cong \Lambda/p\Lambda$. This last $\mathbb{F}_p F_3$-module is irreducible, as mentioned in §13.1, so that $\Gamma + mp\Lambda = m\Lambda$. This implies that

$$\frac{mn}{p}\Lambda = \frac{n}{p}(m\Lambda) = \frac{n}{p}\Gamma + mn\Lambda \subseteq \Gamma,$$

contradicting the minimality of $k = mn$. Consequently, $n = 1$ and $\Gamma = m\Lambda$. $\square$

To complete the proof of Theorem 13.2.1, it remains to establish the existence of a $\mathbb{Z}_2 \circ L_4(5)$-invariant lattice of dimension 248, and this will be done in the next section.

13.5 The representations of $SL_4(q)$ of degree $(q-1)(q^3-1)/2$

We start with the following subsidiary assertion:

Lemma 13.5.1. *Let A be a normal subgroup of prime index r in a finite group B, and suppose that $\xi \in \mathrm{Irr}(B)$. Then the following assertions hold:*

(i) *if $\xi(b) \neq 0$ for some $b \in B \setminus A$, then $\xi|_A \in \mathrm{Irr}(A)$;*

(ii) *if ξ vanishes on $B \setminus A$, then there exists a character $\rho \in \mathrm{Irr}(A)$ such that $\xi = \mathrm{Ind}_A^B(\rho)$.*

Proof. By Clifford's Theorem, $\xi|_A = e\sum_{i=1}^t \rho_i$ for some e, t and some pairwise different irreducible characters ρ_i of A. In particular, $(\xi|_A, \xi|_A)_A = e^2t$ and $et|r$. In case (i) we have

$$1 = (\xi, \xi)_B = \frac{1}{|B|}\sum_{x \in B} |\xi(x)|^2 > \frac{1}{r|A|}\sum_{x \in A} |\xi(x)|^2,$$

that is, $(\xi|_A, \xi|_A)_A < r$. Hence, $et = 1$, $e = t = 1$ and $\xi|_A \in \mathrm{Irr}(A)$. In case (ii) it is clear that $(\xi|_A, \xi|_A)_A = r$ and so $et = e^2t = r$, $e = 1$, $t = r$, $\xi|_A = \rho_1 + \ldots + \rho_r$. Hence, $1 = (\xi|_A, \rho_1)_A = (\xi, \mathrm{Ind}_A^B(\rho_1))_B$ and $\xi = \mathrm{Ind}_A^B(\rho_1)$. □

Throughout this section p is an odd prime number and $q = p^f$. Conjugacy classes and character table of the group $G_q = GL_4(q)$ are described in [Ste 1] (see also [Enn] and [Noz]). We introduce some notation at this point. Let ρ, σ, τ and ω be primitive elements of the fields $\mathbb{F}_q$, $\mathbb{F}_{q^2}$, $\mathbb{F}_{q^3}$ and $\mathbb{F}_{q^4}$ such that $\rho = \sigma^{q+1} = \tau^{q^2+q+1} = \omega^{q^3+q^2+q+1}$ and $\sigma = \omega^{q^2+1}$. Fix an embedding of the multiplicative group $\mathbb{F}_{q^2}^*$ in $\mathbb{C}^*$, and denote the images of ρ and σ in this embedding by ε and η respectively. Set

$$S_q = SL_4(q),\ N_q = (q-1)(q^3-1)/2,$$

$$\mathcal{I}_1 = \mathbb{Z}/(q-1)\mathbb{Z},\ \mathcal{I}_k = \{(x_1, \ldots, x_k) \in \mathcal{I}_1^k | x_i \neq x_j \text{ if } i \neq j\},$$

$$\mathcal{J}_0 = \mathbb{Z}/(q^2-1)\mathbb{Z},\ \mathcal{J}_1 = \{x \in \mathcal{J}_0 | x \not\equiv 0 \pmod{(q+1)}\},$$

$$\mathcal{J}_2 = \mathcal{I}_1 \times \mathcal{J}_1,\ \mathcal{J}_3 = \mathcal{I}_2 \times \mathcal{J}_1,$$

$$\mathcal{J}_4 = \{(x, y) \in \mathcal{J}_1^2 | x \not\equiv y, yq \pmod{(q^2-1)}\},$$

$$\mathcal{R}_0 = \mathbb{Z}/(q^3-1)\mathbb{Z},\ \mathcal{R}_1 = \{x \in \mathcal{R}_0 | x \not\equiv 0 \pmod{(q^2+q+1)}\},\ \mathcal{R}_2 = \mathcal{I}_1 \times \mathcal{R}_1,$$

$$\mathcal{S}_0 = \mathbb{Z}/(q^4-1)\mathbb{Z},\ \mathcal{S}_1 = \{x \in \mathcal{S}_0 | x \not\equiv 0 \pmod{(q^2+1)}\}.$$

The irreducible complex characters χ_k of degree $(q-1)(q^3-1) = 2N_q$ of G_q are parametrized by the elements $k \in \mathcal{J}_1$; moreover, $\chi_k = \chi_{kq}$. Below we give the values of χ_k on representatives of the conjugacy classes of G_q.

Table 13.1

Class	Representative	Parameters	The value of χ_k
$A_1(a)$	$\begin{pmatrix} \rho^a & & & \\ & \rho^a & & \\ & & \rho^a & \\ & & & \rho^a \end{pmatrix}$	$a \in \mathcal{I}_1$	$(q-1)(q^3-1)\varepsilon^{2ak}$
$A_2(a)$	$\begin{pmatrix} \rho^a & & & \\ 1 & \rho^a & & \\ & & \rho^a & \\ & & & \rho^a \end{pmatrix}$	$a \in \mathcal{I}_1$	$-(q-1)\varepsilon^{2ak}$
$A_3(a)$	$\begin{pmatrix} \rho^a & & & \\ 1 & \rho^a & & \\ & & \rho^a & \\ & & 1 & \rho^a \end{pmatrix}$	$a \in \mathcal{I}_1$	$(q^2-q+1)\varepsilon^{2ak}$
$A_4(a)$	$\begin{pmatrix} \rho^a & & & \\ 1 & \rho^a & & \\ & 1 & \rho^a & \\ & & & \rho^a \end{pmatrix}$	$a \in \mathcal{I}_1$	$-(q-1)\varepsilon^{2ak}$
$A_5(a)$	$\begin{pmatrix} \rho^a & & & \\ 1 & \rho^a & & \\ & 1 & \rho^a & \\ & & 1 & \rho^a \end{pmatrix}$	$a \in \mathcal{I}_1$	ε^{2ak}
$A_6(a,b)$	$\begin{pmatrix} \rho^a & & & \\ & \rho^a & & \\ & & \rho^a & \\ & & & \rho^b \end{pmatrix}$	$(a,b) \in \mathcal{I}_2$	0
$A_7(a,b)$	$\begin{pmatrix} \rho^a & & & \\ 1 & \rho^a & & \\ & & \rho^a & \\ & & & \rho^b \end{pmatrix}$	$(a,b) \in \mathcal{I}_2$	0
$A_8(a,b)$	$\begin{pmatrix} \rho^a & & & \\ 1 & \rho^a & & \\ & 1 & \rho^a & \\ & & & \rho^b \end{pmatrix}$	$(a,b) \in \mathcal{I}_2$	0
$A_9(a,b)$	$\begin{pmatrix} \rho^a & & & \\ & \rho^a & & \\ & & \rho^b & \\ & & & \rho^b \end{pmatrix}$	$(a,b) \in \mathcal{I}_2$	$(q-1)^2\varepsilon^{k(a+b)}$
$A_{10}(a,b)$	$\begin{pmatrix} \rho^a & & & \\ 1 & \rho^a & & \\ & & \rho^b & \\ & & & \rho^b \end{pmatrix}$	$(a,b) \in \mathcal{I}_2$	$-(q-1)\varepsilon^{k(a+b)}$
$A_{11}(a,b)$	$\begin{pmatrix} \rho^a & & & \\ 1 & \rho^a & & \\ & & \rho^b & \\ & & 1 & \rho^b \end{pmatrix}$	$(a,b) \in \mathcal{I}_2$	$\varepsilon^{k(a+b)}$

Table 13.1 (continued)

Class	Representative	Parameters	The value of χ_k
$A_{12}(a,b,c)$	$\begin{pmatrix} \rho^a & & & \\ & \rho^a & & \\ & & \rho^b & \\ & & & \rho^c \end{pmatrix}$	$(a,b,c) \in \mathcal{I}_3$	0
$A_{13}(a,b,c)$	$\begin{pmatrix} \rho^a & & & \\ 1 & \rho^a & & \\ & & \rho^b & \\ & & & \rho^c \end{pmatrix}$	$(a,b,c) \in \mathcal{I}_3$	0
$A_{14}(a,b,c,d)$	$\begin{pmatrix} \rho^a & & & \\ & \rho^b & & \\ & & \rho^c & \\ & & & \rho^d \end{pmatrix}$	$(a,b,c,d) \in \mathcal{I}_4$	0
$B_1(a,b)$	$\begin{pmatrix} \rho^a & & & \\ & \rho^a & & \\ & & \sigma^b & \\ & & & \sigma^{bq} \end{pmatrix}$	$(a,b) \in \mathcal{J}_2$	$-(q-1)\varepsilon^{ak}(\eta^{bk}+\eta^{bkq})$
$B_2(a,b)$	$\begin{pmatrix} \rho^a & & & \\ 1 & \rho^a & & \\ & & \sigma^b & \\ & & & \sigma^{bq} \end{pmatrix}$	$(a,b) \in \mathcal{J}_2$	$\varepsilon^{ak}(\eta^{bk}+\eta^{bkq})$
$B_3(a,b,c)$	$\begin{pmatrix} \rho^a & & & \\ & \rho^b & & \\ & & \sigma^c & \\ & & & \sigma^{cq} \end{pmatrix}$	$(a,b,c) \in \mathcal{J}_3$	0
$C_1(a)$	$\begin{pmatrix} \sigma^a & & & \\ & \sigma^{aq} & & \\ & & \sigma^a & \\ & & & \sigma^{aq} \end{pmatrix}$	$a \in \mathcal{J}_1$	$\eta^{2ak}+\eta^{2akq}+(q^2+1)\varepsilon^{-ak}$
$C_2(a)$	$\begin{pmatrix} \sigma^a & & & \\ & \sigma^{aq} & & \\ 1 & & \sigma^a & \\ & 1 & & \sigma^{aq} \end{pmatrix}$	$a \in \mathcal{J}_1$	$\eta^{2ak}+\eta^{2akq}+\varepsilon^{-ak}$
$C_3(a,b)$	$\begin{pmatrix} \sigma^a & & & \\ & \sigma^{aq} & & \\ & & \sigma^b & \\ & & & \sigma^{bq} \end{pmatrix}$	$(a,b) \in \mathcal{J}_4$	$(\eta^{ak}+\eta^{akq})(\eta^{bk}+\eta^{bkq})$
$D_1(a,b)$	$\begin{pmatrix} \rho^a & & & \\ & \tau^b & & \\ & & \tau^{bq} & \\ & & & \tau^{bq^2} \end{pmatrix}$	$(a,b) \in \mathcal{R}_2$	0
$E_1(a)$	$\begin{pmatrix} \omega^a & & & \\ & \omega^{aq} & & \\ & & \omega^{aq^2} & \\ & & & \omega^{aq^3} \end{pmatrix}$	$a \in \mathcal{S}_1$	$\eta^{ak}+\eta^{akq}$

Lemma 13.5.2. *The group $S_q = SL_4(q)$ has characters α and β of degree N_q such that $\chi_k|_{S_q} = \alpha + \beta$, for every k in $\mathcal{J}_1$ with $\frac{q+1}{2}|k$. If k is not divisible by $(q+1)/2$, then $\chi_k|_{S_q} \in \mathrm{Irr}(S_q)$. Finally, if γ is an irreducible character of S_q of degree N_q, then γ is α or β.*

Proof. 1) Suppose first that $\chi = \chi_k$, where $k \in \mathcal{J}_1$ and k is not divisible by $(q+1)/2$. Then χ takes the value $\eta^{ak} + \eta^{akq}$ on the class $E_1(a)$, $a \in \mathcal{S}_1$. Hence, χ cannot vanish simultaneously on two conjugacy classes $E_1(a)$ and $E_1(M-a)$, where $M = (q-1)(q^2+1)/2$. Using this remark, we show that $\chi|_{S_q} \in \mathrm{Irr}(S_q)$. In fact, we have $G_q/S_q \cong \mathbb{Z}_{q-1}$. Decomposing $q-1$ into a product of prime numbers: $q-1 = r_1 r_2 \dots r_t$, we set

$$B_i = \{X \in G_q |\, \det X = \rho^l, r_1 r_2 \dots r_i | l\}.$$

Then

$$B_0 = G_q \rhd B_1 \rhd \dots \rhd B_t = S_q,$$

where $B_{i-1}/B_i \cong \mathbb{Z}_{r_i}$. Furthermore, with a_i denoting $r_1 r_2 \dots r_i$, $E_1(a_i)$ and $E_1(M-a_i)$ are both contained in $B_i \backslash B_{i+1}$; moreover, χ does not vanish on at least one of these classes. Applying Lemma 13.5.1 sequentially to the pairs (B_i, B_{i+1}), $0 \le i \le t-1$, we conclude that $\chi|_{S_q} \in \mathrm{Irr}(S_q)$.

2) Next, suppose that $\chi = \chi_k$, where $k \in \mathcal{J}_1$, $k \equiv 0 \,(\mathrm{mod}\, \frac{q+1}{2})$. Then $k = l(q+1)/2$ for some odd integer l. Set

$$B = \{X \in G_q |\, \det X \in \mathbb{F}_q^{*2} = <\rho^2>\}.$$

Direct verification shows that $\chi \equiv 0$ on $G_q \backslash B$. By Lemma 13.5.1, $\chi|_B$ is a sum of two characters (of degree N_q). Furthermore, $\chi|_{S_q}$ does not depend on $k = l(q+1)/2$, so that there exist S_q-characters α and β of degree N_q such that $\chi|_{S_q} = \alpha + \beta$. Here $\alpha \ne \beta$, since $\alpha(x) + \beta(x) = \chi_k(x) = 1$ for every element x of class $A_5(0)$.

3) Finally, suppose that $\gamma \in \mathrm{Irr}(S_q)$ and $\deg \gamma = N_q$. Decompose $\tilde{\gamma} = \mathrm{Ind}_{S_q}^{G_q}(\gamma) = \delta_1 + \dots + \delta_m$, where $\delta_i \in \mathrm{Irr}(G_q)$. Then

$$1 \le (\delta_i, \tilde{\gamma})_{G_q} = (\delta_i|_{S_q}, \gamma)_{S_q},$$

that is, γ enters $\delta_i|_{S_q}$. By Clifford's Theorem, $\deg \delta_i$ is divisible by N_q; moreover, $\deg \delta_i \le (q-1)N_q$. Using the results of [Ste 1], one can show that $\deg \delta_i = 2N_q$ and $\delta_i = \chi_k$ for some element k of $\mathcal{J}_1$. Furthermore, if k is not divisible by $(q+1)/2$, then $\chi_k|_{S_q}$ is irreducible and so γ does not enter $\chi|_{S_q}$. This means that $k \equiv 0 \,(\mathrm{mod}\, \frac{q+1}{2})$. From the results of 2) it follows that $\delta_i|_{S_q} = \alpha + \beta$. But $\deg \alpha = \deg \beta = \deg \gamma$, and therefore $\gamma \in \{\alpha, \beta\}$. □

Warning. For the present we cannot assert that $\alpha, \beta \in \mathrm{Irr}(S_q)$.

Next, we fix the following standard embeddings $T_q \subset R_q \subset S_q$:

$$T_q = \left\{ \begin{pmatrix} A & 0 \\ 0 & 1 \end{pmatrix} | A \in SL_3(q) \right\} \cong SL_3(q),$$

$$R_q = \left\{ \begin{pmatrix} A & u \\ 0 & 1 \end{pmatrix} | A \in SL_3(q), u \in \mathbb{F}_q^3 \right\} \cong ASL_3(q).$$

Lemma 13.5.3. *In the notation of Lemma* 13.5.2, T_q *admits irreducible characters* $\gamma_1, \dots, \gamma_{(q-1)/2}$ *of degree* $q^3 - 1$ *such that*

$$\alpha|_{T_q} = \beta|_{T_q} = \sum_{1 \le i \le (q-1)/2} \gamma_i.$$

Proof. By Lemma 13.5.2, one can suppose that $\alpha + \beta = \chi|_{S_q}$, where $\chi = \chi_{(q+1)/2}$.

1) We claim that if an element g of S_q has a (3, 1)-block matrix (that is, a matrix of the form $\begin{pmatrix} A & 0 \\ 0 & a \end{pmatrix}$, where $A \in GL_3(q)$ and $a \in \mathbb{F}_q^*$) in some basis $\{e_1, \dots, e_4\}$ of $\mathbb{F}_q^4$, then $\alpha(g) = \beta(g) = \chi(g)/2$. Indeed, the matrices $\begin{pmatrix} E_3 & 0 \\ 0 & b \end{pmatrix}$, $b \in \mathbb{F}_q^*$, centralize g, which means that

$$(C_{G_q}(g) : C_{S_q}(g)) = q - 1 = (G_q : S_q),$$

$$|g^{G_q}| = (G_q : C_{G_q}(g)) = (S_q : C_{S_q}(g)) = |g^{S_q}|,$$

that is, the conjugacy classes g^{G_q} and g^{S_q} are the same. By Clifford's Theorem, α and β are G_q-conjugate, that is, there exists an element c of G_q such that $\beta(x) = \alpha(cxc^{-1})$ for all x in S_q. In particular, $\beta(g) = \alpha(cgc^{-1})$. But $cgc^{-1} \in g^{G_q} = g^{S_q}$, therefore there is an element a in S_q such that $cgc^{-1} = aga^{-1}$. Hence, $\beta(g) = \alpha(cgc^{-1}) = \alpha(aga^{-1}) = \alpha(g) = \chi(g)/2$, as stated.

2) The character table of $T_q = SL_3(q)$ is computed in [SiF]. In particular, the irreducible characters δ_k of degree $q^3 - 1$ of T_q are parametrized by the elements k of $\mathcal{J}_1$; moreover $\delta_k = \delta_{kq}$. We claim that

$$\chi|_{T_q} = \sum_{k \in \mathcal{J}_1, k \equiv 0 \pmod{\frac{q+1}{2}}} \delta_k. \tag{13.5}$$

Clearly, equality (13.5) and item 1) together prove the lemma.

Our calculations are summarized in the following table, where $\tilde{\delta}$ denotes the character from the right-hand side of (13.5). We have preserved the notation for conjugacy classes introduced in [SiF]. Note that the rows $C_1(a)$, $C_2(a)$, $C_3(a, b)$, $a \not\equiv 0\,(3)$, are present only in the case when $3|(q-1)$, and then $\xi = \rho^{(q-1)/3}$. Elements of the classes $C_3(a, b)$, $a \not\equiv 0(3)$, are characterized by the requirements that $\operatorname{Spec} x = \{\xi^a, \xi^a, \xi^a\}$ and $\dim \operatorname{Ker}(x - \xi^a) = 1$, and elements of the classes $C_3(0, b)$ by the similar requirements that $\operatorname{Spec} x = \{1, 1, 1\}$, $\dim \operatorname{Ker}(x - 1) = 1$ and $(x - 1)^2 \neq 0$. Finally, the notation $a \not\equiv 0(b)$ means that the integer b does not divide the integer a.

Table 13.2

Class of x in T_q	x	Class of $X = \begin{pmatrix} x & 0 \\ 0 & 1 \end{pmatrix}$ in S_q	$\tilde{\delta}(x)$, $\chi(X)$
$C_1(0)$	$\begin{pmatrix} 1 & & \\ & 1 & \\ & & 1 \end{pmatrix}$	$A_1(0)$	$(q-1)(q^3-1)$
$C_1(a)$, $a \not\equiv 0(3)$	$\begin{pmatrix} \xi^a & & \\ & \xi^a & \\ & & \xi^a \end{pmatrix}$	$A_6(\frac{a(q-1)}{3}, 0)$	0
$C_2(0)$	$\begin{pmatrix} 1 & & \\ 1 & 1 & \\ & & 1 \end{pmatrix}$	$A_2(0)$	$-(q-1)$
$C_2(a)$, $a \not\equiv 0(3)$	$\begin{pmatrix} \xi^a & & \\ 1 & \xi^a & \\ & & \xi^a \end{pmatrix}$	$A_7(\frac{a(q-1)}{3}, 0)$	0
$C_3(0, b)$	$\begin{pmatrix} 1 & & \\ \rho^b & 1 & \\ & \rho^{2b} & 1 \end{pmatrix}$	$A_4(0)$	$-(q-1)$
$C_3(a, b)$, $a \not\equiv 0(3)$	$\begin{pmatrix} \xi^a & & \\ \rho^b & \xi^a & \\ & \rho^{2b} & \xi^a \end{pmatrix}$	$A_8(\frac{a(q-1)}{3}, 0)$	0
$C_4(a)$, $\frac{q-1}{2} \mid a$, $3a \not\equiv 0(q-1)$	$\begin{pmatrix} \rho^a & & \\ & \rho^a & \\ & & \rho^{-2a} \end{pmatrix}$	$A_9(\frac{q-1}{2}, 0)$	$(-1)^{\frac{q+1}{2}}(q-1)^2$
$C_4(a)$, $a \not\equiv 0(\frac{q-1}{2})$, $3a \not\equiv 0(q-1)$	$\begin{pmatrix} \rho^a & & \\ & \rho^a & \\ & & \rho^{-2a} \end{pmatrix}$	$A_{12}(a, -2a, 0)$	0
$C_5(a)$, $\frac{q-1}{2} \mid a$, $3a \not\equiv 0(q-1)$	$\begin{pmatrix} \rho^a & & \\ 1 & \rho^a & \\ & & \rho^{-2a} \end{pmatrix}$	$A_{10}(\frac{q-1}{2}, 0)$	$(-1)^{\frac{q-1}{2}}(q-1)$
$C_5(a)$, $a \not\equiv 0(\frac{q-1}{2})$, $3a \not\equiv 0(q-1)$	$\begin{pmatrix} \rho^a & & \\ 1 & \rho^a & \\ & & \rho^{-2a} \end{pmatrix}$	$A_{13}(a, -2a, 0)$	0

Table 13.2 (continued)

Class of x in T_q	x	Class of $X = \begin{pmatrix} x & 0 \\ 0 & 1 \end{pmatrix}$ in S_q	$\tilde{\delta}(x)$, $\chi(X)$
$C_6(a, b, c)$, $a \neq b \neq c \neq a$, $abc = 0$	$\begin{pmatrix} \rho^a & & \\ & \rho^b & \\ & & \rho^c \end{pmatrix}$	$A_{12}(a, b, c)$	0
$C_6(a, b, c)$, $a \neq b \neq c \neq a$, $abc \neq 0$	$\begin{pmatrix} \rho^a & & \\ & \rho^b & \\ & & \rho^c \end{pmatrix}$	$A_{14}(a, b, c, 0)$	0
$C_7(a)$, $a \in \mathcal{J}_1$, $a = b(q-1)$	$\begin{pmatrix} \rho^a & & \\ & \sigma^{-a} & \\ & & \sigma^{-aq} \end{pmatrix}$	$B_1(0, -a)$	$-2(q-1)(-1)^b$
$C_7(a)$, $a \in \mathcal{J}_1$, $a \not\equiv 0(q-1)$	$\begin{pmatrix} \rho^a & & \\ & \sigma^{-a} & \\ & & \sigma^{-aq} \end{pmatrix}$	$B_3(0, a, -a)$	0
$C_8(a)$	$\begin{pmatrix} \tau^a & & \\ & \tau^{aq} & \\ & & \tau^{aq^2} \end{pmatrix}$	$D_1(0, a)$	0

This completes the proof of Lemma 13.5.3. □

Before determining the restriction of α and β to the parabolic subgroup R_q of S_q, we describe the projective representations of least degree of $L_2(q)$, following [Tan]. To this end, we set $\mathbb{F} = \mathbb{F}_q$, $\mathbb{F}^* = <\rho>$, $\mathbb{L} = \mathbb{F}_{q^2} = \mathbb{F}(\sqrt{\rho})$, $\mathbb{L}^* = <\sigma>$, $\rho = \sigma^{q+1}$. For $z = x + y\sqrt{\rho} \in \mathbb{L}$, where $x, y \in \mathbb{F}$, set $\bar{z} = x - y\sqrt{\rho}$, $N(z) = z\bar{z}$ and

$$C = \{z \in \mathbb{L} | N(z) = 1\} = <t_0> \cong \mathbb{Z}_{q+1}.$$

Let ν denote a fixed nontrivial character of the additive group of $\mathbb{F}$. Furthermore, π_1 and π_2 denote the characters of $\mathbb{F}^*$ and C given by $\pi_1(\rho) = \pi_2(t_0) = -1$. Set

$$\mathbb{F}^*_{\pm} = \{x \in \mathbb{F}^* | \pi_1(x) = \pm 1\},$$

$$\mathcal{F}^+ = \{f : \mathbb{F}^* \to \mathbb{C} | f(\mathbb{F}^*_-) = 0\},$$

$$\mathcal{F}^- = \{f : \mathbb{F}^* \to \mathbb{C} | f(\mathbb{F}^*_+) = 0\}.$$

Finally, let π be a character of $\mathbb{L}^*$ such that $\pi|_C = \pi_2$. We determine representations $T = T^{\pm}$ of $SL_2(q)$ on $\mathcal{F}^{\pm}$ by the rule:

$$T(g)\varphi(x) = \sum_{y \in \mathbb{F}^*} K(g|x, y)\varphi(y),$$

where $x \in \mathbb{F}^*$, $\varphi \in \mathcal{F}^{\pm}$ and $g = \begin{pmatrix} a & b \\ c & d \end{pmatrix} \in SL_2(q)$. Furthermore,

$$K(g|x, y) = -\frac{1}{q}\nu(\frac{ax+dy}{c}) \sum_{t \in \mathbb{L}, N(t)=yx^{-1}} \nu(-\frac{xt+yt^{-1}}{c})\pi(t)$$

if $c \neq 0$, and

$$K(g|x, y) = \pi(a)\nu(abx)\delta(y - a^2x)$$

if $c = 0$. Here,

$$\delta(x) = \begin{cases} 0 & \text{if} \quad x \in \mathbb{F}^*, \\ 1 & \text{if} \quad x = 0. \end{cases}$$

Then $d(L_2(q)) = (q-1)/2$ (if $q \neq 9$), and $SL_2(q)$ has just two nonequivalent irreducible representations of degree $(q-1)/2$, namely T^+ and T^-.

Consider the split extension $L_q = \mathbb{F}_q.GL_2(q)$ of the additive group of $\mathbb{F}_q$ by $GL_2(q)$ determined as follows:

$$XaX^{-1} = (\det X)^{-1}a$$

for $X \in GL_2(q)$, $a \in \mathbb{F}_q$. Furthermore, if one embeds $\mathbb{F}_p$ in $\mathbb{F}_q$ and

$$SL_2(q).\mathbb{Z}_{p-1} = \{X \in GL_2(q) | (\det X)^{p-1} = 1\}$$

in $GL_2(q)$, then L_q contains the split extension

$$L_p = \mathbb{F}_p.(SL_2(q).\mathbb{Z}_{p-1}).$$

Lemma 13.5.4. *Suppose that $q \geq 5$. Then the representations $T^{\pm}$ of $SL_2(q)$ can be written over $\mathbb{Q}(\zeta)$, $\zeta = \exp(2\pi i/p)$:*

$$\mathcal{F}^{\pm} = V^{\pm} \otimes_{\mathbb{Q}(\zeta)} \mathbb{C}$$

for some $\mathbb{Q}(\zeta)SL_2(q)$-modules V^+ and V^-. Furthermore, if the modules $V^{\pm}$ are viewed as $\mathbb{Q}$-spaces, then they exhaust all the absolutely irreducible L_p-modules of rank $(q-1)(p-1)/2$ with nontrivial action of the subgroup $\mathbb{F}_p = O_p(L_p)$.

Proof. 1) Define the following special basis for the space $\mathcal{F}^+$ (respectively, $\mathcal{F}^-$):

$$\{\Delta_j = \pi(\rho^{-j})\delta(x - \rho^{2j+\mu}) | 0 \leq j < (q-1)/2\},$$

where $\mu = 0$ (respectively, $\mu = 1$). Consider an arbitrary element $g = \begin{pmatrix} a & b \\ c & d \end{pmatrix}$ of $SL_2(q)$, $a = \rho^m$. If $c = 0$, then

$$T(g) : \Delta_j \mapsto \pm\nu(\rho^{2j-m+\mu}b)\Delta_{j-m}$$

(the subscript j of Δ_j is taken modulo $(q-1)/2$). Furthermore if $c \neq 0$, then

$$T(g)\Delta_j = \sum_{0 \le k < (q-1)/2} a_k \Delta_k,$$

where

$$a_k = -\frac{(-1)^{k-j}}{q} \nu\left(\frac{a\rho^{2k+\mu} + d\rho^{2j+\mu}}{c}\right) \times$$

$$\times \sum_{\substack{t=\sigma^m \\ m=2(j-k)+n(q-1) \\ 0 \le n \le q}} (-1)^n \nu\left(-\frac{t\rho^{2k+\mu} + t^{-1}\rho^{2j+\mu}}{c}\right).$$

Recall that ν takes values in $\mathbb{Q}(\zeta)$ only. Therefore, the representations T^+ and T^- are defined over $\mathbb{Q}(\zeta)$ in the basis $\{\Delta_j | 0 \le j < (q-1)/2\}$.

2) Regarding the $\mathbb{Q}(\zeta)SL_2(q)$-module $V = < \Delta_0, \ldots, \Delta_{(q-3)/2} >_{\mathbb{Q}(\zeta)}$ as a $\mathbb{Q}$-space, we set

$$W = < \zeta^i \Delta_j | 0 < i < p, 0 \le j \le (q-3)/2 >_{\mathbb{Q}}$$

(the superscripts $\pm$ are omitted for brevity). Clearly, $SL_2(q)$ preserves V and W. We define operators $\underline{a}$ and $\underline{A}$ on V and W as follows:

$$\underline{a} : v \mapsto \zeta v, \ \forall v \in V; \ \underline{A} = \underline{a}_W : \zeta^i \Delta_j \mapsto \zeta^{i+1} \Delta_j.$$

Furthermore, let Φ be the element of $Gal(\mathbb{Q}(\zeta)/\mathbb{Q})$ such that $\Phi(\zeta) = \zeta^{\theta^{-1}}$, where $\theta = \rho^{(q-1)/(p-1)}$; and define operators $\underline{b}$ and $\underline{B}$ by the rules $\underline{b} : \sum_j a_j \Delta_j \mapsto \sum_j \Phi(a_j)\Delta_j$, where $a_j \in \mathbb{Q}(\zeta)$; $\underline{B} = \underline{b}_W$. Then

(a) $\underline{A}$, $\underline{B}$ preserve W;

(b) $\underline{a}^p = 1_V$, $\underline{A}^p = 1_W$; $\underline{b}^{p-1} = 1_V$, $\underline{B}^{p-1} = 1_W$;

(c) $\underline{A}$ centralizes $SL_2(q)$;

(d) $\underline{B}.\underline{A}.\underline{B}^{-1} = \underline{A}^{\theta^{-1}}$, $\underline{B}\tilde{T}(g)\underline{B}^{-1} = \tilde{T}(JgJ^{-1})$, $\forall g \in SL_2(q)$.

Here $\tilde{T}(g) = T({}^t g^{-1})$ and $J = \begin{pmatrix} \theta & 0 \\ 0 & 1 \end{pmatrix} \in GL_2(q)$. This means that the group $L_p = \mathbb{F}_p.(SL_2(q).\mathbb{Z}_{p-1})$ acts on W, where

$$\mathbb{F}_p := < \underline{A} >, \ \mathbb{Z}_{p-1} := < \underline{B} > .$$

3) Let τ^+ and τ^- be the L_p-characters afforded by W^+ and W^-; β_1 and β_{-1} the $SL_2(q)$-characters afforded by V^+ and V^-. Furthermore, suppose that $\mathrm{Irr}(\mathbb{F}_p) = \{1_{\mathbb{F}_p}, \alpha_1, \ldots, \alpha_{p-1}\}$. One can suppose that the subgroup $L'_p = \mathbb{F}_p \times SL_2(q)$ has characters $\alpha_1\beta_1$ and $\alpha_1\beta_{-1}$ on V^+ and V^-. Then

$$\tau^+|_{L'_p} = \sum_{i=1}^{p-1} \Phi^i(\alpha_1\beta_1) = \alpha_1\beta_{i_1} + \alpha_2\beta_{i_2} + \ldots + \alpha_{p-1}\beta_{i_{p-1}},$$

$$\tau^-|_{L'_p} = \sum_{i=1}^{p-1} \Phi^i(\alpha_1\beta_{-1}) = \alpha_1\beta_{j_1} + \alpha_2\beta_{j_2} + \ldots + \alpha_{p-1}\beta_{j_{p-1}},$$

where $< \Phi >= Gal(\mathbb{Q}(\zeta)/\mathbb{Q})$ as above and $\{i_t, j_t\} = \{1, -1\}$ for $1 \leq t \leq p-1$. Note that $L'_p \lhd L_p$ and that the L_p-orbit of $\alpha_1\beta_{\pm 1}$ has the form $\{\alpha_1\beta_{k_1}, \ldots, \alpha_{p-1}\beta_{k_{p-1}}\}$. So, by Clifford's Theorem, both τ^+ and τ^- are irreducible and, moreover, $\tau^\pm = \mathrm{Ind}_{L'_p}^{L_p}(\alpha_1\beta_{\pm 1})$.

4) Suppose now that $\tau' \in \mathrm{Irr}(L_p)$, $\deg \tau' = (q-1)(p-1)/2$, and τ' is nontrivial when restricted to $\mathbb{F}_p = O_p(L_p)$. Then, by Clifford's Theorem,

$$\tau'|_{\mathbb{F}_p} = \frac{q-1}{2} \sum_{t=1}^{p-1} \alpha_t.$$

As $SL_2(q)$ centralizes $\mathbb{F}_p$, it fixes every α_t-eigensubspace (of dimension $(q-1)/2$) of $\mathbb{F}_p$. Furthermore, if $q \neq 9$ we have $d(L_2(q)) = (q-1)/2$, while $SL_2(9)$ has no nontrivial irreducible representations of degree less than 4 although $d(L_2(9)) = 3$. Thus we have come to the conclusion that $\tau'|_{L'_p} = \sum_{t=1}^{p-1} \alpha_t\beta_{l_t}$, where $l_t \in \{1, -1\}$. Without loss of generality, one can suppose that $l_1 = 1$. Then

$$1 = (\tau'|_{L'_p}, \alpha_1\beta_1)_{L'_p} = (\tau', \mathrm{Ind}_{L'_p}^{L_p}(\alpha_1\beta_1))_{L_p} = (\tau', \tau^+)_{L_p},$$

that is, $\tau' = \tau^+$. □

Lemma 13.5.5. *Suppose that $q \geq 5$. Then the group $R_q = ASL_3(q)$ has either one or two irreducible complex representations of degree $N_q = (q-1)(q^3-1)/2$, and they are rational.*

Proof. 1) Firstly we explain how one can obtain an irreducible character λ of R_q with the properties:

a) $\deg \lambda < (q^2-1)(q^3-1)$;

b) λ is nontrivial when restricted to $E = O_p(R_q) = \mathbb{Z}_p^{3f}$.

Suppose that λ is such a character. Recall that $q = p^f$ and $\zeta = \exp(2\pi i/p)$. The group R_q acts transitively on $E\backslash\{1\}$ and on $\mathrm{Irr}(E)\backslash\{1_E\}$. Set $E^* = \mathrm{Hom}(E, \mathbb{F}_p)$. Then every E-character has the form $\bar{h} : v \mapsto \zeta^{h(v)}$ for some $h \in E^*$. By Clifford's Theorem, $\lambda|_E = l\sum_{h\neq 0}\bar{h}$ for some $l \in \mathbb{N}$, $l < q^2-1$. Correspondingly, the representation space U of λ decomposes as a direct sum $U = \oplus_{h\neq 0}U_h$, where U_h affords the E-character $l\bar{h}$. We fix $h \in E^*\backslash\{0\}$ and set

$$\mathrm{Ker} h =< a_2, \ldots, a_f, b_1, \ldots, b_f, c_1, \ldots, c_f >_{\mathbb{F}_p} \cong \mathbb{Z}_p^{3f-1},$$

where

$$E =< a, b, c >_{\mathbb{F}_q}, \; < a >_{\mathbb{F}_q} =< a = a_1, a_2, \ldots, a_f >_{\mathbb{F}_p},$$

$$< b >_{\mathbb{F}_q} = < b_1, \ldots, b_f >_{\mathbb{F}_p}; \ < c >_{\mathbb{F}_q} = < c_1, \ldots, c_f >_{\mathbb{F}_p}.$$

Then in fact the group $E/\mathrm{Ker} h \cong < a_1 >_{\mathbb{F}_p}$ acts on U_h. Furthermore, the subgroup

$$T^h = \left\{ \begin{pmatrix} A & a \\ 0 & 1 \end{pmatrix} | A \in SL_2(q), a \in \mathbb{F}_q^2 \right\} \cong$$

$$\cong \mathbb{Z}_p^{2f}.SL_2(q)$$

of $T_q \cong SL_3(q)$ also acts on U_h. Since $\dim U_h = l < q^2 - 1$, the subgroup $\mathbb{Z}_p^{2f}$ acts trivially on U_h. This means that if we set $R^h = St_{R_q}(U_h)$, then the kernel of R^h on U_h contains the group $P = \mathbb{Z}_p^{3f-1}.\mathbb{Z}_p^{2f}$ and $\lambda = \mathrm{Ind}_{R^h}^{R_q}(\gamma)$, where γ denotes the character of R^h (and of $R^h/P \cong \mathbb{F}_p \times SL_2(q) \cong L'_p$) afforded by U_h.

Now we set $\widetilde{U_h} = \oplus_{i=1}^{p-1} U_{ih}$. Then the group $\widetilde{R}^h = St_{R_q}(\widetilde{U_h})$ is of the form $E.\widetilde{T}^h$, where

$$\widetilde{T}^h = \left\{ \begin{pmatrix} A & a \\ 0 & \det A^{-1} \end{pmatrix} | a \in \mathbb{F}_q^2, A \in GL_2(q), (\det A)^{p-1} = 1 \right\} \cong$$

$$\cong \mathbb{Z}_p^{2f}.(SL_2(q).\mathbb{Z}_{p-1}).$$

Moreover, the kernel of $\widetilde{R}^h$ on $\widetilde{U_h}$ also contains P. If $\widetilde{\gamma}$ denotes the character of $\widetilde{R}^h$ (and of $\tilde{R}^h/P = \mathbb{F}_p.(SL_2(q).\mathbb{Z}_{p-1}) \cong L_p$) afforded by $\widetilde{U_h}$, then we have again $\lambda = \mathrm{Ind}_{\tilde{R}^h}^{R_q}(\widetilde{\gamma})$.

2) Next, we claim that R_q has a unique irreducible character $\lambda = \lambda_0$ with the properties:

a) $\deg \lambda < N_q$;

b) λ is nontrivial when restricted to E.

Indeed, let $1_{R_q} + \lambda_0$ be the character corresponding to the natural doubly transitive permutation representation of R_q on E. Then λ_0 possesses the indicated properties, and $\deg \lambda_0 = q^3 - 1$. Conversely, let λ' be such a character. According to 1), $\lambda' = \mathrm{Ind}_{R^h}^{R_q}(\gamma)$, where γ is an irreducible character of $L'_p = \mathbb{F}_p \times SL_2(q)$. Since $q \geq 5$ and $\deg \gamma < (q-1)/2 = d(L_2(q))$, γ must be trivial on $SL_2(q)$, that is, $\deg \gamma = 1$. Furthermore, if $\gamma = 1_{R^h}$, then $(\lambda', 1_{R_q})_{R_q} = (\gamma, 1_{R^h})_{R^h} = 1$, and λ' is not irreducible. So γ is one of the $p-1$ nontrivial linear characters $\alpha_1, \ldots, \alpha_{p-1}$ of L'_p. Set $\lambda_i = \mathrm{Ind}_{R^h}^{R_q}(\alpha_i)$. Here the characters $\alpha_1, \ldots, \alpha_{p-1}$ are algebraically conjugate, therefore the characters $\lambda_1, \ldots, \lambda_{p-1}$ are also algebraically conjugate. But one of the characters λ_i coincides with λ_0 and so it is rational. Consequently, $\lambda_1 = \ldots = \lambda_{p-1} = \lambda_0$, that is, $\lambda' = \lambda_0$ as required.

3) Suppose now that a character γ of the group $L'_p = R^h/P$ is of the form $\gamma = \alpha_i \beta_j$, where $1 \leq i \leq p-1$, $j = \pm 1$ and the characters α_i, β_j are as described in the proof of Lemma 13.5.4. Here we show that the character $\lambda = \mathrm{Ind}_{R^h}^{R_q}(\gamma)$ is irreducible. Suppose the contrary: $(\lambda, \nu)_{R_q} > 0$ for some irreducible character ν of

R_q and $\deg\nu < \deg\lambda = N_q$. Then $0 < (\lambda, \nu)_{R_q} = (\gamma, \nu|_{R^h})_{R^h}$, that is, γ enters $\nu|_{R^h}$. Suppose first that $\mathrm{Ker}\nu \supseteq E$. Then $\mathrm{Ker}\gamma \supseteq E$, contradicting the fact that $\gamma = \alpha_i\beta_j$. Therefore the restriction of ν to E is nontrivial. By 2), we have $\nu = \lambda_0$. Hence, $\lambda = \frac{q-1}{2}\lambda_0$ and $\lambda_0|_{R^h} = \frac{q-1}{2}\gamma$. Meanwhile λ_0 is a rational character and the value field $\mathbb{Q}(\gamma)$ of γ contains $\mathbb{Q}(\zeta)$, a contradiction.

4) Finally, let λ be an arbitrary irreducible character of R_q of degree N_q. Since $R_q/E \cong SL_3(q)$ has no irreducible characters of this degree (see [SiF]), the restriction of λ to E must be nontrivial. By the results of 1), we have $\lambda = \mathrm{Ind}_{R^h}^{R_q}(\gamma)$, where γ is an irreducible character of $R^h/P = L'_p = \mathbb{F}_p \times SL_2(q)$ of degree $N_q/(q^3-1) = (q-1)/2$. Moreover, since $E \lhd R_q$ and the restriction of λ to E is nontrivial, it follows that the restriction of γ to E is nontrivial, and so $\gamma = \alpha_i\beta_j$ for some i, j with $1 \le i \le p-1$ and $j = \pm 1$. Thus $\lambda = \lambda_{ij}$, where $\lambda_{ij} = \mathrm{Ind}_{R^h}^{R_q}(\alpha_i\beta_j)$.

Conversely, if $\lambda = \lambda_{ij}$, then λ is irreducible by 3). Furthermore, by 1) we have $\lambda = \mathrm{Ind}_{\widetilde{R}^h}^{R_q}(\widetilde{\gamma})$, where $\widetilde{\gamma}$ is an irreducible character of the group $\widetilde{R}^h/P = L_p$ of degree $(q-1)(p-1)/2$, with nontrivial restriction to $O_p(L_p)$. By Lemma 13.5.4, $\widetilde{\gamma}$ is rational (that is, it is afforded by some rational representation). Hence, λ is also rational. Furthermore, L_p has just two irreducible characters of this kind; therefore, R_q has either one or two irreducible characters λ of degree N_q. □

Proposition 13.5.6. *In the notation of Lemma* 13.5.2, *the characters α and β of the group $S_q = SL_4(q)$ are irreducible and rational, and their kernels are equal to $\Omega_2(Z(S_q)) = \mathbb{Z}_2$. Their restriction to the parabolic subgroup R_q gives two distinct irreducible characters of R_q of degree N_q.*

Proof. When $q = 3$, our assertions follow from [ATLAS]. One can show that R_3 has just three irreducible characters of degree 26 with nontrivial restriction to $O_3(R_3)$, and all of them are rational. So we can assume that $q \ge 5$.

1) Consider the embeddings $S_q \supset R_q \supset T_q$ and the decomposition

$$\alpha|_{R_q} = \sum_{i=1}^{m} \lambda_i, \ \lambda_i \in \mathrm{Irr}(R_q).$$

Suppose that $\alpha|_{R_q}$ is reducible, that is, $m \ge 2$. First, suppose in addition that the restriction of λ_i to $E = O_p(R_q)$ is nontrivial. By item 2) of the proof of Lemma 13.5.5, one can assume that $\lambda_i + 1_{R_q}$ is the character afforded by a permutation representation of R_q on E with point stabilizer $T_q \cong SL_3(q)$. But in that case, the character 1_{T_q} enters $(\lambda_i + 1_{R_q})|_{T_q}$ with multiplicity at least two, that is, 1_{T_q} enters the characters $\lambda_i|_{T_q}$ and $\alpha|_{T_q}$, contradicting Lemma 13.5.3. Hence, $\mathrm{Ker}\lambda_i \supseteq E$ for all i, that is, $\mathrm{Ker}\alpha \supseteq E$. Next, we consider the element $x_1 = \begin{pmatrix} 1 & & & \\ 1 & 1 & & \\ & & 1 & \\ & & & 1 \end{pmatrix}$

of class $A_2(0)$ (in the notation of Table 13.1). Then x_1 is conjugate in S_q to some element of E. Without loss of generality, one can suppose that $x_1 \in E$. Then we have $\alpha(x_1) = N_q$, $\alpha(x_1) + \beta(x_1) = \chi_{(q+1)/2}(x_1) = -(q-1)$, which implies that $|\beta(x_1)| = N_q + (q-1) > \deg \beta$, again a contradiction. Consequently, $\alpha|_{R_q}, \beta|_{R_q} \in \mathrm{Irr}(R_q)$. In particular, $\alpha|_{R_q}$ and $\beta|_{R_q}$ are realized over $\mathbb{Q}$, by Lemma 13.5.5. Since $R_q \subset S_q$, the characters α and β are irreducible.

2) We claim that α and β are rational-valued. For, consider an arbitrary element Φ of $Gal(\mathbb{Q}(\alpha, \beta)/\mathbb{Q})$, where $\mathbb{Q}(\alpha, \beta) = \mathbb{Q}(\{\alpha(x), \beta(x) | x \in S_q\})$ is the value field of the characters α and β. Then α^{Φ} and β^{Φ} are also irreducible characters of S_q of degree N_q. By Lemma 13.5.2, we must have $\{\alpha^{\Phi}, \beta^{\Phi}\} = \{\alpha, \beta\}$. Suppose that $\alpha^{\Phi} = \beta$, and consider the element $x_2 = \begin{pmatrix} 1 & 1 & & \\ & 1 & 1 & \\ & & 1 & 1 \\ & & & 1 \end{pmatrix}$ of class $A_5(0)$ lying in R_q. From Lemma 13.5.5 it follows that $\alpha(x_2) \in \mathbb{Z}$. Hence, $\beta(x_2) = \alpha(x_2)^{\Phi} = \alpha(x_2) = \chi_{(q+1)/2}(x_2)/2 = 1/2$, a contradiction. This means that $\alpha^{\Phi} = \alpha$, $\beta^{\Phi} = \beta$ for all Φ in $Gal(\mathbb{Q}(\alpha, \beta)/\mathbb{Q})$. In other words, $\mathbb{Q}(\alpha) = \mathbb{Q}(\beta) = \mathbb{Q}$, as stated. Here we have shown also that $\alpha(x_2) \neq \beta(x_2)$, that is, $\alpha|_{R_q} \neq \beta|_{R_q}$.

3) Applying Lemma 8.3.1 to the characters α and β, we conclude that both of them are afforded by rational representations. Finally, it is clear that $\mathrm{Ker}\alpha = \mathrm{Ker}\beta = \Omega_2(Z(S_q)) = \mathbb{Z}_2$. □

To complete the proof of Theorem 13.2.1 we need to present an R_5-invariant lattice Λ of rank 248 such that

$$\mathbb{Z}_2 \circ L_4(5) \lhd \mathrm{Aut}(\Lambda) \subseteq \mathbb{Z}_2.\mathrm{Aut}(L_4(5)),$$

where $R_5 = ASL_3(5)$. Embed R_5 in the Lie group $\mathcal{G} = E_8(\mathbb{C})$. In view of Lemma 13.5.5 and Proposition 13.5.6, one can suppose that the character of the adjoint representation of $\mathcal{G}$ (on $\mathcal{L}$) restricted to R_5 is $\alpha|_{R_5}$. By the same Proposition 13.5.6, there exists a 248-dimensional lattice Λ on which $SL_4(5)$ acts with character α. In particular, Λ is R_5-invariant. Setting $\mathbb{G} = \mathrm{Aut}(\Lambda)$ we obtain that $\mathbb{G} \supseteq SL_4(5)/\mathbb{Z}_2 = \mathbb{Z}_2 \circ L_4(5)$. Clearly, $\mathbb{G}$ acts primitively on $\mathcal{L}$. Based on the results of §13.4, one can now conclude that

$$\mathbb{Z}_2 \circ L_4(5) \lhd \mathbb{G} \subseteq \mathbb{Z}_2.\mathrm{Aut}(L_4(5)). \qquad \square$$

By Proposition 13.5.6, we can consider R_q- and S_q-invariant lattices Λ of dimension N_q and determine $\mathbb{G} = \mathrm{Aut}(\Lambda)$. Because our arguments are so similar to those used in §§13.3, 13.4, we shall omit proofs.

Let V be a faithful $\mathbb{Q}(S_q/\mathbb{Z}_2)$-module of rank N_q, Λ an R_q-invariant lattice in V, and set $\mathbb{G} = \mathrm{Aut}(\Lambda)$.

Lemma 13.5.7. *If $q > 3$, the only subgroup of R_q which is absolutely irreducible on V is R_q itself.* □

Note that this lemma fails when $q = 3$.

Proposition 13.5.8. *Suppose that* $\mathbb{G}$ *is an imprimitive linear group on* $V \otimes_{\mathbb{Q}} \mathbb{C}$. *Then* $\mathbb{G}$ *preserves a decomposition* $V = V_1 \oplus \ldots \oplus V_N$, *where*

(i) $\dim V_i = d(q-1)/2$, $(p-1)|d$, $d|(q-1)$;

(ii) $\Lambda_i = \Lambda \cap V_i \cong \Lambda_1$ *is a* $d(q-1)/2$*-dimensional lattice invariant under the subgroup* $\mathbb{F}_q.(SL_2(q).\mathbb{Z}_d)$ *of* L_q;

(iii) V_i *affords the E-character* $\frac{q-1}{2}(\rho_{i1} + \ldots + \rho_{id})$, *where* $\rho_{i1}, \ldots, \rho_{id}$ *are pairwise different;*

(iv) *the permutation action of* $\mathbb{G}$ *on* $\{V_1, \ldots, V_N\}$ *is primitive.* □

Based on this proposition, one can determine the composition structure of $\mathbb{G}$ in the case where $\mathbb{G}$ is imprimitive.

Lemma 13.5.9. *Suppose that* $\mathbb{G}$ *is a primitive linear group on* $V \otimes_{\mathbb{Q}} \mathbb{C}$. *Then* $\mathrm{sol}(\mathbb{G}) = Z(\mathbb{G}) = Z = \mathbb{Z}_2$. *Furthermore, the socle* $L = \mathrm{soc}(\mathbb{G}/Z)$ *is a nonabelian finite simple group.* □

Proposition 13.5.10. *In the notation of Lemma* 13.5.9, *one of the following assertions holds.*

(i) $q = 3$, $L = \mathbb{A}_{27}$ *and* $\mathbb{Z}_2 \times \mathbb{A}_{27} \lhd \mathbb{G} \subseteq \mathbb{Z}_2 \times \mathbb{S}_{27}$.

(ii) $L = L_4(q)$.

(iii) $L = L_m(p^g)$, $m \geq 5$, $jg = 3f$ *and* $m - 1 \geq j > 3(m-1)/4$.

(iv) $L = PSp_{2m}(p^g)$, $m \geq 5$, $jg = 3f$ *and* $m > j > 3m/4$.

(v) $L = \Omega_{2m+1}(p^g)$, $m \geq 3$, $2jg = 3f$ *and* $m - 1 \geq j > 3(m-1)/4$.

(vi) $L = P\Omega^+_{2m}(p^g)$, $m \geq 4$, $2jg = 3f$ *and* $m - 2 \geq j > 3(2m-3)/8$.

(vii) $L = P\Omega^-_{2m}(p^g)$, $m \geq 4$, $2jg = 3f$ *and* $m - 1 \geq j > 3(2m-3)/8$.

(viii) $L = PSU_m(p^g)$, $m \geq 3$, $2jg = 3f$ *and one of the following holds:*

a) $[m/2] \geq j > 3(m-1)/8$,

b) $m - 2 \geq j > 3(m-1)/8$ *and* j *is odd.* □

(Recall that $q = p^f$).

Proposition 13.5.11. *Suppose that* Λ *is an* $SL_4(q)$*-invariant lattice in* V. *Then* $\mathbb{G} = \mathrm{Aut}(\Lambda)$ *is primitive on* $V \otimes_{\mathbb{Q}} \mathbb{C}$. *Furthermore,* $\mathrm{sol}(\mathbb{G}) = Z(\mathbb{G}) = Z = \mathbb{Z}_2$ *and* $L = \mathrm{soc}(\mathbb{G}/Z)$ *satifies one of the following conditions.*

(i) $L = L_{4n}(s)$ *and* $q = s^n$.

(ii) $L = PSp_{2m}(s)$, $q = s^n$ *and* $3n < m < 4n$.

(iii) $L = \Omega_{8n+1}(s)$ *and* $q = s^{2n}$.

(iv) $L = P\Omega^+_{8n+2}(s)$ *and* $q = s^{2n}$.

(v) $L = PSU_{4n}(s)$, $q = s^n$ *and* $n \geq 2$. □

As an immediate consequence of Propositions 13.5.8 and 13.5.10 we obtain:

Corollary 13.5.12. *Suppose that $q = p$, let Λ be an $ASL_3(q)$-invariant lattice in V and set $\mathbb{G} = \mathrm{Aut}(\Lambda)$.*

(i) *If $\mathbb{G}$ is primitive on $V \otimes_{\mathbb{Q}} \mathbb{C}$, then one of the following holds:*

a) $q = 3$ *and* $\mathbb{Z}_2 \times \mathbb{A}_{27} \lhd \mathbb{G} \subseteq \mathbb{Z}_2 \times \mathbb{S}_{27}$,

b) $\mathbb{Z}_2.L_4(p) \lhd \mathbb{G} \subseteq \mathbb{Z}_2.\mathrm{Aut}(L_4(p))$.

(ii) *If $\mathbb{G}$ is imprimitive on $V \otimes_{\mathbb{Q}} \mathbb{C}$, then $\mathbb{G}$ leaves the lattice $\Lambda_1 \oplus \ldots \oplus \Lambda_{p^2+p+1} \cong (\Lambda_1)^{p^2+p+1}$ fixed, where Λ_1 is a $(p-1)^2/2$-dimensional $\mathbb{Z}_p.GL_2(p)$-invariant lattice.* □

All of the statements in Theorem 13.2.2 now follow from Proposition 13.5.6 and Corollary 13.5.12.

Commentary

1) In [Tho 1] the class $\mathcal{T}$ of pairs (Λ, G) satisfying the following assumptions is investigated:

(i) G is a finite group;

(ii) Λ is a torsion-free $\mathbb{Z}G$-module of finite rank;

(iii) $\Lambda/p\Lambda$ is an irreducible $\mathbb{F}_pG$-module for every prime p.

Clearly, in such a pair the Λ is an even unimodular lattice (after a suitable rescaling). Up to that time only two pairs with properties (i) – (iii) were known, namely, $(\Lambda, \mathrm{Aut}(\Lambda))$, where Λ is the root lattice of type E_8, or the Leech lattice (their tensor powers also work). The Thompson-Smith lattice presented in §13.1 gave us a new example. This fact can be regarded as an illustration of properties of the sporadic simple group F_3, all of whose nontrivial irreducible representations over any field have degree not less than 248. It would be interesting to determine the whole of $\mathcal{T}$. However, it took long enough to find a new example! It was not until 1989 that Gow [Gow 3] discovered his new pairs (Λ, G), where G is either a double cover $2^{\pm}\mathbb{S}_n$ of the symmetric group on n symbols with $n = 2m^2 > 8$, or the double cover $2\mathbb{A}_n$ of the alternating group on n symbols, where $n > 8$ is an odd square. Furthermore, Λ is a lattice of dimension $2^{[n/2]-1}$ lying in a basic spin module of G.

Gross [Gro] suggested the notion of *globally irreducible representations*, generalizing the definition of the class $\mathcal{T}$. We shall come back to this question and to the Gow lattices in Chapter 14.

2) It would be very much appreciated if someone could find new ways of looking at the Thompson-Smith lattice, like the Lie approach to the invariant lattice of type G_2 related to $G_2(3)$ suggested by Burichenko and explained in Chapter 8. The same goes for $L_4(5)$-invariant lattices and for possible explanations "in lattice language" of the embedding $L_4(5) \hookrightarrow E_8(4)$.

Apparently, the geometry defined by the internal structure of the group $L_4(5)$ could be the key to constructing an additional automorphism φ such that $<\varphi, \mathrm{Aut}(\mathcal{D})>/\mathbb{Z}_2 = L_4(5)$, where $\mathcal{D}$ is the Borovik decomposition. Such a method was successful in the cases of Co_1 (see §9.7) and F_3 (see §13.1), and also in the construction of the embedding $L_4(5) \hookrightarrow E_8(4)$ given by Cohen, Liebeck, Saxl and Seitz [CLSS].

3) Nearly all examples of rational absolutely irreducible representations (characters) of Chevalley groups known up to now are related to the two following types: the Steinberg module and irreducible constituents of permutation representations. In Section §13.5 we gave an example of rational representations of $SL_4(q)$ of a quite different kind.

Chapter 14

Other lattice constructions

The chapter is devoted to familiarizing readers with several recent lattice constructions. The exposition is of a very conceptual nature, and the proofs usually omitted. For details see the original papers [Bur 3, 4], [Gow 2, 3], [Gro], [Tiep 12, 14, 15].

14.1 A Moufang loop, the Dickson form, and a lattice related to $\Omega_7(3)$

The central object of this section is a particular loop $\mathbb{H}$ of order 3^4, which is used in [Bur 3] to construct

a) a trilinear symmetric form ϑ (the so-called *Dickson form*) on a space $\mathbb{A}$ of dimension 27 with $\mathrm{Aut}(\vartheta)$ isomorphic to the Chevalley group of type E_6;

b) a representation of the central extension $\mathbb{Z}_3 \circ \Omega_7(3)$ over $\mathbb{Z}[\zeta]$, $\zeta = \exp(\frac{2\pi i}{3})$, of dimension 27;

c) the exceptional 27-dimensional Jordan algebra.

As a consequence of this we arrive at the known embeddings

$$L_4(3) \hookrightarrow F_4(2),\ \Omega_7(3) \hookrightarrow {}^2E_6(2)$$

(see [Coo 2], [NoW], [ATLAS]).

A. The loop $\mathbb{H}$ and its automorphism group

$\mathbb{H}$ is the set of ordered quadruples (a, b, c, d), $a, b, c, d \in \mathbb{F}_3$, with the following binary operation:

$$(a, b, c, d).(a', b', c', d') = (a + a', b + b', c + c', d + d' + (c - c')(ab' - a'b)).$$

The first result is readily verified:

Theorem 14.1.1. *Under the binary operation indicated, $\mathbb{H}$ is a commutative Moufang loop, that is, it is a quasigroup with identity element $e = (0, 0, 0, 0)$, and* *inversion operator*

$$x = (a, b, c, d) \mapsto x^{-1} = (-a, -b, -c, -d),$$

and it satisfies the following identities:

$$x(x^{-1}y) = y = (yx^{-1})x, \; xy = yx, \; x.(yz.x) = xy.zx. \qquad \square$$

For any ordered triple (x_1, x_2, x_3), where $x_i = (a_i, b_i, c_i, d_i) \in \mathbb{H}$, $i = 1, 2, 3$, its *associator* is defined to be the element

$$a(x_1, x_2, x_3) = ((x_1x_2).x_3)\,(x_1.(x_2x_3))^{-1},$$

which is equal to $(0, 0, 0, \delta)$, where

$$\delta = \det \begin{pmatrix} a_1 & a_2 & a_3 \\ b_1 & b_2 & b_3 \\ c_1 & c_2 & c_3 \end{pmatrix}.$$

Furthermore, the *associative centre*

$$K = \{w \in \mathbb{H} | \forall x, y \in \mathbb{H}, xw = wx,$$

$$a(w, x, y) = a(x, w, y) = a(x, y, w) = 1\}$$

of $\mathbb{H}$ consists of all elements of the form $(0, 0, 0, d)$, $d \in \mathbb{F}_3$. We shall frequently use the map

$$x = (a, b, c, d) \in \mathbb{H} \mapsto \bar{x} = (a, b, c) \in \bar{\mathbb{H}} = \mathbb{H}/K \cong \mathbb{Z}_3^3,$$

which is a loop epimorphism with kernel K. Conversely, the following statement holds.

Theorem 14.1.2. *Let $\hat{\mathbb{H}}$ be a commutative loop, and $\psi : \hat{\mathbb{H}} \to \mathbb{Z}_3^3$ a loop epimorphism with kernel* $\mathrm{Ker}\psi = \{1, u, u^2\} \cong \mathbb{Z}_3$. *Let D denote the standard determinant defined on $\mathbb{Z}_3^3$. Suppose in addition that the following conditions are satisfied:*

(i) $x^3 = 1$ *for all* $x \in \hat{\mathbb{H}}$;

(ii) $a(x, y, z) = u^\delta$, *where* $\delta = D(\psi(x), \psi(y), \psi(z))$ *for* $x, y, z \in \hat{\mathbb{H}}$;

(iii) $a(u, x, y) = a(x, u, y) = a(x, y, u) = 1$ *and* $xu = ux$ *for all* $x, y \in \hat{\mathbb{H}}$.

Then $\hat{\mathbb{H}} \cong \mathbb{H}$. $\square$

Clearly, every φ in $\mathrm{Aut}(\mathbb{H})$ preserves K, and so induces an automorphism $\bar{\varphi}$ of $\bar{\mathbb{H}}$. Moreover,

$$\varphi(0, 0, 0, 1) = (0, 0, 0, \det \bar{\varphi}).$$

In what follows an essential role is played by the subgroup E of $\mathrm{Aut}(\mathbb{H})$ consisting of the maps

$$n = n_{A,B,C} : (a, b, c, d) \mapsto (a, b, c, Aa + Bb + Cc + d),\ A, B, C \in \mathbb{F}_3.$$

As a consequence of Theorem 14.1.2 we obtain:

Corollary 14.1.3. $\mathrm{Aut}(\mathbb{H})$ *is a split extension of* $E \cong \mathbb{Z}_3^3$ *by* $\mathrm{Aut}(\bar{\mathbb{H}}) \cong GL_3(3)$. *Furthermore, the subgroup*

$$\mathrm{Aut}(\mathbb{H})_0 = \{\varphi \in \mathrm{Aut}(\mathbb{H}) \mid \det \bar{\varphi} = 1\}$$

is the unique extension of E *by* $SL_3(3)$. □

B. The space $\mathbb{A}$, *the Dickson form, and the exceptional Jordan algebra*
We are about to construct forms ϑ and μ on a certain 27-dimensional $\mathrm{Aut}(\mathbb{H})_0$-module $\mathbb{A}$. From now on, let $\mathbb{K}$ be a field of characteristic other than 3 containing a primitive cube root ζ of unity. Furthermore, when we are dealing with the Hermitian form μ and the twisted group ${}^2E_6$ (see below) we shall suppose in addition that $\mathbb{K} = \mathbb{K}_0(\zeta)$ is a quadratic extension of a subfield $\mathbb{K}_0$. Let $\hat{\mathbb{A}} = \, < x | x \in \mathbb{H} >_{\mathbb{K}}$ be the monoid algebra of $\mathbb{H}$ over $\mathbb{K}$, I the ideal of $\hat{\mathbb{A}}$ generated by $(0, 0, 0, 1) - \zeta$, and set

$$\mathbb{A} = \hat{\mathbb{A}}/I.$$

It is clear that $\mathbb{A}$ is a 27-dimensional $\mathrm{Aut}(\mathbb{H})_0$-module. Furthermore, for $a, b, c \in \bar{\mathbb{H}}$ we set

$$\kappa(a, b, c) = \begin{cases} 0 & \text{if} \quad a + b + c \neq 0, \\ 1 & \text{if} \quad a + b + c = 0, a \neq b, \\ -2 & \text{if} \quad a = b = c. \end{cases}$$

We are now able to define the desired trilinear form ϑ on $\mathbb{A}$:

$$\vartheta(x, y, z) = \kappa(\bar{x}, \bar{y}, \bar{z})(xy)z \in \mathbb{K},\ x, y, z \in \mathbb{H}.$$

It is easy to check that ϑ is symmetric and $\mathrm{Aut}(\mathbb{H})_0$-invariant. To construct the desired Hermitian form μ, we denote the automorphism of $\mathbb{K}$ over $\mathbb{K}_0$ taking ζ into ζ^{-1} by τ. Choose a section $\bar{x} \in \bar{\mathbb{H}} \mapsto \tilde{x} \in \mathbb{H}$ of $\bar{\mathbb{H}}$ in $\mathbb{H}$. This means that $\bar{\tilde{x}} = \bar{x}$ for all $\bar{x}$ in $\bar{\mathbb{H}}$. Clearly, $\{\tilde{x} | \bar{x} \in \bar{\mathbb{H}}\}$ forms a basis of $\mathbb{A}$. We define our Hermitian form μ as follows:

$$\mu(\tilde{x}, \tilde{y}) = \delta_{\bar{x},\bar{y}},\ \forall \bar{x}, \bar{y} \in \bar{\mathbb{H}};\ \mu(v, u) = \mu(u, v)^{\tau},\ \forall u, v \in \mathbb{A}.$$

Since

$$\mu((a, b, c, d), (a', b', c', d')) = \delta_{a,a'}\delta_{b,b'}\delta_{c,c'}\zeta^{d-d'},$$

μ is $\mathrm{Aut}(\mathbb{H})_0$-invariant and does not depend on the choice of section $\bar{x} \mapsto \tilde{x}$.

Theorem 14.1.4. *Suppose that $\mathbb{K}$ is an algebraically closed or finite field of characteristic other than 3 containing a primitive cube root ζ of unity. Then ϑ is a Dickson form in the sense that the isometry group $O(\mathbb{A}, \vartheta)$ is isomorphic to the Chevalley group $E_6(\mathbb{K})$. If, in addition, $\mathbb{K} = \mathbb{K}_0(\zeta)$ is a quadratic extension of a finite subfield $\mathbb{K}_0$, then the isometry group $O(\mathbb{A}, \vartheta, \mu)$ is isomorphic to the Chevalley group of type ${}^2E_6$ defined over $\mathbb{K}$.*

Proof. Note that the 27 elements

$$K_i(r, s) = (r, s, 0, 0) + (r, s, 1, i) + (r, s, -1, -i) \text{ with } i, r, s \in \mathbb{F}_3,$$

form a basis of $\mathbb{A}$. Now choose a new basis $\{x_i, x_i', x_{jk} | 1 \leq i \leq 6, 1 \leq j < k \leq 6\}$ of $\mathbb{A}$ by setting

$$x_1 = K_0(0, 0),\ x_2 = K_1(0, 2),\ x_3 = -K_1(1, 0),\ x_4 = -K_1(1, 1),$$

$$x_5 = -K_1(1, 2),\ x_6 = -K_2(0, 1),\ x_1' = K_2(0, 2),\ x_2' = K_0(0, 1),$$

$$x_3' = K_0(1, 2),\ x_4' = K_0(1, 0),\ x_5' = K_0(1, 1),\ x_6' = -K_1(0, 0),$$

$$x_{12} = K_0(0, 2),\ x_{13} = K_0(2, 1),\ x_{14} = K_0(2, 0),\ x_{15} = K_0(2, 2),$$

$$x_{16} = K_2(0, 0),\ x_{23} = K_2(2, 2),\ x_{24} = K_2(2, 1),\ x_{25} = K_2(2, 0),$$

$$x_{26} = -K_1(0, 1),\ x_{34} = K_2(1, 0),\ x_{35} = -K_2(1, 2),\ x_{36} = K_1(2, 0),$$

$$x_{45} = K_2(1, 1),\ x_{46} = K_1(2, 2),\ x_{56} = K_1(2, 1).$$

For convenience we set $x_{ji} = -x_{ij}$ when $1 \leq i < j \leq 6$. Then ϑ takes the value 9 on triples of basic vectors of the form (x_i, x_j', x_{ij}) or $(x_{i_1,i_2}, x_{i_3,i_4}, x_{i_5,i_6})$, whenever the sequence $(i_1, i_2, i_3, i_4, i_5, i_6)$ is an even permutation of the sequence $(1, 2, 3, 4, 5, 6)$. On the remaining triples ϑ is 0. Furthermore,

$$\mu(K_i(r, s), K_{i'}(r', s')) = 3\delta_{i,i'}\delta_{r,r'}\delta_{s,s'}.$$

Our claims were proved by Aschbacher [Asch 4] in the basis $\{x_i, x_i', x_{ij}\}$. □

For $\bar{x}, \bar{y} \in \mathbb{H}$ we set

$$\delta(\bar{x}, \bar{y}) = \begin{cases} 2 & \text{if } \bar{x} = 0 \text{ or } \bar{y} = 0 \text{ or } \bar{x} = -\bar{y}, \\ -2 & \text{if } \bar{x} = \bar{y} \neq 0, \\ 1 & \text{otherwise.} \end{cases}$$

Define a multiplication $\circ$ on $\mathbb{A}$ such that

$$x \circ y = xy\delta(\bar{x}, \bar{y})$$

for all $x, y \in \mathbb{H}$.

Theorem 14.1.5. *There exists just one simple 27-dimensional Jordan algebra* $\mathbb{U}$ *over* $\mathbb{C}$ *such that* $\mathrm{Aut}(\mathbb{U}) \supseteq \mathbb{Z}_3^3.SL_3(3)$. *Namely, if we take* $\mathbb{K} = \mathbb{C}$, *then* $\mathbb{U} \cong (\mathbb{A}, \circ)$. □

C. The group G and its action on $\mathbb{A}$

Our next aim is to construct a finite subgroup G of $O(\mathbb{A}, \vartheta)$ that contains $G_0 = \mathrm{Aut}(\mathbb{H})_0$, whose central factor-group $G/Z(G)$ is isomorphic to the automorphism group $\mathrm{Aut}(\mathcal{D}) = 3^{3+3}.SL_3(3)$ of the unique IOD $\mathcal{D}$ of the Lie algebra of type E_6 (see Chapter 4). Clearly, to each element $n \in E = O_3(G_0)$ there corresponds an element $\bar{n} \in \mathrm{Hom}(\bar{\mathbb{H}}, \mathbb{F}_3)$ such that

$$n(x) = x\zeta^{\bar{n}(\bar{x})}$$

for all $x \in \mathbb{H}$. Denote by T_t the multiplication $x \mapsto tx$, $x \in \mathbb{A}$, and set

$$Z = \langle T_\zeta \rangle \cong \mathbb{Z}_3,\ C = \langle E; T_x | x \in \mathbb{H} \rangle\,,\ G = \langle C, G_0 \rangle\,.$$

For all $x, y \in \mathbb{H}$, $n \in E$, we have $[T_x, n] \in Z$ and $[T_x, T_y] \in E$. In fact, the following statement holds.

Proposition 14.1.6. *The group C is isomorphic to* 3^{1+3+3}, *namely,* $[C, C] = \hat{E} = E \times Z$, $Z(C) = [C, \hat{E}] = Z$, *and* $C/\hat{E} \cong \mathbb{Z}_3^3$. *C acts irreducibly on* $\mathbb{A}$. *Furthermore,* $Z(G) = Z$, *and* $G/C \cong SL_3(3)$. *Finally, under the assumptions of Theorem* 14.1.4, *G preserves both the forms* ϑ *and* μ. □

Clearly, $\bar{G} = G/Z = (C/Z).(G_0/E) \cong 3^{3+3}.SL_3(3)$. Furthermore, $\bar{E} = \hat{E}/Z$ and $\bar{C}/\bar{E} \cong C/\hat{E}$ are irreducible $SL_3(3)$-modules. The isomorphism $\bar{G} \cong \mathrm{Aut}(\mathcal{D})$ now follows from the following simple statement:

Proposition 14.1.7. *There exists just one finite group* $\bar{G}$ *with the following properties:*

(i) $\bar{G} \rhd \bar{C}$, $\bar{G}/\bar{C} \cong SL_3(3)$;

(ii) $\bar{C}/[\bar{C}, \bar{C}]$ *and* $[\bar{C}, \bar{C}]$ *are irreducible* 3*-dimensional modules for* $\bar{G}/\bar{C}$. □

D. The lattice Λ

In this subsection we set $\mathbb{K} = \mathbb{C}$ and $\mathbb{K}_0 = \mathbb{R}$. Clearly, μ is positive definite in this case. The desired lattice Λ is generated (over $\mathbb{Z}$) by the following vectors in $\mathbb{A}$:

$$3x,\ L(x, y, z) = x.(\sum_{\alpha,\beta \in \mathbb{F}_3} y^\alpha z^\beta), \tag{14.1}$$

where $x, y, z \in \mathbb{H}$ and $\bar{y}, \bar{z}$ are linearly independent in $\bar{\mathbb{H}}$.

Clearly, Λ is a free 27-dimensional $\mathbb{Z}[\zeta]$-module in $\mathbb{A}$ invariant under $G = 3^{1+3+3}.SL_3(3)$. We present a further automorphism Φ of this lattice:

$$\Phi(x, y, z, 0) = \begin{cases} (-x, -y, 0, 0) & \text{if } z = 0, \\ \frac{1}{3}(x, y, z, 0)\left(\sum_{\alpha,\beta\in\mathbb{F}_3}(\alpha, \beta, 0, 0)\right) & \text{if } z \neq 0. \end{cases}$$

Theorem 14.1.8. $\Phi \in \mathrm{Aut}(\Lambda)$.

Sketch of Proof. It is readily verified that Φ preserves the Hermitian form μ. Let R denote the subgroup of $\mathbb{H}$ consisting of elements $(a, b, 0, 0)$ with $a, b \in \mathbb{F}_3$. By its definition, Φ centralizes the subgroup

$$P_0 = \{f \in \mathrm{Aut}(\mathbb{H})_0 | f(R) = R\}.$$

Indeed, $\Phi(x) = x^{-1}$ for x in R, and $\Phi(x) = \frac{1}{3}x(\sum_{y\in R} y)$ if x is orthogonal to R. Furthermore, $\Phi^2 = 1$, and $\Phi E \Phi \subseteq P = P_0 E$. The set (14.1) of generating vectors for Λ splits into P-orbits with the following representatives:

$$v_{1,a} = 3\zeta^a = 3(0, 0, 0, a),\ a \in \mathbb{F}_3;$$

$$v_2 = 3(1, 0, 0, 0);\ v_3 = 3(0, 0, 1, 0);$$

$$u_{1,a} = \zeta^a L(1, (1, 0, 0, 0), (0, 1, 0, 0)),\ a \in \mathbb{F}_3;$$

$$u_2 = L((0, 0, 1, 0), (1, 0, 0, 0), (0, 1, 0, 0));$$

$$u_{3,a} = \zeta^a L(1, (1, 0, 0, 0), (0, 0, 1, 0)),\ a \in \mathbb{F}_3;$$

$$u_4 = L((0, 1, 0, 0), (1, 0, 0, 0), (0, 0, 1, 0)).$$

It is obvious that Φ maps the vectors $v_{1,a}$, v_2, and $u_{1,a}$ into vectors of Λ. Furthermore, $\Phi : v_3 \leftrightarrow u_2$, and it is easy to verify that

$$\Phi : u_{3,a} \mapsto u_{3,a},\ u_4 \mapsto L((0, -1, 0, 0), (1, 0, 0, 0), (0, 0, 1, 0)) \in \Lambda.$$

This means that Φ preserves all the representatives chosen above modulo Λ. But Φ normalizes P, and so Φ preserves Λ. □

E. The internal geometry of Λ

Here we deal with algebras $\mathbb{A} = \mathbb{A}_{\mathbb{C}}$ (respectively, $\mathbb{A}_{\mathbb{E}}$) defined over the field $\mathbb{K} = \mathbb{C}$ (respectively, $\mathbb{K} = \mathbb{E} = \mathbb{F}_4$). Correspondingly, $\mathbb{K}_0$ is $\mathbb{R}$ or $\mathbb{F}_2$; and the forms ϑ, μ are denoted by $\vartheta_{\mathbb{C}}$, $\mu_{\mathbb{C}}$, and $\vartheta_{\mathbb{E}}$, $\mu_{\mathbb{E}}$.

It can be verified directly that

$$\vartheta_{\mathbb{C}}(x, y, z) \in \mathbb{Z}[\zeta],\ \mu_{\mathbb{C}}(x, y) \in \mathbb{Z}[\zeta]$$

for all $x, y, z \in \Lambda$, whereas

$$\vartheta_{\mathbb{C}}(3, 3(1,0,0,0), 3(-1,0,0,0)) = 27 \notin 2\mathbb{Z}[\zeta],\ \mu_{\mathbb{C}}(3,3) = 9 \notin 2\mathbb{Z}[\zeta].$$

Hence, the forms $\vartheta_{\mathbb{C}}$ and $\mu_{\mathbb{C}}$ induce nontrivial forms $\bar{\vartheta}$ and $\bar{\mu}$ on the 27-dimensional space $\Lambda/2\Lambda$ over $\mathbb{Z}[\zeta]/2\mathbb{Z}[\zeta] \cong \mathbb{E}$. One can identify $\Lambda/2\Lambda$ with $\mathbb{A}_{\mathbb{E}}$ in such a way that the pair $(\bar{\vartheta}, \bar{\mu})$ is proportional to $(\vartheta_{\mathbb{E}}, \mu_{\mathbb{E}})$.

Note that, unlike G, Φ does not preserve $\vartheta_{\mathbb{C}}$. But Φ preserves $\vartheta_{\mathbb{C}}$ *modulo* 2. Furthermore, Φ preserves $\mu_{\mathbb{C}}$ and a fortiori $\bar{\mu}$. As a consequence, we obtain:

Theorem 14.1.9 [Bur 3]. $< G, \Phi > \subseteq O(\Lambda/2\Lambda, \bar{\vartheta}, \bar{\mu}) \cong 3.{}^2E_6(2)$. □

To get the embedding $\Omega_7(3) \hookrightarrow {}^2E_6(2)$, we have to show that $< G, \Phi >$ is isomorphic to (the non-split central extension) $\mathbb{Z}_3 \circ \Omega_7(3)$. To this end, we introduce two graphs, Γ_1 and Γ_2. The vertex set of Γ_1 is the set of triples $\{v, \zeta v, \zeta^2 v\}$, where v is one of the generating vectors described in (14.1); two distinct vertices $\{v, \zeta v, \zeta^2 v\}$, $\{v', \zeta v', \zeta^2 v'\}$ are adjacent if and only if $\mu(v, v') \neq 0$. Observe that in fact the vectors (14.1) exhaust all the minimal vectors of Λ, hence, Γ_1 is invariant under $\mathrm{Aut}(\Lambda)$. Furthermore, consider a space $W = \mathbb{F}_3^7$ endowed with the standard bilinear form B. The vertex set of Γ_2 is chosen to be the set of lines $< w >_{\mathbb{F}_3}$, where $w \in W$ and $B(w, w) = 1$. Two distinct vertices $< w >_{\mathbb{F}_3}$, $< w' >_{\mathbb{F}_3}$ are adjacent if and only if $B(w, w') = 0$. Clearly, the factor-group $\bar{G} = G/Z \cong 3^{3+3}.SL_3(3)$ acts on Γ_1. We can identify $\bar{G}$ with the group of transformations $f \in SO(W, B) \cong SO_7(3)$ that preserve a fixed maximal totally singular subspace U of W and has determinant 1 on U. Hence, $\bar{G}$ also acts on Γ_2. We list some properties of the action of $\bar{G}$ on $\Gamma = \Gamma_i$, $i = 1, 2$:

(a) $\bar{G}$ has just two orbits $\mathcal{O}_1, \mathcal{O}_2$ on the vertex set of Γ. Furthermore, $|\mathcal{O}_1| = 27$, $|\mathcal{O}_2| = 351$;

(b) the stabilizer $St_{\bar{G}}(v)$ is isomorphic to $\mathbb{Z}_3^3.SL_3(3)$ if $v \in \mathcal{O}_1$, and $3^{1+2}.\mathbb{Z}_3^2.SL_2(3)$ if $v \in \mathcal{O}_2$;

(c) no two distinct vertices v, v' in $\mathcal{O}_1$ are adjacent;

(d) the automorphism group $\mathrm{Aut}(\Gamma)$ acts on the vertex set of Γ as a transitive permutation group with subdegrees 1, 117, and 260.

Our next theorem shows that the graphs Γ_1 and Γ_2 are isomorphic.

Theorem 14.1.10. *Suppose that $\bar{G}$ acts on a graph Γ with action having properties* (a) – (d). *Then Γ is uniquely determined, and* $\mathrm{Aut}(\Gamma) = SO_7(3)$. □

The main result of the section is the following.

Theorem 14.1.11 [Bur 3]. *Let* $\mathbf{t}$ *denote the central involution* $-1_{\mathbb{A}}$. *Then* $\mathrm{Aut}(\Lambda) = < G, \Phi > \times < \mathbf{t} >$, *and* $< G, \Phi >$ *is a non-split central extension of* $Z = \mathbb{Z}_3$ *by* $\Omega_7(3)$.

Proof. Consider the natural homomorphism

$$\Theta : \mathrm{Aut}(\Lambda) \to \mathrm{Aut}(\Gamma_1) \cong SO_7(3).$$

One can show that $\mathrm{Ker}\ \Theta = Z \times <\mathbf{t}> \cong \mathbb{Z}_6$. Furthermore,

$$3^{3+3}.SL_3(3) = \bar{G} \hookrightarrow \mathrm{Im}\ \Theta \subseteq SO_7(3),$$

hence $\mathrm{Im}\ \Theta$ is one of

$$3^{3+3}.SL_3(3),\ 3^{3+3}.GL_3(3),\ \Omega_7(3),\ SO_7(3).$$

In the proof of Theorem 14.1.8, we mentioned that Φ does not normalize the subgroup $\hat{E} = E \times Z$. In other words, $\mathbb{Z}_3^3 = (E.\mathrm{Ker}\ \Theta)/\mathrm{Ker}\ \Theta$ cannot be a normal subgroup of $\mathrm{Im}\ \Theta$. Moreover, if $\mathrm{Im}\ \Theta \supseteq 3^{3+3}.GL_3(3)$, there exists a subgroup $H \cong \mathbb{Z}_3^3.GL_3(3)$ of $\mathrm{Aut}(\Lambda)$ such that $H \supset G_0 = \mathrm{Aut}(\mathbb{H})_0 \cong \mathbb{Z}_3^3.SL_3(3)$. On the other hand, the faithful representation of G_0 on $\mathbb{A}$ cannot be extended to $H = \mathbb{Z}_3^3.GL_3(3)$, which is a contradiction. Thus we have $\mathrm{Im}\ \Theta = \Omega_7(3)$. As $\bar{G} = 3^{3+3}.SL_3(3)$ is a maximal subgroup of $\Omega_7(3)$, we conclude that $\mathrm{Aut}(\Lambda) =< G, \Phi, \mathbf{t} >$. Furthermore, all the automorphisms contained in $< G, \Phi >$ have determinant 1 on $\mathbb{A}$, and $\det \mathbf{t} = -1$. Hence, $\mathrm{Aut}(\Lambda) =< G, \Phi > \times < \mathbf{t} >$, and $< G, \Phi >= Z.\Omega_7(3)$. The extension is non-split, because $\Omega_7(3)$ has no faithful complex representations of degree 27. By the way, note that $\mathbb{Z}_6.SO_7(3)$ acts on Λ as the group of semilinear automorphisms. □

F. Remarks on $L_4(3)$

Note that $G_0 = \mathrm{Aut}(\mathbb{H})_0$ fixes the bilinear form η defined on $\mathbb{A}$ such that

$$\eta(x, y) = \begin{cases} 0 & \text{if } \bar{x} + \bar{y} \neq 0 \\ -xy & \text{if } \bar{x} + \bar{y} = 0 \end{cases}$$

for all $x, y \in \mathbb{H}$. Consider a section

$$\bar{x} \in \bar{\mathbb{H}} \mapsto \tilde{x} \in \mathbb{H}$$

such that $\tilde{y} = \tilde{x}^{-1}$ whenever $\bar{y} = -\bar{x}$. Set

$$\mathbb{A}_0 = \left\{ \sum_{\bar{x} \in \bar{\mathbb{H}}} a_{\bar{x}} \tilde{x} | a_{\bar{x}} \in \mathbb{C},\ a_0 = 0,\ a_{-\bar{x}} = \overline{a_{\bar{x}}} \right\},$$

and $\Lambda_0 = \Lambda \cap \mathbb{A}_0$. Note that the restriction of η to the 26-dimensional real space $\mathbb{A}_0$ is positive definite, and so Λ_0 is a G_0-invariant Euclidean lattice. Furthermore, we have $\vartheta_{\mathbb{C}}(x, y, z) \in \mathbb{Z}$ and $\eta(x, y) \in \mathbb{Z}$ for $x, y, z \in \Lambda_0$. Hence, $\vartheta_{\mathbb{C}}$ and η induce some $\mathbb{F}_2$-valued forms $\bar{\vartheta}$ and $\bar{\eta}$ on the 26-dimensional $\mathbb{F}_2$-space $\Lambda_0/2\Lambda_0$. Based on the results of [Asch 4] and using the arguments of subsection E, one can prove the following results.

Theorem 14.1.12 [Bur 3]. *The lattice* Λ_0 *is invariant under* $G_0 = \mathrm{Aut}(\mathbb{H})_0 \cong \mathbb{Z}_3^3.SL_3(3)$, *and under* Φ. *Furthermore,* $\mathrm{Aut}(\Lambda_0) = \mathbb{Z}_2 \times < G_0, \Phi >$. *Finally,* $< G_0, \Phi > = L_4(3).\mathbb{Z}_2$ *is a subgroup of* $\mathrm{Aut}(L_4(3))$ *that is different from* $PGL_4(3)$ (it is denoted in [ATLAS] by $L_4(3).2_2$). □

Observe that the existence of Λ_0 is predicted by Theorem 13.2.2.

Theorem 14.1.13 [Bur 3]. *The group* $< G_0, \Phi >$ *preserves both the forms* $\bar{\vartheta}$, $\bar{\eta}$. *Furthermore,* $O(\Lambda_0/2\Lambda_0, \bar{\vartheta}, \bar{\eta})$ *is isomorphic to the Chevalley group* $F_4(2)$. □

Corollary 14.1.14 [Coo 2], [NoW]. $L_4(3) \hookrightarrow F_4(2)$. □

14.2 The Steinberg module for $SL_2(q)$ and related lattices

The aim of this section is to study invariant lattices lying in the Steinberg module of the group $G = L_2(q)$. Here $q = p^m$ is a power of a prime number p. If $\mathcal{P} = \mathbb{F}_q \cup \{\infty\}$ denotes the projective line over $\mathbb{F}_q$, then G acts on $\mathcal{P}$ as usual:

$$\begin{pmatrix} a & b \\ c & d \end{pmatrix} z = \frac{az+b}{cz+d}.$$

Consider a space $V_1 = \oplus_{z\in\mathcal{P}}\mathbb{C}e_z$, on which G acts via permutations of the basis vectors e_z. Then the subspace

$$V = \left\{\sum \lambda_z e_z | \sum \lambda_z = 0\right\},$$

which is an irreducible G-submodule, is said to be the *Steinberg module* of G. V affords the unique irreducible complex G-character of degree q. Throughout this section $\sum$ denotes summations over z that runs over $\mathcal{P}$. One can endow V with the standard inner product

$$(\sum \lambda_z e_z | \sum \mu_z e_z) = \sum \lambda_z \mu_z.$$

Set

$$\Lambda_0 = \left\{\sum \lambda_z e_z | \lambda_z \in \mathbb{Z}, \sum \lambda_z = 0\right\}.$$

It is easy to see that every G-invariant lattice Λ in V is similar to some sublattice intermediate between Λ_0 and $(q+1)\Lambda_0$. Clearly, the lattices

$$\Lambda_{q+1,r} = \left\{\sum \lambda_z e_z \in \Lambda_0 | \exists \lambda \in \mathbb{Z}, \forall z \in \mathcal{P}, \lambda_z \equiv \lambda \pmod r\right\},$$

where $r \in \mathbb{N}$, $r|(q+1)$, are examples of such sublattices.

To define other series of invariant lattices we consider the natural permutation $\mathbb{F}_2G$-module $U_1 = \oplus \mathbb{F}_2 u_z$. Of course, U_1 has the G-submodules

$$\ell = < \sum u_z >_{\mathbb{F}_2},\ U = \left\{ \sum c_z u_z | c_z \in \mathbb{F}_2, \sum c_z = 0 \right\}.$$

Moreover, when $q \not\equiv \pm 1 \pmod 8$ the submodules 0, ℓ, U, and U_1 exhaust all the G-submodules of U_1. If $q \equiv \pm 1 \pmod 8$, then just two additional G-submodules $\mathcal{C}_j$, $j = 1, 2$, arise known as *generalized quadratic residue binary codes* of length $q+1$ (see e.g. [McS]). We give a description of these codes for the case $q = p$ following [Shau]. To this end, we identify each element

$$(c_0, \ldots, c_{q-1}, c_\infty) = \sum c_z u_z \in U$$

with the element $c_0 + c_1 x + \ldots + c_{q-1} x^{q-1}$ of $\mathbb{F}_2[x]/(x^q - 1)$. Furthermore, we decompose $x^q - 1$ as a product

$$x^q - 1 = (x-1) \prod_{a \in R^+} (x - \zeta^a) \prod_{a \in R^-} (x - \zeta^a) = (x-1) h^+(x) h^-(x),$$

where ζ is a primitive pth root of unity over $\mathbb{F}_2$, and R^+ (respectively, R^-) denotes the set of quadratic residues (respectively, non-residues) modulo p. Then $\mathcal{C}_1$ and $\mathcal{C}_2$ can be identified with the ideals of $\mathbb{F}_2[x]/(x^q - 1)$ generated by $h^+(x)$ and $h^-(x)$, respectively. If we endow U_1 with the natural G-invariant symmetric form $(\sum c_z u_z; \sum d_z u_z) = \sum c_z d_z$, then the code $\mathcal{C}^*$ *dual* to a code $\mathcal{C} \subseteq U_1$ is defined to be

$$\mathcal{C}^* = \{c \in U_1 | (c; \mathcal{C}) = 0\}.$$

If $q \equiv -1 \pmod 8$, the codes $\mathcal{C}_1$ and $\mathcal{C}_2$ are self-dual; if $q \equiv 1 \pmod 8$ they are dual each to other. For $q \equiv \pm 1 \pmod 8$, $2|r$, $r|(q+1)$, $j = 1, 2$, we define $\Lambda_{q+1,r,j}$ to be the set of elements $\sum \lambda_z e_z \in \Lambda_{q+1,r/2}$ such that

$$\exists C \in \mathcal{C}_j,\ \lambda_z \equiv \lambda_\infty \pmod r \Leftrightarrow z \in C.$$

The first main result is:

Theorem 14.2.1 [Ple 1], [Bur 4]. *The lattices $\Lambda_{q+1,r}$ and $\Lambda_{q+1,r,j}$ introduced above form a complete system of representatives of the similarity classes of G-invariant lattices in the Steinberg module V.*

Sketch of Proof. The study of G-invariant lattices in V reduces to the study of the submodule structure of the G-module $\Lambda_0 / p_i^{\alpha_i} \Lambda_0$, where

$$q + 1 = p_1^{\alpha_i} \ldots p_k^{\alpha_k}$$

is the decomposition of $q+1$ into prime powers. Setting $\mathbb{Z}_m = \mathbb{Z}/m\mathbb{Z}$ for $m \in \mathbb{N}$, one can consider the natural permutation $\mathbb{Z}_m G$-module $V_m^1 = \oplus \mathbb{Z}_m u_z$. Then $\Lambda_0/m\Lambda_0$ is isomorphic to

$$V_m = \left\{ \sum a_z u_z \in V_m^1 | \sum a_z = 0 \right\}$$

as $\mathbb{Z}_m G$-module. If m divides $q+1$, then V_m contains the trivial submodule $L_m =< \sum u_z >_{\mathbb{Z}_m}$, and we set $\bar{V}_m = V_m/L_m$. Further analysis is based on the following statement:

Lemma 14.2.2. *Let r^α be a nontrivial prime-power dividing $q+1$. Then*
(i) *L_r is the unique minimal submodule of V_r;*
(ii) *$\bar{V}_{r^\alpha}$ is decomposable only when $r^\alpha = 2$, $q \equiv \pm 1 \pmod 8$; moreover, in this case we have $\bar{V}_{r^\alpha} = (\mathcal{C}_1/L_r) \oplus (\mathcal{C}_2/L_r)$;*
(iii) *if $r \neq 2$, then $\bar{V}_r$ is irreducible.* □

Suppose now that Λ is a G-invariant lattice in V, and $\Lambda_0 \supseteq \Lambda \supseteq (q+1)\Lambda_0$. One can suppose also that $\Lambda \not\subseteq s\Lambda_0$ for all $s > 1$. Let m be the least natural number such that $\Lambda \supseteq m\Lambda_0$. Clearly, $m|(q+1)$. If $m = 1$, then $\Lambda = \Lambda_0 = \Lambda_{q+1,1}$. Suppose that $m > 1$. Note that $\Lambda_{q+1,m}/m\Lambda_0 = L_m$. Furthermore, when $q \equiv \pm 1 \pmod 8$ and $2|m$ we denote by $Q'_{m,j}$ the submodule of $V_m = \Lambda_0/m\Lambda_0$ that corresponds to $\mathcal{C}_j$ under the identification $(m/2)V_m \rightsquigarrow V_2$, and set $Q_{m,j} = Q'_{m,j} + L_m$. Then $\Lambda_{q+1,m,j}/m\Lambda_0 = Q_{m,j}$. It is shown in [Bur 4] that the only nontrivial submodules of V_m are L_m and $Q_{m,j}$. This completes the proof of Theorem 14.2.1. □

It is not difficult to see that

$$(\Lambda_{q+1,r})^* = \frac{1}{q+1}\Lambda_{q+1,(q+1)/r}.$$

In particular, if $q+1$ is a square greater than 8, then the lattice

$$\frac{1}{\sqrt{q+1}}\Lambda_{q+1,\sqrt{q+1}}$$

is unimodular, with root system of type E_8 (if $q = 8$), and A_q (if $q > 8$). Furthermore, if we set $j^* = j$ for $q \equiv -1 \pmod 8$, and $j^* = 3 - j$ for $q \equiv 1 \pmod 8$, then

$$(\Lambda_{q+1,r,j})^* = \frac{1}{q+1}\Lambda_{q+1,2(q+1)/r,j^*}.$$

In particular, the following statement holds.

Corollary 14.2.3. *If $\frac{1}{8}(q+1)$ is an integer square, then for $j = 1, 2$, the lattices*

$$\frac{1}{\sqrt{q+1}}\Lambda_{q+1,\sqrt{2(q+1)},j}$$

are odd unimodular lattices with minimum norm 1 *if* $q = 7$, *and* 4 *if* $q > 7$. □

The second main result is:

Theorem 14.2.4 [Bur 4]. *If* $(q, r) \neq (7, 2), (7, 4), (8, 3)$, *then*

$$\mathrm{Aut}(\Lambda_{q+1,r}) = \mathbb{Z}_2 \times \mathbb{S}_{q+1}.$$

If $q \equiv \pm 1 \pmod 8$ *and* $q \neq 7$, *then*

$$\mathrm{Aut}(\Lambda_{q+1,r,j}) = \mathbb{Z}_2 \times \mathrm{Aut}(\mathcal{C}_j).$$

In the exceptional cases listed above, the lattices $\Lambda_{8,2}$, $\Lambda_{8,4}$, $\Lambda_{9,3}$, $\Lambda_{8,2,j}$, $\Lambda_{8,4,j}$, and $\Lambda_{8,8,j}$ are isometric to the root lattices of type E_7, E_7^*, E_8, D_7, $7A_1$, and D_7 respectively. Furthermore, modulo the classification of finite simple groups we have (see e.g. [Shau])

$$\mathrm{Aut}(\mathcal{C}_j) = \begin{cases} ASL_3(2) = \mathbb{Z}_2^3.SL_3(2) & \text{if } q = 7, \\ M_{24} & \text{if } q = 23, \\ P\Sigma L_2(q) = L_2(q).Gal(\mathbb{F}_q/\mathbb{F}_p) & \text{otherwise.} \end{cases}$$

We come now to the proof of Theorem 14.2.4. The statement is trivial for the lattices $\Lambda_{q+1,r}$. For the case $\Lambda = \Lambda_{q+1,r,j}$ it is enough to show that $\mathbb{G} = \mathrm{Aut}(\Lambda)$ is contained in $\mathrm{Aut}(\Lambda_0) = \mathbb{Z}_2 \times \mathbb{S}_{q+1}$ (if $q > 7$). Assuming the contrary, we consider a lattice $\Lambda = \Lambda_{q+1,r,j}$ with the least r (for given q) such that $\mathbb{G} \not\subseteq \mathrm{Aut}(\Lambda_0)$.

First, we claim that the only $\mathbb{G}$-invariant lattice Λ_1 with $r\Lambda_0 \subseteq \Lambda_1 \subseteq \Lambda_{q+1,r,j}$ is Λ itself. Indeed, if $\Lambda_1 \neq \Lambda$, then Λ_1 is either $\frac{r}{r_1}\Lambda_{q+1,r_1,j'}$ for some $r_1 > 1$ and some j', or $\frac{r}{r_1}\Lambda_{q+1,r_1}$ for some r_1. In the former case, $\mathrm{Aut}(\Lambda_1) \subseteq \mathrm{Aut}(\Lambda_0)$ by the minimality of r, and so $\mathbb{G}$ is contained in $\mathrm{Aut}(\Lambda_0)$, a contradiction. In the latter case, $\mathrm{Aut}(\Lambda_1)$ is contained in $\mathrm{Aut}(\Lambda_0)$ since $q > 8$, again a contradiction.

Now set $s = r/2$ and consider the $\mathbb{G}$-invariant sublattice

$$\Lambda_2 = \left\langle x \in \Lambda \,|\, (x|x) = 8s^2,\ (x|\Lambda) \subseteq 2s^2\mathbb{Z} \right\rangle_{\mathbb{Z}}$$

of Λ. Clearly, $\Lambda_2 \supseteq r\Lambda_0$. As we have just shown, in this case $\Lambda_2 = \Lambda$. In particular, the norm $(x|x)$ of every vector x in Λ is divisible by r^2. But $x_0 = (1, \ldots, 1, -q) \in \Lambda$, and so $q + 1$ is divisible by r^2. In particular, $8|(q+1)$.

Next, assume that $s > 1$. Since $\Lambda = \Lambda_2$ is generated by the vectors of norm $8s^2$ and $\Lambda \not\subseteq s\Lambda_0$, there exists a vector $x_1 \in \Lambda \backslash s\Lambda_0$ with $(x_1|x_1) = 8s^2$. Clearly, all the coefficients ν_z of $x_1 = \sum \nu_z e_z$ are nonzero, and so $q + 1 \leq 8s^2$. This implies that $q + 1 = 4s^2$ or $q + 1 = 8s^2$. In the former case, we have $q = (2s-1)(2s+1)$ is a prime power, which is impossible if $s > 1$. This means that $q + 1 = 8s^2$. In this case all the coefficients ν_z of x_1 are equal to ± 1, and both signs occur (because $\sum \nu_z = 0$). By the definition of $\Lambda_{q+1,r,j}$, we must have $s = 2$, $q = 31$, $\Lambda = \Lambda_{32,4,j}$.

When $s = 1$, the minimum weight d of codewords of $\mathcal{C}_j$ does not exceed $8s^2 = 8$, of course. This can occur only when $q \leq 31$. Hence, $\Lambda = \Lambda_{24,2,j}$ or $\Lambda_{32,2,j}$.

Finally, the exceptional cases $\Lambda = \Lambda_{24,2,j}$, $\Lambda_{32,2,j}$, and $\Lambda_{32,4,j}$ listed above have been excluded individually (with the use of a computer!). □

14.3 The Weil representations of finite symplectic groups and the Gow-Gross lattices

Let p be an odd prime, and set $S = Sp_{2n}(p)$ for the symplectic group over $\mathbb{F}_p$. In [Gow 2], Gow considered Euclidean integral lattices in the space of the Weil representation of S. More precisely, the Weil representation $\mathcal{W}$ of S is a complex representation of degree p^n that can be obtained from the action of S on an extraspecial group p_+^{1+2n} (as outer automorphism group). See, for example, [Isa 1], [Sei 2], or [War] for a more general approach. $\mathcal{W}$ is a sum of two irreducible representations that have degrees $(p^n - 1)/2$ and $(p^n + 1)/2$. One of these representations, which we shall denote by W_1, is faithful and has even degree, and the kernel of the other representation, W_2, is just the centre $Z(S) \cong \mathbb{Z}_2$ of S. Following [Gow 2], we shall refer to W_1 and W_2 as *Weil representations*.

Suppose now that $p \equiv 3 \pmod 4$. It is shown in [Gow 2] that the characters ψ_i of the W_i, $1 \leq i \leq 2$, each generate the field $\mathbb{Q}(\sqrt{-p})$ over the rational field $\mathbb{Q}$, and have Schur index 1 over $\mathbb{Q}$. Hence, there exists an absolutely irreducible $\mathbb{Q}G$-module V that affords the S-character $\psi + \bar{\psi}$, where $\psi = \psi_1$ or ψ_2, and $G = S.\mathbb{Z}_2 = \mathbb{Z}_2.\mathrm{Aut}(S)$. The goal of this section is to describe the automorphism group $\mathbb{G} = \mathrm{Aut}(\Lambda)$ of any G-invariant (Euclidean integral) lattice lying in V. The main result of [Gow 2] is that when n is even and $\psi = \psi_1$, every such lattice (after suitable renorming the inner product) is even and unimodular.

Later, the same lattices are considered by Gross [Gro] in the context of the so-called *globally irreducible representations* (for the precise definition see §14.5). Thus, when $p \equiv 3 \pmod 4$ and $\psi = \psi_1$, the representation V is globally irreducible. It is shown also that, for p odd, there are two globally irreducible representations V of $Sp_{2n}(p^2)$ of dimension $p^{2n} - 1$ over $\mathbb{Q}$, which are related to the Weil representations and lead to even unimodular Euclidean lattices (of dimension $p^{2n} - 1$). A model for these lattices when $S = SL_2(p^2)$ is due to Elkies (see [Gro]). As to other lattices, only their existence is proved. We have neither a good model nor an estimate for minimal vectors for them. On the other hand, it is not difficult to carry over the methods of this section to these lattices and obtain a description of their automorphism groups.

We shall refer to integral lattices related to the Weil representations of finite symplectic groups as the *Gow-Gross lattices*. From now on we shall suppose that $S = Sp_{2n}(p)$ and $p \equiv 3 \pmod 4$.

In the simplest case $p^n = 9$, $\psi = \psi_1$, the lattices Λ are isometric to the root lattice of type E_8. Furthermore, the representation W_1 realizes the group $\mathbb{Z}_3 \times Sp_4(3)$ as a group of 4-dimensional complex reflections [ATLAS]. More concretely, suppose that $U = \mathbb{C}^4$ is endowed with the standard Hermitian form $u \circ v = \sum_i u_i \bar{v}_i$. Set $\zeta = \exp(\frac{2\pi i}{3})$, $\theta = \zeta - \bar{\zeta}$, and consider the set Ω consisting of the 240 vectors of norm 3 of the form

$$\pm\zeta^i(\theta, 0, 0, 0),\ \pm\zeta^i(0, \theta, 0, 0),\ \pm(0, \zeta^i, \zeta^j, \zeta^k),\ \pm(\zeta^i, 0, \zeta^j, -\zeta^k)$$

(the last three coordinates can be permuted cyclically). Each complex reflection of order 3 of U can be written in the form

$$r_a : u \mapsto u - \frac{(1-\zeta)(u \circ a)}{a \circ a} a$$

for some $a \in U \backslash \{0\}$. The set $\{r_a | a \in \Omega\}$ of 40 distinct reflections generates a group $\mathbb{Z}_3 \times Sp_4(3)$ that acts irreducibly on U. We shall suppose that W_1 is obtained by restricting this action to S. Moreover, $\mathbb{Z}_3 \times S$ preserves the Hermitian lattice $\Gamma =< \Omega >_{\mathbb{Z}}$. One can change Γ into an integral lattice $\Lambda_1 = \Gamma_{\mathbb{Z}}$ by setting

$$(u|v) = \mathrm{Tr}_{\mathbb{Q}(\zeta)/\mathbb{Q}}(u \circ v).$$

Of course, this procedure can be applied to any Hermitian lattice (over $\mathbb{Z}[\zeta]$). In particular, we set $\Lambda_2 = (\theta\Gamma)_{\mathbb{Z}}$. Our first result is:

Theorem 14.3.1 [Tiep 12]. *Up to similarity, there exists just*

(i) *one* $Sp_4(3)$*-invariant* $\mathbb{Z}[\zeta]$*-lattice in* $\mathbb{C}^4$*, namely,* Γ*, and* $\mathrm{Aut}(\Gamma) = \mathbb{Z}_3 \times Sp_4(3)$*;*

(ii) *two* $GSp_4(3)$*-invariant integral lattices in* $\mathbb{C}^8$*, namely,* Λ_1 *and* Λ_2*, and* $\mathrm{Aut}(\Lambda_i) = W(E_8) = \mathbb{Z}_2.O_8^+(2)$.

(Recall that $GSp_4(3) \cong Sp_4(3).\mathbb{Z}_2$).

Proof. (i) Suppose that Γ' is a Hermitian S-invariant lattice in U. Then Γ' is fixed by every reflection r_a, $a \in \Omega$. In particular,

$$\frac{1}{3}(1-\zeta)(u \circ a)a \in \Gamma'$$

for all $u \in \Gamma'$, $a \in \Omega$. From this it follows that Γ' is similar to Γ.

(ii) We describe the action of $G = GSp_4(3)$ on $W = V \otimes_{\mathbb{Q}} \mathbb{C}$. To this end, it is sufficient to identify W with

$$(U_{\mathbb{R}}) \otimes_{\mathbb{R}} \mathbb{C} =< e_i, f_i | 1 \le i \le 4 >_{\mathbb{C}},$$

where

$$U = < e_i = (0, \dots, \underbrace{1}_{i}, \dots, 0) | 1 \leq i \leq 4 >_{\mathbb{C}},$$

and $f_i = \zeta e_i$. If an element $s \in S$ has matrix $A + \zeta B$ in the basis $\{e_i\}$, where $A, B \in M_4(\mathbb{R})$, then $\tilde{s} = (s_{\mathbb{R}})^{\mathbb{C}}$ has matrix $\begin{pmatrix} A & -B \\ B & A - B \end{pmatrix}$ in the basis $\{e_i, f_i\}$. Furthermore, denote by τ the $\mathbb{R}$-linear operator of complex conjugation on U (in the basis $\{e_i\}$), and set $\tilde{\tau} = (\tau_{\mathbb{R}})^{\mathbb{C}}$. Then the matrices $\tilde{s}$, $s \in S$, and $\tilde{\tau}$ define an 8-dimensional irreducible representation of $G = S. < \tau >$ on W.

Suppose now that Λ is a G-invariant lattice in W. Clearly,

$$S = \{f \in \mathbb{Z}_3 \times S | \det f|_U = 1\}.$$

In particular, $\bar{\zeta} r_a \in S$ for all $a \in \Omega$. Set

$$s_a = \left((\bar{\zeta} r_a)_{\mathbb{R}}\right)^{\mathbb{C}},$$

$$\Lambda_a = \left\{v + s_a(v) + s_a^2(v) | v \in \Lambda\right\},$$

$$\overline{\Lambda} = \sum_{a \in \Omega} \Lambda_a, \ \underline{\Lambda} = \sum_{i=1}^{4} \Lambda_{\theta e_i}.$$

One can show that $\overline{\Lambda}$ is similar to Λ_1 or Λ_2. Furthermore,

$$\frac{1}{3}\underline{\Lambda} \supseteq \Lambda \supseteq \overline{\Lambda} \supseteq \underline{\Lambda}, \ [\overline{\Lambda} : \underline{\Lambda}] = 3^2.$$

Hence, Λ is also similar to Λ_1 or Λ_2. □

Our main result is:

Theorem 14.3.2 [Tiep 12]. *Let p be a prime such that $p \equiv 3 \pmod 4$, suppose that $(n, p) \neq (1, 3)$, and let Λ be an integral lattice in V which is invariant under G if $n \geq 2$, and under S if $n = 1$. Set $\mathbb{G} = \mathrm{Aut}(\Lambda)$, $N = \dim \Lambda$. Then $\mathbb{G} = \mathbb{Z}_2.\mathrm{Aut}(PSp_{2n}(p))$, with the following possible exceptions:*

(i) *$p^n = 9$, $N = 8$, and $\mathbb{G} = W(E_8)$;*

(ii) *$p = 3$, $N \geq 10$, and $\mathbb{G} = (\mathbb{Z}_3 \times \mathbb{Z}_2.PSp_{2n}(3)).\mathbb{Z}_2$;*

(iii) *$n = 1$, $N = p + 1$, and either $(p, \mathbb{G}) = (7, W(E_8))$, or $(p, \mathbb{G}) = (23, \mathbb{Z}_2 \circ Co_1)$;*

(iv) *$n = 1$, $N = p + 1$, and $\mathrm{soc}(\mathbb{G}/\mathrm{abel}(\mathbb{G}))$ belongs to the list*

$$\left\{\mathbb{A}_{p+1}, M_{24}\,(p = 23), L_2(p)\,(p \equiv -1 \pmod 8), p > 23)\right\};$$

(v) *$n = 1$, $N = p + 1$, and $\mathrm{soc}(\mathbb{G}/\mathrm{sol}(\mathbb{G}))$ belongs to the list*

$$\{\mathbb{A}_7\,(p = 7), M_{12}\,(p = 11), L_2(p)\,(p \equiv 3 \pmod 8) \text{ or } p = 7, 23)\}.$$

Moreover, the cases (i) – (iv) *actually occur.*

Recall that abel(X) denotes a maximal abelian normal subgroup of X (if it exists). We exclude the case $p^n = 3$ because all 4-dimensional lattices are well known (see [BBNWZ]).

The proof of Theorem 14.3.2 splits into several steps. Set $W = \Lambda \otimes_{\mathbb{Z}} \mathbb{C}$, $Z =< \mathbf{t} : v \mapsto -v > \cong \mathbb{Z}_2$. Clearly, W is a sum of two irreducible S-submodules U and $\bar{U}$ of dimension $N/2$. Let χ be the $\mathbb{G}$-character afforded by Λ. The following two statements are obvious.

Lemma 14.3.3. *One of the following assertions holds.*

(i) $C_{\mathbb{G}}(S) = Z$.

(ii) $C_{\mathbb{G}}(S) = Z \times \mathbb{Z}_3$, $p = 3$. □

Note that possibility (ii) is related to the fact that when $p = 3$ the representations W_i, $i = 1, 2$, can be realized over $\mathbb{Q}(\sqrt{-3})$.

Lemma 14.3.4. *One of the following assertions holds.*

(i) $\mathbb{G}$ *is primitive on* W.

(ii) $\{U, \bar{U}\}$ *is the unique imprimitivity system for* $\mathbb{G}$ *on* W.

(iii) $S = SL_2(p)$, $N = p+1$, $\mathbb{G}$ *has an imprimitivity system* $\{U_1, \ldots, U_{p+1}\}$ *(and then* $\mathbb{G} = A.H$, *where* A *is an abelian normal subgroup,* H *is a faithful doubly transitive permutation group of degree* $p+1$, *and* $L_2(p) \subseteq H \subseteq \mathbb{S}_{p+1}$*).* □

Proposition 14.3.5. *Theorem* 14.3.2 *holds if assertion* 14.3.4(iii) *holds.*

Proof. Suppose that assertion 14.3.4(iii) holds. It is not difficult to see that one of the following cases must occur:

a) $\mathbb{A}_{p+1} \subseteq H \subseteq \mathbb{S}_{p+1}$;

b) $L_2(p) \subseteq H \subseteq PGL_2(p)$;

c) $p = 7$, $H = ASL_3(2)$;

d) $p = 11$, $H = M_{12}$;

e) $p = 23$, $H = M_{24}$.

It is also not difficult to exhibit a unimodular S-invariant lattice Λ with $\mathbb{G} = \mathbb{Z}_2^{p+1}.\mathbb{S}_{p+1}$. To realize the remaining possibilities one can use extended quadratic residue binary codes $\mathcal{C}$ of length $p+1$ (see [Shau]). □

Lemma 14.3.6. *Suppose that* $\mathbb{G}$ *is reducible. Then* $\mathbb{G} = \mathbb{Z}_2.L_2(p)$, *or else* $(p, N, \mathbb{G}) = (7, 8, \mathbb{Z}_2.\mathbb{A}_7)$.

Proof. It is clear that $\chi = \rho + \bar{\rho}$, where $\rho \in \mathrm{Irr}(\mathbb{G})$, $\rho \neq \bar{\rho}$, $S = SL_2(p)$, and $\rho|_S = \psi$. In particular, $\mathrm{Ker}\rho = \mathrm{Ker}\chi = 1$. This means that $\mathbb{G}$ has a faithful irreducible complex character of degree $(p \pm 1)/2$. Furthermore, a Sylow p-subgroup of $\mathbb{G}$

has order p and is not normal in $\mathbb{G}$. Finally, $p \geq 7$ and $Z(\mathbb{G}) = Z$ in accordance with Lemma 14.3.3. It remains to apply the result of [Fer]. □

From now on, we shall assume that $\mathbb{G}$ is absolutely irreducible on V, and that one of the assertions 14.3.4(i), 14.3.4(ii) holds. Using the same arguments as in §12.4, one can show that $Z \subseteq \mathrm{sol}(\mathbb{G}) \subseteq C_{\mathbb{G}}(S)$, and moreover that $L = \mathrm{soc}(\mathbb{G}/\mathrm{sol}(\mathbb{G}))$ is a nonabelian finite simple group. It is shown in [Tiep 12, 18] that

– if L is a finite group of Lie type of characteristic other than p, then the triple (p^n, N, L) is $(9, 8, D_4(2))$ or $(7, 8, D_4(2))$ or $(7, 8, A_3(2) \cong \mathbb{A}_8)$;
– if L is a finite group of Lie type of characteristic p, then $L = PSp_{2n}(p)$;
– if L is an alternating group, then $p^n = 7$, $N = 8$, and $L = \mathbb{A}_7$ or $L = \mathbb{A}_8$;
– if L is one of the 26 sporadic groups, then the triple (p^n, N, L) is $(11, 12, M_{12})$ or $(23, 24, Co_1)$.

This completes the proof of Theorem 14.3.2. □

14.4 The basic spin representations of the alternating groups, the Barnes-Wall lattices, and the Gow lattices

As usual, let $\mathbb{S}_n$ and $\mathbb{A}_n$ denote the symmetric and alternating groups of degree n. Schur showed in [Schur] that for $n > 3$ and $n \neq 6$, $\mathbb{S}_n$ has exactly two non-isomorphic double covers, $2^+\mathbb{S}_n$ and $2^-\mathbb{S}_n$. When $n = 6$ it turns out that there is a single double cover, as $2^+\mathbb{S}_6 \cong 2^-\mathbb{S}_6$. We shall let $2\mathbb{S}_n$ denote either of these two double covers. By definition, $2\mathbb{S}_n$ contains the subgroup $< \mathbf{z} >$ generated by a central involution $\mathbf{z}$ and $2\mathbb{S}_n/ < \mathbf{z} > \cong \mathbb{S}_n$. Given $\tau \in \mathbb{S}_n$, we let τ' be an element of $2\mathbb{S}_n$ that projects onto τ. Choosing τ to be a transposition, it is known that $(\tau')^2 = 1$ in $2^+\mathbb{S}_n$, whereas $(\tau')^2 = \mathbf{z}$ in $2^-\mathbb{S}_n$. The (unique) double cover $2\mathbb{A}_n$ of $\mathbb{A}_n$ is a subgroup of index 2 in both $2^+\mathbb{S}_n$ and $2^-\mathbb{S}_n$.

Of particular importance in the construction of all *spin representations* (that is, faithful irreducible representations of $2\mathbb{S}_n$), are the *basic spin representations*, the spin representations of the least degree $2^{[(n-1)/2]}$. If n is odd, there is a single basic spin representation of $2\mathbb{S}_n$, and its restriction to $2\mathbb{A}_n$ consists of two conjugate irreducible *basic spin representations of* $2\mathbb{A}_n$. If n is even, there are two basic spin representations of $2\mathbb{S}_n$, one being the multiple of the other by the sign character, and these representations both restrict to $2\mathbb{A}_n$ to give the same irreducible *basic spin representation of* $2\mathbb{A}_n$. Thus the basic spin representations of $2\mathbb{A}_n$ are of degree $2^{[n/2]-1}$.

Under some restrictions on n the basic spin representations of $2\mathbb{S}_n$ and $2\mathbb{A}_n$ can be written over the rational field $\mathbb{Q}$. In this case one can consider G-invariant lattices Λ in the $\mathbb{Q}G$-module V that affords basic spin characters of G, where $G =$

$2\mathbb{S}_n$ or $2\mathbb{A}_n$. Moreover, sometimes these lattices possess the following remarkable property highlighted by Thompson [Tho 1]:

$$\Lambda/p\Lambda \text{ is an irreducible } \mathbb{F}_p G\text{-module, for every prime } p. \quad (14.2)$$

Clearly, if $(*|*)$ is a G-invariant symmetric integral form defined on Λ, then after suitable rescaling the lattice $(\Lambda, (*|*))$ is even unimodular [Tho 1]. Only three basic examples of pairs (G, Λ) which satisfy condition (14.2) are known; namely, where Λ is the root lattice of type E_8, the Leech lattice Λ_{24}, or the Thompson-Smith lattice Λ_{248} (see also Chapter 13), together with their tensor powers; and $G = \mathrm{Aut}(\Lambda)$. The aim of [Gow 3] was to construct new examples of such a pair. We reproduce the main results of this paper here.

First, suppose that $n = 2m + 1$ is odd. Then the basic spin character ψ of $2\mathbb{S}_n$ is rational-valued. If θ is an irreducible constituent of the restriction of ψ to $2\mathbb{A}_n$, the field generated over the rational numbers $\mathbb{Q}$ by the values of θ is $\mathbb{Q}(\sqrt{(-1)^m n})$ [Schur]. Thus θ is rational-valued precisely when n is an odd square. Moreover, in this case θ is realized over $\mathbb{Q}$, and the following statement holds.

Theorem 14.4.1 [Gow 3]. *Suppose that $n = 2m + 1$ is an odd square greater than 8. Then the basic spin character of degree 2^{m-1} of $2\mathbb{A}_n$ is afforded by an irreducible $\mathbb{Q}(2\mathbb{A}_n)$-module, V say. Moreover, if Λ is a $2\mathbb{A}_n$-invariant rational integral lattice in V, then $\Lambda/p\Lambda$ is an irreducible $\mathbb{F}_p(2\mathbb{A}_n)$-module for all primes p. In particular, Λ supports an even symmetric positive definite unimodular form that is $2\mathbb{A}_n$-invariant.* □

Next, suppose that $n = 2m$ is even. Let ψ denote either of the basic spin characters of $2\mathbb{S}_n$ (of degree 2^{m-1}), let π be an n-cycle in $\mathbb{S}_n$ and suppose that the element π' of $2\mathbb{S}_n$ projects onto π. The value of ψ at π' is $\pm\sqrt{\varepsilon m}$, where $\varepsilon = \pm 1$. Both values of ε occur, the signs being different for $2^+\mathbb{S}_n$ and $2^-\mathbb{S}_n$. We have $\varepsilon = 1$ if m is odd and we are working in $2^+\mathbb{S}_n$, or if m is even and we are working in $2^-\mathbb{S}_n$; otherwise $\varepsilon = -1$. Thus these characters are rational-valued for $2^-\mathbb{S}_n$ precisely when m is an even square, and rational-valued for $2^+\mathbb{S}_n$ precisely when m is an odd square. Moreover, in this case ψ is realized over $\mathbb{Q}$:

Theorem 14.4.2 [Gow 3]. *Suppose that $n = 2m$, where $m > 1$, and let ψ be either of the basic spin characters of $2\mathbb{S}_n$ of degree 2^{m-1}. Then if m is an odd square, ψ is afforded by an irreducible $\mathbb{Q}(2^+\mathbb{S}_n)$-module, V, say. If Λ is a $2^+\mathbb{S}_n$-invariant rational integral lattice in V, then $\Lambda/p\Lambda$ is an irreducible $\mathbb{F}_p(2^+\mathbb{S}_n)$-module for all primes p. In particular, Λ supports an even symmetric positive definite unimodular form that is $2^+\mathbb{S}_n$-invariant. The same holds for $2^-\mathbb{S}_n$ if m is an even square.* □

From now on, we shall suppose that the basic spin character θ (of degree $2^{[n/2]-1}$) is afforded by an irreducible $\mathbb{Q}G$-module V, where $G = 2\mathbb{A}_n$. In particular,

$n \geq 8$ (see [ATLAS]). The goal of this section is to study the *Gow lattices*, that is, arbitrary G-invariant integral lattices Λ in V, and we find all possibilities for the groups $\mathbb{G} = \mathrm{Aut}(\Lambda)$. We prove also the conjecture stated by Gow in [Gow 3] that there exists a direct connection between these lattices Λ and the even unimodular Barnes-Wall lattices $BW_{2^{2d+1}}$ [BaW]. Namely, when $n = 8k$ is divisible by 8, the Barnes-Wall lattice $BW_{2^{4k-1}}$ occurs among the Gow lattices. As a consequence, we get the rationality and the reducibility modulo 2 of the basic spin character of $2\mathbb{A}_n$ provided $8|n$ – facts mentioned in [Gow 3] as open questions. The main result of the section is the following:

Theorem 14.4.3 [Tiep 15]. *Let Λ be a $2\mathbb{A}_n$-invariant lattice in a faithful $\mathbb{Q}(2\mathbb{A}_n)$-module V of dimension $2^{[n/2]-1}$, $n \geq 8$, and set $\mathbb{G} = \mathrm{Aut}(\Lambda)$. Then one of the following assertions holds.*

(i) *$n = 8, 9$, Λ is isometric to the root lattice of type E_8, and $\mathbb{G} = W(E_8) = \mathbb{Z}_2.O_8^+(2)$.*

(ii) *$n > 9$, $\mathbb{G} = 2\mathbb{A}_n$, and one of the following condition holds:*

a) *n is an odd square;*

b) *n is divisible by 8;*

c) *$n \equiv 2 \pmod 8$, and n is a sum of two squares.*

(iii) *$n > 9$, $n + 1$ is an odd square, and $\mathbb{G} = 2\mathbb{A}_{n+1}$.*

(iv) *$n > 9$, $n = 2\ell^2$ is a double square, and either*

a) *$2|\ell$, and $\mathbb{G} = 2^-\mathbb{S}_n$, or*

b) *$2 \not| \ell$, and $\mathbb{G} = 2^+\mathbb{S}_n$.*

(v) *$n > 9$, $n = 8k$ is divisible by 8, Λ is isometric to the Barnes-Wall lattice $BW_{2^{4k-1}}$, and $\mathbb{G} = 2_+^{1+2(4k-1)} \circ \Omega_{8k-2}^+(2)$.*

(vi) *$n > 9$, $n = 8k$ is divisible by 8, $\mathrm{sol}(\mathbb{G}) = P = 2_+^{1+2(4k-1)}$, and $\mathbb{G}/P \in \{\mathbb{A}_n, \mathbb{S}_n\}$.*

All possibilities, except perhaps case (vi), occur.

We note that a weaker version of Theorem 14.4.3 has been proved by Gow (unpublished). The Barnes-Wall lattices are also considered in [Gro] in the context of *globally irreducible representations* (see §14.5). Furthermore, spin representations are studied in many articles, among which we mention [Mor] and [Naz]. Finally, quite recently using a generalization of the Brauer groups Turull described in [Tur] a remarkable algorithm that settles completety the question as to when a given spin character of a symmetric or alternating group can be realized over $\mathbb{Q}$. In particular, the basic spin characters of $2\mathbb{A}_n$ are rational precisely when n satisfies one of the conditions (ii)a) – c) stated in Theorem 14.4.3.

Before proving Theorem 14.4.3, we extract some consequences from this result.

Corollary 14.4.4 [Tiep 15]. *When n is divisible by 8 the basic spin representation of $2\mathbb{A}_n$ is realized over $\mathbb{Q}$.*

Proof. By Theorem 14.4.3, $2\mathbb{A}_{8k}$ preserves the Barnes-Wall lattice $BW_{2^{4k-1}}$ in its basic spin module. □

Corollary 14.4.5 [Tiep 15]. *Suppose that n is divisible by 8 and that Λ is a faithful $\mathbb{Z}(2\mathbb{A}_n)$-module of dimension $2^{[n/2]-1}$. Then $\Lambda/2\Lambda$ is a reducible $2\mathbb{A}_n$-module.*

Proof. Set $k = n/8$. It is not difficult to see that $\Lambda/2\Lambda$ is a reducible H-module if $\mathrm{Aut}(\Lambda) \supseteq H \rhd P = 2_+^{1+2(4k-1)}$. In particular, this is so when Λ is the Barnes-Wall lattice $\Gamma = BW_{2^{4k-1}}$ and $H = \mathrm{Aut}(\Gamma)$.

Suppose now that $\Lambda/2\Lambda$ is an irreducible $2\mathbb{A}_n$-module for some faithful $\mathbb{Z}(2\mathbb{A}_n)$-lattice of dimension $2^{[n/2]-1}$. By the Deuring-Noether Theorem, one can suppose that $\Gamma \subseteq \Lambda$ but $\Gamma \not\subseteq 2\Lambda$. Since $\Lambda/2\Lambda$ is irreducible, we have $\Gamma + 2\Lambda = \Lambda$. Hence, $\Gamma \cap 2\Lambda = 2\Gamma$, and $\Gamma/2\Gamma \cong \Lambda/2\Lambda$ as $2\mathbb{A}_n$-modules. But in this case the $2\mathbb{A}_n$-module $\Gamma/2\Gamma$ is irreducible, a contradiction. In fact, one can show that $\Lambda/2\Lambda$ is a sum of two absolutely irreducible $\mathbb{F}_2\mathbb{A}_n$-submodules of dimension $2^{[n/2]-2}$. □

Corollary 14.4.6 [Tiep 15]. *Suppose that $8|n$, $n > 8$, Λ is a faithful $\mathbb{Z}(2\mathbb{A}_n)$-lattice of dimension $2^{[n/2]-1}$, and $\mathbb{G} = \mathrm{Aut}(\Lambda)$. Then*

(i) *the $\mathbb{G}$-module $\Lambda/2\Lambda$ is reducible precisely when $\mathbb{G} = 2\mathbb{A}_n$ or $\mathrm{sol}(\mathbb{G}) = 2_+^{1+2(4k-1)}$, where $n = 8k$;*

(ii) *the $\mathbb{G}$-module $\Lambda/2\Lambda$ is irreducible precisely when $n+1$ is a square and $\mathbb{G} = 2\mathbb{A}_{n+1}$, or n is a double square and $\mathbb{G} = 2\mathbb{S}_n$.* □

We now come to the proof of Theorem 14.4.3. The following statement can be found in [Schur] and [Gow 3].

Lemma 14.4.7. *Suppose that $n \geq 7$. The basic spin character θ of G is a primitive complex character. If c' denotes the preimage of order 3 in G of a 3-cycle $c \in \mathbb{A}_n$, then $\theta(c') = -\deg\theta/2$. Moreover, the minimum degree of spin characters is $2^{[(n-1)/2]}$ for $2\mathbb{S}_n$, and $2^{[n/2]-1}$ for $2\mathbb{A}_n$.* □

For the proof of the following statement see [Wag 1] and [KlP].

Lemma 14.4.8. *Suppose that $n > 9$. Then the alternating group $\mathbb{A}_n$ has a unique faithful $\mathbb{F}_2$-module V_2 of least degree. Furthermore, $\dim V_2 = 2[(n-1)/2]$, and the second cohomology group $H^2(\mathbb{A}_n, V_2)$ is trivial.* □

Sketch of Proof of Theorem 14.4.3. The case $n = 9$ is treated in [Gow 3]. Suppose that $n = 8$. By [LPW], the reduction θ_p of θ modulo p is an irreducible Brauer character if p is odd, and a sum of two irreducible Brauer characters of degree 4 if $p = 2$. In particular, if a G-invariant inner product $(*|*)$ on Λ is chosen such that $(\Lambda|\Lambda) \not\subseteq r\mathbb{Z}$ for any $r > 1$, then Λ is an even lattice. Suppose that Λ is not isometric to the root lattice of type E_8. Then $\det\Lambda$ is divisible by some prime number p, and so $\Lambda \supset \Lambda_p \supset p\Lambda$, where

$$\Lambda_p = \{x \in \Lambda | (x|\Lambda) \subseteq p\mathbb{Z}\}.$$

This means that θ_p is reducible, that is, $p = 2$. Furthermore, the assignment

$$(x + \Lambda_2, y + \Lambda_2) \mapsto (x|y) \pmod 2$$

defines a non-degenerate symplectic G-invariant form on Λ/Λ_2. Thus we get a nontrivial homomorphism

$$2\mathbb{A}_8 = G \to O(\Lambda/\Lambda_2) = Sp_4(2) \cong \mathbb{S}_6,$$

a contradiction. Hence, Λ is isometric to the root lattice of type E_8.

In what follows we shall suppose that $n > 9$. Using Lemma 14.4.7, one can show that

$$Z = Z(\mathbb{G}) =< \mathbf{z} : v \mapsto -v >\cong \mathbb{Z}_2$$

is the largest abelian normal subgroup in $\mathbb{G}$. If, in addition, the largest soluble normal subgroup sol($\mathbb{G}$) is just Z, then standard arguments (see §12.4) enable us to state that $L = \mathrm{soc}(\mathbb{G}/Z)$ is a nonabelian finite simple group. Moreover, the full preimage $\tilde{L}$ of L in $\mathbb{G}$ is perfect and contains $G = 2\mathbb{A}_n$. It was shown in [Tiep 15] using the Landázuri-Seitz-Zalesskii lower bound (see Table A7) and Lemma 14.4.7 that $L = \mathbb{A}_{n+1}$ and $n+1$ is an odd square, or else $L = \mathbb{A}_n$. In the former case it follows immediately that $\mathbb{G} = \tilde{L} = 2\mathbb{A}_{n+1}$. In the other case we have $\mathbb{G} = 2\mathbb{S}_n$ and n is a double square, or else $\mathbb{G} = G$.

Suppose now that sol($\mathbb{G}$) $\neq Z$. Then $\mathbb{G}$ normalizes some extraspecial 2-subgroup $P = 2^{1+2m}_{\pm}$. Since the 2-modular representation of $G/Z = \mathbb{A}_n$ on P/Z is faithful, we have $m \geq [(n-1)/2]$ by Lemma 14.4.8. But P acts faithfully on the space V of dimension $2^{[n/2]-1}$, hence $m \leq [n/2] - 1$. We thus conclude that n is even, and $m = [n/2] - 1$. Now the (faithful irreducible) representation of P on V is real, so by computing the Schur-Frobenius indicator we see that $P = 2^{1+2m}_{+}$. If we identify V_2 with P/Z as minimal $\mathbb{F}_2\mathbb{A}_n$-modules (see Lemma 14.4.8), then the map

$$xZ \mapsto x^2 \in Z \cong \mathbb{F}_2,\ x \in P$$

defines a nontrivial $\mathbb{A}_n$-invariant quadratic form of type (+) (that is, with highest dimension of totally singular subspaces equal to m) on V_2. It is not difficult to see that this can occur only when n is divisible by 8. Set $k = n/8$. The action of $\mathbb{G}$ on P via conjugation induces embeddings

$$\mathbb{A}_{8k} \hookrightarrow \mathbb{G}/P \hookrightarrow \mathrm{Out}(P) \cong O^{+}_{8k-2}(2).$$

It is proved in [Dye] that $\mathbb{A}_{8k}$ is a maximal subgroup of Ω^{+}_{8k-2}. Hence, either

$$\mathbb{A}_{8k} \subseteq \mathbb{G}/P \subseteq \mathbb{S}_{8k},$$

or $\mathbb{G}/P \supseteq \Omega^+_{8k-2}$. In the former case sol($\mathbb{G}$) coincides with P, and we come to conclusion (vi) of Theorem 14.4.3. For the analysis of the latter case we need the following three lemmas proved in [Tiep 15].

Lemma 14.4.9. *Suppose that*

(i) Φ *is an irreducible complex representation of a central extension* $A = \mathbb{Z}_2.\mathbb{A}_m$, *where* $m \geq 5$;

(ii) *the operator* $\Phi(c')$ *acts fixed-point-freely, where* c' *is the preimage of order* 3 *in* A *of some* 3-*cycle* $c \in \mathbb{A}_m$.

Then A *is a double cover for* $\mathbb{A}_m$, *and* Φ *is faithful.* □

Lemma 14.4.10. *Suppose that* U *is a quadratic* $\mathbb{F}_2$-*space of type* $(+)$ *of dimension* $2m+2$, $m \geq 1$. *Then the group*

$$\Omega(U) = [O(U), O(U)] \cong \Omega^+_{2m+2}(2)$$

has a unique conjugacy class of cyclic subgroups $<\varphi>$ *of order* 3 *such that* $\mathrm{Ker}(\varphi - 1_U)$ *is a quadratic subspace of type* $(-)$ *of dimension* $2m$. □

The elements φ mentioned in Lemma 14.4.10 will be called 3-*special.*

Lemma 14.4.11. *If* Ψ *denotes an embedding of* $\mathbb{A}_{8k}$ *in* $\Omega^+_{8k-2}(2)$, *then* $\Psi(c)$ *is* 3-*special for every* 3-*cycle* c *in* $\mathbb{A}_{8k}$. □

Let us come back to the case $\mathbb{G} \rhd P$, $\mathbb{G} \supseteq H = P.\Omega^+_{8k-2}(2)$. Without loss of generality one can suppose that $H = \mathrm{Aut}(\Lambda_0)$, and

$$\Lambda_0 \otimes_{\mathbb{Z}} \mathbb{C} = \Lambda \otimes_{\mathbb{Z}} \mathbb{C} = W,$$

where $\Lambda_0 = BW_{2^{4k-1}}$ is the Barnes-Wall lattice of dimension 2^{4k-1}. Furthermore, note that

$$H \supseteq \left(2^{1+2}_{-}.\Omega^-_2(2)\right) * \left(2^{1+2m}_{-}.\Omega^-_{2m}(2)\right) = R^* * H_1,$$

where $m = 4k-2$, R^* is the group of units of the Hurwitz ring R of integral quaternions:

$$R = < i, j, k, \frac{1}{2}(1+i+j+k) | i^2 = j^2 = -1, k = ij = -ji >_{\mathbb{Z}},$$

and $H_1 = 2^{1+2m}_{-}.\Omega^-_{2m}(2)$. This means that Λ is a Hermitian RH_1-lattice. It is shown in [Gro] that Λ is isometric to Λ_0 as a Euclidean lattice. In particular,

$$\mathbb{G} = H = 2^{1+2(4k-1)}_{+} \circ \Omega^+_{8k-2}(2).$$

Conversely, we claim that $\mathbb{G} = H$ contains a double cover of $\mathbb{A}_{8k}$. To this end, we again use the embedding $R^* * H_1 \subseteq H$. Let φ be an element of order 3 in R^*.

Then it is clear that φ acts fixed-point-freely on $W = \Lambda_0 \otimes_{\mathbb{Z}} \mathbb{C}$. Next, consider the subgroup $B = (P/Z).\mathbb{A}_{8k}$ of H/Z. By Lemma 14.4.8, B is split over $O_2(B) = P/Z$. Hence, H contains a subgroup $A = Z.\mathbb{A}_{8k}$. Let c be a 3-cycle in $\mathbb{A}_{8k}$, and c' be the preimage of order 3 of c in A. By Lemma 14.4.11, the image of c' under the projection $H \to H/P$ is 3-special. On the other hand, the image of φ in H/P is also 3-special. By Lemma 14.4.10, the subgroups $<c'>P$ and $<\varphi>P$ are conjugate in H. By Sylow's Theorem, $<c'>$ and $<\varphi>$ are also conjugate. In particular, c' acts fixed-point-freely on W. Applying Lemma 14.4.9, we arrive at the conclusion that A is a double cover of $\mathbb{A}_{8k}$, and that W affords the basic spin character of A.

Thus we have shown that $\mathbb{G}$ satisfies one of the conclusions (ii) – (vi) of Theorem 14.4.3. Taking the Barnes-Wall lattice, we see that possibility (v) is realized. Suppose that $n+1$ is an odd square greater than 10. Then by Theorem 14.4.1, Λ can be chosen to be $2\mathbb{A}_{n+1}$-invariant, so that $\mathbb{G} \supseteq 2\mathbb{A}_{n+1}$. Furthermore, the $\mathbb{G}$-module $\Lambda/p\Lambda$ is irreducible for all primes p. In particular, $O_2(\mathbb{G}) = Z$ (see the proof of Corollary 14.4.5). Hence, cases (v) and (vi) cannot occur. Moreover, $2\mathbb{A}_{n+1} \not\hookrightarrow 2\mathbb{S}_n$. This means that case (iii) must occur. Similarly, when n is a double square greater than 9, possibility (iv) is realized for a suitable invariant lattice Λ. Finally, case (ii) occurs if, for instance, n is an odd square greater than 9.

This completes the proof of Theorem 14.4.3. □

14.5 Globally irreducible representations and some Mordell-Weil lattices

In this section we consider two series of Euclidean integral lattices Λ which were first constructed by Elkies [Elk] as the Mordell-Weil lattices, that is, the groups of points, modulo torsion, on some elliptic curves E over the global function field of some algebraic curve X:

$$\Lambda = \mathrm{Mor}(X, E)/\text{translations} = \mathrm{Hom}(J_X, E), \tag{14.3}$$

where Hom means homomorphisms of abelian varieties and J_X is the Jacobian of X. Here Λ is endowed with the Hermitian form

$$(f, g) \mapsto f \circ {}^t g \in R = \mathrm{End}(E), \tag{14.4}$$

where ${}^t g : E \to J_X$ is the dual homomorphism, and E and J_X are identified with their dual varieties via the standard principal polarizations.

In the first case, X is the hyperelliptic curve of genus $(p-1)/2$ defined over the prime field $\mathbb{F}_p$, $p \equiv 3 \pmod 4$, by the equation $y^2 = x^p - x$. Furthermore, E denotes the elliptic curve with equation $v^2 = u^3 - u$. Clearly, the Frobenius endomorphism F_E of E satisfies the relation $F_E^2 = -p$ in $R = \mathrm{End}(E)$. Moreover, R can be identified with the unique maximal order $\mathbb{Z}[(1+\sqrt{-p})/2]$ of the imaginary quadratic field $\mathbb{K} = \mathbb{Q}(\sqrt{-p})$. The group $G = L_2(p)$ acts on X by the rule

$$\begin{pmatrix} a & b \\ c & d \end{pmatrix} \circ (x, y) = \left(\frac{ax+b}{cx+d}, \frac{y}{(cx+d)^{(p+1)/2}} \right).$$

Hence, Λ is a Hermitian RG-lattice. If we endow Λ with the inner product

$$(f|g) = \mathrm{Tr}_{\mathbb{K}/\mathbb{Q}}(f \circ {}^t g), \tag{14.5}$$

then Λ becomes a Euclidean integral G-invariant lattice.

In the second case, p is an arbitrary prime number and $q = p^f$. Furthermore, $\mathbb{K}$ denotes the quaternion division algebra over $\mathbb{Q}$ ramified at p and ∞, and R is a maximal order in $\mathbb{K}$. Now E is a supersingular elliptic curve over $\mathbb{F}_{q^2}$ with Frobenius endomorphism equal to $-q$ and $\mathrm{End}(E) \cong R$, and X is the Fermat curve of exponent $q+1$ with equation $x^{q+1} + y^{q+1} + z^{q+1} = 0$ over $\mathbb{F}_{q^2}$. One can define a Hermitian and an integral bilinear forms on Λ as in (14.4) and (14.5). Finally, if $\alpha \mapsto \bar{\alpha} = \alpha^q$ is the antilinear involutory automorphism of $\mathbb{F}_{q^2}$, then the groups $S = PSU_3(q)$ and $G = PGU_3(q)$ act in the natural way on X and on Λ.

The same lattices were studied later by Gross [Gro] in the more general context of the so-called *globally irreducible representations*. Recall that in §14.4 we dealt with G-invariant integral lattices Λ that satisfy condition (14.2) enunciated by Thompson in [Tho 1]. In particular, in this case the centralizer algebra

$$\mathbb{K} = \mathrm{End}_G(V) = \{\alpha \in \mathrm{End}_{\mathbb{Q}}(V) | \forall g \in G, g.\alpha = \alpha.g\},$$

where $V = \Lambda \otimes_{\mathbb{Z}} \mathbb{Q}$, is just $\mathbb{Q}$. Gross suggested considering a more general situation, that where $\mathbb{K}$ can be an imaginary quadratic field or a quaternion algebra. More precisely, let G be a finite group, V a finite dimensional $\mathbb{Q}G$-module with character χ, and $\mathbb{K} = \mathrm{End}_G(V)$. Then we have:

Lemma 14.5.1. *The following conditions are equivalent.*

(i) *The $\mathbb{R}G$-module $V \otimes_{\mathbb{Q}} \mathbb{R}$ is irreducible.*

(ii) *The $\mathbb{R}$-algebra $\mathbb{K} \otimes_{\mathbb{Q}} \mathbb{R}$ is a division algebra.*

(iii) $\frac{1}{|G|} \sum_{g \in G} \frac{1}{2}(\chi(g)^2 + \chi(g^2)) = 1.$

(iv) *The space of G-invariant symmetric bilinear forms on $V \otimes_{\mathbb{Q}} \mathbb{R}$ has dimension 1.* □

This means that if one of the conditions (i) – (iv) formulated in Lemma 14.5.1 holds, then the $\mathbb{Q}$-algebra $\mathbb{K}$ is isomorphic to $\mathbb{Q}$, an imaginary quadratic field $\mathbb{Q}(\sqrt{-d})$ with $d \in \mathbb{N}$, or a definite quaternion algebra $\left(\frac{a,b}{\mathbb{Q}}\right)$ with $a, b \in \mathbb{Q}$, $a, b < 0$ (see [Pie]). From now on we assume that $V \otimes_{\mathbb{Q}} \mathbb{R}$ is irreducible. Let R be a maximal order in $\mathbb{K}$ and Λ an RG-lattice in V, that is,

a) Λ is a free $\mathbb{Z}$-submodule of rank $n = \dim_{\mathbb{Q}} V$ of V;
b) Λ is invariant under the natural actions of R and G.

If $\wp$ is a maximal (two-sided) ideal of R, then one can consider the reduced representation

$$V_{\wp} = \Lambda/\wp\Lambda \text{ of } G \text{ over } k_{\wp} = R/\wp R. \tag{14.6}$$

Lemma 14.5.2. *The following assertions are equivalent.*
(i) *For every maximal ideal $\wp$ of R, the representation $V_{\wp}$ of G is irreducible over $k_{\wp}$.*
(ii) *Every RG-lattice in V has the form $\alpha\Lambda$, where α is a fractional ideal of R.* □

Definition 14.5.3 [Gro]. A rational representation V of a finite group G is said to be *globally irreducible* if V satisfies the conditions of Lemmas 14.5.1 and 14.5.2.

As an immediate consequence of this definition we obtain:

Proposition 14.5.4 [Gro]. *If V is a globally irreducible representation of G, then the two-sided class group $Cl(R)$ of R acts simply transitively on the set of isomorphism classes of Euclidean $\mathbb{Z}G$-lattices Λ in V with* $\mathrm{End}_G(\Lambda) = R$. □

To the first series of the Elkies lattices there corresponds a (unique) globally irreducible representation V of degree $p-1$ of $G = L_2(p)$ ($p \equiv 3 \pmod 4$ and prime) with $\mathrm{End}_G(V) = \mathbb{Q}(\sqrt{-p})$. Namely, V affords the G-character $\psi + \bar{\psi}$, where ψ is an irreducible complex character of a $\frac{1}{2}$-discrete series of G [Car].

The second series is related to a rational representation V of degree $2q(q-1)$ of $G = PGU_3(q)$ ($q = p^f$ a prime power) with quaternion algebra $\mathbb{K} = \mathrm{End}_G(V)$. In this case, V affords the G-character 2ψ, where ψ denotes the irreducible cuspidal unipotent complex character of degree $q(q-1)$ of G [Car]. V is globally irreducible if $f \leq 2$. Furthermore, when $q = p^2$ the resulting lattices are even unimodular, for example, when $q = 4$ the lattices Λ are isometric to the Leech lattice Λ_{24}.

As is shown in [Gro], the class of globally irreducible representations includes

– the rational representation of degree $p^n - 1$ of $G = Sp_{2n}(p)$ with $\mathbb{K} = \mathbb{Q}(\sqrt{-p})$ ($p \equiv 3 \pmod 4$, p prime) related to the Weil representations of G and considered in §14.3;

– the rational representations of degree $p^{2n} - 1$ of $G = Sp_{2n}(p^2)$ with $\mathrm{End}_G(V) = \mathbb{K}$, the definite quaternion algebra ramified at p and ∞ (p an odd prime) [Gro];

– the representation V of dimension 2^{n+1} of $G = 2_-^{1+2n}.\Omega_{2n}^-(2)$ on the Barnes-Wall lattices with $\mathrm{End}_G(V) = \mathbb{K}$, the division ring of Hamilton's quaternions [Gro], and, of course, the basic spin representations of $2\mathbb{A}_{(2n+1)^2}$ and $2\mathbb{S}_{2n^2}$ that are considered in §14.4.

The rest of the section is devoted to describing the automorphism groups of the Mordell-Weil lattices discussed by Elkies and Gross.

A. The case $G = L_2(p)$
Recall that p is a prime and $p \equiv 3 \pmod 4$. The following statement holds.

Proposition 14.5.5. *There exists a unique irreducible representation V of $G = L_2(p)$ of dimension $p-1$ over $\mathbb{Q}$ with $\mathrm{End}_G(V) = \mathbb{K} = \mathbb{Q}(\sqrt{-p})$. The representation V is globally irreducible. Furthermore, the reduced representation $V_\wp$ (see (14.6)), where $\wp = \pi R$, $\pi = \sqrt{-p}$, is isomorphic to $\mathrm{Sym}^{(p-3)/2}(U_2)$, where U_2 denotes the standard 2-dimensional representation of $SL_2(p)$ over $\mathbb{F}_p$.* □

Now we fix a G-invariant positive definite inner product $(*|*)$ on V, whose existence follows from Lemma 14.5.1, and assume that Λ is a $\mathbb{Z}G$-lattice in V. As usual, the dual lattice Λ^* is defined to be

$$\Lambda^* = \{x \in V | (x|\Lambda) \subseteq \mathbb{Z}\}.$$

Lemma 14.5.6. *After suitable rescaling of the form $(*|*)$ we have*

$$\Lambda^* = \pi^{-1}\Lambda, \ \det\Lambda = [\Lambda^* : \Lambda] = p^{(p-1)/2}.$$

Moreover, Λ is an even lattice.

Proof. By Lemma 14.5.2, we have $\Lambda^* = \alpha\Lambda$ for some fractional ideal α of R. Since $\wp = \pi R$ is the unique ramified ideal of R, α is a power of $\wp$. After suitable rescaling, one can suppose that $\alpha = R$ or $\alpha = \pi^{-1}R$. It is not difficult to see that the $\mathbb{F}_2G$-module $\Lambda/2\Lambda$ has a composition series with two factors of dimension $(p-1)/2 > 1$. In the meantime, the even part

$$\Lambda^{(2)} = \{x \in \Lambda | (x|x) \in 2\mathbb{Z}\}$$

is a $\mathbb{Z}G$-sublattice of codimension ≤ 1 in Λ. Therefore $\Lambda^{(2)} = \Lambda$, that is, the lattice Λ is even. Because the dimension $\dim\Lambda = p-1 \equiv 2 \pmod 4$ is not divisible by 8, Λ cannot be unimodular. This means that $\alpha = \pi^{-1}R$ and $\Lambda^* = \pi^{-1}\Lambda$. □

The class group $Cl(R)$ has odd order h (see [BoS]). For example, $h = 1$ if and only if $p = 3$, 7, 11, 19, 43, 67, or 163. It is shown in [Gro] that

a) there exists a $\mathbb{K}$-antilinear involutory endomorphism τ of V that normalizes G and preserves $(*|*)$;

b) there exists an RG-lattice Λ in V invariant under τ.

By Proposition 14.5.4, one can index representatives of the isomorphism classes of RG-lattices by elements of $Cl(R)$ in such a way that

$$\tau(\Lambda_1) = \Lambda_1, \ \Lambda_\alpha = \alpha\Lambda_1. \tag{14.7}$$

We shall use this indexing in what follows.

When $p = 3$, Λ is a root lattice of type A_2. As we have mentioned above, $\mathrm{End}(E) = R$ in the general case, where E is the elliptic curve over $\mathbb{F}_p$ with equation $v^2 = u^3 - u$. Furthermore, for each ideal class α of R there is a unique elliptic curve E_α with $\mathrm{End}(E_\alpha) = R$ and $\mathrm{Hom}(E, E_\alpha) \cong \alpha$ as R-module. Finally, Λ_α is obtained from formula (14.3) on replacing E by E_α.

A cyclotomic model for Λ_α is given in [Adl]. Let ζ be a primitive pth root of unity in $\mathbb{C}$, and recall that $\mathbb{K} = \mathbb{Q}(\sqrt{-p})$ is a subfield of the cyclotomic field $\mathbb{Q}(\zeta)$. We may realize the representation V of G on $\mathbb{Q}(\zeta)$ in such a way that the unipotent element $g = \begin{pmatrix} 1 & 1 \\ 0 & 1 \end{pmatrix}$ acts as multiplication by ζ, and the semisimple element $f = \begin{pmatrix} \theta & 0 \\ 0 & \theta^{-1} \end{pmatrix}$, where $\mathbb{F}_p^* =< \theta >$, acts via the element of the Galois group taking ζ to ζ^{θ^2}. If α is an ideal of R, we let $\mathbf{i}(\alpha)$ be the extended ideal $\alpha\mathbb{Z}[\zeta]$. Then

$$\Lambda_\alpha = \mathbf{i}(\alpha).(1 - \zeta)^{-(p-3)/4},$$

with G-invariant inner product

$$(u|v) = \mathrm{Tr}_{\mathbb{Q}(\zeta)/\mathbb{Q}}(u.\bar{v}).$$

In particular, taking $\alpha = \pi R$, so that $\mathbf{i}(\alpha) = (1 - \zeta)^{(p-1)/2}\mathbb{Z}[\zeta]$, we see that Λ_1 *is isometric to the Bondal-Craig lattice* $A_{p-1}^{(p+1)/4} \cong \Gamma_{(p+1)/4}$ considered in §9.6 (if we identify $\mathbb{Z}[\zeta]$ with the ring $\mathbb{Z} < x >$ introduced in §9.1). The following simple statement shows that the two lattice series Λ_α with $\alpha \in Cl(R)$, and Γ_k with $0 \leq k < (p - 1)/2$, intersect only at the point $\Lambda_1 \cong \Gamma_{(p+1)/4}$.

Lemma 14.5.7. *Suppose that* $\mathrm{Aut}(\Gamma_k)$ *has a subgroup* H *such that*
(i) $H \supset \mathbb{Z}_2 \times (\mathbb{Z}_p.\mathbb{Z}_{(p-1)/2}) =< g, f, \mathbf{t} >$, *where* $\mathbf{t}(v) = -v$ *for all* $v \in V$;
(ii) $\mathrm{End}_H(V) \supset \mathbb{Q}$.
Then Γ_k *is isometric to* $\Gamma_{(p+1)/4}$. □

Next, we determine the group $\mathbb{G} = \mathrm{Aut}(\Lambda)$, where $\Lambda = \Lambda_\alpha$ is viewed in the cyclotomic model. Set

$$P =< g >\cong \mathbb{Z}_p,\ Z =< \mathbf{t} >\cong \mathbb{Z}_2,\ B = N_G(P) =< g, f > .$$

Using the same arguments as in §9.6, one can prove the following statements (for details see [Tiep 14]).

1. $C_{\mathbb{G}}(P) = Z \times P$. In particular, $P \in \mathrm{Syl}_p(\mathbb{G})$ and $Z \times B$ is a normal subgroup of index $\ell = 1$ or 2 in $N = N_{\mathbb{G}}(P)$.
2. $L = \Lambda/p\Lambda^*$ is a $\frac{p-1}{2}$-dimensional irreducible faithful $\mathbb{F}_p\mathbb{G}$-module. Moreover, the P-module L is indecomposable, and the element g acts on L with minimal polynomial $(t - 1)^{(p-1)/2}$.
3. Suppose that $P \subseteq H \subseteq \mathbb{G}$. Then one of the following assertions holds.

(i) H has just one composition factor $L_2(p)$, and all other its composition factors are p'-groups.

(ii) $P \lhd H \subseteq N$.

4. If $\ell = 1$, then $\mathbb{G} = Z \times G$.

5. If $\ell = 2$, then $\mathbb{G} = Z \times PGL_2(p)$.

We are now able to prove:

Theorem 14.5.8 [Tiep 14]. *Let p be a prime number such that $p \equiv 3 \pmod 4$, let $R = \mathbb{Z}[(1+\sqrt{-p})/2]$ be the unique maximal order in $\mathbb{K} = \mathbb{Q}(\sqrt{-p})$, and set $h = |Cl(R)|$. Furthermore, let V denote the unique globally irreducible representation of $G = L_2(p)$ of dimension $p-1$ with $\mathrm{End}_G(V) = \mathbb{K}$; and suppose that $\Lambda_1, \ldots, \Lambda_h$ are representatives of the isomorphism classes of RG-lattices in V. Then (after suitable reindexing) we have*

$$\mathrm{Aut}(\Lambda_i) = \begin{cases} \mathbb{Z}_2 \times PGL_2(p) & \text{if} \quad i = 1, \\ \mathbb{Z}_2 \times L_2(p) & \text{if} \quad i > 1. \end{cases}$$

Proof. Recall that we can index Λ_α in such a way that $\tau \in \mathrm{Aut}(\Lambda_1)$ (see (14.7)); moreover, $\mathrm{Aut}(\Lambda_\alpha) \in \{Z \times G, Z \times PGL_2(p)\}$. Hence,

$$\mathrm{Aut}(\Lambda_1) = Z \times PGL_2(p).$$

Conversely, suppose that $\mathbb{G} = \mathrm{Aut}(\Lambda_\alpha) = Z \times PGL_2(p)$. Then $\mathbb{G}$ is absolutely irreducible on V, and so $\mathrm{End}_{\mathbb{G}}(V) = \mathbb{Q}$. Consider an arbitrary element $\varphi \in \mathbb{G} \backslash (Z \times G)$. As $G \lhd \mathbb{G}$, φ normalizes $\mathbb{K} = \mathrm{End}_G(V)$; more precisely, φ induces a nontrivial automorphism of $\mathbb{K}$. This means that $\varphi x \varphi^{-1} = \bar{x}$ for all x in $\mathbb{K}$, that is, φ is $\mathbb{K}$-antilinear. On the one hand, we have some canonical $\mathbb{K}$-antilinear involutory endomorphism τ. "Adjusting" φ by means of G, one can suppose that $\varphi = k\tau$ for some $k \in \mathbb{K}^*$. Hence,

$$\alpha\Lambda_1 = \Lambda_\alpha = \varphi(\Lambda_\alpha) = \varphi(\alpha\Lambda_1) = k\tau(\alpha\Lambda_1) = k\bar{\alpha}\tau(\Lambda_1) = k\bar{\alpha}\Lambda_1,$$

that is, $\alpha = k\bar{\alpha}$. But the order $h = |Cl(R)|$ is odd, therefore we have $(\alpha) = 1$ and $\Lambda_\alpha = \Lambda_1$. □

B. The case $G = PGU_3(q)$

Recall that $q = p^f$ and p is a prime. Henceforth we shall assume that $q > 2$, because when $q = 2$ the lattice arising is isomorphic to the root lattice of type D_4. Furthermore, we set $S = PSU_3(q)$. The following statement is obvious (see Tables A6, A7).

Lemma 14.5.9. $P(S) = q^3 + 1$ *if* $q \neq 5$, *and* $P(S) = 50$ *if* $q = 5$. *Furthermore,* $R_{p'}(S) = q(q-1)$. □

It is known [Car] that the irreducible cuspidal unipotent complex character ψ of dimension $q(q-1)$ of G satisfies $\mathbb{Q}(\psi) = \mathbb{Q}$, $\psi|_S \in \mathrm{Irr}(S)$. For all $\ell \neq p$, a model for a representation with character ψ over the ℓ-adic field $\mathbb{Q}_\ell$ is given by the action of G on the étale cohomology $H^1(X, \mathbb{Q}_\ell)$ of the Fermat curve (of exponent $(q+1)$) over an algebraically closed field of characteristic p. Furthermore, the Schur index $m_{\mathbb{Q}}(\psi)$ of ψ over $\mathbb{Q}$ is 2, therefore G admits an irreducible $\mathbb{Q}$-module V of dimension $2q(q-1)$ with character 2ψ. Then $\mathrm{End}_G(V) = \mathbb{K}$, the quaternion division algebra over $\mathbb{Q}$ ramified at p and ∞.

Proposition 14.5.10. *The representation V of G is irreducible over $\mathbb{R}$, and for all primes $\ell \neq p$ its reduction is irreducible over $k_\ell = M_2(\mathbb{Z}/\ell\mathbb{Z})$. If $\wp$ is the unique ramified ideal in $\mathbb{K}$, the reduction $V_\wp$ is irreducible over $k_\wp = \mathbb{F}_{p^2}$ if and only if $q = p$ or $q = p^2$.* □

In fact, the $\mathbb{R}S$-module V is also irreducible, therefore the centralizer algebra $\mathrm{End}_S(V)$ is a division algebra over $\mathbb{Q}$. In what follows we denote $\psi|_s$ by the same symbol ψ. Furthermore, we consider an arbitrary S-invariant lattice in V and determine the automorphism group $\mathbb{G} = \mathrm{Aut}(\Lambda)$. By Lemma 14.5.1, there exists a unique S-invariant positive definite inner product on Λ. Let χ denote the $\mathbb{G}$-character afforded by Λ. As examples of such a lattice one can take the second series of Elkies lattices.

The first our goal is to distinguish in $\mathbb{G}$ a normal subgroup H with the following properties:

(i) $S \subseteq H$;

(ii) $\mathrm{sol}(H) = Z(H)$.

Recall that $\mathbb{G}$ can be *reducible* on $V \otimes_{\mathbb{Q}} \mathbb{C}$.

Lemma 14.5.11. *Suppose that $\mathbb{G}$ is reducible, or imprimitive and irreducible on $V \otimes_{\mathbb{Q}} \mathbb{C}$. Then there exists a normal subgroup $\mathbb{G}_0$ of index 1 or 2 in $\mathbb{G}$ with the following properties:*

(i) $S \subseteq \mathbb{G}_0$;

(ii) $\mathbb{G}_0$ acts on V with character $\chi_1 + \bar{\chi}_1$, where $\chi_1 \in \mathrm{Irr}(\mathbb{G}_0)$ and $\chi_1|_S = \psi$. Moreover, $Z(\mathbb{G}_0) = C_{\mathbb{G}_0}(S) = \mathbb{Z}_e$, where $e = 2$, 4, or 6.

Proof. If $\mathbb{G}$ is reducible on V, then V is a sum of two irreducible $\mathbb{G}$-submodules V_1 and V_2, and one can take $\mathbb{G}_0 = \mathbb{G}$. If $\mathbb{G}$ is irreducible but imprimitive on V, then $\mathbb{G}$ permutes the components of some decomposition $V = V_1 \oplus V_2$, where V_1 and V_2 afford the S-character ψ. In this case it is enough to set $\mathbb{G}_0 = St_{\mathbb{G}}(V_1, V_2)$. □

In addition to Lemma 14.5.11 we set $\mathbb{G}_0 = \mathbb{G}$ for the case where $\mathbb{G}$ is primitive on V.

Lemma 14.5.12. *Every abelian normal subgroup of* $\mathbb{G}_0$ *is contained in* $Z(\mathbb{G}_0)$. *Furthermore,* $Z(\mathbb{G}_0) \subseteq C_{\mathbb{G}_0}(S)$, *and* $Z(\mathbb{G}_0) = \mathbb{Z}_e$, *where* $e = 2$, 4, *or* 6. □

We note that the group $C_{\mathbb{G}}(S)$ can be nonabelian. For example, in the case $q = 4$, this group coincides with the group $R^* \cong SL_2(3)$ of units, where R is the Hurwitz ring of integral quaternions:

$$R = \left\langle i, j, k, \frac{1}{2}(1+i+j+k) | i^2 = j^2 = -1, k = ij = -ji \right\rangle_{\mathbb{Z}}.$$

However, $C_{\mathbb{G}}(S)$ is a finite subgroup of the 4-dimensional division algebra $\mathrm{End}_S(V)$, so by [Ami] we can state:

Lemma 14.5.13. *Every Sylow r-subgroup* Q *of* $C_{\mathbb{G}}(S)$ *is cyclic or a generalized quaternion group*

$$Q = Q_{2^{n+1}} = \left\langle x, y | x^{2^{n-1}} = y^2, y^4 = 1, yxy^{-1} = x^{-1} \right\rangle$$

of order 2^{n+1}, $n \geq 2$. *Furthermore,* $C_{\mathbb{G}}(S)$ *is isomorphic to one of* $\mathbb{Z}_2$, $\mathbb{Z}_4$, $\mathbb{Z}_6$, Q_8, $G_{6,2}$, *or* $SL_2(3)$, *where*

$$G_{6,2} = \left\langle x, y | x^3 = y^2, y^4 = 1, yxy^{-1} = x^{-1} \right\rangle.$$ □

Using Lemmas 14.5.12 and 14.5.13 one can prove that one of the following statements holds:

(i) $\mathrm{sol}(\mathbb{G}_0) = Z(\mathbb{G}_0)$;

(ii) $\mathbb{G}$ is primitive and contains a normal subgroup P isomorphic to $2_-^{1+2} \cong Q_8$.

We are now in a position to construct the desired normal subgroup H.

Proposition 14.5.14. *The group* $\mathbb{G} = \mathrm{Aut}(\Lambda)$ *has a normal subgroup* H *with the following properties:*

(i) $S \subseteq H$;

(ii) $\mathrm{sol}(H) = Z(H) \in \{\mathbb{Z}_2, \mathbb{Z}_4, \mathbb{Z}_6\}$;

(iii) *either* $\chi|_H$ *is an irreducible primitive complex character, or else* $\chi|_H = \chi_1 + \bar{\chi}_1$ *for some irreducible primitive complex character* χ_1 *of* H *with* $\chi_1|_S = \psi$;

(iv) $\mathbb{G}/H \subseteq \mathbb{Z}_2$ *or* $2_-^{1+2} * H \lhd \mathbb{G} \subseteq (2_-^{1+2} * H).\mathbb{S}_3$.

Proof. When $\mathbb{G}$ is primitive on V and contains a normal subgroup $P \cong Q_8$, we set $H = C_{\mathbb{G}}(P)$. Otherwise one can take $H = \mathbb{G}_0$ and apply Lemmas 14.5.12 and 14.5.13. □

Our further arguments recall those used in §12.4. Firstly, it follows from Proposition 14.5.14 that $L = \mathrm{soc}(H/Z(H))$ is a nonabelian simple group. Secondly,

Lemma 14.5.9 and the classification of finite simple groups are used in [Tiep 14] to show that one of the following assertions holds.

a) $L = S = PSU_3(q)$.

b) $q = 3$ and $L = PSU_4(3)$ or $L = J_2$.

c) $q = 4$ and $L \in \{G_2(4), Suz, Co_1\}$.

We have arrived at the main result of this section:

Theorem 14.5.15 [Tiep 14]. *Let V be a faithful rational representation of $G = PGU_3(q)$ of degree $2q(q-1)$, where $q = p^f > 2$ and p is a prime number. Suppose that Λ is an S-invariant integral lattice in V, where $S = PSU_3(q)$, and set $\mathbb{G} = \mathrm{Aut}(\Lambda)$. Then one of the following assertions holds.*

(i) *$q = 3$, $\mathbb{G} = \mathbb{Z}_3.O_6^-(3)$, and Λ is isometric to the Coxeter–Todd lattice K_{12}.*

(ii) *$q = 4$, $\mathbb{G} = \mathbb{Z}_2 \circ Co_1$, and Λ is isometric to the Leech lattice Λ_{24}.*

(iii) *$\mathbb{G}$ admits a normal subgroup H such that*

a) *$Z(H) \times S \lhd H \subseteq Z(H).\mathrm{Aut}(S)$, $Z(H) \in \{\mathbb{Z}_2, \mathbb{Z}_4, \mathbb{Z}_6\}$;*

b) *$\mathbb{G}/H \subseteq \mathbb{Z}_2$ or $2_-^{1+2} * H \lhd \mathbb{G} \subseteq (2_-^{1+2} * H).\mathbb{S}_3$.*

Sketch of Proof. By Proposition 14.5.14, it is sufficient to consider the exceptional possibilities

$$(q, L) = (3, PSU_4(3)),\ (3, J_2),\ (4, G_2(4)),\ (4, Suz),\ (4, Co_1).$$

1) First, suppose that $(q, L) = (3, PSU_4(3))$. It is known [ATLAS] that L has Schur multiplier $\mathbb{Z}_4 \times \mathbb{Z}_3^2$, and that a central extension $\tilde{L} = \mathbb{Z}_6 \circ L$ has just two irreducible complex characters α, $\bar{\alpha}$ of degree 6; furthermore, $\mathbb{Q}(\alpha) = \mathbb{Q}(\sqrt{-3})$. By [CoS 1], the automorphism group of the *complex Coxeter–Todd lattice*

$$\Gamma = \left\{ \frac{1}{\sqrt{3}}(x_1, \ldots, x_6) \right| $$

$$\left. x_i \in \mathbb{Z}[\zeta],\ x_1 \equiv \ldots \equiv x_6 \ (\mathrm{mod}\ \sqrt{-3}),\ \sum_i x_i \equiv 0 \ (\mathrm{mod}\ 3) \right\},$$

where $\zeta = \exp(\frac{2\pi i}{3})$, is $\tilde{L}.\mathbb{Z}_2$. Here the Hermitian inner product is as usual:

$$u \circ v = \sum_i u_i \bar{v}_i$$

for $u = (u_1, \ldots, u_6)$, $v = (v_1, \ldots, v_6)$. If we regard Γ as a $\mathbb{Z}$-module, and endow Γ with the Euclidean inner product

$$(u|v) = \mathrm{Tr}_{\mathbb{Q}(\zeta)/\mathbb{Q}}(u \circ v),$$

then Γ becomes an integral lattice Λ known as the *Coxeter–Todd lattice* K_{12} (see [CoS 7]). One can prove that

$$\mathrm{Aut}(\Lambda) = \mathbb{Z}_3.O_6^-(3)$$

(recall that $P\Omega_6^-(3) \cong PSU_4(3)$).

Conversely, suppose that Λ' is a 12-dimensional lattice invariant under some central extension of L. According to [ATLAS], non-principal irreducible projective complex characters of L of degree ≤ 12 can be realized as ordinary characters only of $\tilde{L} = \mathbb{Z}_6 \circ L$, and then they coincide with α or $\bar{\alpha}$. Let φ be a central element of order 3 of $\tilde{L}$. Then φ acts fixed-point-freely on $\Lambda' \otimes_{\mathbb{Z}} \mathbb{C}$, and so one can change Λ' into a $\mathbb{Z}[\zeta]$-lattice Γ' by setting

$$\zeta.u = \varphi(u),$$

$$u \circ v = (u|v) + \zeta^2(u|\varphi(v)) + \zeta(u|\varphi^2(v)).$$

It is not difficult to see that Γ' is isometric to Γ (as Hermitian lattices), and so Λ' is isometric to $\Lambda = K_{12}$.

2) Next, suppose that $(q, L) = (3, J_2)$. It is known [ATLAS] that L has only three irreducible projective complex characters of degree ≤ 12, namely, the trivial 1_L, and algebraic conjugates α and β, where $\deg\alpha = \deg\beta = 6$, and $\mathbb{Q}(\alpha) = \mathbb{Q}(\beta) = \mathbb{Q}(\sqrt{5})$. This means that Λ affords the (projective) L-character $\alpha + \beta$. But the Schur-Frobenius indicator of α and β is -1, contrary to Lemma 14.5.1.

3) Finally, suppose that $q = 4$ and $L \neq S$. Since

$$G_2(4) \hookrightarrow Suz \hookrightarrow Co_1,$$

it is sufficient to show that every 24-dimensional lattice Λ' invariant under a central extension $\tilde{L} = \mathbb{Z}_m.L$ of $L = G_2(4)$ is isometric to the Leech lattice. From [ATLAS] it follows that $m = 2$ and that Λ' affords the $\tilde{L}$-character 2ρ, where ρ is the unique non-principal irreducible projective complex L-character of degree not exceeding 24.

Following [Wil 1], we describe a 6-dimensional quaternionic lattice Γ for $\tilde{L}$. To this end, we consider the ring $R =< i, j, k, \omega >_{\mathbb{Z}}$ of integral quaternions in the division algebra

$$\mathbb{H} = \left\langle 1, i, j, k | i^2 = j^2 = -1, k = ij = -ji \right\rangle_{\mathbb{R}},$$

where $\omega = (-1 + i + j + k)/2$. The group R^* of units is generated by i, j, k, ω, and is isomorphic to $SL_2(3)$. It is not difficult to see that every right ideal of R is principal; furthermore, the right ideal $R_2 = (1 + i)R$ is in fact a two-sided ideal, and $R/R_2 \cong \mathbb{F}_4$. Now Γ is defined as a left R-module with basis $\{e_1, \ldots, e_6\}$, where

$$e_1 = (2 + 2i, 0, \ldots, 0),\ e_2 = (2, 2, 0, \ldots, 0),$$

$$e_3 = (0, 2, 2, 0, 0, 0),\ e_4 = (i + j + k, 1, \ldots, 1),$$

$$e_5 = (0, 0, 1 + k, 1 + j, 1 + j, 1 + k),\ e_6 = (0, 1 + j, 1 + j, 1 + k, 0, 1 + k),$$

with the usual inner product:

$$u \circ v = \sum_i u_i \bar{v}_i$$

for $u = \sum_i u_i e_i$, $v = \sum_i v_i e_i$. Clearly, Γ coincides with the *dual* lattice

$$\Gamma^* = \{u | u \circ \Gamma \subseteq 2R_2\}.$$

If we endow Γ with the Euclidean inner product

$$(u|v) = \mathrm{Tr}_{\mathbb{H}/\mathbb{R}}(u \circ v),$$

then Γ becomes an integral lattice Λ isometric to the Leech lattice. Finally, the group

$$\mathrm{Aut}(\Gamma) = \{\varphi \in GL(\Gamma \otimes_R \mathbb{H}) | \varphi(\Gamma) = \Gamma;$$

$$\forall u, v \in \Gamma, \forall r \in R, \varphi(x) \circ \varphi(y) = x \circ y, \varphi(rx) = r\varphi(x)\}$$

is just $\tilde{L} = \mathbb{Z}_2 \circ G_2(4)$ [Wil 1]. Note that right multiplications by units α in R^* are automorphisms of the quaternionic lattice Γ; meanwhile the left multiplications by units α in R^* are automorphisms only of the integral lattice Λ, but not of Γ. This explains the embedding

$$SL_2(3) * 2G_2(4) \hookrightarrow 2Co_1,$$

where mX denotes a non-split central extension $\mathbb{Z}_m \circ X$.

Conversely, any two subgroups of type $G_2(4)$ in Co_1 are conjugate, and so every embedding $G_2(4) \hookrightarrow Co_1$ can be extended to an embedding

$$SL_2(3) * 2G_2(4) \hookrightarrow 2Co_1.$$

Hence, every 24-dimensional $\tilde{L}$-invariant lattice Λ' can be isometrically embedded in Λ. If we choose such an embedding with the least index $[\Lambda : \Lambda']$, then one can show that in fact $\Lambda' = \Lambda$. □

Remark 14.5.16. The Conway sporadic simple group Co_1 has maximal subgroups $3Suz.\mathbb{Z}_2$, $(\mathbb{A}_4 \times G_2(4)).\mathbb{Z}_2$, and $(\mathbb{A}_5 \times J_2).\mathbb{Z}_2$. We have shown above that every 24-dimensional lattice invariant under a central extension of $G_2(4)$, Suz, or Co_1, is isometric to the Leech lattice. But this assertion fails for the case J_2: [Lin] contains the construction of a 24-dimensional J_2-invariant lattice which is not isometric to the Leech lattice.

Commentary

It worth pointing out that the Dickson forms for the Chevalley groups of types F_4, E_6, ${}^2E_6$, are well-studied objects (see [Asch 4]); and the same is true of the Hermitian lattice related to $\Omega_7(3)$ (see [ATLAS]). Moreover, one can prove using Theorem 14.1.10 that the lattice described in [ATLAS] and the lattice Λ expounded in §14.1 are isometric. Nevertheless, it appeared to us that these remarkable objects deserve one further approach, namely, the loop approach demonstrated in §14.1. Unfortunately, attempts to apply this approach to the embedding $Fi_{22} \hookrightarrow {}^2E_6(2)$ did not succeed.

The rationality of the Steinberg representation is well known (see e.g. [Car]). The simplest case, that of $L_2(q)$, was considered in §14.2. It would be interesting to study invariant lattices contained in the Steinberg modules of other finite groups of Lie type.

Mordell-Weil lattices have been used by Elkies [Elk] and Shioda [Shi] to obtain new sphere packings denser than the known ones in dimensions n, $54 \leq n \leq 4096$.

As shown in §§14.3 – 14.5, the concept of globally irreducible representations introduced by Gross turns out to be a very fruitful one. We have mentioned that this approach led to the discovery of new series of even unimodular Euclidean lattices of dimension 2^m (for specific values of m); $p^{2n} - 1$, p an odd prime; and $2p^2(p^2 - 1)$, p a prime. Unfortunately, these lattices are constructed implicitly, and little is known about them. It would be very interesting to describe geometric properties and explicit models of these lattices, and also to find new examples of globally irreducible representations. Some results in this direction have recently been obtained in [Tiep 20 – 25]. Namely, a necessary condition for a given $\mathbb{Q}G$-module (G a finite group) to be globally irreducible has been given in [Tiep 20]. (Recall that a sufficient condition for global irreducibility is indicated in [Gro].) With the use of this criterion, all globally irreducible representations associated with projective representations of finite groups of Lie type (of small rank 1, 2), and of most of the sporadic finite simple groups have been classified in [Tiep 20, 21, 23, 25]. It turns out that all up to now known examples of globally irreducible representations (with a few exceptions) are of the following two sorts: irreducible components of the Weil representations of finite classical groups, and basic spin representations of symmetric and alternating groups. All globally irreducible representations of these two sorts have been classified in [Tiep 22, 24]. In particular, a new series of even unimodular root-free Euclidean lattices of rank $2(p^n - 1)$, where $p \equiv 1 (\mathrm{mod}\, 4)$ is a prime, has been discovered in [Tiep 20]. The question of whether the lattices associated with globally irreducible representations can be realized as sublattices of Mordell-Weil lattices of certain curves in general remains still open; this is closely related to the arithmetic of curves, in particular, the order of the Tate-Shafarevitch group, as it has very recently been shown by Gross and Elkies.

Appendix

In this appendix we list some properties of complex Lie groups and finite simple groups that are used frequently in our book.

1. Jordan subgroups of Lie groups [Ale 1].
Let $\mathcal{G}$ be a complex simple Lie group with trivial centre. Then the list of Jordan subgroups J (up to $\mathcal{G}$-conjugacy) is given in Table A1.

Table A1. Jordan subgroups

$\mathcal{G}$	J	$C_{\mathcal{G}}(J)/J$	$N_{\mathcal{G}}(J)/J$
A_{p^n-1}, p a prime	$\mathbb{Z}_p^{2n}$	1	$Sp_{2n}(p)$
B_n, $\boldsymbol{n} \geq 3$	$\mathbb{Z}_2^{2n}$	1	$\mathbb{S}_{2n+1}$
$C_{2^{n-1}}$, $\boldsymbol{n} \geq 2$	$\mathbb{Z}_2^{2n}$	1	$O_{2n}^{-}(2)$
$D_{2^{n-1}}$, $\boldsymbol{n} \geq 3$	$\mathbb{Z}_2^{2n}$	1	$O_{2n}^{+}(2)$
D_{n+1}, $\boldsymbol{n} \geq 4$	$\mathbb{Z}_2^{2n}$	1	$\mathbb{S}_{2n+2}$
G_2	$\mathbb{Z}_2^3$	1	$SL_3(2)$
F_4	$\mathbb{Z}_3^3$	1	$SL_3(3)$
E_6	$\mathbb{Z}_3^3$	$\mathbb{Z}_3^3$	$\mathbb{Z}_3^3.SL_3(3)$
E_8	$\mathbb{Z}_5^3$	1	$SL_3(5)$
E_8	$\mathbb{Z}_2^5$	$\mathbb{Z}_2^{10}$	$\mathbb{Z}_2^{10}.SL_5(2)$
$D_4.\mathbb{Z}_3$ ($\mathcal{G}$ is not simple)	$\mathbb{Z}_2^3$	$\mathbb{Z}_2^6$	$\mathbb{Z}_2^6.SL_3(2)$

2. Doubly transitive permutation groups with nonabelian socle [Cam 3].
Table A2 lists all the non-abelian finite simple groups L that can arise as socles of doubly transitive permutation groups of degree N. Here the group L is also doubly transitive, except in the case $L = L_2(8)$, $N = 28$.

Further information concerns finite simple groups.

Table A2. Doubly transitive permutation groups with non-abelian socle

L	N	Remarks
$\mathbb{A}_n$, $n \geq 5$	n	two representations, if $n = 6$
$\left\{ \begin{array}{l} L_d(q),\ d \geq 2, \\ (d, q) \neq (2, 2), (2, 3) \end{array} \right.$	$(q^d - 1)/(q - 1)$	two representations (on lines and on hyperplanes), if $d > 2$
${}^2A_2(q)$	$q^3 + 1$	$q > 2$
${}^2B_2(q)$	$q^2 + 1$	$q = 2^{2a+1} > 2$
${}^2G_2(q)$	$q^3 + 1$	$q = 3^{2a+1} > 3$
$B_d(2)$, $d > 2$	$2^{d-1}(2^d \pm 1)$	the actions on quadratic forms
$L_2(11)$	11	two representations
$L_2(8)$	28	
$\mathbb{A}_7$	15	two representations
M_{11}	11, 12	
M_{12}	12	two representations
M_{22}	22	
M_{23}	23	
M_{24}	24	
HS	176	two representations
Co_3	276	

3. Outer automorphisms and the Schur multiplier

If L is a finite simple group, then the group $\mathrm{Out}(L) = \mathrm{Aut}(L)/L$ of outer automorphisms of L is solvable ("Schreier's conjecture") [Gor 2]. More precisely, $\mathrm{Out}(\mathbb{A}_n)$ is equal to $\mathbb{Z}_2$ if $n \neq 6$, and $\mathbb{Z}_2^2$ if $n = 6$. The Schur multiplier $Mult(\mathbb{A}_n)$ is equal to $\mathbb{Z}_2$ if $n \neq 6, 7$, and $\mathbb{Z}_6$ if $n = 6, 7$. The order $|\mathrm{Out}(L)|$ for finite groups L of Lie type is given in Table A3, where $q = p^f$. If L is one of 26 sporadic finite simple groups, then $|\mathrm{Out}(L)|$ is given in Table $A4$ (see [ATLAS]).

If L is a finite simple group other than $\mathbb{A}_n$ not mentioned in Table A5, then (see [KlL 2])

$$Mult(L) = \mathbb{Z}_d,$$

where the parameter d is as in Tables A3, A4.

In Table A4 we have listed also the values of $d(L)$, where L is a sporadic simple group, and $d(X)$ denotes the minimum degree of faithful projective complex representations of a finite group X. In what follows we shall use the following notation for an arbitrary finite group X: $P(X)$ (respectively $P_f(X)$, $P_f^t(X)$) denotes the minimum degree of nontrivial (respectively faithful, faithful transitive) permutation representations of X. Furthermore,

$$R_{\mathbb{F}}(X) = \min\left\{n | X \hookrightarrow PGL_n(\mathbb{F})\right\},$$

where $\mathbb{F}$ is a field;

$$R_p(X) = \min\left\{R_{\mathbb{F}}(X) | \mathrm{char}\mathbb{F} = p\right\},$$

Table A3. Finite simple groups of Lie type

L	d	$\lvert\mathrm{Out}(L)\rvert$	$R_p(L)$
$L_n(q) = A_{n-1}(q)$	$\gcd(\boldsymbol{n}, q-1)$	$\begin{cases} 2df, & \boldsymbol{n} \geq 3 \\ df, & \boldsymbol{n} = 2 \end{cases}$	$\boldsymbol{n}$
$PSU_n(q) = {}^2A_{n-1}(q)$	$\gcd(\boldsymbol{n}, q+1)$	$\begin{cases} 2df, & \boldsymbol{n} \geq 3 \\ df, & \boldsymbol{n} = 2 \end{cases}$	$\boldsymbol{n}$
$PSp_{2m}(q) = C_m(q)$, $(m, q) \neq (2, 2)$	$\gcd(2, q-1)$	$\begin{cases} df, & m \geq 3 \\ 2f, & m = 2 \end{cases}$	$2m$
$\Omega_{2m+1}(q) = B_m(q)$, $2 \nmid q$	2	$2f$	$2m+1$
$P\Omega^+_{2m}(q) = D_m(q)$, $m \geq 3$	$\gcd(4, q^m - 1)$	$\begin{cases} 2df, & m \neq 4 \\ 6df, & m = 4 \end{cases}$	$2m$
$P\Omega^-_{2m}(q) = {}^2D_m(q)$, $m \geq 2$	$\gcd(4, q^m + 1)$	$2df$	$2m$
$G_2(q)$	1	$\begin{cases} f, & p \neq 3 \\ 2f, & p = 3 \end{cases}$	$7 - \delta_{p,2}$
$F_4(q)$	1	$\gcd(2, p).f$	$26 - \delta_{p,3}$
$E_6(q)$	$\gcd(3, q-1)$	$2df$	27
$E_7(q)$	$\gcd(2, q-1)$	df	56
$E_8(q)$	1	f	248
${}^2B_2(q)$, $q = 2^{2a+1}$	1	f	4
${}^2G_2(q)$, $q = 3^{2a+1}$	1	f	7
${}^2F_4(q)$, $q = 2^{2a+1}$	1	f	26
${}^3D_4(q)$	1	$3f$	8
${}^2E_6(q)$	$\gcd(3, q+1)$	$2df$	27

where p is a prime;

$$R_{p'}(X) = \min\{R_\ell(X) \mid \ell \text{ a prime}, \ell \neq p\},$$

$$R(X) = \min\{R_p(X) \mid p \text{ a prime}\}.$$

In particular, $d(X) = R_{\mathbb{C}}(X)$. The last column of Table A3 gives the values of $R_p(L)$ ("the minimum degree of faithful projective representations in natural characteristic") for finite groups of Lie type defined over a field of characteristic p [KlL 2].

4. The minimum degree of projective representations of alternating groups [Schur], [Wag 1 – 3].
For $n = 5, 6, 7, 8$, or $n \geq 9$ we have $R(\mathbb{A}_n) = 2, 2, 3, 4$, or $n - 2$, respectively.

Table A4. Sporadic finite simple groups

L	d	$\lvert \mathrm{Out}(L) \rvert$	$d(L)$
M_{11}	1	1	10
M_{12}	2	2	10
M_{22}	12	2	10
M_{23}	1	1	22
M_{24}	1	1	23
J_1	1	1	56
J_2	2	2	6
J_3	3	2	18
J_4	1	1	1333
HS	2	2	22
Suz	6	2	12
McL	3	2	22
Ru	2	1	28
$He = F_7$	1	2	51
Ly	1	1	2480
$O'N$	3	2	342
Co_1	2	1	24
Co_2	1	1	23
Co_3	1	1	23
Fi_{22}	6	2	78
Fi_{23}	1	1	782
Fi'_{24}	3	2	783
$HN = F_5$	1	2	133
$Th = F_3$	1	1	248
$BM = F_2$	2	1	4371
$M = F_1$	1	1	196883

Table A5. The Schur multiplier of some finite simple groups

L	$Mult(L)$
$L_2(4)$, $L_3(2)$, $L_4(2)$, $PSU_4(2)$, $Sp_6(2)$, $G_2(4)$, $F_4(2)$	$\mathbb{Z}_2$
$G_2(3)$	$\mathbb{Z}_3$
$L_2(9) \cong Sp_4(2)'$, $\Omega_7(3)$	$\mathbb{Z}_6$
$L_3(4)$	$\mathbb{Z}_4 \times \mathbb{Z}_{12}$
$PSU_4(3)$	$\mathbb{Z}_3 \times \mathbb{Z}_{12}$
$PSU_6(2)$, ${}^2E_6(2)$	$\mathbb{Z}_2 \times \mathbb{Z}_6$
$\Omega_8^+(2)$, ${}^2B_2(8)$	$\mathbb{Z}_2^2$
$P\Omega_{2m}^+(q)$, $2 \nmid q$, $2 \mid m$	$\mathbb{Z}_2^2$

Furthermore, if s denotes the number of nonzero terms in the dyadic decomposition of n, then the degree of every faithful representation of a double covering group of $\mathbb{S}_n$ (respectively $\mathbb{A}_n$) in odd characteristics is divisible by $2^{[(n-s)/2]}$ (respectively $2^{[(n-s-1)/2]}$). Finally, if $n \geq 8$, then

$$d(\mathbb{A}_n) = d(\mathbb{S}_n) = n - 1.$$

5. The minimum degree of permutation representations [Coo 1], [KlL 2].
In Table A6 the values of $P(L)$ for the classical finite simple groups, and also for finite twisted groups of types ${}^2B_2(q)$, ${}^2G_2(q)$, are given. It should be pointed out that there are actually two errors in [Coo 1], corrected in [KlL 2].

Table A6. $P(L)$ for classical finite simple groups

L	$P(L)$
$L_n(q)$, $(\boldsymbol{n}, q) \neq (2,5), (2,7), (2,9), (2,11), (4,2)$	$(q^n - 1)/(q - 1)$
$L_2(5)$, $L_2(7)$, $L_2(9)$, $L_2(11)$, $L_4(2)$	$5, 7, 6, 11, 8$
$PSp_{2m}(q)$, $m \geq 2, q > 2, (m, q) \neq (2, 3)$	$(q^{2m} - 1)/(q - 1)$
$Sp_{2m}(2)$, $m \geq 3$	$2^{m-1}(2^m - 1)$
$Sp_4(2)'$, $PSp_4(3)$	$6, 27$
$\Omega_{2m+1}(q)$, $m \geq 3, q > 3, 2 \not| q$	$(q^{2m} - 1)/(q - 1)$
$\Omega_{2m+1}(3)$, $m \geq 3$	$3^m(3^m - 1)/2$
$P\Omega^+_{2m}(q)$, $m \geq 4, q \geq 3$	$(q^m - 1)(q^{m-1} + 1)/(q - 1)$
$\Omega^+_{2m}(2)$, $m \geq 4$	$2^{m-1}(2^m - 1)$
$P\Omega^-_{2m}(q)$, $m \geq 4$	$(q^m + 1)(q^{m-1} - 1)/(q - 1)$
$PSU_3(q)$, $q \neq 5$	$q^3 + 1$
$PSU_3(5)$	50
$PSU_4(q)$	$(q + 1)(q^3 + 1)$
$PSU_n(q)$, $\boldsymbol{n} \geq 5$, $(\boldsymbol{n}, q) \neq (6m, 2)$	$\frac{(q^n-(-1)^n)(q^{n-1}-(-1)^{n-1})}{q^2-1}$
$PSU_n(2)$, $6 \mid \boldsymbol{n}$	$2^{n-1}(2^n - 1)/3$
${}^2B_2(q)$, $q = 2^{2a+1} > 2$	$q^2 + 1$
${}^2G_2(q)$, $q = 3^{2a+1} > 3$	$q^3 + 1$

6. The minimum degree of projective representations of finite groups of Lie type in coprime characteristic [LaS], [SeZ].
Suppose that L is a finite group of Lie type defined over a field of characteristic p. Then $R_{p'}(L) \geq e(L)$, where $e(L)$ ("the Landázuri-Seitz-Zalesskii lower bound") is as in Table A7.

Table A7. The Landázuri-Seitz-Zalesskii lower bound

L	$e(L)$	Exceptions
$L_2(q)$	$(q-1)/\gcd(2,q-1)$	$\begin{cases} R_{2'}(L_2(4)) = 2, \\ R_{3'}(L_2(9)) = 3. \end{cases}$
$L_n(q)$, $n \geq 3$	$(q^n-1)/(q-1)-n$	$\begin{cases} R_{2'}(L_3(2)) = 2, \\ R_{2'}(L_3(4)) = 4, \\ R_{2'}(L_4(2)) = 7, \\ R_{3'}(L_4(3)) = 26. \end{cases}$
$PSp_{2m}(q)$, $m \geq 2$	$\begin{cases} (q^m-1)/2, & 2 \nmid q \\ \frac{(q^m-1)(q^m-q)}{2(q+1)}, & 2 \mid q \end{cases}$	$R_{2'}(Sp_4(2)') = 2$
$PSU_n(q)$, $n \geq 3$	$\begin{cases} (q^n-q)/(q+1), & 2 \nmid n \\ (q^n-1)/(q+1), & 2 \mid n \end{cases}$	$\begin{cases} R_{2'}(PSU_4(2)) = 4, \\ R_{3'}(PSU_4(3)) = 6. \end{cases}$
$P\Omega^+_{2m}(q)$, $m \geq 4$	$\begin{cases} \frac{(q^{m+1}-q)(q^{m-2}+1)}{q^2-1} - m, & q > 3 \\ \frac{(q^m-1)(q^{m-1}-1)}{q^2-1} - 7\delta_{2,p}, & q \leq 3 \end{cases}$	$R_{2'}(\Omega^+_8(2)) = 8$
$P\Omega^-_{2m}(q)$, $m \geq 4$	$\frac{q(q^{2m-2}-1)}{q^2-1} - q^{m-1} - m + 2$	
$\Omega_{2m+1}(q)$, $m \geq 3, 2 \nmid q$	$\begin{cases} \frac{q^{2m}-1}{q^2-1} - m, & q \neq 3 \\ \frac{(q^m-1)(q^m-q)}{q^2-1}, & q = 3 \end{cases}$	$R_{3'}(\Omega_7(3)) = 27$
$E_6(q)$	$q^9(q^2-1)$	
$E_7(q)$	$q^{15}(q^2-1)$	
$E_8(q)$	$q^{27}(q^2-1)$	
$F_4(q)$	$\begin{cases} q^6(q^2-1), & 2 \nmid q \\ q^7(q^3-1)(q-1)/2, & 2 \mid q \end{cases}$	$R_{2'}(F_4(2)) \geq 44$
${}^2E_6(q)$	$q^9(q^2-1)$	
$G_2(q)$	$q(q^2-1)$	$\begin{cases} R_{3'}(G_2(3)) = 14, \\ R_{2'}(G_2(4)) = 12. \end{cases}$
${}^3D_4(q)$	$q^3(q^2-1)$	
${}^2F_4(q)$	$q^4(q-1)(q/2)^{1/2}$	
${}^2B_2(q)$	$(q-1)(q/2)^{1/2}$	$R_{2'}({}^2B_2(8)) = 8$
${}^2G_2(q)$	$q(q-1)$	

Bibliography

[Abd 1] K. S. Abdukhalikov, *On invariant lattices in Lie algebras of type A_{q-1}*, Uspekhi Mat. Nauk 43 (1988), no. 1, 187 – 188; English transl. in Russian Math. Surveys 43 (1988), no. 1, 227 – 228.

[Abd 2] K. S. Abdukhalikov, *On automorphism group of the type A_{2^m-1} Lie algebra's orthogonal decomposition*, Izv. Vyssh. Uchebn. Zaved. Mat. 1991, no. 10, 11 – 14; English transl. in Soviet Math. 35 (1991), no. 10, 9 – 12.

[Abd 3] K. S. Abdukhalikov, *Invariant integral lattices in Lie algebras of type A_{p^m-1}*, Mat. Sb. 184 (1993), no. 4, 61 – 104; English transl. in Russian Acad. Sci. Sb. Math.

[Abd 4] K. S. Abdukhalikov, *Integral lattices associated with the finite affine group,* Mat. Sb. 185 (1994) (to appear, in Russian).

[Ada] J. F. Adams, *2-tori in E_8*, Math. Ann. 278 (1987), no. 1 – 4, 29 – 39.

[Adl] A. Adler, *Some integral representations of $PSL(2, p)$ and their applications*, J. Algebra 72 (1981), 115 – 145.

[Ale 1] A. V. Alekseevskii, *Finite commutative Jordan subgroups of complex simple Lie groups*, Funktsional Anal. i Prilozhen. 8 (1974), no. 4, 1 – 4; English transl. in Functional Anal. Appl. 8 (1974), 277 – 279.

[Ale 2] A. V. Alekseevskii, *Maximal finite subgroups of Lie groups*, Funktsional Anal. i Prilozhen. 9 (1975), no. 3, 79 – 80; English transl. in Functional Anal. Appl. 9 (1975), 248 – 250.

[Ale 3] A. V. Alekseevskii, *Structure of maximal finite primitive subgroups of Lie groups*, Uspekhi Mat. Nauk 30 (1975), no. 5, 197 – 198 (in Russian).

[Ale 4] A. V. Alekseevskii, *Maximal solvable subgroups of Lie groups*, Funktsional Anal. i Prilozhen. 14 (1980), no. 2, 44 – 45; English transl. in Functional Anal. Appl. 14 (1980), 114 – 115.

[Ami] S. A. Amitsur, *Finite subgroups of division rings*, Trans. Amer. Math. Soc. 80 (1955), no. 2, 1 – 386.

[Asch 1] M. Aschbacher, *On the maximal subgroups of the finite classical groups*, Invent. Math. 76 (1984), 469 – 514.

[Asch 2] M. Aschbacher, '*Overgroups of Sylow Subgroups in Sporadic Groups*', Mem. Amer. Math. Soc. 60 (1986), no. 343, 235 pp.

[Asch 3] M. Aschbacher, *Chevalley groups of type G_2 as the group of a trilinear form*, J. Algebra 109 (1987), 193 – 259.

[Asch 4] M. Aschbacher, *The 27–dimensional module for E_6*. I, Invent. Math. 89 (1987), 159 – 195.

[AsK 1] E. F. Assmus Jr. and J. D. Key, *Affine and projective planes*, Discrete Math. 83 (1990), 161 – 187.

[AsK 2] E. F. Assmus Jr. and J. D. Key, *Hadamard matrices and their designs : a coding-theoretic approach*, Trans. Amer. Math. Soc. 330 (1992), 269 – 294.

[ATLAS] J. H. Conway, R. T. Curtis, S. P. Norton, R. A. Parker and R. A. Wilson, '*An ATLAS of finite groups*', Clarendon Press, Oxford, 1985, 252 pp.

[Avr] G. S. Avrunin, *A vanishing theorem for second degree cohomology*, J. Algebra 53 (1978), no. 2, 382 – 388.

[BaB] C. Bachoc and C. Batut, *Étude algorithmique de réseaux construits avec la forme trace* (Preprint).

[BaE 1] R. D. Baker and G. L. Ebert, *Enumeration of two-dimensional flag-transitive planes*, Algebras Groups Geom. 3 (1985), 248 – 257.

[BaE 2] R. D. Baker and G. L. Ebert, *Construction of two-dimensional flag-transitive planes*, Geom. Dedicata 27 (1988), 9 – 14.

[BaI] E. Bannai and T. Itô, '*Algebraic Combinatorics I : Association Schemes*', Benjamin/Cummings, Menlo Park, 1984, 425 pp.

[Ban] E. Bannai, '*Positive Definite Unimodular Lattices with Trivial Automorphism Groups*', Mem. Amer. Math. Soc. 85 (1990), no. 429, 70 pp.

[BaW] E. S. Barnes and G. E. Wall, *Some extreme forms defined in terms of abelian groups*, J. Amer. Math. Soc. 1 (1959), 47 – 63.

[Bay] E. Bayer-Fluckiger, *Definite unimodular lattices having an automorphism of given characteristic polynomial,* Comment. Math. Helvet. 59 (1984), 509 – 538.

[BBNWZ] H. Brown, R. Bülow, J. Neubüser, H. Wondratschek and H. Zassenhaus, '*Crystallographic Groups of Four-dimensional Space*', Wiley, 1978, 443 pp.

[BDDKLS] F. Buekenhout, A. Delandtsheer, J. Doyen, P. B. Kleidman, M. W. Liebeck and J. Saxl, *Linear spaces with flag-transitive automorphism groups*, Geom. Dedicata 36 (1990), no. 1, 89 – 94.

[BKT] A. I. Bondal, A. I. Kostrikin and Pham Huu Tiep, *Invariant lattices, the Leech lattice and its even unimodular analogues in Lie algebra A_{p-1}*, Mat. Sb. 172 (1986), no. 8, 435 – 464; English transl. in Math. USSR-Sb. 58 (1986), 435 – 465.

[Blau] H. I. Blau, *On real and rational representations of finite groups*, J. Algebra 150 (1992), 57 – 72.

[Bon 1] A. I. Bondal, *Invariant lattices in Lie algebras of type A_{p-1}*, Vestnik Moskov. Univ. Ser. I. Mat. Mekh. 1986, no. 1, 52 – 54; English transl. in Moscow Univ. Math. Bull. 41 (1986), no. 1, 49 – 51.

[Bon 2] A. I. Bondal, *Private communication.*

[Bor 1] A. V. Borovik, *Jordan subgroups of simple algebraic groups*, Algebra i logika 28 (1989), no. 2, 144 – 159; English transl. in Algebra and Logic 28 (1989), no. 2, 97 – 108.

[Bor 2] A. V. Borovik, *Structure of finite subgroups of simple algebraic groups*, Algebra i logika 28 (1989), no. 3, 249 – 279; English transl. in Algebra and Logic 28 (1989), no. 3, 163 – 183.

[Bor 3] A. V. Borovik, *Jordan subgroups and orthogonal decompositions*, Algebra i logika 28 (1989), no. 4, 382 – 392; English transl. in Algebra and Logic 28 (1989), no. 4, 248 – 255.

[Bor 4] A. V. Borovik, *Finite subgroups of simple algebraic groups*, Dokl. Akad. Nauk SSSR 309 (1989), no. 4, 784 – 786; English transl. in Soviet Math. Dokl. 40 (1990), no. 3, 570 – 573.

[Bor 5] A. V. Borovik, *A maximal subgroup in the simple finite group $E_8(q)$*, Contemporary Mathematics 131 (1992), Pt. 1, 67 – 79.

[Borc 1] R. E. Borcherds, *Vertex algebras, Kac-Moody algebras and the monster*, Proc. Nat. Acad. Sci. U.S.A. 83 (1986), 3068 – 3071.

[Borc 2] R. E. Borcherds, *Generalized Kac-Moody algebras*, J. Algebra 115 (1988), 501 – 512.

[Borc 3] R. E. Borcherds, *Lattice like the Leech lattice*, J. Algebra 130 (1990), no. 1, 219 – 234.

[Borc 4] R. E. Borcherds, *The monster Lie algebra*, Adv. Math. 83 (1990), no. 1, 30 – 47.

[Borc 5] R. E. Borcherds, *Central extensions of generalized Kac-Moody algebras*, J. Algebra 140 (1991), 330 – 335.

[Borc 6] R. E. Borcherds, *Vertex algebras*, (to appear).

[Borc 7] R. E. Borcherds, *Monstrous moonshine and monstrous Lie superalgebras*, Invent. Math. 109 (1992), 405 – 444.

[BoS] Z. I. Borevich and I. R. Shafarevich, '*Number Theory*', Nauka, Moscow, 1964, 495 pp.; Third ed. 1985; English transl.: Academic Press, 1966.

[BoT] A. Borel and J. Tits, *Éléments unipotents et sous-groupes paraboliques de groupes réductifs*. I, Invent. Math. 12 (1971), no. 2, 95 – 104.

[Bour 1] N. Bourbaki, '*Algèbre*', Chap. 1 - 9, Hermann, Paris.

[Bour 2] N. Bourbaki, '*Groupes et Algèbres de Lie*', Chap. 1 - 8, Actualites Sci. Industr., no. 1337, Hermann, Paris, 1968 – 1975; Chap. 9, Masson, Paris, 1982.

[BrE] M. Broué and M. Enguehard, *Une familie infinie de formes quadratiques entières, leurs groupes d'automorphismes*, Ann. Sci. École Norm. Sup. 6 (1973), no. 1, 17 – 53.

[BuG] M. Busarkin and I. Gortschakov, '*Finite Split Groups*', Nauka, Moscow (in Russian).

[Bur 1] V. P. Burichenko, *Transitive orthogonal decompositions of simple complex Lie algebras of F_4 and E_6 types*, Vestnik Moskov. Univ. Ser. I. Mat. Mekh. 1988, no. 4, 78 – 80; English transl. in Moscow Univ. Math. Bull. 43 (1988), no. 4, 74 – 76.

[Bur 2] V. P. Burichenko, '*Invariant Lattices of Type F_4*', M. Sc. Thesis, Moscow University, 1988 (in Russian).

[Bur 3] V. P. Burichenko, *On a special loop, the Dickson form and the lattice connected with $O_7(3)$*, Mat. Sb. 182 (1991), no. 10, 1408 – 1429; English transl. in Math. USSR-Sb. 74 (1993), 145 – 167.

[Bur 4] V. P. Burichenko, *Invariant lattices in the Steinberg module and their automorphism groups*, Mat. Sb. 184 (1993), no. 12, 145 – 156; English transl. in Russian Acad. Sci. Sb. Math.

[BuT] V. P. Burichenko and Pham Huu Tiep, *Invariant lattices of types F_4 and E_6: the automorphism groups*, Commun. Algebra 21 (1993), no. 12, 4641 – 4677.

[CaE] H. Cartan and S. Eilenberg, '*Homological Algebra*', Princeton Univ. Press, Princeton, New Jersey, 1956, 390 pp.

[CaL] P. J. Cameron and J. H. van Lint, '*Graphs, Codes and Designs*', London Math. Soc. Lecture Note Ser. no. 43, Cambridge Univ. Press, Cambridge, 1980, 147 pp.

[Cam 1] P. J. Cameron, '*Parallelisms of Complete Designs*', London Math. Soc. Lecture Note Ser. no. 23, Cambridge Univ. Press, Cambridge, 1976, 144 pp.

[Cam 2] P. J. Cameron, *Permutation groups on unordered sets*, Higher Combinatorics. NATO Adv. Study Inst. Ser. C: Math. and Phys. Sci. 31 (1977), Reidel, Dordrecht, 217 – 239.

[Cam 3] P. J. Cameron, *Finite permutation groups and finite simple groups*, Bull. London Math. Soc. 13 (1981), 1 – 22.

[Car] R. W. Carter, '*Finite Groups of Lie Type : Conjugacy Classes and Complex Characters*', Wiley-Interscience, 1985, 544 pp.

[Cas] J. W. S. Cassels, '*Rational Quadratic Forms*', Academic Press, New York, 1978, 413 pp.

[CLSS] A. M. Cohen, M. W. Liebeck, J. Saxl and G. M. Seitz, *The local maximal subgroups of exceptional groups of Lie type, finite and algebraic*, Proc. London Math. Soc. (3) 64 (1992), no. 1, 21 – 48.

[CoG] A. M. Cohen and R. L. Griess, *On finite simple subgroups of the complex Lie group of type* E_8, Proc. Sympos. Pure Math. 47 (1987), Pt. 2, 367 – 405.

[Con 1] J. H. Conway, *A characterization of Leech's lattice*, Invent. Math. 7 (1969), 137 – 142.

[Con 2] J. H. Conway, *A group of order 8315553613086720000*, Bull. London Math. Soc. 1 (1969), 79 – 88.

[Coo 1] B. N. Cooperstein, *Minimal degree for a permutation representation of a classical group*, Israel J. Math. 30 (1978), 213 – 235.

[Coo 2] B. N. Cooperstein, *The geometry of root subgroups in exceptional groups.* II, Geom. Dedicata 15 (1983), 1 – 45.

[CoS 1] J. H. Conway and N. J. A. Sloane, *The Coxeter-Todd lattice, the Mitchell group and related sphere packings*, Math. Proc. Cambridge Philos. Soc. 93 (1983), 421 – 440.

[CoS 2] J. H. Conway and N. J. A. Sloane, *Low-dimensional lattices.* I. *Quadratic forms of small determinant*, Proc. Roy. Soc. London Ser. A, 1988, no. 418, 17 – 41.

[CoS 3] J. H. Conway and N. J. A. Sloane, *Low-dimensional lattices.* II. *Subgroups of* $GL(n, \mathbb{Z})$, Proc. Roy. Soc. London Ser. A, 1988, no. 419, 29 – 68.

[CoS 4] J. H. Conway and N. J. A. Sloane, *Low-dimensional lattices.* III. *Perfect forms*, Proc. Roy. Soc. London Ser. A, 1988, no. 418, 43 – 80.

[CoS 5] J. H. Conway and N. J. A. Sloane, *Low-dimensional lattices.* IV. *The mass formula*, Proc. Roy. Soc. London Ser. A, 1988, no. 419, 259 – 286.

[CoS 6] J. H. Conway and N. J. A. Sloane, *Low-dimensional lattices.* V. *Integral coordinates for integral lattices*, Proc. Roy. Soc. London Ser. A, 1989, no. 1871, 211 – 232.

[CoS 7] J. H. Conway and N. J. A. Sloane, '*Sphere Packings, Lattices and Groups*', Springer-Verlag, New York et al, 1988, 663 pp.

[CoS 8] J. H. Conway and N. J. A. Sloane, *A new upper bound for the minimum of an integral lattice of determinant one,* Bull. Amer. Math. Soc. 23 (1990), 383 – 387.

[CoS 9] J. H. Conway and N. J. A. Sloane, *On the minimum of unimodular lattices I: Upper bounds* (in preparation).

[CoS 10] J. H. Conway and N. J. A. Sloane, *On the minimum of unimodular lattices II: Lower bounds* (in preparation).

[CoW] A. M. Cohen and D. B. Wales, *Finite subgroups of* $G_2(\mathbb{C})$, Commun. Algebra 11 (1983), no. 4, 441 – 459.

[CPS] J. H. Conway, V. Pless and N. J. A. Sloane, *The binary self-dual codes of length up to* 32 *: a revised enumeration*, J. Combin. Theory Ser. A, 60 (1992), 183 – 195.

[CSe] A. M. Cohen and G. M. Seitz, *The* $r-$*rank of groups of exceptional Lie type*, Centrum voor Wiskunde en Informatica, Preprint PM-R8607, 1986.

[CuR] C. Curtis and I. Reiner, '*Representation Theory of Finite Groups and Associative Algebras*', Pure and Applied Math. v. 11, Interscience, New York, 1962; Second ed. 1966, 689 pp.

[Cur] R. T. Curtis, *On subgroups of .0.* II. *Local structure*, J. Algebra 63 (1980), no. 2, 413 – 434.

[Cze] T. Czerwinski, *The collineation groups of the translation planes of order* 25, Geom. Dedicata 39 (1991), 125 – 137.

[CzO] T. Czerwinski and D. Oakden, *The translation planes of order twenty-five*, J. Combin. Theory Ser. A, 59 (1992), 193 – 217.

[DeM] D. I. Deriziotis and G. O. Michler, *Character table and blocks of finite simple triality groups* ${}^3D_4(q)$, Trans. Amer. Math. Soc. 303 (1987), 39 – 70.

[Demb] P. Dembowski, '*Finite Geometries*', Springer-Verlag, New York, 1968, 375 pp.

[Demp 1] U. Dempwolff, *On extensions of an elementary group of order* 2^5 *by* $GL(5,2)$, Rend. Sem. Mat. Univ. Padova 48 (1973), 359 – 361.

[Demp 2] U. Dempwolff, *On extensions of elementary abelian* $2-$*groups by* Σ_n, Glasnik Matematicki 14 (1979), 35 – 40.

[Dieu] J. Dieudonné, '*La Géométrie des Groupes Classiques*', Troisieme ed., Springer-Verlag, Berlin, 1971, 129 pp.

[DiM] J. D. Dixon and B. Mortimer, *The primitive permutation groups of degree less than* 1000, Math. Proc. Cambridge Philos. Soc. 103 (1988), 213 – 238.

[Dye] R. H. Dye, *Alternating groups as maximal subgroups of the special orthogonal groups over the field of two elements*, J. Algebra 71 (1981), 472 – 480.

[Ebe] G. L. Ebert, *Translation planes of order* q^2*: Asymptotic estimates*, Trans. Amer. Math. Soc. 238 (1978), 301 – 308.

[Elk] N. Elkies, *On Mordell-Weil lattices,* Max-Planck-Institut f. Mathematik, Arbeitstagung 1990, Bonn.

[Enn] V. Ennola, *On the characters of the finite unitary groups*, Ann. Acad. Scient. Fenn. Ser. AI Math. 1963, no. 323, 3 – 35.

[Ero] V. A. Erokhin, *Automorphism groups of* 24*-dimensional even unimodular lattices*, Zap. Nauchn. Sem. Leningrad. Otdel. Mat. Inst. Steklov. (LOMI) 116 (1982), 68 – 73; English transl. in J. Soviet Math. 26 (1984), 1876 – 1879.

[Fad] D. K. Faddeev, *On complex representations of the full affine group over a finite field*, Dokl. Akad. Nauk SSSR 230 (1976), 295 – 297; English transl. in Soviet Math. Dokl. 17 (1976), no. 5, 1315 – 1318.

[Feit 1] W. Feit, *Groups with a cyclic Sylow subgroup*, Nagoya Math. J. 27 (1966), no. 2, 571 – 584.

[Feit 2] W. Feit, *On integral representations of finite groups*, Proc. London Math. Soc. (3) 29 (1974), 633 – 683.

[Feit 3] W. Feit, *Some lattices over* $\mathbb{Q}(\sqrt{-3})$, J. Algebra 52 (1978), 248 – 263.

[Feit 4] W. Feit, '*The Representation Theory of Finite Groups*', North-Holland Publishing Company, Amsterdam et al, 1982, 502 pp.

[Fer] P. A. Ferguson, *Finite complex linear groups of degree less than* $(2q+1)/3$, Proc. Sympos. Pure Math. 37 (1980), 413 – 417.

[Fis 1] B. Fischer, *Finite groups generated by* 3*-transpositions*, Notes, Mathematics Institute, University of Warwick, 1970.

[Fis 2] B. Fischer, *Finite groups generated by* 3*-transpositions.* I, Invent. Math. 13 (1971), no. 3, 232 – 246.

[FLM 1] I. B. Frenkel, J. Lepowsky and A. Meurman, *A natural representation of the Fischer-Griess Monster with the modular function J as character*, Proc. Nat. Acad. Sci. U.S.A. 81 (1984), 3256 – 3260.

[FLM 2] I. B. Frenkel, J. Lepowsky and A. Meurman, '*Vertex Operator Algebras and the Monster*', Academic Press, Boston et al, 1988, 502 pp.

[Ger] P. Gerardin, *Weil representations associated to finite fields*, J. Algebra 46 (1977), 54 – 101.

[GHJ] F. M. Goodman, P. de la Harpe and V. F. R. Jones, '*Coxeter Graphs and Towers of Algebras*', Springer, New York, 1989, 288 pp.

[Gor 1] D. Gorenstein, '*Finite Groups*', Harper and Row, New York, 1968, 527 pp.

[Gor 2] D. Gorenstein, '*Finite Simple Groups. An Introduction on Their Classification*', Plenum Press, New York and London, 1982, 333 pp.

[Gow 1] R. Gow, *Real representations of the finite orthogonal and symplectic groups of odd characteristic*, J. Algebra 96 (1985), 249 – 274.

[Gow 2] R. Gow, *Even unimodular lattices associated with the Weil representation of the finite symplectic group*, J. Algebra 122 (1989), 510 – 519.

[Gow 3] R. Gow, *Unimodular integral lattices associated with the basic spin representations of* $2A_n$ *and* $2S_n$, Bull. London Math. Soc. 21 (1989), 257 – 262.

[Gri 1] R. L. Griess, *Automorphisms of extraspecial groups and nonvanishing degree 2 cohomology*, Pacific J. Math. 48 (1973), 403 – 422.

[Gri 2] R. L. Griess, *On a subgroup of order* $|2^{15}GL(5,2)|$ *in* $E_8(\mathbb{C})$*, the Dempwolff group and* $Aut(D_8 \circ D_8 \circ D_8)$, J. Algebra 40 (1976), no. 1, 271 – 279.

[Gri 3] R. L. Griess, *The friendly giant*, Invent. Math. 69 (1982), 1 – 102.

[Gri 4] R. L. Griess, *Code loops*, J. Algebra 100 (1986), 224 – 234.

[Gri 5] R. L. Griess, *Sporadic groups, code loops and nonvanishing cohomology*, J. Pure Appl. Algebra 44 (1987), 191 – 214.

[Gri 6] R. L. Griess, *Code loops and a large finite group containing triality for* D_4, Rend. Circ. Mat. Palermo (2) Suppl. no. 19 (1988), 79 – 98.

[Gri 7] R. L. Griess, *A Moufang loop, the exceptional Jordan algebra, and a cubic form in* 27 *variables*, J. Algebra 131 (1990), 281 – 293.

[Gro] B. H. Gross, *Group representations and lattices*, J. Amer. Math. Soc. 3 (1990), 929 – 960.

[HaH] P. Hall and G. Higman, *On the p-length of p-soluble groups and reduction theorems for Burnside's problem*, Proc. London Math. Soc. 6 (1956), no. 21, 1 – 42.

[Hahn] A. Hahn, *The coset lattices of E. S. Barnes and G. E. Wall*, J. Austral. Math. Soc. Ser. A, 49 (1990), 418 – 433.

[HaJ] P. de la Harpe and V. F. R. Jones, *Paires de sous-algèbres semi-simples et graphes fortement réguliers*, C. R. Acad. Sci. Paris Ser. I Math. 311 (1990), 147 – 150.

[Hall 1] M. Hall, '*The Theory of Groups*', Macmillan Company, New York, 1959, 434 pp.

[Hall 2] M. Hall, *Construction of finite simple groups*, in: Computers in Algebra and Number Theory, SIAM-AMS Proceedings, Vol. IV, Providence, R.I., 1971, pp. 109 – 134.

[Her 1] C. Hering, *Eine nicht-desarguessche zweifach transitive affine Ebene der Ordnung* 27, Abh. Math. Sem. Univ. Hamburg 34 (1970), no. 3 – 4, 203 – 208.

[Her 2] C. Hering, *Transitive linear groups which contain irreducible subgroups of prime order*. II, J. Algebra 93 (1985), 151 – 164.

[Her 3] C. Hering, *On linear groups which contain an irreducible subgroup of prime order,* Proc. Inter. Conf. on Projective Planes, Washington State Univ. Press, Pullman, 1973, pp. 99 – 105.

[Hes] W. H. Hesselink, *Special and pure gradings of Lie algebras*, Math. Z. 179 (1982), no. 1, 135 – 149.

[Hup] B. Huppert, '*Endliche Gruppen.*I', Zweiter Nachdruck, Springer-Verlag, Berlin, 1983, 796 pp.

[Isa 1] I. M. Isaacs, *Characters of soluble and symplectic groups*, Amer. J. Math. 95 (1973), 594 – 635.

[Isa 2] I. M. Isaacs, '*Character Theory of Finite Groups*', Academic Press, New York, 1976, 303 pp.

[Iva 1] D. N. Ivanov, *Orthogonal decompositions of Lie algebras of type A_{p^n-1}, and isotropic fiberings*, Uspekhi Mat. Nauk 42 (1987), no. 4, 187 – 188; English transl. in Russian Math. Surveys 42 (1987), no. 4, 141 – 142.

[Iva 2] D. N. Ivanov, *Orthogonal decompositions of semisimple associative algebras*, Vestnik Moskov. Univ. Ser. I Mat. Mekh. 1988, no. 1, 9 – 14; English transl. in Moscow Univ. Math. Bull. 43 (1988), no. 1, 10 – 15.

[Iva 3] D. N. Ivanov, *Orthogonal decompositions of Lie algebras of types A_{p^n-1} and D_n with a finite number of classes of similar invariant sublattices*, Vestnik Moskov. Univ. Ser. I Mat. Mekh. 1989, no. 2, 40 – 43; English transl. in Moscow Univ. Math. Bull. 44 (1989), no. 2, 59 – 64.

[Iva 4] D. N. Ivanov, *A theorem on the subalgebras forming an orthogonal decomposition of an associative algebra*, Uspekhi Mat. Nauk 44 (1989), no. 2, 231 – 232; English transl. in Russian Math. Surveys 44 (1989), no. 2, 283 – 284.

[Iva 5] D. N. Ivanov, *An analogue of the Wagner theorem for orthogonal decompositions of the matrix algebra $M_n(\mathbb{C})$*, Uspekhi Mat. Nauk 49 (1994), no. 1, 215 – 216; English transl. in Russian Math. Surveys.

[Iva 6] D. N. Ivanov, *Orthogonal decompositions of associative algebras. Selected problems* (to appear).

[Iva 7] D. N. Ivanov, *Automorphisms of orthogonal decompositions and group algebras of groups with a partition* (to appear).

[Jac] N. Jacobson, '*Lie Algebras*', Interscience Publishers, John Wiley and Sons, N.Y. and London, 1962, 331 pp.

[Jack] D. Jackson, Ph. D. Thesis, Cambridge, England, 1971.

[Jan] Z. Janko, *A characterization of the simple group $G_2(3)$*, J. Algebra 12 (1969), 360 – 371.

[Jip] Z. Jiping, *On linear groups of degree at most $|P|-1$*, J. Algebra 143 (1991), 307 – 319.

[Kac] V. G. Kac, '*Infinite Dimensional Lie Algebras*', Second ed., Cambridge University Press, Cambridge, New York, 1985, 280 pp.

[Kal] M. J. Kallaher, '*Affine planes with transitive collineation groups*', Elsevier North Holland, N. Y. et al., 1982, 155 pp.

[Kan 1] W. M. Kantor, *k-homogeneous groups*, Math. Z. 124 (1972), 261 – 265.

[Kan 2] W. M. Kantor, *Linear groups containing a Singer cycle*, J. Algebra 62 (1980), 232 – 234.

[Kan 3] W. M. Kantor, *Strongly regular graphs defined by spreads*, Israel J. Math. 41 (1982), no. 4, 298 – 312.

[Kan 4] W. M. Kantor, *Spreads, translation planes and Kerdock sets.* I, SIAM J. Algebraic Discrete Methods 3 (1982), 151 – 165.

[Kan 5] W. M. Kantor, *Spreads, translations planes and Kerdock sets.* II, SIAM J. Algebraic Discrete Methods 3 (1982), 303 – 318.

[Kan 6] W. M. Kantor, *Expanded, sliced and spread spreads*, Lecture Notes in Pure and Appl. Math. 82 (1983), 251 – 261.

[Kan 7] W. M. Kantor, *Flag-transitive planes*, Lecture Notes in Pure and Appl. Math. 103 (1985), 179 – 181.

[Kan 8] W. M. Kantor, *Kerdock codes and related planes*, Discrete Math. 106/107 (1992), 297 – 302.

[Kan 9] W. M. Kantor, *Two families of flag-transitive planes*, Geom. Dedicata 41 (1992), 191 – 200.

[Kan 10] W. M. Kantor, *Note on Lie algebras, finite groups and finite geometries,* in: Groups and Geometries, Proc. Como Conference May 1993, Walter de Gruyter, Berlin 1994.

[KKU 1] A. I. Kostrikin, I. A. Kostrikin and V. A. Ufnarovskii, *Orthogonal decompositions of simple Lie algebras*, Dokl. Akad. Nauk SSSR 260 (1981), no. 3, 526 – 530; English transl. in Soviet Math. Dokl. 24 (1981), no. 2, 292 – 296.

[KKU 2] A. I. Kostrikin, I. A. Kostrikin and V. A. Ufnarovskii, *Orthogonal decompositions of simple Lie algebras (type A_n)*, Trudy Mat. Inst. Steklov 158 (1981), 105 – 120; English transl. in Proc. Steklov Inst. Math. 1983, iss. 4 (158), 113 – 130.

[KKU 3] A. I. Kostrikin, I. A. Kostrikin and V. A. Ufnarovskii, *Multiplicative decompositions of simple Lie algebras*, Dokl. Akad. Nauk SSSR 262 (1982), no. 1, 29 – 33; English transl. in Soviet Math. Dokl. 25 (1982), no. 1, 23 – 27.

[KKU 4] A. I. Kostrikin, I. A. Kostrikin and V. A. Ufnarovskii, *On the question about the uniqueness of orthogonal decompositions of Lie algebras of types A_n and C_n*. I, II. Issledovaniya po algebre i topologii, Mat. Issled. 74 (1983), Shtiinsa, Kishinev (in Russian).

[KKU 5] A. I. Kostrikin, I. A. Kostrikin and V. A. Ufnarovskii, *Invariant lattices of type G_2 and their automorphism groups*, Trudy Mat. Inst. Steklov. 165 (1984), 79 – 97; English transl. in Proc. Steklov Inst. Math. 1985, iss. 3 (165), 85 – 106.

[KKU 6] A. I. Kostrikin, I. A. Kostrikin and V. A. Ufnarovskii, *On decompositions of classical Lie algebras*, Trudy Mat. Inst. Steklov 166 (1984), 107 – 120; English transl. in Proc. Steklov Inst. Math. 1986, iss. 1 (166), 117 – 134.

[KKU 7] A. I. Kostrikin, I. A. Kostrikin and V. A. Ufnarovskii, *Decompositions in simple Lie algebras*, preprint, Kishinev, 1983, 79 pp (in Russian).

[Kle 1] P. B. Kleidman, *The maximal subgroups of $G_2(q)$ (q odd) and of ${}^2G_2(q)$*, J. Algebra 117 (1988), no. 1, 30 – 71.

[Kle 2] P. B. Kleidman, '*The Low-dimensional Finite Classical Groups and Their Subgroups*', Longman Research Notes. Math. Ser. (to appear).

[Kle 3] P. B. Kleidman, *The finite 2-transitive spreads and translation planes*, (manuscript).

[KlL 1] P. B. Kleidman and M. W. Liebeck, *On a theorem of Feit and Tits*, Proc. Amer. Math. Soc. 107 (1989), no. 2, 315 – 322.

[KlL 2] P. B. Kleidman and M. W. Liebeck, '*The Subgroup Structure of the Finite Classical Groups*', London Math. Soc. Lecture Note Ser. no. 129, Cambridge University Press, 1990, 303 pp.

[KlP] A. S. Kleshchev and A. A. Premet, *On second degree cohomology of symmetric and alternating group,* Commun. Algebra 21 (1993), 583 – 600.

[Kne] M. Kneser, *Klassenzahlen definiter quadratischer Formen*, Arch. Math. (Basel) 8 (1957), 241 – 250.

[KoN] H. Koch and G. Nebe, *Extremal even unimodular lattices of rank 32 and related codes,* Math. Nachr. 161 (1993), 309 – 319.

[Kos 1] A. I. Kostrikin, *Invariant lattices in Lie algebras and their automorphism groups*, in: Group Theory, Singapore, 1987, Walter de Gruyter, Berlin, New York, 171 – 181.

[Kos 2] A. I. Kostrikin, *Some new results on orthogonal decompositions of Lie algebras*, Proc. Fifth Internat. Conf. on Hadronic Mechanics and Nonpotential Interactions, August 13 – 17, 1990, Univ. of Northern Iowa, Nova Science Publishers Inc. (to appear).

[Kosh] S. Koshitani, *The Loewy structure of the projective indecomposable modules for $SL(3,3)$ and its automorphism group in characteristic* 3, Commun. Algebra 15 (1987), no. 6, 1215 – 1253.

[KoT 1] A. I. Kostrikin and Pham Huu Tiep, *Irreducible orthogonal decompositions of simple Lie algebras of type A_n*, Dokl. Akad. Nauk SSSR 314 (1990), no. 4, 782 – 786; English transl. in Soviet Math. Dokl. 42 (1991), no. 2, 538 – 542.

[KoT 2] A. I. Kostrikin and Pham Huu Tiep, *Classification of the irreducible orthogonal decompositions of simple complex Lie algebras of type A_n*, Algebra i Analiz 3 (1991), no. 3, 86 – 109; English transl. in St. Petersburg Math. J. 3 (1992), no. 3, 571 – 593.

[KoV 1] H. Koch and B. B. Venkov, *Über ganzzahlige unimodulare euklidische Gitter*, J. reine und angew. Math. 398 (1989), 144 – 168.

[KoV 2] H. Koch and B. B. Venkov, *Über gerade unimodulare Gitter der Dimension 32,* III, Math. Nachr. 152 (1991), 191 – 213.

[Lang] S. Lang, '*Algebra*', Third ed., Addison-Wesley, Reading MA, 1993, 906 pp.

[LaS] V. Landázuri and G. Seitz, *On the minimal degrees of projective representations of the finite Chevalley groups*, J. Algebra 32 (1974), 418 – 443.

[Lee] J. Leech, *Notes on sphere packings*, Canad. J. Math. 19 (1967), 251 – 267.

[Lev] V. I. Levenshtein, *On bounds for packings in n-dimensional Euclidean space*, Dokl. Akad. Nauk SSSR 245 (1979), 1299 – 1303; English transl. in Soviet Math. Dokl. 20 (1979), 417 – 421.

[Lie] M. W. Liebeck, *On the orders of maximal subgroups of the classical groups*, Proc. London Math. Soc. (3) 50 (1985), 426 – 446.

[Lie 2] M. W. Liebeck, *The affine permutation groups of rank three*, Proc. London Math. Soc. (3) 54 (1987), 477 – 516.

[LiN] R. Lidl and H. Niederreiter, '*Finite Fields*', Cambridge University Press, Cambridge, London et al. 1987, 755 pp.

[Lin] J. H. Lindsey, *A new lattice for the Hall-Janko group*, Proc. Amer. Math. Soc. 103 (1988), no. 3, 703 – 709.

[LiS] M. W. Liebeck and J. Saxl, *Primitive permutation groups containing an element of large prime order*, J. London Math. Soc (2) 31 (1985), no. 2, 237 – 249.

[LPS] M. W. Liebeck, C. E. Praeger and J. Saxl, *On the O'Nan-Scott theorem for finite primitive permutation groups*, J. Austral. Math. Soc. Ser. A, 44 (1988), 389 – 396.

[LPW] K. Lux, R. A. Parker and R. A. Wilson, '*Atlas of Finite Groups: Modular Character Tables*', (in preparation).

[Lun] H. Lüneburg, '*Translations Planes*', Springer-Verlag, Berlin a.o., 1980, 278 pp.

[Mag] K. Magaard, '*The Maximal Subgroups of the Chevalley Groups $F_4(F)$, Where F is Finite or Algebraically Closed Field of Characteristic $\neq 2, 3$*', Ph. D. Thesis, California Institute of Technology, Pasadena, California, 1990.

[MaS] G. Mason and E. Shult, *The Klein correspondence and the ubiquity of certain translation planes*, Geom. Dedicata 21 (1986), 29 – 50.

[McK 1] J. H. McKay, *A setting for the Leech lattice*, in: Finite Groups' 72, North-Holland, Amsterdam, 1973, 117 – 118.

[McK 2] J. H. McKay, *The non-abelian simple groups G, $|G| < 10^6$ – character tables*, Commun. Algebra 7 (1977), no. 13, 1407 – 1445.

[McS] F. J. MacWilliams and N. J. A. Sloane, '*The Theory of Error Correcting Codes*', Parts I, II, North-Holland Publ. Comp., Amsterdam, N.Y., 1978, 762 pp.

[Mor] A. O. Morris, *The spin representations of the symmetric group*, Proc. London Math. Soc. (3) 12 (1962), 55 – 76.

[MuW] A. Munemasa and Y. Watatani, *Paires orthogonales de sous-algèbres involutives*, C. R. Acad. Sci. Paris Ser. I Math. 314 (1992), 329 – 331.

[Naz] M. L. Nazarov, *Young's orthogonal form of irreducible projective representations of the symmetric group*, J. London Math. Soc. (2) 42 (1990), 437 – 451.

[Nie] H.-V. Niemeier, *Definite quadratische Formen der Dimension* 24 *und Diskriminante* 1, J. Number Theory 5 (1972), 142 – 178.

[NoW] S. Norton and R. A. Wilson, *The maximal subgroups of $F_4(2)$ and its automorphism group*, Commun. Algebra 17 (1989), no. 11, 2809 – 2824.

[Noz] S. Nozawa, *On the characters of the finite general unitary group $U(4, q^2)$*, J. Fac. Sci. Univ. Tokyo Sec. IA Math. 19 (1972), no. 3, 258 – 293.

[O'Me] O. T. O'Meara, '*Lectures on Linear Groups*', Providence, Amer. Math. Soc., Rhode Island, 1974, 87 pp.

[Ost] T. G. Ostrom, '*Finite Translation Planes*', Lecture Notes in Math. no. 158, Springer, Berlin a.o., 1970, 112 pp.

[PaZ] J. Patera and H. Zassenhaus, *The Pauli matrices in n dimensions and finest gradings of simple Lie algebras of type A_{n-1}*, J. Math. Phys. 29 (1988), no. 3, 665 – 673.

[Pie] R. Pierce, '*Associative Algebras*', Springer-Verlag, New York, 1982, 436 pp.

[Ple 1] W. Plesken, '*Group Rings of Finite Groups over p-adic Integers*', Lecture Notes in Math. no. 1026, Springer, Berlin - Heidelberg - New York, 1983, 151 pp.

[Ple 2] W. Plesken, *Finite unimodular groups of prime degree and circulants*, J. Algebra 97 (1985), 286 – 312.

[Ple 3] W. Plesken, *Some applications of representation theory*, Progress in Math. 95 (1991), 477 – 496.

[PlN] W. Plesken and G. Nebe, '*Finite Rational Matrix Groups*', Mem. Amer. Math. Soc. (to appear).

[PlP 1] W. Plesken and M. Pohst, *On maximal finite irreducible subgroups of $GL(n, \mathbb{Z})$*. I, II, Math. Comp. 31 (1977), no. 138, 536 – 577; III, IV, V, Math. Comp. 34 (1980), no. 149, 245 – 301.

[PlP 2] W. Plesken and M. Pohst, *Constructing integral lattices with prescribed minimum*, Math. Comp. 45 (1985), no. 171, 209 – 221.

[PlP 3] W. Plesken and M. Pohst, *Constructing integral lattices with prescribed minimum*. II, Math. Comp. 60 (1993), no. 202, 817 – 825.

[Pog] B. A. Pogorelov, *Primitive permutation groups of low degree*. I, II, Algebra i Logica 19 (1980), no. 3, 348 – 379, 423 – 457; English transl. in Algebra and Logic 19 (1981), 230 – 254, 278 – 296.

[Pop 1] S. Popa, *Orthogonal pairs of $*$-subalgebras in finite von Neumann algebras*, J. Operator Theory 9 (1983), 253 – 268.

[Pop 2] S. Popa, *Maximal injective subalgebras in factors associated with free groups*, Adv. Math. 50 (1983), 27 – 48.

[Pop 3] S. Popa, *Relative dimension, towers of projections and commuting squares of subfactors*, Pacific J. Math. 137 (1989), 181 – 207.

[Pop 4] S. Popa, *Classification of subfactors: the reduction to commuting squares*, Invent. Math. 101 (1990), 19 – 43.

[Popo] I. Popovici, *Graduations spéciales simples*, Bull. Soc. Roy. Sci. Liège 39 (1970), no. 5 – 6, 218 – 228.

[Pra] C. E. Praeger, *Primitive permutation groups containing an element of order p of small degree, p a prime*, J. Algebra 34 (1975), no. 3, 540 – 546.

[Que 1] H. - G. Quebbemann, *Zur Klassifikation unimodularer Gitter mit Isometrie von Primzahlordnung*, J. reine und angew. Math. 326 (1981), 158 – 170.

[Que 2] H. - G. Quebbemann, *Unimodular lattices with isometries of large prime order*. II, Math. Nachr. 156 (1992), 219 – 224.

[Ras] R. Rasala, *On the minimal degrees of characters of* $\mathbb{S}_n$, J. Algebra 45 (1977), no. 1, 132 – 181.

[SAG] '*Seminar on Algebraic Groups and Related Finite Groups*', Editors: A. Borel et al, Lecture Notes in Math., no. 131, Springer-Verlag, 1970, 321 pp.

[Sah] C. - H. Sah, *Cohomology of split group extensions*. I, J. Algebra 29 (1974), no. 2, 255 – 302; II, J. Algebra 45 (1977), no. 1, 17 – 68.

[San] G. Sandlöbes, *Perfect groups of order less than* 10^4, Commun. Algebra 9 (1981), 477 – 490.

[Schur] I. Schur, *Über die Darstellung der symmetrischen und alternierenden Grüppe durch gebrochene lineare Substitutionen*, Gesammelte Abhandlungen, Bd. 1, 346 – 441.

[Sei 1] G. M. Seitz, *Flag-transitive subgroups of Chevalley groups*, Ann. of Math. (2) 97 (1973), 27 – 56.

[Sei 2] G. M. Seitz, *Some representations of classical groups*, J. London Math. Soc. (2) 10 (1975), 115 – 120.

[Ser 1] J. - P. Serre, '*Lie Algebras and Lie Groups*', Lecture Notes in Math., no. 1500, Springer-Verlag, 1992, 168 pp.

[Ser 2] J. - P. Serre, '*Cours d'arithmétique*', Presses Universitaires de France, Paris, 1970, 188 pp.

[Ser 3] J. - P. Serre, '*Linear Representations of Finite Groups*', Graduate Texts in Math. 42, New York etc., Springer-Verlag, 1977, 170 pp.

[SeS] E. Seah and D. R. Stinson, *On the enumeration of one-factorizations of complete graphs containing prescribed automorphism groups*, Math. Comput. 50 (1988), no. 182, 607 – 618.

[SeZ] G. M. Seitz and A. E. Zalesskii, *On the minimal degrees of projective representations of the finite Chevalley groups*. II, J. Algebra 158 (1993), 233 – 243.

[Shaf] I. R. Shafarevich, '*Basic Algebraic Geometry*', Grundlehren Math. Wiss., Bd. 213, Springer-Verlag, 1974, 439 pp.

[Shau] E. P. Shaughnessy, *Codes with simple automorphism groups*, Arch. Math. (Basel) 22 (1971), no. 5, 459 – 466.

[She] F. A. Sherk, *Translation planes of order* 16, Lecture Notes in Pure and Appl. Math. 82 (1983), 401 – 412.

[Shi] T. Shioda, *Mordell-Weil lattices and sphere packings*, Amer. J. Math. 113 (1991), 931 – 948.

[SiF] W. Simpson and J. S. Frame, *The character tables for* $SL(3,q)$, $SU(3,q^2)$, $PSL(3,q)$, $PSU(3,q^2)$, Canad. J. Math. 25 (1973), 486 – 494.

[Sim] C. C. Sims, *Computational methods in the study of permutation groups*, in: Computational Problems in Abstract Algebra, Pergamon Press, 1970, pp. 169 – 184.

[Smi 1] P. E. Smith, '*On Certain Finite Simple Groups*', Ph. D. Thesis, Cambridge, 1975.

[Smi 2] P. E. Smith, *A simple subgroup of M? and* $E_8(3)$, Bull. London Math. Soc. 8 (1976), no. 9, 161 – 165.

[SSL] '*Séminaire "Sophus Lie". Théorie des Algèbres de Lie. Topologie des Groupes de Lie*', École Norm. Super., Paris, 1954 – 1955.

[Ste 1] R. Steinberg, *The representations of* $GL(3, q)$, $GL(4, q)$, $PGL(3, q)$ *and* $PGL(4, q)$, Canad. J. Math. 3 (1951), 225 – 235.

[Ste 2] R. Steinberg, '*Endomorphisms of Linear Algebraic Groups*', Mem. Amer. Math. Soc. 80 (1968), 108 pp.

[Ste 3] R. Steinberg, '*Lectures on Chevalley Groups*', Lecture Notes, Yale University, 1967 – 1968, 277 pp.

[Sue] C. Suetake, *Flag transitive planes of order* q^n *with a long cycle* l_∞ *as a collineation*, Graphs Combin. 7 (1991), 183 – 195.

[Sun] V. S. Sunder, *On commuting squares and subfactors*, prepublication, Indian Statistical Institute, Bangalore, 1990.

[Sup 1] D. A. Suprunenko, '*Matrix Groups*', Nauka, Moscow, 1972; English transl.: Providence, Rhode Island, 1976, 252 pp.

[Sup 2] I. D. Suprunenko, *Subgroups of* $GL(n, p)$ *containing* $SL(2, p)$ *in an irreducible representation of degree n*, Mat. Sb. 109 (1979), no. 3, 453 – 468; English transl. in Math. USSR-Sb. 37 (1980), 425 – 440.

[Suz 1] M. Suzuki, *On a class of doubly transitive groups*. I, Ann. of Math. (2) 75 (1962), no. 1, 105 – 145; II, Ann. of Math. (2) 79 (1964), no. 3, 514 – 589.

[Suz 2] M. Suzuki, '*Group Theory*', Grundlehren der mathematischen Wissenschaften, Bd. 247, Springer-Verlag, 1982, 434 pp; Bd. 248, Springer-Verlag, 1986, 621 pp.

[Tan] S. Tanaka, *Construction and classification of irreducible representations of special linear groups of the second order over a finite field*, Osaka J. Math. 4 (1967), 65 – 84.

[Tho 1] J. G. Thompson, *Finite groups and even lattices*, J. Algebra 38 (1976), no. 2, 523 – 524.

[Tho 2] J. G. Thompson, *A conjugacy theorem for* E_8, J. Algebra 38 (1976), no. 2, 525 – 530.

[Tho 3] J. G. Thompson, *A simple subgroup of* $E_8(3)$, Finite Groups Symposium, N. Iwahori, Japan Soc. for Promotion of Science, 1976, pp. 113 – 116.

[Tiep 1] Pham Huu Tiep, *One construction of even unimodular lattices*, Vestnik Moskov. Univ. Ser. I Mat. Mekh. 1986, no. 1, 54 – 56; English transl. in Moscow Univ. Math. Bull. 41 (1986), no. 1, 52 – 54.

[Tiep 2] Pham Huu Tiep, *Lattices in Lie algebras of type* A_{p-1}*: The Witt lattice and the Leech lattice*, in: Selected questions of algebra, geometry and discrete mathematics, Proc. Young Scientists' Conf. Algebra, Geometry and Topology, Moscow Univ., 1987, pp. 129 – 134 (in Russian).

[Tiep 3] Pham Huu Tiep, *Invariant sublattices in a Cartan subalgebra*, Vestnik Moskov. Univ. Ser. I Mat. Mekh. 1988, no. 4, 72 – 75; English transl. in Moscow Univ. Math. Bull. 43 (1988), no. 4, 65 – 68.

[Tiep 4] Pham Huu Tiep, *Lattices of root-type in Lie algebras* D_{2^m} *and* B_{2^m-1}, Vestnik Moskov. Univ. Ser. I Mat. Mekh. 1989, no. 1, 100 – 102; English transl. in Moscow Univ. Math. Bull. 44 (1989), no. 1, 92 – 95.

[Tiep 5] Pham Huu Tiep, *Lattices of non-radical type in the Lie algebras B_3 and D_4*, Uspekhi Mat. Nauk 44 (1989), no. 1, 217 – 218; English transl. in Russian Math. Surveys 44 (1989), no. 1, 247 – 248.

[Tiep 6] Pham Huu Tiep, *Small lattices in Lie algebras A_{p-1}*, Vestnik Moskov. Univ. Ser. I Mat. Mekh. 1989, no. 4, 70 – 72; English transl. in Moscow Univ. Math. Bull. 44 (1989), no. 4, 67 – 70.

[Tiep 7] Pham Huu Tiep, *Irreducible orthogonal decompositions in Lie algebras*, Mat. Sb. 180 (1989), no. 10, 1396 – 1414; English transl. in Math. USSR-Sb. 68 (1991), no. 1, 257 – 275.

[Tiep 8] Pham Huu Tiep, *A characteristic property of the multiplicative orthogonal decomposition of the Lie algebra D_4*, Vestnik Moskov. Univ. Ser. I Mat. Mekh. 1990, no. 5, 49 – 53; English transl. in Moscow Univ. Math. Bull. 45 (1990), no. 5, 46 – 49.

[Tiep 9] Pham Huu Tiep, *Irreducible J-decompositions of the Lie algebras A_{p^n-1}*, Mat. Zametki 49 (1991), no. 5, 128 – 134; English transl. in Math. Notes 49 (1991), 531 – 535.

[Tiep 10] Pham Huu Tiep, *On orthogonal decompositions of Lie algebras of type D_p and C_p*, Vestnik Moskov. Univ. Ser. I Mat. Mekh. 1991, no. 3, 13 – 16; English transl. in Moscow Univ. Math. Bull. 46 (1991), no. 3, 13 – 16.

[Tiep 11] Pham Huu Tiep, *A classification of the irreducible orthogonal decompositions of the simple complex Lie algebras of type B_n*, Commun. Algebra 19 (1991), no. 10, 2729 – 2775.

[Tiep 12] Pham Huu Tiep, *Weil representations of finite symplectic groups, and Gow lattices*, Mat. Sb. 182 (1991), no. 8, 1161 – 1183; English transl. in Math. USSR-Sb. 73 (1992), no. 2, 535 – 555.

[Tiep 13] Pham Huu Tiep, *A reduction theorem for invariant lattices of type A_n*, Dokl. Akad. Nauk SSSR 319 (1991), no. 1, 78 – 82; English transl. in Soviet Math. Dokl. 44 (1992), no. 1, 75 – 79.

[Tiep 14] Pham Huu Tiep, *Automorphism groups of some Mordell-Weil lattices*, Izv. Ross. Akad. Nauk Ser. Mat. 56 (1992), no. 3, 509 – 537; English transl. in Russian Acad. Sci. Izv. Math. 40(1993), no. 3, 477 – 501.

[Tiep 15] Pham Huu Tiep, *Basic spin representations of alternating groups, Gow lattices, and Barnes-Wall lattices*, Mat. Sb. 183 (1992), no. 11, 99 – 116; English transl. in Russian Acad. Sci. Sb. Math. 77 (1994), no. 2, 351 – 365.

[Tiep 16] Pham Huu Tiep, *Invariant lattices of type E_8 and their automorphism groups*, Algebra i Analiz 4 (1992), no. 5, 227 – 256; English transl. in St. Petersburg Math. J. 4 (1993), no. 5, 1029 – 1054.

[Tiep 17] Pham Huu Tiep, *A reduction theorem for invariant integral lattices of type A_n*, Nova J. Algebra and Geometry, 1 (1992), no. 3, 261 – 296.

[Tiep 18] Pham Huu Tiep, '*Orthogonal Decompositions and Integral Lattices*', D. Sc. Thesis, Moscow State University, Moscow, 1991.

[Tiep 19] Pham Huu Tiep, *On the Popa conjecture* (preprint).

[Tiep 20] Pham Huu Tiep, *Some globally irreducible representations*, Preprint 26 (1993), Institute for Experimental Mathematics, University of Essen (submitted to J. Algebra).

[Tiep 21] Pham Huu Tiep, *Globally irreducible representations of the finite symplectic group $Sp_4(q)$*, Preprint 4 (1994), Institute for Experimental Mathematics, University of Essen (to appear in Commun. Algebra).

[Tiep 22] Pham Huu Tiep, *Weil representations as globally irreducible representations,* Preprint 5 (1994), Institute for Experimental Mathematics, University of Essen (submitted to Math. Nachr.).

[Tiep 23] Pham Huu Tiep, *Globally irreducible representations of $SL_3(q)$,* Preprint 6 (1994), Institute for Experimental Mathematics, University of Essen.

[Tiep 24] Pham Huu Tiep, *Basic spin representations of $2\mathbb{S}_n$ and $2\mathbb{A}_n$ as globally irreducible representations,* Preprint 7 (1994), Institute for Experimental Mathematics, University of Essen (to appear in Arch. Math. (Basel)).

[Tiep 25] Pham Huu Tiep, *Globally irreducible representations of ${}^2A_2(q)$* (in preparation).

[Tiep 26] Pham Huu Tiep, *On the solvability of the kernel of an orthogonal decomposition* (in preparation).

[Tits] J. Tits, *Quaternions over $\mathbb{Q}(\sqrt{5})$, Leech's lattice and the sporadic group of Hall-Janko,* J. Algebra 63 (1980), 56 – 75.

[Tur] A. Turull, *The Schur index of projective characters of symmetric and alternating groups*, Ann. of Math. 135 (1992), 91 – 124.

[Ufn 1] V. A. Ufnarovskii, *On the computation of a certain automorphism group*, Mat. Issled. 85 (1985), Shtiinsa, Kishinev, 109 – 129 (in Russian).

[Ufn 2] V. A. Ufnarovskii, *On suitable matrices of second order*, Mat. Issled. 90 (1986), Shtiinsa, Kishinev, 113 – 136 (in Russian).

[Ust] V. A. Ustimenko, *Maximality of the group $P\Gamma L_n(q)$ acting on subspaces of dimension m*, Dokl. Akad. Nauk SSSR 240 (1978), no. 6, 1305 – 1308; English transl. in Soviet Math. Dokl. 19 (1978), 769 – 772.

[Ven 1] B. B. Venkov, *On the classification of integral even unimodular* 24*-dimensional quadratic forms*, Trudy Mat. Inst. Steklov 148 (1978), 65 – 76; English transl. in Proc. Steklov Inst. Math. 1980, iss. 4 (148), 63 – 74.

[Ven 2] B. B. Venkov, *Odd unimodular lattices*, Zap. Nauchn. Sem. Leningrad. Otdel. Mat. Inst. Steklov. (LOMI) 86 (1979), 40 – 48; English transl. in J. Soviet Math. 17 (1981), 1967 – 1974.

[Ven 3] B. B. Venkov, *Even unimodular Euclidean lattices in dimension* 32, Zap. Nauchn. Sem. Leningrad. Otdel. Mat. Inst. Steklov. (LOMI) 116 (1982), 44 – 55, 161 – 162; English transl. in J. Soviet Math. 26 (1984), 1860 – 1867.

[Ven 4] B. B. Venkov, *Unimodular lattices and strongly regular graphs*, Zap. Nauchn. Sem. Leningrad. Otdel. Mat. Inst. Steklov. (LOMI) 129 (1983), 30 – 38; English transl. in J. Soviet Math. 29 (1985), 1121 – 1127.

[Ven 5] B. B. Venkov, *Voronoi parallelohedra for certain unimodular lattices*, Zap. Nauchn. Sem. Leningrad. Otdel. Mat. Inst. Steklov. (LOMI) 132 (1983), 57 – 61; English transl. in J. Soviet Math. 30 (1985), 1833 – 1836.

[Ven 6] B. B. Venkov, *Even unimodular Euclidean lattices of dimension* 32. II, Zap. Nauchn. Sem. Leningrad. Otdel. Mat. Inst. Steklov. (LOMI) 134 (1984), 34 – 58; English transl. in J. Soviet Math. 36 (1987), 21 – 38.

[Ven 7] B. B. Venkov, *On even unimodular extremal lattices*, Trudy Mat. Inst. Steklov 165 (1984), 43 – 48; English transl. in Proc. Steklov Inst. Math. 1985, iss. 3 (165), 47 – 52.

[Ven 8] B. B. Venkov, '*Unimodular Euclidean Lattices*', D. Sc. Thesis, St. Petersburg State University, St. Petersburg, 1984.

[VGO] E. B. Vinberg, V. V. Gorbatsevich and A. L. Onishchik, '*Structure of Lie Groups and Lie Algebras*', Encyclopedia of Mathematical Sciences, Editors: R. V. Gamkrelidze, Vol. 41, Moscow, 1990; English transl. in Springer-Verlag.

[ViO] E. B. Vinberg and A. L. Onishchik, '*The Seminar on Lie Groups and Algebraic Groups*', Nauka, Moscow, 1988, 344 pp (in Russian); English transl.: 'Lie Groups and Algebraic Groups', Springer-Verlag, 1990.

[Wag 1] A. Wagner, *The faithful linear representations of least degree of S_n and A_n over a field of characteristic* 2, Math. Z. 151 (1976), 127 – 137.

[Wag 2] A. Wagner, *The faithful linear representations of least degree of S_n and A_n over a field of odd characteristic*, Math. Z. 154 (1977), 103 – 114.

[Wag 3] A. Wagner, *An observation on the degrees of projective representations of the symmetric and alternating groups over an arbitrary field*, Arch. Math. (Basel) 29 (1977), 583 – 589.

[War] H. N. Ward, *Representations of symplectic groups*, J. Algebra 20 (1972), 182 – 195.

[Weil] A. Weil, '*Basic Number Theory*', Springer-Verlag, 1967, 284 pp.

[Wie] H. Wielandt, '*Finite Permutation Groups*', Academic Press, N. Y. and London, 1964, 114 pp.

[Wil 1] R. A. Wilson, *The quaternionic lattice for $2G_2(4)$ and its maximal subgroups*, J. Algebra 77 (1982), 449 – 466.

[Wil 2] R. A. Wilson, *The complex Leech lattice and maximal subgroups of the Suzuki group*, J. Algebra 84 (1983), 151 – 188.

[Wil 3] R. A. Wilson, *The maximal subgroups of Conway's group Co_1*, J. Algebra 85 (1983), no. 1, 144 – 165.

[ZaS] A. E. Zalesskii and I. D. Suprunenko, *Representations of dimension $(p^n \mp 1)/2$ of symplectic group of degree $2n$ over a field of characteristic p*, Vestsi Akad. Navuk BSSR Ser. Fiz.-Mat. Navuk, 1987, no. 6, 9 – 15 (in Russian).

[Zsi] K. Zsigmondy, *Zur Theorie der Potenzreste*, Monath. Math. Phys. 3 (1892), 265 – 284.

Notation

$H \hookrightarrow G$	H isomorphic to a subgroup of G
$H.G$	an extension of H by G
$H \circ G$	a non-split extension of H by G
$H \times G$	the direct product of H and G
$H * G$	a central product of H and G
$H \wr G$	the wreath product of H and G
$G' = [G, G]$	the commutator subgroup of G
G^{∞}	the last member of the derived series of G
$\mathrm{Syl}_p(G)$	the set of Sylow p-subgroups of G
$O_p(G)$, $O_{p'}(G)$	the largest normal p- (p'-subgroup) of G
$\vert < g > \vert$	the order of an element g
$\vert X\vert$	the cardinality of a finite set X
$Z(G)$	the centre of G
$\exp(G)$	the exponent of G
$\mathrm{soc}(G)$	the product of all minimal normal subgroups of G
$\mathrm{abel}(G)$	a maximal abelian normal subgroup of G
$\mathrm{sol}(G)$	the largest soluble normal subgroup of G
$\mathrm{Irr}(G)$	the set of irreducible complex characters of G
$Mult(G)$	the Schur multiplier of G
$P(G)$	the least degree of nontrivial permutation representations of G
$P_f(G)$	the least degree of faithful permutation representations of G
$P_f^t(G)$	the least degree of faithful transitive permutation representations of G
$d(G)$	the least degree of faithful projective complex representations of G
$R_p(G)$	the least degree of faithful projective representations of G over fields of characteristic p
$R_{p'}(G)$	the least degree of faithful projective representations of G over fields of characteristic $\neq p$
$R(G)$	the least degree of faithful projective representations of G
$e(G)$	the Landázuri-Seitz-Zalesskii lower bound for a finite Chevalley group G
$e_0(G)$	the least degree of nontrivial projective complex representations of G
$e_p(G)$	the least degree of nontrivial projective representations of G over fields of characteristic p
$e_{p'}(G)$	the least degree of nontrivial projective representations of G over fields of characteristic $\neq p$

χ_H, $\chi\|_H$	the restriction of a character χ to a subgroup H
$\mathrm{Ind}_H^G(\chi)$	the character induced from a character of a subgroup H of G
$\mathbb{Z}_n$	the cyclic group of order n
$\mathbb{Z}_p^n$	the elementary abelian group of order p^n
$p_{\pm}^{1+2n}$	an extraspecial group of order p^{1+2n}
p^{m+n}	an extension of $\mathbb{Z}_p^m$ by $\mathbb{Z}_p^n$
$\mathbb{S}_n$, $\mathbb{A}_n$	the symmetric and alternating groups on n symbols
$\mathrm{Sym}(X)$	the permutation group on a set X
$L_n(q)$	the projective special linear group of degree n over $\mathbb{F}_q$
$PSU_n(q)$	the projective special unitary group of degree n over $\mathbb{F}_q$
$Sp_{2n}(q)$	the symplectic group of degree $2n$ over $\mathbb{F}_q$
$O_n^{\pm}(q)$	the orthogonal group (of type $\pm$) of degree n over $\mathbb{F}_q$
$\Omega_n^{\pm}(q)$	the unique nonabelian composition factor of $O_n^{\pm}(q)$
$Chev(p)$	the set of finite simple groups of Lie type defined over fields of characteristic p
$Chev(p')$	the set of finite simple groups of Lie type defined over fields of characteristic $\neq p$
$\mathbb{F}_q$	the field consisting of q elements
$\mathbb{F}^*$	the multiplicative group of a field $\mathbb{F}$
$\mathrm{char}\mathbb{F}$	the characteristic of a field $\mathbb{F}$
$\mathrm{Tr}_{\mathbb{F}/\mathbb{F}_0}$	the trace ($\mathbb{F}_0$ is a subfield of $\mathbb{F}$)
$N_{\mathbb{F}/\mathbb{F}_0}$	the norm ($\mathbb{F}_0$ is a subfield of $\mathbb{F}$)
$Gal(\mathbb{F}/\mathbb{F}_0)$	the Galois group of $\mathbb{F}$ over a subfield $\mathbb{F}_0$
$\mathrm{Tr}X$	the trace of a matrix X
${}^tX$	the matrix transposed to X
gcd	greatest common divisor
Λ^*	the lattice dual to Λ
V^*	the space dual to V
$S^n(V) = \mathrm{Sym}^n(V)$, $\wedge^n(V)$	the symmetric and exterior powers of V
$W(R)$, $A(R)$	the Weyl group and the isometry group of a root system R
$\mathrm{Aut}(X)$	the automorphism group of X
$\mathrm{Inn}(X)$	the inner automorphism group of X
$\mathrm{Out}(X)$	the outer automorphism group of X

Author Index

Subject Index